The Physics of Semiconductors

Modern fabrication techniques have made it possible to produce semiconductor devices whose dimensions are so small that quantum-mechanical effects dominate their behavior. This book describes the key elements of quantum mechanics, statistical mechanics, and solid-state physics that are necessary in understanding these modern semiconductor devices. Theoretical results are illustrated with reference to real devices such as photodiodes, flat-panel displays, and metal-oxide–semiconductor field-effect transistors.

The author begins with a review of elementary quantum mechanics and then describes more advanced topics, such as multiple quantum wells. He then discusses equilibrium and nonequilibrium statistical mechanics. Following this introduction, he provides a thorough treatment of solid-state physics, covering electron motion in periodic potentials, electron–phonon interactions, and recombination processes. The final four chapters are devoted exclusively to actual applications, ranging from simple junctions to the latest electroluminescent devices.

The book contains many homework exercises and is suitable as a textbook for electrical engineering, materials science, or physics students taking courses in solid-state device physics. It will also be a valuable reference for practicing engineers in optoelectronics and related areas.

Further material related to this book can be found on the worldwide web at: http://www.ece.gatech.edu/research/labs/comp_elec/

Kevin Brennan received his B.S. degree from the Massachusetts Institute of Technology and his M.S. and Ph.D. degrees from the University of Illinois, Urbana-Champaign. He is a Professor and Institute Fellow in the School of Electrical and Computer Engineering at the Georgia Institute of Technology. He was the recipient of a National Science Foundation Presidential Young Investigator Award and is the author of more than 110 technical articles. Professor Brennan has served as a consultant to various industrial and governmental organizations and holds several U.S. patents for his work on avalanche photodiodes.

The Physics of Semiconductors

with applications to
optoelectronic devices

KEVIN F. BRENNAN

CAMBRIDGE
UNIVERSITY PRESS

PUBLISHED BY THE PRESS SYNDICATE OF THE UNIVERSITY OF CAMBRIDGE
The Pitt Building, Trumpington Street, Cambridge, United Kingdom

CAMBRIDGE UNIVERSITY PRESS
The Edinburgh Building, Cambridge CB2 2RU, UK   http://www.cup.cam.ac.uk
40 West 20th Street, New York, NY 10011-4211, USA   http://www.cup.org
10 Stamford Road, Oakleigh, Melbourne 3166, Australia

First published 1999

Typeset in Sabon 10.25/13pt. and Antique Olive in LaTeX 2_ε [TB]

A catalog record for this book is available from the British Library

Library of Congress Cataloging in Publication data
Brennan, Kevin F., 1956–
    The physics of semiconductors : with applications to
optoelectronic devices / Kevin F. Brennan.
        p.   cm.
    Includes bibliographical references.
    ISBN 0-521-59350-6
    1. Semiconductors.   2. Optoelectronic devices.   I. Title.
QC611.B75   1999
621.3815'2 – dc21   98-29503
                                               CIP

ISBN 0 521 59350 6   hardback
ISBN 0 521 59662 9   paperback

Transferred to digital printing 2003

To my mother and the memory of my father

Contents

Chapter 13
Optoelectronic Emitters   673

Chapter 14
Field-Effect Devices   709

†Section can be omitted without loss of continuity.

Preface

The maturation of epitaxial crystal growth capabilities, such as molecular-beam epitaxy and chemical beam epitaxy, has enabled the realization of a host of new ultrasmall semiconductor devices. Aside from the feature size reduction of conventional semiconductor devices, particularly transistors, a totally new class of semiconductor devices has been invented. These structures, called superlattices/multiple-quantum-well devices, consist of alternating layers of different semiconductor materials, often measuring only a few atomic layers thick. These new semiconductor devices operate well within the range in which quantum-mechanical phenomena become prevalent. As a consequence, most new semiconductor devices behave according to quantum-mechanical effects rather than classical effects. Therefore the understanding of these new device types requires a firm grounding in the basics of quantum mechanics. It is the purpose of this book to introduce the engineering student, particularly those interested in studying solid-state devices, to the principles of quantum mechanics, statistical mechanics, and solid-state physics. Following this introduction, the physics of semiconductors and various device structures is examined.

The book contains fourteen chapters in total. The first four chapters are concerned with the standard principles of quantum mechanics for a one-particle system. I have attempted to condense the vast literature on this subject into just four chapters that will present the salient features of quantum mechanics. I have included a few topics, most notably a short presentation on relativistic quantum mechanics, for completeness. The instructor may elect to skip different sections as he or she sees fit. Sections in the Table of Contents marked by a † are optional, in the sense that material following that section does not directly depend on the material within that section. The first three chapters deal with problems that can be solved exactly in quantum mechanics. I feel that it makes the most sense to show first all the systems that can be solved exactly, including relativistic ones, before jumping to more complicated yet more interesting problems. Chapter 4 presents approximation methods, most notably perturbation theory, both time dependent and time independent. Armed with these principles, the student has most of the essential features of quantum mechanics needed to understand the workings of semiconductor devices.

Beginning with Chapter 5, the basic principles of statistical mechanics are presented. Equilibrium statistical mechanics is covered in Chapter 5, while nonequilibrium statistical mechanics is discussed in Chapter 6. In Chapter 6 the Boltzmann equation is derived and the basic modeling equations for semiconductor device simulation, the drift-diffusion and the Poisson equations, are developed. The chapter also includes a brief discussion of superconductivity.

Again, one may elect to include or omit this section. I would strongly urge the instructor to present this material since I have repeatedly found that the students' imagination is greatly sparked by this material. I personally believe that every solid-state engineer should have some understanding of the workings of super-conductors. This is especially true after the recent discovery of high-temperature superconductors.

Chapter 7 begins the introduction to solids. Chapter 7 focuses on multielec-tron systems, molecular formation, energy-band formation through the tight binding approach, and crystalline symmetries. The behavior of electrons in a periodic potential is the subject of Chapter 8. Topics such as effective-mass the-ory, the Brillouin zone, the nearly-free-electron model, and cellular methods are included. Lattice vibrations and phonons are presented in Chapter 9. Chapter 10 concludes the presentation of the underlying physics of semiconductors by fo-cusing on generation and recombination processes.

The balance of the book, Chapters 11–14, is devoted to device applications. Specifically, I discuss junctions in Chapter 11, detailing the workings of the most important junction types in both equilibrium and nonequilibrium. Photonic de-tectors and detection are the basis of Chapter 12. Photonic emitters are presented next in Chapter 13. The book concludes with a discussion of field-effect devices in Chapter 14. Obviously the choice of topics is not exhaustive. Many impor-tant topics have been omitted, particularly bipolar junction transistors, which are briefly discussed only in Chapter 12. My omission of many topics is due to the vastness of the subject matter. I have chosen to focus on junctions, detectors, emitters, and field-effect transistors since these are the most important devices in compound semiconductor device applications.

In writing such a book, I have obviously borrowed from those whom have gone before, and I am certainly indebted to many references. I have tried to include a full list of all the references that I have contacted at the end of the book.

From a pedagogic point of view, I have developed this book from notes I have written for a three-quarter first-year graduate-level course given in the School of Electrical Engineering at the Georgia Institute of Technology. Typically, I teach virtually all the material in Chapters 1–4 in the first quarter. In the second quarter, all of Chapters 5–8 and Section 9.1 of Chapter 9 are covered. The material in Chapters 10–14 is typically covered in a third-quarter class on semiconductor devices that follows the previous two quarters.

I would like to thank my many colleagues at Georgia Tech and elsewhere for their interest and helpful insight. I am particularly indebted to Dr. W. R. Callen, Jr., for his many suggestions throughout the writing of this book. I am also indebted to the many students at Georgia Tech who have taken the courses from which this book was developed. Their incisive criticisms, both during and after the courses, have been most helpful. In particular, I would like to thank Ali Adibi for his careful and critical reading of versions of the manuscript. I would also like to thank Joel Jackson for electronically coding all the figures, Enrico Bellotti for revising the computer codes and generating many figures, and William Darkwah for his work on translating the computer codes into C. I am also grateful to Maziar Farahmand for his help in some of the mathematical derivations in the text.

Finally, I would like to thank my friends and family for their enduring support.

Atlanta, January 1998

1

Basic Concepts
in Quantum Mechanics

Quantum mechanics forms the basis of modern physics. In a sense it is the parent theory about which we construct our view of the physical world. Briefly, quantum mechanics is the theory by which we describe the behaviors of subatomic and atomic particles, such as electrons, of which the macroscopic world is made. Although it is not necessary to treat macroscopic objects by use of quantum mechanics, the laws of quantum mechanics and their implications are completely consistent with Newton's Laws of Motion, which we know are applicable to most macroscopic objects. As we will see below, Newton's Laws of Motion are a special subset of quantum mechanics; quantum mechanics reduces to Newton's Laws at macroscopic dimensions.

Before we begin our study of quantum mechanics, it is of interest to explain why quantum mechanics is of importance in the study of modern electrical engineering. Many new areas of electrical engineering are based on developments that can be understood only through the use of quantum mechanics. Among these are the broad areas of

1. semiconductors and solid-state electronic devices,
2. electro-optics and lasers,
3. superconductors.

It would be fair to say that in the study of each of the above areas some knowledge of quantum mechanics is essential. In this book, some basic concepts in quantum mechanics are presented that are necessary in the study of the above-mentioned disciplines.

1.1 Introduction

The concept most basic toward the understanding of quantum mechanics is the concept of measurement. In classical physics, the range in which Newton's Laws of Motion apply, the question of measurement rarely is of fundamental importance. We tacitly assume that a measurement of a physical observable, say length, can be made unequivocally. In other words, we assume that everyone will obtain the same measured value of the length of an object at all times. In fact, this seems so obvious that we rarely discuss it at all. However, in quantum mechanics, there is no guarantee that one will always measure the same value of a physical observable at all times. The physical observables commonly measured – microscopic particles, energy, momentum, angular momentum, etc. – can be altered by the mere action of the measurement itself. As a result, one cannot

1

definitively state what value of a particular observable a quantum-mechanical particle will have until a measurement of that observable has been performed. In addition, the measured value may be very different from one measurement to the next.

Let us first consider a simple example of measurement that can help us understand this uncertainty. Let us define the two sides of a coin as the only two possible states of the coin. The coin will always be in its final state in one of only two possible states, heads up or tails up (heads or tails). If the coin is flipped we expect then that the coin will end up either heads or tails after the coin has come to rest. In a sense, we have measured the state of the coin by examining what side is up after it has been flipped. The experiment that we will perform then is something like this. We start with a two-sided coin, heads and tails. The coin is in some initial state, either heads up or tails up. The coin is flipped, and it comes to rest with either heads up or tails up. Of course, we do not know which side is up or down until we examine the coin. By looking at the coin, we measure the state of the coin and decide if it is heads or tails.

Let us consider a variation on the above experiment. As we will see in Chapter 4, an electron also has two possible states that it can be in. These states are related to the spin angular momentum of the electron. As we will see, the spin angular momentum of an electron can be oriented in one of two possible ways. For convenience we call these orientations up or down and the state of the electron either spin up or spin down. A similar experiment to the flipping coin can be performed with an electron. Consider the case in which an electron is created through a beta decay of a nucleus. If we know nothing about the nature of the decay, we would expect to find the emitted electron in either the spin-up or the spin-down state with equal probability, just as we expect to find a flipped coin in either the heads or the tails state. It is important to recognize the fact that we do not know what state the electron is in, though, until we measure it. This is identical to the case of the coin: we do not know whether the coin is heads or tails until we look at it.

What, though, is the state the electron is in before we measure it? Intuitively, we would argue that it must be either spin up or spin down and just assume that its state is well defined at all times and that when we measured it we simply observed what state it was in. Quantum mechanics tells us that this is not correct. The electron is much like the flipped coin. One could ask the question, what state is the coin in before it comes to rest? In other words, is the coin heads or tails while it is spinning in the air? Such a question is ridiculous to ask about the spinning coin because the coin is not in a definite state, heads or tails, until after it has come to rest. The coin is in a definite state only after we have measured what state it is in. Quantum mechanics tells us that to ask the same question about the electron state before measuring it is also meaningless. The electron is not in a definite state until we have measured it. What seems strange about this argument is that we have grown used to the idea in classical physics that the external world is independent of ourselves. To the classical mind, the presence or the absence of the experimenter has no consequence on the results of the experiment. Quantum mechanics tells us that this is not true. The mere action of measurement alters the state of the system, changing it from an indefinite state (coin spinning in the air, electron moving through space) into a definite state (coin comes to rest in either the heads or the tails configuration; electron measured as having either spin up or spin down).

Although the analogy between the spinning coin and an electron is somewhat attractive, there is an important difference between these two situations. One can argue that although the state of the coin is unknown while it is flipping in the air, its final state, either heads or tails, is predictable based on its initial condition and the force applied to set it spinning. Although such a calculation may prove difficult, it is, at least in principle, tractable. Therefore the uncertainty in the final state of the spinning coin is due only to ignorance and not to something fundamental. However, for the electron this is not the case. As we discuss below, knowledge of the electron's state is inherently uncertain until a measurement is made.

The thought experiment described above indicates that there is an inherent uncertainty in the behaviors of microscopic particles. Essentially, the results of our thought experiment indicate that the electron is in an indefinite state until a measurement is made on it and that as a consequence there is an inherent uncertainty in the state of the electron before measurement. Let us carefully discuss this point. This implies that not only does the experimenter not know the state of the electron before the measurement but the electron itself does not know. In other words, there is no way one can know in advance precisely in what state the electron will be found on measurement. As mentioned above, this is completely foreign to classical physics. It would be akin to stating that the orbit of the Earth is uncertain and not predictable.

What proof do we have that the uncertainty in the state of the electron is fundamental and is not just a consequence of our own ignorance of the state of the electron? After all, one could argue that the electron is in a well-defined state unknown to us until we measure it. The uncertainty in that case is due only to ignorance on the part of the observer, not to any fundamental uncertainty in the nature of the electron's state itself. This is the uncertainty we ascribe to the flipping coin, that is, in principle we can predict with certainty which state it will finally be in but it is extremely difficult to do that in practice. To understand what is meant by uncertainty as applied to a quantum-mechanical entity, let us more clearly define quantum-mechanical uncertainty. The well-known Heisenberg Uncertainty Principle of quantum mechanics states that certain sets of conjugate observables cannot be simultaneously known with infinite precision. The Uncertainty Principle is most frequently written in terms of the uncertainties in the real space position Δx and linear momentum Δp_x of the particle as

$$(\Delta x)(\Delta p_x) \geq \hbar/2.$$

The above equation implies that the product in the uncertainties of the two physical observables, position x and linear momentum p_x, must be always greater than or equal to Planck's constant divided by 2π, written as $\hbar$. Physically, this states that the two observables, position and linear momentum, cannot be simultaneously known with infinite precision. The Uncertainty Principle states that the uncertainty in the knowledge of the values of canonical observables is fundamental.

We still have not clearly made an argument that would indicate that the uncertainty is fundamental. One could argue that there are hidden variables that preset the value of a physical observable. For example, let us assume that there is a set of unobservable variables that determines precisely what values the position and the momentum of the particle will have at any given time. Although we cannot measure the values of both the momentum and the position precisely, the hidden variables would tell us precisely what the position and the momentum of the particle are at any given time if we were able to read them. Each value of x corresponds to a specific set of values of the hidden variables and, as a result, the collection of all values of the hidden variables can be regarded as a set of distinct and clearly defined subensembles. Below it is argued that such an arrangement is not consistent with quantum-mechanical interference, specifically as applied to that of one particle.

To understand the failings of the hidden-variable theory and the necessity of concluding that quantum-mechanical uncertainty is fundamental, let us consider the following experiment. In optics there is an experiment called Young's experiment or the two-slit experiment. The experimental arrangement consists of a sheet in which two slits are cut. Light is shone incident upon the surface, and the transmitted beam falls onto a screen set up behind the surface, as shown in Figure 1.1.1. The resulting interference pattern attained resembles that sketched in Figure 1.1.2.a. There is a central maximum of light intensity surrounded by a minimum on either side. Arranged around the central maximum are different maxima of varying intensity, as shown in the diagram. As can be clearly seen, the interference pattern illustrates that total cancellation of the intensity occurs at different locations across the screen. Owing to the fact that the beams of light emerging from the two different slits have

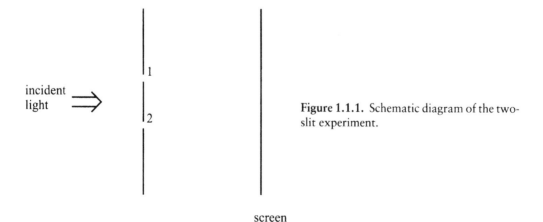

Figure 1.1.1. Schematic diagram of the two-slit experiment.

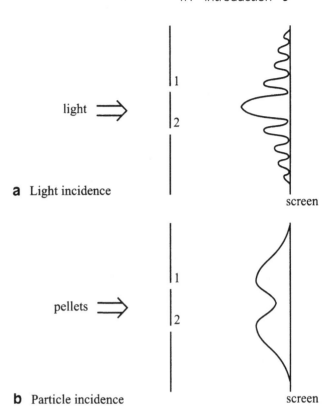

a Light incidence

b Particle incidence

Figure 1.1.2. Interference patterns in a two-slit experiment under light and particle incidence conditions.

different optical path lengths, they arrive at the screen with a phase difference. A maximum in the intensity occurs if the optical path difference is equal to an integral multiple of the wavelength of the light. A minimum occurs if the optical path difference is equal to a half-integral number of wavelengths. In this case, cancellation of the light intensity incident from each of the two slits leads to an intensity minimum. It is important to recognize that the interference pattern generated from incident light is due to the fact that the light travels as a wave and, as such, one wave front can propagate through both slits simultaneously. The resulting interference pattern arises from the subsequent cancellations and reinforcements of the wave front's emerging from either of the two slits onto the screen.

If a collection of particles, such as small-shot pellets, is incident upon the sheet, a different pattern forms on the screen. In this case, the shot can pass through only one slit at a time. One shot cannot pass through both slits simultaneously. As a result, the pattern generated on the screen resembles that sketched in Figure 1.1.2.b. In this case, the pattern on the screen represents the locations at which shots were detected. The peaks imply that these locations are the most likely points at which the shot would be collected. The decaying tails in the pattern indicate that there is a lower probability of finding the particles in these regions. Note that no cancellation in the pattern occurs. There are no interference effects in this case. This is obvious in classical mechanics since a particle cannot interfere with itself; one particle incident upon the sheet can pass through only one of the slits at a time, and no interference can thus occur.

What happens if the sheet is bombarded by quantum-mechanical particles? Let us consider what happens if we use electrons as the incident particles and let us assume that the slits are very small. There are two different experimental conditions that we will consider. The first is exactly as that described above: the electrons fall incident upon the sheet, pass through the slits, and reach the screen. In the second condition, slightly behind each slit we place charge detectors that will record the passage of an electron through them on their flight to the screen. What is observed is truly remarkable. The two cases lead to two different results. In the first case, the pattern observed on the screen would resemble that of Figure 1.1.2.a, which corresponds to wave motion. In the second case, the pattern observed on the screen resembles that shown in Figure 1.1.2.b, which corresponds to particle motion. In the first case, a sheet with two slits in it with nothing but a screen behind it leads to an interference pattern like that arising from waves. The only difference between the first and the second cases is the presence of the charge detectors behind each slit to record the passage of the electrons through them. In the second case, the observed pattern resembles that due to particles. One would obtain exactly the same results if only one electron were incident upon the sheet. In other words, one electron incident upon the sheet would give rise to an interference pattern resembling that from waves if no detectors were present, while it would give rise to a pattern resembling that from particles if the detectors were present. Under certain conditions then, an electron is found to interfere with itself as if it were a wave, while under other conditions an electron behaves as if it were a localized particle. How can we interpret these results?

To understand the findings of the two-slit experiment as applied to quantum-mechanical particles, it is important to reiterate that particles are highly localized in space while waves are highly unlocalized in space. An ideal wave extends throughout space while an ideal particle is compressed to a point in space. The presence of the charge detectors behind the slits in case two somehow changes the result of the experiment. As one may recall, an electron contains a fundamental unit of electric charge. Smaller quantities of electronic charge have not been observed. It is an experimental fact that if one attempts to measure electronic charge, only an integral multiple of the fundamental unit of charge can ever be observed; the electron cannot subdivide itself into parts, each carrying a fraction of the electronic charge. Therefore, if we place charge detectors behind the slits, we would expect to observe only discrete units of the electronic charge. In other words, the charge will pass through only one slit at a time in its entirety for one electron. As a result, the pattern emerging on the screen should be one similar to that for particles. This is indeed what is seen experimentally.

How, though, can we resolve the fact that when the detectors are not present a wavelike interference pattern emerges, even if only one electron is present? This would imply that the electron interferes with itself and as such must pass through both slits simultaneously, subdividing itself in contradiction to what is stated above. How can this be so? The electron does act as a wave in this case and as a particle in the presence of the detectors, since an electron, as all quantum-mechanical particles, displays a wave–particle duality. If an experiment is designed to measure an electron as a particle, as for example the use of charge detectors in the above case, then the electrons will be observed to behave as particles. If an experiment is designed to illustrate the wave properties of an electron, as in the interference experiment without the detectors, the electron will behave as a wave. Therefore the nature of the experiment determines the behavior of what is observed! An electron behaves as a particle if we observe it in such a manner as to bring out its particlelike aspects. An electron behaves as a wave if we observe it in such a manner as to bring out its wavelike aspects. The electron contains mutually exclusive aspects, particlelike and wavelike properties. When the wavelike properties are observed, the particlelike properties are hidden. When the particlelike properties are observed, the wavelike properties are hidden. It depends on the nature of the measurement as to which one is observed at any given time. When the electron is measured in the two-slit experiment, we say that its wave function is collapsed into either a particlelike or a wavelike state. The measurement acts to collapse the electronic state into one of the two possibilities, wavelike or particlelike behavior, much like the catching of the spinning coin collapses the state of the coin into one of its two possibilities, heads or tails.

The analogy to the spinning coin is helpful in understanding the nature of the electron. The coin also contains mutually exclusive aspects, heads and tails. The coin cannot be in both states simultaneously: it has either heads up or tails up, but both cannot be up simultaneously. As for the electron, when the coin is collapsed into heads up, the tails aspect is hidden, and when the coin is measured with tails up, the heads aspect is missing. We all agree that it is meaningless to ask what state the coin is in while spinning. Similarly, it is meaningless to ask what state the electron is in until it is measured.

Finally, let us return to the hidden-variable argument. In the above discussion we stated that an electron can interfere with itself; a single electron in a two-slit experiment can give rise to an interference pattern similar to that of waves. According to the hidden-variable theory, the electron should still pass through only one of the two slits, even though we do not know or cannot know in advance which one it will pass through. As we recall, the hidden-variable theory states that a physical observable, in this case which slit the electron will pass through, is preset by the action of a set of hidden variables, even though these variables are incapable of being observed. The action of the hidden variables must be independent of the experiment since they must act in advance, presetting its behavior. But if this is the case, then the electron's state in the two-slit experiment can never differ depending on the experimental situation since the detectors are placed behind the slits. The knowledge of the detectors' presence would have to be known to the electron in advance of its passage through the slits, a highly unreasonable assumption. Instead, we conclude that there are no hidden variables and that the electron has no preset state. All states are available to it until some action of measurement is applied, after which the electron is collapsed into a definite state. This implies that quantum-mechanical uncertainty is a fundamental property of matter and is not simply due to the ignorance of the observer. Therefore the process of measuring the state of a quantum-mechanical particle results in determining its state!

There are many particles that can be treated as quantum mechanical. Aside from electrons, the particles of light, called photons, are also quantum mechanical. In fact, light also behaves in a fashion that is nearly identical to that of electrons. Under some conditions it behaves as a wave and under other conditions it behaves as a particle. Again, depending on the experimental setup, either property can be observed but never can both wave and particle behaviors be observed concurrently. In Section 1.2 evidence that light has both wavelike and particlelike behaviors is discussed.

1.2 Evidence for a Quantum-Mechanical Picture of Radiation and Matter

Before we discuss the rudiments of quantum mechanics, it is important to understand the experimental underpinnings of the theory. In Section 1.1, we argued through a thought experiment that there is an inherent uncertainty associated with the knowledge of an electron's state. To understand the peculiar circumstances surrounding the two-slit experiment, it is necessary to postulate that an electron has a wave–particle duality. Sometimes an electron will exhibit wavelike properties while under other conditions an electron will exhibit particlelike properties. The nature of the electron's state depends on the experimental situation. As we discussed at the end of Section 1.1, the wave–particle duality is not confined to electrons only, but is exhibited by photons as well. In fact, all quantum-mechanical particles exhibit a wave–particle duality. In this section, experimental evidence that confirms the wave–particle duality of the electron and light is discussed. By inference it is concluded that all quantum-mechanical particles exhibit a wave–particle duality.

The classical theory of light indicates that light can be described as an electromagnetic wave and, as such, exhibits wavelike behavior. Common experiences indicate that light is a wave, that is, diffraction of light by a grating and the refraction of light at an interface between two different media. One of the great vindications of Maxwell's theory of electromagnetism was to show that Maxwell's equations gave rise to wave propagation. It is just these electromagnetic waves that form the classical description of light.

That light is an electromagnetic wave can be seen quite readily from Maxwell's equations as follows. The four Maxwell equations in mks units are

$$\vec{\nabla} \cdot \vec{D} = \rho, \qquad \vec{\nabla} \times \vec{E} = \frac{-\partial \vec{B}}{\partial t},$$

$$\vec{\nabla} \cdot \vec{B} = 0, \qquad \vec{\nabla} \times \vec{H} = \vec{J} + \frac{\partial \vec{D}}{\partial t}, \qquad (1.2.1)$$

where the constitutive relations between the microscopic and the macroscopic fields,

$$\vec{D} = \epsilon \vec{E}, \qquad \vec{B} = \mu \vec{H}, \tag{1.2.2}$$

are assumed. A scalar potential Φ can be defined in terms of the electric field $\vec{E}$ as

$$\vec{E} = -\vec{\nabla}\Phi, \tag{1.2.3}$$

and a vector potential $\vec{A}$ can be defined in terms of the $\vec{B}$ field as

$$\vec{B} = \vec{\nabla} \times \vec{A}. \tag{1.2.4}$$

Substituting Eq. (1.2.4) for $\vec{B}$ into the third Maxwell equation gives

$$\vec{\nabla} \times \vec{E} = -\partial/\partial t(\vec{\nabla} \times \vec{A}),$$

$$\vec{\nabla} \times \left[\vec{E} + \frac{\partial \vec{A}}{\partial t} \right] = 0. \tag{1.2.5}$$

The curl of the quantity in the brackets above vanishes. From the theorems of vector calculus, if the curl of a vector function vanishes, then it must be equal to the gradient of a scalar function. As a result, the expression in the brackets of the second of Eqs. (1.2.5) can be rewritten as

$$\vec{E} + \frac{\partial \vec{A}}{\partial t} = -\vec{\nabla}\Phi, \tag{1.2.6}$$

or equivalently as

$$\vec{E} = -\vec{\nabla}\Phi - \frac{\partial \vec{A}}{\partial t}, \tag{1.2.7}$$

where the vector potential $\vec{A}$ is related to $\vec{B}$ through Eq. (1.2.4).

Let us consider the solution of the Maxwell equations in free space, assuming that a charge density ρ is present. The divergence of $\vec{D}$ can then be written as

$$\vec{\nabla} \cdot \epsilon \vec{E} = \rho. \tag{1.2.8}$$

Substituting for the electric field $\vec{E}$ Eq. (1.2.7), Eq. (1.2.8) becomes

$$\vec{\nabla} \cdot \epsilon \left[-\vec{\nabla}\Phi - \frac{\partial \vec{A}}{\partial t} \right] = \rho, \tag{1.2.9}$$

and applying the divergence operator gives

$$\nabla^2 \Phi + \partial/\partial t(\vec{\nabla} \cdot \vec{A}) = \frac{-\rho}{\epsilon}. \tag{1.2.10}$$

Similarly we can write the equation for the curl of $\vec{H}$ by using the constitutive relations:

$$\vec{\nabla} \times \vec{H} = \vec{J} + \frac{\partial \vec{D}}{\partial t},$$

$$\vec{\nabla} \times \frac{\vec{B}}{\mu} = \vec{J} + \frac{\partial \epsilon \vec{E}}{\partial t}. \tag{1.2.11}$$

Substituting into the second of Eqs. (1.2.11) the expressions given by Eqs. (1.2.4) and (1.2.7) for $\vec{B}$ and $\vec{E}$ yields

$$1/\mu \vec{\nabla} \times \vec{\nabla} \times \vec{A} = \vec{J} + \epsilon \partial/\partial t \left[-\vec{\nabla}\Phi - \frac{\partial \vec{A}}{\partial t} \right]. \tag{1.2.12}$$

Equation (1.2.12) can be simplified with the vector formula

$$\vec{\nabla} \times \vec{\nabla} \times \vec{A} = \vec{\nabla}(\vec{\nabla} \cdot \vec{A}) - \nabla^2 \vec{A} \tag{1.2.13}$$

to

$$1/\mu[-\nabla^2 \vec{A} + \vec{\nabla}(\vec{\nabla} \cdot \vec{A})] = \vec{J} - \epsilon \left[\vec{\nabla} \frac{\partial \Phi}{\partial t} + \frac{\partial^2 \vec{A}}{\partial t^2} \right],$$

$$1/\mu[-\nabla^2 \vec{A} + \vec{\nabla}(\vec{\nabla} \cdot \vec{A})] + \epsilon \left[\vec{\nabla} \frac{\partial \Phi}{\partial t} + \frac{\partial^2 \vec{A}}{\partial t^2} \right] = \vec{J},$$

$$-\nabla^2 \vec{A} + \vec{\nabla}(\vec{\nabla} \cdot \vec{A}) + \mu\epsilon \left[\vec{\nabla} \frac{\partial \Phi}{\partial t} + \frac{\partial^2 \vec{A}}{\partial t^2} \right] = \mu\vec{J},$$

$$\nabla^2 \vec{A} - \vec{\nabla}(\vec{\nabla} \cdot \vec{A}) - \mu\epsilon \left[\vec{\nabla} \frac{\partial \Phi}{\partial t} + \frac{\partial^2 \vec{A}}{\partial t^2} \right] = -\mu\vec{J},$$

$$\nabla^2 \vec{A} - \mu\epsilon \frac{\partial^2 \vec{A}}{\partial t^2} - \vec{\nabla}(\vec{\nabla} \cdot \vec{A}) - \mu\epsilon \vec{\nabla} \frac{\partial \Phi}{\partial t} = -\mu\vec{J}. \tag{1.2.14}$$

The last result above can be further simplified by use of a gauge transformation. A gauge transformation can be understood as follows. The potentials Φ and $\vec{A}$ are related to the electric field $\vec{E}$ and the magnetic field $\vec{B}$ through differential vector operators as

$$\vec{E} = -\vec{\nabla}\Phi, \qquad \vec{B} = \vec{\nabla} \times \vec{A}.$$

The gradient of any scalar function can be added to $\vec{A}$ without changing the value of $\vec{B}$ since the curl of the grad of any scalar function is always zero. Hence the value of $\vec{A}$ can be altered when $\vec{\nabla}\Lambda$ is added, where Λ is a scalar function, leaving $\vec{B}$ invariant:

$$\vec{A} \Rightarrow \vec{A} + \vec{\nabla}\Lambda. \tag{1.2.15}$$

We find $\vec{B}$ by taking the curl of the new value of $\vec{A}$,

$$\vec{B} = \vec{\nabla} \times (\vec{A} + \vec{\nabla}\Lambda) = \vec{\nabla} \times \vec{A} + \vec{\nabla} \times \vec{\nabla}\Lambda = \vec{\nabla} \times \vec{A}, \qquad (1.2.16)$$

which simply gives the original value of $\vec{B}$. $\vec{B}$ is said to be invariant under the above transformation of $\vec{A}$. This transformation is called a gauge transformation. $\vec{A}$ alone cannot be transformed without altering Φ since $\vec{E}$ is defined by Eq. (1.2.7) in terms of both Φ and $\vec{A}$ as

$$\vec{E} = -\vec{\nabla}\Phi - \frac{\partial \vec{A}}{\partial t}.$$

Therefore, when $\vec{A}$ is transformed, the scalar potential Φ must be redefined in order that the value of $\vec{E}$ also remain invariant. $\vec{E}$ will remain invariant provided that the term $-\partial\Lambda/\partial t$ is added to Φ:

$$\Phi \Rightarrow \Phi - \partial\Lambda/\partial t. \qquad (1.2.17)$$

After both transformations of Φ and $\vec{A}$ are made, $\vec{E}$ is given as

$$\vec{E} = -\vec{\nabla}\Phi + \frac{\partial \vec{\nabla}\Lambda}{\partial t} - \frac{\partial \vec{A}}{\partial t} - \frac{\partial \vec{\nabla}\Lambda}{\partial t}, \qquad (1.2.18)$$

which, of course, just yields the original value of $\vec{E}$:

$$\vec{E} = -\vec{\nabla}\Phi - \frac{\partial \vec{A}}{\partial t}.$$

In summary, the value of the potentials can be transformed through a gauge transformation such as to leave the fields invariant.

A gauge transformation enables us to choose freely a set of potentials $\vec{A}$ and Φ without changing the values of $\vec{E}$ and $\vec{B}$. We can choose $\vec{A}$ and Φ that satisfy

$$\vec{\nabla}\cdot\vec{A} + \mu\epsilon\frac{\partial \Phi}{\partial t} = 0, \qquad (1.2.19)$$

recognizing that there exists some gauge transformation that involves the potential Λ that will leave $\vec{E}$ and $\vec{B}$ invariant. This specific choice of gauge is called the Lorentz gauge and the above condition is called the Lorentz condition. With this gauge, the equation for $\vec{J}$ reduces to

$$\nabla^2 \vec{A} - \mu\epsilon\frac{\partial^2 \vec{A}}{\partial t^2} - \vec{\nabla}\left[\vec{\nabla}\cdot\vec{A} + \mu\epsilon\frac{\partial \Phi}{\partial t}\right] = -\mu\vec{J},$$

$$\nabla^2 \vec{A} - \mu\epsilon\frac{\partial^2 \vec{A}}{\partial t^2} = -\mu\vec{J}. \qquad (1.2.20)$$

Similarly, Eq. (1.2.10) can be rewritten as

$$\nabla^2 \Phi + \partial/\partial t(\vec{\nabla} \cdot \vec{A}) = -\rho/\epsilon, \qquad \vec{\nabla} \cdot \vec{A} = -\mu\epsilon \frac{\partial \Phi}{\partial t}. \tag{1.2.21}$$

Substituting the second of Eqs. (1.2.21) for the divergence of $\vec{A}$ into the first of Eqs. (1.2.21), we find that

$$\nabla^2 \Phi - \mu\epsilon \frac{\partial^2 \Phi}{\partial t^2} = -\rho/\epsilon. \tag{1.2.22}$$

Examination of the above equations for the potentials Φ and $\vec{A}$ shows that both obey classical wave equations. If the source terms $\vec{J}$ and ρ are assumed to be zero, then the wave equations for the potentials become

$$\nabla^2 \vec{A} - \mu\epsilon \frac{\partial^2 \vec{A}}{\partial t^2} = 0,$$

$$\nabla^2 \Phi - \mu\epsilon \frac{\partial^2 \Phi}{\partial t^2} = 0. \tag{1.2.23}$$

The propagation velocity of the wave is simply given by the square root of the negative reciprocal of the coefficient in front of the time-derivative term. Hence the wave propagates with a velocity of

$$c = \frac{1}{\sqrt{\mu\epsilon}}, \tag{1.2.24}$$

which is exactly the speed of light in free space. We find that the vector and the scalar electromagnetic potentials propagate at the speed of light in free space. Given that the potentials have this property, the fields do as well. Therefore, from Maxwell's equations (the fundamental result of classical electromagnetism), it is found that light is an electromagnetic wave, and as such will exhibit wavelike properties.

Several experiments also indicate that light is a wave. The most familiar observations that can be explained only if we assume that light behaves as a wave are

1. the refraction of light in passing from one medium into another,
2. the diffraction of light from a single narrow slit,
3. the interference of two separate beams of light, as in Young's experiment described in Section 1.1.

All the above-mentioned effects are readily explained if it is assumed that light is a wave. The evidence for treating light as an electromagnetic wave is compelling; the wave nature of light can be readily derived from first principles by use of only Maxwell's equations, and familiar experience–refraction and diffraction–indicates that light is a wave. Therefore one would be tempted to conclude that

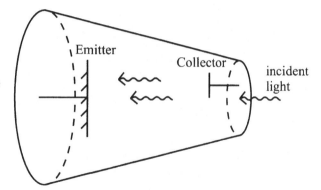

Figure 1.2.1. Schematic drawing of the experimental setup used in the photoelectric effect measurement.

light is simply a classical electromagnetic wave. This was indeed the case until the end of the nineteenth century.

The wave picture of light came into question around 1900. Several experiments were performed that could not be readily reconciled with the wave theory of light. Among these are the photoelectric effect and the Compton effect.

1. The Photoelectric Effect

The photoelectric effect is based on a simple experiment in which incident light is shone onto the surface of a metal enclosed within a vacuum tube, as shown in Figure 1.2.1. Under illumination, electrons are observed to be emitted from the surface of the metal after absorbing the incident light. The emitted electrons are collected by the metal contact, labeled collector in Figure 1.2.1, and a current flows in the external circuit. Application of a negative voltage to the collector acts to retard the motion of the ejected electrons, reducing the number collected, which in turn lowers the measured photocurrent. It is found that there is a maximum potential, called V_s, at and above which no photocurrent is measured. This maximum potential is just that amount of energy needed to counteract the maximum kinetic energy of the electrons emitted from the metal surface. The cutoff potential V_s then represents the maximum kinetic energy K_{max} of all of the emitted electrons. Hence,

$$K_{max} = qV_s. \tag{1.2.25}$$

The experimental measurements show that K_{max} is independent of the incident-light intensity. It is further observed that there is a cutoff frequency v_0 below which no photoelectrons are emitted. It is also observed that there is no appreciable lag between the time that the light is absorbed and the time that the photoelectrons are emitted. The experimental observations are summarized in Table 1.2.1.

Are these observations consistent with the wavelike picture of light? Let us examine each of these observations in detail and evaluate them in the context of

---

Table 1.2.1
Summary of the Photoelectric Effect

Observations:
1) K_{max} is independent of the light intensity.
2) There exists a cutoff frequency, υ_0, below which no electrons are emitted.
3) There is no time lag for electron emission.

Classical Theory:
1) Energy of the light is directly dependent upon the light intensity.
2) No dependence on the frequency of the light is expected. There should be no cutoff frequency.
3) Absorption of energy takes a finite length of time.

Quantum Picture:
1) As the intensity doubles, the number of photons doubles but the energy content of each photon remains the same. Therefore, the amount of energy an electron can absorb remains fixed.
2) The energy of a photon is $h\upsilon$. Energy transfer occurs only in a discrete manner, not continuously. After a photon is absorbed by an electron its energy changes by a discrete amount. If the photon's energy is not large enough for an electron to be freed from the metal, no absorption occurs. Therefore, there exists a cutoff frequency.
3) The absorption of a photon is instantaneous so no time lag occurs.

---

the classical theory of light. From the definition of the Poynting vector, the energy flow per unit area in a plane electromagnetic wave is determined from the cross product of the $\vec{E}$ and the $\vec{H}$ fields. The energy content is proportional to the amplitude of the fields and subsequently the intensity of the light. Therefore, as the intensity increases, the total energy content of the light wave increases. One would then expect that as more energy is incident upon the surface with increasing intensity, more energy will be imparted to the electrons. Since more energy is imparted to the electrons, it would seem that the maximum kinetic energy of the electrons would increase with increasing light intensity. This, however, is not observed.

Based on the classical picture of light, photoelectric emission should occur for any frequency of incident light, provided that the intensity is sufficiently large for photoemission. In the formula for the Poynting vector, the energy content of the light is independent of its frequency and instead depends on the amplitudes of the field and, as such, the intensity. The experimental observations indicate that there is a well-defined cutoff frequency below which there is no photoemission independent of the incident-light intensity in direct contradiction to the classical theory.

Finally, according to the classical theory, there should be a measurable time lag between the time the light is first shone onto the metal and the time the first electron is emitted from the surface. This follows from the fact that it takes time for an electron to absorb sufficient energy from the incident wave for it to escape. However, no time lag has ever been observed. The electrons are essentially emitted instantaneously from the metal surface.

From the above discussion it is apparent that the experimental observations made in the photoelectric effect are not consistent with explanations derived from classical theory, that light is simply an electromagnetic wave. The resolution of this apparent paradox was first made by Einstein, who postulated that light consists of packets of energy, called photons. The energy of each photon is assumed to depend on only the frequency of the light υ as

$$E = h\upsilon, \tag{1.2.26}$$

where h is Planck's constant, given as 4.134×10^{-15} eV s.

The physics of the photoelectric effect can then be understood as follows. The entire energy of a photon is transferred to a single electron in the metal during a photoabsorption event. The energy transfer occurs then only in discrete quantities of energy; on the absorption of a photon, an electron changes its energy by an amount equal to $h\upsilon$. These discrete quantities of energy are called quanta. When an electron absorbs a quantum of energy (a photon in this case), the change in the electron's energy is abrupt. This is different from the case of absorption of energy from a wave, wherein the absorption of energy is a continuous process and takes a finite length of time to be accomplished.

By using the photon picture of light, we can now resolve the difficulties raised by the photoelectric effect observations. As discussed above, it is experimentally observed that the maximum electron kinetic energy is independent of the light intensity. If it is assumed that light comprises photons, then if the intensity is doubled, this implies that the number of incident photons is doubled but the amount of energy each photon supplies remains unchanged. Since an electron can absorb energy in only discrete amounts through the absorption of an entire photon, the change in the electron's kinetic energy is always just the energy of the photon $h\upsilon$. The photoemitted electrons with the highest energy content to begin with, those least bound within the metal, will have the greatest kinetic energy of the emitted carriers, which can be no larger than the initial energy plus the photon energy. The maximum kinetic energy of the electrons emitted from the metal must then remain unchanged. This is true independent of the light intensity. An increase in light intensity leads to an increase in only the number of electrons emitted from the metal and not an increase in the energy of each emitted electron. Hence K_{max} is expected to be independent of the light intensity in agreement with experimental observation.

The existence of a cutoff frequency υ_0 below which no electrons are photoemitted from the metal is readily explained since the energy of each photon is fixed by the frequency of the light. Assuming that there is a minimum energy that an electron must absorb in order to be emitted from the metal, if the energies of the incident photons are each less than this amount of energy, no photoemission can occur. Therefore there is a cutoff frequency υ_0 below which the energy of each incident photon is less than that needed to eject an electron from the metal. Under these conditions, no photocurrent is observed independent of the number of photons present.

EXAMPLE 1.2.1 Photoelectric Emission

The work function of a material is defined as the energy needed to remove an electron from the metal. (As will be discussed in Chapter 11, the work function is the energy difference between the vacuum level and what is called the Fermi level.) The work function of Al, Φ, is given as 4.25 eV.

a. **Does the photoelectric effect occur for incident light of a 6800-Å wavelength?**

To answer this question, all that is needed is to determine what the energy E of a photon of this wavelength is. If E is greater than Φ, then photoemission will occur. Otherwise it will not.

$$E = hc/\lambda = \frac{(4.134 \times 10^{-15} \text{eV s})(3.0 \times 10^{10} \text{ cm/s})}{6800 \times 10^{-8} \text{ cm}},$$

which calculates to $E = 1.82 \text{ eV}$. Therefore $E < \Phi$, and no photoemission occurs.

b. **What is the cutoff wavelength for Al?**

$$\lambda = \frac{hc}{\Phi}.$$

Substituting in for hc and Φ into the above equation, we find that

$$\lambda = \frac{12{,}400 \text{ eV Å}}{4.25 \text{ eV}} = 2917 \text{ Å}.$$

2. The Compton Effect

Another important experiment that contradicts the wave picture of light is the Compton effect experiment. The Compton effect involves the scattering of incident radiation, x rays particularly, off of a solid target, such as graphite. In the Compton effect experiment, it is observed that two beams of different wavelengths, λ and λ', are reflected from the target. The wavelength of one of the reflected beams, λ, is equal to the wavelength of the incident beam. The wavelength of the second reflected beam, λ', is shifted in magnitude to a greater value,

$$\lambda' = \lambda + \Delta\lambda, \tag{1.2.27}$$

where the quantity $\Delta\lambda$ is called the Compton shift.

The existence of a shifted wavelength is not consistent with the expectations of classical theory. From the classical wave picture of radiation, it is expected that the scattered radiation will be of the same wavelength as the incident radiation. An incident electromagnetic wave sets a charge oscillating at the same frequency as the frequency of the wave. As is well known from classical electromagnetics, an accelerated charge radiates at a frequency equal to that of the oscillation. The reradiated electromagnetic wave from the oscillating charge should have the same frequency as the incident wave. The presence of a reflected beam shifted in wavelength to lower energy is contradictory to classical theory.

To explain the wavelength shift in the reflected beam, Compton postulated that the incident light must consist of photons. The incident photons collide with both free electrons and bound electrons in the metal. If the photon–electron scattering is imagined to be similar to that of billiard balls, then the incident particles transfer some kinetic energy to the target particles and emerge with a lower energy than before. (In reality, a single photon is first absorbed by an electron, and a photon of different energy is emitted. However, we can treat the Compton effect collisions like hard-sphere scatterings for simplicity and still retain the essential physics.)

As in all collisions, both energy and momentum must be conserved within the system. When these conservation laws are applied to the collision, the magnitude of the Compton shift can be determined. The collision process is sketched in Figure 1.2.2 for both before and after the interaction. The particle energies are

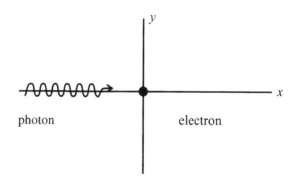

Figure 1.2.2. Compton effect scattering showing the system both before and after the collision.

a Before the collision

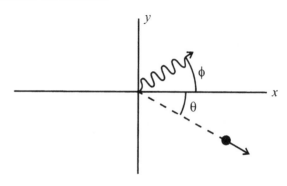

b After the collision

defined as

$h\upsilon$: incident-photon energy,
$h\upsilon'$: scattered-photon energy,
mc^2: total electron energy,
$m_0 c^2$: electron rest-mass energy.

The kinetic energy of the electron before the interaction is assumed to be zero, while after the interaction it is given as the difference between the total energy and the rest-mass energy (this result is derived in detail in Chapter 3, Section 3.5, on relativity and relativistic quantum mechanics):

$$K = (m - m_0)c^2, \tag{1.2.28}$$

where the mass m is defined as

$$m = \frac{m_0}{\sqrt{\left(1 - \frac{v^2}{c^2}\right)}}. \tag{1.2.29}$$

Conservation of energy requires that the initial energy E_i of the system before the collision must be equal to the final energy E_f of the system after the collision. Initially the electron is at rest and a photon of energy $h\upsilon$ is incident. Therefore E_i is simply equal to $h\upsilon$. After the collision occurs, the electron gains kinetic energy from the incident photon. The final energy of the electron and the photon is then

$$E_f = h\upsilon' + (m - m_0)c^2. \tag{1.2.30}$$

The conservation of energy yields

$$h\upsilon = h\upsilon' + (m - m_0)c^2. \tag{1.2.31}$$

The linear momentum of the system must also be conserved. The momentum of a photon can be expressed using what is known as the de Broglie relation:

$$p = h/\lambda. \tag{1.2.32}$$

Momentum is a vector quantity. All components of the momentum must be conserved before and after the collision. If it is assumed that the incident photon moves along only the x axis initially, as shown in Figure 1.2.2.a, the scattered photon and the electron travel off in only the x and the y directions after the interaction (Figure 1.2.2.b). It is clear that only two dimensions need to be considered to describe the collision. Therefore it is necessary to write conservation equations of the linear momentum in the x and the y directions. Conservation of the x component of momentum gives

$$h/\lambda = h/\lambda' \cos\phi + \frac{m_0 v \cos\theta}{\sqrt{\left(1 - \frac{v^2}{c^2}\right)}}, \tag{1.2.33}$$

where ϕ is the angle between the scattered photon and the x axis and θ is the angle between the scattered electron and the x axis. Conservation of the y component of the momentum yields

$$0 = h/\lambda' \sin\phi - \frac{m_0 v \sin\theta}{\sqrt{\left(1 - \frac{v^2}{c^2}\right)}}. \qquad (1.2.34)$$

The Compton shift is readily found when these equations are combined (see Problem 8 at the end of this chapter):

$$\Delta\lambda = (\lambda' - \lambda) = \frac{h}{m_0 c}(1 - \cos\phi). \qquad (1.2.35)$$

Comparison with experimental measurements shows good agreement with the above formula. Thus the observed shift in wavelength of the scattered beam to lower energy is explained on the assumption that the radiation consists of photons that scatter off of free electrons.

The unmodified line in the Compton experiment is due to photon scatterings with bound electrons. In this case, the electron and the atom recoil after the collision. The effective mass within the Compton formula should then be replaced by that of the electron plus the host atom M_0, as opposed to just the electron mass. Generally the atomic mass is far greater than the electron mass, typically by ~ 4 orders of magnitude, so the electron mass is neglected and M_0 is then used in the Compton formula in place of m_0. The Compton shift arising from the scatterings with the atoms is roughly 4 orders of magnitude smaller than that from free electrons. As a result, the scattered wavelength is essentially unchanged from the incident wavelength, giving rise to an unmodified line in the reflected spectrum.

Both the Compton effect and the photoelectric effect indicate that light cannot be simply viewed as a classical wave. Instead, these experiments indicate that light has corpuscular properties as well. The basic question raised then is how can we reconcile the Compton effect and the photoelectric effect observations, that light is composed of particles (photons), with the result from classical electromagnetics and diffraction and refraction experiments that indicate that light is a wave? After all, particles and waves are two completely different, and essentially opposite, descriptions. The wave and the particle pictures are mutually exlusive; how can an entity be both a particle and a wave? The answer to this dilemma lies in the duality of nature, as discussed in Section 1.1. The behavior of light and all quantum-mechanical particles depends on the particulars of the experiment used to observe it. If an experiment is designed to illustrate the wave nature of light, such as a diffraction experiment or Young's experiment, that is what is observed. If an experiment is designed to test the particle nature of light, such as the Compton or the photoelectric effects, then the light will behave as a collection of particles. Measurement of the state of the electron through the action of the specific experiment causes the electron to collapse into a definite state, just as the catching of the spinning coin causes it to collapse into either heads or tails up. The state of the electron, as the state of the spinning coin, is unknown until a measurement is made, whereupon the system collapses into a definite state.

It is important to recognize that light will never appear to be both a particle and a wave simultaneously. These are mutually exclusive states of the system and cannot occur at the same time, much like a flipped coin cannot be both heads and tails simultaneously after it has come to rest. When light is observed as a photon, it loses all its wavelike properties and vice versa. The wavelike and the particlelike pictures are said to be complementary. This is known as the Principle of Complementarity.

Similarly, matter exhibits properties akin to both a particle and a wave description. For example, electrons can be localized, appearing as particles as in the Millikan oil-drop experiment. Alternatively, electrons can be diffracted by crystals (see Chapter 8) and thus appear as waves. As in the situation for light, an electron never exhibits both wavelike and particlelike properties simultaneously.

The question becomes, how do we treat matter and light, given that they exhibit wavelike and particlelike behaviors? How can we mathematically formulate the dynamics of matter and light, taking into account the fact that they have an inherent duality? Sometimes they behave as particles and under different observation conditions they behave as waves. How this is treated is the subject of the first several chapters of this book.

1.3 Wave Packets and Free-Particle Motion

The discussion presented in Sections 1.1 and 1.2 indicates that there is strong evidence that all quantum-mechanical particles exhibit a wave–particle duality. It appears that the only way that we can understand experimental observations that indicate wavelike behavior under some conditions and particlelike behavior under different conditions is to postulate that electrons, photons, and similar quantum-mechanical particles have an inherent wave–particle duality that collapses into definite wavelike or particlelike states after a measurement is performed. The question is, how are the particlelike aspects represented in conjunction with the wavelike aspects for a quantum-mechanical particle? Clearly, if we try to represent matter by plane waves then there is no description of the particlelike aspects. Conversely, if we picture matter as a classical point-like particle, the wavelike description is lost. To describe a quantum-mechanical particle mathematically, including its wave–particle dualism, a new technique is necessary.

It should be stressed that when a measurement is performed to localize the position of an electron, the electron collapses into a state with definite position (a point particle). If a measurement is performed to localize the electron's wavelength [or equivalently its momentum from Eq. (1.2.32)] the electron collapses into a state with a definite wavelength (a plane wave). We define the wave function as the mathematical description of the electron. After a measurement then, it is said that the wave function collapses into a state of definite position, momentum, or whatever is physically being examined. Therefore the wave function must collapse into a plane-wave state if an experiment is performed in which the electron's wavelength is localized. Alternatively, the wave function must collapse into a state with definite position if an experiment is performed in which the electron is localized in space. What, though, is the form of the wave function before a measurement?

The most useful construction of the wave function is that of a superposition of waves that form a wave packet whose spatial width Δx and momentum width Δp are consistent with the Uncertainty Principle. Therefore the wave function represents an electron with an inherent uncertainty in its spatial position and simultaneously an inherent uncertainty in its momentum. The wave packet is constructed from the linear superposition of plane waves of various k such that they interfere constructively in only a small region Δx that forms the spatial extent of the electron and interferes destructively elsewhere. Let us define the wave function to be $\Psi(x, y, z, t)$. From the above definition of a wave packet, Ψ can be described by a linear superposition of two or more waves, $\Psi_1(x, y, z, t)$ and $\Psi_2(x, y, z, t)$. Ψ is written then as $\Psi = \Psi_1 + \Psi_2$. Mathematically the decomposition of Ψ into the sum of two or more wave functions implies that an arbitrary wave function can be determined from a linear superposition of different wave functions. This is called the Principle of Linear Superposition. As we see below, the Principle of Linear Superposition is of great importance.

Let us consider the two-slit experiment again. As we know from the classical theory of waves, the intensity of a wave is described by the square of the amplitude of the wave. Let $|\Psi_1|^2$ represent the intensity on the screen made behind the sheet containing the two slits if slit 2 is blocked. Let $|\Psi_2|^2$ represent the intensity on the screen made if slit 1 is blocked. What would be the resulting intensity on the screen if both slits are open? In accordance with the Principle of Superposition, the amplitude on the screen must be equal to the sum of the amplitudes from each of the two slits, 1 and 2. The intensity is given by the square of the sum of the amplitudes. As a result, the intensity on the screen with both slits open must be given as

$$|\Psi|^2 = |\Psi_1 + \Psi_2|^2, \tag{1.3.1}$$

which is precisely what is expected for self-interference. From the arguments presented in Section 1.1, it was determined that a single electron would interfere with itself if it were incident upon the two slits, provided no attempt was made to localize the position of the electron (no detectors were present behind the slits). Therefore the intensity given by the square of the sum of the amplitudes is consistent with self-interference and with the basic tenets of quantum mechanics. It is important to recognize that the intensity on the screen in the two-slit experiment without detectors is not given by

$$|\Psi|^2 = |\Psi_1|^2 + |\Psi_2|^2, \tag{1.3.2}$$

since this is not consistent with self-interference.

As discussed above, the wave function must be constructed so as to be consistent with the Uncertainty Principle as well as reduce to a definite state of momentum or position after a measurement of either property is performed. It is relatively simple to construct a wave function that is in a definite momentum state. From Eq. (1.2.32), known as the de Broglie relation, the momentum of a quantum-mechanical particle is related to its wavelength as $p = h/\lambda$ or $p = \hbar k$ where $\hbar$ is defined as Planck's constant h divided by 2π. Therefore a particle has definite momentum; by this we mean that the particle has a single value of its momentum if it has a single value of λ. A state that has only one value of λ is a plane-wave state and can be represented as

$$\Psi = A e^{i(kx - \omega t)}, \tag{1.3.3}$$

where A is the amplitude, k is $2\pi/\lambda$, ω is $2\pi\upsilon$, and t is the time. In this case, the momentum of the electron is definite: it is h/λ or, equivalently, $\hbar k$. However, the real-space position of such a particle is competely unknown. Note that this particular plane wave extends infinitely throughout the x direction and as such has no definite position. This result is consistent with the Uncertainty Principle since Δp is zero (the electron is in a definite state of momentum) and Δx is ∞. Similarly, if the electron is localized in real space, Ψ can be expressed as

$$\Psi = A\delta(x - x_0), \tag{1.3.4}$$

where $\delta(x - x_0)$ is a Dirac delta function and has value of 1 when x is equal to x_0 and is zero elsewhere, precisely what is meant by real-space localization. In this case, Δx is zero while Δp is ∞.

From the above discussion, it is clear that the position and the momentum variables form a Fourier conjugate pair, much like frequency and time do in circuit theory. As the reader may recall from elementary circuit theory, it is often useful to formulate circuit problems in the frequency domain as opposed to the time domain. It is easy to move from one domain to the other through a Fourier transform. These transformations are given for time as

$$f(t) = \frac{1}{\sqrt{2\pi}} \int g(\omega) e^{i\omega t} \, d\omega \tag{1.3.5}$$

and for frequency as

$$g(\omega) = \frac{1}{\sqrt{2\pi}} \int f(t) e^{-i\omega t} \, dt, \tag{1.3.6}$$

where ω is the circular frequency, defined as $2\pi\upsilon$. Recall from circuit theory that if a signal has a sharp, singular frequency then if it is transformed back into the time domain, the result is a horizontal line that covers all times, as shown in Figure 1.3.1. Similarly, if the signal can be characterized as occurring at a precise instant of infinitesimal duration, then it can be represented as a vertical spike in the time domain and a horizontal line in the frequency domain, as shown in Figure 1.3.2.

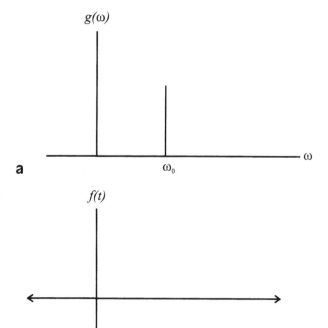

Figure 1.3.1. a. Frequency function corresponding to a signal which has only one frequency component ω_0. b. Corresponding time function. Note that $f(t)$ extends over all values of t.

In a similar fashion, position and momentum form a Fourier conjugate pair. If the particle is completely localized in real space (vertical spike in the real-space plot), then its corresponding momentum is completely unknown, as shown in Figure 1.3.3. If the particle's momentum is completely certain, then its position is totally unknown. This follows from the physical nature of particles and waves.

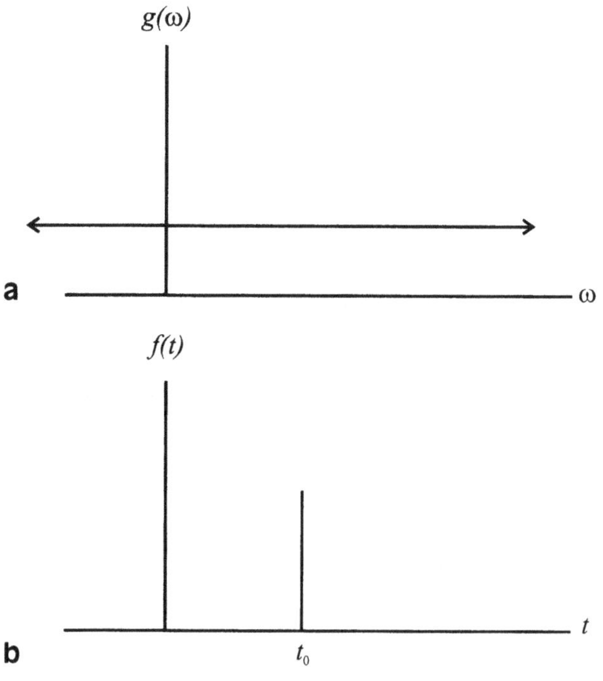

Figure 1.3.2. a. Frequency function for a system which has a singular time value $t = t_0$. b. Time function for a system which vanishes everywhere except at $t = t_0$.

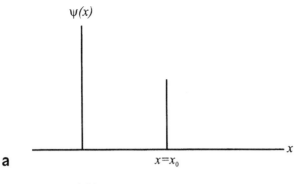

a

$x=x_0$

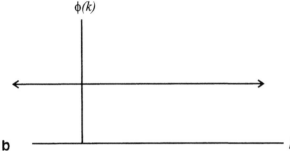

b

Figure 1.3.3. a. $\psi(x)$ for a completely localized particle. b. Corresponding momentum function $\phi(k)$.

If a quantum-mechanical particle is found to exhibit wavelike properties after a measurement, it loses all its particlelike attributes in accordance with the Principle of Complementarity. If, on the other hand, a quantum-mechanical particle is found to exhibit particlelike properties after a measurement, it loses all its wavelike attributes. Mathematically these relationships can be expressed through use of Fourier transforms for position,

$$\psi(x) = \frac{1}{\sqrt{2\pi}} \int \phi(p) e^{ipx/\hbar} \, dp, \tag{1.3.7}$$

and for momentum,

$$\phi(p) = \frac{1}{\sqrt{2\pi}} \int \psi(x) e^{-ipx/\hbar} \, dx, \tag{1.3.8}$$

where $\psi(x)$ and $\phi(p)$ are the position and the momentum functions, respectively. k can be used in place of p, if we choose, since they are simply related: $p = \hbar k$.

Let us check that these definitions are consistent with the fact that when the position is sharply defined the momentum is indefinite and when the momentum is sharply defined the position is indefinite. First, let us more fully define the delta function introduced above. A delta function is nonzero for only one precise value and is defined as

$$\int \delta(x - x_0) f(x) \, dx = f(x_0), \tag{1.3.9}$$

where it is assumed that the range of integration includes x_0. If $f(x)$ is chosen to be 1, then the delta function is defined as

$$\int \delta(x - x_0) \, dx = 1 \qquad (1.3.10)$$

if the range of integration includes the point x_0 and zero if it does not. A state that is completely localized in terms of position can be represented as

$$f(x) = \delta(x - x_0) \qquad (1.3.11)$$

since it will have a nonzero value only at the point $x = x_0$ and have a zero value elsewhere.

Consider now the Fourier transform of this function. It is given as

$$\phi(p) = \frac{1}{\sqrt{2\pi}} \int \delta(x - x_0) e^{-ipx/\hbar} \, dx = \frac{e^{-ipx_0/\hbar}}{\sqrt{2\pi}}, \qquad (1.3.12)$$

which states that a delta function in real space transforms into a plane-wave state in k or momentum space. This is comparable with the result that a definite state in momentum space is a plane-wave state in real space. In other words, a state that has a precise value of momentum [$\phi(p)$ would be a delta function] is represented in real space as a plane wave [$\psi(x)$ is a plane wave].

The question, though, is, how do we treat an electronic state that is localized neither in real space nor in momentum space? In general, we have found that until a measurement is performed on the electron its state is unknown. It exists as some wave packet that has a finite extent in real and momentum space in accordance with the Uncertainty Principle. Such a state is diagrammatically sketched in Figure 1.3.4.

A wave packet can be thought of as consisting of a superposition of many waves that destructively interfere everywhere except within a small region, as shown in the figure. How is this wave packet constructed? It should be recalled

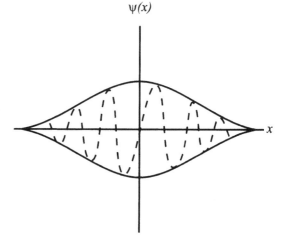

Figure 1.3.4. Sketch of an electron wave packet in real space x.

that any periodic function can be decomposed into a sum of sine and cosine waves in a Fourier series. This is a fundamental result in Fourier analysis and, as we shall see in Section 1.6, a fundamental result in mathematics. Additionally, an aperiodic function can be represented by Fourier transforms. The only difference between the aperiodic case and the periodic case is that the wave packet must be composed of waves with a continuous range of wavelengths in the aperiodic case rather than a discrete set of values as in the periodic case. Therefore a wave packet can be formed with the Fourier transforms defined above for x and p.

As an example of a wave packet, consider a function characterized in real space as

$$\psi(x) = e^{\frac{-x^2}{2b^2}}, \tag{1.3.13}$$

which is simply a Gaussian function. The corresponding momentum-space function $\phi(p)$ can be found from the Fourier transform of $\psi(x)$. This can be accomplished as follows. $\phi(p)$ is given as

$$\phi(p) = \frac{1}{\sqrt{2\pi}} \int e^{\frac{-x^2}{2b^2}} e^{-ipx/\hbar} \, dx, \tag{1.3.14}$$

which can be rewritten as

$$\phi(p) = \frac{1}{\sqrt{2\pi}} \int e^{-\left(\frac{x^2}{2b^2} + ipx/\hbar\right)} \, dx. \tag{1.3.15}$$

Completing the square in the exponent, we obtain

$$\phi(p) = \frac{1}{\sqrt{2\pi}} \int e^{\frac{-1}{2b^2}(x^2 + i2b^2 px/\hbar)} \, dx,$$

$$\phi(p) = \frac{1}{\sqrt{2\pi}} \int e^{\frac{-1}{2b^2}\left(x^2 + 2ib^2 px/\hbar - \frac{b^4 p^2}{\hbar^2} + \frac{b^4 p^2}{\hbar^2}\right)} \, dx,$$

$$\phi(p) = \frac{1}{\sqrt{2\pi}} \int e^{\frac{-1}{2b^2}(x + ib^2 p/\hbar)^2} e^{\frac{-b^4 p^2}{2b^2 \hbar^2}} \, dx, \tag{1.3.16}$$

since

$$(x + ib^2 p/\hbar)^2 = x^2 + 2ib^2 px/\hbar - b^4 p^2/\hbar^2.$$

Therefore the integral becomes

$$\phi(p) = \frac{2}{\sqrt{2\pi}} e^{\frac{-b^4 p^2}{2b^2 \hbar^2}} \int e^{-\frac{1}{2b^2}(x + ib^2 p/\hbar)^2} \, dx. \tag{1.3.17}$$

Renaming variables as

$$u = x + ib^2 p/\hbar; \qquad du = dx, \tag{1.3.18}$$

we obtain,

$$\phi(p) = \frac{2}{\sqrt{2\pi}} e^{-\frac{b^4 p^2}{2b^2 \hbar^2}} \int e^{-\frac{u^2}{2b^2}} \, du, \tag{1.3.19}$$

which simply integrates out with bounds from 0 to ∞ to

$$\phi(p) = b \, e^{-\frac{b^2 p^2}{2\hbar^2}}. \tag{1.3.20}$$

Note that $\phi(p)$ is peaked at $p = 0$. As b increases, $\phi(p)$ becomes more compressed and $\psi(x)$ becomes more spread out. Physically the state becomes more spread out in real space whereas it becomes more localized in momentum space. This is precisely the meaning of the Uncertainty Principle; as the momentum is localized, the spatial extent of the state is delocalized and there is a greater probability of finding the particle in a larger spatial region. Thus, to construct a wave packet that is highly localized in space, $\psi(x)$ is sharply peaked and b must be small, resulting in a large range of values of the momentum required for making $\phi(p)$. Therefore the use of Fourier transforms in the construction of the wave packet is consistent with the Uncertainty Principle and is of use in describing the state of the electron. In Section 1.4 the physical meaning behind the wave function is discussed.

1.4 **Probability and Quantum Mechanics**

In Section 1.3 we have determined that a quantum-mechanical particle can be represented mathematically by what we call a wave function. It was further found that a wave function is constructed from a linear superposition of plane waves so as to yield a function consistent with the Uncertainty Principle, that is, it has a finite extension in real and momentum spaces. The question is, though, what does the wave function represent physically? We have said it is useful in describing a quantum-mechanical particle, but what does it mean and how is it of use in computing the physically measurable properties of the particle? In this section, we discuss the physical meaning of the wave function.

From our discussion in Section 1.1, we recognize that until a measurement is performed, the precise state of the quantum-mechanical particle is unknown. It is unknown not only to us, the observers, but is inherently unknown until it is measured. The action of measurement, we have said, acts to collapse the wave function, meaning that the state of the particle (the electron will be commonly used as an example of a quantum-mechanical particle) becomes definite. It would be reasonable then to say that before a measurement, one can talk about only the probability that an electron will be found in a certain state on measurement. This is similar to the case of the spinning coin. In that case, we talk of the probability of finding the coin as either heads up or tails up. In the example of the coin, there is a 50% chance of finding the coin with heads up and a 50% chance of finding it with tails up. Once the coin is caught, it is in a definite state, heads or tails, and there is no longer any uncertainty. The electron behaves in essentially the same way. Before making a measurement of the state of the electron, all we can say is that there is a certain probability of finding the electron in a certain definite state. Once a measurement is made, the electron wave function collapses into a definite state of the system and there is no longer any uncertainty. The electron wave function then is associated with the probability of finding the electron in a given state on measurement.

We define the wave function as the probability amplitude and its modulus square as the probability density. The integral of the probability density over a spatial volume is interpreted as the probability of finding the particle within that volume. If the volume is taken as all space (infinite dimensions), then the probability of finding the particle must be 100%; the particle must be somewhere within the universe. Assigning the probability to be 1 for 100%, we say that the wave function is normalized if its integral over all space is 1. Mathematically, then,

$$|\Psi|^2 \equiv \text{probability density.} \tag{1.4.1}$$

$|\Psi|^2 \, \mathrm{d}^3 r$ is the probability of finding the particle within the volume $\mathrm{d}^3 r$, and the normalization condition is given as

$$\int |\Psi|^2 \, \mathrm{d}^3 r = 1. \tag{1.4.2}$$

From these definitions, it is possible to determine the probability of finding an electron in a certain state.

Let us quickly review the basics of probability theory that we will need before we proceed. Let $E(k)$ be defined as an event in a set of occurrences or events labeled $E(1)$, $E(2)$, $E(3)$, etc., with a probability of occurrence $P(1)$, $P(2)$, $P(3)$, etc. It is assumed that the probability is normalized; in other words the sum of the probabilities of all of the possible occurrences must be 1:

$$\sum P(k) = 1. \tag{1.4.3}$$

For simplicity, let us again use the example of a coin. There are two possible events, heads and tails. Let the event heads be called $E(1)$ and the event tails be called $E(2)$. The probabilities of each event occuring are $P(1)$ and $P(2)$, respectively. In this case, each event occurs with an equal probability of 0.5. Let us define a random variable x that takes on value $x(1)$ if $E(1)$ occurs, $x(2)$ if $E(2)$ occurs, etc. An example of this would be a simple game in which I am given one point for the event heads occurring and three points for the event tails occurring. Then $x(1) = 1$ and $x(2) = 3$. The expectation value of x is defined as the weighted sum $\sum x(k)P(k)$. In this specific case, the expectation value of x, often written as $\langle x \rangle$, is

$$\langle x \rangle = 1(0.5) + 3(0.5) = 2, \tag{1.4.4}$$

the sum of the probabilities of occurrence (0.5 in each case) times the value of x (1 and 3). Therefore we find the expectation value of a quantity by summing the product of the probability of occurrence of each value times its value, provided that the probability is properly normalized.

Next consider a large number of trials N occurring. $N(1)$ leads to event $E(1)$, $N(2)$ leads to event $E(2)$, etc. The probability that event k occurs then is given by

$$\frac{N(k)}{N} = P(k) \tag{1.4.5}$$

for large-number N. The average value of the variable x is then equal to the expectation value of x since

$$\frac{\sum x(k)N(k)}{N} = \langle x \rangle. \tag{1.4.6}$$

The variance of x is defined as

$$(\Delta x)^2 = \langle (x - \langle x \rangle)^2 \rangle = \sum [x(k) - \langle x \rangle]^2 P(k) = \langle x^2 \rangle - \langle x \rangle^2. \tag{1.4.7}$$

In words, the variance measures the departure of the distribution from the mean, and the variance is said to equal the mean of the square minus the square of the mean.

If the distribution is assumed to be continuous, the expectation value is defined with an integral in place of the sum. Therefore the expectation value of x is

$$\langle x \rangle = \int x\rho(x)\,dx, \tag{1.4.8}$$

where $\rho(x)$ is defined as the probability density. From our discussion of the meaning of the wave function, we recognize that the probability density in quantum mechanics is simply the product of $\Psi^*\Psi$. Therefore $\langle x \rangle$ in quantum mechanics is given as

$$\langle x \rangle = \frac{\int \Psi^* x \Psi \, d^3 r}{\int \Psi^* \Psi \, d^3 r}, \tag{1.4.9}$$

where Ψ^* represents the complex conjugate of Ψ and the denominator is used to ensure proper normalization of the wave functions. Generally the expectation value of any physical observable $f(r)$ is found from the following prescription,

$$\langle f(r) \rangle = \int \Psi^* f(r) \Psi \, d^3 r, \tag{1.4.10}$$

provided that the state Ψ is properly normalized. Physically, this means that the expected value of the physical observable $f(r)$ is simply the mean value of this observable in the state Ψ. The value obtained, $\langle f(r) \rangle$, is the expected value of the physical observable in the state Ψ after a measurement is made. If, for example,

$f(r)$ is equal to x, the position of the particle $\langle x \rangle$ represents the expected value of the position of the particle after a measurement is made of the position when the particle is characterized by the wave function Ψ.

Before we end this section, let us turn our attention briefly to the description of the dynamics of the electron. Up to this point, it has been mentioned that the state of an electron can be described by a wave packet or a wave function that consists of the superposition of many Fourier components such that total destructive interference occurs everywhere except within a small region of finite extent. This ensures that the wave function is consistent with the Uncertainty Principle. How, though, does this wave function evolve in space and time? What governs its dynamics? The Schroedinger equation has been put forth to describe the dynamics of quantum particles. In Section 1.5 we will see the origins of the Schroedinger equation. For now, it is simply introduced. The Schroedinger equation is defined as

$$\frac{-\hbar^2}{2m}\nabla^2\Psi + V\Psi = E\Psi = i\hbar\frac{\partial\Psi}{\partial t}, \tag{1.4.11}$$

where V is the potential in which the particle moves, ∇^2 is the differential operator,

$$\nabla^2 = d^2/dx^2 + d^2/dy^2 + d^2/dz^2, \tag{1.4.12}$$

and m is the mass of the particle. We will explore the solution of this equation throughout the course of this book. Before we do so, it is necessary to define operators and their relationship to quantum mechanics.

1.5 Dynamical Variables and Operators

In order to learn the origin and the meaning of the Schroedinger equation it is useful to first introduce operators. There is a differential operator associated with every physical observable in quantum mechanics. For example, associated with the linear momentum p there is a differential operator. As is shown below, this operator in one dimension is given as $\hbar/i\,\partial/\partial x$. We begin our discussion of operators by deriving the expression for the momentum operator.

We start from the interpretation of $\Psi^*\Psi$. From the discussion in Section 1.4, $\Psi^*\Psi$ represents the probability density. If we define ρ as the probability density $\Psi^*\Psi$ and $\vec{j}$ as the probability current density, then if the probability is assumed to be conserved, there must be a continuity equation for the probability current density. This has the same form as any current continuity equation, namely,

$$\vec{\nabla}\cdot\vec{j} + \partial\rho/\partial t = 0. \tag{1.5.1}$$

Physically, Eq. (1.5.1) means that the probability current-density flux through a Gaussian surface must be equal to the time rate of change of the probability density within the volume enclosed by the Gaussian surface. In other words, the probability density is conserved; any leakage of probability outside of an enclosed volume must be equal to the loss of probability density within the volume. The student may recall that in electromagnetics we again encounter a continuity equation. In this case, the outward flux of electrical current density across a Gaussian surface must be equal to the time rate of change of the charge density enclosed within the volume.

The Schroedinger equation for Ψ is simply given as

$$-\hbar^2/2m\nabla^2\Psi + V\Psi = E\Psi = i\,\hbar\partial\Psi/\partial t. \tag{1.5.2}$$

Multiply both sides of the Schroedinger equation by Ψ^*, the complex conjugate of Ψ. Next we take the complex conjugate of the resulting equation. The two equations obtained are

$$\frac{-\hbar^2}{2m}\Psi^*\nabla^2\Psi + \Psi^*V\Psi = i\,\hbar\Psi^*\partial\Psi/\partial t, \tag{1.5.3}$$

$$\frac{-\hbar^2}{2m}\Psi\nabla^2\Psi^* + \Psi V\Psi^* = -i\,\hbar\Psi\partial\Psi^*/\partial t. \tag{1.5.4}$$

Subtract Eq. (1.5.4) from Eq. (1.5.3) to yield

$$\frac{-\hbar^2}{2m}[\Psi^*\nabla^2\Psi - \Psi\nabla^2\Psi^*] = i\,\hbar[\Psi^*\partial\Psi/\partial t + \Psi\partial\Psi^*/\partial t]. \tag{1.5.5}$$

Expanding out the derivative of $\Psi^*\Psi$ with respect to time gives

$$i\,\hbar\partial/\partial t(\Psi^*\Psi) = i\,\hbar[\Psi^*\partial\Psi/\partial t + \Psi\partial\Psi^*/\partial t]. \tag{1.5.6}$$

Therefore we can replace the right-hand side of Eq. (1.5.6) with $i\,\hbar\partial/\partial t(\Psi^*\Psi)$, obtaining,

$$\frac{\hbar^2}{2m}[\Psi^*\nabla^2\Psi - \Psi\nabla^2\Psi^*] + i\,\hbar\partial/\partial t(\Psi^*\Psi) = 0. \tag{1.5.7}$$

Equation (1.5.7) can be simplified through use of the following vector identity. Given an arbitrary vector $\vec{a}$ multiplied by the scalar Ψ, the divergence of this quantity is

$$\vec{\nabla}\cdot(\Psi\vec{a}) = \vec{a}\cdot\vec{\nabla}\Psi + \Psi\vec{\nabla}\cdot\vec{a}. \tag{1.5.8}$$

Therefore, if $\vec{a}$ is $\vec{\nabla}\Psi$, then

$$\vec{\nabla}\cdot(\Psi^*\vec{\nabla}\Psi - \Psi\vec{\nabla}\Psi^*) = \vec{\nabla}\Psi\cdot\vec{\nabla}\Psi^* + \Psi^*\nabla^2\Psi - \vec{\nabla}\Psi^*\cdot\vec{\nabla}\Psi - \Psi\nabla^2\Psi^*. \tag{1.5.9}$$

As a result, the first term in Eq. (1.5.7) simplifies to

$$\vec{\nabla}\cdot[\Psi^*\vec{\nabla}\Psi - \Psi\vec{\nabla}\Psi^*] = \Psi^*\nabla^2\Psi - \Psi\nabla^2\Psi^*. \tag{1.5.10}$$

Combining these results together, we obtain

$$\partial/\partial t(\Psi^*\Psi) + \frac{\hbar}{2mi}\vec{\nabla}\cdot[\Psi^*\vec{\nabla}\Psi - \Psi\vec{\nabla}\Psi^*] = 0. \tag{1.5.11}$$

This has precisely the form of the continuity equation defined above. Given that the probability density ρ is defined as the product of $\Psi^*\Psi$, then, in order to agree with the continuity equation, the probability current density must be defined as

$$\vec{j} = \frac{\hbar}{2mi}(\Psi^*\vec{\nabla}\Psi - \Psi\vec{\nabla}\Psi^*). \tag{1.5.12}$$

With this definition of the probability current density, Eq. (1.5.11) becomes

$$\vec{\nabla}\cdot\vec{j} = -\frac{\partial\rho}{\partial t}. \tag{1.5.13}$$

Next let us consider the time derivative of $\langle x \rangle$, the expectation value of x. Classically, we expect that the time rate of change of the position of a particle is simply equal to its velocity. To see what happens quantum mechanically, we note that the time rate of change of the expectation value of x in one dimension is written mathematically as

$$\frac{\partial}{\partial t}\langle x \rangle = \frac{\partial}{\partial t}\int \Psi^* x\Psi\,dx = \int x\left(\frac{\partial\rho}{\partial t}\right)dx. \tag{1.5.14}$$

The time derivative operates on only the wave functions and not on the operator x. This is because the only time-dependent quantities are the wave functions. As we will see in Chapter 4, the time dependence of the problem can be folded completely into either the wave functions or the operators, leading to two different, although equivalent, pictures or formulations of quantum mechanics. These pictures are called the Schroedinger and the Heisenberg pictures, respectively. The time rate of change of ρ can be reexpressed as the divergence of $\vec{j}$ by use of Eq. (1.5.12). Substituting Eq. (1.5.13) into Eq. (1.5.14) yields

$$d/dt\langle x \rangle = -\int x(\vec{\nabla}\cdot\vec{j})\,d^3r. \tag{1.5.15}$$

Integrating by parts and simplifying to one dimension (for mathematical convenience, it is easier to work with only one component; no loss of generality occurs), we obtain

$$u \equiv x, \qquad v \equiv j_x,$$

$$du \equiv dx, \qquad dv \equiv \frac{dj_x}{dx}dx,$$

$$d/dt\langle x \rangle = -xj_x|_{\text{surface}} + \int j_x\,dx. \tag{1.5.16}$$

The integration is over all space. The surface at which the current density is evaluated in the first term above is the surface of the Gaussian sphere that encloses all space. The probability current density at infinity must vanish since the probability of finding the electron outside of the universe must be zero. The probability current density approaches zero faster than x approaches infinity. As a result, the product of x and j_x is zero. Subsequently the surface term vanishes. Therefore the time rate of change of $\langle x \rangle$ is given as

$$d/dt\langle x \rangle = \int j_x \, dx. \tag{1.5.17}$$

Extending this to three dimensions yields

$$d/dt\langle \vec{r} \rangle = \int \vec{j} \, d^3r. \tag{1.5.18}$$

If we now substitute

$$\vec{j} = \frac{\hbar}{2mi}(\Psi^* \vec{\nabla} \Psi - \vec{\nabla} \Psi^* \Psi) \tag{1.5.19}$$

in for $\vec{j}$, $d/dt\langle \vec{r} \rangle$ becomes

$$d/dt\langle \vec{r} \rangle = \frac{\hbar}{2mi} \int (\Psi^* \vec{\nabla} \Psi - \vec{\nabla} \Psi^* \Psi) \, d^3r. \tag{1.5.20}$$

Multiplying by the mass m results in

$$m \, d/dt\langle \vec{r} \rangle = \frac{\hbar}{2i} \int \Psi^* \vec{\nabla} \Psi \, d^3r - \frac{\hbar}{2i} \int \vec{\nabla} \Psi^* \Psi \, d^3r. \tag{1.5.21}$$

Integrating the last term above by parts, again by first considering only one dimension, we obtain

$$u \equiv \Psi, \qquad\qquad v \equiv \Psi^*,$$
$$du \equiv \frac{d\Psi}{dx}dx, \qquad dv \equiv \frac{d\Psi^*}{dx}dx,$$

$$-\frac{\hbar}{2i} \int \frac{d\Psi^*}{dx} \Psi \, dx = -\frac{\hbar}{2i} \left[\Psi\Psi^*|_{\text{surface}} - \int \Psi^* \frac{d\Psi}{dx} dx \right]. \tag{1.5.22}$$

Substituting this back into the equation for $m \, d/dt\langle x \rangle$, we obtain

$$m \, d/dt\langle x \rangle = \frac{\hbar}{2i} \int \Psi^* \frac{d\Psi}{dx} dx - \frac{\hbar}{2i} \Psi\Psi^*|_{\text{surface}} + \frac{\hbar}{2i} \int \Psi^* \frac{d\Psi}{dx} dx. \tag{1.5.23}$$

Combining the first and the last terms on the right-hand side of Eq. (1.5.23) and recognizing that the surface term must vanish again, we obtain

$$m \, d/dt\langle x \rangle = \frac{\hbar}{i} \int \Psi^* \frac{d\Psi}{dx} dx = \int \Psi^* \frac{\hbar}{i} \frac{d\Psi}{dx} dx. \tag{1.5.24}$$

Note that the left-hand side of Eq. (1.5.24) is simply the mass times the average velocity. From Newtonian mechanics, we recognize that the product of the mass and the average velocity is simply the mean linear momentum p. If the Correspondence Principle is used (the laws of quantum mechanics must reduce to those of classical mechanics in the range in which classical mechanics holds), then $m\,d/dt\langle x\rangle$ must be equal to the mean linear momentum along the x direction in quantum mechanics as well. Since the right-hand side of Eq. (1.5.24) is a quantum-mechanical expression, it must give the quantum-mechanical description of the mean linear momentum or its expectation value. In other words, the expectation value of the linear momentum in quantum mechanics must be given as (generalizing to three dimensions)

$$\langle \vec{p}\,\rangle = \int \Psi^* \frac{\hbar}{i}\vec{\nabla}\Psi\, d^3r. \tag{1.5.25}$$

From our previous definition of the expectation value of any physical observable $f(r)$,

$$\langle f(r)\rangle = \int \Psi^* f(r)\Psi\, d^3r, \tag{1.5.26}$$

the linear momentum in quantum mechanics must correspond to the differential operator $\hbar/i\vec{\nabla}$. We say that the term $\hbar/i\vec{\nabla}$ is the quantum-mechanical linear-momentum operator.

Before we discuss the nature of quantum-mechanical operators in general, it is useful at this point to examine the value of $d/dt\langle \vec{p}\,\rangle$. In classical mechanics, the time rate of change of the linear momentum is equal to the net external force acting on the object. For conservative forces, this can be written as the negative gradient of the potential energy. Therefore, when we evaluate $d/dt\langle p\rangle$ in quantum mechanics, we expect to obtain something that resembles the negative gradient of the potential energy. Consider then

$$d/dt\langle \vec{p}\,\rangle = d/dt \int \Psi^* \frac{\hbar}{i}\vec{\nabla}\Psi\, d^3r. \tag{1.5.27}$$

The Schroedinger equation and its complex conjugate are

$$i\hbar \frac{d}{dt}\Psi = -\frac{\hbar^2}{2m}\nabla^2\Psi + V\Psi, \tag{1.5.28}$$

$$-i\hbar \int \frac{d}{dt}\Psi^* = -\frac{\hbar}{2m}\nabla^2\Psi^* + V\Psi^*. \tag{1.5.29}$$

For simplicity, consider only the x component of p. No loss of generality occurs since a cyclic permutation of the components of p can be made to give the result in three dimensions. Hence,

$$d/dt\langle p_x\rangle = \int \left(\frac{d}{dt}\Psi^*\right)\frac{\hbar}{i}\frac{d}{dx}\Psi\, d^3r + \int \Psi^* \frac{\hbar}{i}\frac{d}{dx}\left(\frac{d\Psi}{dt}\right)d^3r$$

$$= -i\hbar \int \frac{d\Psi^*}{dt}\frac{d\Psi}{dx}\, d^3r - i\hbar \int \Psi^* \frac{d}{dx}\left(\frac{d\Psi}{dt}\right)d^3r. \tag{1.5.30}$$

We can use the Schroedinger equation and its complex conjugate to substitute in for the time derivatives of Ψ and Ψ^*. After these substitutions are made, the time rate of change of the x component of the momentum is

$$d/dt\langle p_x \rangle = -\frac{\hbar^2}{2m} \int \nabla^2 \Psi^* \frac{d\Psi}{dx} d^3r + \frac{\hbar^2}{2m} \int \Psi^* \frac{d}{dx} \nabla^2 \Psi \, d^3r$$
$$+ \int V\Psi^* \frac{d\Psi}{dx} d^3r - \int \Psi^* \frac{d}{dx}(V\Psi) d^3r. \tag{1.5.31}$$

Combining terms, we obtain

$$d/dt\langle p_x \rangle = -\frac{\hbar^2}{2m} \int \left[\nabla^2 \Psi^* \frac{d\Psi}{dx} - \Psi^* \nabla^2 \frac{d\Psi}{dx} \right] d^3r$$
$$+ \int \left[V\Psi^* \frac{d\Psi}{dx} - \Psi^* \frac{d(V\Psi)}{dx} \right] d^3r. \tag{1.5.32}$$

To simplify the above expression further we need to recall Green's theorem from advanced calculus or electromagnetics. According to Green's theorem, a surface integral can be transformed into a volume integral and vice versa by use of the following prescription,

$$\int_{\text{surface}} (u\vec{\nabla}v - v\vec{\nabla}u) \cdot d\vec{S} = \int_{\text{volume}} (u\nabla^2 v - v\nabla^2 u) \, d^3r. \tag{1.5.33}$$

Green's theorem can be used to simplify the expression for the time rate of change of p_x by making the assignments

$$u \equiv \frac{d\Psi}{dx}, \qquad v \equiv \Psi^*. \tag{1.5.34}$$

Then the first term on the right-hand side of Eq. (1.5.32) is

$$\int \left[\nabla^2 \Psi^* \frac{d\Psi}{dx} - \Psi^* \nabla^2 \frac{d\Psi}{dx} \right] d^3r = \int [u\nabla^2 v - v\nabla^2 u] \, d^3r, \tag{1.5.35}$$

which must be equal to the surface integral

$$\int_{\text{surface}} \left(\frac{d\Psi}{dx} \vec{\nabla}\Psi^* - \Psi^* \vec{\nabla} \frac{d\Psi}{dx} \right) \cdot d\vec{S} = 0 \tag{1.5.36}$$

by Green's theorem. The surface is the boundary at infinity that encloses all space. The wave function and its first derivative must be bounded, otherwise this would imply that there is some probability of finding the particle outside of the bounds of the universe. Subsequently the values of Ψ and $d\Psi/dx$ must vanish along the surface. Therefore the only surviving terms in the expression for the derivative of p_x are those that involve V. Noting this and expanding out the derivative with respect to x of $V\Psi$, we find that Eq. (1.5.32) becomes

$$d/dt\langle p_x \rangle = \int \left(V\Psi^* \frac{d\Psi}{dx} - \Psi^* V \frac{d\Psi}{dx} - \Psi^* \Psi \frac{dV}{dx} \right) d^3r. \tag{1.5.37}$$

The first two terms on the right-hand side of Eq. (1.5.37) subtract out, leaving

$$d/dt\langle p_x \rangle = -\int \Psi^* \frac{dV}{dx} \Psi \, d^3r = -\langle \nabla_x V \rangle = \langle F_x \rangle. \tag{1.5.38}$$

Equation (1.5.38) is just Newton's Second Law of Motion; the time rate of change of the linear momentum is equal to the mean or the expectation value of the gradient of the potential energy or the force. Therefore we have shown that the quantum-mechanical description of the expectation value of the linear momentum is consistent with the classical laws of motion. In addition, we have shown that the classical observable p, the linear momentum, is replaced in quantum mechanics by a differential operator $\hbar/i\nabla$ and that use of this formalism leads to the correct physics.

The above discussion can be generalized in the following way. For each physically observable quantity there exists a differential operator associated with it. Furthermore, we determine the expectation value of the physical observable by applying the operator onto the wave function, multiplying by the complex conjugate of the wave function, and integrating over all space; we find the expectation value from the usual prescription given by Eq. (1.4.10). The function $f(r)$ in our definition of the expectation value [Eq. (1.4.10)] is replaced by its associated differential operator. In the case of linear momentum, the operator associated with p is simply $\hbar/i\nabla$. The same relationship holds between the quantum-mechanical operators as that between the corresponding classical physical quantities. In other words, if a physical observable, the kinetic energy for example, can be represented in terms of other physical quantities, in this case momentum and mass $p^2/2m$, then the equivalent quantum-mechanical operator for the kinetic energy is found by use of the same relationship. This is required from the Correspondence Principle. We are led then to a general prescription: In going from classical mechanics to quantum mechanics, we first express the physical observable of interest in terms of observables for which the corresponding quantum-mechanical operator is known. Next we substitute in the quantum-mechanical expressions for these variables. The resulting expression is the quantum-mechanical operator for the particular observable of interest.

Using as an example the kinetic energy, we can determine its equivalent quantum-mechanical operator as follows. We start by expressing the kinetic energy in terms of observables whose quantum-mechanical operators are known, in this case $p^2/2m$. We substitute in the quantum-mechanical operator $\hbar/i\nabla$ for p. The resulting quantum-mechanical operator expression for the kinetic energy

T is

$$-\frac{\hbar^2}{2m}\nabla^2. \tag{1.5.39}$$

The quantum-mechanical operator associated with the position r is simply itself. In one dimension then the operator associated with x is x. In this book, the potentials we consider are only a function of position. Therefore the potential $V(x)$ transforms quantum mechanically into $V(x)$. With this result, the total energy H of a particle moving in a potential $V(x)$ is given as the sum of the kinetic and the potential energies:

Table 1.5.1
Classical Physical Observables and Their Associated Quantum-Mechanical Operators

Physical Quantity	Operator
Momentum p	$\hbar/i \nabla$
Position r	r
Kinetic energy T	$-\hbar^2/2m \, \nabla^2$
Potential energy V	$V(r)$
Total energy H	$-\hbar^2/2m \, \nabla^2 + V(r)$
Total energy E	$i\hbar\partial/\partial t$

$$H = T + V; \qquad H = -\frac{\hbar^2}{2m}\nabla^2 + V(x). \tag{1.5.40}$$

A small list of physical observables and their corresponding quantum-mechanical operator expressions is presented in Table 1.5.1.

1.6 **Properties of Operators**

Having defined what we mean by an operator in quantum mechanics, let us next consider their mathematical properties. First let us consider the properties of an operator that corresponds to a real physical observable. Physically, the expectation value of any real physical quantity must be a real number. This is because the expectation value is the mean value of the observable measured in a particular quantum-mechanical state. The quantity measured must have physical meaning and as such must therefore be real. Consider the expectation value of the quantity F. This is given by

$$\langle F \rangle = \int \Psi^* F \Psi \, d^3r. \tag{1.6.1}$$

Given that F is a physical observable, then $\langle F \rangle$ is a real number. If $\langle F \rangle$ is a real number, then it must be equal to its complex conjugate. Hence we must have

$$\langle F \rangle = \langle F \rangle^*. \tag{1.6.2}$$

Substituting in the integral expressions for $\langle F \rangle$ and $\langle F \rangle^*$, we obtain

$$\langle F \rangle = \int \Psi^*(F\Psi)\, d^3r, \qquad \langle F \rangle^* = \int (F\Psi)^*\Psi \, d^3r. \tag{1.6.3}$$

From the equality of the expectation value of F and its complex conjugate,

$$\int \Psi^*(F\Psi)\, d^3r = \int (F\Psi)^*\Psi \, d^3r \tag{1.6.4}$$

must be satisfied for any operator F that corresponds to a real physical observable. If an operator F satisfies this condition we say that it is an Hermitian operator, and the property reflected in Eq. (1.6.5) is called hermiticity.

An operator $F^\dagger$ is said to be the adjoint operator of F if $F^\dagger$ has the following property:

$$\int \Psi^*(F\Psi)\, d^3r = \int (F^\dagger \Psi)^* \Psi\, d^3r, \tag{1.6.5}$$

where the complex conjugate is taken of the entire quantity within the parentheses. If $F^\dagger = F$, then F is said to be self-adjoint. Note that if an operator is self-adjoint it must also be Hermitian. In fact we say that a Hermitian operator is self-adjoint. The importance of knowing that an operator is self-adjoint or Hermitian is that that operator then must correspond to a real physical observable. Conversely, if an operator corresponds to a real physical observable, it must be Hermitian.

Box 1.6.1 Generalization of Hermiticity

The definition of hermiticity can be generalized from the definition provided by Eq. (1.6.4) in the following way. If Ψ_1 and Ψ_2 are two admissible wave functions that are not necessarily the same, then the definition of hermiticity can be generalized as

$$\int_{-\infty}^{+\infty} (A\Psi_1)^* \Psi_2\, dx = \int_{-\infty}^{+\infty} \Psi_1^* A\Psi_2\, dx.$$

This assertion can be proved as follows. Let Ψ be an arbitrary wave function that is formed by a linear combination of Ψ_1 and Ψ_2 as

$$\Psi = a_1 \Psi_1 + a_2 \Psi_2,$$

where a_1 and a_2 are coefficients. From the definition of hermiticity, a Hermitian operator A must satisfy

$$\int (A\Psi)^* \Psi\, dx = \int \Psi^*(A\Psi)\, dx$$

in the state Ψ. Substituting the linear combination given above in for Ψ, we find that the left-hand side of the above equation becomes

$$\int [A(a_1\Psi_1 + a_2\Psi_2)]^*(a_1\Psi_1 + a_2\Psi_2)\, dx$$

$$= \int a_1^2 (A\Psi_1)^* \Psi_1\, dx + \int a_2^2 (A\Psi_2)^* \Psi_2\, dx + \int a_1^* a_2 (A\Psi_1)^* \Psi_2\, dx$$

$$+ \int a_1 a_2^* (A\Psi_2)^* \Psi_1\, dx.$$

The right-hand side is

$$\int (a_1\Psi_1 + a_2\Psi_2)^* [A(a_1\Psi_1 + a_2\Psi_2)]\, dx$$

$$= \int a_1^2 \Psi_1^* A\Psi_1\, dx + \int a_2^2 \Psi_2^* A\Psi_2\, dx + \int a_1^* a_2 \Psi_1^* A\Psi_2\, dx + \int a_1 a_2^* \Psi_2^* A\Psi_1\, dx.$$

The left- and the right-hand sides of the above equations are equal since A is Hermitian.

Note that the terms involving a_1^2 and a_2^2 subtract out because of the hermiticity of A. The resulting expression after rearranging is

$$a_1^* a_2 \left[\int (A\Psi_1)^* \Psi_2 \, dx - \int \Psi_1^* A\Psi_2 \, dx \right] = a_1 a_2^* \left[\int (A\Psi_2)^* \Psi_1 \, dx - \int \Psi_2^* A\Psi_1 \, dx \right].$$

But the left-hand side is equal to the complex conjugate of the right-hand side. The coefficients a_1 and a_2 are arbitrary. The phase of each of the right- and the left-hand sides is arbitrary. As a result, the only way in which the equality can hold is if each side is identically equal to zero. Subsequently the general property of hermiticity is proved.

As an example of hermiticity, let us show that the linear momentum is Hermitian by showing that $p = p^\dagger$. We consider only one dimension. The general result in three dimensions can be obtained by a simple permutation of variables. The linear-momentum operator in one dimension is defined as

$$p = \frac{\hbar}{i} \frac{d}{dx}. \tag{1.6.6}$$

To demonstrate that p is Hermitian, p must satisfy Eq. (1.6.5). Therefore, for p to be Hermitian, the following equation must be true:

$$\int \Psi^* \left(\frac{\hbar}{i} \frac{d}{dx} \Psi \right) dx - \int \left(\frac{\hbar}{i} \frac{d}{dx} \Psi \right)^* \Psi \, dx. \tag{1.6.7}$$

To show this equality, let us consider the right-hand side. Taking the complex conjugate of $\hbar/i$ gives $-\hbar/i$. The right-hand side is then

$$-\frac{\hbar}{i} \int \frac{d\Psi^*}{dx} \Psi \, dx. \tag{1.6.8}$$

Integrating this expression by parts using the assignments

$$u = \Psi, \qquad\qquad v = \Psi^*,$$
$$du = \frac{d\Psi}{dx} \, dx, \qquad dv = \frac{d\Psi^*}{dx} \, dx,$$

yields

$$= -\frac{\hbar}{i} \Psi \Psi^* \Big|_{\text{surface}} + \frac{\hbar}{i} \int \Psi^* \frac{d\Psi}{dx} \, dx. \tag{1.6.9}$$

However, the surface term vanishes as before, since the wave function must be bounded at $\pm\infty$. With this condition, the left- and the right-hand sides of Eq. (1.6.7) become

$$\frac{\hbar}{i} \int \Psi^* \frac{d\Psi}{dx} = \frac{\hbar}{i} \int \Psi^* \frac{d\Psi}{dx}, \tag{1.6.10}$$

showing the equivalence of p and $p^\dagger$.

We next ask the question, what must be the form of the state Ψ such that it has a precise, well-defined, sharp value of the physical observable corresponding to the operator A? In other words, we seek the nature of a quantum-mechanical state that has a definite value of a particular physical observable corresponding to the operator A. When we state that Ψ is in a definite state of A, this means that it has no variance. Therefore, from our definition of the variance, the variance of A is

$$(\Delta A)^2 = \langle (A - \langle A \rangle)^2 \rangle = \langle A^2 \rangle - \langle A \rangle^2. \tag{1.6.11}$$

In this case, the variance $(\Delta A)^2$ is simply zero. Therefore

$$\int \Psi^*(A - \langle A \rangle)^2 \Psi \, d^3r = 0. \tag{1.6.12}$$

If A is Hermitian (which it must be if A corresponds to a physical observable), then, by using the property stated by Eq. (1.6.5), we obtain

$$\int [(A - \langle A \rangle)\Psi]^* \, [(A - \langle A \rangle)\Psi] \, d^3r = 0. \tag{1.6.13}$$

Therefore

$$\int |(A - \langle A \rangle)\Psi|^2 \, d^3r = 0. \tag{1.6.14}$$

For the above to equal zero in general, the expression within the verticals must be zero. This yields

$$(A - \langle A \rangle)\Psi = 0, \qquad A\Psi = \langle A \rangle \Psi. \tag{1.6.15}$$

Since $\langle A \rangle$ is the expectation value of the operator A in the state Ψ, it is a number and is often just written as A'. Therefore the state Ψ must satisfy

$$A\Psi = A'\Psi \tag{1.6.16}$$

to be in a definite state of the operator A. The number A' is a real number and is called the eigenvalue of A, which is simply equal to the expectation value of A, provided that the wave function Ψ is an eigenstate of A. The state Ψ is said to be an eigenfunction of the operator A and is in a sharp, well-defined state of A.

The operator equation $A\Psi = A'\Psi$ is called an eigenvalue equation. Physically, it means the following. As we have seen above, if the state Ψ satisfies $A\Psi = A'\Psi$, then the state Ψ is in a sharp state of A, and the value of the observable corresponding to the operator A in the state Ψ is A'. For example, if A is assumed to be p (the operator is $\hbar/i \, d/dx$), the linear-momentum operator, and Ψ is taken equal to e^{ikx}, then the resulting eigenvalue equation is

$$\frac{\hbar}{i} \frac{d}{dx} e^{ikx} = p' e^{ikx}, \tag{1.6.17}$$

where p' is the eigenvalue of e^{ikx}. p' can be determined when the operator is applied to e^{ikx} as

$$\frac{\hbar}{i}\frac{d}{dx}e^{ikx} = \hbar k\, e^{ikx}, \tag{1.6.18}$$

which yields $\hbar k$ as the eigenvalue of p. Note that $\hbar k$ is simply a number and it physically represents the value of the momentum in the state e^{ikx}. Therefore the state e^{ikx} has a sharp well-defined value of the momentum $\hbar k$ and is an eigenfunction of p. This is not surprising since e^{ikx} is a plane-wave state and has a definite wavelength (and therefore momentum).

We can think of an eigenvalue equation as the mathematical representation of a measurement. The operator A, when applied onto a state Ψ, measures the value of the physical observable that corresponds to A in the state Ψ. The value of A in the state Ψ is the eigenvalue of A and physically represents the value of that observable when it is measured in the state Ψ. If a number is obtained after A is applied onto Ψ, an eigenvalue is obtained; then the state Ψ is a definite state of A.

From the definition of an eigenvalue equation, we can construct the Schroedinger equation. The Schroedinger equation can be understood as an eigenvalue equation for the total energy of a quantum-mechanical particle. The total energy operator is given from Table 1.5.1 as $H = T + V$, where T is the kinetic energy and V is the potential energy. T is defined as $p^2/2m$. Substituting the operator form of p in for p^2 in T, we obtain the total energy operator:

$$H = -\frac{\hbar^2}{2m}\nabla^2 + V. \tag{1.6.19}$$

If the eigenvalue of H is E, then the eigenvalue equation for the total energy is

$$H\Psi = E\Psi,$$
$$-\frac{\hbar^2}{2m}\nabla^2\Psi + V\Psi = E\Psi, \tag{1.6.20}$$

which is precisely the Schroedinger equation defined in Eq. (1.4.11).

The Schroedinger equation can be interpreted in the following way. If the state Ψ satisfies the condition $H\Psi = E\Psi$, then Ψ is in a definite state of H with a well-defined energy E. The energy E is the eigenvalue of H and is the actual value of the energy that is obtained when the energy of the state Ψ is measured. Therefore the Schroedinger equation can be looked at as mathematically stating the condition under which a state Ψ has a definite value of the total energy. As we have seen from the above, we obtain the differential form of Schroedinger's equation by substituting in the quantum mechanical operator forms for T and V following the rules outlined in Section 1.5. Let us next consider some properties of eigenfunctions and eigenfunction equations.

Two different eigenfunctions Ψ_1 and Ψ_2 that correspond to distinct eigenvalues A'_1 and A'_2 of a single Hermitian operator A must be orthogonal. This statement can be proved as follows. If Ψ_1 and Ψ_2 are eigenfunctions of A, then

$$A\Psi_1 = A_1'\Psi_1,$$
$$A\Psi_2 = A_2'\Psi, \tag{1.6.21}$$

where A_1' and A_2' are eigenvalues of A. Multiplying the first equation of Eqs. (1.6.21) by Ψ_2^* and taking the complex conjugate of the second equation and multiplying the result by Ψ_1 yields

$$\Psi_2^* A\Psi_1 = \Psi_2^* A_1' \Psi_1,$$
$$\Psi_1(A\Psi_2)^* = \Psi_1(A_2'\Psi_2)^*. \tag{1.6.22}$$

But the complex conjugate of the eigenvalue A_2' is simply equal to itself, since A_2' must be a real number. Hence

$$(A_2'\Psi_2)^* = A_2'\Psi_2^*. \tag{1.6.23}$$

Subtracting the second equation of Eqs. (1.6.22) from the first and making use of the hermiticity of A yields

$$\int [\Psi_2^* A\Psi_1 - \Psi_2^* A\Psi_1]\, d^3r = \int [(A_1' - A_2')\Psi_1\Psi_2^*]\, d^3r. \tag{1.6.24}$$

Clearly the left-hand side is equal to zero. Therefore the right-hand side must also be equal to zero. This implies that either A_1' is equal to A_2', in other words the eigenvalues are not distinct, in contradiction to the original assumption, or that

$$\int \Psi_2^*\Psi_1\, d^3r = 0, \tag{1.6.25}$$

which simply states that Ψ_2 and Ψ_1 are orthogonal. In summary, what has been demonstrated is that eigenfunctions corresponding to two different eigenvalues of the same operator must be orthogonal. This is an important theorem because it states that all the eigenfunctions, with unique eigenvalues, of a Hermitian operator (one that corresponds to a real, physical observable) must be linearly independent. No one of the eigenfunctions can be expressed as a linear combination of the others, because each eigenfunction is orthogonal to all the others. Therefore if there are n eigenfunctions of a particular operator, these eigenfunctions are all mutually orthogonal. If these eigenfunctions completely span the space, then they form a basis set. A set of functions that completely spans the space is said to be a complete set.

The eigenfunctions in quantum mechanics are defined to be square-integrable functions in configuration space. What this means is that the eigenfunctions are defined in a space that possesses certain properties. As discussed in Section 1.7, the space is linear and complete. By complete we mean that the full set of eigenfunctions spans the space. Since the eigenfunctions of a Hermitian operator are mutually orthonormal and they are complete, they must form a basis set. Therefore, the complete set of eigenfunctions of any Hermitian operator is guaranteed to form a basis set.

To see the importance of this result, let us reiterate the mathematical definition of a basis set. A basis set is defined as a complete set of vectors that completely span the space and are mutually orthonormal. The most common example of a basis set is the set of unit vectors $\hat{x}$, $\hat{y}$, and $\hat{z}$ used in describing three-dimensional space. These three vectors completely span three-dimensional space and are orthonormal. Therefore any three-dimensional vector can be written as a linear combination of these three basis vectors. We are not limited to just the three vectors $\hat{x}$, $\hat{y}$, and $\hat{z}$, however, in describing a general vector in three-dimensional space. Other choices, such as the spherical unit vectors $\hat{r}$, $\hat{\theta}$, and $\hat{\phi}$ can be used as well.

Another example of a basis set that is familiar to us is the set of Fourier functions. As we know from elementary circuit theory, any periodic function can be expanded as a linear combination of the Fourier functions; any periodic function can be expanded in terms of a Fourier series. The Fourier functions form a complete set and are all mutually orthogonal. As such they form a basis set from which any periodic function can be expanded. The use of orthogonal functions, such as the Fourier series, in the expansion of an arbitrary function is the basis of digital signal processing. For example, an auditory signal can be expanded in terms of its Fourier components. In general, any arbitrary waveform can be digitized if it is reduced into a sum of discrete components or frequencies through use of an orthonormal Fourier series expansion.

One might now ask the question, how do we treat a state that is not an eigenstate of a specific operator? From our above discussion, we have found that a wave function that satisfies an eigenvalue equation for an operator A is in a definite state of A; it has a definite value of the physical observable that corresponds to A. However, how do we describe a state that is not in an eigenstate of an operator, one that does not obey an eigenvalue equation? Let us consider a state Ψ that does not satisfy the eigenvalue equation for A. From our above discussion, it is clear that the state Ψ can be expanded in terms of the eigenfunctions of A since these functions form a basis set. The complete set of eigenfunctions of any Hermitian operator forms a basis. Therefore any arbitrary function Ψ can be expressed as a linear combination of the complete set of eigenfunctions of any Hermitian operator. Assuming that the eigenfunctions Ψ_i of the operator A are known and they are orthonormal, Ψ can be expanded then as

$$\Psi = \sum c_i \psi_i, \tag{1.6.26}$$

where c_i are the coefficients in the expansion. The expectation value of A in the state Ψ can then be determined from the definition of the expectation value as

$$\langle A \rangle = \int \Psi^* A \Psi \, d^3r = \int \left(\sum c_i^* \psi_i^* \right) A \sum c_i \Psi_i \, d^3r. \tag{1.6.27}$$

But the coefficients, c_i and c_i^* are simply numbers and can be moved outside of the integral. The expectation value of A becomes

$$\langle A \rangle = \int \psi_i^* A \psi_i \, d^3r \sum c_i^* c_i. \tag{1.6.28}$$

We can simplify this further by recognizing that since all the ψ_i are eigenfunctions of A, each of these functions satisfies $A\psi_i = A_i' \psi_i$. Making this substitution, we can express the expectation value of A as

$$\langle A \rangle = \sum |c_i|^2 A_i' \int \psi_i^* \psi_i \, d^3r. \tag{1.6.29}$$

Since the eigenfunctions ψ_i are all orthonormalized, the integral simply gives 1 for each of the ith wave functions. $\langle A \rangle$ then is

$$\langle A \rangle = \sum |c_i|^2 A_i'. \tag{1.6.30}$$

Equation (1.6.30) states that the expectation value of an operator within an arbitrary state is given by the sum over all the eigenvalues times the square of the coefficients in the expansion.

How, though, are the coefficients in the expansion of Ψ determined? To find the c_i's in the expansion of Ψ we make use of the orthogonality of the basis eigenfunctions. From the definition of orthogonality,

$$\int \psi_i^* \psi_j \, d^3r = \delta_{ij}, \tag{1.6.31}$$

where $\delta_{ij} = 0$ if $i \neq j$ and is 1 if $i = j$. Therefore we can find the jth coefficient in the expansion of Ψ by multiplying the wave function by the jth eigenstate,

$$\psi_j^* \Psi = \sum c_i \psi_j^* \psi_i \tag{1.6.32}$$

and integrating over all space. This leads to

$$\int \psi_j^* \Psi \, d^3r = \sum \int c_i \psi_j^* \psi_i \, d^3r = \sum c_i \delta_{ij}, \tag{1.6.33}$$

where the orthogonality condition has been used on the right-hand side. The sum over i of $c_i \delta_{ij}$ is simply c_j. Therefore

$$c_j = \int \psi_j^* \Psi \, d^3r. \tag{1.6.34}$$

What though is the meaning of c_j? c_j was defined above as a coefficient in the expansion of an arbitrary wave function Ψ in terms of the eigenfunctions ψ_i of a Hermitian operator A. The key point in understanding the meaning of the coefficients c_j is to recognize that when a measurement is performed on any given arbitrary state, then that state must collapse into a definite eigenstate of the operator. In other words, when a measurement is performed, only a definite state can be observed, even if the initial state is not a definite state. The initial state must collapse into a definite state. This can be understood from the spinning coin. While the coin is spinning (an arbitrary state), it is neither heads nor tails. However, once the coin is caught (a measurement is made), then the state of the coin collapses into one of the two possibilities, heads or tails. In the language of quantum mechanics, heads and tails are the two eigenstates of the flipping operator. While the coin is in the air it is in an arbitrary state. Once the coin is caught it is found in only one of the two possible states with 100% certainty; it has collapsed into one of the two eigenstates of the flipping operator.

To see how this relates to the meaning of the coefficients c_i in the expansion, consider the following. If the wave function Ψ is normalized, then $\Psi^*\Psi$ yields

$$\int \Psi^*\Psi \, d^3r = 1 = \int \sum_i c_i^*\psi_i^* \sum_i c_i\psi_i \, d^3r = \sum_i |c_i|^2 \int \psi_i^*\psi_i \, d^3r. \quad (1.6.35)$$

But the eigenfunctions ψ_i are orthonormal. Therefore the integral of $\psi_i^*\psi_i$ is simply 1. As a result,

$$\sum_i |c_i|^2 = 1. \quad (1.6.36)$$

The sum of the squares of the coefficients in the expansion of Ψ is normalized. Therefore each squared coefficient represents some sort of probability. Writing out the product of $\Psi^*\Psi$ in terms of each of the eigenfunctions gives

$$\int \Psi^*\Psi \, d^3r = \int \left[c_1^2\psi_1^2 + c_2^2\psi_2^2 + \cdots + c_n^2\psi_n^2 \right] d^3r. \quad (1.6.37)$$

Recall that $\Psi^*\Psi$ physically represents a probability density. This implies in turn that the sum

$$c_1^2\psi_1^2 + c_2^2\psi_2^2 + \cdots + c_n^2\psi_n^2 \quad (1.6.38)$$

is also a probability density. Each of the squared coefficients represents the probability of finding Ψ in one of the possible eigenstates ψ_i.

Perhaps a better understanding of the meaning of the squared coefficients c_i^2 is obtained by an analogy to digital signal processing. In digital signal processing, an arbitrary signal is decomposed into its Fourier components. The squares of the coefficients in the decomposition represent how much of each Fourier component is present in the signal. Provided that the sum of the squares of the coefficients in the decomposition is normalized, each squared coefficient represents the relative proportion of that Fourier component present in the signal. In the quantum-mechanical decomposition considered here, the squares of the coefficients represent the probability of finding the state Ψ in any one of the eigenstates ψ_i after a measurement has been made on Ψ.

The above discussion leads to a more general statement known as the Fundamental Expansion Postulate. **The Fundamental Expansion Postulate states that**

Every physical observable can be represented by a Hermitian operator with eigenfunctions, $\psi_1, \psi_2, \ldots, \psi_n$, and every physical state Ψ can be expanded as a linear combination of these eigenstates as $\sum c_i \psi_i$, where each coefficient is given as

$$c_j = \int \psi_j^* \Psi \, d^3 r. \tag{1.6.39}$$

The set of eigenfunctions ψ_i forms a complete set.

What are the conditions under which two or more observables can be simultaneously known? Specifically, what is the condition of the operators if the physical observables to which they correspond can be simultaneously measured to be in sharp, well-defined states? In order for a state Ψ to be in a sharp, well-defined state of an operator A or B it must be an eigenstate of A or B. Therefore Ψ must be a simultaneous eigenstate of the operators A and B. As such, Ψ must satisfy

$$A\Psi = A'\Psi, \qquad B\Psi = B'\Psi \tag{1.6.40}$$

simultaneously. Applying B to the first equation of Eqs. (1.6.40) and A to the second yields

$$BA\Psi = BA'\Psi = A'B\Psi,$$
$$AB\Psi = AB'\Psi = B'A\Psi. \tag{1.6.41}$$

Subtracting the second equation of Eqs. (1.6.41) from the first equation, we obtain

$$BA\Psi - AB\Psi = A'B\Psi - B'A\Psi,$$
$$(BA - AB)\Psi = A'B'\Psi - B'A'\Psi, \tag{1.6.42}$$

where we have made use of the eigenvalue equations for A and B in simplifying the right-hand side. A' and B' are simply numbers and must therefore commute: $A'B' = B'A'$. As a result, the right-hand side vanishes, leaving

$$(BA - AB)\Psi = 0. \tag{1.6.43}$$

Consequently, for two observables to be simultaneously known (have simultaneous eigenvalues), the operators corresponding to those observables must satisfy the condition

$$AB = BA. \tag{1.6.44}$$

Any two operators that satisfy the above condition are said to commute, and the observables corresponding to these operators can be simultaneously known.

Consider two Hermitian operators A and B that do not commute. In general, the commutation relation between noncommuting operators can be written as

$$AB - BA = iC. \tag{1.6.45}$$

The product of the variances of A and B obeys the following relation (a proof of this is given in Merzbacher 1970):

$$(\Delta A)^2 \cdot (\Delta B)^2 \geq \frac{1}{4}\langle C \rangle^2. \tag{1.6.46}$$

The uncertainty relation between x and p_x can be obtained by application of these results. Let A be x and B be p_x. The commutation relation between these operators is then

$$xp_x - p_x x = iC. \tag{1.6.47}$$

The number C can be found by evaluation of the left-hand side of the commutator relation. To determine what the left-hand side of Eq. (1.6.47) is, assume that it acts on a function f. In other words, evaluate

$$(xp_x - p_x x)f. \tag{1.6.48}$$

Substituting $\hbar/i \, d/dx$ in for p_x yields

$$
\begin{aligned}
xp_x - p_x x &= \left(x\frac{\hbar}{i}\frac{d}{dx} - \frac{\hbar}{i}\frac{d}{dx}x \right)f \\
&= x\frac{\hbar}{i}\frac{d}{dx}f - \frac{\hbar}{i}\frac{d}{dx}xf.
\end{aligned} \tag{1.6.49}
$$

The differential operator d/dx acts on everything to its right. Therefore d/dx differentiates f in the first term and the product of xf in the second term. The commutator is then

$$x\frac{\hbar}{i}\frac{df}{dx} - \frac{\hbar}{i}f - \frac{\hbar}{i}x\frac{df}{dx}. \tag{1.6.50}$$

But the first and the last terms subtract out, leaving $i\hbar f$. The function f is arbitrary, so the commutator is

$$(xp_x - p_x x) = i\hbar. \tag{1.6.51}$$

Therefore the number C is equal to $\hbar$ in this case. Using this result and the fact that the product of the uncertainties in A and B must be greater than $1/4\langle C \rangle^2$, we find that the product of the uncertainties is

$$(\Delta A)^2(\Delta B)^2 \geq \left(\frac{\hbar}{2}\right)^2. \tag{1.6.52}$$

Substituting Δx and Δp_x in for ΔA and ΔB yields

$$\Delta x \Delta p_x \geq \frac{\hbar}{2}, \tag{1.6.53}$$

which is precisely the Uncertainty Principle. Some properties of commutators are collected in Table 1.6.1.

Table 1.6.1
Properties of Commutators

Definition: $[A, B] = AB - BA$

1. $[A, B] + [B, A] = 0$
2. $[A, A] = 0$
3. $[A, B + C] = [A, B] + [A, C]$
4. $[A + B, C] = [A, C] + [B, C]$
5. $[A, BC] = [A, B]C + B[A, C]$
6. $[AB, C] = [A, C]B + A[B, C]$
7. $(A, [B, C]) + (C, [A, B]) + (B, [C, A]) = 0$
8. $e^A = 1 + A/1! + A^2/2! + A^3/3! + \cdots$

1.7 Linear Vector Spaces and Quantum Mechanics

In the above sections, the basic rudiments of quantum mechanics that are most useful in engineering have been presented. However, as a means of enhancing our understanding and appreciation of the properties of operators and electronic states, it is helpful to consider the wave function Ψ as a vector quantity, called a state vector. In this way, many of the properties introduced in the previous sections will seem more natural and more easily understood.

Let us first review the basic definitions of vector spaces and basis vectors. A vector space is defined as a set of vectors that is closed under both addition and scalar multiplication. Given any two vectors u and v in the space, then $u + v$ and αu, where α is a scalar, are also in the space. The basis vectors of the vector space comprise a set of vectors that fully spans the space and the vectors are all linearly independent. For example, one set of basis vectors of the simple three-dimensional real space comprises the vectors $\hat{x}$, $\hat{y}$, and $\hat{z}$. These three vectors fully span the space; any three-dimensional vector can be expressed as a linear combination of the above three basis vectors. Similarly, the vectors $\hat{r}$, $\hat{\theta}$, and $\hat{\phi}$ also form a basis set for the three-dimensional real space. For example, a real-space vector within a two-dimensional space can be written as a linear combination of the two-dimensional vectors $\hat{x}$ and $\hat{y}$.

A similar situation exists for a state vector in quantum mechanics. Consider the state function Ψ. Ψ can be written as a linear combination of the basis vectors α and β, which are assumed to span the vector space. Clearly, any arbitrary state vector can be written as a linear combination of the basis vectors that span the vector space, just as any arbitrary real-space vector can be written as a linear combination of the real-space basis vectors. This, though, is precisely what is meant by the Fundamental Expansion Postulate introduced in Section 1.6: every physical state can be written as a linear combination of the complete set of eigenfunctions of any real Hermitian operator. It is important to note that the squares of the coefficients in the expansion of the state Ψ in terms of the basis vectors represent the probability of measuring the arbitrary state Ψ in one of the eigenstates or basis states of the Hamiltonian operator. Therefore the projection of a quantum-mechanical state vector into any basis vector enables the determination of the probability of finding the system in the state corresponding to the basis vector. The probability is given simply by the square of the value of the projection.

The quantum-mechanical vector space has exactly the same properties as any vector space; it is closed under addition and scalar multiplication. The properties of the space are summarized as follows:

Given that Ψ_a and Ψ_b are specific vectors in a vector space and μ and λ are scalars,

1. $\Psi_a + \Psi_b = \Psi_b + \Psi_a,$
2. $\Psi_a + (\Psi_b + \Psi_c) = (\Psi_a + \Psi_b) + \Psi_c,$
3. $\mu(\lambda\Psi_a) = (\mu\lambda)\Psi_a,$
4. $\lambda(\Psi_a + \Psi_b) = \lambda\Psi_a + \lambda\Psi_b,$ (1.7.1)
5. Ψ_0 exists such that $\Psi_a + \Psi_0 = \Psi_a.$
6. The vectors Ψ_a, Ψ_b, etc., are all linearly independent if the following relationship does not hold except for the trivial case in which all the coefficients are zero:

$$\lambda_1\Psi_1 + \lambda_2\Psi_2 + \cdots + \lambda_n\Psi_n = 0. \tag{1.7.2}$$

7. The basis vectors, or basis, are formed by the set of n linearly independent vectors in an n-dimensional vector space. These vectors both span the space and are linearly independent. As mentioned above, an arbitrary vector can always be expanded in terms of the basis vectors Ψ_i as

$$\Psi = \sum_i a_i \psi_i, \tag{1.7.3}$$

where a_i are called the components of Ψ.

As for vectors in a real vector space, a scalar product of state vectors can be formed. The scalar product formed between any two state vectors is defined as

$$\int \Psi_a^* \Psi_b \, d^3r = \int \sum_i a_i^* \Psi_i^* \sum_j a_j \Psi_j \, d^3r = \sum_i \sum_j a_i^* b_j \int \Psi_i^* \Psi_j \, d^3r. \tag{1.7.4}$$

Provided that the vectors Ψ_i and Ψ_j are orthonormal, the vector product becomes

$$\int \Psi_a^* \Psi_b \, d^3r = \sum_i \sum_j a_i^* b_j \int \Psi_i^* \Psi_j \, d^3r = \sum_i \sum_j a_i^* b_j \delta_{ij} = \sum_i a_i^* b_i. \tag{1.7.5}$$

How, though, are operators treated in the vector space? An operator A is essentially a means of mapping every vector Ψ in the space into some other vector Ψ' in the space. This can be represented mathematically as

$$\Psi' = A\Psi. \tag{1.7.6}$$

The operators of interest to us are mostly linear. This means that they have the following properties:

$$A(\Psi_a + \Psi_b) = A\Psi_a + A\Psi_b,$$
$$A(\lambda\Psi_a) = \lambda A\Psi_a, \tag{1.7.7}$$

where λ is a scalar. There is also an identity operator 1 that satisfies the condition,

$$\Psi = 1\Psi. \tag{1.7.8}$$

We have also found in Section 1.6 that, in general, no two operators commute. In other words, if A and B are operators, then

$$AB \neq BA \tag{1.7.9}$$

unless the physical observables corresponding to the two different operators A and B can be simultaneously measured.

The above properties of the operators and the state vectors are consistent with the behavior of matrices. It is well known from linear algebra that matrices obey the distributive laws given by the conditions stated above for both the state vectors and the operators. In addition, matrices have inverses and an identity operator and do not generally commute. It is natural then to think of reformulating quantum mechanics by use of a matrix representation for the state vectors and the operators. To see how this can be done, let us make the following definitions. The state vector Ψ can always be written as a linear combination of some basis wave functions. Let the basis functions of interest be ψ_i. Ψ can then be written as

$$\Psi = \sum_i a_i \psi_i. \tag{1.7.10}$$

The above can be reexpressed as the product of two matrices of the form

$$\Psi = [a_1 \quad a_2 \quad a_3 \quad \cdots \quad a_n] \begin{bmatrix} \psi_1 \\ \psi_2 \\ \psi_3 \\ \vdots \\ \psi_n \end{bmatrix}, \tag{1.7.11}$$

where the row matrix represents the coefficients a_i in the expansion of Ψ in the basis vectors ψ_i.

In quantum mechanics, it is possible that a state vector Ψ can be related to a different state vector Ψ' through the application of an operator A as

$$\Psi = A\Psi'. \tag{1.7.12}$$

From the above discussion, it is clear that the state vectors Ψ and Ψ' can be expanded in terms of a basis set of wave vectors ψ_i. The expansions are given as

$$\Psi = \sum_i a_i \psi_i, \qquad \Psi' = \sum_i a_i' \psi_i. \tag{1.7.13}$$

Substituting into Eq. (1.7.12) the expression for Ψ' above yields

$$\Psi = A\sum_i a_i' \psi_i. \tag{1.7.14}$$

Multiplying each side by ψ_j^* and integrating over all space, we obtain

$$\int \psi_j^* \Psi \, d^3r = \sum_i \int \psi_j^* A a_i' \psi_i \, d^3r. \tag{1.7.15}$$

Substituting in for Ψ the expression given in Eq. (1.7.13) on the left-hand side of Eq. (1.7.15) yields

$$\int \psi_j^* \sum_i a_i \psi_i \, d^3r = \sum_i \int a_i \psi_j^* \psi_i \, d^3r = \sum_i a_i \delta_{ij} = a_j. \tag{1.7.16}$$

Therefore a_j can be written as

$$a_j = \sum_i A_{ji} a_i', \tag{1.7.17}$$

where A_{ji} is defined as

$$A_{ji} = \int \psi_j^* A \psi_i \, d^3r \tag{1.7.18}$$

and is called the matrix representation of the operator A in the basis defined by the vectors ψ_i. Equation (1.7.17) can be written as matrices:

$$\begin{bmatrix} a_1 \\ a_2 \\ \vdots \\ a_n \end{bmatrix} = \begin{bmatrix} A_{11} & A_{12} & \cdots & A_{1n} \\ A_{21} & A_{22} & \cdots & A_{2n} \\ \vdots & & & \vdots \\ A_{n1} & A_{n2} & \cdots & A_{nn} \end{bmatrix} \begin{bmatrix} a_1' \\ a_2' \\ \vdots \\ a_n' \end{bmatrix}. \tag{1.7.19}$$

Therefore the operator A is represented by the matrix A_{ji}.

Let us consider what happens when the operator A is applied to a set of state vectors that form the eigenfunctions of the operator A. From the discussion in Section 1.6, if the eigenfunctions of A are ϕ_n, then

$$A\phi_n = a_n\phi_n. \tag{1.7.20}$$

The corresponding matrix representation of A in the basis that comprises the eigenfunctions of A, ϕ_n, is then

$$A_{ji} = \int \phi_i^* A\phi_j \, d^3r \tag{1.7.21}$$

but,

$$A\phi_j = a_j\phi_j. \tag{1.7.22}$$

Substituting Eq. (1.7.22) into Eq. (1.7.21) becomes

$$A_{ji} = \int \phi_i^* a_j\phi_j \, d^3r = a_j\delta_{ij}. \tag{1.7.23}$$

In matrix form Eq. (1.7.23) becomes

$$A = \begin{bmatrix} a_1 & 0 & 0 & \cdots & 0 \\ 0 & a_2 & 0 & \cdots & 0 \\ \cdot & & & & \\ \cdot & & & & \\ \cdot & & & & \\ 0 & 0 & 0 & \cdots & a_n \end{bmatrix}. \tag{1.7.24}$$

Hence we see that the matrix representation of the operator A is diagonal when it is expressed in terms of the basis corresponding to the eigenfunctions of A. Note that the diagonal elements correspond to the eigenvalues of the operator A. Therefore the problem of finding the eigenvalues of the operator A reduces to the problem of diagonalizing the matrix representation of the operator A. The diagonal elements of the matrix A then are simply the eigenvalues of A.

We can use the above property to solve eigenvalue equations. For example, the Schroedinger equation can be written as an eigenvalue equation involving the operator H and the eigenvalues E. This can then be written with matrices. As we have found above, the Schroedinger equation is given as

$$H\Psi = E\Psi. \tag{1.7.25}$$

Given that Ψ is a column vector and that H can be represented as a matrix, the Schroedinger equation can be solved as a matrix eigenvalue equation. As the student may recall from linear algebra, a matrix eigenvalue equation (such as the Schroedinger equation above) can be written as

$$(H - E)\Psi = 0. \tag{1.7.26}$$

Let us consider two cases. The first case is the trivial case in which the wave function Ψ is an eigenstate of the Hamiltonian H. Let us define the eigenstates of H to be the set of wave vectors ψ_i. Then,

$$H\psi_i = E_i\psi_i. \tag{1.7.27}$$

We can write Eq. (1.7.27) in matrix form by noting that the functions ψ_i are column vectors and H is a square matrix. To find the elements of H, multiply each side by a particular eigenstate ψ_j and integrate over all space. This yields

$$\int \psi_j^* H\psi_i \, d^3r = \int \psi_j^* E_i\psi_i \, d^3r. \tag{1.7.28}$$

The left-hand side of Eq. (1.7.28) is simply equal to the matrix element H_{ji}, while the right-hand side is $E_j\delta_{ij}$. Therefore Eq. (1.7.28) reduces to

$$H_{ji} = E_j\delta_{ij} \tag{1.7.29}$$

or

$$H_{ii} = E_i, \qquad H_{ij}(i \neq j) = 0, \tag{1.7.30}$$

implying that the matrix H is diagonal and the diagonal elements are equal to the eigenvalues of H. Therefore, for the simple case in which the wave functions are eigenstates of H, the matrix is diagonal.

A more interesting case is when the wave function Ψ is not an eigenstate of H. In this case, we proceed as in Section 1.6 and expand Ψ in terms of the eigenstates of H, the ψ_i. The Schroedinger equation can then be written as

$$H\Psi = H\sum_i c_i\psi_i = E\sum_i c_i\psi_i. \tag{1.7.31}$$

Again, multiplying both sides by ψ_j^* and integrating over all space yields

$$\int \psi_j^* H\sum_{i_i} c_i\psi_i \, d^3r = \int \psi_j^* E\sum_{i_i} c_i\psi_i \, d^3r. \tag{1.7.32}$$

Simplifying yields

$$\sum_i c_i H_{ji} = \sum_i Ec_i\delta_{ji},$$

$$\sum_i c_i(H_{ji} - E\delta_{ji}) = 0. \tag{1.7.33}$$

In matrix form, this becomes

$$
\begin{bmatrix}
(H_{11} - E) & \cdots & H_{1n} \\
\cdot & \cdots & \cdot \\
\cdot & \cdots & \cdot \\
H_{n1} & \cdots & (H_{nn} - E)
\end{bmatrix}
\begin{bmatrix}
c_1 \\
\cdot \\
\cdot \\
c_n
\end{bmatrix} = 0,
\tag{1.7.34}
$$

where the matrix elements of H are defined as above as

$$
H_{ij} = \int \psi_i^* H \psi_j \, d^3 r.
\tag{1.7.35}
$$

If the wave vectors ψ_i are eigenfunctions of H, then the matrix H is diagonal, as we discussed above. Then the eigenvalues E_i of the operator H are simply equal to the matrix elements:

$$
H_{ii} = \int \psi_i^* H \psi_i \, d^3 r = E_i.
\tag{1.7.36}
$$

The key then to solving the Schroedinger equation in the matrix representation is to diagonalize the Hamiltonian.

From linear algebra, the solution of the matrix Schroedinger equation is obtained if the determinant of the matrix $(H - E)$ vanishes. Under these conditions, the nontrivial solution is obtained. When the determinant of the matrix $(H - E)$ is set equal to zero, the eigenvalues of the system can be obtained. From knowledge of the eigenvalues of H, the eigenfunctions ψ_i can also be determined. An example of this procedure is provided in Example 1.7.1.

EXAMPLE 1.7.1

Consider a system whose Hamiltonian can be described by a 3×3 matrix. The matrix elements are evaluated to be

$$
\begin{bmatrix}
3 & -1 & 0 \\
-1 & 2 & -1 \\
0 & -1 & 3
\end{bmatrix}.
$$

a. **What are the possible results of a measurement of the energy made on this system?**

From the discussion in Section 1.6, the possible values of a measurement of the energy of the system are given by the possible values of the energy eigenvalues. Therefore it is necessary to find the eigenvalues of the above matrix. To do this we consider the solution of the equation

$$
(H - E)\Psi = 0.
$$

The resulting form of the matrix $(H - E)$ is

$$
\begin{bmatrix}
(3 - E) & -1 & 0 \\
-1 & (2 - E) & -1 \\
0 & -1 & (3 - E)
\end{bmatrix}.
$$

To have the nontrivial solution for the energy eigenvalues, the determinant of the above matrix must vanish. Therefore

$$\text{Det}\begin{bmatrix} (3-E) & -1 & 0 \\ -1 & (2-E) & -1 \\ 0 & -1 & (3-E) \end{bmatrix} = 0.$$

Expanding out the above determinant by minors yields

$$(3-E)\begin{vmatrix} (2-E) & -1 \\ -1 & (3-E) \end{vmatrix} + \begin{vmatrix} -1 & 0 \\ -1 & (3-E) \end{vmatrix} = 0.$$

Simplifying yields

$$(3-E)[E^2 - 5E + 4] = 0,$$
$$(3-E)(1-E)(4-E) = 0.$$

Therefore the possible energy eigenvalues are

$$E = 3, 1, 4.$$

b. What are the eigenvectors of the system?

To find the eigenvectors, consider each eigenvalue separately. First consider the eigenvalue $E = 1$. Substituting into $(H - E)$ for $E = 1$ and for the state vector Ψ, the equation $(H - E)\Psi = 0$ becomes

$$\begin{bmatrix} (3-1) & -1 & 0 \\ -1 & (2-1) & -1 \\ 0 & -1 & (3-1) \end{bmatrix}\begin{bmatrix} x_1 \\ x_2 \\ x_3 \end{bmatrix} = 0,$$

$$\begin{bmatrix} (2) & -1 & 0 \\ -1 & (1) & -1 \\ 0 & -1 & (2) \end{bmatrix}\begin{bmatrix} x_1 \\ x_2 \\ x_3 \end{bmatrix} = 0.$$

The resulting set of linear equations is

$$2x_1 - x_2 = 0,$$
$$-x_1 + x_2 - x_3 = 0,$$
$$-x_2 + 2x_3 = 0.$$

Therefore the eigenvector corresponding to the eigenvalue $E = 1$ is

$$\begin{bmatrix} 1 \\ 2 \\ 1 \end{bmatrix}.$$

The eigenvectors corresponding to the other eigenvalues can be found in a similar manner as above. The results are found as follows:

$$E = 3: \quad \begin{bmatrix} 0 & -1 & 0 \\ -1 & -1 & -1 \\ 0 & -1 & 0 \end{bmatrix} \begin{bmatrix} x_1 \\ x_2 \\ x_3 \end{bmatrix} = 0,$$

$$-x_2 = 0,$$
$$-x_1 - x_2 - x_3 = 0,$$
$$-x_2 = 0.$$

Therefore $x_2 = 0$ and $x_1 = -x_3$. The eigenvector associated with $E = 3$ is then

$$E = 3, \qquad \begin{bmatrix} 1 \\ 0 \\ -1 \end{bmatrix}.$$

The eigenvector corresponding to $E = 4$ is

$$E = 4, \qquad \begin{bmatrix} 1 \\ -1 \\ 1 \end{bmatrix}.$$

c. If a state is described by the state vector

$$\Psi = \begin{bmatrix} 2 \\ 3 \\ -2 \end{bmatrix},$$

what are the possible results of a measurement of the energy made on this state and what are their relative probabilities?

A measurement of the energy for the system governed by the Hamiltonian specified in part a can give only 3, 1, or −4. The probabilities are determined from the square of the coefficients in the expansion of Ψ in terms of the eigenvectors of H. To find the coefficients in the expansion of the state vector Ψ, it is necessary to evaluate the overlap of Ψ with each of the eigenvectors ψ_i determined in part b. Therefore we need to evaluate the following overlap integral,

$$\int \Psi^* \Psi_1 \, d^3r = C_1$$

to determine the coefficient in front of the first eigenvector ψ_1 in the expansion of Ψ. C_1 is thus evaluated as follows. First the eigenvectors of the system and the state vector itself need to be normalized. This is accomplished in the usual way from the magnitude of the vector. C_1 can be found as

$$C_1 = \frac{1}{\sqrt{17}}[2 \quad 3 \quad -2]\frac{1}{\sqrt{6}}\begin{bmatrix} 1 \\ 2 \\ 1 \end{bmatrix} = \sqrt{\frac{6}{17}}.$$

Similarly, C_2 and C_3 are

$$C_2 = \frac{2\sqrt{2}}{\sqrt{17}}, \qquad C_3 = \frac{-\sqrt{3}}{\sqrt{17}}.$$

Hence the probability of measuring the system with each energy eigenvalue is given as

$$E = 1, \qquad C_1^2 = \frac{6}{17},$$

$$E = 3, \qquad C_2^2 = \frac{8}{17},$$

$$E = 4, \qquad C_3^2 = \frac{3}{17}.$$

It should be finally noted that the state vector Ψ collapses, on measurement, into one of the possible eigenstates of the system.

Before we end this section, it is useful to introduce a different notation, called the Dirac notation, which is of great importance in quantum mechanics. The principal advantage of Dirac notation is that it is a basis-free notation. By this we mean that each state is described independently of the coordinate system. If we have a wave function ψ_a and a different wave function ψ_b, then the overlap integral between these two states can be written as

$$\int \Psi_a^* \Psi_b \, d^3r = \langle a|b \rangle, \tag{1.7.37}$$

where $\langle a|$ represents Ψ_a^* and $|b\rangle$ represents Ψ_b. We call $\langle a|$ a bra vector and $|b\rangle$ a ket vector. The integral between the states is given then by the expression $\langle a|b \rangle$ or the bracket of a and b. This is the nature of the name of the bra and the ket vectors.

Finally, the matrix element

$$\int \Psi_b^* A \Psi_a \, d^3r, \tag{1.7.38}$$

where A is an arbitrary operator, can be written as $\langle b|A|a \rangle$. An eigenvalue equation can be written in Dirac notation as

$$A|A'\rangle = A'|A'\rangle, \tag{1.7.39}$$

where A' is the eigenvalue of the operator A and $|A'\rangle$ is the corresponding eigenstate.

PROBLEMS

1. Let A and B be self-adjoint operators. Show that
   $$\frac{AB - BA}{i}, \qquad AB + BA$$
   are self-adjoint.

2. A particle is confined in a box of length, L. Given that the nth state eigenfunctions are
   $$\psi_n = \sqrt{\frac{2}{L}} \sin\left(\frac{n\pi x}{L}\right),$$
   calculate
   a. the mean value of the position,
   b. the mean value of x^2,
   c. the mean value of the momentum squared,
   d. the mean value of $p^2/2m$.

3. Given that the ground state of the hydrogen atom is expressed as
   $$\Psi_0 = \frac{1}{\sqrt{\pi}}\left(\frac{1}{a_0}\right)^{\frac{3}{2}} e^{\frac{-r}{a_0}},$$
   where a_0 is a constant, calculate
   a. the location at which the radial probability is a maximum,
   b. the expectation value for r in this state,
   c. explain the results.

4. Evaluate the following commutators:
   a. $[p, x^2]$,
   b. $[p^2, x]$,
   c. $[p^2, x^2]$,
   d. $[p^2, f(x)]$.

5. Show that $\vec{L} = \vec{r} \times \vec{p}$ is Hermitian.

6. Prove the following theorem:
   $$e^A B e^{-A} = B + [A, B] + 1/2![A, [A, B]] + \cdots$$

7. Prove the following theorem:
   $$V^{-1} e^W V = e^{V^{-1}WV},$$
   where V^{-1} is the inverse operator of V.

8. Derive the Compton effect formula:

$$\Delta\lambda = \frac{h}{m_0 c}(1 - \cos\phi).$$

9. Obtain uncertainty relations for the following products of physical observables:

    a. $\Delta x \Delta E$,

    b. $\Delta p \Delta E$.

    Assume that the potential $V(x)$ is given as $V(x) = \sum a_n x^n$.

10. What is the value of $[A, B]$ if A and B have the same eigenfunctions?

11. What is the maximum kinetic energy with which an electron can escape from cesium (work function $= 1.9$ eV) when light of $\lambda = 4000$ Å shines onto it?

12. Consider a harmonic oscillator where H, the Hamiltonian operator, is defined as

$$H = \frac{p^2}{2m} + \frac{1}{2}kx^2$$

    with eigenfunctions $\Psi_m(x)$, $H\Psi_m = E_m\Psi_m$. By evaluating the quantity

$$\int \Psi_m^*(Hpx - pxH)\Psi_m \, dx,$$

    show that

$$\left\langle \frac{p^2}{2m} \right\rangle = \left\langle \frac{kx^2}{2} \right\rangle.$$

    This is known as the quantum-mechanical virial theorem for the harmonic oscillator.

13. Evaluate the following matrix element:

$$\langle k| p_x^2 [p_x, x] |k'\rangle,$$

    where $|k'\rangle$ and $\langle k|$ represent states with sharp values of k' and k, respectively. Recall that the momentum eigenvalues are given as $\hbar k$ and $\hbar k'$ for definite k states k and k'.

14. Determine the possible values of a measurement and the probability of measuring each value in the state Ψ if the Hamiltonian of the system is given as

$$H = \begin{bmatrix} 1 & -2 & 0 \\ -2 & 0 & 2 \\ 0 & 2 & -1 \end{bmatrix}$$

    and Ψ is given as

$$\Psi = \frac{1}{3\sqrt{5}} \begin{bmatrix} -3 \\ 6 \\ 0 \end{bmatrix}.$$

15. If the time operator is assumed to be simply equal to t, then determine an expression for the Uncertainty Relation between time and energy. Do not just cite the answer but derive an expression by using the relationship between the variances of each of these observables stated in the text.

16. Consider the matrix representation of the following system. If the diagonal matrix elements H_{ii} are all equal to 2 and the off-diagonal matrix elements describing the interaction are given as $H_{ij} = -1$ for $ij = 12, 21, 23$, and 32 and $H_{ij} = 0$ otherwise for a 3×3 matrix, determine the possible values of a measurement and

the probability of measuring each value in the state Ψ if Ψ is given as

$$\Psi = \begin{bmatrix} 2 \\ -3\sqrt{2} \\ 0 \end{bmatrix}.$$

17. If a system is described by the following Hamiltonian determine

   a. the possible results of a measurement of the energy of the system,

   b. the eigenvectors:

   $$H = \begin{bmatrix} 4 & 0 & 0 \\ 0 & 2 & 1 \\ 0 & 1 & 2 \end{bmatrix}.$$

   If a state is described by the following wave vector, determine the probability of finding it in each of the possible eigenstates of the system:

   $$Y = \frac{1}{\sqrt{14}} \begin{bmatrix} 2 \\ 1 \\ 3 \end{bmatrix}.$$

2

One-Dimensional
Potential Problems

We next turn our attention to solving the Schroedinger equation for one-dimensional potentials. There are only a few potentials for which the Schroedinger equation can be solved exactly. These are the free-electron problem, the infinite one-dimensional square-well potential, the finite square-well potential, and the harmonic oscillator. The three-dimensional Coulomb potential can also be solved exactly but its discussion is delayed until Chapter 3. In this chapter, we consider each of the one-dimensional potentials in turn.

The first problem that we consider is the free-electron problem. A free particle moves in the absence of a potential; $V(x) = 0$. Under this condition, the Schroedinger equation,

$$H\Psi = E\Psi, \qquad \left[\frac{p^2}{2m} + V(x)\right] = E\Psi, \qquad (2.0.1)$$

reduces to

$$-\frac{\hbar^2}{2m}\nabla^2\Psi = E\Psi, \qquad (2.0.2)$$

where the operator expression for the momentum has been substituted in and the potential has been set to zero. The solution of this differential equation is given in general as a linear combination of plane waves:

$$\Psi = Ae^{ikx} + Be^{-ikx}, \qquad (2.0.3)$$

where k is defined as

$$k \equiv \sqrt{\frac{2mE}{\hbar^2}}. \qquad (2.0.4)$$

Physically the solutions of the free-electron problem correspond to plane-wave states of wave vector k. In Chapter 1, it was stated that plane-wave states have a definite momentum $p = \hbar k$. To see this, let us determine the momentum for the free-particle states.

To evaluate the momentum of a state, we follow the prescription given in Chapter 1; apply the momentum operator onto the state and ascertain if it satisfies an eigenvalue equation. Mathematically, the momentum operator p is applied to the state Ψ to see if it satisfies

$$p\Psi = p'\Psi. \qquad (2.0.5)$$

Substituting its operator form in for p, we find that the eigenvalue equation for the momentum is

$$\frac{\hbar}{i}\frac{d}{dx}\Psi = p'\Psi. \tag{2.0.6}$$

Consider now each part of the expression for Ψ separately, Ae^{ikx} and Be^{-ikx}. Applying the momentum operator onto each term separately yields

$$\frac{\hbar}{i}\frac{d}{dx}Ae^{ikx} = \hbar k Ae^{ikx}, \qquad \frac{\hbar}{i}\frac{d}{dx}Be^{-ikx} = -\hbar k Be^{-ikx}. \tag{2.0.7}$$

Inspection of the above results indicates that the momentum eigenvalues are $\pm\hbar k$. Therefore the plane-wave states Ae^{ikx} and Be^{-ikx} have definite values of momentum $\pm\hbar k$. This result is consistent with de Broglie's hypothesis that plane-wave states have sharp, definite values of momentum and are completely uncertain spatially.

2.1 Infinite Square-Well Potential

A simple example of a one-dimensional potential for which the Schroedinger equation can be readily solved is the infinite one-dimensional square-well potential. The infinite square-well potential is defined as a structure in which the potential is infinite at the boundaries and everywhere outside but is zero everywhere within. The structure is sketched in Figure 2.1.1. There are three regions to consider, which are marked as I, II, and III in the figure. In regions I and III and at the boundaries the potential is infinite. A particle cannot exist within these regions and still maintain a finite energy. Subsequently, the particle's motion is restricted to within the region from $-a/2$ to $+a/2$. Everywhere within region II the potential is constant and, for simplicity, is defined as zero. To obtain the solution of this problem, the Schroedinger equation is solved within region II and the boundary conditions are applied at $x = \pm a/2$. First, write the Schroedinger equation in region II, often referred to as the well or the well region. Since the

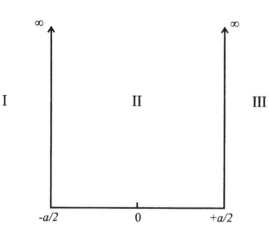

Figure 2.1.1. Infinite square-well potential in one dimension.

potential is zero within the well, the Schroedinger equation is precisely the same as that given by Eq. (2.0.2). The solution is again given by

$$\Psi = A'e^{ikx} + B'e^{-ikx},$$ (2.1.1)

where k is

$$k = \sqrt{\frac{2mE}{\hbar^2}}.$$ (2.1.2)

It is more convenient to rewrite the solution for Ψ in terms of sines and cosines as

$$\Psi = A\cos kx + B\sin kx.$$ (2.1.3)

To obtain the complete solution of the problem, the boundary conditions must be applied. At $x = +a/2$, the wave function must vanish since the potential at this boundary goes abruptly to infinity. Likewise, at $x = -a/2$, the wave function again vanishes. The wave function must go to zero at the boundary since the potential is infinite there. As argued above, the particle cannot exist in a region of infinite potential since it would then have infinite energy, a physical impossibility. The wave function must be zero everywhere that the particle is excluded from since the wave function represents the probability amplitude of locating the particle at a certain point. Making use of the condition that the wave function vanishes at $x = \pm a/2$ yields

$$A\cos\frac{ka}{2} + B\sin\frac{ka}{2} = 0,$$ (2.1.4)

$$A\cos\frac{ka}{2} - B\sin\frac{ka}{2} = 0.$$ (2.1.5)

Adding Eqs. (2.1.4) and (2.1.5), we obtain

$$2A\cos\frac{ka}{2} = 0.$$ (2.1.6)

If the trivial solution $A = 0$ is disregarded, then

$$\cos\frac{ka}{2} = 0.$$ (2.1.7)

For the above result to be true in general, the argument of the cosine must be equal to an odd multiple of $\pi/2$. Therefore

$$\frac{ka}{2} = \frac{\pi}{2}, \frac{3\pi}{2}, \frac{5\pi}{2}, \ldots.$$ (2.1.8)

Solving for k in terms of a gives

$$k = \frac{n\pi}{a}, \quad n = 1, 3, 5 \ldots.$$ (2.1.9)

The resulting solution for the wave function Ψ can be expressed then as

$$\Psi_n^{(1)} = A_n \cos \frac{n\pi x}{a}, \quad n = 1, 3, 5, \ldots. \tag{2.1.10}$$

This is called the class 1 solution for the infinite square well. Note that the class 1 solution involves only cosine functions.

An alternative solution to the problem can be obtained when Eq. (2.1.5) is subtracted from Eq. (2.1.4) to yield

$$2B \sin \frac{ka}{2} = 0. \tag{2.1.11}$$

Again if the trivial solution $B = 0$ is disregarded, the sine must satisfy

$$\sin \frac{ka}{2} = 0. \tag{2.1.12}$$

For the above relation to be true in general, the argument of the sine must then be equal to an integral multiple of π:

$$\frac{ka}{2} = \pi, 2\pi, 3\pi, \ldots. \tag{2.1.13}$$

The general expression for k is then

$$k = \frac{n\pi}{a}, \quad n = 2, 4, 6, \ldots. \tag{2.1.14}$$

The resulting expression for the second class of solutions is

$$\Psi_n^{(2)} = B_n \sin \frac{n\pi x}{a}, \quad n = 2, 4, 6 \ldots. \tag{2.1.15}$$

Both sets of solutions, class 1 and class 2, are solutions to the infinite square-well problem. Before we discuss the nature of these solutions, let us calculate the energy for each case.

The energy eigenvalues of the infinite square well can be readily determined from the values of k. The general expression for k in both classes of solutions is given as

$$k = \sqrt{\frac{2mE}{\hbar^2}}. \tag{2.1.16}$$

Therefore the energy can be expressed in terms of k as

$$E = \frac{\hbar^2 k^2}{2m}. \tag{2.1.17}$$

However k is restricted to certain multiples of π^2/a^2. These are

$$k^2 = \frac{n^2 \pi^2}{a^2}, \quad n = 1, 3, 5, \ldots, \tag{2.1.18}$$

for the class 1 solutions and

$$k^2 = \frac{n^2\pi^2}{a^2}, \quad n = 2, 4, 6, \ldots, \tag{2.1.19}$$

for the class 2 solutions. The energy eigenvalues for the class 1 and the class 2 solutions are then

$$E_n^{(1)} = \frac{\hbar^2}{2m}\frac{n^2\pi^2}{a^2}, \quad n = 1, 3, 5, \ldots, \tag{2.1.20}$$

$$E_n^{(2)} = \frac{\hbar^2}{2m}\frac{n^2\pi^2}{a^2}, \quad n = 2, 4, 6, \ldots. \tag{2.1.21}$$

When these two results are combined, the energy eigenvalues can be written as

$$E_n = \frac{n^2\pi^2\hbar^2}{2ma^2}, \quad n = 1, 2, 3, \ldots, \tag{2.1.22}$$

where the odd integers correspond to the class 1 solutions and the even integers correspond to the class 2 solutions.

Let us examine each solution. The $n = 1$ solution is plotted in Figure 2.1.2. As can be seen from this figure, the wave function goes smoothly to zero at each boundary ($x = \pm a/2$) in accordance with the boundary conditions. The $n = 1$ solution has no nodes, that is, no points that intersect the x axis. The $n = 1$ solution is precisely the same as the fundamental excitation of a vibrating string held fixed at both ends. The amplitude of the wave for a vibrating string is zero at the ends since the string cannot move at those points. The fundamental mode of vibration is then one in which the string vibrates with no nodes present, points of no vibrational amplitude, except at the ends. The energy and the mathematical representation of the wave function accompany the diagram. The $n = 2$ wave function and its energy eigenvalue are plotted in Figure 2.1.3. In the $n = 2$ case, note that the wave function has a node at $x = 0$. The wave function goes through a complete cycle within the well. Hence one full wavelength is enclosed within the well. The $n = 3$ case is shown in Figure 2.1.4. In this case, the wave function

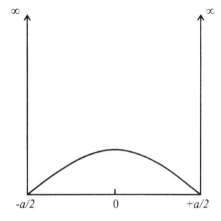

Figure 2.1.2. $n=1$ solution of the infinite square-well potential problem: $E_1 = [(\pi^2\hbar^2)/(2ma^2)]$, $\Psi_1 = A_1 \cos[(\pi x)/a]$.

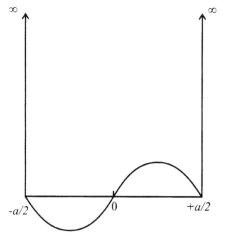

Figure 2.1.3. $n=2$ solution of the infinite square-well potential problem: $E_2 = [(4\pi^2\hbar^2)/(2ma^2)]$, $\Psi_2 = B_1 \sin[(2\pi x)/a]$.

has two nodes. This leads to the general statement that the wave function has $n-1$ nodes.

The two classes of solutions determined above are said to have different parity, evenness or oddness, under reflection about the origin. The class 1 solutions have even parity; on reflection about the midpoint of the well, they map into themselves. The class 1 solutions are cosine functions and are even under change of sign of the argument. The class 2 solutions are odd since they change sign on reflection about the midpoint of the well. The class 2 solutions are sine functions and thus change sign when the argument changes sign. These results are summarized below.

Class 1:

$$\Psi(-x) = \Psi(x), \qquad \Psi_n^{(1)} = A_n \cos\frac{n\pi x}{a} \quad \text{(even parity).} \tag{2.1.23}$$

Class 2:

$$\Psi(-x) = -\Psi(x), \qquad \Psi_n^{(2)} = B_n \sin\frac{n\pi x}{a} \quad \text{(odd parity).} \tag{2.1.24}$$

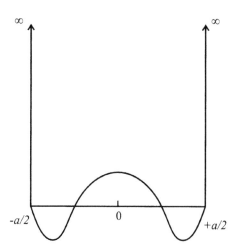

Figure 2.1.4. $n=3$ solution of the infinite square-well potential problem: $E_3 = [(9\pi^2\hbar^2)/(2ma^2)]$, $\Psi_3 = A_3 \cos[(3\pi x)/a]$.

Finally, to complete the solution of the infinite square well, it is necessary to determine the coefficients A_n and B_n. To do this, the wave functions must be properly normalized. Generally the wave function must be integrated over all space. In the case of the infinite square well, the potential is infinity and the wave function vanishes everywhere outside of the well. Therefore the only nonzero value of the wave function occurs within the well. Hence the integration is restricted to values of x between $-a/2$ and $+a/2$. This implies that

$$\int_{-\frac{a}{2}}^{+\frac{a}{2}} A^2 \cos^2 \frac{n\pi x}{a} dx = 1. \tag{2.1.25}$$

When the integration is performed, A is found as

$$A = \sqrt{\frac{2}{a}}. \tag{2.1.26}$$

Similarly, B is

$$B = \sqrt{\frac{2}{a}}. \tag{2.1.27}$$

Therefore the complete solutions of the infinite square-well problem are

$$\Psi_n^{(1)} = \sqrt{\frac{2}{a}} \cos \frac{n\pi x}{a}, \quad n = 1, 3, 5, \ldots, \text{(even parity)}, \tag{2.1.28}$$

$$\Psi_n^{(2)} = \sqrt{\frac{2}{a}} \sin \frac{n\pi x}{a}, \quad n = 2, 4, 6, \ldots, \text{(odd parity)}, \tag{2.1.29}$$

$$E_n = \frac{n^2 \pi^2 \hbar^2}{2ma^2}. \tag{2.1.30}$$

2.2 One-Dimensional Potential Barrier

Next we consider a one-dimensional potential barrier of infinite extent. Such a structure is shown in Figure 2.2.1. The structure extends spatially to infinity along the positive and the negative x directions but is of finite height. It is interesting to consider the dynamics of a quantum-mechanical particle that is incident from the left-hand side of the diagram onto the barrier. We consider two cases, one in which the carrier has energy greater than the potential barrier and the other in which the carrier has energy less than the potential barrier. Before we consider the quantum-mechanical case, let us first consider the classical case.

A classical particle incident upon the barrier with an energy greater than the barrier height is simply transmitted. If the particle's kinetic energy is less than the potential barrier height, it is reflected. In either case, there is no chance that the particle will be partially reflected or partially transmitted; the particle is either transmitted completely with 100% probability or reflected completely with 100% probability.

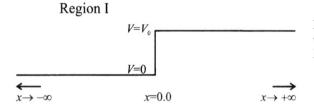

Region II

Region I

$V=V_0$

$V=0$

$x \to -\infty$ $x=0.0$ $x \to +\infty$

Figure 2.2.1. One-dimensional potential barrier of infinite extent and finite height.

The case for a classical wave is quite different. As is well known from classical optics, if a wave is incident from one medium into another (the two mediums differing by the index of refraction), the incident wave may be partially reflected and partially transmitted across the interface. This corresponds to the case in which the incident wave has an energy greater than the potential barrier. Partial reflection and transmission occur for a wave incident from one transmission line into another that differ in their impedances. This is similar to the case of two mediums differing in their indices of refraction. If the transmission lines are impedance matched (they have identical characteristic impedances) then the wave is fully transmitted. If the impedances are not identical, then the wave is partially reflected at the interface.

Quantum-mechanical particles do not behave as classical particles following incidence onto the potential barrier. As in the case of optics and transmission lines, the quantum-mechanical particle has a nonzero probability of being reflected when its energy exceeds the barrier height. What is different between the classical wave behavior and the quantum-mechanical behavior is the interpretation of the transmitted and the reflected waves. In classical physics, the reflected and the transmitted waves at an interface represent a division of the incident-wave power into two separate components representing energy transmission and reflection across the interface. Quantum-mechanical partial reflection and transmission do not imply partial transmission and reflection of the particle. In other words, the particle does not subdivide into smaller parts, part of which is transmitted and part of which is reflected across the boundary. Instead, the wave function is partially reflected and partially transmitted, implying that there is a probability that the particle will be reflected and a probability that the particle will be transmitted at the boundary. After a measurement is made, the particle is found in its entirety to be either transmitted across the boundary or reflected from the boundary; it does not subdivide. The transmitted and the reflected components of the wave function represent the relative probability of finding the particle on either side of the interface.

Let us now consider mathematically the case of a quantum-mechanical particle incident upon the potential barrier whose incident energy exceeds the barrier height. In this case, $E > V_0$. The time-independent Schroedinger equation in region I is

$$\left[-\frac{\hbar^2}{2m}\frac{\partial^2}{\partial x^2} + V(x) \right] \Psi = E\Psi. \tag{2.2.1}$$

But in region I the potential $V(x)$ is taken as zero.

Therefore the Schroedinger equation reduces to that for a free particle:

$$-\frac{\hbar^2}{2m}\frac{\partial^2}{\partial x^2}\Psi = E\Psi. \tag{2.2.2}$$

Solving for Ψ in region I gives

$$\Psi_1 \sim e^{ik_1 x}, \quad k_1 = \sqrt{\frac{2mE}{\hbar^2}}. \tag{2.2.3}$$

In general, Ψ_1 is given as

$$\Psi_1 = A e^{ik_1 x} + B e^{-ik_1 x}, \tag{2.2.4}$$

which is a linear combination of traveling waves to the right and to the left.

In region II, the region to the right of the barrier, the potential $V(x)$ is given as V_0. The Schroedinger equation in this region is then simply

$$\left(-\frac{\hbar^2}{2m}\frac{\partial^2}{\partial x^2} + V_0\right)\Psi = E\Psi. \tag{2.2.5}$$

Solving for Ψ, we obtain

$$\Psi_2 \sim e^{ik_2 x}, \quad k_2 = \sqrt{\frac{2m(E - V_0)}{\hbar^2}}. \tag{2.2.6}$$

Again, the solution generally involves linear combinations of traveling waves. Ψ_1 in region II can then be written as

$$\Psi_2 = C e^{ik_2 x} + D e^{-ik_2 x}. \tag{2.2.7}$$

Given the general solutions for the wave function in each of the two regions, we apply the boundary conditions at the interface $x = 0$. As discussed in Section 2.1, the wave function and its first derivative must be continuous across the interface. There is an alternative boundary condition to the continuity of the first derivative at the interface that reduces to the same condition for most cases of interest. This alternative boundary condition is the conservation of probability current density across the interface. When the conservation of probability current density is used, the most general condition can be obtained. For completeness, let us pause here and examine the implications of this condition.

Given that there are no sources or sinks of probability, then the probability current density must obey a continuity equation. Specifically, the probability current density passing through a Gaussian surface must be equal to the time rate of change of the probability density. This is precisely what was determined in Chapter 1.5:

$$\vec{\nabla} \cdot \vec{j} = \frac{-\partial \rho}{\partial t}. \tag{2.2.8}$$

Consequently, if the Gaussian surface encloses the interface, the time rate of change of the probability current density is zero so the flux of probability current density from one region to the other must be equal. Hence $j_1 = j_2$. The current densities j_1 and j_2 are given assuming that the mass of the particle is the same on either side of the interface (in Section 8.1 we consider the issue of effective mass, in which the mass of the carrier is not in general equal to the free-electron mass) as [Eq. (1.5.10)]

$$\vec{j_1} = \frac{\hbar}{2mi}(\Psi_1^* \vec{\nabla} \Psi_1 - \Psi_1 \vec{\nabla} \Psi_1^*),$$

$$\vec{j_2} = \frac{\hbar}{2mi}(\Psi_2^* \vec{\nabla} \Psi_2 - \Psi_2 \vec{\nabla} \Psi_2^*). \tag{2.2.9}$$

Substituting the expression for Ψ_1 into the expression for j_1 yields, on simplification to one dimension,

$$j_{1x} = \frac{\hbar}{2mi}[(A^* e^{-ik_1 x} + B^* e^{ik_1 x})(ik_1 A e^{ik_1 x} - iBk_1 e^{-ik_1 x})$$
$$- (A e^{ik_1 x} + B e^{-ik_1 x})(-ik_1 A^* e^{-ik_1 x} + ik_1 B^* e^{ik_1 x})]. \tag{2.2.10}$$

The cross terms vanish on expansion, leaving

$$j_{1x} = \frac{\hbar}{2mi}[ik_1 A^2 - ik_1 B^2 + ik_1 A^2 - ik_1 B^2]. \tag{2.2.11}$$

Simplifying yields

$$j_{1x} = \frac{\hbar k_1}{m}[A^2 - B^2]. \tag{2.2.12}$$

Similarly for j_{2x}, we set $D = 0$, implying no incidence from the right-hand side:

$$j_{2x} = \frac{\hbar}{2mi}[C^2 ik_2 - (-ik_2)C^2], \tag{2.2.13}$$

which reduces to

$$j_{2x} = \frac{\hbar k_2}{m}C^2. \tag{2.2.14}$$

But $j_{1x} = j_{2x}$. Therefore, equating the above two expressions for the current densities, we obtain

$$\frac{\hbar k_1}{m}[A^2 - B^2] = \frac{\hbar k_2}{m}C^2,$$

$$[A^2 - B^2] = \frac{k_2}{k_1}C^2,$$

$$1 - \frac{B^2}{A^2} = \frac{k_2}{k_1}\frac{C^2}{A^2}. \tag{2.2.15}$$

But the boundary condition of the continuity of the wave function across the interface gives

$$A + B = C, \qquad \frac{C}{A} = \frac{B}{A} + 1. \tag{2.2.16}$$

Substituting Eq. (2.2.16) for C/A into the last of Eqs. (2.2.15) yields

$$1 - \frac{B^2}{A^2} = \frac{k_2}{k_1} \left(1 + \frac{B}{A}\right)^2,$$

$$\left(1 - \frac{B}{A}\right)\left(1 + \frac{B}{A}\right) = \frac{k_2}{k_1} \left(1 + \frac{B}{A}\right)^2. \tag{2.2.17}$$

Simplifying, we obtain

$$\left(1 - \frac{B}{A}\right) = \frac{k_2}{k_1} \left(1 + \frac{B}{A}\right). \tag{2.2.18}$$

The solution for B/A is

$$\frac{B}{A} = \frac{\left(1 - \frac{k_2}{k_1}\right)}{\left(1 + \frac{k_2}{k_1}\right)}. \tag{2.2.19}$$

The reflection coefficient is given as the square of B/A. Hence the reflection coefficient R is

$$R = \left|\frac{B}{A}\right|^2 = \left(\frac{1 - \frac{k_2}{k_1}}{1 + \frac{k_2}{k_1}}\right)^2. \tag{2.2.20}$$

Provided that the incident-wave function is normalized, the transmission coefficient T is simply given as the difference between 1 and R. Hence T is

$$T = 1 - R. \tag{2.2.21}$$

Substituting in the above expression for R, we find that T is calculated to be

$$T = \frac{4\frac{k_2}{k_1}}{\left(1 + \frac{k_2}{k_1}\right)^2}. \tag{2.2.22}$$

Equation (2.2.22) is the transmission coefficient for an incident quantum-mechanical particle onto a finite potential barrier of infinite spatial extent whose total energy exceeds the potential barrier height.

The second case of interest to us is when the particle has incident energy less than the potential barrier height. Classically, the particle must be forbidden to enter region II since its resulting kinetic energy would necessarily have to be negative. Hence, from a classical point of view, the particle must be reflected from the potential boundary. Quantum mechanically, the situation is quite different. To find what happens, it is necessary to solve the Schroedinger equation throughout the structure once again. In this case, the Schroedinger equation in region I is

$$\frac{-\hbar^2}{2m}\frac{\partial^2}{\partial x^2}\Psi = E\Psi. \tag{2.2.23}$$

Again, the solution is simply a linear combination of traveling waves:

$$\Psi = Ae^{ikx} + Be^{-ikx}, \tag{2.2.24}$$

where k is

$$k = \sqrt{\frac{2mE}{\hbar^2}}. \tag{2.2.25}$$

The Schroedinger equation in region II is

$$\frac{-\hbar^2}{2m}\frac{\partial^2\Psi}{\partial x^2} + V_0\Psi = E\Psi. \tag{2.2.26}$$

In this case, the energy of the particle is less than V_0. Solving for Ψ yields

$$\frac{\partial^2\Psi}{\partial x^2} = \frac{2m(V_0 - E)}{\hbar^2}\Psi,$$
$$\Psi = Ce^{-Kx} + De^{Kx}, \tag{2.2.27}$$

where K is

$$K = \sqrt{\frac{2m(V_0 - E)}{\hbar^2}}. \tag{2.2.28}$$

Again the boundary conditions must be applied. As x approaches positive infinity, the wave function Ψ must go to zero. This implies that the coefficient D must be zero, otherwise Ψ would grow exponentially with increasing x. The wave function within region II is then

$$\Psi = Ce^{-Kx}. \tag{2.2.29}$$

At the interface $x = 0.0$, the wave function and the first derivative of the wave function must be continuous. Applying these two conditions at the interface gives

$$\begin{aligned} A + B &= C &&\text{(continuity of the wave function)},\\ ikA - ikB &= -KC &&\text{(continuity of the first derivative)}. \end{aligned} \tag{2.2.30}$$

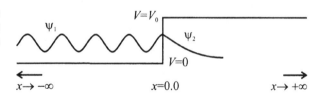

Figure 2.2.2. Sketch of the wave function of a particle whose energy E is less than the potential barrier height V_0. Note that the wave function decays exponentially within region II.

Substituting in $A + B$ for C into the last equation of Eqs. (2.2.30) and dividing through by A yields

$$ik(1 - B/A) = -K(1 + B/A). \tag{2.2.31}$$

Solving for the ratios of B to A and C to A, we obtain

$$\frac{B}{A} = \frac{(ik + K)}{(ik - K)}, \quad \frac{C}{A} = \frac{2ik}{(ik - K)}. \tag{2.2.32}$$

From the above, the magnitude of $|B/A|^2$ is readily seen to be 1. Therefore the reflection from the boundary is 100%.

Nevertheless, the wave function within region II is not zero since the coefficient C is not zero. This implies that the probability amplitude leaks into the potential barrier. In Figure 2.2.2, the wave function is sketched within both regions I and II. As can be seen from this figure, the wave function decays exponentially within the barrier while it is wavelike within region I.

How, though, can the probability amplitude be nonzero in a region, like that of region II, in which the total energy of the particle is less than the potential energy $(E < V_0)$? From the Correspondence Principle we expect that quantum mechanics must not violate classical mechanics in fundamental ways; a particle should still have only a positive value of its kinetic energy. Therefore the particle can never be measured to be within the potential barrier with a negative value of its kinetic energy. How is this reconciled with the quantum-mechanical result that the wave function does not vanish within the potential barrier? First it is instructive to calculate the probability current density within region II. The definition of $\vec{j}$ made in Chapter 1 is

$$\vec{j} = \frac{\hbar}{2mi}[\Psi^* \vec{\nabla} \Psi - \Psi \vec{\nabla} \Psi^*]. \tag{2.2.33}$$

The wave function within region II was found to be

$$\Psi = C e^{-Kx}. \tag{2.2.34}$$

Since the wave function is real, it is equal to its complex conjugate $\Psi = \Psi^*$. Subsequently the probability current density j for a real wave function vanishes.

This implies that no probability current density leaks into the barrier. In other words, there is no net flux of probability across the interface. In other words, if the wave is initially confined within region I to begin with, no probability density flows across the interface, implying 100% reflection back into the incident medium.

This result can be understood by an analogy to total internal reflection in optics. If light is incident from a medium of a higher index of refraction into a medium with a smaller index of refraction and if the angle of incidence is sufficiently large, the light can be totally refracted back into the initial medium. Such a condition is called total internal reflection. When total internal reflection occurs, no optical power is transmitted across the interface; the time-averaged value of the Poynting vector vanishes within the second medium. However, Maxwell's equations require that the tangential component of the electric field must be continuous across an interface. Therefore, although no power is transmitted across the interface between the two mediums, the tangential component of the electric-field vector is continuous across them. Similarly, in quantum mechanics, no probability current density is transmitted across the interface when the total energy of the particle is less than the barrier height, but the wave function is continuous across the interface.

Physically, what happens, though, when the particle's energy is less than the barrier height? The coefficient C in the expression for the wave function in region II is nonzero. Therefore, from our definition of the probability density, $\Psi\Psi^*$ is not zero. What does this mean? The wave function penetrates into the classically forbidden region a finite distance. At first glance, it appears that there is a fundamental paradox; the particle cannot be measured to be within the potential barrier since this would require it to have a negative kinetic energy; however, its probability density is nonzero in this region. How can this be reconciled? Consider the fact that the wave function penetrates only a small distance into the potential barrier. Therefore the probability is appreciable only near $x = 0$, in a range, say Δx, where Δx is the penetration distance. Since the wave function in region II is given as

$$\Psi \sim e^{-Kx} \tag{2.2.35}$$

with K equal to

$$K = \sqrt{\frac{2m(V_0 - E)}{\hbar^2}}, \tag{2.2.36}$$

the wave function goes rapidly to zero when x is larger than $1/K$. Δx can then be approximated as $1/K$. Hence,

$$\Delta x \sim \frac{1}{K} = \frac{\hbar}{\sqrt{2m(V_0 - E)}}. \tag{2.2.37}$$

From the Uncertainty Principle,

$$\Delta x \Delta p_x \geq (\hbar/2), \quad \Delta x \Delta p_x \sim \hbar. \tag{2.2.38}$$

Therefore

$$\Delta p_x \sim \frac{\hbar}{\Delta x} \sim \sqrt{2m(V_0 - E)}. \tag{2.2.39}$$

The uncertainty in the energy of the particle ΔE is given as

$$\Delta E = \frac{(\Delta p_x)^2}{2m} \sim (V_0 - E). \tag{2.2.40}$$

The uncertainty in the particle's energy is equal to the difference between the potential and the energy, $V_0 - E$. Therefore it is impossible to determine definitively if the energy of the particle is less than the potential barrier when it is localized within the penetration distance of region II. As a result, there is no contradiction with classical mechanics. In quantum mechanics, the act of measuring or attempting to localize the particle within the penetration distance changes its energy such that it is impossible to determine whether its energy is less than the barrier height.

Before we conclude this section, it is helpful to summarize the above results. It is found that for quantum-mechanical particle energies in excess of the potential barrier height there is a nonzero reflection coefficient at the interface. In other words, there is a nonzero probability that an incident particle whose kinetic energy exceeds that of the potential barrier will be reflected. This is different from the classical particle-mechanics result in which the particle is always transmitted across the interface if its energy is greater than the potential barrier height. However, the quantum-mechanical result is consistent with what is found for waves. A wave incident from one region into a different region generally undergoes some reflection as well as transmission at the interface. Examples of this are microwave reflection at the boundary between two different transmission lines, optical reflection at an interface formed between materials of different indices of refraction, and mechanical wave reflection of a wave on a string at an interface formed between two strings of different mass density.

A different result occurs when the incident particle has energy less than the barrier height. Classically, such a particle would be reflected at the barrier. Quantum mechanically, the particle still has a 100% probability density current reflection coefficient, but the particle's wave function does not vanish at the interface. Instead, the wave function decays exponentially to zero within the potential barrier. This is similar to the case of total internal reflection in optics; the incident wave has a 100% power reflection coefficient, yet the tangential component of the electric field does not vanish at the interface. The electric field decays exponentially to zero within the nonpropagating medium. We consider this issue in more detail in subsequent sections.

Box 2.2.1 Equivalence of the Continuity of j_x and the First Derivative of Ψ with Respect to x

---

The boundary condition of the continuity of j_x is equivalent to requiring that the product of $1/m\, d\Psi/dx$ be continuous across the interface. This is easily seen from the following. If different masses are assumed present in the two regions, from Eqs. (2.2.12) and (2.2.14) we find that

$$\frac{k_1}{m_1}[A^2 - B^2] = \frac{k_2}{m_2}C^2.$$

The above results from use of the continuity of the current density across the interface. When the form of the wave functions given in Eq. (2.2.24) is used, the continuity of $1/m\, d\Psi/dx$ leads to

$$\frac{1}{m_1}(ik_1 A - ik_1 B) = \frac{1}{m_2}ik_2 C,$$

$$\frac{k_1}{m_1}(A - B) = \frac{k_2}{m_2}C.$$

But $(A + B) = C$ from the continuity of Ψ. Therefore multiplying both sides by $(A + B)$ yields

$$\frac{k_1}{m_1}(A^2 - B^2) = \frac{k_2}{m_2}C^2,$$

which is precisely the same result obtained by requiring the continuity of j across the boundary as found above. Typically the boundary conditions used in solving resonant tunneling problems in solid-state physics are taken as

1. the continuity of the wave function across the boundary:

   $$\Psi_1 = \Psi_2;$$

2. the continuity of $1/m\, d\Psi/dx$ across the boundary:

   $$\frac{1}{m_1}\frac{d\Psi_1}{dx} = \frac{1}{m_2}\frac{d\Psi_2}{dx}.$$

2.3 Finite Square-Well Potential

The next problem of interest to us is the finite square-well potential. This problem is of keen practical interest since many important semiconductor devices utilize finite potential wells. For example, there are semiconductor quantum-well lasers. These devices exploit spatial quantization effects in order to increase the efficiency as well as alter the lasing transition energy of the laser. We discuss such a structure below.

The finite potential well structure is sketched in Figure 2.3.1. As can be seen from this diagram, the potential barrier is finite at the edges of the well as opposed to the infinite energy condition in the structure considered in Section 2.1. The structure is partitioned into three separate regions, denoted as regions I, II, and III in the diagram. As in the previous examples, it is necessary to solve the Schroedinger equation within each of the three regions and then match the

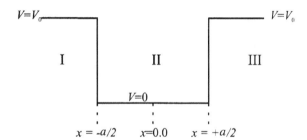

Figure 2.3.1. Sketch of the finite square-well potential diagram showing three distinct regions, I, II, and III.

boundary conditions across all the interfaces. Provided that the potential height is the same in regions I and III, the solution of the Schroedinger equation in these regions is then the same. Two different cases need to be considered. The first is when the particle's energy is less than the potential barrier height, $E < V_0$. The second case is when the particle's energy is greater than the barrier height. Let us consider the first case.

The solution in regions I and III, where the potential height is V_0, can then be found from,

$$-\frac{\hbar^2}{2m}\frac{d^2\Psi}{dx^2} + V_0\Psi = E\Psi. \tag{2.3.1}$$

The Schroedinger equation is readily solved in both regions to yield for $E < V_0$

$$\Psi_1 \sim e^{k_1 x}, \qquad \Psi_3 \sim e^{-k_1 x}, \tag{2.3.2}$$

where k_1 is given as

$$k_1 = \sqrt{\frac{2m(V_0 - E)}{\hbar^2}}. \tag{2.3.3}$$

Next the Schroedinger equation is solved in region II, where the potential is zero. The solution is precisely the same as that found in Sections 2.1 and 2.2, with traveling waves to the right and to the left:

$$\Psi_2 = C'e^{ik_2 x} + D'e^{-ik_2 x}, \tag{2.3.4}$$

where k_2 is

$$k_2 = \sqrt{\frac{2mE}{\hbar^2}}. \tag{2.3.5}$$

It is more convenient, though, to rewrite the exponential functions in terms of sines and cosines. Ψ_2 is given then as

$$\Psi_2 = C\cos k_2 x + D\sin k_2 x. \tag{2.3.6}$$

The boundary conditions at the interfaces between regions I and II and regions II and III are that the wave function and its first derivative must be continuous. Applying these conditions at $x = \pm a/2$, we obtain

$$x = -\frac{a}{2}, \qquad A e^{\frac{-k_1 a}{2}} = C \cos k_2 \frac{a}{2} - D \sin k_2 \frac{a}{2},$$

$$x = \frac{a}{2}, \qquad B e^{-k_1 \frac{a}{2}} = C \cos k_2 \frac{a}{2} + D \sin k_2 \frac{a}{2},$$

$$x = -\frac{a}{2}, \qquad k_1 A e^{\frac{-k_1 a}{2}} = k_2 C \sin k_2 \frac{a}{2} + k_2 D \cos k_2 \frac{a}{2},$$

$$x = \frac{a}{2}, \qquad -k_1 B e^{\frac{-k_1 a}{2}} = -k_2 C \sin k_1 \frac{a}{2} + k_2 D \cos k_2 \frac{a}{2}. \qquad (2.3.7)$$

Adding and subtracting the first two equations and the last two equations of Eqs. (2.3.7) yields

$$(A + B) e^{\frac{-k_1 a}{2}} = 2 C \cos k_2 \frac{a}{2},$$

$$(B - A) e^{\frac{-k_1 a}{2}} = 2 D \sin k_2 \frac{a}{2},$$

$$k_1 (A - B) e^{\frac{-k_1 a}{2}} = 2 k_2 D \cos k_2 \frac{a}{2},$$

$$k_1 (A + B) e^{\frac{-k_1 a}{2}} = 2 k_2 C \sin k_2 \frac{a}{2}. \qquad (2.3.8)$$

Dividing out the exponentials from the above set of equations leads to the following two transcendental equations:

$$k_1 = k_2 \tan \frac{k_2 a}{2},$$

$$-k_1 = k_2 \cot \frac{k_2 a}{2}. \qquad (2.3.9)$$

The only unknown in Eqs. (2.3.9) is the energy eigenvalue E (note that the k's are functions of E and V_0). Owing to the transcendental nature of these two equations, the energy E cannot be solved for in closed form. Therefore it is necessary to solve these equations numerically or graphically. At this point, the problem is essentially solved as far as the physics is concerned. What remains is mathematics. To proceed, we use a numerical approach to solve these equations. Fortran 77 and C programs, which solve for the energy eigenvalues of the finite square-well problem, are available electronically at the book web site. The student is encouraged to retrieve these routines for use in his/her personal computer and to solve for various finite square-well geometries.

As in the infinite square-well problem, only discrete energy levels arise. Therefore electrons or any quantum-mechanical particle confined to a finite square-well potential can have only discrete energy values. The appearance of only discrete energy eigenvalues in a confined system is called spatial quantization.

The effects of spatial quantization are manifest in new semiconductor device structures, called single-quantum-well and multiquantum-well systems. In these device structures, two or more different materials are grown sequentially on top of one another in very narrow layers, typically less than 200 Å. As we will see in Chapter 8, different semiconductor materials have differing energy bandgaps. The energy bandgap is defined as the energy difference between the valence band (energy band in which the electrons occupy localized levels and are not generally free to move in response to an electric field) and the conduction band (energy band in which the electrons are free to move under the application of an electric field). If two materials of different bandgaps are grown on top of one another, a discontinuity in the conduction and the valence bands occurs at the interface between the two materials. The discontinuity in the two bands occurs because of the discontinuity in the energy gap. Depending on how much of the discontinuity occurs within the conduction band or the valence band, a potential discontinuity at the interface can be formed. If a material of smaller bandgap is sandwiched between two layers of material of greater bandgap (for example, GaAs sandwiched between two layers of AlGaAs, as shown in Figure 2.3.2.a) a finite potential well can be formed. This is illustrated in Figure 2.3.2.a. If the width of the well is made sufficiently small that spatial quantization effects occur, then the well is said to be a single quantum well. The resulting energy-band structure of these devices resembles a finite square-well potential. As a result, electrons confined within the well can have only discrete energies. Such structures are made routinely in the laboratory and are even used to make

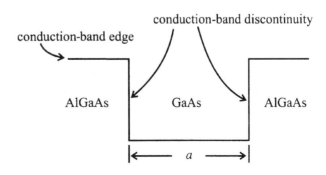

Figure 2.3.2. Sketch of the conduction bands of GaAs and AlGaAs showing a, a single quantum-well system, b, a multiquantum-well system.

a Example of a single quantum well formed by the GaAs/AlGaAs materials system.

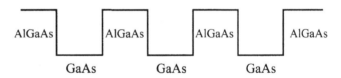

b Multiquantum-well AlGaAs/GaAs system.

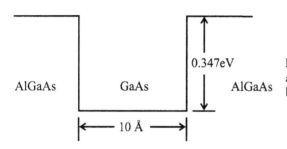

Figure 2.3.3. Finite square well formed when a thin 10-Å-wide layer of GaAs is embedded between two thicker AlGaAs layers.

commercial devices. One of the most important devices made with quantum wells is the quantum-well laser. In many quantum-well lasers, multiple wells are formed. These devices are called multiquantum-well structures. A schematic drawing of a multiquantum-well device is shown in Figure 2.3.2.b.

Let us consider an example problem of a semiconductor quantum-well structure. The structure we will consider is shown in Figure 2.3.3. The well width is chosen to be 10.0 Å and the well depth is assumed to be 0.347 eV. The electron mass within the well is assumed to be the same as the free-electron mass. In Chapter 8 we will discuss the effective mass of an electron in a solid. For now, we avoid such complications. Making use of the computer program, we calculate the allowed energies of the electron in this well as

$$n = 1, \qquad E_1 = 0.12660 \text{ eV}; \qquad n = 2, \qquad E_2 = 0.3413 \text{ eV}.$$

EXAMPLE 2.3.1

Consider the solution of an asymmetric finite square well. In this case, it is assumed that identical masses are present in all three regions of the structure. The potential diagram is sketched in Figure 2.3.4. To solve the problem, it is first necessary to solve the Schroedinger equation in each of the three regions. Next, the boundary conditions and the continuity of the wave function and its first derivative are applied at the intersections of regions I, II, and III. The potential is assumed to equal U_1 within region I, U_3 within region III, and zero in region II.

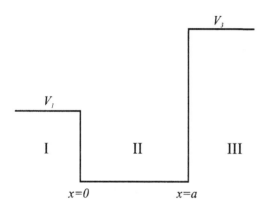

Figure 2.3.4. Asymmetric finite square-well potential.

In this problem, there are several cases to consider in general. The first case is that in which the energy is less than both U_1 and U_3. This case is the most interesting and we consider it here. The other cases, in which the energy is greater than U_1 but less than U_2 and in which the energy is greater than both U_1 and U_2, are left as exercises for the reader.

The solution in region I, $x < 0.0$, is given as

$$\Psi_1 = C_1 e^{K_1 x}, \quad K_1 \equiv \sqrt{\frac{2m(U_1 - E)}{\hbar^2}}. \tag{2.3.10}$$

The solution in region III, $x > a$, also corresponds to an evanescent wave. In region III, the wave function is given as

$$\Psi_3 = C_3 e^{-K_3 x}, \quad K_3 \equiv \sqrt{\frac{2m(U_3 - E)}{\hbar^2}}. \tag{2.3.11}$$

Inside the well, the solution of the Schroedinger equation is once again given by that of a free particle. When the solution is expressed in terms of sines and cosines, Ψ_2 is given as

$$\Psi_2 = C_2 \sin(kx + \delta), \tag{2.3.12}$$

where k is defined as

$$k \equiv \sqrt{\frac{2mE}{\hbar^2}}. \tag{2.3.13}$$

The quantity δ is a phase term. Recall that the only difference between the sine and the cosine functions is a phase angle. Therefore the general method of writing a sine or a cosine function is to choose a sine function with a phase angle within the argument. Applying the boundary conditions at $x = 0.0$ yields

$$C_1 = C_2 \sin \delta,$$
$$K_1 C_1 = k C_2 \cos \delta. \tag{2.3.14}$$

Solving for K_1 yields

$$K_1 = k \cot \delta. \tag{2.3.15}$$

At $x = a$, application of the continuity of the wave function and its first derivative gives

$$C_3 e^{-K_3 a} = C_2 \sin(ka + \delta),$$
$$-K_3 C_3 e^{-K_3 a} = k C_2 \cos(ka + \delta). \tag{2.3.16}$$

Dividing the two equations, we obtain

$$-K_3 = k\cot(ka + \delta). \tag{2.3.17}$$

Therefore we obtain two transcendental equations. Substitute in for K_1 and k into the first of the two transcendental equations (that corresponding to $x = 0$). This yields

$$\cot\delta = \sqrt{\frac{m(U_1 - E)}{mE}},$$

$$\cot^2\delta = \frac{(U_1 - E)}{E} = \frac{1}{\sin^2\delta} - 1. \tag{2.3.18}$$

Substituting $\hbar^2 k^2/2m$ in for E, we can solve $\sin\delta$ as

$$\sin\delta = \frac{\hbar k}{\sqrt{(2mU_1 - \hbar^2 k^2) + \hbar^2 k^2}}. \tag{2.3.19}$$

The terms $\hbar^2 k^2$ in the denominator subtract out, leaving

$$\sin\delta = \pm\frac{\hbar k}{\sqrt{2mU_1}}. \tag{2.3.20}$$

Note that either the positive or the negative roots are feasible solutions for the first case (that corresponding to the first of the two possible transcendental equations). The positive root corresponds to a solution within the first quadrant while the negative root corresponds to a solution within the third quadrant. Equations (2.3.18) imply that $\cot\delta$ is positive since

$$\cot\delta = \sqrt{\frac{U_1 - E}{E}} > 0. \tag{2.3.21}$$

As such, the solution for δ is given as

$$\delta = n\pi + \sin^{-1}\frac{\hbar k}{\sqrt{2mU_1}}. \tag{2.3.22}$$

Next consider the second transcendental equation. The second transcendental equation is given above as

$$-K_3 = k\cot(ka + \delta), \quad K_3 \equiv \sqrt{\frac{2m(U_3 - E)}{\hbar^2}}. \tag{2.3.23}$$

Substituting in for K_3 and k and rearranging yields

$$-\frac{\sqrt{m(U_3 - E)}}{\sqrt{mE}} = \cot(ka + \delta). \tag{2.3.24}$$

Again making use of the trigonometric identity relating the sine and the cotangent, we can solve the expression for $\sin(ka + \delta)$ as

$$\sin(ka + \delta) = \frac{-\hbar k}{\sqrt{2mU_3}}. \tag{2.3.25}$$

The cotangent function is equal to a negative term. Because $\cot(ka+\delta)$ is negative, then the argument $(ka + \delta)$ must be in either the second or the fourth quadrant. Given this condition, the solution for $(ka + \delta)$ is

$$ka + \delta = m\pi + \sin^{-1}\left(\frac{-\hbar k}{2mU_3}\right). \tag{2.3.26}$$

When the following substitutions are made, the conditions on δ can be stated as follows:

$$\sin \delta = \frac{\hbar k}{\sqrt{2mU_1}} \equiv V_1,$$

$$\sin(ka + \delta) = -\frac{\hbar k}{\sqrt{2mU_3}} \equiv -V_2. \tag{2.3.27}$$

Therefore the phase angle δ can be solved in terms of V_1 from the first transcendental equation as

$$\delta = \sin^{-1} V_1 + n\pi. \tag{2.3.28}$$

When the second transcendental equation is used, the expression for $(ka + \delta)$ gives

$$(ka + \delta) = m\pi - \sin^{-1} V_2. \tag{2.3.29}$$

Hence ka can be solved as

$$ka = (ka + \delta) - \delta = m\pi - \sin^{-1} V_2 - (n\pi + \sin^{-1} V_1). \tag{2.3.30}$$

Since both m and n are integers, their difference is also an integer. The final solution then is

$$ka = m\pi - \sin^{-1} V_2 - \sin^{-1} V_1. \tag{2.3.31}$$

The above result constitutes the solution for the case in which the energy is less than both potential barriers.

The solution of this problem can be found graphically as well as by the method sketched out above. The functions are plotted together in Figure 2.3.5. The cotangent functions are periodic in π. A different solution is obtained then for any multiple of π, as is clear from inspection of Figure 2.3.5. The solution is obtained from the intersection of the two curves as shown in the figure. Alternatively, the solution can be found numerically.

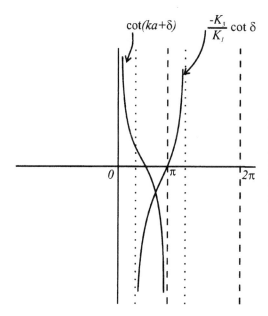

Figure 2.3.5. Rough sketch of $\cot(ka + \delta)$ and $-K_3/K_1 \cot \delta$. Note that there is a difference in phase between the two plots that is due to the different arguments. The two functions are periodic in π and hence will have solutions as multiples of π.

EXAMPLE 2.3.2 Semiconductor Quantum-Well Lasers

One of the most important developments in semiconductor device technology was the invention of semiconductor lasers. These devices provide a very low-power, coherent light source that is useful in communications systems, eye surgery, bar-code scanners, and compact disc players. Of the existing semiconductor laser structures, the quantum-well devices are of the most interest here as examples of a realistic application of finite quantum wells. A quantum-well semiconductor laser is made by the growth of two different semiconductor materials of different energy bandgaps on top of one another. As we will see in Chapter 8, the bandgap energy is the energy difference between nonconducting, localized electronic states called valence-band states and conduction-band states. As the name implies, electrons within the conduction band are free to propagate through the crystal in response to an applied electric field. Electrons within the conduction band carry the current in a solid.

Consider the structure sketched in Figure 2.3.6. The narrow-gap semiconductor layer is chosen here to be GaAs, while the wider-gap semiconductor material is AlGaAs. AlGaAs is a ternary semiconductor alloy that can be made with variable concentrations of Al and Ga. The most general way of writing AlGaAs is $Al_xGa_{1-x}As$, where x represents the mole fraction of Al present in the alloy. x can vary between 0 (pure GaAs) and 1 (pure AlAs). For the example considered here, we choose an Al concentration of 45%, $x = 0.45$. If the GaAs layer is made sufficiently small, $< \sim 150$ Å, spatial quantization effects occur. Under these conditions, quantum levels appear in both the conduction- and the valence-band wells, as sketched in Figure 2.3.6. Therefore the lowest energy that an electron can thermalize to is that of the $n = 1$ state in the conduction-band well. In other words, if an electron is injected into the GaAs layer, the lowest

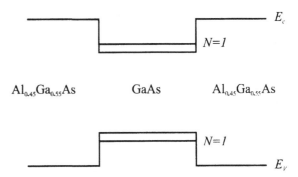

Figure 2.3.6. Example GaAs finite square well formed when a thin (≤ 150-Å) GaAs layer is sandwiched between two AlGaAs layers.

energy that it can attain is equal to that of the $n = 1$ level. In the valence band, a vacancy, commonly called a hole, carries the current. As we will discuss in Chapter 8, a hole is defined as a positively charged particle that is used to model the behavior of a vacancy in an otherwise filled band. Hole energies are defined to increase the deeper the hole moves into the valence band. Therefore the hole will have its minimum energy when it is at the top of the valence band. The lowest energy that a hole injected into the GaAs layer can have then is that of the $n = 1$ state in the valence-band well.

In a semiconductor laser, light is emitted when an electron recombines with a hole in the valence band. In Chapter 10, we will discuss in detail the processes that lead to light emission and absorption in a solid. For now we assume that photon emission occurs when an electron and a hole recombine. The photon of light created by this transition has an energy equal to the difference between the initial electron and hole energies. In the present case of a quantum-well laser, if we neglect any complications because of selection rules, this energy is equal to the sum of the $n = 1$ conduction-band level, the energy bandgap, and the $n = 1$ valence-band level, as can be seen from Figure 2.3.7.

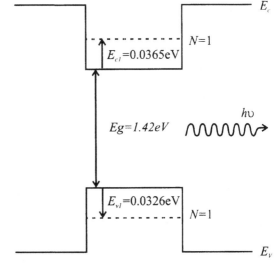

Figure 2.3.7. Sketch of the energy bands for a heterostructure semiconductor system showing the effects of spatial quantization levels on the energy of a photon emitted following an electron–hole recombination event.

To calculate the energy of the photon emitted during an electron–hole recombination event in a quantum-well structure, it is necessary to determine the energy levels of the electron and the hole states in the quantum well. The sum of these energies and the bandgap energy must equal the energy of the emitted photon. As an example, let us calculate the $n = 1$ levels for the electron and the hole in a GaAs/Al$_{0.45}$Ga$_{0.55}$As single quantum well 25 Å wide. If the conduction-band-edge discontinuity forming the electron well is assumed to be 0.345 eV and the valence-band-edge discontinuity forming the hole well is 0.213 eV, the $n = 1$ levels can then be computed numerically. Again, the mass of the electron is assumed to be equal to that of a free-space electron in both layers.

Treating the conduction- and the valence-band structures as separate quantum wells of different depths, 0.347 eV and 0.213 eV, respectively, the energy of the $n = 1$ states can be determined numerically. The $n = 1$ levels are found to be

conduction band $\quad n = 1 \quad E_1 = 0.0365$ eV,
valence band $\qquad n = 1 \quad E_1 = 0.0326$ eV.

The bandgap energy in GaAs is 1.42 eV. Therefore the energy of the emitted photon is simply given as the sum of the $n = 1$ levels in the conduction- and the valence-band quantum wells and the bandgap as $0.0365 + 0.0326 + 1.42 = 1.4891$ eV. Hence a semiconductor laser of this design will emit light of energy 1.489 eV. This corresponds to a wavelength of 834.0 nm.

2.4 One-Dimensional Finite Potential Barrier

In Section 2.2, we discussed the solution for a quantum-mechanical particle incident upon a finite potential barrier of infinite extent. Two different cases were considered: the incident particle's energy less than the potential barrier height and the incident particle's energy greater than the potential barrier height. In the situation in which the incident particle's energy is less than the potential barrier height, it was found that the particle's wave function decays exponentially into the barrier. In addition, no probability current density flows across the interface. All the probability density is reflected back into the incident region. What would happen, though, if the potential barrier was of finite extent? Specifically, it is interesting to consider what would happen if the potential barrier is sufficiently small that the wave function does not decay completely to zero before it reaches the second interface. Classically, there should be no difference in the result if the potential barrier is of finite or infinite spatial extent. In either case, if the particle's energy is less than the potential barrier height, it should be reflected by the barrier. However, something interesting occurs in the quantum-mechanical case. This is the topic of this section.

To solve the problem of a one-dimensional potential barrier, the Schroedinger equation in each region (to the left of the barrier, within the barrier, and to the

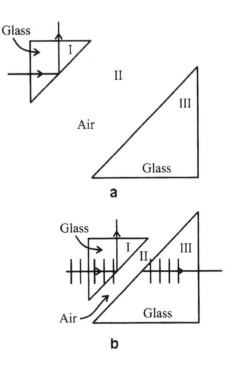

Figure 2.4.1. Illustration of an optical analogy to tunneling through a potential barrier by use of a two-prism system. a. When the two prisms are separated by a relatively large distance, light totally internally reflected in prism I does not propagate into prism III. b. When the spatial separation between the two prisms is shortened, light totally internally reflected in prism I can propagate in prism III even though they do not physically touch. Evanescent modes between the two prisms do not fully decay, leading to repropagation within prism III.

right of the barrier) needs to be solved. As in the previous problems considered in this chapter, after the general solution of the Schroedinger equation is obtained in each region, the boundary conditions at the interfaces must be applied. Before we actually solve the finite potential barrier problem, it is instructive to examine what happens for an optical analog of this problem.

Consider the setup sketched in Figure 2.4.1. A beam of light is incident upon a glass prism such as to be totally internally reflected at the other side, as shown in Figure 2.4.1.a. Hence no power flows out of the prism in the direction of incidence. The electromagnetic power is completely reflected back into region I (the initial prism). Even though the electromagnetic power transmitted into region II is zero, the electric field is not zero everywhere within region II (air in this case). This is because Maxwell's equations require that the tangential component of the electric field must be continuous across any boundary. Therefore, since the electric field is not zero within region I (the initial glass prism), then it must be nonzero within region II, at least near the interface. The field decays exponentially into region II. Thus, very far away from the interface the field will have decayed to zero. This is completely analogous to the situation studied in Section 2.3 for a quantum-mechanical particle incident upon a finite potential barrier of infinite spatial extent; no probability current density is transmitted across the potential barrier, yet the wave function remains continuous across the interface. As a result the wave function is not zero everywhere within the potential barrier but has a finite penetration distance.

What happens, though, if another prism is placed close to, but not in contact with, the first prism? As is well known in optics, a ray of light is seen to emerge from the second prism, as shown in Figure 2.4.1.b. Even though the two prisms are not in contact, power can flow from the first prism to the second. This is due to the fact that the electric-field component has not completely decayed to zero within the thin region of air separating the two prisms. Therefore the field can repropagate within the second prism, and power ultimately flows from region I into region III.

Based on this result in optics, the result in quantum mechanics can be understood. Consider what happens if the finite potential barrier is made sufficiently small that the wave function does not decay to zero before it reaches the third region. As in the case for optics, the wave function will reemerge on the other side of the barrier. This implies that there is a probability of finding the particle on the right-hand side of the barrier given that it is incident from the left-hand side when its energy is less than the barrier height! In other words, there is a probability that the particle can tunnel through the potential barrier. Such a surprising result occurs from the wave nature of the electron. Let us demonstrate tunneling through a finite potential barrier by solving the Schroedinger equation.

A sketch of the finite potential barrier problem that we are interested in is presented in Figure 2.4.2. The structure is partitioned into three separate regions as marked in the figure. The potential barrier has height V_0 and width b, as shown in the diagram. The particle is assumed to be incident only from the left-hand side (region I). The Schroedinger equation and its solution in this region, where the potential is zero, are given as

$$-\frac{\hbar^2}{2m}\frac{\partial^2}{\partial x^2}\Psi = E\Psi,$$

$$\Psi_1 = e^{ik_1 x} + re^{-ik_1 x}, \quad k_1 = \sqrt{\frac{2mE}{\hbar^2}}. \tag{2.4.1}$$

In the above, it is assumed that the incident-wave function is normalized. As a result, the coefficient in front of the first term in the expression for Ψ_1 is 1. The solution within the barrier, assuming that the energy is less than the potential in

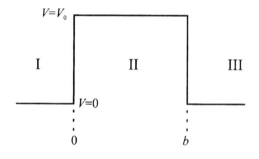

Figure 2.4.2. One-dimensional finite potential barrier of height V_0 and width b.

region II, is given as

$$-\frac{\hbar^2}{2m}\frac{\partial^2\Psi}{\partial x^2} + V_0\Psi = E\Psi,$$

$$\Psi_2 = C_2^+ e^{k_2 x} + C_2^- e^{-k_2 x}, \quad k_2 = \sqrt{\frac{2m(V_0 - E)}{\hbar^2}}. \qquad (2.4.2)$$

Note that the electron wave function has a nonzero value within the potential barrier. It is important to recognize that when the energy of the particle is less than the potential barrier height, the presence of the particle within the barrier is classically forbidden. If the barrier width is not too large, the wave function will be appreciably nonzero at the second interface. The solution of the Schroedinger equation within the third region is equal to that within the first region since the potential is exactly the same therein. Therefore the wave function within region III is given as

$$\Psi_3 = t e^{ik_1 x}. \qquad (2.4.3)$$

Applying the boundary conditions of the continuity of the wave function and of the probability current density (which is equivalent to the continuity of the first derivative of the wave function; see Box 2.2.1) at the interfaces $x = 0$ and $x = b$ leads to

$$1 + r = C_2^+ + C_2^-,$$

$$ik_1 - ik_1 r = k_2 C_2^+ - k_2 C_2^-$$

for $x = 0.0$, and

$$C_2^+ e^{k_2 b} + C_2^- e^{-k_2 b} = t e^{ik_1 b},$$

$$k_2 C_2^+ e^{k_2 b} - k_2 C_2^- e^{-k_2 b} = ik_1 t e^{ik_1 b} \qquad (2.4.4)$$

for $x = b$. Equations (2.4.4) can be written in terms of the four unknowns C_2^+, C_2^-, t, and r. These equations can be solved by elimination or by matrix methods. The solution of these equations is left as an exercise for the reader. The solutions for the transmission and the reflection amplitudes are obtained as

$$t = \frac{4i k_1 k_2 \, e^{(k_2 - ik_1)b}}{(k_2 + ik_1)^2 - (k_2 - ik_1)^2 e^{2k_2 b}},$$

$$r = \frac{(k_2^2 + k_1^2)(1 - e^{-2k_2 b})}{(k_2 + ik_1)^2 e^{-2k_2 b} - (k_2 - ik_1)^2}. \qquad (2.4.5)$$

The transmission and the reflection coefficients are real numbers and are given by the squares of the amplitudes t and r. These are

$$T = \left[1 + \frac{V_0^2 \sinh^2 k_2 b}{4E(V_0 - E)}\right]^{-1},$$

$$R = \left[1 + \frac{4E(V_0 - E)}{V_0^2 \sinh^2 k_2 b}\right]^{-1}. \tag{2.4.6}$$

The above results are correct for incident particles whose energy is less than the barrier height. As is obvious from these results, there is a nonzero probability that a particle incident with energy less than the potential barrier height will reemerge from the barrier. This is called quantum-mechanical tunneling and has great importance in solid-state devices. Before we discuss tunneling in multiple barrier structures, let us consider the situation in which the incident particle has energy greater than the barrier height.

At incident-electron energies greater than the barrier height, the transmissivity is not always 100%. Again, because of the wave nature of the electron, there are only certain incident energies for which the electron is always transmitted. This is of course in variance with what is expected from classical mechanics. In classical mechanics, any particle whose energy exceeds the potential barrier height must be transmitted across it. However, waves are not necessarily transmitted without reflection across a boundary. An elementary example of this is a wave incident from a string of one mass density into another string with a different mass density. As is well known, depending on the conditions, the incident wave may be partially reflected and partially transmitted at the boundary.

A more familiar example to electrical engineers is that of a wave propagating from one transmission line into another. If a wave is incident from one transmission line to another, partial reflection at the interface will occur unless the lines are impedance matched. The problem can be represented as a transmission line with characteristic impedance Z_0, terminated by a load with impedance Z_L, as shown in Figure 2.4.3. If the origin of our coordinate system is taken at the load, then the voltage on the line is simply given by the linear combination of a wave traveling to the right, $V_- e^{ikz}$, and a wave traveling to the left, $V_+ e^{-ikz}$. The voltage on the line can be written then as

$$V = V_+ e^{-ikz} + V_- e^{ikz}. \tag{2.4.7}$$

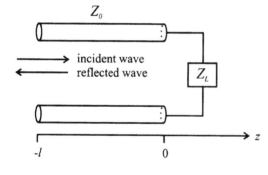

Figure 2.4.3. Transmission line model for incident and reflected waves terminated by a load Z_L.

An alternative expression for the potential can be written in terms of the reflection coefficient Γ as

$$V = V_+(e^{-ikz} + \Gamma e^{ikz}), \tag{2.4.8}$$

where Γ is defined as

$$\Gamma = \frac{V_-}{V_+}. \tag{2.4.9}$$

The current in the line is determined from the derivative of the voltage with respect to the position z:

$$\frac{dV}{dz} = -i\omega L I, \tag{2.4.10}$$

where L and ω are the inductance per unit length and the circular frequency, respectively. Defining k and Z_0 as

$$k \equiv \omega\sqrt{LC}, \qquad Z_0 \equiv \sqrt{\frac{L}{C}}, \tag{2.4.11}$$

we can write the current as

$$I = \frac{1}{Z_0} V_+(e^{-ikz} - \Gamma e^{ikz}). \tag{2.4.12}$$

The impedance at any point along the line is simply defined as the ratio of the voltage to the current at that point. Hence $Z(z)$ is

$$Z(z) = \frac{V(z)}{I(z)} = Z_0 \frac{e^{-ikz} + \Gamma e^{ikz}}{e^{-ikz} - \Gamma e^{ikz}}. \tag{2.4.13}$$

The impedance at the origin $z = 0$ is then given as

$$Z_L = Z_0 \frac{(1 + \Gamma)}{(1 - \Gamma)}, \tag{2.4.14}$$

since $z(0)$ is simply the impedance of the load Z_L. The reflection coefficient can now be easily found in terms of the impedance as

$$\Gamma = \frac{(Z_L - Z_0)}{(Z_L + Z_0)}. \tag{2.4.15}$$

Note that when the numerator vanishes in the expression for the reflection coefficient above, the wave is not reflected at the load. This condition, in which the impedance of the load and the impedance of the line are equal such that no reflection occurs, is called impedance matching. In other words, when two transmission lines are impedance matched, no reflection will occur at the interface between the two lines. Otherwise, in general, there will always be a reflected wave at the interface.

The physics of an electron incident upon a potential barrier with energy in excess of the barrier height is precisely the same as that of a wave incident from one transmission line into another transmission line, as discussed above. In general, there will be a reflected wave amplitude at the interface in either case. However, just as in the case of transmission lines, there are conditions under which the transmission coefficient of an electron onto the barrier is 100%; there is no reflected wave. In order to observe this effect, let us calculate the transmission and the reflection coefficients for a one-dimensional barrier for an electron whose incident energy exceeds the barrier height, $E > V_0$.

The simplest means of obtaining the transmission and the reflection coefficients for $E > V_0$ is to simply substitute ik_2 for k_2 into the expressions for T and R given above. k_2 becomes

$$k_2 = \sqrt{\frac{[2m(E - V_0)]}{\hbar^2}}. \tag{2.4.16}$$

With this substitution, the transmission and the reflection coefficients become

$$T = \left[1 + \frac{V_0^2 \sin^2 k_2 b}{4E(E - V_0)}\right]^{-1},$$

$$R = \left[1 + \frac{4E(E - V_0)}{V_0^2 \sin^2 k_2 b}\right]^{-1}. \tag{2.4.17}$$

Clearly, when $\sin^2 k_2 b$ is zero, the transmission coefficient becomes equal to 1 and the reflection coefficient becomes equal to zero. This is equivalent to the impedance-matching condition discussed for transmission lines.

As an example of transmission of an electron over a one-dimensional potential barrier, consider a barrier of height 0.347 eV and width 20.0 Å. The transmission coefficient as a function of incident-electron energy is sketched in Figure 2.4.4. The transmissivity reaches a peak value at an incident energy of 0.44 eV. The transmissivity, although small, is nonetheless nonzero at electron energies below that of the barrier height. Interestingly, the transmissivity drops significantly above 0.44 eV. The incident energy of 0.44 eV corresponds to the impedance-matching condition for this particular barrier width and height.

As a further example, consider the transmission of a barrier of 10.0 Å in width and height of 1.0 eV. The transmissivity as a function of energy is plotted in Figure 2.4.5. Again note that the transmission coefficient peaks to unity and then drops off with increased energy.

2.5 Multiple Quantum Wells

In the previous sections of this chapter, we have discussed the infinite and the finite square-well and the potential barrier problems. A natural extension of these problems is a multiple-quantum-well system, as shown in Figure 2.3.2. Such

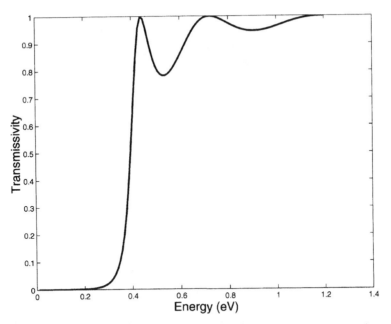

Figure 2.4.4. Transmissivity as a function of incident-electron energy for a single barrier of width 20 Å and height 0.347 eV.

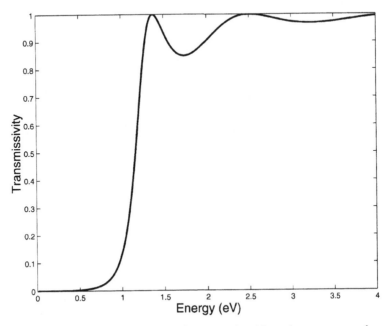

Figure 2.4.5. Transmissivity as a function of incident-electron energy for a single barrier of width 10 Å and height of 1.0 eV.

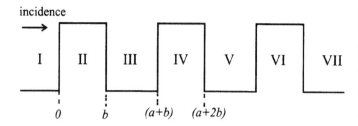

incidence

I II III IV V VI VII

0   b   (a+b)   (a+2b)

Figure 2.5.1. Multibarrier potential structure for a general number of barriers and wells. Each potential well is assumed to be the same and of width a and height V_0. The barriers are all assumed to be identical of width b.

structures are of importance in semiconductor devices, particularly in lasers. One of the most important semiconductor laser designs uses a series of multiple quantum wells. In this section, we consider the physics of a multiple-quantum-well system.

The specific structure that we consider is sketched in Figure 2.5.1. The barrier heights are taken to be V, while the widths of the barriers and the wells are b and a, respectively. The different regions of the solution are apparent from inspection of Figure 2.5.1. We want to find a general solution of the problem for an arbitrary number of wells and barriers as well as for arbitrary potential heights and layer widths. Again, there is more than one case to consider. The first case is when the incident electrons have energy less than the barrier height. Another case is when the incident electrons have energy greater than the barrier height. The first situation is considered here, and the second case is left as an exercise for the reader.

In region I, the Schroedinger equation and its solution are

$$\frac{-\hbar^2}{2m}\frac{d^2\Psi_1}{dx^2} = E\Psi_1,$$ (2.5.1)

$$\Psi_1 = e^{ik_1 x} + R e^{-ik_1 x}, \quad k_1 = \sqrt{\frac{2mE}{\hbar^2}},$$ (2.5.2)

where it is assumed that the particle is incident from only the left and is normalized. In region II, the Schroedinger equation and its solution, assuming that the energy is less than the potential barrier height, are readily found as

$$-\frac{\hbar^2}{2m}\frac{d^2\Psi_2}{dx^2} + V\Psi_2 = E\Psi_2,$$ (2.5.3)

$$\Psi_2 = C_2^+ e^{k_2 x} + C_2^- e^{-k_2 x}, \quad k_2 = \sqrt{\frac{2m(V-E)}{\hbar^2}}.$$ (2.5.4)

Both the exponentially increasing and decreasing functions are included in the solution. In the case of the finite potential barrier of infinite extent, the coefficient of the exponentially increasing term is zero. This was due to the boundary condition at $x = +\infty$; the wave function must go to zero as x approaches infinity. In the case of a finite barrier of finite spatial extent, the exponentially increasing term does not necessarily vanish. Therefore both terms are retained in the solution. The solution within region II can be written in an alternative fashion with the definition of the hyperbolic functions. Ψ_2 can be written then as

$$\Psi_2 = C_2^+ \cosh k_2 x + C_2^- \sinh k_2 x. \tag{2.5.5}$$

Similarly, the Schroedinger equation and its solution in region III are

$$-\frac{\hbar^2}{2m}\frac{d^2\Psi_3}{dx^2} = E\Psi_3, \tag{2.5.6}$$

$$\Psi_3 = C_3^+ \cos k_1 x + C_3^- \sin k_1 x, \quad k_1 = \sqrt{\frac{2mE}{\hbar^2}}. \tag{2.5.7}$$

Since no field is applied to the structure, it is translationally symmetric. As a result, the solution remains the same in each similar region, well or barrier layer, respectively, independent of which specific well or barrier is considered. If a field is applied to the structure, the translational symmetry of the structure will be lifted. The electron energy then depends on its location within the structure. For example, the action of the electric field acts to change the potential energy of the structure as a function of position. Therefore the potential energy changes necessarily from point to point on the application of an electric field.

The imposition of the boundary conditions at the interfaces, the continuity of the wave function, and the probability current density lead to a specific solution. At the first interface the boundary conditions require that

$$e^{ik_1 x} + R e^{-ik_1 x} = C_2^+ \cosh k_2 x + C_2^- \sinh k_2 x, \tag{2.5.8}$$

$$\frac{ik_1}{m}e^{ik_1 x} - \frac{ik_1}{m}R e^{-ik_1 x} = \frac{k_2}{m}C_2^+ \sinh k_2 x + \frac{k_2}{m}C_2^- \cosh k_2 x. \tag{2.5.9}$$

In the above, it is assumed that the mass is the same everywhere in the structure and is simply equal to m. Substituting $x = 0.0$ in for x, we can rearrange the above equations in matrix form as

$$\begin{bmatrix} 1 & 1 \\ \frac{ik_1}{m} & \frac{-ik_1}{m} \end{bmatrix} \begin{bmatrix} 1 \\ R \end{bmatrix} = \begin{bmatrix} 1 & 0 \\ 0 & \frac{k_2}{m} \end{bmatrix} \begin{bmatrix} C_2^+ \\ C_2^- \end{bmatrix}. \tag{2.5.10}$$

Similarly, at the second interface, the boundary conditions give

$$C_2^+ \cosh k_2 b + C_2^- \sinh k_2 b = C_3^+ \cos k_1 b + C_3^- \sin k_1 b,$$

$$\frac{k_2}{m}C_2^+ \sinh k_2 b + \frac{k_2}{m}C_2^- \cosh k_2 b = -\frac{k_1}{m}C_3^+ \sin k_1 b + \frac{k_1}{m}C_3^- \cos k_1 b.$$

$$\tag{2.5.11}$$

Rewriting in matrix form, we obtain

$$
\begin{bmatrix} \cosh k_2 b & \sinh k_2 b \\ \frac{k_2}{m}\sinh k_2 b & \frac{k_2}{m}\cosh k_2 b \end{bmatrix} \begin{bmatrix} C_2^+ \\ C_2^- \end{bmatrix} = \begin{bmatrix} \cos k_1 b & \sin k_1 b \\ \frac{-k_1}{m}\sin k_1 b & \frac{k_1}{m}\cos k_1 b \end{bmatrix} \begin{bmatrix} C_3^+ \\ C_3^- \end{bmatrix}.
$$

$$(2.5.12)$$

The column matrix for the coefficients C_2^+ and C_2^- can be solved for in Eq. (2.5.12) and then substituted into Eq. (2.5.10). Solving for C_2^+ and C_2^- yields

$$
\begin{bmatrix} C_2^+ \\ C_2^- \end{bmatrix} = \frac{m}{k_2}\begin{bmatrix} \frac{k_2}{m}\cosh k_2 b & -\sinh k_2 b \\ \frac{-k_2}{m}\sinh k_2 b & \cosh k_2 b \end{bmatrix} \begin{bmatrix} \cos k_1 b & \sin k_1 b \\ -\frac{k_1}{m}\sin k_1 b & \frac{k_1}{m}\cos k_1 b \end{bmatrix} \begin{bmatrix} C_3^+ \\ C_3^- \end{bmatrix},
$$

$$(2.5.13)$$

where the matrix involving the hyperbolic sine and the hyperbolic cosine has been inverted. Substituting the above expression for the column matrix for C_2^+ and C_2^- yields

$$
\begin{bmatrix} 1 & 1 \\ \frac{ik_1}{m} & \frac{-ik_1}{m} \end{bmatrix}\begin{bmatrix} 1 \\ R \end{bmatrix} = \begin{bmatrix} 1 & 0 \\ 0 & \frac{k_2}{m} \end{bmatrix}\begin{bmatrix} \cosh k_2 b & -\frac{m}{k_2}\sinh k_2 b \\ -\sinh k_2 b & \frac{m}{k_2}\cosh k_2 b \end{bmatrix}
$$
$$
\times \begin{bmatrix} \cos k_1 b & \sin k_1 b \\ -\frac{k_1}{m}\sin k_1 b & \frac{k_1}{m}\cos k_1 b \end{bmatrix}\begin{bmatrix} C_3^+ \\ C_3^- \end{bmatrix}.
$$

$$(2.5.14)$$

Applying the boundary conditions at the third interface, those corresponding to $x = (a+b)$, yields

$$
C_3^+ \cos k_1(a+b) + C_3^- \sin k_1(a+b)
$$
$$
= C_4^+ \cosh k_2(a+b) + C_4^- \sinh k_2(a+b),
$$

$$(2.5.15)$$

$$
\frac{-k_1}{m}C_3^+ \sin k_1(a+b) + \frac{k_1}{m}C_3^- \cos k_1(a+b)
$$
$$
= \frac{k_2}{m}C_4^+ \sinh k_2(a+b) + \frac{k_2}{m}C_4^- \cosh k_2(a+b).
$$

$$(2.5.16)$$

Again, we rewrite the above equations in matrix form, solving for C_3^+ and C_3^- as

$$
\begin{bmatrix} C_3^+ \\ C_3^- \end{bmatrix} = \frac{m}{k_1}\begin{bmatrix} \frac{k_1}{m}\cos k_1(a+b) & -\sin k_1(a+b) \\ \frac{k_1}{m}\sin k_1(a+b) & \cos k_1(a+b) \end{bmatrix}
$$
$$
\times \begin{bmatrix} \cosh k_2(a+b) & \sinh k_2(a+b) \\ \frac{k_2}{m}\sinh k_2(a+b) & \frac{k_2}{m}\cosh k_2(a+b) \end{bmatrix}\begin{bmatrix} C_4^+ \\ C_4^- \end{bmatrix}.
$$

$$(2.5.17)$$

Substituting the result for the column matrix C_3^+ and C_3^- given by Eq. (2.5.17) into Eq. (2.5.14) yields

$$
\begin{bmatrix} 1 & 1 \\ \frac{ik_1}{m} & \frac{-ik_1}{m} \end{bmatrix} \begin{bmatrix} 1 \\ R \end{bmatrix}
$$

$$
= \begin{bmatrix} 1 & 0 \\ 0 & \frac{k_2}{m} \end{bmatrix} \begin{bmatrix} \cosh k_2 b & -\frac{m}{k_2}\sinh k_2 b \\ -\sinh k_2 b & \frac{m}{k_2}\cosh k_2 b \end{bmatrix}
$$

$$
\times \begin{bmatrix} \cos k_1 b & \sin k_1 b \\ -\frac{k_1}{m}\sin k_1 b & \frac{k_1}{m}\cos k_1 b \end{bmatrix} \begin{bmatrix} \cos k_1(a+b) & -\frac{m}{k_1}\sin k_1(a+b) \\ \sin k_1(a+b) & \frac{m}{k_1}\cos k_1(a+b) \end{bmatrix}
$$

$$
\times \begin{bmatrix} \cosh k_2(a+b) & \sinh k_2(a+b) \\ \frac{k_2}{m}\sinh k_2(a+b) & \frac{k_2}{m}\cosh k_2(a+b) \end{bmatrix} \begin{bmatrix} C_4^+ \\ C_4^- \end{bmatrix}. \tag{2.5.18}
$$

The calculation continues iteratively in this fashion until the carrier crosses through all the interfaces. Each matrix that describes the boundary conditions at each interface is called a transfer matrix. In general, if the product of all the transfer matrices that couple the incoming wave vector to the outgoing wave vector through the heterostructure stack is taken as M, then the equation can be reduced into a simpler form. The reflection column matrix can be written in terms of the net transfer matrix M and the final transmission matrix (transmission into the propagating region to the far right-hand side of the stack, region VII in Figure 2.5.1) as

$$
\begin{bmatrix} 1 \\ R \end{bmatrix} = \frac{1}{2ik} \begin{bmatrix} ik & 1 \\ ik & -1 \end{bmatrix} \begin{bmatrix} M_{11} & M_{12} \\ M_{21} & M_{22} \end{bmatrix} \begin{bmatrix} 1 & 1 \\ ik & -ik \end{bmatrix} \begin{bmatrix} \tau \\ 0 \end{bmatrix}. \tag{2.5.19}
$$

In Eq. (2.5.19), it is assumed that the last region is physically identical to the incident region that results in the same k vector's appearing in each region. This is true if no bias is applied across the structure such that it remains translation-ally symmetric. The M matrix represents the result of multiplying the transfer matrices for the heterostructure stack. Finally, the last column matrix has only one nonzero entry, τ, which corresponds to the coefficient of the transmitted wave. The zero entry appears since it is assumed that no wave is incident from the right-hand side. We can find the transmission coefficient for the stack by multiplying the matrices on the right-hand side of Eq. (2.5.19):

$$
\begin{bmatrix} 1 \\ R \end{bmatrix} = \begin{bmatrix} A & B \\ C & D \end{bmatrix} \begin{bmatrix} \tau \\ 0 \end{bmatrix}. \tag{2.5.20}
$$

Multiplying the top row yields

$$
1 = A\tau. \tag{2.5.21}
$$

The transmission coefficient T is found from the square of the transmission amplitude τ as

$$\tau\tau^* = T = \left(\frac{1}{A}\right)\left(\frac{1}{A}\right)^*, \tag{2.5.22}$$

where A is given as

$$A = \frac{1}{2ik}[ikM_{11} - k^2M_{12} + M_{21} + ikM_{22}]. \tag{2.5.23}$$

Note that once the net transfer matrix for the multiple-quantum-well system has been determined, the transmission coefficient can be readily evaluated.

The above result provides a prescription for determining the transmission coefficient of an electron through a multiple finite square well or a multiple-quantum-well structure under zero applied external bias. The solution of the tunneling probability through a multiple-quantum-well structure is of importance in new semiconductor devices. Recently, new crystal-growth techniques, such as molecular-beam epitaxy (MBE), metal-organic chemical-vapor deposition (MOCVD), and chemical-beam epitaxy (CBE), have enabled semiconductor device engineers to make multilayered structures with atomic layer dimensions. Typical structures are made with alternating layers of GaAs and $Al_xGa_{1-x}As$. As we discussed in Section 2.3, the GaAs layers form the wells and the Al-GaAs layers form the barriers in the structure. MBE and CBE are so exacting that crystals can be grown consisting of monolayers of each atomic species. In other words, a structure can be made that consists of alternating layers of GaAs and AlGaAs, each one atom wide! Depending on the the number of alternating layers grown, various multiple-quantum-well geometries can be formed. If the quantum-mechanical tunneling transmission coefficient is nonzero throughout the structure, all the wells are said to be coupled. A coupled multiple-quantum-well structure is given a special name, a superlattice. If the wells are not coupled, the structure is called a multiple-quantum-well device. More is written about superlattices and multiple-quantum-well devices in Chapters 7, 12, and 13.

The transmission coefficient for a multiquantum-well structure is in general algebraically tedious to determine. Therefore it is useful to solve the problem on a computer. Again, a computer code is available at the book web site for this purpose. As an example, let us consider the numerical solution for the transmission coefficient of a multiple-well system.

EXAMPLE 2.5.1

Calculate the transmission coefficient of an electron incident upon a two-well, three-barrier multiple-quantum-well structure. Let us consider a realistic system that can be readily fabricated in the laboratory with MBE, MOCVD, or CBE crystal-growth techniques. The most typical materials system used in growing multiquantum-well structures is the GaAs/AlGaAs system. Specifically, let us choose a 45% Al composition, $Al_{0.45}Ga_{0.55}As$. The difference in the energy bandgaps between the GaAs and the AlGaAs layers in this case is roughly

0.56 eV. The difference in the energy gaps between the two materials must be accounted for in the conduction and the valence bands. Since the energy gaps are not the same, there must be a discontinuity in the conduction and the valence bands to account for the energy-gap difference. In the GaAs/AlGaAs materials system, the energy-gap difference is equal to the sum of the conduction- and the valence-band discontinuities. Experimental results indicate that the conduction-band-edge discontinuity is roughly equal to 62% of the energy-gap difference. Using 62% for the energy-gap difference, the conduction-band edge discontinuity is calculated to be 0.347 eV.

As we will see in Chapter 8, the electron mass within a solid is typically different from its free-space mass. The electron mass in a solid is called the effective mass. In GaAs, the effective mass is 0.067 times the free-electron mass. In $Al_{0.45}Ga_{0.55}As$ the effective mass is 0.1087 times the free-electron mass. The transmission coefficient as a function of incident-electron energy can be determined numerically. Let us consider two different conditions.

a. Determine the peak in the transmission coefficient of an electron if the wells are 50 Å and the barriers are 50 Å in width.

The logarithm of the transmission coefficient of the structure at various incident-electron energies is shown in Figure 2.5.2. Inspection of the figure shows that the peak occurs $\sim E = 0.082$ eV. Note that only one peak appears at low energy in the transmission coefficient. There is only one energy for which the

Figure 2.5.2. Logarithm of the transmissivity versus incident-electron energy for a two-well, three-barrier device. The barriers are assumed to be 50 Å in width and 0.347 eV in height. The wells are assumed for be 50 Å in width.

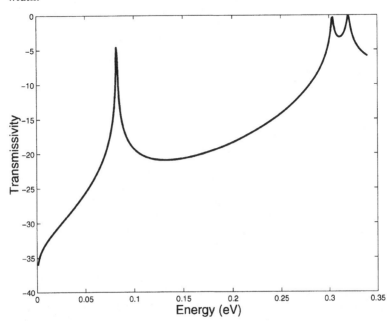

multiquantum-well system appears transparent to an incoming electron at low energy. At energies away from the peak energy, the transmission coefficient is relatively small, and, as such, there is little chance that the electron will be able to penetrate through the entire structure and emerge on the right-hand side. Therefore we see an important result; the probability that an electron, with incident energy less than the barrier height, can penetrate through the entire structure is a strong function of its incident energy.

What, though, is so special about the energy that corresponds to the peak in the transmissivity curve? To answer this question it is instructive to examine a one-well structure.

Consider a single-quantum-well structure made with a 30-Å-wide GaAs well sandwiched between two $Al_{0.45}Ga_{0.55}As$ layers. As in the above example, the well depth is 0.347 eV. For simplicity, let us assume that the electron masses are simply equal to the free-space electron mass in either layer. The eigenenergies of the well are calculated using Eqs. (2.3.9) to be

$$n = 1, \qquad E_1 = 0.0273 \text{ eV};$$
$$n = 2, \qquad E_2 = 0.109 \text{ eV};$$
$$n = 3, \qquad E_3 = 0.236 \text{ eV}.$$

Alternatively, if we calculate the transmission coefficient as a function of incident-electron energy (see Figure 2.5.3), the peaks in the transmissivity as a function

Figure 2.5.3. Logarithm of the transmissivity versus incident-electron energy for a single 30-Å well, GaAs/AlGaAs structure with a 0.347-eV barrier height. Note that three peaks appear in the transmissivity curve corresponding to the three eigenenergies of the well.

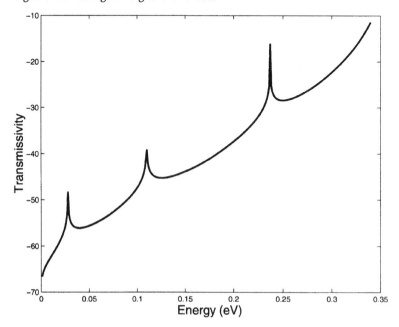

of energy occur roughly at incident energies of:

$E_1 = 0.028$ eV,
$E_2 = 0.110$ eV,
$E_3 = 0.237$ eV.

Comparison with the above values calculated with the finite square-well formula shows that the peaks in the transmission curve correspond to the confined-state eigenenergies. Therefore we conclude that the peaks in the transmissivity curves occur precisely at the eigenenergies of the well. Another way to interpret this result is that the peaks in the transmission coefficient occur at the resonances of the well. Hence, if an electron is incident at an energy that is a resonant energy of the structure, it will be transmitted through the structure. Otherwise, it most likely will be reflected from the structure. To check the validity of this assumption, the wave functions are plotted for each of the three peaks in Figures 2.5.4, 2.5.5, and 2.5.6, respectively. As can be clearly seen from these figures, the first peak corresponds to the $n = 1$ state (note there are no nodes in the plot in Figure 2.5.4), the second peak corresponds to the $n = 2$ state (Figure 2.5.5), while the third peak corresponds to the $n = 3$ state (Figure 2.5.6).

From the above result, it is clear that the eigenenergies of a quantum-well system can be determined from the transmissivity of the structure. By calculating the peaks in the transmission coefficient versus the incident-energy curve, the eigenenergies of the structure can be found.

Figure 2.5.4. Plot of the wave function versus position corresponding to the first peak in Figure 2.5.3 calculated with the transmissivity approach. Note that this wave function closely resembles that for the $n = 1$ state of an infinite square-well potential.

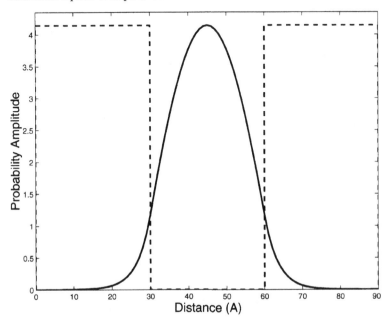

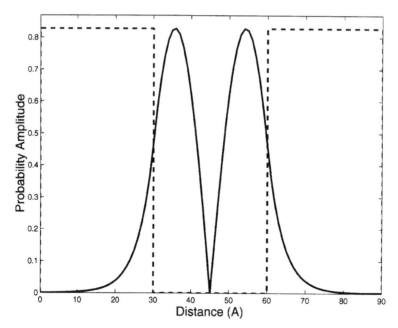

Figure 2.5.5. Plot of the wave function versus position corresponding to the second peak in Figure 2.5.3 calculated with the transmissivity approach. Note that the wave function resembles that of a sine wave, similar to that for the $n = 2$ state of an infinite square-well potential.

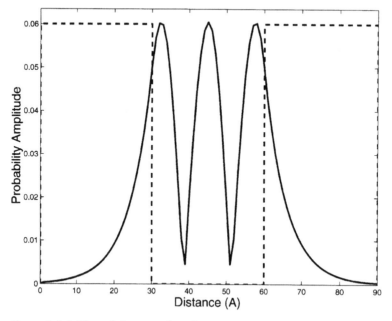

Figure 2.5.6. Plot of the wave function versus position corresponding to the third peak in Figure 2.5.3 calculated with the transmissivity approach. Note that the wave function has two nodes resembling those for the $n = 3$ state of an infinite square-well potential.

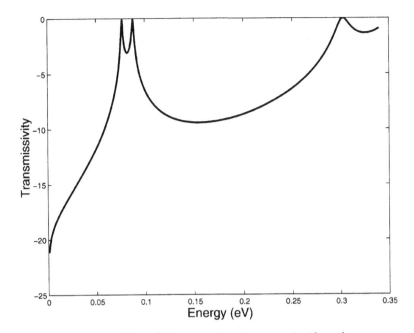

Figure 2.5.7. Logarithm of the transmissivity versus incident-electron energy for a two-well structure, similar to that shown in Figure 2.5.2, except that the barrier widths have been reduced to 25 Å. Note that two peaks now appear at around 0.10 eV. This is due to the interaction of the electron wave function between the two wells of the structure.

b. Determine the transmission coefficient in a two well, three-barrier structure whose dimensions are

well 1 and well 2 = 50 Å each,
barriers 1, 2, and 3 = 25 Å each.

This structure is similar to the two-well structure examined in part a above. Both structures have exactly the same well widths. The only significant difference between the two structures is that the barriers are smaller in the present case than in the first case. The transmission coefficient as a function of incident-electron energy is presented in Figure 2.5.7. It is interesting to compare the calculated transmissivity for this structure with that shown in Figure 2.5.2. Note that for the structure shown in Figure 2.5.2, only one peak in the transmissivity occurs at low energy while two peaks appear at low energy in Figure 2.5.7. Apparently the smaller barrier widths lead to a splitting of the allowed energy eigenvalues in the structure. As the barrier widths decrease, there is a greater interaction of the electron wave functions between each well. As we will see in Chapter 7, as the electron wave functions interact more, the peak in the transmissivity curve splits apart more. This has important ramifications in the study of energy bands in crystals. We will return to this problem in Chapter 7 and discuss its implications in greater detail.

Before we end this section on tunneling, it is important to consider another approach known as the WKB approximation. The WKB approximation is useful in cases in which the potential changes slowly with respect to the de Broglie wavelength of the electron. There are several examples in semiconductor devices in which the WKB approximation can be made to estimate the tunneling current density. One of the most important examples is the tunnel diode. The tunnel diode is famous for being one of the first direct manifestations of quantum-mechanical effects at a macroscopic level. Basically, the tunnel diode is a reverse-biased diode in which tunneling of electrons from within the valence band into the conduction band occurs. As a result, a substantial current can flow under reverse-bias conditions.

Let us consider the WKB approximation in one dimension. The Schroedinger equation is

$$-\frac{\hbar^2}{2m}\frac{d^2\Psi}{dx^2} + V(x) = E\Psi.$$

(2.5.24)

Assume a general solution of the form

$$\Psi(x) = A(x)e^{\frac{iS(x)}{\hbar}},$$

(2.5.25)

which is a generalized plane-wave state with amplitude $A(x)$ and phase $S(x)$. The term involving the second derivative in the Schroedinger equation becomes, on substitution for $\Psi(x)$,

$$-\frac{\hbar^2}{2m}\left[\frac{d^2A}{dx^2} + \frac{2i}{\hbar}\frac{dA}{dx}\frac{dS}{dx} + \frac{i}{\hbar}A\frac{d^2S}{dx^2} - \frac{A}{\hbar^2}\left(\frac{dS}{dx}\right)^2\right]e^{\frac{iS}{\hbar}}.$$

(2.5.26)

Combining the second-derivative term with the terms involving the energy and the potential yields

$$A\left[\frac{1}{2m}\left(\frac{dS}{dx}\right)^2 + V(x) - E\right] - \frac{i\hbar}{2m}\left[2\frac{dA}{dx}\frac{dS}{dx} + A\frac{d^2S}{dx^2}\right] - \frac{\hbar^2}{2m}\frac{d^2A}{dx^2} = 0.$$

(2.5.27)

If $\hbar$ is regarded as a parameter in smallness, then the term involving $\hbar^2$ can be neglected. The first term is then of the order of unity while the second term is of the order of $\hbar$. Because the two terms are of highly different orders of magnitude, to preserve the equality, they must each be separately equal to zero. Once this assignment is made, the first term immediately yields

$$\frac{dS}{dx} = \sqrt{2m(E-V)},$$

(2.5.28)

while the second term is

$$2\frac{dA}{dx}\frac{dS}{dx} + A\frac{d\left(\frac{dS}{dx}\right)}{dx} = 0.$$

(2.5.29)

The derivative of S with respect to x can be associated with the momentum as

$$\frac{dS}{dx} = p. \tag{2.5.30}$$

When this assignment is made, the second term becomes

$$\frac{d}{dx}(A^2 p) = 0. \tag{2.5.31}$$

A is clearly

$$A = \frac{C}{\sqrt{p}}, \tag{2.5.32}$$

where C is an arbitrary constant. We can find the solution for $\Psi(x)$ using Eq. (2.5.32) and by determining $S(x)$. We determine $S(x)$ by integrating out $dS/dx = p(x)$:

$$S(x) = \int p(x)\,dx. \tag{2.5.33}$$

Therefore $\Psi(x)$ is

$$\Psi(x) = A(x)e^{\frac{iS(x)}{\hbar}} = \frac{C}{\sqrt{p}}e^{i\int p(x)\,dx}. \tag{2.5.34}$$

The WKB approximation can be used to determine the form of the wave function when the potential changes slowly with respect to the particle's de Broglie wavelength.

The case of greatest interest to us is that of an electron tunneling through a potential barrier. The tunneling probability can be found from the above form of the wave function as

$$T_t \sim e^{-\frac{2}{\hbar}\int p(x)\,dx}. \tag{2.5.35}$$

This result is of great importance in calculating the tunneling current in many semiconductor devices.

2.6   **One-Dimensional Harmonic Oscillator**

The last problem that we consider in this chapter is that of the one-dimensional harmonic oscillator. As the reader may recall from classical mechanics, a simple harmonic oscillator obeys the equation of motion:

$$m\frac{d^2x}{dt^2} = -kx. \tag{2.6.1}$$

The right-hand side is a restoring force that acts in the direction opposite to the displacement. The solution of this differential equation is oscillatory with a circular frequency ω, given by

$$\omega = \sqrt{\frac{k}{m}}. \tag{2.6.2}$$

The potential associated with harmonic motion is obtained by integration of the force $-kx$:

$$V(x) = \frac{1}{2}kx^2. \tag{2.6.3}$$

In this section, we consider the quantum-mechanical behavior of a particle that moves under the influence of the potential $1/2kx^2$.

The problem then is to solve the Schroedinger equation for a particle in the presence of the potential $1/2kx^2$. The Schroedinger equation is

$$\left[-\frac{\hbar^2}{2m}\frac{d^2}{dx^2} + \frac{1}{2}kx^2 \right]\Psi = E\Psi. \tag{2.6.4}$$

When k is replaced with Eq. (2.6.2), the Schroedinger equation is

$$\left[-\frac{\hbar^2}{2m}\frac{d^2}{dx^2} + \frac{1}{2}m\omega^2x^2 \right]\Psi = E\Psi. \tag{2.6.5}$$

Equation (2.6.5) can be solved in two ways. Both approaches are discussed here at length. The first approach we use is the power-series method.

To simplify the arithmetic involved in the power-series solution, we define some new, dimensionless variables. The position variable x is replaced with the dimensionless variable y:

$$y \equiv \sqrt{\frac{m\omega}{\hbar}}x, \qquad y^2 = \frac{m\omega}{\hbar}x^2. \tag{2.6.6}$$

With this definition, the second-derivative term becomes

$$\frac{d^2}{dy^2} = \frac{\hbar}{m\omega}\frac{d^2}{dx^2}. \tag{2.6.7}$$

We can rewrite the Schroedinger equation by multiplying through by $2m/\hbar^2$ to give

$$\frac{d^2}{dx^2}\Psi - \frac{m^2\omega^2x^2}{\hbar^2}\Psi + \frac{2mE}{\hbar^2}\Psi = 0. \tag{2.6.8}$$

Multiplying the above equation by $\hbar/m\omega$ gives

$$\frac{\hbar}{m\omega}\frac{d^2\Psi}{dx^2} - \frac{m\omega x^2}{\hbar}\Psi + \frac{2E\Psi}{\hbar\omega} = 0. \tag{2.6.9}$$

Substituting in for x in terms of y, we can rewrite the Schroedinger equation completely in terms of y as

$$\frac{d^2\Psi}{dy^2} + \left(\frac{2E}{\hbar\omega} - y^2\right)\Psi = 0. \tag{2.6.10}$$

In the asymptotic limit as y approaches infinity in either the positive or the negative direction, the energy E is very much less than y^2: $E \ll y^2$. In this case Eq. (2.6.10) reduces to

$$\frac{d^2\Psi}{dy^2} - y^2\Psi = 0, \tag{2.6.11}$$

whose solution behaves as

$$\Psi(y) = e^{\pm\frac{y^2}{2}}u(y), \tag{2.6.12}$$

where $u(y)$ is a function of y. The positive exponential term is not admissible since the wave function must be bounded at infinity. Therefore the solution is of the form

$$\Psi(y) = u(y)e^{-\frac{y^2}{2}}. \tag{2.6.13}$$

We can obtain the form of $u(y)$ by substituting the expression for the wave function into the Schroedinger equation. When this substitution is made, then the second-derivative term becomes

$$\frac{d^2}{dy^2}\left(ue^{-\frac{y^2}{2}}\right) = \frac{d}{dy}\left[e^{-\frac{y^2}{2}}\frac{du}{dy} - ye^{-\frac{y^2}{2}}u\right]$$

$$= -ye^{-\frac{y^2}{2}}\frac{du}{dy} + e^{-\frac{y^2}{2}}\frac{d^2u}{dy^2} - e^{-\frac{y^2}{2}}u + y^2e^{-\frac{y^2}{2}}u - ye^{-\frac{y^2}{2}}\frac{du}{dy}. \tag{2.6.14}$$

The Schroedinger equation is then

$$-ye^{-\frac{y^2}{2}}\frac{du}{dy} + e^{-\frac{y^2}{2}}\frac{d^2u}{dy^2} - ue^{-\frac{y^2}{2}} + y^2ue^{-\frac{y^2}{2}} - y\frac{du}{dy}e^{-\frac{y^2}{2}}$$

$$+ \frac{2E}{\hbar\omega}\left(ue^{-\frac{y^2}{2}}\right) - y^2\left(ue^{-\frac{y^2}{2}}\right) = 0. \tag{2.6.15}$$

The fourth and the last terms in Eq. (2.6.15) can be subtracted out, leaving

$$-2ye^{-\frac{y^2}{2}}\frac{du}{dy} + e^{-\frac{y^2}{2}}\frac{d^2u}{dy^2} - ue^{-\frac{y^2}{2}} + \frac{2E}{\hbar\omega}ue^{-\frac{y^2}{2}} = 0. \tag{2.6.16}$$

The exponential function is never equal to zero. Therefore it can be divided out within the equation to give

$$\frac{d^2u}{dy^2} - 2y\frac{du}{dy} + \frac{2E}{\hbar\omega}u - u = 0,$$

$$\frac{d^2u}{dy^2} - 2y\frac{du}{dy} + \left(\frac{2E}{\hbar\omega} - 1\right)u = 0. \qquad (2.6.17)$$

The potential is symmetric in position: $V(x) = 1/2kx^2$. It can be shown that the eigenfunctions of the Schroedinger equation for a particle that experiences a symmetric potential have definite evenness or oddness (see Problem 13 at the end of this chapter). A wave function that has definite evenness or oddness with respect to position x is said to have definite parity. The even-parity states must be symmetric with respect to x. In other words, the even-parity eigenfunctions must obey the following condition:

$$u(y) = u(-y). \qquad (2.6.18)$$

The even states can be expressed in terms of a power series that involves only the even powers of y:

$$u(y) = \sum a_s y^{2s}, \qquad (2.6.19)$$

where a_s are the coefficients in the expansion over y. The first and the second derivatives with respect to y of $u(y)$ are

$$\frac{du}{dy} = \sum 2sa_s y^{2s-1}, \qquad \frac{d^2u}{dy^2} = \sum 2s(2s-1)a_s y^{2s-2}. \qquad (2.6.20)$$

Substituting these expressions into Eqs. (2.6.17) yields

$$\sum 2s(2s-1)a_s y^{2s-2} - 2y\sum 2sa_s y^{2s-1} + \left(\frac{2E}{\hbar\omega} - 1\right)\sum a_s y^{2s} = 0, \qquad (2.6.21)$$

which can be rearranged to

$$\sum_{s=0} 2s(2s-1)a_s y^{2s-2} + \sum_{s=0}\left(\frac{2E}{\hbar\omega} - 1 - 4s\right)a_s y^{2s} = 0. \qquad (2.6.22)$$

The sums in Eq. (2.6.22) are over the variable s, which is called a dummy index since it disappears when the sum is written out in full. Any variable can be used in place of s in the sum with the same result. The index in the sum in the first term in Eq. (2.6.22) can be changed without altering the equation. Let s be replaced with $s + 1$ in the first term. Therefore $2s - 2$ becomes $2s$. With these changes, Eq. (2.6.22) becomes

$$\sum_{s=0} 2(s+1)[2(s+1) - 1]a_{s+1}y^{2s+2-2} + \sum_{s=0}\left(\frac{2E}{\hbar\omega} - 1 - 4s\right)a_s y^{2s} = 0,$$

$$\sum_{s=0} 2(s+1)(2s+1)a_{s+1}y^{2s} + \sum_{s=0}\left(\frac{2E}{\hbar\omega} - 1 - 4s\right)a_s y^{2s} = 0. \qquad (2.6.23)$$

In Eqs. (2.6.23), the sums are over s in both terms and start with the same value, 0. In the original equation, the first value in the sum in the first term is equal to zero since the sum starts at zero. The coefficient s gives zero for $s = 0$. The first term that contributes to the sum in the original first term is $s = 1$. If the index in the sum of the first term is redefined, all that is accomplished is that the first term in the original equation, identically zero to begin with, is removed from the sum. Since the nonzero sums in the two terms in Eqs. (2.6.23) start at the same value of s, they can be simply combined to give

$$\sum_{s=0} \left[2(s + 1)(2s + 1)a_{s+1} + \left(\frac{2E}{\hbar\omega} - 1 - 4s \right) a_s \right] y^{2s} = 0. \tag{2.6.24}$$

In general, y^{2s} is not equal to zero so it can be divided through to yield

$$2(s + 1)(2s + 1)a_{s+1} = a_s \left(4s + 1 - \frac{2E}{\hbar\omega} \right). \tag{2.6.25}$$

Equation (2.6.25) is called a recursion relation, since it relates the succeeding coefficients in terms of the preceding coefficient. The $s + 1$ coefficient can be related to the sth coefficient as

$$a_{s+1} = \frac{\left(4s + 1 - \frac{2E}{\hbar\omega} \right)}{2(s + 1)(2s + 1)} a_s. \tag{2.6.26}$$

For example, for $s = 0$ the recursion relation gives

$$a_1 = \frac{\left(1 - \frac{2E}{\hbar\omega} \right)}{2} a_0, \tag{2.6.27}$$

and for $s = 1$,

$$a_2 = \frac{1}{24} \left(5 - \frac{2E}{\hbar\omega} \right) \left(1 - \frac{2E}{\hbar\omega} \right) a_0. \tag{2.6.28}$$

When the recursion relation is used, a series representation for the even-parity solution of the one-dimensional harmonic oscillator can be obtained.

To aid us in determining the energy eigenvalues for the even-parity solution, it is useful to examine the solution as it approaches infinity. The value for $u(y)$ increases at large s. In the limit as s approaches infinity, the ratio of the coefficients in the sum becomes

$$\frac{a_{s+1}}{a_s} \sim \frac{1}{s}. \tag{2.6.29}$$

This is precisely the same as that for the coefficients in the power-series expansion of an exponential function. Therefore, for large values, $u(y)$ diverges like an exponential function as

$$u(y) \sim e^{y^2}. \tag{2.6.30}$$

As a result, the total wave function, given by the product of $u(y)$ and the exponential function of y^2, results in the wave function's diverging at large values:

$$\Psi \sim e^{\frac{y^2}{2}}.$$
(2.6.31)

In order that the wave function does not diverge at large values, the series over s must terminate at some maximum term. If we call the maximum index in the sum r, then the series must end after the rth term. In other words, the coefficient a_{r+1} and all succeeding coefficients must be zero.

The recursion relation can now be used to determine the possible values of the energy eigenvalues. In general, the $r + 1$ coefficient can be written in terms of the rth coefficient as

$$a_{r+1} = a_r \frac{\left(4r + 1 - \frac{2E}{\hbar\omega}\right)}{2(r + 1)(2r + 1)}.$$
(2.6.32)

Assuming that the rth coefficient is the last term in the sum, a_{r+1} is zero. For a_{r+1} to be zero, the numerator in Eq. (2.6.32) must vanish. Hence,

$$4r + 1 - \frac{2E}{\hbar\omega} = 0.$$
(2.6.33)

Solving for E yields

$$E = \frac{\hbar\omega}{2}(4r + 1).$$
(2.6.34)

If the series has only one term, $r = 0$, then the energy and the a_1 coefficient are

$$E = \frac{\hbar\omega}{2}, \qquad a_1 = 0.$$
(2.6.35)

The eigenfunction $\Psi(y)$ is given in general by the product of the exponential term and $u(y)$ as

$$\Psi(y) = u(y)e^{\frac{-y^2}{2}}.$$
(2.6.36)

If the series has only one term, $u(y)$ is

$$u = \sum a_s y^{2s}, \quad u = a_0.$$
(2.6.37)

The eigenfunction $\Psi(y)$ is then

$$\Psi = a_0 e^{\frac{-y^2}{2}}$$
(2.6.38)

with an eigenvalue of

$$E = \frac{\hbar\omega}{2}.$$
(2.6.39)

As a second example, consider the case in which the solution has two terms, $r = 1$. The sum then runs over $s = 0$ and $s = 1$. The term a_2 is zero. The resulting eigenvalue and eigenfunction are

$$E = \frac{\hbar\omega}{2}(4+1) = \frac{5}{2}\hbar\omega,$$
$$\Psi = a_0 e^{\frac{-y^2}{2}} + a_1 y^2 e^{\frac{-y^2}{2}}, \tag{2.6.40}$$

where a_1 is

$$a_1 = a_0 \frac{\left(1 - \frac{2E}{\hbar\omega}\right)}{2},$$
$$a_1 = \frac{a_0}{2}(1-5) = -2a_0. \tag{2.6.41}$$

Therefore the eigenfunction is

$$\Psi = a_0 e^{\frac{-y^2}{2}}(1 - 2y^2). \tag{2.6.42}$$

When $r = 2$, the series has only three terms; the solutions are

$$E = \frac{9}{2}\hbar\omega,$$
$$\Psi = a_0 e^{\frac{-y^2}{2}} - 2a_0 y^2 e^{\frac{-y^2}{2}} + a_2 y^4 e^{\frac{-y^2}{2}}. \tag{2.6.43}$$

a_2 can be determined from the recursion relation as

$$a_2 = a_1 \frac{(5-9)}{2 \times 2 \times 3} = \frac{2}{3}a_0. \tag{2.6.44}$$

Therefore the eigenfunction is

$$\Psi = a_0 e^{\frac{-y^2}{2}}\left(1 - 4y^2 + \frac{2}{3}y^4\right). \tag{2.6.45}$$

We can find the coefficient a_0 by normalizing the wave function.

Let us next construct the solution for wave functions that have odd parity. Odd-parity wave functions obey

$$u(y) = -u(-y). \tag{2.6.46}$$

Since these functions have odd parity, they are formed by summation over only the odd powers in a power-series expansion. Hence $u(y)$ can be expanded as

$$u(y) = \sum a_s y^{2s+1}, \tag{2.6.47}$$

where the sum over s runs over only the odd numbers. The recursion relation relating successive coefficients in the expansion in terms of the preceding coefficients is found in a manner similar to that described above. The result is

$$a_{s+1} = a_s \frac{4s + 3 - \frac{2E}{\hbar\omega}}{2(s + 1)(2s + 3)}. \tag{2.6.48}$$

Again the series must terminate after some finite number of terms, say, r terms. Therefore the last term in the series corresponds to $s = r$. The coefficient a_{r+1} must be equal to zero. To ensure this condition, the numerator in the above equation must vanish. Hence,

$$4r + 3 - \frac{2E}{\hbar\omega} = 0. \tag{2.6.49}$$

Solving for E yields

$$E = \hbar\omega \left(2r + \frac{3}{2} \right), \quad r = 0, 1, 2, \ldots. \tag{2.6.50}$$

If the series terminates after only one term, then $r = 0$ and the energy eigenvalue and the corresponding eigenvector are

$$E = \frac{3}{2}\hbar\omega, \quad \Psi = a_0 y e^{\frac{-y^2}{2}}. \tag{2.6.51}$$

If the series terminates after two terms, then $r = 1$ and the energy eigenvalue and eigenvector are

$$E = \frac{7}{2}\hbar\omega, \quad \Psi = a_0 y \left(1 - \frac{2}{3}y^2 \right) e^{-\frac{y^2}{2}}. \tag{2.6.52}$$

In general the complete solutions for the energy and the wave funtion to the harmonic oscillator problem can be expressed as

$$E_n = \left(n + \frac{1}{2} \right) \hbar\omega, \quad n = 0, 1, 2, 3 \ldots,$$

$$\Psi_n(x) = \frac{C_n}{2^{\frac{n}{2}}} h_n(y) e^{-\frac{y^2}{2}}, \tag{2.6.53}$$

where

$$C_n \equiv \left[\frac{\sqrt{\frac{m\omega}{\pi\hbar}}}{n!} \right]^{\frac{1}{2}}, \quad y \equiv \sqrt{\frac{m\omega}{\hbar}} x.$$

The functions $h_n(y)$ used in the above expressions are called the Hermite polynomials. The first few Hermite polynomials are collected below:

$$h_0(y) = 1,$$
$$h_1(y) = 2y,$$
$$h_2(y) = 4y^2 - 2,$$
$$h_{n+1}(y) = 2yh_n(y) - h'_n(y). \tag{2.6.54}$$

The above constitute the solutions of the one-dimensional harmonic oscillator. We can derive these solutions in an alternative way by using two different operators, called the raising and the lowering operators. The operator method is an excellent means of illustrating the power and the utility of operators in solving problems in quantum mechanics. It is also of great importance in quantum field theory. Because of its importance, it is worthwhile to digress here and illustrate how the harmonic oscillator problem can be solved with algebraic manipulations of operators. We define the operators a and $a\dagger$ in the following way:

$$a = \frac{1}{\sqrt{2}}\left(y + \frac{d}{dy}\right), \qquad a\dagger = \frac{1}{\sqrt{2}}\left(y - \frac{d}{dy}\right), \tag{2.6.55}$$

where again y is defined in terms of x as

$$y \equiv \sqrt{\frac{m\omega}{\hbar}}x. \tag{2.6.56}$$

Consider the product of a and $a\dagger$. This is given as

$$\begin{aligned} aa\dagger &= \frac{1}{2}\left(y + \frac{d}{dy}\right)\left(y - \frac{d}{dy}\right) \\ &= \frac{1}{2}\left[y^2 - \frac{d^2}{dy^2} - y\frac{d}{dy} + \frac{d}{dy}y\right] \\ &= \frac{1}{2}y^2 - \frac{1}{2}\frac{d^2}{dy^2} - \frac{1}{2}y\frac{d}{dy} + \frac{1}{2}\frac{d}{dy}y. \end{aligned} \tag{2.6.57}$$

$1/2\, d/dy\, y$ can be simplified since the operator d/dy acts on everything to its right. Hence,

$$\frac{1}{2}\frac{d}{dy}(y) = \frac{1}{2} + \frac{1}{2}y\frac{d}{dy}. \tag{2.6.58}$$

With this result, the product of a and $a\dagger$ becomes

$$aa\dagger = \frac{1}{2}y^2 - \frac{1}{2}\frac{d^2}{dy^2} - \frac{1}{2}y\frac{d}{dy} + \frac{1}{2} + \frac{1}{2}y\frac{d}{dy},$$
$$aa\dagger = \frac{1}{2}y^2 - \frac{1}{2}\frac{d^2}{dy^2} + \frac{1}{2}. \tag{2.6.59}$$

When H' is defined as

$$H' \equiv \frac{1}{2}y^2 - \frac{1}{2}\frac{d^2}{dy^2}, \tag{2.6.60}$$

the product of a and $a\dagger$ can be rewritten as

$$aa\dagger = H' + \frac{1}{2}, \qquad a\dagger a = H' - \frac{1}{2}. \tag{2.6.61}$$

We obtain the second relation above by noting that the commutation relation between a and $a\dagger$ is

$$aa\dagger - a\dagger a = 1, \qquad [a, a\dagger] = 1. \tag{2.6.62}$$

Recognizing that a and $a\dagger$ can be expressed in terms of y and that y can be reexpressed in terms of x, we can then write the operators a and $a\dagger$ in terms of x through the following:

$$y = \sqrt{\frac{m\omega}{\hbar}}x, \qquad \frac{d}{dy} = \sqrt{\frac{\hbar}{m\omega}}\frac{d}{dx}, \qquad \frac{d^2}{dy^2} = \frac{\hbar}{m\omega}\frac{d^2}{dx^2},$$

$$a = \frac{1}{\sqrt{2}}\left(y + \frac{d}{dy}\right),$$

$$a = \frac{1}{\sqrt{2}}\left(\sqrt{\frac{m\omega}{\hbar}}x + \sqrt{\frac{\hbar}{m\omega}}\frac{d}{dx}\right). \tag{2.6.63}$$

The momentum operator p_x is simply given as

$$p_x = \frac{\hbar}{i}\frac{d}{dx}, \qquad \frac{d}{dx} = \frac{ip_x}{\hbar}. \tag{2.6.64}$$

Substituting in the last of Eqs. (2.6.63) Eq. (2.6.64) for d/dx in terms of p_x, we find that a and $a\dagger$ can be expressed as

$$a = \frac{1}{\sqrt{2}}\left[\sqrt{\frac{m\omega}{\hbar}}x + \sqrt{\frac{\hbar}{m\omega}}\frac{ip_x}{\hbar}\right],$$

$$a = \frac{1}{\sqrt{2}}\left[\sqrt{\frac{m\omega}{\hbar}}x + i\sqrt{\frac{1}{m\hbar\omega}}p_x\right],$$

$$a\dagger = \frac{1}{\sqrt{2}}\left[\sqrt{\frac{m\omega}{\hbar}}x - i\sqrt{\frac{1}{m\hbar\omega}}p_x\right]. \tag{2.6.65}$$

The product of a and $a\dagger$ can now be formed in terms of x as

$$aa\dagger = \frac{m\omega}{2\hbar}x^2 + \frac{1}{2m\hbar\omega}p_x^2 + i\sqrt{\frac{m\omega}{2\hbar}}\sqrt{\frac{1}{2m\hbar\omega}}p_x x - i\sqrt{\frac{m\omega}{2\hbar}}\sqrt{\frac{1}{2m\hbar\omega}}xp_x.$$

$$(2.6.66)$$

The last two terms above can be combined to yield

$$aa\dagger = \frac{m\omega}{2\hbar}x^2 + \frac{1}{2m\hbar\omega}p_x^2 + \frac{i}{2\hbar}(p_x x - xp_x). \qquad (2.6.67)$$

We can again simplify the last term by recognizing that it is simply the commutator of x and p_x. $aa\dagger$ is then

$$aa\dagger = \frac{1}{\hbar\omega}\left[\frac{p_x^2}{2m} + \frac{1}{2}m\omega^2 x^2\right] + \frac{i}{2\hbar}[p_x, x]. \qquad (2.6.68)$$

But

$$[p_x, x] = \frac{\hbar}{i}, \qquad U(x) = \frac{1}{2}m\omega^2 x^2, \qquad (2.6.69)$$

and

$$H = \frac{p_x^2}{2m} + U(x). \qquad (2.6.70)$$

Therefore the product $aa\dagger$ can be written as

$$aa\dagger = \frac{H}{\hbar\omega} + \frac{1}{2}. \qquad (2.6.71)$$

In Eqs. (2.6.61), H' was defined as

$$H' = aa\dagger - \frac{1}{2}. \qquad (2.6.72)$$

Therefore, by comparing Eqs. (2.6.71) and (2.6.72), we find that H' must be equal to $H/\hbar\omega$. H' is equal to the normalized Hamiltonian. From the above definitions, the Schroedinger equation can be written as

$$\left(aa\dagger - \frac{1}{2}\right)\hbar\omega\Psi = E\Psi,$$

$$\left(aa\dagger - \frac{1}{2}\right)\Psi = \frac{E}{\hbar\omega}\Psi. \qquad (2.6.73)$$

Alternatively, the Schroedinger equation can be written with the commutation relation between a and $a\dagger$. $(aa\dagger - 1/2)$ can be rewritten as

$$aa\dagger - \frac{1}{2} = (1 + a\dagger a) - \frac{1}{2} = a\dagger a + \frac{1}{2}. \qquad (2.6.74)$$

Since $(aa\dagger - 1/2)$ is equal to H' then, $(a\dagger a + 1/2)$ is also equal to H' from the above result. The Schroedinger equation can then be written in an alternative form as

$$\left(a\dagger a + \frac{1}{2} \right) \Psi = \frac{E}{\hbar\omega} \Psi. \tag{2.6.75}$$

To summarize, the Schroedinger equation can be written in two different forms by use of the operators a and $a\dagger$. These forms are

$$\left(aa\dagger - \frac{1}{2} \right) \Psi = \frac{E}{\hbar\omega} \Psi,$$
$$\left(a\dagger a + \frac{1}{2} \right) \Psi = \frac{E}{\hbar\omega} \Psi. \tag{2.6.76}$$

Using the Schroedinger equation written in the forms given above, we can find their solution from strictly algebraic manipulations as follows. Applying $a\dagger$ to the first of these two equations gives

$$a\dagger \left(aa\dagger - \frac{1}{2} \right) \Psi = a\dagger \left(\frac{E}{\hbar\omega} \right) \Psi. \tag{2.6.77}$$

The order of the operators is important, as we discussed in Chapter 1, since they do not commute. Being careful then to keep the correct order of the operators, we find that $a\dagger$ acts through Eq. (2.6.77) resulting in

$$\left(a\dagger aa\dagger - \frac{1}{2}a\dagger \right) \Psi = \frac{E}{\hbar\omega}a\dagger\Psi,$$
$$\left(a\dagger a - \frac{1}{2} \right) a\dagger\Psi = \frac{E}{\hbar\omega}a\dagger\Psi. \tag{2.6.78}$$

Adding $a\dagger\Psi$ to each side of the last equation above gives

$$\left(a\dagger a + \frac{1}{2} \right) a\dagger\Psi = \frac{(E + \hbar\omega)}{\hbar\omega}a\dagger\Psi. \tag{2.6.79}$$

But the second of the two equations in Eqs. (2.6.76) is

$$\left(a\dagger a + \frac{1}{2} \right) \psi = \frac{E}{\hbar\omega}\Psi. \tag{2.6.80}$$

Therefore, if Ψ is an eigenvector of $(aa\dagger+1/2)$, so is $a\dagger\Psi$, but with an eigenvalue of $(E+\hbar\omega)/(\hbar\omega)$ instead of $E/(\hbar\omega)$. Therefore the operator $a\dagger$ raises the energy of the state by one quantum since its application onto an eigenstate results in creating another eigenstate with a higher-energy eigenvalue. In fact, the new state is higher in energy by precisely one quantum of energy, $\hbar\omega$. By virtue of this behavior, $a\dagger$ is called a creation or raising operator.

States of lower energy can be constructed with the operator a. To see how the operator a leads to states with a lower energy, start with the Schroedinger equation in the form

$$\left(a\dagger a + \frac{1}{2}\right)\Psi = \frac{E}{\hbar\omega}\Psi. \tag{2.6.81}$$

The operator a is applied to both sides of Eq. (2.6.81) to yield

$$a\left(a\dagger a + \frac{1}{2}\right)\Psi = \frac{E}{\hbar\omega}a\Psi,$$

$$\left(aa\dagger + \frac{1}{2}\right)a\Psi = \frac{E}{\hbar\omega}a\Psi. \tag{2.6.82}$$

Subtracting $a\Psi$ from both sides results in

$$\left(aa\dagger - \frac{1}{2}\right)a\Psi = \frac{(E - \hbar\omega)}{\hbar\omega}a\Psi. \tag{2.6.83}$$

The first of the two equations in Eqs. (2.6.76) is

$$\left(aa\dagger - \frac{1}{2}\right)\Psi = \frac{E}{\hbar\omega}\Psi. \tag{2.6.84}$$

Clearly, if Ψ is an eigenstate of the operator $(aa\dagger - 1/2)$ with an eigenvalue of $E/\hbar\omega$, then $a\Psi$ is an eigenstate of $(aa\dagger - 1/2)$ with an eigenvalue of $(E - \hbar\omega)/(\hbar\omega)$. The operator a is called an annihilation or lowering operator since it lowers the energy of the state by one quantum, $\hbar\omega$.

The energy of a harmonic oscillator can never be less than zero. Therefore there must be a ground state below which the energy cannot be lowered. Subsequently, if the lowering operator a operates on the ground state it must give zero. Calling the ground state of the harmonic oscillator Ψ_0 and its energy eigenvalue E_0, we find that the Schroedinger equation for the state $a\Psi_0$ is

$$\left(aa\dagger - \frac{1}{2}\right)a\Psi_0 = \frac{(E_0 - \hbar\omega)}{\hbar\omega}a\Psi_0. \tag{2.6.85}$$

Unless $a\Psi_0$ vanishes, Eq. (2.6.85) implies that $a\Psi_0$ is an eigenstate of the Hamiltonian operator with an energy eigenvalue of $(E_0 - \hbar\omega)$ in contradiction to the original definition that E_0 is the minimum energy of the system. Therefore $a\Psi_0$ must be equal to zero, implying that

$$\left(a\dagger a + \frac{1}{2}\right)\Psi_0 = \frac{E_0}{\hbar\omega}\Psi_0,$$

$$a\dagger a\Psi_0 + \frac{1}{2}\Psi_0 = \frac{E_0}{\hbar\omega}\Psi_0. \tag{2.6.86}$$

But $a\Psi_0 = 0$, so $a\dagger a\Psi_0 = 0$ as well. The first term in Eqs. (2.6.86) is then equal to zero. The equation then simplifies to

$$\frac{1}{2}\Psi_0 = \frac{E_0}{\hbar\omega}\Psi_0. \tag{2.6.87}$$

Solving for E_0 yields

$$E_0 = \frac{1}{2}\hbar\omega. \tag{2.6.88}$$

The eigenvectors of all the other states corresponding to solutions of the Schroedinger equation for the harmonic oscillator are determined by successive applications of the operator $a\dagger$ onto the ground state Ψ_0. Therefore the nth state is given as

$$\Psi_n = C_n(a\dagger)^n\Psi_0. \tag{2.6.89}$$

The energy eigenvalues are easily found from the action of the raising operator onto a state. From the result obtained above, $a\dagger$ acts to raise the energy of a state by one quantum, $\hbar\omega$. Therefore the energy of the nth state must be simply equal to the ground-state energy $1/2\hbar\omega$, plus n times $\hbar\omega$. This results in a general expression for the energy eigenvalue of the nth state as

$$E_n = \left(n + \frac{1}{2}\right)\hbar\omega. \tag{2.6.90}$$

The form of the ground-state eigenfunction Ψ_0 can be found in the following way. The application of the lowering operator a onto the ground state must give zero. Therefore

$$a\Psi_0 = 0. \tag{2.6.91}$$

Substituting

$$a = \frac{1}{\sqrt{2}}\left(y + \frac{d}{dy}\right) \tag{2.6.92}$$

in the expression for a yields

$$\left(y + \frac{d}{dy}\right)\Psi_0 = 0,$$

$$\frac{d\Psi_0}{dy} = -y\Psi_0,$$

$$\Psi_0 = Ae^{-\frac{y^2}{2}}, \tag{2.6.93}$$

where A is a normalization constant. The first excited state Ψ_1 is found from the ground state Ψ_0 by application of the raising operator $a\dagger$. Hence,

$$\Psi_1 = a\dagger\Psi_0,$$

$$\Psi_1 = \frac{1}{\sqrt{2}}\left(y - \frac{d}{dy}\right)\Psi_0 = 2Aye^{-\frac{y^2}{2}},$$

$$\Psi_1 = A_1 ye^{-\frac{y^2}{2}}, \tag{2.6.94}$$

etc. Note that exactly the same solutions are obtained as were found with the power-series method.

It is important to realize that the ground-state energy of the harmonic oscillator is not zero. There is a zero-point energy for a quantum-mechanical oscillator. In other words, a quantum-mechanical oscillator has a nonzero energy and motion, even at absolute zero degrees. How can this be? The explanation of the origin of the zero-point energy lies in the Uncertainty Principle. Since a particle cannot have simultaneously sharp values of its momentum and position, if it has a well-defined momentum, as is the case for an oscillator, it must undergo zero-point motion. In other words, a particle in the ground state of an oscillator cannot be at rest since its momentum is well defined in this state, which would violate the Uncertainty Principle.

EXAMPLE 2.6.1 Parabolic Energy Bands

Crystal-growth technologies, such as MOCVD and CBE, not only can achieve exacting layer widths but can also be used to grade the composition of a semi-conductor alloy such as $Al_xGa_{1-x}As$ continuously. By doing so, the bandgap energy of the system changes continuously with real-space position. With these techniques, different materials can be grown with various energy-band structures.

Consider the parabolic energy-band structure shown in Figure 2.6.1. If the equations for the parabolas describing this structure are

$$E = ax^2, \quad \text{conduction band,}$$
$$E = bx^2, \quad \text{valence band,}$$

find the energy of a photon emitted from a transition of an electron out of the $n = 2$ quantum level in the conduction band to the $n = 2$ level in the valence band.

To solve this problem, it is important to note that both the conduction-band and the valence-band structures have a parabolic shape. The band structures are essentially potential energy diagrams. Hence the allowed energy levels, provided that the parabolic well is sufficiently small such that spatial quantization effects occur, will have the same form as those for a harmonic oscillator. This is obvious since the form of the potential is the same in each case. In other words, if we solve the Schroedinger equation for a harmonic oscillator or for the potential-energy bands given above, in either case the potentials used in the Schroedinger

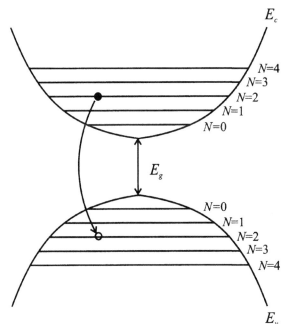

Figure 2.6.1. Sketch of the conduction and the valence bands of a continuously graded semiconductor material showing the spatial quantization energy levels.

equation are the same. Since the equations have the same form, the solution must be essentially the same. The confined quantum levels in each potential well are evenly spaced, as they are in the harmonic oscillator problem. The resulting energy levels are sketched in Figure 2.6.1.

The energy levels are given as

$$E = \left(n + \frac{1}{2}\right)\hbar\omega,$$

where the frequency ω is given by

$$\omega = \sqrt{\frac{k}{m}}.$$

The force constant k in this case is found from the potential-energy expressions $E = ax^2$ and $E = bx^2$. From the definition of the force as the negative gradient of the potential energy, the force can be found as

$-2ax$, conduction band,
$-2bx$, valence band.

For a harmonic oscillator, the equation of motion is simply $F = -kx$, where F is the force. Hence the force constant is simply $2a$ and $2b$ for the conduction and the valence bands, respectively. The frequencies ω are then given as

$$\omega = \sqrt{\frac{2a}{m}}, \qquad \omega = \sqrt{\frac{2b}{m}},$$

where it is assumed that the electron and the hole masses are the same and are both equal to the free-electron mass. Under these conditions, the quantum levels in the conduction and the valence bands are

$$E_c = \left(n + \frac{1}{2}\right) \hbar \sqrt{\frac{2a}{m}}, \tag{2.6.95}$$

$$E_v = \left(n + \frac{1}{2}\right) \hbar \sqrt{\frac{2b}{m}}. \tag{2.6.96}$$

To find the energy of a photon emitted from a recombination of an electron in the $n = 2$ state with a hole in the $n = 2$ state, it is necessary to add the confined quantum state energies for the electron and the hole to the bandgap energy. The result is

$$E = E_g + \left(2 + \frac{1}{2}\right) \hbar \sqrt{\frac{2a}{m}} + \left(2 + \frac{1}{2}\right) \hbar \sqrt{\frac{2b}{m}}.$$

Depending on both a and b, the photon energy can be much greater than the energy gap of the material. In this way, novel semiconductor lasers can be made to emit light at wavelengths different from those possible when only bulk material is used.

EXAMPLE 2.6.2 Molecules

A simplisitic treatment of a binary molecule can be made with the results of Chapter 2. A molecule generally undergoes two different types of motion in addition to translation; these are rotation and vibration. Let us consider each type of motion separately for two simple molecules. As the first example, consider an HO molecule, as shown in Figure 2.6.2. For simplicity, let us first assume that the mass of the oxygen atom is much larger than the mass of the hydrogen atom. With this assumption, the oxygen atom remains stationary and the only movement is that of the hydrogen atom.

Figure 2.6.2. Diagrammatic sketch of an HO molecule showing rotational motion.

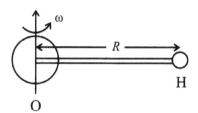

a. **Neglect any vibration and determine the allowed rotational energies of the molecule if the hydrogen atom is a fixed distance R from the oxygen atom and has mass m.**

The Hamiltonian in cylindrical coordinates is given as

$$H = -\frac{\hbar^2}{2m}\left[\frac{d^2}{dr^2} + \frac{1}{r}\frac{d}{dr} + \frac{1}{r^2}\frac{d^2}{d\theta^2} + \frac{d^2}{dz^2}\right].$$

If the rotation is completely confined within the xy plane, then there is no motion along z. Hence the second derivative with respect to z must be zero. By design, the hydrogen atom remains at a fixed distance R from the oxygen atom. Therefore the derivatives with respect to r must also vanish. The Schroedinger equation for the molecule is then

$$H\Psi = E\Psi,$$

where H simplifies to

$$H = -\frac{\hbar^2}{2m}\frac{1}{r^2}\frac{d^2}{d\theta^2}.$$

The moment of inertia for the system is

$$I = mR^2,$$

and so the Schroedinger equation becomes

$$-\frac{\hbar^2}{2mR^2}\frac{d^2}{d\theta^2}\Psi = E\Psi,$$

$$-\frac{\hbar^2}{2I}\frac{d^2\Psi}{d\theta^2} = E\Psi.$$

Solving for Ψ yields

$$\Psi = Ae^{\pm in\theta}, \quad n = \sqrt{\frac{2IE}{\hbar^2}}.$$

We obtain the coefficient A by normalizing the wave function. In this case, the normalization is given from integration over the full range of values of θ as

$$1 = A^2\int_0^{2\pi}\Psi^*\Psi\,d\theta = A^2\int_0^{2\pi}e^{-in\theta}e^{in\theta}\,d\theta = A^2\int_0^{2\pi}d\theta.$$

Therefore A is

$$A = \sqrt{\frac{1}{2\pi}}.$$

Figure 2.6.3. Diagrammatic sketch of an HO molecule showing vibrational motion and its corresponding potential energy diagram.

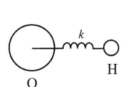

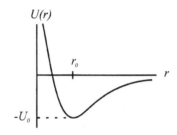

The wave function is then

$$\Psi = \sqrt{\frac{1}{2\pi}} e^{in\theta} \quad \text{or} \quad \sqrt{\frac{1}{2\pi}} e^{-in\theta}.$$

The energy eigenvalues are

$$E = \frac{n^2 \hbar^2}{2I},$$

where single valuedness implies that n is an integer.

b. Let us next consider the case in which the H atom vibrates about the oxygen atom, as in Figure 2.6.3.

If we again assume that the mass of the oxygen atom is much larger than the mass of the hydrogen atom, then the problem can be simplified to that of a single mass on a spring. A typical potential-energy diagram for a molecule is shown in Figure 2.6.3. The potential has a minimum value at $U = U_0$, which occurs at the separation distance r_0. The energy U_0 corresponds to the binding energy of the molecule, and the distance of separation r_0 is the bond length. The potential energy increases toward infinity for r approaching zero because of the nuclear–nuclear repulsion. The potential approaches zero as r approaches infinity because of the decreasing strength of the interaction with distance. We will discuss molecular formation in greater detail in Chapter 7.

The effective spring constant can be determined from the potential-energy curve as follows. The potential $U(r)$ can be expanded in a power series as

$$U(r) = U(r_0) + (r - r_0) \frac{dU}{dr}\bigg|_{r=r_0} + \frac{1}{2}(r - r_0)^2 \frac{d^2U}{dr^2}\bigg|_{r=r_0} + \cdots,$$

where higher-order terms greater than 2 have been neglected. The first derivative of U with respect to r is zero at $r = r_0$ since this is the minimum energy point in the curve. Therefore the potential becomes

$$U(r) = U(r_0) + \frac{1}{2}(r - r_0)^2 \frac{d^2U}{dr^2}.$$

This is essentially the harmonic oscillator problem with a potential of

$$U(x) = U_0 + \frac{1}{2}kx^2.$$

When the above equations are compared, the effective spring constant for the molecule can be identified as

$$k = \frac{d^2 U}{dr^2}\Bigg|_{r=r_0}.$$

The frequency of oscillation then of the H atom is

$$w = \sqrt{\frac{\frac{d^2 U}{dr^2}\big|_{r=r_0}}{m}},$$

and the corresponding vibrational energy of the hydrogen atom is

$$E = \left(n + \frac{1}{2}\right)\hbar\omega,$$

where ω is defined above.

c. Finally, let us consider a molecule consisting of atoms of equal mass.

For example, consider the H_2 molecule. In this case, the assumption that one of the masses remains fixed is no longer valid. Both atoms are free to move. The system is sketched in Figure 2.6.4. Let r represent the separation between the two masses, $r = r_1 - r_2$. r_0 is defined as the equilibrium length of the spring. The equations of motion of the system are

$$m_1 \ddot{r}_1 = k(r - r_0),$$

$$m_2 \ddot{r}_2 = -k(r - r_0).$$

Subtracting the first equation above from the second gives

$$\ddot{r}_2 - \ddot{r}_1 = \ddot{r} = -\frac{k}{\mu}(r - r_0),$$

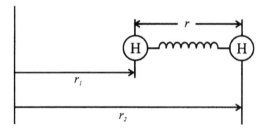

Figure 2.6.4. Diagrammatic sketch of an H_2 molecule showing the relative coordinates.

where μ is the reduced mass that is defined as

$$\mu \equiv \frac{m_1 m_2}{m_1 + m_2}.$$

The problem is again that of a simple harmonic oscillator with a frequency that, in this case, is

$$\omega = \sqrt{\frac{\frac{d^2 U}{dr^2}\big|_{r=r_0}}{\mu}}.$$

PROBLEMS

1. Consider a one-dimensional finite square-well potential. Let it be symmetric about $x = 0$ of width $2a$ and depth V_0. The potential structure is sketched in Figure Pr. 2.1. Calculate the ground-state energy eigenvalue, where $V_0 = 0.347$ eV and $2a = 100$ Å. Let $m = 0.067$ times the free-electron mass.

2. If a quantum-mechanical particle is tethered to a rotating rod, as shown in Figure Pr. 2.2, and moves in a horizontal plane of constant radius R at all times, what must the normalized eigenfunctions be? What are the energy eigenvalues?

3. Semiconductor quantum-well lasers made from the GaAs/AlGaAs material system operate under stimulated emission of photons arising from electron–hole recombination events between confined quantum state levels.

   a. What type of quantum-well profile will give rise to a series of levels in both the conduction and the valence bands such that the levels are equally spaced in energy?

   b. The bandgap energy difference (difference in energy between the conduction- and the valence-band edges) is 1.42 eV. Assume that the electron and the hole masses are 0.067 and 0.50 times the free-electron mass, respectively. The band profile found in part a gives rise to a force on the carriers of $F = -k(x - x_0)$. If we assume that $k = 1.6 \times 10^{12}$ eV/cm^2, find the energy of the lasing transition between the $n = 1$ levels of the system.

4. Use the raising and the lowering operators to construct the $n = 2$ harmonic oscillator wave function from the ground-state wave function Ψ_0. Do not just cite the answer. Show your work.

Figure Pr.2.1. One-dimensional finite square-well potential.

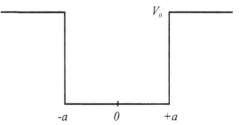

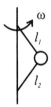

Figure Pr.2.2. Quantum-mechanical particle tethered to a rotating rod.

5. Consider the dimensionless harmonic oscillator Hamiltonian, $H = p^2/2 + x^2/2$, where $p = 1/i \, d/dx$.

a. Show that the wave function

$$\Psi_0 = e^{-\frac{x^2}{2}}$$

is an eigenfunction and determine its eigenvalue.

b. Show that the wave function

$$\Psi_1 = xe^{-\frac{x^2}{2}}$$

is an eigenfunction of H and determine its eigenvalue.

c. By proper choice of α make the wave function

$$\Psi_2 = (1 + \alpha x^2)e^{-\frac{x^2}{2}}$$

orthogonal to Ψ_0 and Ψ_1. Is Ψ_2 an eigenfunction of H? If so, find its eigenvalue.

6. Consider an infinite square well of width a. Determine the probability distribution for various values of the momentum corresponding to the normalized eigenstates.

7. Consider a dimensionless, harmonic oscillator moving only in the x direction described by the following Hamiltonian:

$$H = -\frac{1}{2}\frac{d^2}{dx^2} + \frac{1}{2}x^2.$$

Using the raising and lowering operators a and $a\dagger$, defined as

$$a\dagger = \frac{1}{\sqrt{2}}\left(x - \frac{d}{dx}\right), \qquad a = \frac{1}{\sqrt{2}}\left(x + \frac{d}{dx}\right)$$

and the normalized ground state Ψ_0, show that

$$\int (a\dagger\Psi_0)^* a\dagger\Psi_0 \, dx = \int \Psi_0^* a a\dagger\Psi_0 \, dx = 1,$$

$$\int (a\dagger a\dagger\Psi_0)^* a\dagger a\dagger\Psi_0 \, dx = \int \Psi_0^* a a a\dagger a\dagger\Psi_0 \, dx = 2.$$

Demonstrate the above by using only the commutator relation between a and $a\dagger$ and the property of a acting on the ground state.

$$[a, a^\dagger] = 1; \qquad a\Psi_0 = 0.$$

8. Show that the energy of the harmonic oscillator is always greater than zero. Use the Schroedinger equation in the following form:

$$\frac{\hbar\omega}{2}\left[-\frac{d^2}{d\rho^2} + \rho^2\right]\Psi = E\Psi,$$

where

$$\rho = \sqrt{\frac{m\omega}{\hbar}}x.$$

Figure Pr.2.9. Device schematic of a doping superlattice structure. The unlabeled layers are space-charge neutral.

9. There are essentially two types of artificial, multilayered semiconductor structures known as superlattices, that is, compositional and doping superlattices. In a doping superlattice in equilibrium, the electron potential energy as a function of real-space distance is modulated by a region of positive charge (from ionized donors) and a region of negative charge (from ionized acceptors). The device scheme then is something like that shown in Figure Pr. 2.9, in which the empty cells contain no net electric charge. If the positively charged cell contains exactly the same amount of charge as that of the negatively charged cell, determine

   a. what the potential energy versus the real-space curve is. (Draw a diagram.) Assume that the dimensions and charge concentrations are

	width	charge concentration
positive charge	d_n	N_d^+
negative charge	d_p	N_a^-
uncharged layer	d	0

   b. If the layers are thin, spatial quantization results. Determine the energy eigenvalues of the quantum wells. Hint: Simply define a quantum of energy as done in the harmonic oscillator problem. Do not try to determine its value.

10. Consider an infinite square well of width $2a$. The wave function of a particle trapped inside the well is given as

$$\Psi = C \left[\cos \frac{\pi x}{2a} + \sin \frac{3\pi x}{a} + \frac{1}{4} \cos \frac{3\pi x}{2a} \right].$$

   a. Calculate the coefficient C.

   b. If a measurement of the total energy is made, what are the possible results of such a measurement and what is the probability of measuring each value?

11. Consider the potential well shown in Figure Pr. 2.11. Sketch the approximate behavior of the ground state and the first excited-state wave functions for $E < V_0$. Explain your answer.

Figure Pr.2.11. Infinite one-dimensional potential well containing a small barrier potential.

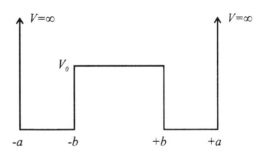

12. Given that the annihilation and the creation operators are

$$a = \frac{1}{\sqrt{2}}\left(y + \frac{d}{dy}\right), \qquad a\dagger = \frac{1}{\sqrt{2}}\left(y - \frac{d}{dy}\right),$$

determine the value of the matrix element,

$$\langle 1|[a, a\dagger]|0\rangle,$$

where $\langle 1|$ is the first excited state and $|0\rangle$ is the ground state of the harmonic oscillator.

13. Prove that the eigenfunctions of a symmetric potential have definite parity if they are nondegenerate.

14. Consider a violin string attached firmly at both ends such that the vertical displacement of the string at each end is zero. Let the string have length L and mass m. The wave velocity of the string is c, and ω is its angular frequency of vibration. Determine the general expression for the modes of vibration $y(x)$.

15. An electron is confined within an infinite square well of width a with a wave function of the form

$$\Psi = A\left[\cos\frac{\pi x}{a} + 0.5\cos\frac{3\pi x}{a}\right].$$

   a. Determine the coefficient A.

   b. Determine the possible momentum eigenvalues for this state.

   c. Determine the possible energy eigenvalues for this state.

   d. What is the probability of measuring each energy eigenvalue?

16. A particle exists in the nth energy level of a one-dimensional potential well with infinite potential and length L. From the solution for the wave function, calculate

   a. the average value of the momentum of the particle $\langle p_x \rangle$,

   b. the average value of the square of the momentum of the particle $\langle p_x^2 \rangle$,

   c. the standard deviation in its momentum

$$\sigma_{p_x} = \sqrt{\langle p_x^2 \rangle - \langle p_x \rangle^2},$$

   d. the average value of the position of the particle $\langle x \rangle$,

   e. the average value of the square of the position of the particle $\langle x^2 \rangle$,

   f. the standard deviation in its position

$$\sigma_x = \sqrt{\langle x^2 \rangle - \langle x \rangle^2},$$

   g. $\Delta x \Delta p_x = \sigma_x \sigma_p$.

17. An H_2 molecule can be modeled as a rigid rod connecting two masses, as shown in Figure Pr.2.17. If the moment of inertia of the model is $I = 2mL^2$, where the mass of the rod is neglected, determine the energy of a photon emitted when the rod undergoes a rotational transition from the first to the ground state. Assume that the rod rotates in the xy plane.

$$H = -\frac{\hbar^2}{2m}\left[\frac{d}{dr^2} + \frac{1}{r}\frac{d}{dr} + \frac{1}{r^2}\frac{d^2}{d\theta^2} + \frac{d^2}{dz^2}\right].$$

Figure Pr.2.17. Diagrammatic sketch of a H$_2$ molecule showing rotational motion.

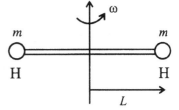

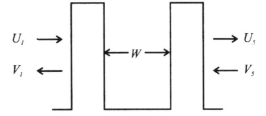

Figure Pr.2.18. Sketch of a two barrier, one well system. The well width is W.

18. **a.** Consider the double barrier structure sketched below. If an electron is assumed to be incident only from the left, determine the transmission coefficient in terms of the transfer matrix elements,

$$M_{11}, M_{12}, M_{11}^*, M_{12}^*, e^{\pm ikw}$$

$$\begin{bmatrix} U_1 \\ V_1 \end{bmatrix} = \begin{bmatrix} M_{11} & M_{12} \\ M_{12}^* & M_{11}^* \end{bmatrix} \begin{bmatrix} e^{-ikw} & 0 \\ 0 & e^{ikw^*} \end{bmatrix} \begin{bmatrix} M_{11} & M_{12} \\ M_{12}^* & M_{11}^* \end{bmatrix} \begin{bmatrix} U_5 \\ V_5 \end{bmatrix}$$

   b. What would be the minimum resonant energy of the double barrier structure if it is assumed for ease in computation that the potential heights can be treated as infinite?

19. Given that a particle obeys the equation,

$$\frac{d^2 x}{dt^2} + \omega^2 x = 0$$

   where ω is defined as,

$$\omega = \sqrt{\frac{k}{m}}$$

   Determine the minimum value of the expectation value of E using the Uncertainty Principle and the fact that

$$\langle p^2 \rangle \geq (\Delta p)^2; \qquad \langle x^2 \rangle \geq (\Delta x)^2.$$

3

Three-Dimensional Problems

In this chapter, three-dimensional problems in quantum mechanics are examined. The most important problem treated here is that of spherical symmetry: motion in a spherically symmetric potential. These types of problems are called central-force problems since the potential and hence the force depends only on r and not on the angle.

3.1 Angular Momentum and Central Forces

In a central-force problem there is no net torque exerted by the potential. This can be readily seen from the definition of torque, the vector product of the radial vector r and the force F. Since the force is given as the gradient of the potential $-\vec{\nabla} V(r)$, the force and the moment arm are collinear. Therefore the potential exerts no torque on the particle. If no net external torque is applied, the angular momentum of a system remains unchanged in accordance with the law of conservation of angular momentum. The angular momentum of a particle that moves in the presence of a central force then must be conserved. This is an important result that applies generally to many different physical problems.

Before we analyze the central-force problem in detail from a quantum-mechanical perspective, it is instructive to pause and look at the symmetry of central-force motion. Every conservation law in physics is essentially a reflection of a symmetry property. For example, the law of conservation of linear momentum is simply a statement about the translational symmetry of a system. If a particle has the same velocity at a point B that it had earlier at a point A, then the linear momentum of the system is said to have been conserved. Another way to look at this is that under the translation from A to B, the system remains invariant; the properties of the system remain the same. In other words, the properties of the system are invariant under a linear translation in space. The law of conservation of linear momentum then is simply a statement concerning the translational spatial symmetry of a system.

Similarly, the law of conservation of energy can be viewed as a statement concerning symmetry in time. A system that is invariant under a translation in time (the properties of the system are the same at a later time compared with those at an earlier time) is one in which the same amount of energy is present at $t = 0$ as at a later time t.

In the situation of interest to us here, that of central-force motion, the conserved quantity is angular momentum. The law of conservation of angular momentum is a statement about the rotational symmetry of a system. If the angular momentum of a system is conserved, it is also rotationally symmetric; the state of the system is preserved under a rotation in space.

What happens, though, when a conservation law is violated? For example, if there is an external torque applied to a system, its angular momentum is not conserved. The angular momentum of the system changes with time because of the presence of the torque. How can this be interpreted in terms of symmetry principles? When an external torque or field is present, there is now a preferred direction in space, a direction that is different from all other directions. The presence of the external field acts to break the symmetry of space. The same situation occurs in the conservation of energy. In a case in which the mechanical energy of a system is not conserved, the system has more mechanical energy initially than it has at a later time. Some of the mechanical energy has been changed into thermal energy and the system is distinguishable from its previous situation. Clearly, the system is asymmetric in time.

Modern physics is concerned with the study of symmetries and the underlying causes that lead to symmetry breaking. In fact, most of the work done currently in unifying the basic force laws – electromagnetic, gravitational, weak, and strong nuclear forces – is concerned with finding an underlying symmetry common to all four interactions and a mechanism by which this symmetry was broken, leading to the appearance of four separate physical interactions. We will see in Section 3.5 how the inclusion of space–time symmetry leads to a natural inclusion of the spin of an electron. As a result, a complete and highly accurate description of the hydrogen atom can be attained.

Let us now turn to calculating the angular momentum in a central-force problem. Recall first the method of transforming from classical to quantum mechanics. The classical dynamical variable of interest is constructed from the position and the momentum variables. In this case, the angular momentum can be expressed as

$$\vec{L} = \vec{r} \times \vec{p}. \tag{3.1.1}$$

Next substitute in the quantum-mechanical expressions for the classical dynamical variables; $\vec{r}$ becomes simply $r(x, y, z)$ (an ordered triple of the real-space vector) and $\vec{p}$ is

$$\vec{p} = \frac{\hbar}{i} \left[\frac{d}{dx}, \frac{d}{dy}, \frac{d}{dz} \right]. \tag{3.1.2}$$

The angular momentum $\vec{L}$ can then be found from constructing the cross product as

$$\vec{L} = \vec{r} \times \frac{\hbar}{i} \vec{\nabla}. \tag{3.1.3}$$

The cross product can be evaluated through use of determinants. As the reader may recall, a simple way in which to evaluate the cross product between two vectors is to write the 3×3 determinant in which the first row is that of the unit vectors in Cartesian coordinates and the second and the third rows are filled with the components of the two vectors. In this case, $\vec{L}$ can be constructed from

$$\vec{L} = \frac{\hbar}{i} \begin{bmatrix} \hat{i} & \hat{j} & \hat{k} \\ x & y & z \\ \dfrac{d}{dx} & \dfrac{d}{dy} & \dfrac{d}{dz} \end{bmatrix}. \tag{3.1.4}$$

Evaluating the determinant gives

$$\vec{L} = \frac{\hbar}{i} \left(y\frac{d}{dz} - z\frac{d}{dy} \right)\hat{i} + \frac{\hbar}{i} \left(z\frac{d}{dx} - x\frac{d}{dz} \right)\hat{j} + \frac{\hbar}{i} \left(x\frac{d}{dy} - y\frac{d}{dx} \right)\hat{k}. \tag{3.1.5}$$

Making use of the definition of p_x, p_y, and p_z, we can rewrite $\vec{L}$ as

$$\vec{L} = (yp_z - zp_y)\hat{i} + (zp_x - xp_z)\hat{j} + (xp_y - yp_x)\hat{k},$$
$$\vec{L} = L_x\hat{i} + L_y\hat{j} + L_z\hat{k}. \tag{3.1.6}$$

It is useful to consider the commutation relations among the different components of $\vec{L}$ and $\vec{p}$. First consider the commutator of L_x and y. This is

$$[L_x, y] = [(yp_z - zp_y), y] = -z[p_y, y] = +z(\hbar i) = i\hbar z. \tag{3.1.7}$$

Other commutators are

$$[L_x, p_y] = i\hbar p_z,$$
$$[L_x, x] = 0,$$
$$[L_x, p_x] = 0. \tag{3.1.8}$$

Using the above results, we can evaluate the commutator of L_x and L_y. Before we actually do the calculation, it is interesting to pause and reflect on what we expect to find. One would expect that each component of the angular momentum can be known simultaneously. As we will see below, L_x and L_y do not commute. From the definition of the commutator, two operators that do not commute cannot be simultaneously known. It seems surprising that different components of the angular momentum cannot be simultaneously determined. This point is discussed in greater detail below. Let us now consider the commutator of L_x and L_y:

$$[L_x, L_y] = [L_x, zp_x - xp_z] = [L_x, z]p_x - x[L_x, p_z]. \tag{3.1.9}$$

The commutators of L_x and z and L_x and p_z can be evaluated as follows:

$$[L_x, z] = [yp_z - zp_y, z]$$
$$= y[p_z, z] - [zp_y, z]. \tag{3.1.10}$$

But the commutator $[zp_y, z]$ is equal to zero. Therefore the commutator becomes

$$= y\frac{\hbar}{i}\left(\frac{\partial}{\partial z}z - z\frac{\partial}{\partial z}\right) = \frac{\hbar}{i}y = -i\hbar y. \tag{3.1.11}$$

The commutator of L_x and p_z is given as

$$[L_x, p_z] = [yp_z - zp_y, p_z] = [yp_z, p_z] - p_y[z, p_z]. \tag{3.1.12}$$

The commutator $[yp_z, p_z]$ is equal to zero since the commutator of a function on itself and an orthogonal function is always zero. This results in

$$-p_y\left(z\frac{\hbar}{i}\frac{\partial}{\partial z} - \frac{\hbar}{i}\frac{\partial}{\partial z}z\right) = p_y\frac{\hbar}{i} = -i\hbar p_y. \tag{3.1.13}$$

Using the above results, we can finally determine the commutator between L_x and L_y. Substituting into Eq. (3.1.9) the above results yields

$$[L_x, L_y] = -i\hbar yp_x - \frac{\hbar}{i}p_y x = -i\hbar yp_x + i\hbar p_y x$$
$$= -i\hbar yp_x + i\hbar p_y x = i\hbar(xp_y - yp_x). \tag{3.1.14}$$

But the last term above can be written in terms of the z component of the angular momentum L_z since

$$L_z = (xp_y - yp_x). \tag{3.1.15}$$

Therefore the commutator of L_x and L_y becomes

$$[L_x, L_y] = i\hbar L_z. \tag{3.1.16}$$

Clearly, L_x and L_y do not commute. Similar relations are found for the other components of $\vec{L}$. The commutators involving $\vec{L}$ are listed below:

$$[L_y, L_z] = i\hbar L_x, \qquad [L_z, L_x] = i\hbar L_y, \qquad [L^2, L] = 0, \tag{3.1.17}$$

where L^2 is

$$L^2 = L_x^2 + L_y^2 + L_z^2. \tag{3.1.18}$$

Interestingly, L^2 commutes with each of the components of $\vec{L}$, but the separate components of $\vec{L}$ do not commute with each other. Physically, this implies that the value of the total angular momentum can be known along with one component of the angular momentum at a time. However, no two components of $\vec{L}$ can be known simultaneously. From a classical perspective, this is a strange result.

Because of the commutation relations, only one component of the angular momentum can be measured at a time. Any one of the three components can

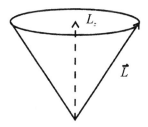

Figure 3.1.1. Sketch of the angular-momentum vector L and its orientation along the z axis.

be chosen to be measured but no two can be simultaneously known. If a wave function is an eigenfunction of L_z, it is also not an eigenfunction of L_x and L_y. Generally, we choose to write the wave function in terms of the eigenfunctions of L_z. In effect then the z axis is defined as the axis along which the measurement is performed. In other words, the z axis represents a preferred direction in space along which an external field, say, is applied. Recall that a physical measurement must be made to the system in order to determine its state. Therefore some field must be applied in order to measure the angular-momentum component. The direction of that field is chosen to be along the z axis. Physically, all that can be known about the angular momentum at any one time is its total magnitude L and one additional component L_z. Figure 3.1.1 sketches visually the angular-momentum vector. As can be seen from the figure, $\vec{L}$ lies somewhere on a cone whose axis is the z axis.

If the particle moves in a central potential, $\vec{L}$ must then commute with the potential energy $V(r)$. If $\vec{L}$ also commutes with the kinetic-energy operator T, then $\vec{L}$ is a constant of the motion; it is a conserved quantity. Looking at it differently, if $\vec{L}$ commutes with the Hamiltonian operator H, the sum of the potential- and the kinetic-energy operators, then the angular momentum and the energy can be simultaneously known. Therefore the energy and the angular momentum can be ascribed to the motion of the particle at all times. The eigenvalues corresponding to these observables are then said to be good quantum numbers. Let us show that $\vec{L}$ is a constant of the motion, given that it commutes with H. Before we consider $\vec{L}$, it is helpful to consider the general case first.

Let A be a time-independent operator. Then the time rate of change of the matrix element of A on the state $\Psi(t)$ is given as

$$i\hbar\frac{d}{dt}[\Psi^*(t)A\Psi(t)] = i\hbar\left[\Psi^*(t)A\frac{d\Psi(t)}{dt}\right] + i\hbar\left[\frac{d\Psi^*(t)}{dt}A\Psi(t)\right]. \qquad (3.1.19)$$

The time-dependent Schroedinger equation is simply

$$i\hbar\frac{d\Psi}{dt} = H\Psi, \qquad -i\hbar\frac{d\Psi^*}{dt} = H\Psi^*. \qquad (3.1.20)$$

Substituting into Eq. (3.1.19) for the time rates of change of Ψ and Ψ^* yields

$$i\hbar\Psi^*A\frac{d\Psi}{dt} + i\hbar\frac{d\Psi^*}{dt}A\Psi = \Psi^*AH\Psi - H\Psi^*A\Psi. \qquad (3.1.21)$$

We can rewrite the right-hand side of Eq. (3.1.21), making use of the hermiticity of H, as

$$\Psi^*AH\Psi - H\Psi^*A\Psi = \Psi^*AH\Psi - \Psi^*HA\Psi$$
$$= \Psi^*(AH - HA)\Psi = \Psi^*[A, H]\Psi. \qquad (3.1.22)$$

Combining Eqs. (3.1.21) and (3.1.22) leads to

$$i\hbar\frac{d}{dt}(\Psi^* A\Psi) = \Psi^*(AH - HA)\Psi. \tag{3.1.23}$$

Integrating each side over all space gives

$$i\hbar\frac{d}{dt}\int(\Psi^* A\Psi)\,d^3r = \int\Psi^*(AH - HA)\Psi\,d^3r. \tag{3.1.24}$$

But

$$\int(\Psi^* A\Psi)\,d^3r = \langle A\rangle, \tag{3.1.25}$$

which is the expectation value of A. When this identification is made, Eq. (3.1.24) becomes

$$i\hbar\frac{d}{dt}\langle A\rangle = \langle AH - HA\rangle. \tag{3.1.26}$$

We see then that the time rate of change of the expectation value of the operator A is equal to the expectation value of the commutator of H and A. For a time-independent operator A, if it commutes with H then the expectation value of A is a constant of the motion; its time derivative is zero and A is time invariant. This is a general result. If any time-independent operator commutes with the Hamiltonian, it is time invariant or a constant of the motion.

In general, A can also be time dependent. In this case the time rate of change of the expectation value of A becomes

$$\frac{d}{dt}\langle A\rangle = \frac{\langle AH - HA\rangle}{i\hbar} + \frac{\partial A}{\partial t}. \tag{3.1.27}$$

The time rate of change of the expectation value depends on both its commutator with H and its partial derivative with respect to time.

In the case of greatest interest here, that of the orbital angular momentum $\vec{L}$, we know that $\vec{L}$ is explicitly time independent. From the result above, if $\vec{L}$ and H commute, then the time rate of change of the expectation value of $\vec{L}$ must be zero and $\vec{L}$ is a constant of the motion. As we will see, the quantity L^2 is more useful to us than $\vec{L}$ itself. Therefore it is of greater use to show that L^2 commutes with H. By a similar argument, $\vec{L}$ also commutes with H, since $\vec{L}$, like L^2 as shown below, commutes with any radial derivative as well as $V(r)$.

We can determine L^2 by forming the cross product between $\vec{r}$ and $\vec{p}$ by using spherical coordinates. To do this we first write the momentum in terms of r, θ, and ϕ, as

$$\vec{p} = \frac{\hbar}{i}\vec{\nabla} = \frac{\hbar}{i}\left(\hat{r}\frac{\partial}{\partial r} + \hat{\phi}\frac{1}{r\sin\theta}\frac{\partial}{\partial\phi} + \hat{\theta}\frac{1}{r}\frac{\partial}{\partial\theta}\right), \tag{3.1.28}$$

where the unit vectors $\hat{r}, \hat{\theta}$, and $\hat{\phi}$ are written in terms of $\hat{i}, \hat{j}$, and $\hat{k}$ as

$$
\begin{aligned}
\hat{r} &= \sin\theta\cos\phi\,\hat{i} + \sin\theta\sin\phi\,\hat{j} + \cos\theta\hat{k}, \\
\hat{\phi} &= -\sin\phi\,\hat{i} + \cos\phi\,\hat{j}, \\
\hat{\theta} &= \cos\theta\cos\phi\,\hat{i} + \cos\theta\sin\phi\,\hat{j} - \sin\theta\hat{k}.
\end{aligned}
\tag{3.1.29}
$$

To find the spherical coordinate representation of the angular momentum, the cross product of $\vec{r}$ and $\vec{p}$ is formed. This yields

$$
\vec{L} = \vec{r} \times \vec{p} = r\hat{r} \times \frac{\hbar}{i}\vec{\nabla} = \frac{\hbar}{i}\left(\hat{\phi}\frac{\partial}{\partial\theta} - \hat{\theta}\frac{1}{\sin\theta}\frac{\partial}{\partial\phi}\right).
\tag{3.1.30}
$$

When the Cartesian forms are substituted in for the unit vectors, $\hat{r}, \hat{\theta}$, and $\hat{\phi}$, $\vec{L}$ becomes

$$
\begin{aligned}
\vec{L} = \frac{\hbar}{i}\Bigg[&(-\sin\phi\,\hat{i} + \cos\phi\,\hat{j})\frac{\partial}{\partial\theta} \\
&- \frac{1}{\sin\theta}(\cos\theta\cos\phi\,\hat{i} + \cos\theta\sin\phi\,\hat{j} - \sin\theta\hat{k})\frac{\partial}{\partial\phi}\Bigg],
\end{aligned}
\tag{3.1.31}
$$

which simplifies, by a combination of like terms, to

$$
\begin{aligned}
\vec{L} = \frac{\hbar}{i}&\left[-\sin\phi\frac{\partial}{\partial\theta} - \cot\theta\cos\phi\frac{\partial}{\partial\phi}\right]\hat{i} \\
&+ \frac{\hbar}{i}\left[\cos\phi\frac{\partial}{\partial\theta} - \cot\theta\sin\phi\frac{\partial}{\partial\phi}\right]\hat{j} + \frac{\hbar}{i}\frac{\partial}{\partial\phi}\hat{k}.
\end{aligned}
\tag{3.1.32}
$$

Hence L_x, L_y, and L_z are

$$
\begin{aligned}
L_x &= \frac{\hbar}{i}\left[-\sin\phi\frac{\partial}{\partial\theta} - \cot\theta\cos\phi\frac{\partial}{\partial\phi}\right], \\
L_y &= \frac{\hbar}{i}\left[\cos\phi\frac{\partial}{\partial\theta} - \cot\theta\sin\phi\frac{\partial}{\partial\phi}\right], \\
L_z &= \frac{\hbar}{i}\frac{\partial}{\partial\phi}.
\end{aligned}
\tag{3.1.33}
$$

L^2 can now be found simply from the dot vector product of $\vec{L}$ onto itself. This is

$$
L^2 = \vec{L}\cdot\vec{L} = -\hbar^2\left[\frac{1}{\sin^2\theta}\frac{\partial^2}{\partial\phi^2} + \frac{1}{\sin\theta}\frac{\partial}{\partial\theta}\left(\sin\theta\frac{\partial}{\partial\theta}\right)\right].
\tag{3.1.34}
$$

Inspection of the above result shows clearly that L^2 is independent of r. Hence L^2 commutes with any function or derivative of r. In a central-force motion problem, the potential is strictly a function of r. Therefore L^2 commutes with

$V(r)$. $\vec{L}$, as given by Eqs. (3.1.33), is also independent of r. Therefore $\vec{L}$ must also commute with $V(r)$.

As we mentioned above, to show that L^2 is a constant of the motion, it is necessary to show that it commutes with H. H is the sum of the potential-energy and the kinetic-energy operators. L^2 has been shown above to commute with $V(r)$. Therefore L^2 and $\vec{L}$ will be constants of the motion if they commute with the kinetic-energy operator T. To show that this indeed happens, it is useful to rewrite the Hamiltonian in terms of L^2.

The Hamiltonian can be written in terms of L^2 through the following. Above we expressed L^2 as equal to the dot vector product of $\vec{L}$ on itself. Substituting $\vec{r} \times \vec{p}$ in place of $\vec{L}$ gives then

$$L^2 = (\vec{r} \times \vec{p}) \cdot (\vec{r} \times \vec{p}),$$
$$L^2 = r^2 p^2 - r(\vec{r} \cdot \vec{p}) \cdot \vec{p} + 2i\hbar \vec{r} \cdot \vec{p}, \qquad (3.1.35)$$

but

$$\vec{r} \cdot \vec{p} = \frac{\hbar}{i} r \frac{\partial}{\partial r}. \qquad (3.1.36)$$

When this substitution is made, L^2 becomes

$$L^2 = r^2 p^2 + \hbar^2 \frac{\partial}{\partial r}\left(r^2 \frac{\partial}{\partial r}\right)$$

$$= r^2 p^2 + \hbar^2 r^2 \frac{\partial^2}{\partial r^2} + 2\hbar^2 r \frac{\partial}{\partial r}. \qquad (3.1.37)$$

The kinetic-energy operator T is simply equal to $p^2/2m$. Solving for p^2 in terms of L^2 gives

$$p^2 = \frac{L^2}{r^2} - \frac{\hbar^2}{r^2} \frac{\partial}{\partial r}\left(r^2 \frac{\partial}{\partial r}\right). \qquad (3.1.38)$$

The kinetic-energy operator T is then

$$T = \frac{p^2}{2m} = \frac{L^2}{2mr^2} - \frac{\hbar^2}{2mr^2} \frac{\partial}{\partial r}\left(r^2 \frac{\partial}{\partial r}\right), \qquad (3.1.39)$$

which is clearly only a function of r and L^2. Given that the potential is only radially dependent, the Hamiltonian is only a function of r, its derivatives, and L^2. L^2 commutes with itself and $\vec{L}$. Since L^2 is independent of r it commutes then with any radial function and its derivatives. As a result, L^2 and H must commute. In addition, $\vec{L}$ and H must commute. Therefore

$$[H, L^2] = 0, \qquad [H, L] = 0. \qquad (3.1.40)$$

Physically, H and $\vec{L}$ have simultaneous eigenstates and can thus be simultaneously known.

Since H and L^2 commute, they satisfy eigenvalue equations for the same eigenstates. If we choose the eigenvalues of L^2 to be $\lambda\hbar^2$, then the eigenvalue equations for H and L^2 can be written as

$$H\Psi = E\Psi, \qquad L^2\Psi = \lambda\hbar^2\Psi, \tag{3.1.41}$$

where Ψ is the simultaneous eigenstate of H and L^2. H is given as

$$H = \left[\frac{L^2}{2mr^2} - \frac{\hbar^2}{2mr^2}\frac{\partial}{\partial r}\left(r^2\frac{\partial}{\partial r}\right) + V(r)\right]. \tag{3.1.42}$$

The eigenvalue equation for H is then

$$\left[\frac{L^2}{2mr^2} - \frac{\hbar^2}{2mr^2}\frac{\partial}{\partial r}\left(r^2\frac{\partial}{\partial r}\right) + V(r)\right]\Psi = E\Psi. \tag{3.1.43}$$

Note that Eq. (3.1.43) is separable since L^2 involves only θ and ϕ, while the remaining terms are only functions of r. Therefore the wave function can be separated into the product of a radial part and an angular part as

$$\Psi(r, \theta, \phi) = R(r)Y(\theta, \phi), \tag{3.1.44}$$

where $Y(\theta, \phi)$ is a solution of the L^2 eigenvalue equation.

When L^2 is applied onto Ψ, L^2 operates on only the radial-independent part $Y(\theta, \phi)$ since it commutes with the radial function $R(r)$. This leads to the following result:

$$L^2\Psi = L^2 R(r)Y(\theta, \phi) = R(r)L^2 Y(\theta, \phi) = R(r)\lambda\hbar^2 Y(\theta, \phi) = \lambda\hbar^2\Psi. \tag{3.1.45}$$

Therefore, in the eigenvalue equation for H, the term involving $L^2/2mr^2$ can be replaced with $(\lambda\hbar^2)/(2mr^2)$. When this replacement is made, the eigenvalue equation for H becomes

$$\left[\frac{\lambda\hbar^2}{2mr^2} - \frac{\hbar^2}{2mr^2}\frac{\partial}{\partial r}\left(r^2\frac{\partial}{\partial r}\right) + V(r)\right]\Psi = E\Psi. \tag{3.1.46}$$

Replacing Ψ with the product of $R(r)$ and $Y(\theta, \phi)$ gives

$$-\frac{\hbar^2}{2mr^2}\frac{\partial}{\partial r}\left(r^2\frac{\partial}{\partial r}\right)RY + \frac{\lambda\hbar^2}{2mr^2}RY + VRY = ERY. \tag{3.1.47}$$

Note that Eq. (3.1.47) has no explicit dependence on θ and ϕ. The angular dependence has been removed by use of the eigenvalue equation for L^2. Therefore the Y dependence can be divided out, leaving only a radially dependent equation. Hence we obtain two different equations. One is radially dependent and the other is not. These equations are

$$L^2 Y = \lambda\hbar^2 Y \tag{3.1.48}$$

and

$$-\frac{\hbar^2}{2mr^2}\frac{\partial}{\partial r}\left(r^2\frac{\partial}{\partial r}\right)R + \frac{\lambda\hbar^2}{2mr^2}R + VR = ER. \tag{3.1.49}$$

Let us consider the solution of each of the above equations separately. The radially independent equation (3.1.48) becomes, on substitution in Eq. (3.1.34) for L^2,

$$-\hbar^2\left[\frac{1}{\sin^2\theta}\frac{\partial^2}{\partial\phi^2} + \frac{1}{\sin\theta}\frac{\partial}{\partial\theta}\left(\sin\theta\frac{\partial}{\partial\theta}\right)\right]Y = \hbar^2\lambda Y. \tag{3.1.50}$$

Dividing through by $\hbar^2$, we obtain the eigenvalue equation for L^2:

$$\left[\frac{1}{\sin^2\theta}\frac{\partial^2}{\partial\phi^2} + \frac{1}{\sin\theta}\frac{\partial}{\partial\theta}\left(\sin\theta\frac{\partial}{\partial\theta}\right)\right]Y = -\lambda Y. \tag{3.1.51}$$

The resulting equation is again separable. This time it can be separated into equations involving θ and ϕ. Therefore Y can be written as the product of theta- and phi-dependent functions as

$$Y(\theta,\phi) = \Theta(\theta)\Phi(\phi). \tag{3.1.52}$$

Substituting Eq. (3.1.52) into Eq. (3.1.51) and simplifying produces

$$-\frac{1}{\Phi}\frac{d^2\Phi}{d\phi^2} = \frac{\sin^2\theta}{\Theta}\left[\frac{1}{\sin\theta}\frac{d}{d\theta}\left(\sin\theta\frac{d\Theta}{d\theta}\right) + \lambda\Theta\right]. \tag{3.1.53}$$

Since each side of Eq. (3.1.53) is a function of only one variable, both the left- and the right-hand sides must be separately equal to a constant. We choose the constant to be m^2. The phi-dependent part can then be written as

$$-\frac{1}{\Phi}\frac{d^2\Phi}{d\phi^2} = m^2, \qquad \frac{d^2\Phi}{d\phi^2} + m^2\Phi = 0, \tag{3.1.54}$$

and the theta-dependent part is

$$\frac{1}{\sin\theta}\frac{d}{d\theta}\left(\sin\theta\frac{d\Theta}{d\theta}\right) + \lambda\Theta = \frac{m^2\Theta}{\sin^2\theta}. \tag{3.1.55}$$

The phi-dependent equation is readily solved to yield

$$\Phi \sim e^{im\phi}; \tag{3.1.56}$$

Φ must be single valued. This follows from the inherent rotational symmetry of the problem. Once the system is rotated in a complete circle, it must go back into itself. In other words, the system must be invariant under a full rotation by 2π, otherwise one could tell that the system has been rotated around in a complete circle, violating the rotational symmetry property. The rotational symmetry of the wave function implies that

$$e^{im\phi} = e^{im(\phi+2\pi)}. \tag{3.1.57}$$

Dividing through by $e^{im\phi}$ yields

$$1 = e^{im2\pi}. \tag{3.1.58}$$

To ensure that the exponential function $e^{im2\pi}$ is equal to 1, m must be equal to an integer, $0, \pm 1, \pm 2, \pm 3, \ldots$. When the above results are combined, the phi-dependent eigenstates can be summarized as

$$\Phi \sim e^{im\phi}, \qquad m = 0, \pm 1, \pm 2, \pm 3, \ldots. \tag{3.1.59}$$

The phi-dependent wave functions are eigenvectors of the L_z operator. The L_z operator was found above to be

$$L_z = \frac{\hbar}{i}\frac{d}{d\phi}. \tag{3.1.60}$$

Applying this operator onto the phi-dependent eigenfunctions gives

$$L_z\Phi = \frac{\hbar}{i}\frac{\partial}{\partial\phi}e^{im\phi} = m\hbar e^{im\phi}. \tag{3.1.61}$$

Clearly, $\Phi(\phi)$ are eigenstates of L_z with eigenvalues of $m\hbar$. This result implies that the z component of the angular momentum of a particle undergoing central-force motion is quantized; only discrete values of the z component of the angular momentum are allowed. This is a general result since it does not depend on the form of the potential; it must simply be a function of r.

The theta-dependent equation can also be solved directly. Rewriting it, including the separation constant m^2, gives

$$\frac{\sin^2\theta}{\Theta}\left[\frac{1}{\sin\theta}\frac{d}{d\theta}\left(\sin\theta\frac{d\Theta}{d\theta}\right) + \lambda\Theta\right] = m^2,$$

$$\frac{1}{\sin\theta}\frac{d}{d\theta}\left(\sin\theta\frac{d\Theta}{d\theta}\right) + \lambda\Theta - \frac{m^2}{\sin^2\theta}\Theta = 0. \tag{3.1.62}$$

Equations (3.1.62) can be solved through a change of variables. When the following substitutions are made,

$$\xi = \cos\theta, \qquad d\xi = -\sin\theta\, d\theta, \qquad F(\xi) = \Theta(\theta), \tag{3.1.63}$$

Eqs. (3.1.62) can be transformed into

$$\frac{d}{d\xi}\left[(1-\xi^2)\frac{dF}{d\xi}\right] - \frac{m^2F}{1-\xi^2} + \lambda F = 0. \tag{3.1.64}$$

The solution of Eq. (3.1.64) is well known. The solutions are the associated Legendre polynomials. Before we consider the general solution, it is instructive to consider a simpler solution, that corresponding to $m = 0$. When $m = 0$, Eq. (3.1.64) becomes

$$\frac{d}{d\xi}\left[(1 - \xi^2)\frac{dF}{d\xi}\right] + \lambda F = 0, \tag{3.1.65}$$

which can be recognized as Legendre's equation. The general solution to Legendre's equation is found with a power series in ξ:

$$F(\xi) = \sum a_k \xi^k. \tag{3.1.66}$$

When Eq. (3.1.66) is substituted into Eq. (3.1.65) (Legendre's equation) a recursion relation is obtained. The recursion relation is

$$a_{k+2} = a_k \frac{k(k + 1) - \lambda}{(k + 1)(k + 2)}, \tag{3.1.67}$$

where k is a positive integer. k has values ranging from 0 to infinity. The series must terminate at some finite value of k. If the series does not terminate, the ratio a_{k+2}/a_k approaches $k/(k + 2)$ as k goes to infinity. As a result, the series behaves like $\sum(1/n)$, which diverges for values of $\theta = 0$ or π. In other words, the solution has singularities at certain values of θ that are not acceptable eigenfunctions of L^2. We choose $a_{l+2} = 0$ such that l is the last term in the series. The recursion relation becomes then

$$0 = a_l \frac{l(l + 1) - \lambda}{(l + 1)(l + 2)}. \tag{3.1.68}$$

In order for Eq. (3.1.68) to be true in general,

$$l(l + 1) = \lambda. \tag{3.1.69}$$

When $m = 0$, the theta-dependent equation is

$$\frac{1}{\sin\theta}\frac{d}{d\theta}\left(\sin\theta\frac{d\Theta}{d\theta}\right) = -\lambda\Theta = -l(l + 1)\Theta. \tag{3.1.70}$$

From Eq. (3.1.34), L^2 is given as

$$L^2 = -\hbar^2\left[\frac{1}{\sin^2\theta}\frac{d^2}{d\phi^2} + \frac{1}{\sin\theta}\frac{d}{d\theta}\left(\sin\theta\frac{d}{d\theta}\right)\right]. \tag{3.1.71}$$

The application of L^2 on θ gives only theta-dependent terms as

$$L^2 = -\frac{\hbar^2}{\sin\theta}\frac{d}{d\theta}\left(\sin\theta\frac{d}{d\theta}\right). \tag{3.1.72}$$

A comparison of Eqs. (3.1.70) and (3.1.72) shows that the left-hand side of

Table 3.1.1
Values of l and Their Corresponding Eigenvalues

l	Eigenvalue	Spectroscopic State
0	0	s
1	$2\hbar^2$	p
2	$6\hbar^2$	d
3	$12\hbar^2$	f
.	.	
.	.	alphabetic
.	.	from
l	$l(l+1)\hbar^2$	here

Eq. (3.1.70) is equal to L^2 divided by $-\hbar^2$. As a result,

$$\frac{L^2}{-\hbar^2}\Theta = -\lambda\Theta = -l(l+1)\Theta.$$

(3.1.73)

Multiplying through by $-\hbar^2$, we obtain

$$L^2\Theta = \hbar^2\lambda\Theta = \hbar^2 l(l+1)\Theta.$$

(3.1.74)

Inspection of Eq. (3.1.74) indicates that the functions Θ are eigenfunctions of L^2 with eigenvalues $\hbar^2 l(l+1)$, where l is an integer.

l is called the orbital angular-momentum quantum number. Historically, a spectroscopic notation was adopted for the different values of l. The values of l can assume any positive integer and zero, and the corresponding spectroscopic states are collected in Table 3.1.1.

The actual polynomial solutions to Eq. (3.1.60) are known as the Legendre polynomials. The normalized solutions are written in a special way as

$$P_l(\xi) = \frac{1}{2^l l!}\frac{d^l}{d\xi^l}(\xi^2 - 1)^l.$$

(3.1.75)

The $P_l(\xi)$ are determined from the power-series solution $F(\xi)$, given by Eq. (3.1.61). The first few terms of the Legendre polynomials are

$$P_0(\xi) = 1,$$

$$P_1(\xi) = \xi,$$

$$P_2(\xi) = \frac{1}{2}(3\xi^2 - 1),$$

(3.1.76)

where $\xi \equiv \cos\theta$. It is important to note that the Legendre polynomials form an orthonormal set. When integrated over all values of theta, which ranges from 0 to π, the orthogonality condition is

$$\int_0^\pi P_l(\cos\theta)P_{l'}(\cos\theta)\sin\theta\,d\theta = 0, \qquad l' \neq l.$$

(3.1.77)

The normalization condition for the Legendre polynomials is

$$\int_{-1}^{+1}|P_l(\xi)|^2\,d\xi = \frac{2}{2l+1}.$$

(3.1.78)

When m is not equal to zero, the solution of the angular equation involves both θ and ϕ. The general solutions in this case are the associated Legendre polynomials. The normalized forms are given as

$$P_l^m(\xi) = (1 - \xi^2)^{\frac{m}{2}} \frac{d^m P_l(\xi)}{d\xi^m}$$

$$= \frac{1}{2^l l!} (1 - \xi^2)^{\frac{m}{2}} \frac{d^{l+m}}{d\xi^{l+m}} (\xi^2 - 1)^l. \tag{3.1.79}$$

The associated Legendre polynomials form the solution to Eq. (3.1.64) and obey orthonormality conditions similar to those of the Legendre polynomials. The associated Legendre polynomials are orthogonal,

$$\int_{-1}^{+1} P_l^m(\xi) P_{l'}^{m'}(\xi) \, d\xi = 0, \qquad l \neq l', \tag{3.1.80}$$

and they also are normalized as

$$\int_{-1}^{+1} \left[P_l^m(\xi) \right]^2 d\xi = \frac{2}{2l+1} \frac{(l+m)!}{(l-m)!}. \tag{3.1.81}$$

Often it is useful to redefine the solutions to Eq. (3.1.64) by normalizing them with respect to an integration over the entire solid angle. The resulting functions are slightly different from the associated Legendre polynomials and are called the spherical harmonics. The spherical harmonics $Y(\theta, \phi)$ are defined in the following way:

$$Y_{lm}(\theta, \phi) = \sqrt{\frac{2l+1}{4\pi} \frac{(l-m)!}{(l+m)!}} (-1)^m e^{im\phi} P_l^m(\cos\theta). \tag{3.1.82}$$

The subscripts l and m represent the l and the m angular-momentum quantum numbers. As discussed above, both l and m are integers. l has values ranging from 0 to infinity while m is given from Eq. (3.1.59) as $0, \pm 1, \pm 2$, etc. The maximum value of m depends on l. From the nature of the spherical harmonics, the value of m is restricted to lie between $-l$ and $+l$. In other words, m can have any integer value lying between $-l$ and $+l$. The first few spherical harmonics are given as

$$Y_{00} = \frac{1}{\sqrt{4\pi}},$$

$$Y_{10} = \sqrt{\frac{3}{4\pi}} \cos\theta,$$

$$Y_{1,\pm 1} = \mp \sqrt{\frac{3}{8\pi}} e^{\pm i\phi} \sin\theta. \tag{3.1.83}$$

As in the case of the Legendre polynomials, the spherical harmonics form a complete basis set. They are mutually orthogonal and completely span the space. In addition, they are normalized. The orthogonality condition can be expressed as

$$\int_0^{2\pi} \int_0^\pi Y_{lm}^*(\theta, \phi) Y_{l',m'}(\theta, \phi) d\Omega = \delta_{ll'} \delta_{mm'}, \tag{3.1.84}$$

where m is an integer such that

$$-l \leq m \leq +l. \tag{3.1.85}$$

It is important to recognize that the spherical harmonics are solutions to the eigenvalue equations for the operators L^2 and L_z. Hence the Y_{lm}'s satisfy

$$L^2 Y_{lm} = \hbar^2 l(l+1) Y_{lm},$$
$$L_z Y_{lm} = m\hbar Y_{lm} \tag{3.1.86}$$

and are clearly the eigenfunctions of the operators L^2 and L_z.

Finally, it is useful to define two new operators, the angular-momentum raising and lowering operators. These operators, L_+ and L_-, are defined as

$$L_+ = L_x + i L_y, \qquad L_- = L_x - i L_y. \tag{3.1.87}$$

Making use of Eqs. (3.1.33), we can combine the components L_x and L_y to give L_+ and L_- as

$$L_+ = \hbar e^{i\phi} \left(\frac{\partial}{\partial\theta} + i \cot\theta \frac{\partial}{\partial\phi} \right),$$
$$L_- = -\hbar e^{-i\phi} \left(\frac{\partial}{\partial\theta} - i \cot\theta \frac{\partial}{\partial\phi} \right). \tag{3.1.88}$$

The operators L_+ and L_- behave in a similar fashion to the raising and lowering operators determined for the harmonic oscillator. In Chapter 2, we found raising and lowering operators that, when applied onto an eigenstate, either increased or decreased its corresponding energy by one quantum. In the case of the angular-momentum raising and lowering operators, the z component of the angular momentum is either increased or decreased by one. Mathematically, the operation of L_+ and L_- can be summarized as

$$L_+ Y_{lm} = \hbar\sqrt{(l-m)(l+m+1)} \, Y_{l,m+1}(\theta, \phi),$$
$$L_- Y_{lm} = \hbar\sqrt{(l+m)(l-m+1)} \, Y_{l,m-1}(\theta, \phi). \tag{3.1.89}$$

Before we end this section, it is important to review the results obtained. For a spherically symmetric potential, that is, a potential that depends only on r, the angular momentum is a constant of the motion. Quantum mechanically, we have found that although the angular momentum is a conserved quantity in a central-force motion problem, only one component of the angular momentum and the magnitude of the angular momentum can be specified at any one instant in time. Generally, the z component of the angular momentum L_z is chosen to be measured along with the total angular momentum L. The equation that describes the angular motion can be separated into two independent parts, one part that depends on θ and the other that depends on ϕ. The solution of the phi-dependent part is also an eigenfunction of the L_z operator. The solution of the theta-dependent part is also an eigenfunction of the operator L^2. Owing to the fact that L^2 commutes with L, the theta-dependent part is an eigenfunction of L as well. Hence the angular solution specifies the z component of the angular momentum and the orbital angular momentum itself. As we discussed above, only these two properties of the angular momentum can be known simultaneously. The eigenvalues of the L_z and the L^2 operators are specified by two integers m and l. m has only integer values lying between and including $-l$ and $+l$. l has integer values ranging from 0 to infinity. In Section 3.2 we will find what the maximum restriction on the value of l is.

Since both L^2 and L_z commute with one another and with the Hamiltonian H, then the physical observables corresponding to these operators can be simultaneously determined. Hence, the energy, orbital angular momentum, and the z component of the orbital angular momentum can all be known simultaneously in a central-force motion problem. The motion of a quantum-mechanical particle in a central force can be completely specified by the quantum numbers corresponding to its energy, angular momentum, and z component of its orbital angular momentum. As we will see in Section 3.2, this has important ramifications in the behavior of an electron in a hydrogen atom.

3.2 Hydrogen Atom

We next turn our attention to solving the radial part of the three-dimensional Schroedinger equation in spherical coordinates. The equation of interest is Eq. (3.1.49). In Section 3.1, the general solution for the angular part, Eq. (3.1.50), was obtained. Inspection of Eq. (3.1.49) shows that the potential $V(r)$ needs to be specified to obtain a specific solution of the radial equation.

The problem of greatest interest to us here is that of the hydrogen atom, in which the potential $V(r)$ is proportional to $1/r$. As we will see below, the states of the hydrogen atom will be characterized by three quantum numbers since the problem is a three-dimensional one. Each dimension, r, θ, and ϕ, introduces an independent degree of freedom in the problem and hence three independent coordinates are necessary to describe fully the motion of the particle. As we found in Chapter 2, in quantum mechanics the motion of a particle in a bound system is restricted to only discrete states. Hence the motion of the particle in the presence of the potential $V(r)$ will be quantized along each direction. Each coordinate will be specified by an independent quantum number.

It is first interesting to solve Eq. (3.1.49) for the case in which $V(r) = 0$. This problem is essentially that of an infinite three-dimensional well. The solution, for which Cartesian coordinates x, y, and z are used, is presented in Box 3.2.1. This approach is equivalent to extending the one-dimensional infinite square-well problem discussed in Chapter 2 to three dimensions.

Box 3.2.1 Three-Dimensional Rectangular Potential Well

Consider the motion of an electron within a three-dimensional rectangular box. The potential at the edges of the box and everywhere outside of the box is assumed to be infinite. The potential everywhere within the box is assumed to be zero. As in the case of the one-dimensional infinite square well, the wave function must vanish at the boundaries of the box. The wave function is then completely confined within the box. Given that the problem has rectangular symmetry, the Schroedinger equation can be readily solved as follows.

In three dimensions in rectangular coordinates the Schroedinger equation is

$$-\frac{\hbar^2}{2m}\left[\frac{\partial^2}{\partial x^2} + \frac{\partial^2}{\partial y^2} + \frac{\partial^2}{\partial z^2}\right]\Psi(xyz) + V(xyz)\Psi(xyz) = E\Psi(xyz).$$

The potential $V(xyz)$ is zero everywhere within the box and infinite at the edges and outside of the box. The Schroedinger equation within the box is

$$-\frac{\hbar^2}{2m}\left[\frac{\partial^2}{\partial x^2} + \frac{\partial^2}{\partial y^2} + \frac{\partial^2}{\partial z^2}\right]\Psi = E\Psi.$$

The equation is completely separable in x, y, and z. Therefore the solution for Ψ is given as the product of three separate, independent functions: $X(x)$, $Y(y)$, and $Z(z)$. Substituting in the product of $X(x)Y(y)Z(z)$ for Ψ into the Schroedinger equation and dividing through by Ψ yields

$$-\frac{\hbar^2}{2m}\left[\frac{1}{X}\frac{\partial^2 X}{\partial x^2} + \frac{1}{Y}\frac{\partial^2 Y}{\partial y^2} + \frac{1}{Z}\frac{\partial^2 Z}{\partial z^2}\right] = E.$$

Each term in the above equation must be separately equal to a constant. Choosing the separation constants as k_x^2, k_y^2, and k_z^2, we can write each of the terms as

$$\frac{1}{X}\frac{\partial^2 X}{\partial x^2} = -k_x^2,$$

$$\frac{1}{Y}\frac{\partial^2 Y}{\partial y^2} = -k_y^2,$$

$$\frac{1}{Z}\frac{\partial^2 Z}{\partial z^2} = -k_z^2.$$

Substituting each of these relations into the above equation gives

$$-\frac{\hbar^2}{2m}\left[-k_x^2 - k_y^2 - k_z^2\right] = E.$$

But,

$$k^2 = k_x^2 + k_y^2 + k_z^2.$$

Therefore the energy E can be written as

$$E = \frac{\hbar^2 k^2}{2m}.$$

The solution $\Psi(xyz)$ is given as the product of the functions $X(x)$, $Y(y)$, and $Z(z)$. The equations for x, y, and z are immediately solved to give

$$X(x) = C_1^- \sin k_x x + C_1^+ \cos k_x x,$$

$$Y(y) = C_2^- \sin k_y y + C_2^+ \cos k_y y,$$

$$Z(z) = C_3^- \sin k_z z + C_3^+ \cos k_z z.$$

The boundary conditions imply that the wave function must vanish along all the faces of the box. In other words, along the faces described as $x = 0$ and $x = L_x$, $Y = 0$ and $y = L_y$, and $z = 0$ and $z = L_z$, the wave function must be zero. Mathematically, these conditions can be expressed as

$$\Psi(0, y, z) = 0, \qquad \Psi(x, 0, z) = 0, \qquad \Psi(x, y, 0) = 0,$$

$$\Psi(L_x, y, z) = 0, \qquad \Psi(x, L_y, z) = 0, \qquad \Psi(x, y, L_z) = 0.$$

To satisfy the boundary conditions at $x = 0$, $y = 0$, and $z = 0$, the coefficients for the cosine terms must vanish. This implies that

$$C_1^+ = C_2^+ = C_3^+ = 0.$$

Only the sine terms remain in the expressions for $X(x)$, $Y(y)$, and $Z(z)$. The wave function Ψ is given by the product of these three functions. When all the normalization constants are grouped together into one normalization constant A, the wave function has the form

$$\Psi = A \sin k_x x \sin k_y y \sin k_z z.$$

Application of the boundary conditions at $x = L_x$, $y = L_y$, and $z = L_z$ to the functions $X(x)$, $Y(y)$, and $Z(z)$ yields

$$\sin k_x L_x = 0, \qquad \sin k_y L_y = 0, \qquad \sin k_z L_z = 0.$$

To recover the nontrivial solution, the arguments of the sine functions must obey

$$k_x = \frac{n_x \pi}{L_x}, \qquad k_y = \frac{n_y \pi}{L_y}, \qquad k_z = \frac{n_z \pi}{L_z},$$

where n_x, n_y, and n_z are integers. The choice of these values for k_x, k_y, and k_z guarantees that the sine functions will vanish at the edges of the box. Substituting the above expressions into the expression for the total energy yields

$$E = \frac{\hbar^2}{2m} \left[\frac{n_x^2 \pi^2}{L_x^2} + \frac{n_y^2 \pi^2}{L_y^2} + \frac{n_z^2 \pi^2}{L_z^2} \right].$$

As can be seen from the above equation, the energy is quantized. The energy eigenvalues depend on three separate quantum numbers, n_x, n_y, and n_z. These quantum numbers specify the motion of the particle in each of the three dimensions, x, y, and z. The eigenfunctions are given by

$$\Psi = A \sin \frac{n_x \pi x}{L_x} \sin \frac{n_y \pi y}{L_y} \sin \frac{n_z \pi z}{L_z}.$$

In the hydrogen atom, the potential is spherically symmetric, $V(r)$. The potential in this case depends inversely on the separation of the charges, $\sim 1/r$. Therefore the angular solution is precisely the same as that determined in Section 3.1. The only difference between the hydrogen atom problem and the results presented in Section 3.1 is the presence of the radially symmetric potential $1/r$ in the radial equation. This is what we must solve. To solve the radial equation including $V(r)$, it is useful to first rewrite it with a radial-momentum operator p_r, which is defined as

$$p_r = \frac{\hbar}{i}\left(\frac{d}{dr} + \frac{1}{r}\right). \tag{3.2.1}$$

The kinetic-energy term is then

$$\frac{p_r^2}{2m}. \tag{3.2.2}$$

The general solution of the Schroedinger equation will be of the form $R_{nl}(r)$ $Y_{l,m}(\theta,\phi)$. $R(r)$ is a function of a polynomial of r. In the case in which the potential is $1/r$, the form of $R(r)$ is $u(r)/r$, where $u(r)$ is a polynomial in r. The choice of $u(r)/r$ is principally motivated by the solution of the spherical potential well. In that case, it can be shown that the solution is of the form $1/r\, u(r)$. Applying the p_r^2 operator given above to this choice for $R(r)$ gives

$$p_r^2 R = p_r^2 \frac{u}{r} = p_r \frac{\hbar}{i}\left(\frac{\partial}{\partial r} + \frac{1}{r}\right)\frac{u}{r} = p_r \frac{\hbar}{i}\left[\frac{u'}{r} - \frac{u}{r^2} + \frac{u}{r^2}\right]$$

$$= p_r \frac{\hbar}{i}\frac{u'}{r} = p_r \frac{\hbar}{i}\frac{1}{r}\frac{\partial u}{\partial r}$$

$$= \left(\frac{\hbar}{i}\right)^2 \left(\frac{\partial}{\partial r} + \frac{1}{r}\right)\frac{u'}{r},$$

$$\left(\frac{\hbar}{i}\right)^2\left[\frac{u''}{r} - \frac{u'}{r^2} + \frac{u'}{r^2}\right] = \left(\frac{\hbar}{i}\right)^2\left(\frac{u''}{r}\right) = \left(\frac{\hbar}{i}\right)^2 \frac{1}{r}\frac{\partial^2 u}{\partial r^2} = -\frac{\hbar^2}{r}\frac{\partial^2 u}{\partial r^2}. \tag{3.2.3}$$

It is important to first demonstrate that the application of p_r^2 is equivalent to the differential operators in the kinetic-energy term in H, which is

$$-\hbar^2 \frac{1}{r^2}\frac{\partial}{\partial r}\left(r^2 \frac{\partial}{\partial r}\right) = -\hbar^2\left(\frac{\partial^2}{\partial r^2} + \frac{2}{r}\frac{\partial}{\partial r}\right). \tag{3.2.4}$$

To show this, consider p_r^2. Applying this operator onto the state Ψ yields

$$p_r^2 \Psi = -\hbar^2\left(\frac{\partial}{\partial r} + \frac{1}{r}\right)\left(\frac{\partial}{\partial r} + \frac{1}{r}\right)\Psi = -\hbar^2\left[\frac{\partial^2}{\partial r^2} - \frac{1}{r^2} + \frac{1}{r^2} + \frac{1}{r}\frac{\partial}{\partial r} + \frac{1}{r}\frac{\partial}{\partial r}\right]\Psi$$

$$= -\hbar^2\left(\frac{\partial^2}{\partial r^2} + \frac{2}{r}\frac{\partial}{\partial r}\right)\Psi. \tag{3.2.5}$$

Comparing Eqs. (3.2.4) and (3.2.5) clearly shows that they are equal. Hence use of $p_r^2/2m$ for the kinetic-energy term recovers the same result as before.

The Schroedinger equation can then be written with p_r as

$$\left[\frac{p_r^2}{2m} + \frac{\hbar^2 l(l+1)}{2mr^2} + V(r)\right]\Psi = E\Psi. \tag{3.2.6}$$

Substituting u/r in for Ψ yields

$$-\frac{\hbar^2}{2mr}\frac{\partial^2 u}{\partial r^2} + \frac{\hbar^2 l(l+1)}{2mr^2}\left(\frac{u}{r}\right) + V(r)\left(\frac{u}{r}\right) = E\left(\frac{u}{r}\right), \tag{3.2.7}$$

where use has been made of Eq. (3.2.3) in simplifying the first term. The factor $1/r$ is common to all four terms and can thus be multiplied out, leaving

$$\left[-\frac{\hbar^2}{2m}\frac{\partial^2}{\partial r^2} + \frac{\hbar^2 l(l+1)}{2mr^2} + V(r)\right]u = Eu. \tag{3.2.8}$$

As r approaches zero, u approaches zero since $rR = u$. The potential $V(r)$ is given (in cgs units) as

$$V(r) = \frac{-Zq^2}{r}, \tag{3.2.9}$$

where Z is the atomic number, q is the electron charge, and r the distance of separation between the charge and the force center. Equation (3.2.8) can be simplified if some substitutions are made to make it dimensionless. We define the following quantities:

$$r = a_0\rho, \qquad E = E_0\epsilon. \tag{3.2.10}$$

Substituting Eqs. (3.2.10) into Eq. (3.2.8) for r and E yields

$$\left[-\frac{\hbar^2}{2m a_0^2}\frac{1}{\partial\rho^2}\frac{\partial^2}{} + \frac{\hbar^2 l(l+1)}{2m a_0^2 \rho^2}\right]u - \frac{Zq^2}{a_0\rho}u = E_0\epsilon u. \tag{3.2.11}$$

Since the constants a_0 and E_0 are currently undefined, let us define them such that

$$\frac{Zq^2}{a_0} = \frac{\hbar^2}{m a_0^2} = E_0. \tag{3.2.12}$$

Solving in terms of a_0 and E_0 gives

$$a_0 = \frac{\hbar^2}{mq^2 Z}, \qquad E_0 = \frac{Z^2 q^4 m}{\hbar^2}. \tag{3.2.13}$$

With these assignments, Eq. (3.2.11) becomes

$$\left[-\frac{1}{2} \frac{\partial^2}{\partial \rho^2} + \frac{l(l+1)}{2\rho^2} - \frac{1}{\rho} \right] u = \epsilon u, \tag{3.2.14}$$

which is dimensionless as desired. We can solve Eq. (3.2.14) in the usual manner by assuming a power-series solution. We guess a solution of the form

$$u = A_0 \rho^n e^{-\frac{\rho}{n}} + \cdots + A_q \rho^n \rho^q e^{-\frac{\rho}{n}}. \tag{3.2.15}$$

Substituting the first term in the expansion into the second-derivative term in Eq. (3.2.14) yields

$$\frac{\partial^2}{\partial \rho^2} (\rho^n e^{-\frac{\rho}{n}}) = \frac{1}{n^2} \rho^n e^{-\frac{\rho}{n}} - \frac{2n\rho^{n-1}e^{-\frac{\rho}{n}}}{n} + n(n-1)\rho^{n-2}e^{-\frac{\rho}{n}}. \tag{3.2.16}$$

Substituting in the above result along with the first term for u into Eq. (3.2.14) gives then

$$-\frac{1}{2}\left[\frac{1}{n^2}\rho^n e^{-\frac{\rho}{n}} - \frac{2\rho^n e^{-\frac{\rho}{n}}}{\rho} + \frac{n(n-1)\rho^n}{\rho^2}e^{-\frac{\rho}{n}} \right]$$
$$+ \frac{l(l+1)\rho^n}{2\rho^2}e^{-\frac{\rho}{n}} - \frac{\rho^n e^{-\frac{\rho}{n}}}{\rho} = \epsilon\rho^n e^{-\frac{\rho}{n}}. \tag{3.2.17}$$

Like terms of the same power of ρ must be equal. Hence equating the terms of power ρ^n gives

$$-\frac{1}{2n^2}\rho^n e^{-\frac{\rho}{n}} = \epsilon\rho^n e^{-\frac{\rho}{n}}. \tag{3.2.18}$$

When simplified, Eq. (3.2.18) results in an expression for ϵ as

$$\epsilon = -\frac{1}{2n^2}. \tag{3.2.19}$$

The total energy E is then determined from the definition of ϵ from the second equation of Eqs. (3.2.10). From Eqs. (3.2.10) and (3.2.12) E is found to be

$$E = E_0\epsilon = -\frac{1}{2n^2}\frac{Z^2 q^4 m}{\hbar^2}. \tag{3.2.20}$$

Note that the energy is found in terms of known physical constants, the electric charge e, the electron mass m, the atomic number Z, and $\hbar$. Substituting in numerical values for each of the physical constants q, m and $\hbar$, and assuming that $Z = 1$, we find that the energy levels of the hydrogen atom are

$$E = \frac{-13.6}{n^2}\text{eV}. \tag{3.2.21}$$

The ground state of the hydrogen atom corresponds to $n = 1$. From Eq. (3.2.21) this state has an energy of $-13.6\,\mathrm{eV}$ in close agreement with experiment.

Equating terms of order ρ^{n-2} in Eq. (3.2.17) leads to a relationship between l and n. This can be seen from

$$\frac{1}{2}n(n-1)\frac{\rho^{n}e^{-\frac{\rho}{n}}}{\rho^{2}} = \frac{l(l+1)}{2}\frac{\rho^{n}e^{-\frac{\rho}{n}}}{\rho^{2}}. \tag{3.2.22}$$

Dividing out the common factors above leads to a simple relation between l and n as

$$\frac{n(n-1)}{2} = \frac{l(l+1)}{2}. \tag{3.2.23}$$

Therefore, for Eq. (3.2.23) to be satisfied, n must be an integer. Solving for l in terms of n yields

$$l(l+1) - n(n-1) = 0,$$
$$l^{2} + l - n(n-1) = 0. \tag{3.2.24}$$

Below we can simplify the term under the square root by recognizing that

$$l = \frac{-1 \pm \sqrt{1 + 4n(n-1)}}{2},$$
$$l = \frac{-1 \pm \sqrt{1 + 4n^{2} - 4n}}{2},$$

which factor as

$$4n^{2} - 4n + 1 = (2n-1)(2n-1). \tag{3.2.25}$$

Subsequently, l can be determined then as

$$l = \frac{-1 \pm (2n-1)}{2},$$
$$l = -\frac{1}{2} \pm \frac{2n-1}{2}. \tag{3.2.26}$$

l must be positive so only the positive root is chosen. l can be determined in terms of n then as

$$l = \frac{-1 + 2n - 1}{2} = \frac{2n-2}{2},$$
$$l = n - 1. \tag{3.2.27}$$

l ranges in value from 0 up to $n-1$. Therefore, if n is equal to 1, then l has only value 0. If n is equal to 2, l has values 0 and 1, etc.

It is instructive at this point to review the possible quantum numbers and their relationships in the solution of the radially symmetric hydrogen atom problem. Because of the spherical symmetry, the angular solution to the hydrogen atom is precisely the same as that obtained in Section 3.1. In that case, the angular solution Y_{lm} involved two different variables, θ and ϕ. Motion involving these coordinates was found to be quantized, being specified by the quantum numbers l and m. The equations describing the motion in θ and ϕ are Eqs. (3.1.50) and (3.1.49), respectively. Further analysis showed that m and l specify the eigenvalues of the z component and the total angular momentum, respectively. These relations can be summarized as

$$-\frac{1}{\Phi}\frac{d^2\Phi}{d\phi^2} = m^2,$$

$$\frac{\sin^2\theta}{\Theta}\left[\frac{1}{\sin\theta}\frac{d}{d\theta}\left(\sin\theta\frac{d\Theta}{d\theta}\right) + l(l+1)\Theta\right] = m^2,$$

$$L_z = \frac{\hbar}{i}\frac{\partial}{\partial\phi}, \qquad L_z\Phi = m\hbar\Phi,$$

$$L^2 Y = l(l+1)\hbar^2 Y. \tag{3.2.28}$$

The hydrogen atom problem is a three-dimensional problem. Therefore its solution requires the specification of motion along three different independent directions. These three directions are specified by three different spatial coordinates. Because of the spherical symmetry of the problem, the coordinates chosen are spherical coordinates r, θ, and ϕ. The motion of the particle requires the specification of each of these three variables as found above. Since the motion of the particle in the hydrogen atom is quantized, the solution requires the specification of three independent quantum numbers that correspond to the three independent spatial directions. The three quantum numbers are n, l, and m, which specify the motion along the three directions r, θ, and ϕ. Each of these quantum numbers is restricted to integer values.

It is important to recognize that the energy of the hydrogen atom depends on only the quantum number n. This is not surprising since n specifies the radial solution and as such specifies the radial position of the electron with respect to the force center. Since the potential and hence the force depend on only the separation distance of the electron and the force center (in the case of the hydrogen atom the force center lies at the center of the nucleus, a positively charged proton), the energy of the electron within the hydrogen atom potential should depend on only its radial position.

The quantum number l specifies the solution in θ as well as the total angular momentum of the state L. l was found above to range in values from 0 to $n - 1$. The last quantum number needed to specify the solution of the hydrogen atom is m. The value of m determines the z component of the orbital angular momentum as well as the solution in ϕ. m is an integer whose values range over the complete set of integers, $0, \pm 1, \pm 2, \ldots$, subject to the restriction that $-l \le m \le +l$ for each l value. The allowed values of the three quantum numbers n, l, and m are summarized in Table 3.2.1.

What, though, are the hydrogen atom wave functions? The higher-order wave functions can be generated from the power-series solution given by Eq. (3.2.15). This series can be rewritten in a more convenient form as

$$u = \rho^s \sum_{q=0}^{\infty} A_q \rho^q e^{-\frac{\rho}{n}}, \qquad (3.2.29)$$

where A_q are the coefficients in the sum. u is the solution to Eq. (3.2.8):

$$\left[-\frac{\hbar^2}{2m} \frac{d^2}{dr^2} + \frac{\hbar^2 l(l+1)}{2mr^2} + V(r) \right] u = Eu.$$
$$(3.2.30)$$

Table 3.2.1
Allowed Values of the Quantum Numbers in the Hydrogen Atom Problem

n	l	m
1	0	0
2	0	0
	1	$-1, 0, +1$
3	0	0
	1	$-1, 0, +1$
	2	$-2, -1, 0, +1, +2$
4	0	0
	1	$-1, 0, +1$
	2	$-2, -1, 0, +1, +2$
	3	$-3, -2, -1, 0, +1, +2, +3$
etc.		

For simplicity, in solving Eq. (3.2.8) with the above series it is easier to lump all the powers of ρ into a general function of r, which we call $g(r)$. u can then be reexpressed as

$$u = g(r) e^{-\frac{r}{na_0}}. \qquad (3.2.31)$$

The first and the second derivatives of u are obtained by the differentiation of Eq. (3.2.31) to give

$$\frac{du}{dr} = g' e^{-\frac{r}{na_0}} - \frac{1}{na_0} g e^{-\frac{r}{na_0}},$$
$$\frac{d^2u}{dr^2} = g'' e^{-\frac{r}{na_0}} - \frac{2}{na_0} g' e^{-\frac{r}{na_0}} + \frac{1}{na_0^2} e^{-\frac{r}{na_0}}. \qquad (3.2.32)$$

Substituting Eqs. (3.2.32) into the Schroedinger equation, Eq. (3.2.30), leads to

$$-\frac{\hbar^2}{2m} \left[g'' e^{-\frac{r}{na_0}} - \frac{2}{na_0} g' e^{-\frac{r}{na_0}} + \frac{1}{(na_0)^2} g e^{-\frac{r}{na_0}} \right]$$
$$+ \frac{\hbar^2 l(l+1)}{2mr^2} g e^{-\frac{r}{na_0}} - \frac{Zq^2}{r} g e^{-\frac{r}{na_0}} = Eg e^{-\frac{r}{na_0}}. \qquad (3.2.33)$$

Equation (3.2.33) can be simplified through use of Eq. (3.2.30). The energy E is given by

$$E = \frac{-Z^2 q^4 m}{2n^2 \hbar^2}. \qquad (3.2.34)$$

Substituting the above expression for E into Eq. (3.2.33), the right-hand side of Eq. (3.2.33) becomes

$$-\frac{mq^4 Z^2}{2\hbar^2 n^2} g \, e^{-\frac{r}{na_0}}. \tag{3.2.35}$$

We can simplify the third term on the left-hand side of Eq. (3.2.33) by making use of the definition of a_0 [see Eqs. (3.2.13)]:

$$a_0 = \frac{\hbar^2}{mq^2 Z}. \tag{3.2.36}$$

The coefficient of the third term is

$$-\frac{\hbar^2}{2m} \frac{1}{(na_0)^2}. \tag{3.2.37}$$

Substituting Eq. (3.2.36) for a_0, we find that the coefficient of the third term of Eq. (3.2.33) becomes

$$-\frac{\hbar^2}{2m} \frac{m^2 q^4 Z^2}{n^2 \hbar^4} = -\frac{mq^4 Z^2}{2\hbar^2 n^2}. \tag{3.2.38}$$

Hence the third term on the left-hand side is precisely the same as the term on the right-hand side of Eq. (3.2.33). Subtracting out both terms Eq. (3.2.33) becomes

$$-\frac{\hbar^2}{2m}\left[g'' e^{-\frac{r}{na_0}} - \frac{2}{na_0} g' e^{-\frac{r}{na_0}}\right] + \frac{\hbar^2 l(l+1)}{2mr^2} g \, e^{-\frac{r}{na_0}} - \frac{Zq^2}{r} g \, e^{-\frac{r}{na_0}} = 0. \tag{3.2.39}$$

To ensure that Eqs. (3.2.29) and (3.2.31) are equal, $g(r)$ must be defined as

$$g(r) = r^s \sum_q A_q r^q. \tag{3.2.40}$$

Using this form for $g(r)$, we can compute the first and the second derivatives of $g(r)$, g' and g''. g' and g'' are given as

$$g' = s A_0 r^{s-1} + (s+1) A_1 r^s + \cdots + (s+q) A_q r^{s+q-1} + \cdots,$$

$$g'' = s(s-1) A_0 r^{s-2} + (s+1) s A_1 r^{s-1} + \cdots$$

$$+ (s+q)(s+q-1) A_q r^{s+q-2} + \cdots. \tag{3.2.41}$$

The first and the second derivatives can be substituted into Eq. (3.2.39) and the exponential term divided out to give

$$-\frac{\hbar^2}{2m}\left[g'' - \frac{2}{na_0}g'\right] + \frac{\hbar^2 l(l+1)}{2mr^2}g - \frac{Ze^2}{r}g = 0,$$

$$g'' - \frac{2}{na_0}g' - \frac{l(l+1)}{r^2}g - \frac{2mZe^2}{\hbar^2 r}g = 0. \qquad (3.2.42)$$

If the equation is made dimensionless through use of Eqs. (3.2.10), the second of Eqs. (3.2.42) becomes

$$\frac{d^2g}{d\rho^2} - \frac{2}{n}\frac{dg}{d\rho} - \frac{l(l+1)}{\rho^2}g - \frac{2}{\rho}g = 0. \qquad (3.2.43)$$

Substituting the derivatives given by Eqs. (3.2.41), Eq. (3.2.43) is

$$s(s-1)A_0\rho^{s-2} + (s+1)sA_1\rho^{s-1} + \cdots + (s+q)(s+q-1)A_q\rho^{s+q-2} + \cdots$$
$$- l(l+1)[A_0\rho^{s-2} + A_1\rho^{s-1} + \cdots + A_q\rho^{s+q-2} + \cdots]$$
$$- \frac{2}{na_0}[sA_0\rho^{s-1} + (s+1)A_1\rho^s + \cdots + (s+q)\rho^{s+q-1}A_q + \cdots]$$
$$+ \frac{2mZe^2}{\hbar^2}[A_0\rho^{s-1} + A_1\rho^s + A_2\rho^{s+1} + \cdots + A_q\rho^{s+q-1} + \cdots]$$
$$= 0. \qquad (3.2.44)$$

Equation (3.2.44) must equal zero for all ρ. The indicial equation for like powers of ρ^{s-2} is, as before,

$$s(s-1) - l(l+1) = 0. \qquad (3.2.45)$$

The coefficient of the general term, that with ρ^{s+q-1}, is

$$(s+q+1)(s+q)A_{q+1} - \frac{2}{n}(s+q)A_q + 2A_q - l(l+1)A_{q+1} = 0. \qquad (3.2.46)$$

But from Eq. (3.2.45), s is equal to $l+1$. Making this substitution and solving for A_{q+1} yields

$$A_{q+1} = A_q\left[\frac{\frac{2}{n}(l+1+q) - 2}{(l+q+2)(l+q+1) - l(l+1)}\right], \qquad (3.2.47)$$

Table 3.2.2
Some Hydrogen Atom Wave Functions

Quantum Numbers

n	l	m_1	Eigenfunctions
1	0	0	$\frac{1}{\sqrt{\pi}}\left(\frac{Z}{a_0}\right)^{3/2}e^{-\frac{Zr}{a_0}}$
2	0	0	$\frac{1}{4\sqrt{2\pi}}\left(\frac{Z}{a_0}\right)^{3/2}$ $\left(2-\frac{Zr}{a_0}\right)e^{-\frac{Zr}{2a_0}}$
2	1	0	$\frac{1}{4\sqrt{2\pi}}\left(\frac{Z}{a_0}\right)^{3/2}$ $\left(\frac{Zr}{a_0}\right)e^{-\frac{Zr}{2a_0}}\cos\theta$
2	1	± 1	$\frac{1}{8\sqrt{2\pi}}\left(\frac{Z}{a_0}\right)^{3/2}$ $\left(\frac{Zr}{a_0}\right)e^{-\frac{Zr}{2a_0}}\sin\theta$ $e^{\pm i\phi}$

which can be reexpressed as

$$A_{q+1} = A_q\left[\frac{\frac{2}{n}(l+1+q)-2}{(q+1)(q+2l+2)}\right].$$

From the above recursion relation, it is possible to generate the hydrogen atom eigenfunctions. For example, for $l \neq 0$, the solution is

$$u = \rho^s \sum_{q=0}^{n} A_q\rho^q e^{-\rho},$$

where $R = u/r$ and $\rho = Zr/a_0 n$, neglecting corrections for the reduced mass (see Problem 1).

When the above relations are used, higher-order wave functions can be found through use of the recursion relation and power-series definition. Some of the hydrogen atom wave functions are collected in Table 3.2.2.

3.3 The Periodic Table of the Elements

In section 3.2 we determined the solution of the Schroedinger equation for a Coulomb potential between two oppositely charged point charges. This corresponds to a simple hydrogen atom: an electron in orbit about a positively charged nucleus consisting of a proton. For simplicity, the effect of the finite mass of the proton has been neglected (see Problem 1). The solution involves the specification of three quantum numbers, n, l, and m. m is commonly denoted as m_l and is referred to in this manner henceforth. As discussed in Section 3.2, n, l, and m_l describe the motion of the electron in r, θ, and ϕ. Motion in these three directions is quantized. Only certain discrete values of n, l, and m_l are allowed. Therefore only discrete energy states are possible within the hydrogen atom. Each state is specified by the allowed values of n, l and m_l. Several of these states have been collected in Table 3.2.1. The energy of each of these states depends on only n and is given by Eq. (3.2.34).

The energy levels of hydrogenlike atoms, those that consist of only one electron but may have more than one proton within the nucleus (for example, ionized helium, doubly ionized lithium, etc.), are found in a manner similar to finding the energy levels of hydrogen. The only significant difference is the value of the atomic number Z in the expression for the energy. These atoms have a scheme of energy levels similar to that of hydrogen.

It is interesting to consider the number of allowed states at a given energy in the hydrogen atom. We can do this by counting the number of states possible for a given value of n. Inspection of Table 3.2.1 shows that for $n = 1$, the ground state of the hydrogen atom, there is only one possible state. The ground state has quantum numbers $n = 1$, $l = 0$, and $m_l = 0$. The first excited state of the hydrogen atom corresponds to $n = 2$. The possible $n = 2$ states are found from inspection of Table 3.2.1. As can be seen from the table, for $n = 2$ there are two possible values of l, $l = 0$ and $l = 1$. Corresponding to the $l = 0$ case, there is only one possible value of m_l, $m_l = 0$. For $l = 1$, there are three possible values of m_l, $m_l = 0$, ± 1. Hence there are four possible states for $n = 2$. Further inspection of Table 3.2.1 shows that there are nine possible states corresponding to $n = 3$ and sixteen for $n = 4$. The number of states corresponding to greater values of n can be found in a similar manner.

It is important to realize that for a given value of n, there are only certain discrete states available in which electrons can be added. Consider then, what happens if the atom consists of more than one electron. Neglecting any electron–electron interaction (we will discuss this topic in detail in Chapter 9), each electron can be placed into only one possible state. As we will discuss in great detail in Chapters 5 and 7, no two electrons can simultaneously occupy the exact same quantum state. In other words, only one electron can be in any particular quantum state at any given time. Therefore, if an additional electron is added to a hydrogen atom, it must occupy a different state from that of the first electron. This is a general principle of nature and is commonly called the Pauli principle. Subsequently, in the ground state of a multielectron atom, only one electron occupies the $n = 1, l = 0$, and $m_l = 0$ state. The other electrons occupy the $n = 2, n = 3$, etc., states until all the electrons are assigned an electronic state.

As the reader may recall from basic chemistry, the elements differ from one another by the number of electrons and protons present. Hydrogen, the simplest element, consists of only one electron and proton. The next element is helium, which consists of two electrons and two protons. The third element is lithium, which has three electrons and three protons, etc. The Periodic Table of the Elements consists of the arrangement of the elements in accordance to increasing number of electrons and protons. The Periodic Table is arranged in a special way (see Figure 3.3.1). Elements within each column have similar chemical properties, while the elements in each row form a progression. As the reader may recall from basic chemistry, the last atom in each row is chemically inert and corresponds to an atom in which its outermost electronic shell of acceptable states is completely filled. Another electron added to an atom with a completely filled outermost shell must be placed into the next energy orbital.

As mentioned above, the elements are grouped into vertical columns in accordance with their chemical properties. In other words, all the elements in Column IA have a similar chemistry. For example all of them will react with Column VIIA elements to form ionic substances. All the elements in Column VIIIA again have similar properties. In this case, all these elements are inert; they do not participate in chemical reactions. Thus it is important to recognize that the ordering of the Periodic Table arises from the periodicity in the chemical properties of the elements. The question is, of course, how can this observed periodicity in the chemical properties of the elements be explained?

1 H																	2 He
3 Li	4 Be											5 B	6 C	7 N	8 O	9 F	10 Ne
11 Na	12 Mg											13 Al	14 Si	15 P	16 S	17 Cl	18 Ar
19 K	20 Ca	21 Sc	22 Ti	23 V	24 Cr	25 Mn	26 Fe	27 Co	28 Ni	29 Cu	30 Zn	31 Ga	32 Ge	33 As	34 Se	35 Br	36 Kr
37 Rb	38 Sr	39 Y	40 Zr	41 Nb	42 Mo	43 Tc	44 Ru	45 Rh	46 Pd	47 Ag	48 Cd	49 In	50 Sn	51 Sb	52 Te	53 I	54 Xe
55 Cs	56 Ba	57 La	72 Hf	73 Ta	74 W	75 Re	76 Os	77 Ir	78 Pt	79 Au	80 Hg	81 Tl	82 Pb	83 Bi	84 Po	85 At	86 Rn
87 Fr	88 Ra	89 Ac															

58 Ce	59 Pr	60 Nd	61 Pm	62 Sm	63 Eu	64 Gd	65 Tb	66 Dy	67 Ho	68 Er	69 Tm	70 Yb	71 Lu
90 Th	91 Pa	92 U	93 Np	94 Pu	95 Am	96 Cm	97 Bk	98 Cf	99 Es	100 Fm	101 Md	102 No	103 Lr

Figure 3.3.1. The Periodic Table of the Elements.

The arrangement of atoms in the Periodic Table can be understood from knowledge of the possible quantum states in an atom. Once all the possible states corresponding to a given value of n are filled, any additional electrons must be placed in higher energy states, those with a greater value of n, in accordance with the Pauli principle. The number of allowed states for each value of n determines the number of states in each shell. Therefore, once all the states for a given value of n are filled, the resulting atom should be chemically inert and thus be found within Column VIIIA. As more electrons are added to the atom, they are promoted to the next-highest n-level shell and begin filling states for $l = 0, l = 1$, etc. Since the outermost configuration of the atoms become the same, the chemical properties will repeat themselves.

Now let us see if our results obtained in Section 3.2 predict the correct periodicity of the atoms. In Section 3.2 the number of states corresponding to the $n = 2$ case is found to be equal to 4. Hence there can be up to four electrons within the $n = 2$ shell. For $n = 1$ there is only one allowed state. As a result, there can be only one electron within the $n = 1$ shell. The number of electrons that can be accommodated within each shell should correspond to the number of atoms within each row of the Periodic Table. Therefore, based on the results of Section 3.2, there should be one atom in the first row, four atoms in the second row, nine atoms in the third row, etc. Inspection of the Periodic Table (see Figure 3.3.1) indicates that there are actually two atoms in the first row, hydrogen and helium, eight atoms in the second row, lithium through neon, etc. In fact, it appears that there are exactly twice as many allowed energy states for each value of n, as predicted by the results from

Section 3.2. Clearly, the solution of the Schroedinger equation for the hydrogen atom obtained in Section 3.2 fails to account adequately for the empirical evidence provided by the Periodic Table. The solution obtained seems to be incomplete since it fails to predict the twofold multiplicity of each electronic state.

Apparently there must be an additional degree of freedom present in the hydrogen atom problem that accounts for the twofold multiplicity in the number of allowed energy levels in the problem. What, though, could this additional degree of freedom be? After all, the Schroedinger equation specifies the motion of the electron in all three spatial directions, so it should give a complete description of the motion of the particle. However, empirical evidence clearly shows that the solution of the Schroedinger equation leads to an incomplete solution since it cannot account for a twofold multiplicity in the number of allowed energy states. To see the reason why there is an additional degree of freedom in the problem, the Schroedinger equation must be generalized to be consistent with the special theory of relativity. In Section 3.4 we will see that a new degree of freedom arises when space and time are taken into account. This additional degree of freedom is a property of the electron itself. It is an independent degree of freedom and is related to the spin angular momentum of the electron. As we will see, the spin can have two possible orientations, which results in a multiplicity in the number of allowed states by a factor of 2.

3.4 Basic Concepts in Relativity Theory

To understand the origin of the electron spin and its importance to the Periodic Table, it is first necessary to learn the special relativity theory. The theory of relativity was developed by Einstein before Schroedinger wave mechanics at the beginning of the twentieth century. The basic principle of the theory of relativity is that the laws of physics must be identical in all inertial reference frames. Since this is a general requirement, any fundamental physical theory must then be relativistically invariant. As such, quantum mechanics must be consistent with relativity.

Soon after the development of quantum mechanics, Paul Dirac extended the theory so as to be relativistically invariant. Dirac discovered that the electron spin and its value are immediately predicted from relativistic quantum mechanics. The existence of electron-spin angular momentum and its magnitude, previously postulated to be $\hbar/2$, follows directly from the requirement of relativistic invariance. The existence of electron spin is necessary to ensure that the electron's dynamics are consistent with relativity. This discovery was one of the great accomplishments of the quantum theory.

In this section and in Section 3.5 we discuss the basics of the relativity theory and the relativistic extension of quantum mechanics. This material completes all the simple problems in quantum mechanics that can be solved exactly. In Chapter 4, we will discuss how to solve problems by using approximation techniques.

As mentioned above, the special theory of relativity states that all inertial reference frames are physically equivalent. An inertial reference frame is defined as a frame in which no accelerations are present. In an inertial reference frame any observer cannot determine whether he/she is in motion relative to another inertial reference frame by doing an experiment completely confined within his/her own reference frame. For example, if a passenger is flying in a jetliner at constant velocity while experiencing no turbulence, a passenger cannot distinguish whether he/she is moving and the clouds (or ground) are stationary or whether the clouds are moving and he/she is stationary. (Of course, we know from experience that we are moving.) In fact, if a passenger sits in the middle of the plane and cannot look out of the windows, if the plane travels at constant velocity without turbulence, it is impossible to tell whether the plane is actually moving or not. In other words, everything seems the same as if the passenger were sitting in the plane at rest on the runway. Under such circumstances there is no way of telling whether the plane is moving or not.

However, when the plane is accelerating or decelerating, as in take-off or landing, one can easily distinguish motion. During an acceleration, a passenger on the plane feels a force on his/her body and is gently (if the plane accelerates smoothly) forced back into his/her seat. This force arises from the acceleration of the plane and is not a true force in the sense that it does not arise from the gradient of a potential. Forces such as these are called fictitious forces since they arise within only an accelerated reference frame. An inertial observer watching the acceleration of the plane (in this case such an observer cannot be on board but is standing on the ground) describes the force on the passenger's body as simply due to the plane accelerating into the passenger. The plane accelerates first and pushes the people on board along with it. It is this pushing that arises from the acceleration that accounts for the fictitious force. Therefore, as far as the observer on the ground (the inertial observer) is concerned, there is no force acting on the passenger, only an acceleration.

Although the passenger and the ground observer may disagree as to the cause of the "force" on the passenger, both will agree that the passenger is moving while the ground observer is not. Clearly, in the case of accelerations, it is easy to distinguish between which frame is in motion and which is not. In the above example, both observers, the one on the plane and the one on the ground, agree that the plane is in motion not the ground.

A different result occurs if we consider two planes moving with constant velocity in the absence of all turbulence. Let us also assume that the planes are above a uniform cloud cover such that there are no noticeable features around them. From the above discussion, a passenger in either plane cannot tell that he/she is moving. Everything seems the same as if the passenger were sitting quietly at home. The passenger feels no sensation of motion and sees no motion. If the two planes approach one another, then a passenger in one plane, call it plane 1, will see the other plane, plane 2, move relative to himself/herself. A passenger in plane 2 will see plane 1 move relative to himself/herself but in the opposite way as the first passenger did. In either case, each passenger will state that the other plane was moving but in opposite directions. Each observer is convinced that he is at rest while the other is moving. Under the described conditions, there is no way that either observer can prove that he/she is moving or not moving, only that there is a relative movement of his/her plane with respect to the other. This is what is meant by relative motion.

The question is, is it possible to distinguish between inertial reference frames? Putting it more simply, is there some experiment that either passenger in planes 1 and 2 can perform that will enable him/her to determine his/her absolute motion? Let us first consider our own experiences on commercial airliners. While in the air, again assuming no turbulence, one can move about the cabin, put the folding tray table down, and place objects on the table without worrying whether they will move off it. All these actions can be made equally as well when the plane is in constant motion or at rest. If we confine our observations to only within the cabin, we cannot tell then if the plane is moving or not. However, when the plane is accelerating or decelerating, if we put the tray table down and put a cup on it, that cup will no longer stay in place. It will slide off the tray table. Thus we can determine unequivocally that the plane is in motion from observations completely confined to the plane itself.

Einstein provided the first general answer to the question concerning motion of inertial frames. He recognized that it is impossible to ascertain the motion of an inertial reference frame by performing an experiment completely confined within it. He further recognized that light behaves in a rather strange way in terms of motion. At the end of the nineteenth century, James Maxwell succeeded in unifying the theories of electricity and magnetism into one theory of electromagnetism. Maxwell's equations lead directly to a wave equation for light, as was shown in Chapter 1. From this result, it was concluded that light is an electromagnetic wave that travels through vacuum in all directions with a constant speed c. Inspection of the result obtained in Chapter 1.2 shows that there is no account taken of the motion of the source of the radiation or that of the observer. Maxwell's equations predict that the velocity of light is always c in a vacuum, independent of the motion of the source or the observer. Consequently, this is in contradiction to what we would expect based on our discussion above of relative motion.

For example, consider the case of a boy throwing a baseball from the back of a truck with velocity (relative to him) of v. If the truck is moving forward with velocity V, relative to an observer stationary on the ground, then the velocity of the ball with respect to the stationary observer is $V - v$. This is what is meant by classical or Galilean relativity. We would expect then that the speed of light would behave in a similar manner. If, instead of a baseball, the boy turned on a flashlight at the back of the truck according to Galilean relativity, the magnitude of the velocity of the light emitted by the flashlight relative to the stationary observer would be $|V - c|$. However, when the speed of light is measured in the laboratory in a similar type of experiment, its velocity is always seen to be c! The speed of light is always observed to be the same in all inertial reference frames, independent of the motion of the source or observer. How can this be reconciled?

Einstein resolved this puzzle by realizing the validity of two fundamental facts. These facts he termed postulates and they are as follows:

Postulate 1: All inertial frames of reference are equivalent. If two frames of reference differ by only uniform motion (constant velocity) then they are physically equivalent, and observers in both frames must concur in their descriptions of physics.

Postulate 2: The speed of light is constant in all equivalent inertial reference frames.

The above two postulates, if pushed to their logical extremes, as Einstein did, lead to some surprising results. The most important of these results is that simultaneity is relative. In other words, two events judged to be simultaneous in one reference frame are not necessarily simultaneous in another. This has tremendous ramifications in measurements, such as length and time intervals, since measurements involve simultaneous events. Let us now explore the implications of the theory of relativity.

Consider two different inertial reference frames. Given the two frames $S(x, t)$ and $S'(x', t')$, we can construct a means of mapping from one frame of reference to the other. Such a mapping provides a transformation from one set of coordinates to another. For example, x' and t' as measured in the frame S' can be transformed into x and t, as measured in the frame S. The transformation connecting these two coordinates must be linear. x' and t' must be expressed as linear functions of x and t and vice versa. If the transformations were not linear then uniform motion in one frame of reference would not be seen as linear uniform motion in the other frame of reference in violation of Postulate 1. Therefore the position as measured in S, x, can be related to the position and time as measured in S' as

$$x = ax' + bt', \tag{3.4.1}$$

and similarly,

$$x' = ax - bt, \tag{3.4.2}$$

where a and b are constants. The above relationships are consistent with the simple Galilean transformations, which are given as

$$x' = x - vt, \qquad x = x' + vt', \tag{3.4.3}$$

where v is the relative velocity between the two frames. We define the motion of the origin of S as measured in S' by putting $x = 0$ into Eq. (3.4.1). When this substitution is made Eq. (3.4.1) yields

$$\frac{x'}{t'} = -\frac{b}{a}. \tag{3.4.4}$$

The motion of the origin of S' as measured in S is defined also by the substitution of $x' = 0$ into Equation (3.4.2) above. This yields

$$\frac{x}{t} = \frac{b}{a}. \tag{3.4.5}$$

The velocity of separation between the two frames must be equal and opposite. Thus the velocity of frame 2 as measured by frame 1 must be opposite and equal to the velocity of frame 1 as measured by frame 2. Hence,

$$-\frac{x'}{t'} = \frac{x}{t} = v. \tag{3.4.6}$$

This implies that

$$\frac{b}{a} = v. \tag{3.4.7}$$

Consider next the behavior of a light signal as measured in each frame. Let the signal originate at the origin of both coordinate systems. Again let the origins coincide when the signal is emitted. This is done for simplicity in the mathematics. The physical result is independent of how we perform the experiment. In the frame S, the position of the expanding wave front is given by

$$x = ct, \tag{3.4.8}$$

while in S' the position of the expanding wave front is

$$x' = ct', \tag{3.4.9}$$

where c is the velocity of light. It is important to recognize that the speed of light is the same, c, in both reference frames. Substituting the value of x and x' into Eqs. (3.4.1) and (3.4.2) yields

$$ct = ax' + bt',$$
$$ct' = ax - bt, \tag{3.4.10}$$

but

$$ct = act' + bt',$$
$$ct' = act - bt. \tag{3.4.11}$$

Simplifying Eqs.(3.4.11) yields

$$ct = (ac + b)t',$$
$$ct' = (ac - b)t. \tag{3.4.12}$$

Eliminating t and t' from Eqs. (3.4.12) gives

$$ct = (ac + b)(ac - b)\frac{t}{c},$$
$$c^2 = a^2c^2 - b^2. \tag{3.4.13}$$

We can simplify this further by using the relationship between b/a and v given by Eq. (3.4.7). Then

$$b = av, \qquad b^2 = a^2v^2. \tag{3.4.14}$$

Therefore

$$c^2 = a^2(c^2 - v^2),$$
$$a^2 = \frac{c^2}{c^2 - v^2} = \frac{1}{1 - \frac{v^2}{c^2}},$$
$$a = \frac{1}{\sqrt{\left(1 - \frac{v^2}{c^2}\right)}}, \qquad b = \frac{v}{\sqrt{1 - \frac{v^2}{c^2}}}. \tag{3.4.15}$$

The position x can be found in terms of x' and t' then as

$$x = ax' + bt',$$
$$x = \frac{x'}{\sqrt{1 - \frac{v^2}{c^2}}} + \frac{vt'}{\sqrt{1 - \frac{v^2}{c^2}}},$$
$$x = \frac{(x' + vt')}{\sqrt{1 - \frac{v^2}{c^2}}}, \tag{3.4.16}$$

while x' can be determined from x and t as

$$x' = \frac{x}{\sqrt{1 - \frac{v^2}{c^2}}} - \frac{vt}{\sqrt{1 - \frac{v^2}{c^2}}},$$
$$x' = \frac{x - vt}{\sqrt{1 - \frac{v^2}{c^2}}}. \tag{3.4.17}$$

How do the coordinates y and z transform? For simplicity, we consider here motion along only the x direction. In this case, since space is isotropic, all directions perpendicular to the unique direction specified by the motion must be equivalent. Therefore the coordinates y and z transform into themselves.

The time as measured by frame S, t is not the same as that measured by frame S', t'. This is somewhat surprising since in classical mechanics time is assumed to be the same for every observer. Time intervals in special relativity are different between two different observers because of the relativity of two simultaneous events. In other words, observers in different frames of reference will disagree

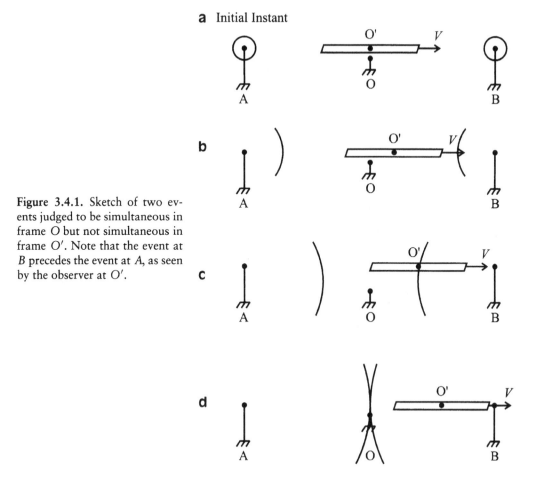

Figure 3.4.1. Sketch of two events judged to be simultaneous in frame O but not simultaneous in frame O'. Note that the event at B precedes the event at A, as seen by the observer at O'.

as to whether an event judged to be simultaneous in one frame is seen to be simultaneous in another. This can be understood as follows.

Consider two frames of reference, S and S'. The two frames move relative to one another with velocity v. At an initial instant the origins of the two reference frames coincide, as shown in Figure 3.4.1. At precisely this instant, a light wave originates at both A and B, as shown in Figure 3.4.1. A and B are equidistant from the origin of S, O. At this instant, A and B are equidistant from point O in S, which in turn are equidistant from the origin of S', O'. The question is, is the emission of light from A and B judged to be a simultaneous event in both reference frames? To answer this question, let us first consider an observer at the origin O in frame S. Let us take the case in which this observer judges the two events to be simultaneous. Will the observer at O' in frame S' also judge the events as being simultaneous? S' as seen from S moves to the right at velocity v. If the observer in S determines that the emission events at A and B are simultaneous, then the light-wave fronts originating from A and B must reach O at precisely the same time. However, since it takes a finite amount of time for the wave fronts to reach O from A and

a Initial Instant

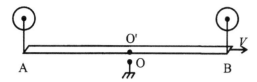

b

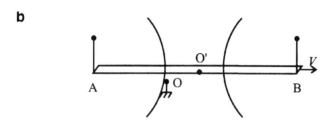

Figure 3.4.2. Sketch of two events judged to be simultaneous in O' but not simultaneous in O. Note that the event originating at A precedes the event originating at B in O.

c

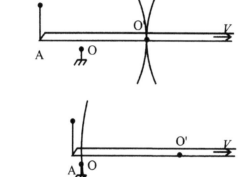

d

B, O' is no longer aligned with O on the arrival of the wave fronts. O' has moved to the right. Therefore O' intersects the wave front emanating from B before it intersects the wave front originating from A. Hence, the events are simultaneous in S while the event at B precedes the event at A, as seen in frame S'.

Of course, the situation could be reversed. Let us consider what happens if the observer at O' in S' judges the events to be simultaneous. In this case (see Figure 3.4.2), the light signals from A and B arrive simultaneously at O' rather than at O. From the point of view of S', S moves to the left with velocity v. Therefore, in order that the wave fronts arrive simultaneously at O', the wave front from A must arrive earlier at O than the wave front from B, as shown in Figure 3.4.2. Hence the events are simultaneous in S' while the event at A precedes the event at B, as seen from S.

From the above discussion, it is clear that simultaneity is a relative concept. What is judged to be simultaneous in one reference frame is not judged to be simultaneous in another. Therefore measurements such as length, which requires the location of the end points of an object simultaneously, will vary between frames of reference. Time intervals also will vary between different frames of reference.

The time transformations between frames are readily determined from the transformations for x and x' as follows. Let

$$\gamma \equiv \frac{1}{\sqrt{1 - \frac{v^2}{c^2}}}. \tag{3.4.18}$$

With this definition, the transformations for x and x' can be written as

$$x = \gamma(x' + vt'), \qquad x' = \gamma(x - vt). \tag{3.4.19}$$

Substituting in for x within the second equation of Eqs. (3.4.19) from the first equation gives

$$x' + \gamma vt = \gamma^2(x' + vt'), \tag{3.4.20}$$

which can be rearranged to

$$\gamma vt = \gamma^2 x' + \gamma^2 vt' - x'. \tag{3.4.21}$$

Therefore t can be solved as

$$t = \frac{1}{\gamma v}[\gamma^2 vt' + x'(\gamma^2 - 1)],$$

$$t = \gamma t' + \frac{1}{\gamma}\frac{x'}{v}(\gamma^2 - 1). \tag{3.4.22}$$

But $(\gamma^2 - 1)$ is equal to

$$\frac{1}{1 - \frac{v^2}{c^2}} - 1 = \frac{1 - \left(1 - \frac{v^2}{c^2}\right)}{\left(1 - \frac{v^2}{c^2}\right)} = \frac{\frac{v^2}{c^2}}{1 - \frac{v^2}{c^2}}. \tag{3.4.23}$$

The equation for t is then given as

$$t = \gamma t' + \frac{1}{\gamma}\left(\frac{x'}{v}\right)\frac{\frac{v^2}{c^2}}{1 - \frac{v^2}{c^2}}, \tag{3.4.24}$$

which simplifies to

$$t = \gamma t' + \gamma\frac{vx'}{c^2}. \tag{3.4.25}$$

The transformations between the frames S and S' are called the Lorentz

Table 3.4.1
Lorentz Transformations in One
Dimension

$x' = \gamma(x - vt)$	$x = \gamma(x' + vt')$
$y' = y$	$y = y'$
$z' = z$	$z = z'$
$t' = \gamma\left(t - \frac{vx}{c^2}\right)$	$t = \gamma\left(t' + \frac{vx'}{c^2}\right)$

transformations. These transformations for one dimension are compiled in Table 3.4.1.

As mentioned above, because of the relativity of simultaneous events, observers in two different reference frames will disagree about the length of an object. An object's length is measured to be greatest when the body is at rest relative to the observer. When it moves with velocity v relative to an observer, its measured length is contracted in the direction of motion by the factor

$$\sqrt{1 - \frac{v^2}{c^2}}, \tag{3.4.26}$$

while its dimensions perpendicular to the direction of its motion are unaffected. This effect is known as the Lorentz contraction principle. To understand the origin of this effect, consider a rod at rest in the S' frame of length $L' = x'_2 - x'_1$. The length of the rod in frame S' is called its proper length, since it is measured at rest with respect to the observer. Such a measurement is what we typically make when we measure the length of an object. We seek to answer the question, what is the length of the rod as measured by an observer in frame S that moves with velocity v relative to S'? In classical physics, the length of the rod is assumed to be invariant; it has the same length in either reference frame. However, we have seen above that the relativity theory indicates that two events that are judged to occur simultaneously in one frame are not necessarily judged to be simultaneous in another frame. Therefore any process that involves making two simultaneous events, such as locating the two ends of a rod simultaneously, will give different results depending on the nature of the observer.

Let us now use the Lorentz transformations to determine what the measured length in the frame S will be. Let the proper length of the rod be L', as measured in S'. It is important to recognize that to measure the length of the rod in S' the positions of the end points of the rod as measured by an observer in S', call them x'_2 and x'_1, must be found at the same time t'. Similarly, the length of the rod as measured in S requires locating the end points of the rod, x_2 and x_1, at the same time t. The length of the rod in frame S is then given as $x_2 - x_1$. The measured lengths of the rod in frames S and S' can be compared as follows. The proper length L' is given as

$$L' = x'_2 - x'_1. \tag{3.4.27}$$

But x'_2 and x'_1 can be found in terms of the variables measured in S by use of the Lorentz transformations as

$$x'_2 = \frac{x_2 - vt_2}{\sqrt{1 - \frac{v^2}{c^2}}}, \qquad x'_1 = \frac{x_1 - vt_1}{\sqrt{1 - \frac{v^2}{c^2}}}. \tag{3.4.28}$$

Therefore L' is given in terms of the coordinates in S as

$$L' = \frac{(x_2 - x_1) - v(t_2 - t_1)}{\sqrt{1 - \frac{v^2}{c^2}}}. \tag{3.4.29}$$

The separation $x_2 - x_1$ measured in S is the length of the rod as measured in S. However, the end points of the rod must be located simultaneously at time $t_2 = t_1$. The proper length L' is related to the length measured in the frame S, L, as

$$L' = \frac{L}{\sqrt{1 - \frac{v^2}{c^2}}}, \qquad L = L'\sqrt{1 - \frac{v^2}{c^2}}. \tag{3.4.30}$$

Inspection of Eqs. (3.4.30) shows that, given that L' is the proper length of the rod, the length of the rod as measured by an observer in a frame moving with respect to S' with velocity v is contracted by a factor of $(1 - v^2/c^2)^{1/2}$.

Let us next consider the measurement of time intervals between two different reference frames. The passage of a time interval requires a simultaneous measurement. As a result, we expect that the duration of a time interval will be different depending on the motion of the observer. What is found is that a clock is measured to go at its fastest rate when it is at rest relative to the observer. When a clock moves at a constant velocity v with respect to the observer, it is measured to move at a slower rate. The amount by which the clock slows is given by the factor

$$\sqrt{1 - \frac{v^2}{c^2}}. \tag{3.4.31}$$

This effect, that a moving clock runs slowly, is called time dilation.

To see the origin of the time dilation effect, consider a clock at rest at position x' in the frame S'. An interval, $t'_2 - t'_1$, as measured in S', is measured in S as the difference between the two times t_2 and t_1. The clock is stationary in frame S', so its position remains the same at all times. Therefore x'_2, the position of the clock at time t'_2, must equal its position at time t'_1, x'_1. However, as seen from S, the clock is not stationary since S moves with respect to S' with velocity v. Therefore the position of the clock as measured by an observer in S at different times is not the same. x_2 does not equal x_1. t_1 and t_2 must be found with the S' coordinates, since the clock is stationary in this frame. Therefore

$$t_1 = \frac{t'_1 + \frac{vx'_1}{c^2}}{\sqrt{1 - \frac{v^2}{c^2}}}, \qquad t_2 = \frac{t'_2 + \frac{vx'_1}{c^2}}{\sqrt{1 - \frac{v^2}{c^2}}}, \tag{3.4.32}$$

where use has been made of the fact that $x'_1 = x'_2$. The difference between t_1

and t_2 is

$$t_2 - t_1 = \frac{t_2' - t_1'}{\sqrt{1 - \frac{v^2}{c^2}}}. \tag{3.4.33}$$

The time interval measured in the rest frame, in this case S', is $t_2' - t_1'$. This interval is called the proper time interval. The time interval measured in a different frame S, which moves with respect to S' with velocity v, is given as $t_2 - t_1$. From the above relation it is clear that the measured time interval is least in the rest frame than in any other frame. This is the origin of the statement that "moving clocks run slow."

At this point, the reader may wonder if the above results have ever been experimentally probed. An interesting experiment has been made that verifies the time dilation effect. This experiment involves the measurement of the decay time of a subatomic particle (for a full discussion see Frisch and Smith 1963). The particle analyzed is a muon that is formed from cosmic-ray showers in the Earth's upper atmosphere. A muon has a finite lifetime before it decays into a different particle. In the rest frame of the muons the decay rate is observed to be, say, τ seconds (the actual value is not important to us, only the relative difference measured between the two frames). The experiment involves comparison of the lifetime of the muons as measured in the rest frame versus measurement of the lifetime in a frame that moves relative to the muons with a high velocity v. The measurement in the latter reference frame can be made as follows. Muons formed in the upper atmosphere move at a velocity close to the speed of light as they descend toward the ground. Therefore an observer on the Earth moves relative to the muons with a velocity close to the speed of light. If the number of muons is counted at a mountaintop and then compared with the number of muons that survive at the bottom of the mountain, the average lifetime of the muons in the moving reference frame can be assessed. Comparison is then made between the lifetime in the moving frame and the lifetime in the stationary frame. When the experiment is performed, it is found that many more muons survive in transiting the distance from the top of the mountain to the bottom than would be predicted from the rest frame lifetime. Few muons are expected to survive if the time taken for the muons to travel from the top of the mountain to the bottom, call it t, is larger than the average lifetime of the particles τ. The only way that many muons can survive is if their clocks run slowly with respect to the fixed observers on the Earth. In other words, the natural clock of the muons, their rate of decay, is measured as slower in the moving reference frame than in the stationary frame.

One might argue, however, that if the observer rides along with the muons, their lifetimes must be the same as in the laboratory rest frame. This is of course true; an observer at rest with respect to the muons will always measure the same average lifetime. However, to an observer at rest in the frame of the muons, the height of the mountain is seen to be foreshortened because of the Lorentz contraction principle. Therefore, if an observer travels with the muons, the average lifetime that that observer would measure is the proper lifetime, but the distance traveled by the muons would be measured to be much shorter than that measured by the fixed observers on the Earth. There are numerous puzzles and apparent paradoxes in special relativity. Some of these puzzles are included in the problems at the end of the chapter. For a more thorough examination of the kinematics of relativity the reader is referred to the books of French (1968), Resnick (1972), and Taylor and Wheeler (1963).

The above discussion concerned the kinematics of special relativity. Of even greater importance is the dynamics of special relativity. Postulate 1 implies that all the laws of physics are the same in all inertial reference frames. Therefore momentum and energy must be conserved during collisions in any reference frame. An observer in any inertial reference frame must judge that the energy and the momentum of a system undergoing a collision are conserved. Although observers in different frames may disagree as to the measured values of the momentum and the energy, they must nevertheless agree that these quantities are conserved during the interaction. Consider a collision between two particles A and B, as observed from two different reference frames S and S'. Let S move relative to S' with velocity v. For the two different observers to agree that the momentum of the system is conserved, the mass of the particles must be redefined as

$$m = \frac{m_0}{\sqrt{1 - \frac{v^2}{c^2}}},$$
(3.4.34)

where m_0 is the proper mass, that measured at rest with respect to the particle, and m is the mass measured by an observer moving with velocity v relative to the particle. The corresponding values of the momentum components as measured in the moving reference frame are

$$p_x = \frac{m_0 v_x}{\sqrt{1 - \frac{v^2}{c^2}}}, \qquad p_y = \frac{m_0 v_y}{\sqrt{1 - \frac{v^2}{c^2}}}, \qquad p_z = \frac{m_0 v_z}{\sqrt{1 - \frac{v^2}{c^2}}},$$
(3.4.35)

where v_x, v_y, and v_z are the components of the velocity of separation v. The details of the above derivation are standard and are left as an exercise (see Problem 16).

Using Eqs. (3.4.35), the general form of the kinetic energy can be readily obtained. We can determine an expression for the kinetic energy simply by calculating the work done by an external force in increasing the speed of a particle from rest to a final velocity u. The kinetic-energy difference due to the action of an applied electric field is then

$$K = \int \vec{F} \cdot d\vec{x}.$$
(3.4.36)

Let us first calculate the kinetic energy classically. To do this we recognize that the force can be expressed as the time rate of change of the momentum:

$$F = m_0 \frac{dv}{dt}. \tag{3.4.37}$$

Classically, the kinetic energy then is simply given as

$$K = \int m_0 \frac{dv}{dt} \, dx, \tag{3.4.38}$$

where the force is assumed to act along only the x direction. Evaluating the integral yields

$$K = \int m_0 \frac{dv}{dx} \frac{dx}{dt} \, dx = \int m_0 \, dv \frac{dx}{dt} = \int_0^v m_0 v \, dv,$$

$$K = \frac{1}{2} m_0 v^2, \tag{3.4.39}$$

which is the well-known classical result.

The expression for the kinetic energy in relativistic mechanics can be found by integration of the relativistic expression for the force instead. The force is still simply given by the time rate of change of the momentum but, in this case, the mass is not time independent. Therefore K is

$$K = \int F \, dx = \int \frac{d}{dt}(mv) \, dx = \int \frac{d(mv)}{dx} \frac{dx}{dt} \, dx = \int d(mv) \frac{dx}{dt},$$

$$K = \int (m \, dv + v \, dm)v, \tag{3.4.40}$$

since both m and v depend on t. The above integral can be evaluated with the following result. The mass is simply

$$m^2 c^2 - m^2 v^2 = m_0^2 c^2,$$

$$m = \frac{m_0}{\sqrt{1 - \frac{v^2}{c^2}}},$$

$$m^2 \left(1 - \frac{v^2}{c^2}\right) = m_0^2. \tag{3.4.41}$$

Taking differentials of each side yields

$$2mc^2 \, dm - m^2 2v \, dv - v^2 2m \, dm = 0. \tag{3.4.42}$$

Dividing both sides by $2m$ yields

$$mv \, dv + v^2 \, dm = c^2 \, dm. \tag{3.4.43}$$

But this is exactly the integrand in Eq. (3.4.40). Therefore the kinetic energy can be written as

$$K = \int_{m = m_0}^{m = m} c^2 \, dm, \tag{3.4.44}$$

which readily integrates to

$$K = mc^2 - m_0 c^2,$$

$$K = m_0 c^2 \left[\frac{1}{\sqrt{1 - \frac{v^2}{c^2}}} - 1 \right]. \tag{3.4.45}$$

The total energy E must be equal to the sum of the kinetic and the rest energies. The rest energy is simply $m_0 c^2$. E can be expressed using the first of Eqs. (3.4.45) as

$$E = K + m_0 c^2 = mc^2 - m_0 c^2 + m_0 c^2 = mc^2, \tag{3.4.46}$$

which is the well-known result from the relativity theory.

Finally, let us determine the relationship between E and p, the total energy and the momentum of the particle. From the definition of the relativistic mass, one can easily write

$$m^2 c^2 - m^2 v^2 = m_0^2 c^2. \tag{3.4.47}$$

The momentum p, though, is simply the product of the mass and the velocity: $p = mv$. Substituting p in for mv, in Eq. (3.4.47) we obtain

$$m^2 c^2 - p^2 = m_0^2 c^2. \tag{3.4.48}$$

Multiplying both sides by c^2 yields

$$m^2 c^4 - p^2 c^2 = m_0^2 c^4. \tag{3.4.49}$$

The total energy E was found above to be equal to mc^2. Substituting E^2 in then for $m^2 c^4$ yields

$$E^2 - p^2 c^2 = \left(m_0 c^2 \right)^2. \tag{3.4.50}$$

If the rest energy is defined as

$$E_0 = m_0 c^2, \tag{3.4.51}$$

the relation between the momentum and the energy becomes

$$E^2 = p^2 c^2 + E_0^2. \tag{3.4.52}$$

In Section 3.5 we will use this relationship to find relativistically invariant expressions in quantum mechanics.

3.5 **Relativistic Quantum Mechanics**

There are two different approaches that are commonly used in developing a relativistically invariant formulation of quantum mechanics. Both approaches are briefly discussed here. The first method leads directly to the Klein–Gordon equation. The second approach we will take leads to the Dirac equation, which describes the motion of an electron. It is the Dirac equation that predicts that the electron must have a spin angular momentum of $1/2\,\hbar$ in order that the total angular momentum of the electron be preserved. Our primary goal in this section is to demonstrate, through use of the Dirac equation, that an electron must have a spin angular momentum.

1. The Klein–Gordon Equation

In Section 3.4 it was determined that the energy and the momentum of a particle in special relativity theory are related by

$$E^2 = (pc)^2 + \left(m_0 c^2\right)^2. \tag{3.5.1}$$

If the rest mass m_0 of the particle is zero, as is the case for a photon, then Eq. (3.5.1) reduces to

$$E = pc. \tag{3.5.2}$$

From the de Broglie relationship,

$$p = \frac{h}{\lambda}. \tag{3.5.3}$$

The energy E of a massless particle can be reexpressed as

$$E = \frac{hc}{\lambda} = h\upsilon, \tag{3.5.4}$$

which is consistent with the result determined in Chapter 1 for a photon.

It is tempting to simply quantize Eq. (3.5.1) following the rules stated in Chapter 1. Equation (3.5.1) can thus be quantized when each variable is replaced with its quantum-mechanical operator. The operator for the energy is

$$E = i\hbar\frac{\mathrm{d}}{\mathrm{d}t}, \tag{3.5.5}$$

while the operator for p is

$$\vec{p} = -i\hbar\vec{\nabla}. \tag{3.5.6}$$

Therefore E^2 and p^2 can be expressed as

$$E^2 = -\hbar^2\frac{\mathrm{d}^2}{\mathrm{d}t^2}, \qquad p^2 = -\hbar^2\nabla^2. \tag{3.5.7}$$

Once these substitutions are made, Eq. (3.5.1) becomes

$$-\hbar^2 \frac{d^2 \Psi}{dt^2} = -\hbar^2 c^2 \nabla^2 \Psi + m_0^2 c^4 \Psi, \tag{3.5.8}$$

which is a quantum relativistic equation. Equation (3.5.8) is called the Klein–Gordon equation. It turns out that the Klein–Gordon equation describes the relativistic behavior of a spinless particle. For example, it can be used to describe the motion of a pi meson. The behavior of such particles does not concern us here. Our goal is to show that the spin of the electron naturally arises from relativistic invariance. However, before we leave the Klein–Gordon equation, it is interesting to pause and consider its implications.

It is important to note that the Klein–Gordon equation describes a scalar wave function. This can be observed in the following way. Note that the Klein–Gordon equation can be rewritten as

$$\left[\frac{1}{c^2} \frac{d^2}{dt^2} - \nabla^2 + \left(\frac{m_0 c}{\hbar} \right)^2 \right] \Psi = 0. \tag{3.5.9}$$

Inspection of Eq. (3.5.9) shows that it resembles a classical wave equation. The operator

$$\frac{1}{c^2} \frac{d^2}{dt^2} - \nabla^2 \tag{3.5.10}$$

has exactly the same form in any inertial reference frame since it transforms into itself under the Lorentz transformations. The constant term in Eq. (3.5.9) is also of course invariant under a Lorentz transformation. Subsequently Eq. (3.5.9) is Lorentz invariant and its solutions Ψ are invariant. This implies that

$$\Psi(r', t') = \Psi(r, t). \tag{3.5.11}$$

The solutions of the Klein–Gordon equation are plane waves. These are given as

$$\Psi \sim e^{i(\vec{k} \cdot \vec{r} - \omega t)}, \tag{3.5.12}$$

where Ψ is an eigenfunction of both E and p. This can be shown as follows. Applying the operator for E to Ψ yields

$$i\hbar \frac{\partial \Psi}{\partial t} = i\hbar \frac{\partial}{\partial t} \left[e^{i(\vec{k} \cdot \vec{r} - \omega t)} \right] = i\hbar(-i\omega) e^{i(\vec{k} \cdot \vec{r} - \omega t)} = \hbar\omega \, e^{i(\vec{k} \cdot \vec{r} - \omega t)}. \tag{3.5.13}$$

Therefore

$$H\Psi = E_{\text{op}} \Psi = \hbar\omega \Psi, \tag{3.5.14}$$

giving an energy eigenvalue of $\hbar\omega$. Similarly, p acting on Ψ yields

$$
\begin{aligned}
\vec{p}\,\Psi &= -i\hbar\vec{\nabla}e^{i(\vec{k}\cdot\vec{r}-\omega t)} \\
&= -i\hbar\, ik\, e^{i(\vec{k}\cdot\vec{r}-\omega t)} \\
&= \hbar k\, e^{i(\vec{k}\cdot\vec{r}-\omega t)}.
\end{aligned}
\tag{3.5.15}
$$

Therefore

$$
p_{\mathrm{op}}\Psi = \hbar k \Psi
\tag{3.5.16}
$$

with the momentum eigenvalue of $\hbar k$. With these results, the Klein–Gordon equation can be solved as follows. Starting with

$$
-\hbar^2\frac{\partial^2\Psi}{\partial t^2} = -\hbar^2 c^2\nabla^2\Psi + m_0^2 c^4\Psi,
\tag{3.5.17}
$$

substitutions for the differential operators can be made as

$$
\frac{\partial^2\Psi}{\partial t^2} = -\omega^2\Psi, \qquad \nabla^2\Psi = -k^2\Psi,
\tag{3.5.18}
$$

which gives

$$
\hbar^2\omega^2\Psi = \left(\hbar^2 k^2 c^2 + m_0^2 c^4\right)\Psi.
\tag{3.5.19}
$$

Solving for $\hbar\omega$ yields

$$
\hbar\omega = \pm\sqrt{\hbar^2 k^2 c^2 + m_0^2 c^4}.
\tag{3.5.20}
$$

Note that the energy is multivalued! The Klein–Gordon equation allows for negative-energy eigenvalues! This bizarre property implies the existence of antiparticles that have been experimentally observed in accordance with the above prediction. The positive-energy states correspond to particles while the negative-energy states correspond to antiparticles. Further discussion of the Klein–Gordon equation is beyond the level of this book. The interested reader is referred to the references.

2. The Dirac Equation

One can construct an alternative relativistic extension of the Schroedinger equation. Dirac started with the classical Schroedinger equation and again with the relativistic relation between the energy and the momentum,

$$
E^2 = p^2 c^2 + m^2 c^4
\tag{3.5.21}
$$

and rewrote the Hamiltonian as

$$
H = \sqrt{p^2 c^2 + m^2 c^4},
\tag{3.5.22}
$$

yielding

$$i\hbar\frac{\partial\Psi}{\partial t} = \sqrt{p^2c^2 + m^2c^4}\,\Psi. \tag{3.5.23}$$

Substituting in the quantum-mechanical operator for p yields

$$i\hbar\frac{\partial\Psi}{\partial t} = \sqrt{-\hbar^2c^2\nabla^2 + m^2c^4}\,\Psi. \tag{3.5.24}$$

This expression is not valid since it is not symmetric with respect to the time and the space derivatives. The time and the space derivatives must have the same order. Dirac found that the simplest such scheme can be attained if it is assumed that the derivatives are linear in both time and space. With this assumption, the Hamiltonian operator can be constructed to be

$$H = c\vec{\alpha}\cdot\vec{p} + \beta mc^2. \tag{3.5.25}$$

The relativistic equation then is

$$H\Psi = E\Psi,$$
$$(c\vec{\alpha}\cdot\vec{p} + \beta mc^2)\Psi = E\Psi$$

or

$$(E - c\vec{\alpha}\cdot\vec{p} - \beta mc^2)\Psi = 0. \tag{3.5.26}$$

Substituting the expressions for the quantum-mechanical operators for p and E into Eq. (3.5.26) leads to

$$\left(i\hbar\frac{\partial}{\partial t} - c\vec{\alpha}\cdot\frac{\hbar}{i}\vec{\nabla} - \beta mc^2\right)\Psi = 0,$$
$$\left(i\hbar\frac{\partial}{\partial t} + i\hbar c\vec{\alpha}\cdot\vec{\nabla} - \beta mc^2\right)\Psi = 0. \tag{3.5.27}$$

The question is, what are α and β? Equations (3.5.27) must be valid for a free particle. There can be no terms in H (say α and β) that depend on either t or p (time and space coordinates). Otherwise the derivatives of these quantities would give rise to forces. α and β must be independent of r, t, p, and E. Consequently, α and β must commute with each of these quantities.

Any solution of Equations (3.5.27) must also be a solution of the relativistic energy equation. This follows from the fact that in the absence of external fields, the free-particle solution must always be recovered. Under these conditions, the quantum-mechanical wave functions must resemble the motion of classical particles. The solutions must then obey

$$E^2 = p^2c^2 + m^2c^4. \tag{3.5.28}$$

To show that the solutions of Eq. (3.5.26) satisfy Eq. (3.5.28), it is useful to multiply Eq. (3.5.26) by

$$(E + c\vec{\alpha} \cdot \vec{p} + \beta mc^2).\tag{3.5.29}$$

This yields

$$
\begin{aligned}
E^2 - c^2\big[&\alpha_x^2 p_x^2 + \alpha_y^2 p_y^2 + \alpha_z^2 p_z^2 + (\alpha_x\alpha_y + \alpha_y\alpha_x)p_x p_y + (\alpha_y\alpha_z + \alpha_z\alpha_y)p_y p_z \\
&+ (\alpha_z\alpha_x + \alpha_x\alpha_z)p_z p_x\big] - m^2 c^4 \beta^2 - mc^3\big[(\alpha_x\beta + \beta\alpha_x)p_x \\
&+ (\alpha_y\beta + \beta\alpha_y)p_y + (\alpha_z\beta + \beta\alpha_z)p_z\big] = 0,
\end{aligned}\tag{3.5.30}
$$

where we are careful to note that E and p are differential operators. The free-particle solution can be recovered from Eq. (3.5.30) if the following conditions hold:

1. $\alpha_x^2 = \alpha_y^2 = \alpha_z^2 = \beta^2 = 1$,

2. $\alpha_x\alpha_y + \alpha_y\alpha_x = \alpha_y\alpha_z + \alpha_z\alpha_y = \alpha_z\alpha_x + \alpha_x\alpha_z = 0$,

3. $\alpha_x\beta + \beta\alpha_x = \alpha_y\beta + \beta\alpha_y = \alpha_z\beta + \beta\alpha_z = 0.$ (3.5.31)

When these conditions are used, Eq. (3.5.30) becomes

$$E^2 - c^2 p^2 - m^2 c^4 = 0,\tag{3.5.32}$$

which is precisely the free-particle solution. Note that conditions 1 and 2 above imply that α_x, α_y, α_z, and β all anticommute in pairs. Also their squares are all unity.

In general two operators that anticommute cannot be numbers or scalars but must be matrices. For example, any two numbers cannot anticommute. Therefore both α and β can be written as matrices. H is Hermitian so α and β must also be Hermitian. From condition 1 in Eqs. (3.5.31) the squares of each matrix are unity so their eigenvalues must be either 1 or -1. The matrix β is chosen, somewhat arbitrarily, to be diagonal. Hence it must consist of either $+1$'s or -1's along the diagonal. β can be constructed then as

$$\beta = \begin{bmatrix} 1 & 0 & 0 & 0 \\ 0 & 1 & 0 & 0 \\ 0 & 0 & -1 & 0 \\ 0 & 0 & 0 & -1 \end{bmatrix}.\tag{3.5.33}$$

The anticommutation relation between β and each component of α found above implies that $\alpha_x\beta + \beta\alpha_x = 0$. This condition, though, is the product of matrices and can be written as (assuming for the moment that the matrices are 2×2)

$$\begin{bmatrix} \alpha_{11} & \alpha_{12} \\ \alpha_{21} & \alpha_{22} \end{bmatrix}\begin{bmatrix} 1 & 0 \\ 0 & -1 \end{bmatrix} + \begin{bmatrix} 1 & 0 \\ 0 & -1 \end{bmatrix}\begin{bmatrix} \alpha_{11} & \alpha_{12} \\ \alpha_{21} & \alpha_{22} \end{bmatrix} = 0.\tag{3.5.34}$$

Multiplying out yields

$$\begin{bmatrix} \alpha_{11} & -\alpha_{12} \\ \alpha_{21} & -\alpha_{22} \end{bmatrix} + \begin{bmatrix} \alpha_{11} & \alpha_{12} \\ -\alpha_{21} & -\alpha_{22} \end{bmatrix} = 0,$$

$$\begin{bmatrix} (\alpha_{11} + \alpha_{11}) & (-\alpha_{12} + \alpha_{12}) \\ (\alpha_{21} - \alpha_{21}) & (-\alpha_{22} - \alpha_{22}) \end{bmatrix} = 0. \tag{3.5.35}$$

In general, the jl component is given as

$$(\alpha_x)_{jl}(\beta_j + \beta_l) = 0, \tag{3.5.36}$$

where β_j and β_l represent the possible eigenvalues of the β matrix ± 1. Note that if β_j is equal to β_l, then the only way in which Eq. (3.5.36) equals zero is if $(\alpha_x)_{jl} = 0$. However, if $\beta_j \neq \beta_l$, then β_j must be equal to $-\beta_l$ and α_x need not be zero. Inspection of the result of multiplying the α_x matrix by the β matrix indicates that the off-diagonal terms of α_x need not be zero. Hence α_x can be written as

$$\alpha_x = \begin{bmatrix} 0 & \alpha_{x1} \\ \alpha_{x2} & 0 \end{bmatrix}. \tag{3.5.37}$$

Note that if α and β are chosen to be 2×2 matrices, then the anticommutation relationships of Eqs. (3.5.31) are indeed satisfied. This is readily seen from

$$\begin{bmatrix} 0 & \alpha_{x1} \\ \alpha_{x2} & 0 \end{bmatrix}\begin{bmatrix} 1 & 0 \\ 0 & -1 \end{bmatrix} + \begin{bmatrix} 1 & 0 \\ 0 & -1 \end{bmatrix}\begin{bmatrix} 0 & \alpha_{x1} \\ \alpha_{x2} & 0 \end{bmatrix}$$

$$= \begin{bmatrix} 0 & -\alpha_{x1} \\ \alpha_{x2} & 0 \end{bmatrix} + \begin{bmatrix} 0 & \alpha_{x1} \\ -\alpha_{x2} & 0 \end{bmatrix} = 0, \tag{3.5.38}$$

which is precisely the anticommutation requirement.

In addition, the square of the each component of α must be one. Therefore $\alpha_x^2 = 1$. To ensure that this requirement is satisfied, it is necessary that the product of the matrices be

$$\begin{bmatrix} 0 & \alpha_{x1} \\ \alpha_{x2} & 0 \end{bmatrix}\begin{bmatrix} 0 & \alpha_{x1} \\ \alpha_{x2} & 0 \end{bmatrix} = \begin{bmatrix} \alpha_{x1}\alpha_{x2} & 0 \\ 0 & \alpha_{x2}\alpha_{x1} \end{bmatrix} = \begin{bmatrix} 1 & 0 \\ 0 & 1 \end{bmatrix}. \tag{3.5.39}$$

Therefore

$$\alpha_{x1}\alpha_{x2} = 1, \tag{3.5.40}$$

$$\alpha_{x2}\alpha_{x1} = 1. \tag{3.5.41}$$

If α_{x1} has n rows and m columns and α_{x2} has m rows and n columns, then it is not possible to satisfy the above condition unless both α_{x1} and α_{x2} are square matrices of equal dimension. First choose the minimum dimension of two. Such a choice gives three general 2×2 matrices. The matrices chosen must also be linearly independent and of course must satisfy conditions (3.5.31). There are many 2×2 matrices that are linearly independent. For convenience, we choose three matrices, called the Pauli matrices, which are linearly independent 2×2 matrices. These are

$$\sigma_x = \begin{bmatrix} 0 & 1 \\ 1 & 0 \end{bmatrix}, \qquad \sigma_y = \begin{bmatrix} 0 & -i \\ i & 0 \end{bmatrix}, \qquad \sigma_z = \begin{bmatrix} 1 & 0 \\ 0 & -1 \end{bmatrix}. \tag{3.5.42}$$

A fourth matrix that anticommutes with the above three matrices cannot be found. This is because the above three matrices and the identity matrix I are all linearly independent and completely span the space of 2×2 matrices. From our definition of a basis set, the Pauli matrices and the identity matrix form a basis set in the space of 2×2 matrices. As such, all other 2×2 matrices can be written as a linear combination of these matrices. This result implies then that in order to satisfy the conditions specified in Eqs. (3.5.31), matrices of a higher rank than two are required. Subsequently, we are led to examine 4×4 matrices.

Let us consider the following set of 4×4 matrices. These have the form

$$\beta = \begin{bmatrix} 1 & 0 & 0 & 0 \\ 0 & 1 & 0 & 0 \\ 0 & 0 & -1 & 0 \\ 0 & 0 & 0 & -1 \end{bmatrix}, \qquad \alpha_x = \begin{bmatrix} 0 & 0 & 0 & 1 \\ 0 & 0 & 1 & 0 \\ 0 & 1 & 0 & 0 \\ 1 & 0 & 0 & 0 \end{bmatrix},$$

$$\alpha_y = \begin{bmatrix} 0 & 0 & 0 & -i \\ 0 & 0 & i & 0 \\ 0 & -i & 0 & 0 \\ i & 0 & 0 & 0 \end{bmatrix}, \qquad \alpha_z = \begin{bmatrix} 0 & 0 & 1 & 0 \\ 0 & 0 & 0 & -1 \\ 1 & 0 & 0 & 0 \\ 0 & -1 & 0 & 0 \end{bmatrix}. \tag{3.5.43}$$

The above matrices can be abbreviated for convenience in writing as

$$\beta = \begin{bmatrix} 1 & 0 \\ 0 & -1 \end{bmatrix} \qquad \alpha_{xyz} = \begin{bmatrix} 0 & \delta_{xyz} \\ \delta_{xyz} & 0 \end{bmatrix}, \tag{3.5.44}$$

where

$$\delta_x = \sigma_x, \qquad \delta_y = \sigma_y, \qquad \delta_z = \sigma_z. \tag{3.5.45}$$

Of course, other 4×4 matrices can be found that satisfy the conditions in Eqs. (3.5.31) and are linearly independent. However, the choice above is convenient for our purposes and is that which Dirac used so as to be consistent with Pauli's spin formulation.

With the above definitions of the operators for α and β, let us next consider the behavior of an electron in the presence of a central-force potential $V(r)$ by using the Dirac equation. The Dirac equation is given as

$$[c\vec{\alpha} \cdot \vec{p} + \beta mc^2 + V(r)]\Psi = i\hbar \frac{\partial \Psi}{\partial t}, \tag{3.5.46}$$

where Ψ is a 4×1 matrix and $V(r)$ is equal to $V(r)I$. For a central-force potential, the total angular momentum of the electron must be conserved as discussed above. If the total angular momentum of an electron in a hydrogen atom is given by $\vec{L}$, the orbital angular momentum, then $\vec{L}$ must be a constant of the motion or a conserved quantity. This must be true since in any central-force motion problem there is no net torque exerted by the potential and the total angular momentum must be conserved. Subsequently, if $\vec{L}$ is the total angular momentum of the electron in an atom, $d\vec{L}/dt$ must be zero. From Eq. (3.1.26), the time rate of change of $\vec{L}$ is found from the commutator of $\vec{L}$ and H. The Dirac Hamiltonian supercedes the Schroedinger Hamiltonian. Therefore, to determine if $\vec{L}$ is a conserved quantity, it is necessary to examine whether $\vec{L}$ commutes with the Dirac Hamiltonian. The time rate of change of $\vec{L}$ is then

$$i\hbar \frac{d\vec{L}}{dt} = [\vec{L}, H] = -[H, \vec{L}] = -[(c\vec{\alpha} \cdot \vec{p} + \beta mc^2 + V), \vec{L}]. \tag{3.5.47}$$

Simplifying to one component of L, L_x, yields

$$i\hbar \frac{dL_x}{dt} = -[H, L_x] = -[c(\vec{\alpha} \cdot \vec{p}), L_x] - [\beta mc^2, L_x] - [V, L_x]. \tag{3.5.48}$$

Clearly, $[\beta mc^2, L_x] = 0$ since βmc^2 is a constant. In addition, the commutator of the potential V and L_x vanishes since the potential is a function of only r and L_x is a function of only angle. Therefore, in order that the orbital angular momentum of the electron be conserved in a central-force motion problem, it remains to show that the commutator

$$-[c\vec{\alpha} \cdot \vec{p}, L_x]$$

must vanish. Let us focus our attention then on evaluating this commutator.
Consider then

$$i\hbar \frac{dL_x}{dt} = -[c\vec{\alpha} \cdot \vec{p}, L_x] = -c(\vec{\alpha} \cdot \vec{p} L_x - L_x \vec{\alpha} \cdot \vec{p}). \tag{3.5.49}$$

Clearly α commutes with p and L_x from our earlier discussion since it commutes with r and p. But does p commute with L_x? Expanding out the right-hand side yields

$$i\hbar \frac{dL_x}{dt} = -c\vec{\alpha} \cdot (\vec{p} L_x - L_x \vec{p}) = c\vec{\alpha} \cdot (L_x \vec{p} - \vec{p} L_x),$$

$$L_x = (yp_z - zp_y),$$

$$i\hbar \frac{dL_x}{dt} = c\vec{\alpha} \cdot [(yp_z - zp_y)\vec{p} - \vec{p}(yp_z - zp_y)]. \tag{3.5.50}$$

In Chapter 1 the commutators of the spatial and the momentum components were evaluated to be

$$[x, p_x] = [y, p_y] = [z, p_z] = i\hbar, \tag{3.5.51}$$

and

$$[x, p_y] = 0, \qquad [x, p_z] = 0, \tag{3.5.52}$$

etc. The expression for the time rate of change of the x component of the orbital angular momentum becomes then

$$
\begin{aligned}
&= c\vec{\alpha} \cdot [yp_z p_x \hat{i} + yp_z p_y \hat{j} + yp_z p_z \hat{k} - zp_y p_x \hat{i} - zp_y p_y \hat{j} - zp_y p_z \hat{k} \\
&\quad - p_x yp_z \hat{i} - p_y yp_z \hat{j} - p_z yp_z \hat{k} + p_x zp_y \hat{i} + p_y zp_y \hat{j} + p_z zp_y \hat{k}] \\
&= c\vec{\alpha} \cdot [p_z(yp_y - p_y y)\hat{j} - p_y(zp_z - p_z z)\hat{k}] \\
&= c\vec{\alpha} \cdot [p_z[y, p_y]\hat{j} - p_y[z, p_z]\hat{k}] \\
&= c[\alpha_y p_z[y, p_y] - \alpha_z p_y[z, p_z]].
\end{aligned} \tag{3.5.53}
$$

We can simplify the above by making use of the commutators

$$[y, p_y] = i\hbar, \qquad [z, p_z] = i\hbar \tag{3.5.54}$$

to

$$i\hbar\frac{dL_x}{dt} = -i\hbar c[\alpha_z p_y - \alpha_y p_z]. \tag{3.5.55}$$

From the above result it is clear that $\vec{L}$ does not commute with the relativistic expression for H. Subsequently the time rate of change of $\vec{L}$ is not zero, implying that $\vec{L}$ is not a constant of the motion. Nevertheless, the original statement that the total angular momentum of the electron must be conserved in a central-force motion problem, like that considered here, must still be true. It must be concluded then that $\vec{L}$ is not the total angular momentum of the electron since it is apparently not a conserved quantity. There must then be an additional component of the angular momentum that, when it is added to $\vec{L}$, gives the total angular momentum of the electron. This additional term must be such that its commutator with H is just the negative of Eq. (3.5.55). After such a term is added to $\vec{L}$ then, the total angular momentum of the electron will commute with H.

Before we determine the magnitude of this additional component of the angular momentum, let us consider an analogous situation in classical mechanics. In the Earth–Sun system, the Earth revolves around the Sun in an elliptical orbit, as shown in Figure 3.5.1. The orbital motion of the Earth around the Sun has an orbital angular momentum associated with it, as shown in the diagram marked

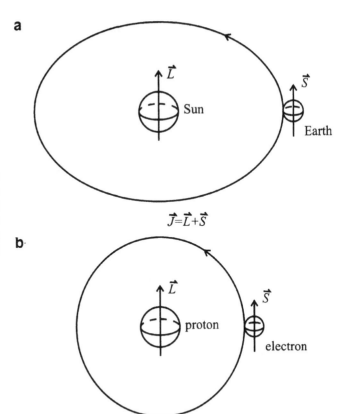

Figure 3.5.1. a. Sketch of the Earth–Sun system showing the orbital L and spin S angular momentum. b. Sketch of the electron–proton system showing the orbital and the spin motions of the particles.

as $\vec{L}$. The total angular momentum of the Earth, though, has an additional component due to the spin motion of the Earth about its own axis. Neglecting any tidal friction effects, the spin angular momentum of the earth is essentially independent of the orbital angular momentum of the Earth–Sun system. Therefore the spin angular momentum can be pictured as an independent degree of freedom in the Earth–Sun system. The total angular momentum $\vec{J}$ of the Earth is then found as the vector sum of the orbital angular momentum $\vec{L}$ and the spin angular momentum $\vec{S}$, as $\vec{J} = \vec{L} + \vec{S}$.

In a similar manner, the motion of an electron in an atom can be pictured as the result of both an orbital motion and a spin motion. In direct analogy to our experience with classical bodies, such as the Earth and the Sun, we can envision that the electron has an inherent angular momentum associated with it. For lack of a better description, the most natural way to think of this is that the electron spins and as such has a spin angular momentum. The resulting total angular momentum of the electron in an atom is then found as the vector sum of the orbital and the spin angular momentum. However, the rules governing how the angular-momentum vectors are summed are quite different from those for the simple classical case. The basics of the addition of angular momentum are discussed in Section 3.7.

Let us now construct the form of $\vec{J}$, the total angular momentum. What we seek is a term that when added to $\vec{L}$ will ensure that the commutator with H vanishes. In this way, the conservation of the total angular momentum of the electron will be guaranteed. Consider then the following. From Eqs. (3.5.44), α can be abbreviated as

$$\alpha = \begin{bmatrix} 0 & \delta_{xyz} \\ \delta_{xyz} & 0 \end{bmatrix}, \tag{3.5.56}$$

where

$$\delta_x = \sigma_x, \qquad \delta_y = \sigma_y, \qquad \delta_z = \sigma_z. \tag{3.5.57}$$

Let us define σ' as

$$\sigma' \equiv \begin{bmatrix} \delta & 0 \\ 0 & \delta \end{bmatrix} \tag{3.5.58}$$

such that σ'_x satisfies

$$i\hbar \frac{d\sigma'_x}{dt} = \sigma'_x H - H\sigma'_x. \tag{3.5.59}$$

Substituting in the relativistic expression for H, we obtain

$$i\hbar \frac{d\sigma'_x}{dt} = \sigma'_x(c\vec{\alpha} \cdot \vec{p} + \beta mc^2 + V) - (c\vec{\alpha} \cdot \vec{p} + \beta mc^2 + V)\sigma'_x. \tag{3.5.60}$$

Recognizing that βmc^2 and V commute with σ'_x, we can write Eq. (3.5.60) as

$$i\hbar \frac{d\sigma'_x}{dt} = \sigma'_x(c\vec{\alpha} \cdot \vec{p}) - c(\vec{\alpha} \cdot \vec{p})\sigma'_x. \tag{3.5.61}$$

σ'_x commutes with p but not with α. The right-hand side of Eq. (3.5.61) can be evaluated as follows. Expanding out the dot product and making use of the commutation of σ'_x and p and p and α yields

$$i\hbar \frac{d\sigma'_x}{dt} = c\vec{p} \cdot [\sigma'_x\alpha_x\hat{i} + \sigma'_x\alpha_y\hat{j} + \sigma'_x\alpha_z\hat{k} - \alpha_x\sigma'_x\hat{i} - \alpha_y\sigma'_x\hat{j} - \alpha_z\sigma'_x\hat{k}]. \tag{3.5.62}$$

Each of the components in Eq. (3.5.62) can be evaluated by use of the matrix forms of σ'_x and α_x and α_y and α_z. Consider the $\hat{i}$ component:

$$\sigma'_x\alpha_x - \alpha_x\sigma'_x, \qquad \sigma'_x = \begin{bmatrix} \sigma_x & 0 \\ 0 & \sigma_x \end{bmatrix}, \qquad \alpha_x = \begin{bmatrix} 0 & \sigma_x \\ \sigma_x & 0 \end{bmatrix}. \tag{3.5.63}$$

Therefore, forming the product of σ'_x and α_x yields

$$\sigma'_x\alpha_x = \begin{bmatrix} \sigma_x & 0 \\ 0 & \sigma_x \end{bmatrix}\begin{bmatrix} 0 & \sigma_x \\ \sigma_x & 0 \end{bmatrix} = \begin{bmatrix} 0 & \sigma_x^2 \\ \sigma_x^2 & 0 \end{bmatrix} \tag{3.5.64}$$

and the product of α_x and σ'_x yields

$$\alpha_x \sigma'_x = \begin{bmatrix} 0 & \sigma_x \\ \sigma_x & 0 \end{bmatrix} \begin{bmatrix} \sigma_x & 0 \\ 0 & \sigma_x \end{bmatrix} = \begin{bmatrix} 0 & \sigma_x^2 \\ \sigma_x^2 & 0 \end{bmatrix}. \tag{3.5.65}$$

Hence the difference is zero and the $\hat{i}$ component vanishes.

The $\hat{j}$ component,

$$\sigma'_x \alpha_y - \alpha_y \sigma'_x \tag{3.5.66}$$

can be evaluated similarly with the matrix forms of σ'_x and α_y. These are

$$\sigma'_x = \begin{bmatrix} \sigma_x & 0 \\ 0 & \sigma_x \end{bmatrix}, \qquad \alpha_y = \begin{bmatrix} 0 & \sigma_y \\ \sigma_y & 0 \end{bmatrix}. \tag{3.5.67}$$

The $\hat{j}$ component can then be found from

$$
\begin{aligned}
\sigma'_x \alpha_y - \alpha_y \sigma'_x &= \begin{bmatrix} \sigma_x & 0 \\ 0 & \sigma_x \end{bmatrix} \begin{bmatrix} 0 & \sigma_y \\ \sigma_y & 0 \end{bmatrix} - \begin{bmatrix} 0 & \sigma_y \\ \sigma_y & 0 \end{bmatrix} \begin{bmatrix} \sigma_x & 0 \\ 0 & \sigma_x \end{bmatrix} \\
&= \begin{bmatrix} 0 & \sigma_x \sigma_y \\ \sigma_x \sigma_y & 0 \end{bmatrix} - \begin{bmatrix} 0 & \sigma_y \sigma_x \\ \sigma_y \sigma_x & 0 \end{bmatrix} \\
&= \begin{bmatrix} 0 & \sigma_x \sigma_y - \sigma_y \sigma_x \\ \sigma_x \sigma_y - \sigma_y \sigma_x & 0 \end{bmatrix}.
\end{aligned} \tag{3.5.68}
$$

But the commutator $[\sigma_x, \sigma_y] = \sigma_x \sigma_y - \sigma_y \sigma_x$ is given as

$$
\begin{aligned}
\sigma_x \sigma_y &= \begin{bmatrix} 0 & 1 \\ 1 & 0 \end{bmatrix} \begin{bmatrix} 0 & -i \\ i & 0 \end{bmatrix} = \begin{bmatrix} i & 0 \\ 0 & -i \end{bmatrix}, \\
\sigma_y \sigma_x &= \begin{bmatrix} 0 & -i \\ i & 0 \end{bmatrix} \begin{bmatrix} 0 & 1 \\ 1 & 0 \end{bmatrix} = \begin{bmatrix} -i & 0 \\ 0 & i \end{bmatrix}.
\end{aligned} \tag{3.5.69}
$$

But σ_z is defined as

$$\sigma_z = \begin{bmatrix} 1 & 0 \\ 0 & -1 \end{bmatrix}. \tag{3.5.70}$$

Therefore

$$\sigma_x \sigma_y = i\sigma_z, \qquad \sigma_y \sigma_x = -i\sigma_z. \tag{3.5.71}$$

When Eqs. (3.5.71) are combined, the commutator of σ_x and σ_y is $2i\sigma_z$. With this result, the $\hat{j}$ component becomes

$$\sigma'_x \alpha_y - \alpha_y \sigma'_x = \begin{bmatrix} 0 & 2i\sigma_z \\ 2i\sigma_z & 0 \end{bmatrix} = 2i \begin{bmatrix} 0 & \sigma_z \\ \sigma_z & 0 \end{bmatrix} = 2i\alpha_z. \tag{3.5.72}$$

Similarly, the $\hat{k}$ component can be determined with the matrix forms of σ_x' and α_z. The result is

$$\sigma_x'\alpha_z - \alpha_z\sigma_x' = -2i\alpha_y. \tag{3.5.73}$$

With the above expressions for the $\hat{j}$ and the $\hat{k}$ components, Eq. (3.5.62) can be written as

$$
\begin{aligned}
i\hbar\frac{d\sigma_x'}{dt} &= c\vec{p} \cdot \left[\left(\sigma_x'\alpha_y - \alpha_y\sigma_x'\right)\hat{j} + \left(\sigma_x'\alpha_z - \alpha_z\sigma_x'\right)\hat{k}\right] \\
&= c\vec{p} \cdot [2i\alpha_z\hat{j} - 2i\alpha_y\hat{k}] \\
&= c[p_y 2i\alpha_z - 2i\alpha_y p_z] \\
&= 2ic[p_y\alpha_z - \alpha_y p_z].
\end{aligned} \tag{3.5.74}
$$

It is important to note that Eq. (3.5.74) is essentially the same as Eq. (3.5.55), which is equal to the time rate of change of L_x. Recalling Eq. (3.5.55), we find that the time rate of change of L_x is

$$i\hbar\frac{dL_x}{dt} = -i\hbar c[\alpha_z p_y - \alpha_y p_z]. \tag{3.5.75}$$

But Eq. (3.5.74),

$$i\hbar\frac{d\sigma_x'}{dt} = 2ic[\alpha_z p_y - \alpha_y p_z], \tag{3.5.76}$$

is essentially equal to dL_x/dt to within a constant. If $i\hbar\,d\sigma_x'/dt$ is multiplied by $\hbar/2$ and added to $i\hbar\,dL_x/dt$, then the sum vanishes. This is clearly seen from

$$
\begin{aligned}
i\hbar\frac{dL_x}{dt} + \frac{i\hbar^2}{2}\frac{d\sigma_x'}{dt} &= -i\hbar c(\alpha_z p_y - \alpha_y p_z) + i\hbar c(\alpha_z p_y - \alpha_y p_z) = 0, \\
i\hbar\frac{dL_x}{dt} + \frac{i\hbar^2}{2}\frac{d\sigma_x'}{dt} &= 0.
\end{aligned} \tag{3.5.77}
$$

Therefore the expression given by Eq. (3.5.74) multiplied by $\hbar/2$ is precisely what is needed to add to Eq. (3.5.55) to ensure that the total angular momentum of the electron vanishes. The addition of this extra term to L_x satisfies the requirement that the net angular momentum vanish in a central-force motion problem.

We define J_x, the x component of the total angular momentum then as

$$J_x = L_x + \frac{\hbar}{2}\sigma_x'. \tag{3.5.78}$$

With this definition, the total angular momentum of the electron is guaranteed to vanish as

$$i\hbar\frac{d}{dt}\left[L_x + \frac{\hbar}{2}\sigma'_x\right] = 0,$$

$$i\hbar\frac{d}{dt}J_x = 0. \tag{3.5.79}$$

We define the total angular momentum of the electron to be J. From the above, the x component of the total angular momentum J_x, is found to be conserved in a central-force potential. The above results can be summarized as follows. Originally it was determined that L_x was not a conserved quantity in the relativistic description. Therefore, since we know that the total angular momentum of an electron must be conserved in a central-force potential, L must not be the total angular momentum of an electron. An additional term has to be added to L_x to ensure that the total angular momentum of the electron is conserved. This additional term was found to be

$$\frac{\hbar}{2}\sigma'_x. \tag{3.5.80}$$

We call this additional term the x component of the spin angular momentum of the electron. The orbital angular momentum is specified by the allowed values of L, determined in accordance with the results found in Section 3.1, while the spin angular momentum has magnitude $1/2\hbar\sigma'$. It is important to recognize that the spin angular momentum is a natural consequence of the extension of the conservation of angular momentum to a relativistic system.

As we will see in Section 3.6, the spin angular momentum of the electron has two possible orientations. It can be either up or down, leading to two different possible states of the electron. In other words, the spin of an electron gives an additional multiplicity of 2 to the total number of allowed states. For example, an electron in the $n = 1$ state must have $l = 0$ and $m_l = 0$ from the rules determined in Section 3.3. This would imply that only one state exists for $n = 1$. However, the multiplicity of 2, because of the two possible orientations of the spin angular momentum of an electron, provides for two possible electronic states for $n = 1$.

$$n = 1, l = 0, m_l = 0, m_s = +1/2; \quad n = 1, l = 0, m_l = 0, m_s = -1/2, \tag{3.5.81}$$

where m_s is called the spin quantum number and represents the two possible states of the spin angular momentum. Given that the spin produces a multiplicity in the number of states of a factor of 2, it is now possible to account for the Periodic Table of the Elements. As we discussed in Section 3.3, the allowed states of the electron predicted by Schroedinger's equation are a factor of 2 too small to account for the known elements. By introduction of the spin angular momentum from the requirement that the total angular momentum of the electron must vanish in a relativistic system, that factor of 2 necessary to explain the Periodic Table naturally appears. In Section 3.6 the experimental evidence for the existence of the electron's spin angular momentum and its measurement is discussed.

3.6 Experimental Evidence of Spin Angular Momentum

Stern and Gerlach first experimentally measured the spin angular momentum of an electron. Their experiment preceded the theory developed by Dirac by roughly ten years. The experiment itself actually measures the magnetic moment of the electron rather than the spin angular momentum directly. As we will see below, the spin angular momentum and the magnetic moment are intimately linked. Thus the measurement of one leads to the determination of the other. In this section, we discuss the experimental evidence provided by the Stern–Gerlach experiment for the existence of the spin angular momentum of an electron.

To understand the basis of the Stern–Gerlach experiment it is important to first review the motion of a dipole in an external field. As mentioned above, the Stern–Gerlach experiment was devised to measure the orientation of the magnetic dipole moment of an electron in an external field. As we will see, the electron has an internal magnetic moment that is antiparallel to the spin angular-momentum vector. We can obtain the simplest picture of a magnetic dipole by modeling the atom as an electron moving in a circular orbit around the nucleus. The magnetic field $\vec{B}$ can be written as the curl of the vector potential $\vec{A}$

$$\vec{B} = \vec{\nabla} \times \vec{A}, \tag{3.6.1}$$

since the divergence of $\vec{B}$ is zero (there are no magnetic monopoles, at least not confirmed at this writing). The magnetic field can be determined from the current density $\vec{J}$ and the electric field $\vec{E}$ as

$$\vec{\nabla} \times \vec{B} = \frac{4\pi}{c} \vec{J} + \frac{1}{c} \frac{d\vec{E}}{dt}. \tag{3.6.2}$$

If the fields are time independent, then the time rate of change of the electric field is zero, $d\vec{E}/dt = 0$, and Eq. (3.6.2) reduces to

$$\vec{\nabla} \times \vec{B} = \frac{4\pi}{c} \vec{J}. \tag{3.6.3}$$

Substituting the curl of $\vec{A}$ for $\vec{B}$ gives

$$\vec{\nabla} \times \vec{\nabla} \times \vec{A} = \frac{4\pi}{c} \vec{J}. \tag{3.6.4}$$

We next apply the vector identity

$$\vec{\nabla} \times \vec{\nabla} \times \vec{A} = \vec{\nabla}(\vec{\nabla} \cdot \vec{A}) - \nabla^2 \vec{A}. \tag{3.6.5}$$

As we discussed in Chapter 1, a gauge transformation can be performed in which the value of $\vec{A}$ is chosen such that the divergence of $\vec{A}$ vanishes. As we determined in Chapter 1, we are free to make any choice of gauge since it alters only the form of the potentials $\vec{A}$ and Φ (the electrostatic potential). With this choice of gauge, Eq. (3.6.4) becomes

$$\vec{\nabla} \times \vec{\nabla} \times \vec{A} = -\nabla^2 \vec{A},$$
$$-\nabla^2 \vec{A} = \frac{4\pi}{c} \vec{J}. \tag{3.6.6}$$

The solution of Eq. (3.6.6) is, in general,

$$\vec{A} = \frac{1}{C} \int \frac{\vec{J}(x_2, y_2, z_2)}{r_{12}} d^3 r_2, \tag{3.6.7}$$

where r_{12} is the distance between the current loop and the point of observation. If the charge orbits at a constant angular velocity then the current is constant. This is due to the fact that the current of one electron moving in a circle is simply given by the quotient of its charge q and the period of its revolution T as $i = q/T$. The period of revolution of the electron can be expressed as

$$T = \frac{2\pi r}{v}. \tag{3.6.8}$$

Therefore the current is

$$i = \frac{qv}{2\pi r}. \tag{3.6.9}$$

The integration of the current density of the revolving electron $\vec{J}$ over the volume is equal to the integrated product of the current and the differential path length $d\vec{l}$. When these results are combined with the expression for the vector potential given by Eq. (3.6.7), $\vec{A}$ becomes

$$\vec{A} = \frac{I}{c} \int_{\text{loop}} \frac{d\vec{l}_2}{r_{12}}. \tag{3.6.10}$$

We can evaluate the vector potential by performing the path integral over the current loop. The current loop and the coordinate system are sketched in Figure 3.6.1.a. As can be seen from the diagram, the current loop is within the xy plane while the observation point P_1 is located at the coordinates $(0, y_1, z_1)$ and is separated from the current loop by a distance r_{12}. Since the observation point lies within the y–z plane the diagram can be redrawn as Figure 3.6.1.b. If we make the approximation that the point P_1 is very far away from the origin

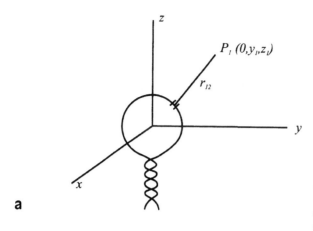

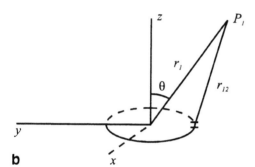

Figure 3.6.1. a. Current loop and observation point P_1. b. Sketch of the geometry of the current loop and the observation point to show that P_1 lies within the yz plane.

and that this distance of separation is very much larger than the radius of the current loop, then the two vectors r_1 and r_{12} can be assumed to be parallel near the origin. This is sketched in Figure 3.6.2.a. When this approximation is made, the distance separating the observation point P_1 and the current loop r_{12} can be expressed in terms of r_1 as

$$r_{12} = r_1 - y_2 \sin\theta, \tag{3.6.11}$$

where y_2 is defined from Figure 3.6.2.b.

Before we proceed, it is useful to consider the symmetry of the problem. The vector $\vec{A}$ at point P will not have a y component. The observation point is confined within the yz plane and therefore has no x component. Two vectors of equal length connecting the loop to the observation point must be symmetrically placed with respect to the yz plane (see Figure 3.6.3). Therefore the y components of the differential path elements compensate for one another, and there is no net value of the vector potential $\vec{A}$ in the y direction at point P_1. The x component does not vanish, however, since, when the r_{12} vectors are equal in length, the x components do not compensate for one another. The vector potential will also not have a z component since the current does not have a component in this direction. The current loop is assumed to lie completely

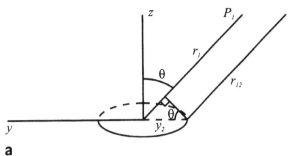

Figure 3.6.2. a. Observation point P_1 is assumed to lie very far away from the origin such that r_1 is parallel to r_{12}. b. Sketch of the current loop confined entirely within the xy plane showing the differential elements used in the calculation.

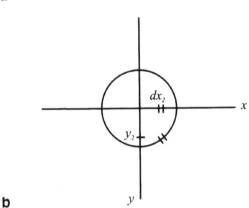

Figure 3.6.3. Sketch of the current loop showing that the y components of $\vec{A}$ compensate one another while the x components do not.

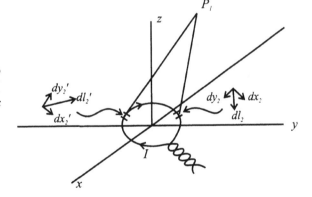

within the xy plane. The only component of $\vec{A}$ that does not vanish then is the x component. Subsequently, $\vec{A}$ is given as

$$\vec{A} = \frac{I}{c} \int \frac{dx_2}{r_{12}} \hat{\imath},\qquad(3.6.12)$$

where we have made use of the fact that the differential path element $d\vec{l}_2$ is equal to

$$d\vec{l}_2 = dx_2\,\hat{\imath} + dy_2\,\hat{\jmath}.\qquad(3.6.13)$$

From Figure 3.6.2, r_{12} can be written as

$$r_{12} \sim r_1 - y_2 \sin \theta. \tag{3.6.14}$$

The reciprocal of r_{12} is

$$\frac{1}{r_{12}} \sim \frac{1}{r_1 - y_2 \sin \theta} \sim \frac{1}{r_1} \left[\frac{1}{1 - \frac{y_2 \sin \theta}{r_1}} \right]. \tag{3.6.15}$$

We can express the expression on the right-hand side above as

$$\frac{1}{r_1} \left[1 + \frac{y_2 \sin \theta}{r_1} \right]. \tag{3.6.16}$$

Finally, $1/r_{12}$ can be approximated as

$$\frac{1}{r_{12}} \sim \frac{1}{r_1} \left[1 + \frac{y_2 \sin \theta}{r_1} \right]. \tag{3.6.17}$$

With this substitution, the vector potential becomes

$$\vec{A} = \hat{i} \frac{I}{cr_1} \int_{\text{loop}} \left(1 + \frac{y_2 \sin \theta}{r_1} \right) dx_2, \tag{3.6.18}$$

where r_1 and θ are constant since r_1 does not change as we move around the loop. Recognizing this, we find that the integral for A is

$$\vec{A} = \hat{i} \frac{I \sin \theta}{cr_1^2} \int_{\text{loop}} y_2 \, dx_2 + \frac{I}{cr_1} \int_{\text{loop}} dx_2 \, \hat{i}. \tag{3.6.19}$$

The integral of the path element alone around a closed loop is zero. Hence the last term in Eq. (3.6.19) vanishes, leaving

$$\vec{A} = \hat{i} \frac{I \sin \theta}{cr_1^2} \int_{\text{loop}} y_2 \, dx_2. \tag{3.6.20}$$

The integral of $y_2 \, dx_2$ around the closed loop is simply the area of the loop. This is obvious from inspection of Figure 3.6.2.b. The magnitude of the vector potential can be readily evaluated as

$$A = I \frac{\sin \theta a}{cr^2}, \tag{3.6.21}$$

where a is the area of the current loop and r is the distance from the center of the loop to the observation point. $\vec{A}$ is perpendicular to the plane formed by the vector $\vec{r}$ and the normal vector to the current loop.

It is convenient to reexpress the vector potential in terms of the magnetic-moment vector μ. The magnetic moment is defined as

$$\mu = \frac{Ia}{c}. \tag{3.6.22}$$

In mks units, μ is simply given as the product of the current and the area of the loop Ia. The direction of μ is along the direction normal to the plane of the current loop. With this definition, the vector potential can be rewritten as

$$\vec{A} = \frac{\vec{\mu} \times \vec{r}}{r^2}. \tag{3.6.23}$$

The magnetic moment can be related to the angular-momentum vector. What we desire is to relate the spin angular momentum to the electron magnetic moment. Rather than do this directly, it is easier and perhaps more informative to illustrate how the magnetic moment and the angular momentum are related by use of a simpler scheme. Consider an electron in a circular orbit. The magnitude of its orbital angular momentum is given as

$$L = mvr, \tag{3.6.24}$$

where m is the mass of the particle, v is its velocity, and r is the radius of its orbit. The vector potential $\vec{A}$ associated with the moving charge is perpendicular to the vector normal to the current loop $\hat{n}$, as we found above. The magnetic-moment vector $\vec{\mu}$ is parallel to $\hat{n}$. The angular-momentum vector $\vec{L}$ is also parallel or antiparallel to $\hat{n}$, which is readily seen from the definition of $\vec{L}$ ($\vec{L} = \vec{r} \times \vec{p}$) and inspection of Figure 3.6.4. Assume that $\vec{L}$ is parallel to $\hat{n}$. The current flows in the opposite direction from that of the electron. Hence the current flow is in the negative direction with respect to the electron velocity. Combining the expression for the current of a single electron moving in a circular orbit, Eq. (3.6.9), with the definition of the magnetic moment, Eq. (3.6.22), we can write the magnitude of the magnetic moment as

$$\mu = \frac{qvr}{2c} \text{ (cgs)}, \qquad \mu = \frac{qvr}{2} \text{ (mks)}. \tag{3.6.25}$$

But the magnitude of the angular momentum $\vec{L}$ is simply equal to mvr. Substituting then L/m for vr, we find that the magnitude of the magnetic moment becomes

$$\mu = \frac{qL}{2mc} \text{ (cgs)}, \qquad \mu = \frac{qL}{2m} \text{ (mks)}. \tag{3.6.26}$$

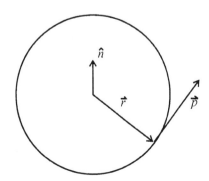

Figure 3.6.4. Sketch showing an electron in a circular orbit of radius r. Note that $\vec{L} = \vec{r} \times \vec{p}$ is parallel to $\hat{n}$.

In mks units the magnetic moment of an electron in a circular orbit is related to its angular momentum by the factor $q/2m$. Given Eqs. (3.6.26), once either the magnetic moment or the angular momentum of the electron is determined, the other is immediately known.

In general, the magnitude of the ratio of the magnetic moment to the angular momentum of a particle is written as

$$\frac{\mu}{L} = \frac{g_l \mu_b}{\hbar},$$ (3.6.27)

where μ_b is the Bohr magneton and is defined as

$$\mu_b = \frac{q\hbar}{2m}.$$ (3.6.28)

Therefore the magnitude of the ratio of μ to L is

$$\frac{\mu}{L} = g_l \frac{q}{2m},$$ (3.6.29)

where g_l is called the Lande g factor. In the case of an electron, g_l is simply one.

Although the magnitude of the magnetic moment is related to the magnitude of the orbital angular momentum through Eq. (3.6.29), how are the directions related? From the above discussion, it is clear that since the current moves in the opposite direction to the electron's velocity, the magnetic moment must point in the opposite direction to the angular-momentum vector. Hence $\vec{\mu}$ and $\vec{L}$ are antiparallel and $\vec{\mu}$ can be written in terms of $\vec{L}$ as

$$\vec{\mu} = -g_l \frac{\mu_b}{\hbar} \vec{L}.$$ (3.6.30)

It should now be clear that once the magnetic moment is determined the orbital angular momentum can also be determined. Hence a measurement of $\vec{\mu}$ leads directly to a determination of $\vec{L}$. However, we are interested in not only the relationship between $\vec{L}$ and $\vec{\mu}$ but also the relationship between the spin angular momentum and the internal magnetic moment of the electron. A similar relationship to that in Eq. (3.6.30) holds between the spin angular momentum of an electron and its internal magnetic moment. The magnitude of the spin magnetic moment can be expressed in terms of the spin angular momentum S as

$$\mu_s = -g_s \frac{\mu_b S}{\hbar},$$ (3.6.31)

where S was determined in Section 3.5 to be $S = \hbar/2\,\sigma'$ with $S_z = \hbar/2\,\sigma'_z$. g_s in this case is simply equal to 2 and μ_b is again the Bohr magneton (see Box 3.6.1).

Box 3.6.1 Spin Angular Momentum

The physics of spin angular momentum follows exactly the same rules as those for orbital angular momentum. Earlier in this chapter we found that the orbital angular momentum $\vec{L}$ and its z component L_z can be simultaneously known. The same is true for spin; $\vec{S}$ and S_z can be simultaneously known. From Eq. (3.1.74) it was found that the eigenvalue of the operator L^2 is $\hbar^2 l(l+1)$, where l is an integer. l represents the possible values of the orbital angular momentum. Similarly, the eigenvalue of S^2 is $\hbar^2 s(s+1)$, where s is again an integer and it represents the magnitude of the spin angular-momentum vector. The eigenvalue of the operator L_z is given by Eq. (3.1.61) as $m\hbar$, where m is an integer. Similarly, the eigenvalue of S_z is $m_s \hbar$. The primary difference between the allowed values of S_z and L_z, though, is that the only allowed values of m_s are $\pm 1/2$. These values of m_s are the only possible values since a beam of $\vec{L} = 0$ electrons splits into two beams and only two beams. Therefore the spin can have only two orientations within a magnetic field. The two orientations are specified by the two values of S_z and thus m_s.

When Eq. (3.6.31) is used, the z component of the spin magnetic moment can be found as

$$\mu_{sz} = -\frac{g_s \mu_b}{\hbar} \hbar m_s,$$
$$\mu_{sz} = -g_s \mu_b m_s.$$

Since the beam splits into only two beams, the product of g_s and m_s must have two values. m_s has two values, $\pm 1/2$. Therefore g_s must be equal to 2 to give a product of ± 1.

The point of the Stern–Gerlach experiment is that an apparatus (explained below) is used to determine the possible orientations of the magnetic moment of certain select atoms. Classically, one would expect that the magnetic moment of an atom can assume any orientation with respect to an external field. However, from the above discussion, the magnetic moment of the electron depends on both its orbital and spin angular-momentum values. From our discussion in Section 3.1, we know that the orbital angular momentum is quantized; it can assume only certain discrete values. From the relationship between L and μ given by Eq. (3.6.30), the magnetic moment can assume only discrete values and orientations as well. We can assess the influence of the spin angular momentum by choosing to examine an atom in which the net orbital angular momentum is zero. Such an atom would either have a completely filled outer shell or have only a single electron in the outermost s orbital. Stern and Gerlach chose to examine just such an atom and measured the orientations of its magnetic moment. Before we discuss what they observed let us examine their experiment.

To understand how the Stern–Gerlach experiment works, it is necessary to understand how a dipole reacts to an externally applied field. For ease in

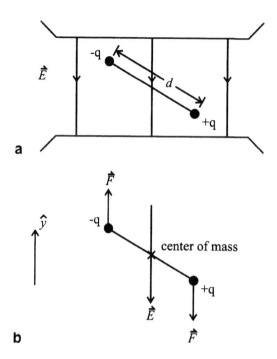

Figure 3.6.5. a. Electric dipole in the presence of a uniform electric field E. b. Free-body diagram showing the forces acting on the dipole from the external field.

understanding, let us consider an electric dipole in the presence of an external electric field E. Although we use an electric dipole, the result is equally applicable to magnetic dipoles in a magnetic field as well. Consider an electric dipole in a uniform electric field applied between two plates, as shown in Figure 3.6.5.a. As can be seen from the diagram, the dipole consists of two opposite but equal charges separated by a distance d. The forces acting on the dipole are shown in the free-body diagram sketched in Figure 3.6.5.b. The net force on the charge marked $-q$ is $-qFy$. The net force on the charge marked $+q$ is $+qFy$. Therefore the net force on the dipole is zero. The center of mass of the dipole does not move. Although there is no net force on the dipole, there is a torque. The net torque about the center of mass of the dipole is

$$\vec{\tau} = \vec{r} \times \vec{F}, \tag{3.6.32}$$

where r is equal to $d/2$, half the length of the dipole and the force is simply qF. The torque is then

$$\vec{\tau} = 2qr F \sin\theta \hat{z},$$
$$\vec{\tau} = qd F \sin\theta \hat{z}, \tag{3.6.33}$$

where θ is defined as the angle between F and r. Recognizing that the electric dipole moment p is defined as qd, we find that the torque is simply given by the vector product of p and F:

$$\vec{\tau} = \vec{p} \times \vec{F}. \tag{3.6.34}$$

A similar relationship holds for the magnetic moment. The torque on a magnetic dipole from an external magnetic field is

$$\vec{\tau} = \vec{\mu} \times \vec{B}. \tag{3.6.35}$$

The potential energy of orientation of a magnetic dipole in a magnetic field is given by the dot product between the magnetic moment and the external field:

$$\Delta E = -\vec{\mu} \cdot \vec{B}. \tag{3.6.36}$$

As was found above, a uniform field will not displace the center of mass of a dipole. It exerts only a torque on the dipole. However, a nonuniform field will exert a net force on a dipole. In a nonuniform field, the charges will experience different levels of force, depending on which is closest to the strongest field gradient. The net force acting on the dipole in this case is

$$\vec{F} = \vec{\nabla}(\vec{\mu} \cdot \vec{B}). \tag{3.6.37}$$

Stern and Gerlach sent a beam of silver atoms through an inhomogenous magnetic field. The nonuniform field will exert a net force on each atom, depending on the orientation of its magnetic moment. As mentioned above, classically it is expected that the magnetic moments of the atoms can assume any orientation and as such the beam of silver atoms should emerge spread out continuously from the maximum allowed value to the minimum value. Stern and Gerlach observed a different result. An incident beam of silver atoms emerged from the nonuniform field split into two and only two sections. This implies that the orientation of the magnetic moment of a silver atom is quantized and can assume only two orientations.

Given our discussion of quantum mechanics to this point, it is not surprising that the magnetic moment of the silver atoms is quantized since m is related to L (a quantity that we found is quantized in all atoms) by Eq. (3.6.30). The question is, though, is the observation of only two beams consistent with the rules for the quantization of orbital angular momentum? Only the magnitude of L and one component can be known simultaneously, typically L_z. The z direction is defined as being along the probing field direction. As discussed in Section 3.1, the z component of L can have only certain discrete values. These are given as

$$L_z = m_l \hbar. \tag{3.6.38}$$

L has $(2l + 1)$ possible orientations since m_l and thus L_z have $(2l + 1)$ possible values. Therefore, the magnetic moment μ must also have $(2l + 1)$ possible orientations from Eq. (3.6.30). Since l is a positive integer from 0 to n, $2l + 1$ is always an odd number. Subsequently, the orbital angular momentum always has an odd number of possible orientations and as a result, the magnetic moment associated with orbital motion must also have an odd number of possible orientations.

The experimental observation is, however, that the beam of silver atoms splits into two separate beams by the action of the nonuniform field. To understand this result, let us first examine the electron configuration in silver. Silver has forty-seven electrons and its electronic configuration is

$$1s^2\, 2s^2\, 2p^6\, 3s^2\, 3p^6\, 4s^2\, 3d^{10}\, 4p^6\, 5s^1\, 4d^{10}. \tag{3.6.39}$$

There is only one electron in the outer shell occupying the $5s$ orbital. Since all the other shells are completely filled, the net angular momentum of the atom is equal to that of the unfilled shell, the $5s$ level. The orbital angular momentum of an s orbital is zero. Hence the orbital angular momentum of the silver atom is zero. Accordingly, a beam of silver atoms would not split at all from the interaction of the orbital magnetic moment and the nonuniform field. Therefore there must be another effect for a single electron that accounts for the two possible orientations of the magnetic moment. From Eq. (3.6.31), there is a magnetic moment associated with the spin angular momentum of the electron. The experimental evidence provided by the Stern–Gerlach experiment indicates that there must be two possible orientations of the spin magnetic moment and as such there must then be only two orientations of the spin angular momentum. S must have only two possible states. These two states we call spin up or spin down, and they are commonly represented as ↑ for spin up and ↓ for spin down. In summary then, the Stern–Gerlach experiment indicates that there are only two possible spin states for an electron, either spin up or spin down. An electron can be in one or the other spin state and as such there is a twofold degeneracy in the number of allowed electronic states due to the spin of the electron.

Finally, we might ask, how does the spin affect the energy of the electron in an atom? From Eq. (3.6.36) the energy of the magnetic moment in the presence of an external field is given by the dot vector product of the moment and the field

$$\Delta E = -\vec{\mu}_s \cdot \vec{B}. \tag{3.6.40}$$

Subsequently the energy of the electron will be different if the spin magnetic moment is parallel or antiparallel to the field. In an atom, the orbital motion of the electron about the nucleus of the atom generates an orbital magnetic moment. Therefore the spin and the orbital magnetic moments can interact, leading to different energy states for the electron. Depending on the strength of the interaction between the orbital and the spin magnetic moments, the electron will have a different energy. This is called the spin–orbit interaction.

The magnitude of the spin–orbit interaction can be estimated in the following way.

The spin–orbit interaction energy is given by the dot product of the spin magnetic moment and the magnetic field:

$$\Delta E = \frac{g_s \mu_b \vec{S} \cdot \vec{B}}{\hbar},$$
(3.6.41)

where B is the magnetic field produced by the orbital motion of the electron. Equation (3.6.41) is valid if evaluated in a frame of reference in which the electron is at rest. In the case of interest, we have to be careful since the field B arises from the motion of the electron itself. To determine the interaction energy properly, we need to transform into the frame of reference in which the nucleus is at rest and the electron is in motion. Use must be made of the Lorentz transformations. The details of this transformation can be found in the references and are omitted here. Once this transformation is made, an additional factor of 1/2 is introduced into the expression for the interaction energy. The result for the interaction energy is

$$\Delta E = \frac{1}{2} \frac{g_s \mu_b}{\hbar} \vec{S} \cdot \vec{B}.$$
(3.6.42)

The magnetic field arising from the orbital motion of the electron can be determined from the Biot–Savart Law:

$$\vec{B} = \frac{\mu_0}{4\pi} \frac{\vec{j} \times \vec{r}}{r^3} = -\frac{Zq\mu_0}{4\pi} \frac{\vec{v} \times \vec{r}}{r^3} \text{ (mks)}.$$
(3.6.43)

The electric field is (in mks units)

$$\vec{E} = \frac{Zq}{4\pi \epsilon_0} \frac{\vec{r}}{r^3}.$$
(3.6.44)

The magnetic field is then

$$\vec{B} = -\mu_0 \vec{v} \times \frac{Zq}{4\pi} \frac{\vec{r}}{r^3},$$
$$\vec{B} = -\epsilon_0 \mu_0 \vec{v} \times \vec{E}.$$
(3.6.45)

Since

$$c^2 = \frac{1}{\epsilon_0 \mu_0},$$
(3.6.46)

$\vec{B}$ is given as

$$\vec{B} = \frac{1}{c^2} \vec{v} \times \vec{E}.$$
(3.6.47)

The electric field can be reexpressed as the gradient of the potential as

$$\vec{E} = \frac{1}{q}\frac{dV(r)}{dr}\frac{\vec{r}}{r}.$$

(3.6.48)

With this substitution, $\vec{B}$ is

$$\vec{B} = -\frac{1}{qc^2}\frac{1}{r}\frac{dV(r)}{dr}\vec{v}\times\vec{r},$$

$$\vec{B} = -\frac{1}{mqc^2}\frac{1}{r}\frac{dV(r)}{dr}\vec{p}\times\vec{r},$$

(3.6.49)

since $p = mv$. But the orbital angular momentum $\vec{L}$ is simply equal to $\vec{r}\times\vec{p}$. Therefore $\vec{L} = -\vec{p}\times\vec{r}$. Making this substitution, we have finally

$$\vec{B} = \frac{1}{mqc^2}\frac{1}{r}\frac{dV}{dr}\vec{L}.$$

(3.6.50)

Plugging the above expression for B back into Eq. (3.6.42) yields

$$\Delta E = \frac{g_s\mu_b}{2\hbar mqc^2 r}\frac{dV}{dr}\vec{S}\cdot\vec{L}$$

(3.6.51)

for the spin–orbit interaction energy. Examination of Eq. (3.6.51) shows that the spin–orbit energy depends on the dot product of $\vec{S}$ and $\vec{L}$. $\vec{L}$ has different values depending on which orbital state the electron is in. If the electron is in an s orbital, $\vec{L}$ is zero. In this case, the spin–orbit interaction vanishes. If the electron is in a state such as the p or the d orbitals, then the spin–orbit interaction is not zero and the original energy eigenvalues for the electronic states found in Section 3.2 need to be modified by this correction. It should be noted that the spin–orbit interaction is typically a small effect, but nevertheless can be important in many atomic systems.

3.7 Addition of Angular Momentum

In classical mechanics, the addition of the orbital and spin angular momentum to obtain the total angular momentum presents no challenge. The total angular momentum is simply the vector sum of the orbital and spin components. For a composite system, such as the planetary system about the Sun, the total angular momentum can be found in several ways. To first approximation, it can be assumed that all the orbital and spin components are independent. The total angular momentum is given by the addition of the orbital angular momentum of the planets together and then the addition of their total spin angular momentum. Alternatively, the same result is obtained by the addition of the orbital and spin components for each planet together first and then the addition of all these together. The manner in which the momenta are added does not matter in classical mechanics. However, the summation of the angular momentum in quantum mechanics is not so simple. The primary problem is that the total angular momentum, just like the orbital angular momentum, cannot be simultaneously measured along with all three of its components. Subsequently, there are rules that govern the allowed values of $\vec{J}$ and J_z. In this section, no attempt is made to give a complete description of the addition of angular momentum, but rather the subject is simply introduced to provide the reader with some basic information on which to later build if necessary.

To understand how $\vec{L}$ and $\vec{S}$ are added to obtain the total angular momentum in quantum mechanics, it is necessary first to consider what the possible values of $\vec{J}$, the total angular momentum, are. We can obtain the possible values of $\vec{J}$ by examining the properties of the operator J. Some of the basic properties of the angular momentum operator are as follows.

The total angular momentum vector $\vec{J}$ includes the spin and orbital components. $\vec{J}$ has three independent spatial components, J_x, J_y, and J_z. The commutation relations among each of these components are exactly the same as those for the orbital angular momentum components discussed in Section 3.1. The corresponding commutation relations for the components of $\vec{J}$ are

$$J_x J_y - J_y J_x = i\hbar J_z, \qquad J_y J_z - J_z J_y = i\hbar J_x, \qquad J_z J_x - J_x J_z = i\hbar J_y. \tag{3.7.1}$$

Raising and lowering operators J_+ and J_- can be defined in direct analogy to those for the orbital angular momentum as

$$J_+ = J_x + iJ_y, \qquad J_- = J_x - iJ_y. \tag{3.7.2}$$

The square of the total angular momentum is given as

$$J^2 = J_x^2 + J_y^2 + J_z^2. \tag{3.7.3}$$

J^2, of course, commutes with $\vec{J}$. J^2 also commutes with J_z.

The eigenstates of J^2 and J_z are defined as

$$J_z|\lambda, m\rangle = m\hbar|\lambda, m\rangle, \qquad J^2|\lambda, m\rangle = \lambda\hbar^2|\lambda, m\rangle. \tag{3.7.4}$$

There is a definite relationship between λ and $m: \lambda \geq m^2$. Given this relation, the possible values of m given j can be determined with the raising and the lowering operators as follows. $J^2 - J_z^2$ can be written as

$$J^2 - J_z^2 = J_x^2 + J_y^2, \tag{3.7.5}$$

which is simply equal to

$$J^2 - J_z^2 = 1/2(J_+J_- + J_-J_+) = 1/2(J_+J_+^\dagger + J_+^\dagger J_+). \tag{3.7.6}$$

But any operator of the form $AA^\dagger$ has only nonnegative expectation values. Therefore

$$\langle \lambda m | J^2 - J_z^2 | \lambda m \rangle = (\lambda - m^2)\hbar^2 \geq 0. \tag{3.7.7}$$

The operators J_+ and J_- are raising and lowering operators, respectively. In likewise manner to the definition of the orbital angular-momentum raising and lowering operators J_+ and J_-, operating on $|\lambda m\rangle$ gives

$$J_+|\lambda, m\rangle = C_+(\lambda, m)\hbar|\lambda, m+1\rangle, \qquad J_-|\lambda, m\rangle = C_-(\lambda, m)\hbar|\lambda, m-1\rangle. \tag{3.7.8}$$

The condition $\lambda \geq m^2$ limits the value of m for any λ. Therefore there must be a maximum value of m for any given λ. Let this maximum value of m be j. Then,

$$J_+|\lambda, j\rangle = 0. \tag{3.7.9}$$

Consider the operation of J_-J_+ on $|\lambda, j\rangle$. This becomes

$$J_-J_+|\lambda, j\rangle = (J^2 - J_z^2 - \hbar J_z)|\lambda, j\rangle = (\lambda - j^2 - j)\hbar^2|\lambda, j\rangle = 0, \tag{3.7.10}$$

where we have made use of the fact that $J_-J_+ = J_x^2 + J_y^2 - \hbar J_z$, which of course is simply equal to $J^2 - J_z^2 - \hbar J_z$. Proof of this relationship is given in Box 3.7.1. The quantity in Eq. (3.7.10) must be equal to zero because J_+ operating on

---

Box 3.7.1 Relationship of J_+ and J_- to J^2

First note that, in direct analogy to the case for the orbital angular-momentum operators, the raising and the lowering operators J_+ and J_- are defined as

$$J_+ = J_x + iJ_y, \qquad J_- = J_x - iJ_y.$$

Since $J^2 = J_x^2 + J_y^2 + J_z^2$, then

$$J^2 - J_z^2 - \hbar J_z = J_x^2 + J_y^2 - \hbar J_z.$$

Also J_-J_+ is

$$J_-J_+ = (J_x - iJ_y)(J_x + iJ_y) = J_x^2 + J_y^2 + i[J_x, J_y].$$

But $[J_x, J_y] = i\hbar J_z$. With this substitution, we obtain the desired result,

$$J_-J_+ = J_x^2 + J_y^2 - \hbar J_z.$$

$|\lambda, j\rangle$ must be zero since j is the maximum value. Therefore $\lambda = j^2 + j = j(j+1)$.

Similarly, there must also be a lowest value of $m, m = j'$, such that $J_-|\lambda, j'\rangle = 0$. In an analogous manner to the above,

$$J_+J_-|\lambda j'\rangle = (J^2 - J_z^2 + \hbar J_z)|\lambda, j'\rangle = (\lambda - j'^2 + j')\hbar^2|\lambda, j'\rangle = 0. \qquad (3.7.11)$$

Therefore $\lambda = j'(j' - 1)$, so $j' = -j$. Clearly m has values ranging from $-j$ to $+j$.

The possible allowed values of j are specified as follows. j must be either a nonnegative integer or half integer. j has values then of

$$j = 0, \frac{1}{2}, 1, \frac{3}{2}, 2, \ldots . \qquad (3.7.12)$$

The corresponding eigenvalues of J_z are $m\hbar$. There are $2j+1$ total eigenvalues of J_z corresponding to each value of j, since m ranges from $-j$ to $+j$. The possible values of m, the z component of the angular momentum for the composite system, are then

$$m = -j, -j + 1, -j + 2, \ldots, +j. \qquad (3.7.13)$$

The rules for the addition of angular momentum can now be determined. These rules are rather subtle. As an example of the system for which the angular momentum is added, let us envision an atom with multiple electrons. The question we would like to answer is, what is the total angular momentum of the collection of electrons in the atom? In other words, how are the individual particles' angular momenta added to give the total angular momentum of the composite system? The first question one would ask is, how is the total angular momentum of the system $\vec{J}$ determined from the orbital and spin angular-momentum values of each of the component electrons? It is not obvious how this is done. Depending on the atomic system, the total angular momentum is formed in different ways.

In general, $\vec{J}$ is found from the vector sum of $\vec{L}$ and $\vec{S}$ as

$$\vec{J} = \vec{L} + \vec{S}. \qquad (3.7.14)$$

Let us consider two electrons, one with total angular momentum $\vec{J}_1$ and the other with total angular momentum $\vec{J}_2$. There are two distinct ways in which the total angular momentum $\vec{J}$, the sum of $\vec{J}_1$ and $\vec{J}_2$, can be formed.

The first approach is given by summing the orbital contributions $\vec{L}_1$ and $\vec{L}_2$ and the spin contributions $\vec{S}_1$ and $\vec{S}_2$ separately. In this approach, the total orbital and spin components of each individual particle are determined first and are added together to give the total orbital and spin values. The total orbital angular momentum $\vec{L}$ and total spin angular momentum $\vec{S}$ are, respectively,

$$\vec{L} = \vec{L}_1 + \vec{L}_2, \qquad \vec{S} = \vec{S}_1 + \vec{S}_2. \tag{3.7.15}$$

Once the total orbital and spin angular momenta are determined, the total angular momentum of the system $\vec{J}$ is given as

$$\vec{J} = \vec{L} + \vec{S}. \tag{3.7.16}$$

This is known as the LS coupling scheme for the addition of angular momentum. It is applicable to systems in which both $\vec{L}$ and $\vec{S}$ are good quantum numbers, those systems in which $\vec{L}$ and $\vec{S}$ are well defined for each state. The LS coupling scheme is of greatest importance in light elements for which the spin–orbit interaction is weak.

The second method is known as JJ coupling. In this scheme, the total angular momentum of each particle $\vec{J}_1$ and $\vec{J}_2$ is found first as

$$\vec{J}_1 = \vec{L}_1 + \vec{S}_1, \qquad \vec{J}_2 = \vec{L}_2 + \vec{S}_2. \tag{3.7.17}$$

The total angular momentum J is in this scheme:

$$\vec{J} = \vec{J}_1 + \vec{J}_2. \tag{3.7.18}$$

JJ coupling is most applicable for heavy elements in which the spin–orbit interaction is strong.

The actual number and the nature of the achievable composite angular-momentum states can be determined from a simple rule. This rule is called the **vector addition theorem for angular momenta**. When a system of angular momenta j_1 is added to a system with angular momenta j_2, the resulting possible values of the total angular momentum j ranges from the maximum value $|j_1 + j_2|$ in integer steps to the minimum value $|j_1 - j_2|$. The complete set of possible values of the total angular momentum is then

$$|j_1 + j_2|, |j_1 + j_2 - 1|, |j_1 + j_2 - 2|, \ldots, |j_1 - j_2|. \tag{3.7.19}$$

For each composite state of total angular momentum j, there is a z component of the angular-momentum eigenstate m. The possible values of m (that for the composite system) range from $+j$ to $-j$ for each state. This gives a total number of states for each value of j of $(2j + 1)$. Note that these results are consistent with the rules determined above for the values of the z component of the total angular momentum. However, it is not immediately obvious why the values for j range from $|j_1 - j_2|$ to $|j_1 + j_2|$. To see how the restrictions on j occur, consider the following.

The largest possible value for m must be $j_1 + j_2$ and this must be the largest possible value for j. For $m = j_1 + j_2 - 1$, there are two independent states possible. These independent states are constructed by the combination of $m_1 = j_1, m_2 = j_2 - 1$ or $m_1 = j_1 - 1, m_2 = j_2$. One of these combinations belongs to the $j_1 + j_2$ state, while the other belongs to a different j state, $j = j_1 + j_2 - 1$.

For $m = j_1 + j_2 - 2$, there are three possible independent states. This continues until the state where $m = |j_1 - j_2|$ is reached. At this point, the maximum number of independent states has been reached and this is the smallest value of j needed to account for all the states. Therefore the values of j range from $|j_1 + j_2|$ to $|j_1 - j_2|$.

To illustrate the above result, consider the following example. Let the first electron have $j_1 = 2$ and the second electron have $j_2 = 1$. Then the first electron has $2j_1 + 1$ values for m_1 while the second electron has $2j_2 + 1$ values for m_2. According to the rules mentioned above, the total angular momentum j should range from $|j_1 + j_2|$ to $|j_1 - j_2|$. For this example then j should have values of 3, 2, or 1. The corresponding values of m should be

Table 3.7.1
Possible Ways of Adding m_1 and m_2

m_1	m_2	Total m	Total j
+2	+1	+3	3
+2	0	+2	3 or 2
+1	+1	+2	3 or 2
+2	−1	+1	3, 2, or 1
+1	0	+1	3, 2, or 1
0	+1	+1	3, 2, or 1
+1	−1	0	3, 2, or 1
0	0	0	3, 2, or 1
−1	+1	0	3, 2, or 1
0	−1	−1	3, 2, or 1
−1	0	−1	3, 2, or 1
−2	+1	−1	3, 2, or 1
−1	−1	−2	3 or 2
−2	0	−2	3 or 2
−2	−1	−3	3

for $j = 3, m = -3, -2, -1, 0, 1, 2, 3$;
for $j = 2, m = -2, -1, 0, 1, 2$;
for $j = 1, m = -1, 0, 1$.

This gives a total of fifteen states.

Alternatively we can construct the possible states with the help of Table 3.7.1. In this table, the total number of values of m and j are determined given m_1 and m_2 by simple enumeration of the possibilities.

When the result of the table and the result from the rules are compared, it is clear that for the present example both methods agree. The reader can check this result for other combinations.

Although our discussion here is rather brief, the above presentation should provide the reader with sufficient background to enable further study. The interested reader is referred for more detail to the many excellent texts on quantum mechanics listed in the references.

PROBLEMS

1. Consider a positronium atom that consists of an electron in orbit around a positron. A positron has the same mass as an electron but has a positive unit charge. If the reduced mass of the system is defined as

$$\mu = \frac{m_1 m_2}{m_1 + m_2},$$

where m_1 and m_2 are the electron and the positron masses, respectively, determine the wavelength of the electronic transition from the $n = 3$ to $n = 2$ state.

Hint: In the text, the hydrogen atom problem is solved assuming that the mass of the proton is infinite. In other words, the center of mass of the problem is at the center of the proton. In this problem, the center of mass is not at the positron. When the motion of the electron is formulated about the center of mass, the reduced mass results. The reduced mass is the effective mass of the electron moving about the center of mass. Solve the Schroedinger equation then for a particle of mass μ.

2. An electron is in a state whose wave function is

$$\Psi_1 = (x + iy) = \sin\theta e^{i\phi} r.$$

a. Given that the operator corresponding to the z component of the angular momentum L_z is

$$L_z = -\frac{\hbar}{i}\frac{\partial}{\partial\phi},$$

find the eigenvalue for the state Ψ_1.

b. If the electron has a wave function Ψ_2,

$$\Psi_2 = (y + iz),$$

determine the possible values of L_z for this state.

c. Using the results in part b, determine the probability of each eigenvalue occurring after a measurement is performed on the electron initially in state Ψ_2.

3. a. Calculate the expectation value of the potential energy U in the $n = 2, l = 0$ state of the hydrogen atom.

b. Calculate the expectation value of the potential energy U in the $n = 2, l = 1$ state of the hydrogen atom.

Discuss the results in terms of E and the expectation value of the kinetic energy K.

4. A small bead of mass m slides freely along a circular ring of radius R. Using the concept of a wavelike aspect of the bead and the fact that the corresponding wave function is single valued, determine

a. the allowed energies of the bead,

b. the allowed angular-momentum values.

Hint: In this case, it is easiest if you assume that the bead is a wave and determine the conditions for which an integral number of half-wavelengths can fit around the loop. There is no need to solve the Schroedinger equation in this case.

5. A three-dimensional harmonic oscillator of mass m and force constant k is in a stationary state of $n_x = 0, n_y = 0$, and $n_z = 2$.

a. Does this state have definite L^2? If so, give its value.

b. Does this state have definite L_z? If so, give its value.

c. Give the possible results of a simultaneous measurement of L^2 and L_z in this state and give the probabilities for finding each of them.

6. It has been suggested that U^{235} can be separated from U^{238} by laser excitation in the following way. The energy levels in the U^{235} isotope are slightly different from those in the U^{238} isotope since the nuclear mass is different. Consequently, if one tunes a laser to the U^{235} ionization energy, only U^{235} atoms will ionize and be collected

by an applied electric field. To demonstrate the validity of this concept, consider the hydrogen atom. Compare the $n = 2$ with $n = 1$ transition energies in the following systems:

a. hydrogen atom,

b. deuterium atom (one proton, one neutron, and one electron),

c. $(He^3)^+$ ion (two protons, one neutron, and one electron).

7. Determine the value of the matrix element

$$\langle l', m' | [L_+, L_-] | l, m \rangle,$$

where

$$L_+ = L_x + i L_y, \qquad L_+ Y_{lm} = \hbar \sqrt{(l - m)(l + m + 1)} Y_{l,m+1},$$

$$L_- = L_x - i L_y, \qquad L_- Y_{lm} = \hbar \sqrt{(l + m)(l - m + 1)} Y_{l,m-1}.$$

8. Determine the energy levels for a particle with angular momentum $l = 0$ in a spherical potential well such that

$$U(r) = -U_0 \quad \text{for } r \le a, \qquad U(r) = 0 \quad \text{for } r \ge a.$$

Note: the solution of

$$\frac{d^2}{dr^2}(r R) - k^2 (r R) = 0$$

is

$$R \sim \frac{2 \sin kr}{r}, \qquad r < a.$$

Simply set up the transcendental equation. Do not attempt to solve it.

9. State the four quantum numbers of the electronic states in bromine.

10. The transition rate is defined as the probability per second that an atom in a certain energy level will make a transition to another level. Given that the transition probability for an electric dipole transition from state 2 to state 1 is proportional to $\langle 1 | (-er) | 2 \rangle$, determine the relative transition probability between the $n = 1$ and $n = 2, l = 0$ states in a hydrogen atom.

11. The commutators of L_x, L_y, and L_z are formed from cyclic permutation of each, where

$$[L_x, L_y] = i\hbar L_z.$$

If Ψ is an eigenstate of L^2 and L_z, show that

$$\phi = (L_x + i L_y)\Psi$$

is also an eigenstate of L^2 and L_z.

12. Consider the motion of an electron in a screened Coulomb potential, described as

$$V(r) = \frac{q^2}{4\pi\epsilon} \frac{e^{-\beta|r_1 - r_2|}}{|r_1 - r_2|}.$$

a. Write expressions by using both Dirac notation and an integral representation for two electrons that interact through this potential. Assume that the electrons are in plane-wave states of wave vectors k_1 and k_2 initially and scatter into states of wave vectors k_1' and k_2' finally.

b. If the complete Fourier transform of the screened Coulomb interaction,

$$\frac{e^{-\beta u}}{u} \qquad (u \equiv r_1 - r_2),$$

is

$$\frac{1}{\gamma^2 + \beta^2},$$

where γ is the Fourier transform variable conjugate to u. Evaluate the integral in part a.

13. An electron is in the following state:

$$\Psi(r, \theta, \phi) = A(1 - 2\alpha r^2 \cos^2 \theta) e^{\frac{-\alpha r^2}{2}}.$$

a. Does this state have definite L^2? If so, give its value.

b. Does this state have definite L_z? If so, give its value.

c. Give the possible results of a simultaneous measurement of L^2 and of L_z in this state and the probabilities for measuring each of them.

14. What is the value of the matrix element

$$\langle l + 1 | [L_+, L_-] | l \rangle ?$$

15. A particle of mass m moves in the xy plane in a circular orbit of radius $R = 1$. If the Hamiltonian in cylindrical coordinates is

$$H = -\frac{\hbar^2}{2m} \left[\frac{\partial^2}{\partial r^2} + \frac{1}{r} \frac{\partial}{\partial r} + \frac{1}{r^2} \frac{\partial^2}{\partial \theta^2} + \frac{\partial^2}{\partial z^2} \right].$$

a. What are the possible results of a measurement of the total energy if the particle has the wave function

$$\Psi = C \left[\cos 2\theta + \frac{1}{2} \cos 4\theta + \frac{1}{2} \sin 4\theta \right] ?$$

b. What is the probability of measuring each value?

16. Using the conservation of linear momentum in a two-body elastic collision, show that the relativistic mass expression is a necessary consequence of this collision. Use the velocity transformations

$$u_x' = \frac{u_x - v}{1 - \frac{u_x v}{c^2}} \qquad u_y' = \frac{u_y \sqrt{1 - \frac{v^2}{c^2}}}{1 - \frac{u_x v}{c^2}},$$

where the velocity v is along the x direction.

17. Consider a relativistic electron moving in a magnetic field B. Determine the radius of the particle's orbit if its kinetic energy is 10 MeV (10^7 eV) and if it moves in a magnetic field of 2.0 Wb/m^2.

18. The mean lifetime of cosmic-ray muons observed from the Earth is measured to be 1.6×10^{-5} s. The mean lifetime of the muons in the rest frame of the muons is 2.3×10^{-6} s. What is the speed of the cosmic-ray muons with respect to the Earth?

19. State the four quantum numbers of all the filled electronic states in yttrium, which has atomic number 39.

20. Estimate the value of the spin–orbit interaction in the $n = 2, l = 1$ state of the hydrogen atom.

21. In a spontaneous decay of a particle, if two equal-mass daughter particles are emitted in opposite directions, each with velocity $0.67\,c$ with respect to the center of mass of the initial particle, what is the correct speed of one of the new particles with respect to the other?

22. Start with the Dirac equation with $V = 0$,

$$[c\vec{\alpha} \cdot \vec{p} + \beta m_0 c^2]\Psi = E\Psi,$$

and determine the values of the momentum of a photon.

23. What do each of the four quantum numbers used in the description of an electronic state of an atom correspond to in terms of spatial coordinates?

24. A meson moving in the x direction with a kinetic energy as measured in a frame S equal to its rest energy decays into two photons. In frame S', at rest with respect to the meson, the photons are emitted in the positive and the negative y' directions. Find the energies of the photons in frame S'. Also determine the velocity of separation between frames S and S'.

Approximation Methods in Quantum Mechanics

In the previous chapters, we have solved a series of problems that all have an exact solution. However, there are only a select few problems in quantum mechanics that provide exact analytical solutions. Those problems that can be solved exactly are the free particle, the one-dimensional barrier potential, the finite and the infinite square wells, the infinite triangular well, the harmonic oscillator, and the hydrogen atom, including both its nonrelativistic and relativistic solutions. However, there are many problems different from the above-mentioned few as well as problems that cannot be approximated by these solutions. What happens then if we want to solve a different problem, such as the helium atom or the hydrogen atom in an applied electric field? To solve these problems it is necessary to use approximation techniques. In this chapter, we consider some of these techniques, the most important being the time-independent and the time-dependent perturbation theories. We begin with a discussion of time-independent disturbances in a nondegenerate system.

4.1 Time-Independent Perturbation Theory: The Nondegenerate Case

We seek to determine the solution to a problem in which a small disturbance is applied to an otherwise determinable system. Perturbation theory is used most often in cases for which the solution to the undisturbed (unperturbed) system is known and the perturbation is small. For example, perturbation theory is useful in solving for the energy levels in a hydrogen atom in the presence of a small external electric field. The solution of the hydrogen atom alone is known precisely as we found in Chapter 3. Given that the applied field is relatively small, the disturbance or perturbation is also small. Therefore the solution for the energy levels of the system should not be very different from the unperturbed situation.

Consider a system whose Hamiltonian can be dissociated into two separate parts, one part corresponding to the unperturbed system and the other part representing the perturbation. We can write the total Hamiltonian then as

$$H = H_0 + gV, \tag{4.1.1}$$

where H_0 represents the unperturbed part of the Hamiltonian whose eigenfunctions and eigenvalues are assumed to be known and V is a small perturbative potential of the order of g in smallness. The factor g is introduced here as a bookkeeping tool that will directly indicate the order of each term. We want to solve the general problem

$$H\Psi_n = E_n \Psi_n, \tag{4.1.2}$$

assuming that the unperturbed problem

$$H_0 \Psi_n^{(0)} = E_n^{(0)} \Psi_n^{(0)} \tag{4.1.3}$$

has already been solved. In the example mentioned above, that of a hydrogen atom in an external electric field, H_0 is the Hamiltonian describing the unperturbed hydrogen atom with eigenfunctions $\Psi_n^{(0)}$ and gV is the perturbation corresponding to the applied field.

From our supposition that the perturbation is small, it is reasonable to assume that the energy eigenvalues for the complete problem (perturbed and unperturbed Hamiltonians) can be written as

$$E_n = E_n^{(0)} + g E_n^{(1)} + g^2 E_n^{(2)} + \cdots, \tag{4.1.4}$$

where $E_n^{(0)}$ is the unperturbed energy eigenvalue and $g E_n^{(1)}$ is the first-order correction to the energy, $g^2 E_n^{(2)}$ is the second-order correction to the energy, etc. The wave functions can also be expanded in powers of the perturbation parameter g as

$$\Psi_n = \Psi_n^{(0)} + g \Psi_n^{(1)} + g^2 \Psi_n^{(2)} + \cdots. \tag{4.1.5}$$

By substituting Eq. (4.1.5) for the electronic wave function and the energy into the Schroedinger equation, we can obtain the first-order correction as well as all other higher-order corrections to the energy and the wave function. Of greatest interest are the first-order corrections to the energy and the wave functions and the second-order correction to the energy. By definition, higher-order corrections should be unimportant since the perturbation must be small in order for this approach to be valid. Consider then

$$H\Psi_n = E_n \Psi_n,$$
$$(H_0 + gV)\big[\Psi_n^{(0)} + g\Psi_n^{(1)} + g^2 \Psi_n^{(2)} + \cdots\big]$$
$$= \big[E_n^{(0)} + g E_n^{(1)} + g^2 E_n^{(2)} + \cdots\big]\big[\Psi_n^{(0)} + g\Psi_n^{(1)} + g^2 \Psi_n^{(2)} + \cdots\big]. \tag{4.1.6}$$

Expanding out both sides of the above yields

$$H_0 \Psi_n^{(0)} + g H_0 \Psi_n^{(1)} + g^2 H_0 \Psi_n^{(2)} + \cdots + g V \Psi_n^{(0)} + g^2 V \Psi_n^{(1)} + g^3 V \Psi_n^{(2)} + \cdots$$
$$= E_n^{(0)} \Psi_n^{(0)} + E_n^{(0)} g \Psi_n^{(1)} + \cdots + g E_n^{(1)} \Psi_n^{(0)} + g^2 E_n^{(1)} \Psi_n^{(1)} + g^3 E_n^{(1)} \Psi_n^{(2)} + \cdots$$
$$+ g^2 E_n^{(2)} \Psi_n^{(0)} + g^3 E_n^{(2)} \Psi_n^{(1)} + g^4 E_n^{(2)} \Psi_n^{(2)} + \cdots. \tag{4.1.7}$$

The factor g reflects the order of approximation in the expansion of the eigenvalues, eigenvectors, and the potential. Zeroth order corresponds to $g = 0$, the unperturbed solution. First order corresponds to g, second order to g^2, etc. In order for the equality to hold, terms of the same order must be equal. Therefore, equating terms of equal order, which are easily grouped since the power of g in each term determines its order, yields

$$H_0\Psi_n^{(0)} = E_n^{(0)}\Psi_n^{(0)} \tag{4.1.8}$$

for zeroth order,

$$H_0 g\Psi_n^{(1)} + gV\Psi_n^{(0)} = E_n^{(0)} g\Psi_n^{(1)} + gE_n^{(1)}\Psi_n^{(0)} \tag{4.1.9}$$

for first order, and

$$g^2 H_0\Psi_n^{(2)} + g^2 V\Psi_n^{(1)} = g^2 E_n^{(0)}\Psi_n^{(2)} + g^2 E_n^{(1)}\Psi_n^{(1)} + g^2 E_n^{(2)}\Psi_n^{(0)} \tag{4.1.10}$$

for second order. Higher-order terms of course can be considered also. However, for our purposes, only the first three orders, zeroth, first, and second, are of interest. Note that the zeroth-order term is true by design. It is simply Schroedinger's equation for the unperturbed system.

The first-order term leads to an expression for the first-order correction to the energy eigenvalues. This can be determined as follows. Dividing out the factor of g from Eq. (4.1.9) yields

$$H_0\Psi_n^{(1)} + V\Psi_n^{(0)} = E_n^{(0)}\Psi_n^{(1)} + E_n^{(1)}\Psi_n^{(0)}. \tag{4.1.11}$$

Multiplying through on each side by the complex conjugate of $\Psi_n^{(0)}$ and integrating over all space, we find that Eq. (4.1.11) becomes

$$\int \Psi_n^{(0)*} H_0\Psi_n^{(1)} \, d^3r + \int \Psi_n^{(0)*} V\Psi_n^{(0)} \, d^3r$$
$$= \int \Psi_n^{(0)*} E_n^{(0)}\Psi_n^{(1)} \, d^3r + \int \Psi_n^{(0)*} E_n^{(1)}\Psi_n^{(0)} \, d^3r. \tag{4.1.12}$$

Making use of the hermiticity of H_0, we can write the first term on the left-hand side of Eq. (4.1.12) as

$$\int \Psi_n^{(0)*} H_0\Psi_n^{(1)} \, d^3r = \int \left[H_0\Psi_n^{(0)}\right]^* \Psi_n^{(1)} \, d^3r = \int E_n^{(0)}\Psi_n^{(0)*}\Psi_n^{(1)} \, d^3r. \tag{4.1.13}$$

Substituting Eq. (4.1.13) back into Eq. (4.1.12) yields

$$E_n^{(0)} \int \Psi_n^{(0)*}\Psi_n^{(1)} \, d^3r + \int \Psi_n^{(0)*} V\Psi_n^{(0)} \, d^3r$$
$$= E_n^{(0)} \int \Psi_n^{(0)*}\Psi_n^{(1)} \, d^3r + \int \Psi_n^{(0)*} E_n^{(1)}\Psi_n^{(0)} \, d^3r. \tag{4.1.14}$$

The first terms on the left- and the right-hand sides are equal. As such, we are left with

$$\int \Psi_n^{(0)*} V\Psi_n^{(0)} \, d^3r = E_n^{(1)} \int \Psi_n^{(0)*}\Psi_n^{(0)} \, d^3r. \tag{4.1.15}$$

Therefore the expression for $E_n^{(1)}$, the first-order correction to the energy, is

$$E_n^{(1)} = \frac{\int \Psi_n^{(0)*} V \Psi_n^{(0)} \, d^3 r}{\int \Psi_n^{(0)*} \Psi_n^{(0)} \, d^3 r}. \tag{4.1.16}$$

Inspection of the above result shows that the first-order correction to the energy is equal to the matrix element of the perturbation between the unperturbed eigenstates $\Psi_n^{(0)}$.

Next let us find the first-order correction to the wave functions Ψ_n, the eigenstates of the complete Hamiltonian H. To do this we need to consider the first-order term

$$[H_0 - E_{n'}^{(0)}] \Psi_{n'}^{(1)} = [E_{n'}^{(1)} - V] \Psi_{n'}^{(0)}, \tag{4.1.17}$$

where n' labels the state. From the expansion postulate, discussed in Chapter 1, any arbitrary wave function can be expanded in terms of a basis set of eigenstates. Therefore the first-order correction $\Psi_{n'}^{(1)}$ can be expanded in terms of a complete set of eigenfunctions. Since H_0 is Hermitian, its eigenstates form a complete basis set. $\Psi_{n'}^{(1)}$ can then be expanded in terms of the $\Psi_n^{(0)}$ as

$$\Psi_{n'}^{(1)} = \sum_n A_n \Psi_n^{(0)}. \tag{4.1.18}$$

Substituting expansion (4.1.18) into Eq. (4.1.17) gives

$$[H_0 - E_{n'}^{(0)}] \sum_n A_n \Psi_n^{(0)} = [E_{n'}^{(1)} - V] \Psi_{n'}^{(0)},$$

$$\sum_n H_0 A_n \Psi_n^{(0)} - E_{n'}^{(0)} A_n \Psi_n^{(0)} = [E_{n'}^{(1)} - V] \Psi_{n'}^{(0)}. \tag{4.1.19}$$

But the states $\Psi_n^{(0)}$ are eigenstates of H_0. Therefore

$$\sum_n H_0 A_n \Psi_n^{(0)} = \sum_n A_n E_n^{(0)} \Psi_n^{(0)}. \tag{4.1.20}$$

Substituting Eq. (4.1.20) into Eq. (4.1.19) yields

$$\sum_n A_n [E_n^{(0)} - E_{n'}^{(0)}] \Psi_n^{(0)} = [E_{n'}^{(1)} - V] \Psi_{n'}^{(0)}. \tag{4.1.21}$$

Multiplying each side of Eq. (4.1.21) by $\Psi_m^{(0)*}$ and integrating over all space, we obtain

$$\int \sum_n A_n \Psi_m^{(0)*} E_n^{(0)} \Psi_n^{(0)} \, d^3 r - \int \sum_n A_n \Psi_m^{(0)*} E_{n'}^{(0)} \Psi_n^{(0)} \, d^3 r$$

$$= \int \Psi_m^{(0)*} E_{n'}^{(1)} \Psi_{n'}^{(0)} \, d^3 r - \int \Psi_m^{(0)*} V \Psi_{n'}^{(0)} \, d^3 r. \tag{4.1.22}$$

The wave functions $\Psi_m^{(0)}$ and $\Psi_n^{(0)}$ are orthogonal since they are separate eigenfunctions of H_0. Using this property, we can simplify Eq. (4.1.22) as follows. Each term in Eq. (4.1.22) can be written as

$$\int \sum_n A_n \Psi_m^{(0)*} E_n^{(0)} \Psi_n^{(0)} \, d^3r = A_m E_m^{(0)},$$

$$\int \sum_n A_n \Psi_m^{(0)*} E_{n'}^{(0)} \Psi_n^{(0)} \, d^3r = A_m E_{n'}^{(0)},$$

$$\int \Psi_m^{(0)*} E_{n'}^{(1)} \Psi_{n'}^{(0)} \, d^3r = 0, \qquad m \neq n' \text{ (See Liboff 1992, p. 651)},$$

$$\int \Psi_m^{(0)*} V \Psi_{n'}^{(0)} \, d^3r = \langle m | V | n' \rangle. \tag{4.1.23}$$

Replacing each term in Eq. (4.1.22) with Eqs. (4.1.23)

$$A_m \left[E_m^{(0)} - E_{n'}^{(0)} \right] = -\langle m | V | n' \rangle. \tag{4.1.24}$$

If $m = n'$, then $E_m^{(0)} - E_{n'}^{(0)} = 0$, since the eigenvalues must then be the same because the eigenfunctions are identical. As a result, A_m is not determinable from the above. If $m \neq n'$, then

$$A_m(m \neq n') = \frac{-\langle m | V | n' \rangle}{\left[E_m^{(0)} - E_{n'}^{(0)} \right]}, \tag{4.1.25}$$

which, in coordinate representation, is

$$A_m(m \neq n') = -\frac{\int \Psi_m^{(0)*} V \Psi_{n'}^{(0)} \, d^3r}{\left[E_m^{(0)} - E_{n'}^{(0)} \right]}. \tag{4.1.26}$$

Using this expression for the coefficients, we can find the first-order correction to the wave functions as

$$\Psi_{n'}^{(1)} = \sum_n A_n \Psi_n^{(0)},$$

$$\Psi_{n'}^{(1)} = -\sum_{m \neq n'} \frac{\int \Psi_m^{(0)*} V \Psi_{n'}^{(0)} \, d^3r}{\left[E_m^{(0)} - E_{n'}^{(0)} \right]} \Psi_m^{(0)}. \tag{4.1.27}$$

It is important to note that the sum excludes the term $m = n'$ since the denominator vanishes under this condition. In other words, the first-order correction to the wave function involves the sum over all the zeroth-order wave functions, excluding the one for which the denominator vanishes. The numerator is simply the matrix element of the perturbation V between the n'th state and all the other eigenstates of H_0. Equations (4.1.27) can be written in terms of Dirac notation as

$$\Psi_{n'}^{(1)} = -\sum_{m \neq n'} \frac{\langle m | V | n' \rangle}{\left[E_m^{(0)} - E_{n'}^{(0)} \right]} \Psi_m^{(0)}, \tag{4.1.28}$$

and in another useful shorthand notation as

$$\Psi_{n'}^{(1)} = -\sum_{m \neq n'} \frac{V_{mn'}}{\left[E_m^{(0)} - E_{n'}^{(0)} \right]} \Psi_m^{(0)}. \tag{4.1.29}$$

The first-order correction to the energy is

$$E_n^{(1)} = V_{nn} = \langle n | V | n \rangle = \int \Psi_n^{(0)*} V \Psi_n^{(0)} \, d^3 r. \tag{4.1.30}$$

The first-order correction to both the energy and the wave function is sufficient for most problems. However, sometimes the first-order correction to the energy in perturbation theory vanishes, and as such the perturbation must be taken to second order. Therefore it is important to determine the general expression for the second-order correction to the energy in perturbation theory.

To find the second-order correction to the energy, we start with the second-order equation, Eq. (4.1.10):

$$H_0 \Psi_n^{(2)} + V \Psi_n^{(1)} = E_n^{(0)} \Psi_n^{(2)} + E_n^{(1)} \Psi_n^{(1)} + E_n^{(2)} \Psi_n^{(0)}. \tag{4.1.31}$$

Multiply Eq. (4.1.31) on each side by $\Psi_n^{(0)*}$ and integrate over all space to yield

$$\int \Psi_n^{(0)*} H_0 \Psi_n^{(2)} \, d^3 r + \int \Psi_n^{(0)*} V \Psi_n^{(1)} \, d^3 r$$
$$= E_n^{(0)} \int \Psi_n^{(0)*} \Psi_n^{(2)} \, d^3 r + E_n^{(1)} \int \Psi_n^{(0)*} \Psi_n^{(1)} \, d^3 r + E_n^{(2)} \int \Psi_n^{(0)*} \Psi_n^{(0)} \, d^3 r. \tag{4.1.32}$$

Using the hermiticity property of H_0 and the fact that the $\Psi_n^{(0)}$'s are eigenfunctions of H_0, we can simplify Eq. (4.1.32) as

$$\int \Psi_n^{(0)*} H_0 \Psi_n^{(2)} \, d^3 r = \int \left[H_0 \Psi_n^{(0)} \right]^* \Psi_n^{(2)} \, d^3 r = E_n^{(0)} \int \Psi_n^{(0)*} \Psi_n^{(2)} \, d^3 r, \tag{4.1.33}$$

which cancels two terms in Eq. (4.1.31). Equation (4.1.31) then becomes

$$\int \Psi_n^{(0)*} V \Psi_n^{(1)} \, d^3 r = E_n^{(1)} \int \Psi_n^{(0)*} \Psi_n^{(1)} \, d^3 r + E_n^{(2)} \int \Psi_n^{(0)*} \Psi_n^{(0)} \, d^3 r. \tag{4.1.34}$$

The first-order correction to the wave functions can be expanded in terms of the zeroth-order wave functions as

$$\Psi_n^{(1)} = \sum_m A_m \Psi_m^{(0)}, \tag{4.1.35}$$

where $m \neq n$. Substituting in Eq. (4.1.35) for $\Psi_n^{(1)}$ into the right-hand side of Eq. (4.1.34) yields

$$\int \Psi_n^{(0)*} V \Psi_n^{(1)} \, \mathrm{d}^3 r = E_n^{(1)} \int \Psi_n^{(0)*} \sum_m A_m \Psi_m^{(0)} \, \mathrm{d}^3 r + E_n^{(2)} \int \Psi_n^{(0)*} \Psi_n^{(0)} \, \mathrm{d}^3 r.$$

(4.1.36)

Since the sum over m excludes the nth term, then the first term on the right-hand side vanishes. The second-order correction to the energy in perturbation theory is easily solved to yield

$$E_n^{(2)} = \frac{\int \Psi_n^{(0)*} V \Psi_n^{(1)} \, \mathrm{d}^3 r}{\int \Psi_n^{(0)*} \Psi_n^{(0)} \, \mathrm{d}^3 r}.$$

(4.1.37)

From the above it is clear that the second-order correction to the energy depends on the first-order correction to the wave functions. Substituting in for $\Psi_n^{(1)}$, Eq. (4.1.29), we can rewrite $E_n^{(2)}$ as

$$E_n^{(2)} = \frac{\int \Psi_n^{(0)*} V \left[\sum_{m \neq n} \frac{V_{mn}}{E_n^{(0)} - E_m^{(0)}} \Psi_m^{(0)} \right] \mathrm{d}^3 r}{\int \Psi_n^{(0)*} \Psi_n^{(0)} \, \mathrm{d}^3 r},$$

(4.1.38)

where

$$V_{mn} = \langle m | V | n \rangle = \int \Psi_m^{(0)*} V \Psi_n^{(0)} \, \mathrm{d}^3 r.$$

(4.1.39)

Therefore, with the above definition for V_{mn}, $E_n^{(2)}$ can be written as

$$E_n^{(2)} = \sum_{m \neq n} \frac{|V_{mn}|^2}{E_n^{(0)} - E_m^{(0)}},$$

(4.1.40)

assuming that the zeroth-order eigenfunctions are properly normalized.

Let us consider a sample problem to illustrate the use of time-independent, nondegenerate perturbation theory. An interesting example is that of a hydrogen atom in a weak electric field. The effect of the electric field on the energy levels of the atom is called the Stark effect. The applied electric field is assumed to be constant and is applied along the z axis. The choice of the z direction is of course simply arbitrary. The energy of an electron in a uniform electric field F_0 is given as

$$U = q F_0 z,$$

(4.1.41)

where z is the distance over which the field acts. More generally, in vector form, the perturbation is given as

$$U = q \vec{F}_0 \cdot \vec{r}.$$

(4.1.42)

The total Hamiltonian of the system is the sum of the unperturbed Hamiltonian, that of the hydrogen atom itself and that due to the action of the applied electric field. The solution of the Schroedinger equation is precisely known for the hydrogen atom and was given in Chapter 3. The total Hamiltonian of the system then can be written as

$$H = H_0 + H', \tag{4.1.43}$$

where H_0 describes the atom by itself and H' describes the perturbation due to the applied field. Since the applied field is small, it is assumed that the perturbation H' is also small. Therefore we are justified in using perturbation theory in solving the problem. We will solve for the correction to the energy eigenvalues to second order. As we will see below, this is necessary since the first-order correction to the energy vanishes.

The energy for the total system can be written to the second-order correction as

$$E_n = E_n^{(0)} + E_n^{(1)} + E_n^{(2)}, \tag{4.1.44}$$

where $E_n^{(1)}$ and $E_n^{(2)}$ are given from Eqs. (4.1.30) and (4.1.38). The expression for the energy is then

$$E_n = E_n^{(0)} + \langle n| V |n\rangle + \sum_{m \neq n} \frac{|V_{mn}|^2}{E_n^{(0)} - E_m^{(0)}}. \tag{4.1.45}$$

The first-order correction to the energy is determined by substituting the perturbation given by Eq. (4.1.42) in for V in Eq. (4.1.45). This becomes

$$\langle n| V |n\rangle = q\vec{F} \cdot \int \Psi_n^{(0)*} \vec{r} \, \Psi_n^{(0)} \, d^3r \equiv r_{nn}, \tag{4.1.46}$$

where r_{nn} represents the matrix element of the position vector r between the nth zero-order eigenstates of H_0. The matrix element V_{mn} can be written with the same notation as

$$V_{mn} = q\vec{F} \cdot \int \Psi_m^{(0)*} \vec{r} \, \Psi_n^{(0)} \, d^3r = q\vec{F} \cdot \vec{r}_{mn}, \tag{4.1.47}$$

where $\vec{r}_{mn}$ is defined as

$$\vec{r}_{mn} \equiv \int \Psi_m^{(0)*} \vec{r} \, \Psi_n^{(0)} \, d^3r. \tag{4.1.48}$$

The total energy of the system, the hydrogen atom, and the applied electric field is

$$E_n = E_n^{(0)} + q\vec{F} \cdot \vec{r}_{nn} + q^2 \sum_{m \neq n} \frac{(\vec{F} \cdot \vec{r}_{mn})(\vec{F} \cdot \vec{r}_{nm})}{E_n^{(0)} - E_m^{(0)}}. \tag{4.1.49}$$

Consider each term in Eq. (4.1.49). The first term, $E_n^{(0)}$, is the zeroth-order energy that is equal to the unperturbed energy of the hydrogen atom. The first-order correction is given by the second term. This term can be evaluated once the unperturbed state of the hydrogen atom is known. For example, consider the case in which the electron is in the ground state of the unperturbed hydrogen atom. The unperturbed eigenfunction is, in this case,

$$\Psi_1^{(0)} = \frac{1}{\sqrt{\pi}} \frac{1}{a_0^{\frac{3}{2}}} e^{-\frac{r}{a_0}}. \tag{4.1.50}$$

Note that $\Psi_1^{(0)}$ has full rotational symmetry. It remains invariant under any rotation. Since r is a vector quantity, the mean value of the vector r in a spherically symmetric state must be zero. Subsequently, the first-order correction to the energy must vanish in a spherically symmetric state. To first order then, the energy of the $n = 1$ state of the hydrogen atom is not altered by the application of an electric field.

Physically, the expectation value of $-e\vec{r}$ represents the expectation value of the electric dipole moment. This follows from the definition of the electric dipole moment as the expectation value of the position vector $\vec{r}$ in terms of the probability density ρ as

$$\vec{p} = -q \int \rho \vec{r} \, d^3 r. \tag{4.1.51}$$

ρ, to zeroth order, is given as

$$\rho = \Psi_n^{(0)*} \Psi_n^{(0)}. \tag{4.1.52}$$

Therefore the expectation value of the dipole moment vanishes to first order in states with spherical symmetry. The expectation value of the dipole moment does not necessarily vanish, though, for states that are not spherically symmetric, such as the $2p$ states. In those cases, the first-order correction to the energy is not necessarily zero.

The second-order correction to the energy does not vanish, however, for the $n = 1$ state. If the field is assumed to be applied along the z direction, the dot product of the field and r is simply equal to $F_0 Z_{nm}$. Therefore the second-order correction to the energy is

$$E_n^{(2)} = q^2 F_0^2 \frac{\sum_{n \neq m} |Z_{nm}|^2}{E_n^{(0)} - E_m^{(0)}}. \tag{4.1.53}$$

What, though, is the value of Z_{nm}? To determine Z_{nm}, it is necessary to evaluate the first-order correction to the eigenfunctions. The first-order correction to the eigenfunctions can be expressed in this case as

$$\Psi_n^{(1)} = \sum_{n \neq m} \frac{V_{mn}}{E_n^{(0)} - E_m^{(0)}} \Psi_m^{(0)}$$

$$= \sum_{n \neq m} \frac{q\vec{F} \cdot \vec{r}_{mn}}{E_n^{(0)} - E_m^{(0)}} \Psi_m^{(0)}. \tag{4.1.54}$$

To determine r_{mn}, it is useful to evaluate the electric dipole moment by using the wave functions expanded to first order. The probability density ρ to first order is given as

$$\rho = \left| \Psi_n^{(0)} + \Psi_n^{(1)} \right|^2, \tag{4.1.55}$$

where $\Psi_n^{(1)}$ is the first-order correction to the wave function. The expectation value of the electric dipole moment is given by Eq. (4.1.51). Substituting into Eq. (4.1.51), Eq. (4.1.55) we calculate the electric dipole moment as

$$\vec{p} = -q \int \vec{r} \left| \Psi_n^{(0)} \right|^2 d^3r - q^2 \int \Psi_n^{(0)*} \vec{r} \sum_{m \neq n} \frac{\vec{F} \cdot \vec{r}_{mn}}{E_n^{(0)} - E_m^{(0)}} \Psi_m^{(0)} d^3r$$

$$- q^2 \int \sum_{m \neq n} \frac{\vec{F} \cdot \vec{r}_{nm}}{E_n^{(0)} - E_m^{(0)}} \Psi_m^{(0)*} \vec{r} \Psi_n^{(0)} d^3r$$

$$+ a \text{ squared term in } \Psi_n^{(1)}. \tag{4.1.56}$$

Neglecting the squared term in $\Psi_n^{(1)}$, we can express the electric dipole moment as

$$\vec{p} = -q \int \left| \Psi_n^{(0)} \right|^2 \vec{r} \, d^3r - q^2 \sum_{m \neq n} \frac{\vec{r}_{nm} \vec{r}_{mn} \cdot \vec{F}}{E_n^{(0)} - E_m^{(0)}} - q^2 \sum_{m \neq n} \frac{\vec{r}_{mn} \vec{r}_{nm} \cdot \vec{F}}{E_n^{(0)} - E_m^{(0)}}. \tag{4.1.57}$$

If the first term above is defined as the zeroth-order value of the dipole moment p_0 and the second and the third terms are combined under one sum, the electric dipole moment to first order can be expressed as

$$\vec{p} = \vec{p}_0 - q^2 \sum_{m \neq n} \frac{(\vec{r}_{nm} \vec{r}_{mn} + \vec{r}_{mn} \vec{r}_{nm}) \cdot \vec{F}}{\left[E_n^{(0)} - E_m^{(0)} \right]}. \tag{4.1.58}$$

Therefore the electric dipole moment can be written to first order as

$$\vec{p} = \vec{p}_0 + \overset{\leftrightarrow}{\alpha} \cdot \vec{F}, \tag{4.1.59}$$

where $\vec{\overset{\leftrightarrow}{\alpha}}$ is defined as the polarizability and has the value

$$\overset{\leftrightarrow}{\alpha} = -q^2 \sum_{m \neq n} \frac{(\vec{r}_{nm}\,\vec{r}_{mn} + \vec{r}_{mn}\,\vec{r}_{nm}) \cdot \vec{F}}{\left[E_n^{(0)} - E_m^{(0)}\right]}. \tag{4.1.60}$$

The term $\overset{\leftrightarrow}{\alpha} \cdot \vec{F}$ is the induced electric dipole moment. The first term p_0 is the permanent dipole moment of the atom. Physically, Eq. (4.1.59) can be interpreted as follows. For states of spherical symmetry, the permanent electric dipole moment vanishes. This is because the mean value of the vector r is zero when averaged over a spherically symmetric state. As a result, if the outermost electrons are all in spherically symmetric states (s orbitals) then the atom has no permanent electric dipole moment. However, if an electric field is applied, an induced electric dipole moment is produced.

In the presence of an external field, an electron in a spherical orbital is attracted by the field. The presence of the external field produces a preferred direction in space along which the electron becomes aligned. As a result, the electron is no longer equally likely to be found in any direction around the atom. Hence the atom becomes polarized, producing a weak electric dipole moment.

The energy of the atom in an external electric field F is given in general by the dot vector product of the dipole moment and the applied electric field, $-\vec{p} \cdot \vec{F}$. When the result obtained above for the electric dipole moment is used, the energy of the system is given as

$$-\vec{p} \cdot \vec{F} = -\vec{p}_0 \cdot \vec{F} - \vec{F} \cdot \overset{\leftrightarrow}{\alpha} \cdot \vec{F} + \cdots = E_n, \tag{4.1.61}$$

where E_n is the total energy of the nth state. Since the total energy is given as the sum of the zeroth-, first-, second-, etc. order corrections,

$$E_n = E_n^{(0)} + E_n^{(1)} + E_n^{(2)} + \cdots, \tag{4.1.62}$$

it is clear that the second-order correction to the energy of the atom is given by the interaction of the induced electric dipole moment and the electric field.

4.2 Degenerate Perturbation Theory

In Section 4.1 it was found that the first-order correction to the wave functions in perturbation theory contains a term in the denominator involving the difference between the zeroth-order energy eigenvalues of the system. The sum over the states excludes the case of $m = n$, since the coefficient is assumed zero to remove its arbitrariness. Therefore no singularity is encountered when the first-order correction to the wave functions is evaluated. What would happen, though, if any two zeroth-order eigenstates had the same energy? In other words, how do we treat the case of degeneracy? Inspection of Eq. (4.1.28) clearly shows that if two different eigenstates, call them m and n, have exactly the same energy,

$E_m^{(0)} = E_n^{(0)}$, then the denominator is zero and the expression for the first-order correction to the wave function diverges. The approach developed in Section 4.1 is no longer valid if the system is degenerate. In this section we address this issue and develop an alternative perturbation formulation that can treat degenerate systems.

The case of degeneracy can be stated as follows. Let us assume that the zeroth-order solution of the Schroedinger equation has several, call them g, solutions, all of the same energy. These g states are then all degenerate. Let these degenerate states be written as

$$\Psi_{n,m}^{(0)}, \quad m = 1, 2, 3, \dots, g, \tag{4.2.1}$$

where the superscript (0) represents the order of correction to the wave function (in this case, the wave function is to zeroth order). The first subscript n represents the state of the electron. The second subscript m represents which one of the g degenerate states is being considered. m ranges in value from 1 to g. Each of the states $\Psi_{n,m}^{(0)}$ is an eigenstate of H_0 with precisely the same eigenvalue $E_n^{(0)}$. From the discussion in Chapter 1, any linear combination of the g degenerate eigenstates is also an eigenstate. Linear combinations of the degenerate eigenstates can be constructed as

$$\Psi_n^{(0)\alpha} = \sum_{m=1}^{g} b_m^\alpha \Psi_{n,m}^{(0)}, \tag{4.2.2}$$

where the coefficients are labeled as b_m^α. The new wave functions $\Psi_n^{(0)\alpha}$ are constructed as a linear combination of the old degenerate states. Our aim is to construct the new wave functions from the original degenerate wave functions such that only diagonal matrix elements are retained, thereby eliminating the singular coefficients. By using the new wave functions, then, we can use the original approach for nondegenerate perturbation theory to solve the problem. The issue is then, how do we determine the coefficients in the expansion given above such that the matrix elements for the degenerate states are diagonal?

The coefficients b_m^α must be chosen such that the terms in the first-order correction to the wave function vanish whenever the denominator is zero. Recall that the first-order correction to the wave functions is given as

$$\Psi_n^{(1)} = -\sum_{m \neq n} \frac{V_{mn}}{E_n^{(0)} - E_m^{(0)}} \Psi_m^{(0)}, \tag{4.2.3}$$

where V_{mn} is defined as

$$V_{mn} = \int \Psi_m^{(0)*\alpha} H' \Psi_n^{(0)} \, d^3r. \tag{4.2.4}$$

The new wave functions $\Psi_{n,m}^{(0)\alpha}$ must be constructed such that V_{mn} vanishes whenever $E_n^{(0)} = E_m^{(0)}$. Under this condition then, no singularity occurs in the first-order correction to the wave functions. Therefore the coefficients b_m^α are chosen such that

$$\int \Psi_n^{(0)*\alpha} H' \Psi_n^{(0)\beta} \, d^3r \equiv H'_{\alpha,\beta} = 0. \qquad (4.2.5)$$

The problem of choosing the new wave functions such that the matrix elements $H'_{\alpha\beta}$ vanish for all $\alpha \neq \beta$ is known as diagonalizing the $g \times g$ matrix, $H'_{\alpha\beta}$. To construct the diagonalization we start with the Schroedinger equation for the total Hamiltonian:

$$(H_0 + H')\Psi_n = E_n \Psi_n. \qquad (4.2.6)$$

Let us consider the case in which the ground state is assumed to be g-fold degenerate. The new wave function formed from a linear combination of the degenerate wave functions can be written as

$$\Psi_0^{(0)\alpha} = \sum_{m=1}^{g} b_m \Psi_{0,m}^{(0)}. \qquad (4.2.7)$$

The above expansion shows that the new ground-state wave function is found from a linear combination of the g-fold degenerate ground-state wave functions. The original problem consists of g linearly independent $n = 0$ states, all with the same energy. As discussed above, it is necessary to form new states, the Ψ_0^α for which the matrix element $V_{mn} = 0$ when $E_{0,m}^{(0)} - E_{0,n}^{(0)} = 0$. The matrix elements must vanish when the new wave functions are not identical, $\alpha \neq \beta$. To make the construction, we start with the Schroedinger equation for the entire system with the new wave functions. This is given then as

$$(H_0 + H')\Psi_0^{(0)\alpha} = E_0 \Psi_0^{(0)\alpha}, \qquad (H_0 - E_0)\Psi_0^{(0)\alpha} = -H'\Psi_0^{(0)\alpha}. \qquad (4.2.8)$$

Substituting the expansion in terms of the degenerate eigenfunctions given by Eq. (4.2.2) in for $\Psi_0^{(0)\alpha}$ yields

$$(H_0 - E_0)\left[\sum_{m=1}^{g} b_m \Psi_{0,m}^{(0)} \right] = -\sum_{m=1}^{g} b_m H' \Psi_{0,m}^{(0)}. \qquad (4.2.9)$$

But the zeroth-order wave functions are eigenstates of H_0. Therefore the application of H_0 onto the wave function on the left-hand side of Eq. (4.2.9) above simplifies to

$$\sum_{m=1}^{g} b_m H_0 \Psi_{0,m}^{(0)} = \sum_{m=1}^{g} E_0^{(0)} b_m \Psi_{0,m}^{(0)}. \qquad (4.2.10)$$

With the above substitution, Eq. (4.2.9) becomes

$$\sum_{m=1}^{g} \left[E_0^{(0)} - E_0 \right] b_m \Psi_{0,m}^{(0)} = -\sum_{m=1}^{g} b_m H' \Psi_{0,m}^{(0)}. \qquad (4.2.11)$$

Multiplying Eq. (4.2.11) by $\Psi_{0,p}^{(0)*}$ and integrating over all space gives

$$\int \sum_{m=1}^{g} \Psi_{0,p}^{(0)*} [E_0^{(0)} - E_0] b_m \Psi_{0,m}^{(0)} \, d^3r = \int \sum_{m=1}^{g} -\Psi_{0,p}^{(0)*} b_m H' \Psi_{0,m}^{(0)} \, d^3r. \qquad (4.2.12)$$

We can simplify the left-hand side of Eq. (4.2.12) by recognizing that all the factors sandwiched between the two wave functions are numbers. Therefore the difference between the energies and the coefficients can be factored out, leaving a sum over the zeroth-order wave functions. This can be simplified as follows. Since the zeroth-order wave functions are orthonormalized, the integral on the left-hand side can be reduced to

$$[E_0^{(0)} - E_0] \sum_{m=1}^{g} \int \Psi_{0,p}^{(0)*} b_m \Psi_{0,m}^{(0)} \, d^3r = b_p [E_0^{(0)} - E_0], \qquad (4.2.13)$$

since

$$\int \Psi_{0,p}^{(0)*} \Psi_{0,m}^{(0)} \, d^3r = \delta_{m,p}. \qquad (4.2.14)$$

Therefore Eq. (4.2.12) becomes

$$[E_0^{(0)} - E_0] b_p = -\sum_{m=1}^{p} b_m H'_{pm}, \qquad (4.2.15)$$

where H'_{pm} is given as

$$H'_{pm} \equiv \int \Psi_{0,p}^{(0)*} H' \Psi_{0,m}^{(0)} \, d^3r. \qquad (4.2.16)$$

If E'_0 is defined as

$$-E'_0 = [E_0^{(0)} - E_0], \qquad E_0 = E_0^{(0)} + E'_0, \qquad E'_0 = E_0^{(1)}, \qquad (4.2.17)$$

then Eq. (4.2.16) can be written as

$$\sum_{m=1}^{g} H'_{pm} b_m = E'_0 b_p. \qquad (4.2.18)$$

It is important to recognize that E'_0 is the first-order correction to the energies as shown by Eqs. (4.2.17). The indices in Eq. (4.2.18), p and m, are free and dummy indices, respectively. A dummy index, m in Eq. (4.2.18) above, is a repeated index that disappears whenever the summation is written out in full. A free index, like p above, is one that appears only once in each term of the equation. p has values from 1 to g. Equation (4.2.18) can be conveniently rewritten as

$$\sum_{m=1}^{g} (H'_{pm} - E'_0 \delta_{mp}) b_m = 0. \qquad (4.2.19)$$

Box 4.2.1   Matrix Representation of Degenerate Perturbation Theory

We can recast the discussion using the matrix formulation of quantum mechanics presented in Section 1.7. It is perhaps helpful to do this to aid in our understanding of degenerate perturbation theory. As in Section 1.7, we seek the eigenvalues of the Hamiltonian H' when the wave function is not necessarily an eigenstate of H'. The matrix of the perturbation Hamiltonian H' in the states $1 - n$ with the first g state degenerate is

$$H' = \begin{bmatrix} H_{11} & H_{12} & . & H_{1g} & . & . & H_{1n} \\ . & . & . & . & . & . & . \\ . & . & . & . & . & . & . \\ H_{g1} & . & . & H_{gg} & . & . & . \\ . & . & . & . & . & . & . \\ . & . & . & . & . & . & . \\ H_{n1} & . & . & . & . & . & H_{nn} \end{bmatrix}.$$

The $g \times g$ submatrix corresponds to the g-fold degenerate states of the system. From the discussion in the text, the coefficient in the expansion of the first-order correction to the eigenfunctions for these states vanishes when the energy is degenerate. This situation must be avoided. To prevent this, the eigenstates must be reexpressed in terms of a new basis such that only the diagonal elements are retained, thereby eliminating the singular coefficients. The new basis comprises states formed from a linear combination of the original degenerate states. This new basis is defined as $\Psi^{(0),\alpha}$. Now if the new basis is chosen such that the off-diagonal elements in the $g \times g$ submatrix vanish, then no singularity occurs. Hence in the new basis the $g \times g$ submatrix is diagonal and the matrix H' becomes

$$H' = \begin{bmatrix} H_{\alpha\alpha} & . & . & 0 & . & H_{1n} \\ . & H_{\beta\beta} & . & . & . & . \\ . & . & . & . & . & . \\ 0 & . & . & H_{gg} & . & . \\ . & . & . & . & . & . \\ H_{n1} & . & . & . & . & H_{nn} \end{bmatrix},$$

where all the elements in the $g \times g$ matrix are zero except those along the diagonal. From the discussion in Section 4.1 on nondegenerate perturbation theory, the diagonal elements $H'_{\alpha\alpha}$, etc., are simply the first-order corrections to the energy eigenvalues due to the presence of the perturbation. Therefore the goal in degenerate perturbation theory is to diagonalize the degenerate submatrix. The diagonal elements correspond then to the first-order corrections to the eigenenergies. To diagonalize the submatrix, a suitable basis set must be selected. The selection of this basis follows the arguments presented in the text.

The solutions of Eq. (4.2.19), known as the secular equation, for the coefficients b_p generate the correct zeroth-order linear combination of the ground-state wave functions, which satisfy the condition that the matrix element vanishes whenever the denominator in Eq. (4.2.3) is zero.

The problem is, how do we now find the coefficients in the secular equation? We obtain the nontrivial solution for the coefficients b_m by setting the determinant of the factor multiplying them to zero. In other words, to obtain the nontrivial solution to Eq. (4.2.19),

$$\det\left(H'_{pm} - E'_0 \delta_{mp}\right) = 0. \tag{4.2.20}$$

Therefore, by setting the determinant equal to zero, we can obtain the energy eigenvalues that correspond to the new wave functions. Backsubstituting into Eq. (4.2.18) for the eigenvalues, we can obtain the coefficients for the new wave functions. Generally, the perturbation acts such that the new eigenfunctions will no longer be degenerate and the resulting eigenvalues obtained from Eq. (4.2.20) are different.

The energy values E' determined from the solution of Eq. (4.2.20) are indeed the energy eigenvalues for the new linear combinations of the original degenerate eigenstates. This is obvious from Eqs. (4.2.17), wherein the energy E'_0 is shown to be precisely the first-order correction to the energies. Note that the action of the perturbation acts to lift the degeneracy of the original states. In the presence of the perturbation, the degeneracy is removed.

It is useful to demonstrate that the resulting eigenfunctions $\Psi_0^{(0)\alpha}$ are orthogonal to one another. Given that the new eigenfunctions are orthogonal and are no longer degenerate, they will not lead to a divergence in Eq. (4.2.3). To demonstrate the orthogonality of the wave functions, consider

$$\int \Psi_0^{(0)\alpha*} \Psi_0^{(0)\beta} \, d^3r = \int \sum_{p=1}^{g} b_p^{\alpha*} b_p^{\beta} \Psi_{0,p}^{(0)*} \Psi_{0,p}^{(0)} \, d^3r. \tag{4.2.21}$$

To evaluate the above matrix element, it is necessary to determine what the product of the two eigenfunctions is. Consider the following. The secular equation was found to be

$$\sum_{m=1}^{g} H_{pm} b_m = E b_p. \tag{4.2.22}$$

Multiplying each side by $b_p^{\beta*}$ and summing over p yields

$$\sum_p b_p^{\beta*} \sum_m H_{pm} b_m^{\alpha} = \sum_p E^{\alpha} b_p^{\alpha} b_p^{\beta*}, \tag{4.2.23}$$

where the sums still range from 1 to g. An equivalent expression can be obtained by multiplying by $b_p^{\alpha*}$, giving

$$\sum_p b_p^{\alpha*} \sum_m H_{pm} b_m^{\beta} = \sum_p E^{\beta} b_p^{\beta} b_p^{\alpha*}. \tag{4.2.24}$$

Take the complex conjugate of Eq. (4.2.24) and interchange the indices p and m. The indices can be switched without altering the value of the expression. This becomes

$$\sum_p b_p^{\beta*} \sum_m H_{mp}^* b_m^\alpha = \sum_p E^{\beta*} b_p^\alpha b_p^{\beta*}. \tag{4.2.25}$$

However, $H_{mp}^* = H_{pm}$ because of the hermiticity of H. Subtracting Eq. (4.2.25) from Eq. (4.2.23) yields

$$(E^\alpha - E^{\beta*}) \sum_p b_p^\alpha b_p^{\beta*} = 0. \tag{4.2.26}$$

Of course $E^{\beta*}$ is equal to E^β since the energy is real. If $\alpha = \beta$, then the energy eigenvalues must of course be equal. However, if the energy eigenvalues are different, then the sum over the coefficients must vanish. Therefore the coefficients for the new wave functions are orthogonal, and of course the new wave functions are orthogonal.

It is helpful to examine a few worked examples to understand how to use degenerate perturbation theory. Two worked examples follow.

EXAMPLE 4.2.1

Consider a doubly degenerate state. We seek to evaluate the general expression for the energy eigenvalues in the presence of a perturbation H'. Use of first-order perturbation theory is prohibited since there is a degeneracy and hence a singularity appears in the equations. Therefore we need to use degenerate perturbation theory to find the energy eigenvalues from the action of the perturbation. Start with the secular equation, Eq. (4.2.19):

$$\sum_{m=1}^g (H_{pm}' - E'\delta_{mp}) b_m^\alpha = 0, \tag{4.2.27}$$

where $g = 2$. Remember that g specifies the degeneracy of the system. In this case it is 2. The new nondegenerate eigenfunctions are

$$\Psi_0^{(0)\alpha} = \sum_{p=1}^2 b_p^\alpha \Psi_{0,p}^{(0)}. \tag{4.2.28}$$

We obtain the corrected energy eigenvalues from the secular equation by insisting that the determinant of the prefactor vanish. This becomes then

$$\det(H_{pm}' - E'\delta_{mp}) = 0. \tag{4.2.29}$$

Expanding out the secular equation yields

$$\begin{aligned} (H_{11}' - E')b_1^\alpha + H_{12}' b_2^\alpha &= 0, \\ H_{21}' b_1^\alpha + (H_{22}' - E')b_2^\alpha &= 0. \end{aligned} \tag{4.2.30}$$

In matrix form,

$$\begin{bmatrix} (H'_{11} - E') & H'_{12} \\ H'_{21} & (H'_{22} - E') \end{bmatrix} \begin{bmatrix} b_1^\alpha \\ b_2^\alpha \end{bmatrix} = 0. \tag{4.2.31}$$

For the nontrivial solution to exist, the determinant of the coefficients must vanish. Expanding out the determinant above yields

$$(H'_{11} - E')(H'_{22} - E') - H'_{21} H'_{12} = 0. \tag{4.2.32}$$

Solving for the corrected energy eigenvalues, we obtain

$$(E')^2 - E'(H'_{11} + H'_{22}) + (H'_{11} H'_{22} - H'_{21} H'_{12}) = 0,$$

$$E' = \frac{(H'_{11} + H'_{22})}{2} \pm \frac{\sqrt{[(H'_{11} + H'_{22})^2 - 4H'_{11} H'_{22} + 4|H'_{12}|^2]}}{2}. \tag{4.2.33}$$

The product of H'_{12} and H'_{21} can be written as the square of the absolute value of H'_{12} since H' is Hermitian. Note that there are two eigenvalues E'_+ and E'_-. If there were n degenerate states we would obtain n corrected eigenvalues. Expanding out the term under the square root, the eigenvalues are given as

$$E' = \frac{1}{2} \left[(H'_{11} + H'_{22}) \pm \sqrt{H'^2_{11} + H'^2_{22} + 2H'_{11} H'_{22} - 4H'_{11} H'_{22} + 4|H'_{12}|^2} \right],$$

$$E' = \frac{1}{2} \left[(H'_{11} + H'_{22}) \pm \sqrt{H'^2_{11} + H'^2_{22} - 2H'_{11} H'_{22} + 4|H'_{12}|^2} \right],$$

$$E' = \frac{1}{2} \left[(H'_{11} + H'_{22}) \pm \sqrt{(H'_{11} - H'_{22})^2 + 4|H'_{12}|^2} \right]. \tag{4.2.34}$$

Solving for the corrected energy eigenvalues, E'_+ and E'_-, gives

$$E'_+ = \frac{1}{2} \left[(H'_{11} + H'_{22}) + \sqrt{(H'_{11} - H'_{22})^2 + 4|H_{12}|^2} \right],$$

$$E'_- = \frac{1}{2} \left[(H'_{11} + H'_{22}) - \sqrt{(H'_{11} - H'_{22})^2 + 4|H_{12}|^2} \right]. \tag{4.2.35}$$

Note that the corrected energy eigenvalues are different. The presence of the perturbation acts to lift the degeneracy and the resulting eigenstates within the perturbation are no longer degenerate.

The coefficients for the corrected eigenfunctions can be determined as follows. From the secular equation, b_2^α can be written in terms of b_1^α as

$$b_2^\alpha = -\frac{(H'_{11} - E'_+)}{H'_{12}} b_1^\alpha,$$

$$b_2^\alpha = -\frac{b_1^\alpha}{H'_{12}} \left[\frac{1}{2}(H'_{11} - H'_{22}) - \frac{1}{2}\sqrt{(H'_{11} - H'_{22})^2 + 4|H'_{12}|^2} \right], \tag{4.2.36}$$

where the above expression for E'_+ has been substituted in. The second set of coefficients can be determined similarly as

$$b_2^\beta = -\frac{(H'_{11} - E'_-)}{H'_{12}} b_1^\beta,$$

$$b_2^\beta = -\frac{b_1^\beta}{H'_{12}} \left[\frac{1}{2}(H'_{11} - H'_{22}) + \frac{1}{2}\sqrt{(H'_{11} - H'_{22})^2 + 4|H'_{12}|^2} \right]. \tag{4.2.37}$$

The corrected eigenfunctions can be written with the coefficients then as

$$\Psi_0^{(0)\alpha} = b_1^\alpha \Psi_{0,1}^{(0)} + b_2^\alpha \Psi_{0,2}^{(0)},$$

$$\Psi_0^{(0)\beta} = b_1^\beta \Psi_{0,1}^{(0)} + b_2^\beta \Psi_{0,2}^{(0)}, \tag{4.2.38}$$

where the above expressions for the b's can be substituted in.

Finally, it is useful to demonstrate that the resulting corrected wave functions are orthogonal. This can be shown from the definition of orthogonality as

$$\int \Psi_0^{(0)\alpha *} \Psi_0^{(0)\beta} \, d^3 r = \int \left(b_1^{\alpha *} \Psi_{0,1}^{(0)*} + b_2^{\alpha *} \Psi_{0,2}^{(0)*} \right) \cdot \left(b_1^\beta \Psi_{0,1}^{(0)} + b_2^\beta \Psi_{0,2}^{(0)} \right) d^3 r. \tag{4.2.39}$$

Multiplying out each term on the right-hand side of Eq. (4.2.39) gives four terms. These are

$$= \int \left[b_1^{\alpha *} b_1^\beta \Psi_{0,1}^{(0)*} \Psi_{0,1}^{(0)} + b_1^{\alpha *} b_2^\beta \Psi_{0,1}^{(0)*} \Psi_{0,2}^{(0)} + b_2^{\alpha *} b_1^\beta \Psi_{0,2}^{(0)*} \Psi_{0,1}^{(0)} \right.$$

$$\left. + b_2^{\alpha *} b_2^\beta \Psi_{0,2}^{(0)*} \Psi_{0,2}^{(0)} \right] d^3 r. \tag{4.2.40}$$

The second and the third terms in Eq. (4.2.40) integrate out to zero since the degenerate states $\Psi_{0,1}^{(0)}$ and $\Psi_{0,2}^{(0)}$ are orthogonal. As a result, the integral is simply

$$\int \Psi_0^{(0)*\alpha} \Psi_0^{(0)\beta} \, d^3 r = b_1^{\alpha *} b_1^\beta + b_2^{\alpha *} b_2^\beta. \tag{4.2.41}$$

Substituting in for $b_1^{\alpha *}$, b_1^β, etc., we find that the overlap integral is

$$= b_1^{\alpha *} b_1^\beta - \frac{b_1^{\alpha *}}{H'_{12}} \left[\frac{1}{2}(H'_{11} - H'_{22}) - \frac{1}{2}\sqrt{(H'_{11} - H'_{22}) + 4|H'_{12}|^2} \right]$$

$$* \left(-\frac{b_1^\beta}{H'_{12}} \right) \left[\frac{1}{2}(H'_{11} - H'_{22}) + \frac{1}{2}\sqrt{(H'_{11} - H'_{22})^2 + 4|H'_{12}|^2} \right]. \tag{4.2.42}$$

When the last two terms above are multiplied together, the cross terms vanish, leaving

$$= b_1^{\alpha*} b_1^{\beta} + \frac{b_1^{\alpha*} b_1^{\beta}}{|H_{12}'|^2} \left[\frac{1}{4} (H_{11}' - H_{22}')^2 - \frac{1}{4} \left[(H_{11}' - H_{22}')^2 + 4|H_{12}'|^2 \right] \right]. \qquad (4.2.43)$$

Simplification of the terms within the brackets above gives

$$= b_1^{\alpha*} b_1^{\beta} + \frac{b_1^{\alpha*} b_1^{\beta}}{|H_{12}'|^2} \left(-|H_{12}'|^2 \right), \qquad (4.2.44)$$

which is clearly equal to zero. Therefore the corrected eigenfunctions are found to be orthogonal, consistent with our proposed construction.

Let us consider a second example of the use of degenerate perturbation theory.

EXAMPLE 4.2.2

Consider a system in which the perturbation is defined as

$$H' = \begin{bmatrix} 0 & \lambda \\ \lambda & 0 \end{bmatrix}. \qquad (4.2.45)$$

The elements of the perturbation matrix H' are then

$$H_{11}' = H_{22}' = 0, \qquad H_{12}' = H_{21}' = \lambda. \qquad (4.2.46)$$

The total Hamiltonian for the system is defined as the sum of the perturbed and unperturbed parts H' and H_0 as

$$H = H_0 + H'. \qquad (4.2.47)$$

Let there be two degenerate eigenfunctions of H_0 called U_1 and U_2. By definition then

$$H_0 U_1 = E_0 U_1, \qquad (4.2.48)$$
$$H_0 U_2 = E_0 U_2, \qquad (4.2.49)$$

where the degenerate energy eigenvalue is defined as E_0. The problem is then to find the correct first-order energy eigenvalues and eigenvectors in the presence of the perturbation H'. We start with the secular equation, Eq. (4.2.19):

$$\sum_{p=1}^{2} H_{mp}' b_p = E' b_m, \qquad m = 1, 2. \qquad (4.2.50)$$

Summing over the free index m yields

$$\begin{aligned} m = 1: \quad & H_{11}' b_1 + H_{12}' b_2 = E' b_1, && (4.2.51) \\ m = 2: \quad & H_{21}' b_1 + H_{22}' b_2 = E' b_2. && (4.2.52) \end{aligned}$$

But the matrix elements are defined above. Substituting into Eqs. (4.2.51) and (4.2.52) the values for each matrix element from Eqs. (4.2.46) yields

$$\lambda b_2 = E' b_1, \tag{4.2.53}$$

$$\lambda b_1 = E' b_2. \tag{4.2.54}$$

The energy eigenvalues can be determined from the determinant of the above system of equations. This gives

$$\det \left[\begin{bmatrix} 0 & \lambda \\ \lambda & 0 \end{bmatrix} - \begin{bmatrix} E' & 0 \\ 0 & E' \end{bmatrix} \right] = 0,$$

$$\det \begin{bmatrix} -E' & \lambda \\ \lambda & -E' \end{bmatrix} = 0. \tag{4.2.55}$$

Expanding out the determinant yields

$$E'^2 - \lambda^2 = 0,$$

$$E' = \pm \lambda. \tag{4.2.56}$$

If $E' = +\lambda$, then b_1 must be equal to b_2. If $E' = -\lambda$, then b_1 is equal to $-b_2$. Therefore the normalized eigenfunctions and their corresponding eigenvalues are given as

$$U_+ = \frac{1}{\sqrt{2}} U_1 + \frac{1}{\sqrt{2}} U_2 = \frac{1}{\sqrt{2}} (U_1 + U_2), \qquad E = E_0 + \lambda, \tag{4.2.57}$$

$$U_- = \frac{1}{\sqrt{2}} U_1 - \frac{1}{\sqrt{2}} U_2 = \frac{1}{\sqrt{2}} (U_1 - U_2), \qquad E = E_0 - \lambda. \tag{4.2.58}$$

Comparison of the above result to that obtained in example 4.2.1 shows that they are essentially the same, as expected. Note that the perturbation H' acts to lift the degeneracy of the states. One state goes to a lower energy while the other is raised to a higher energy. This result is of great importance to the study of degenerate systems, like atoms. We will discuss this point in great detail in Chapter 7. However, it is interesting to pursue the point a bit here to demonstrate the power of degenerate perturbation theory and its implications.

Using the above result, we can understand qualitatively how molecules form. Consider the case of two hydrogen atoms that are both in the ground state. If the atoms are initially very far apart, the electrons then have identical energy eigenvalues and wave functions. However, if the two atoms are brought close together the electron wave functions begin to overlap between the two atoms. As a result, the one electron solutions for the energy eigenvalues and eigenfunctions from the one atom problem are no longer valid. Instead, it is necessary to treat the system as a multiparticle problem.

The energies of the separate electronic states within the atoms are degenerate when the atoms are spatially separated. When the two atoms are brought close together the coupling of the wave functions leads to a splitting of the degeneracy as we have found in the example problems above. Two different energy eigenvalues occur. One corresponds to an electronic state of higher energy than the noninteracting state while the other corresponds to a lower energy than the noninteracting state. The lower-energy state forms the molecular bonding level while the higher-energy state forms the molecular antibonding level. It is important to recognize that a state is formed whose energy is less than that of an electron in an isolated atom. This is the reason why it is energetically favorable for molecules to form. We will discuss this point in great detail in Chapter 7.

4.3   Time-Dependent Perturbation Theory

We now turn our attention to time-dependent perturbations of a system. Phenomena such as electromagnetic transitions and phonon scatterings can be determined from relationships derived with time-dependent perturbation theory. In this and the following sections, we will derive the physics that underlies the study of transitions between states in a quantum-mechanical system. To obtain an understanding of transitions it is first necessary to consider how a system develops in time quantum mechanically.

A time-dependent wave function can be written in Dirac notation as

$$\Psi(t) = |a, t\rangle, \qquad \Psi(t_0) = |a, t_0\rangle, \tag{4.3.1}$$

where t represents the time and a indicates all other quantum numbers. The question is, how does $\Psi(t)$ depend on $\Psi(t_0)$, the wave function at time, $t = t_0$? We define a linear operator $T(t, t_0)$ as a time-translation operator. $T(t, t_0)$ translates the system in time from $t = t_0$ to $t = t$. $T(t, t_0)$ is defined quantitatively as

$$\Psi(t) = |a, t\rangle = T(t, t_0)|a, t_0\rangle = T(t, t_0)\Psi(t_0). \tag{4.3.2}$$

Similarly, the time translation of a state from $t = t_0$ to $t = t_2$ can be written as a translation in time from t_0 to t_2. Since the final time is arbitrary, t_2 could just as easily be t_1. Hence going from t_0 to t_2 can be accomplished first by translation from t_0 to t_1 and from t_1 to t_2. These results quantitatively are written as

$$\Psi(t_2) = T(t_2, t_1)\Psi(t_1) = T(t_2, t_1)T(t_1, t_0)\Psi(t_0) = T(t_2, t_0)\Psi(t_0). \tag{4.3.3}$$

Equation (4.3.3) shows that the time-translation operator $T(t, t_0)$ is linear. In summary, $T(t, t_0)$ satisfies the condition that a total translation in time can be subdivided into a progression of smaller translations in time. In other words,

$$T(t_2, t_0) = T(t_2, t_1)T(t_1, t_0). \tag{4.3.4}$$

Note that the identity time-translation operator $T(t, t)$ is defined as

$$T(t, t) = 1, \tag{4.3.5}$$

and the inverse time-translation operator is

$$T(t, t_0)T(t_0, t) = 1. \tag{4.3.6}$$

Equation (4.3.6) states that when the system is translated first from t_0 to t and then from t back to t_0, no net time translation occurs.

The question is then, what is $T(t, t_0)$ mathematically? The time-translation operator can be determined with help from the time-dependent Schroedinger equation. The time-dependent Schroedinger equation is

$$i\hbar \frac{d\Psi(t)}{dt} = H(t)\Psi(t). \tag{4.3.7}$$

The time dependence of $\Psi(t)$ can be determined in a general way by integration of the time-dependent Schroedinger equation. This can be performed as follows:

$$\frac{d\Psi(t)}{\Psi(t)} = \frac{-iH(t)}{\hbar} dt. \tag{4.3.8}$$

Integrating with respect to time, we can write the above equation as

$$\int_{\Psi(t_0)}^{\Psi(t)} \frac{d\Psi(t)}{\Psi(t)} = -\frac{i}{\hbar} \int_{t_0}^{t} H \, dt. \tag{4.3.9}$$

Integrating and substituting in the bounds on the left-hand side yields, if H is time independent,

$$\ln\left[\frac{\Psi(t)}{\Psi(t_0)}\right] = -iH\frac{(t - t_0)}{\hbar}, \tag{4.3.10}$$

which reduces to,

$$\Psi(t) = \Psi(t_0)e^{\frac{-iH(t-t_0)}{\hbar}}. \tag{4.3.11}$$

Hence the exponential term generates a translation of the wave function in time from $t = t_0$ to $t = t$. Therefore we can associate $T(t, t_0)$ with the exponential function as

$$T(t, t_0) = e^{\frac{-iH(t-t_0)}{\hbar}}. \qquad (4.3.12)$$

For small time increments $(t - t_0) \sim \epsilon$ the time-translation operator can be expressed by a Taylor series expansion of the exponent as

$$T(t + \epsilon, t) = 1 - \frac{iH\epsilon}{\hbar}, \qquad (4.3.13)$$

where the expansion is taken to only first order in smallness. Note that the time-translation operator satisfies all the qualities discussed above, that is, the exponential operator is linear, has an inverse, and has an identity operator.

With the above definition of the time-translation operator, the time development of a state function can be determined. What we seek in general is the solution of the Schroedinger equation when the state functions are time dependent. Specifically, we seek the solution to the following physical problem. Consider a system that is initially in a well-defined state characterized by an unperturbed hamiltonian H_0 whose eigenstates are considered known. At some specific time, say t_0, a perturbation is applied to the system that may or may not be time dependent. The question is, in what state will the system be at a later time t? In other words, how does the action of the perturbation change the state of the system?

Physically, we are suggesting that a perturbation, when applied to a system in some initial state, can cause the system to make a transition into another state. The physical problem that we seek to examine is that of a system, for example, a hydrogen atom, whose electron is initially in some specific state. A perturbation is then applied, for example an electromagnetic field. At some later time t, after which the perturbation has been applied, the system is reexamined. It is possible then that the electron within the hydrogen atom has changed its state under the influence of the applied electromagnetic field. This effect is said to be an electromagnetic transition. It is our purpose now to determine how these transitions can occur.

To determine how a perturbation can cause a transition from one state to another, it is useful to write the total Hamiltonian H as the sum of two separate terms, a time-independent part H_0 and a perturbation part V. The eigenfunctions of H_0 are assumed to be known and are represented as $\Psi_n^{(0)}$. The eigenfunctions then satisfy

$$H_0 \Psi_n^{(0)} = E_n^{(0)} \Psi_n^{(0)}. \qquad (4.3.14)$$

It is important to recognize that the eigenstates of H_0 form a basis set. Therefore, from the completeness theorem, any arbitrary state can be expanded in terms of these basis states. Consequently the time-dependent wave functions can then be expanded in terms of the time-independent wave functions as

$$\Psi(t_0) = \sum_n C_n e^{\frac{-iH_0 t_0}{\hbar}} \Psi_n^{(0)}, \tag{4.3.15}$$

where the coefficients in the expansion are C_n. In general, the time-dependent wave functions at any given time t are written as

$$\Psi(t) = \sum_n C_n(t) e^{\frac{-iH_0 t}{\hbar}} \Psi_n^{(0)}, \tag{4.3.16}$$

where the coefficients are also time dependent. The coefficients $C_n(t)$ can be found from the expansion in the usual way by multiplying by $\Psi_m^{(0)*} e^{iH_0 t/\hbar}$ and integrating over all space. The resulting expression is

$$\int \Psi_m^{(0)*} e^{\frac{iH_0 t}{\hbar}} \Psi(t) \, d^3 r = \sum_n C_n(t) \int \Psi_m^{(0)*} \Psi_n^{(0)} \, d^3 r. \tag{4.3.17}$$

When the orthogonality property of the time-independent wave functions is used, the coefficients are given as

$$C_m(t) = \int \Psi_m^{(0)*} e^{\frac{iH_0 t}{\hbar}} \Psi(t) \, d^3 r. \tag{4.3.18}$$

Physically, the square of the coefficients $C_m(t)$ can be interpreted as the probability of finding the particle in the mth state at the time t. This follows from our discussion in Chapter 1, wherein we determined that the squares of the coefficients in the expansion of an arbitrary wave function Ψ in terms of the eigenstates Ψ_i of a Hermitian operator are the relative probabilities of finding the state Ψ in a particular eigenstate after a measurement has been made.

The expression above for $C_m(t)$ can be further simplified when the exponential operator is applied. To see how the exponential operator acts, first consider the Hermitian property of the Hamiltonian:

$$H_0 \Psi_m^{(0)*} = E_m^{(0)*} \Psi_m^{(0)*} = E_m^{(0)} \Psi_m^{(0)*}, \tag{4.3.19}$$

since the eigenvalues of a Hermitian operator are real and

$$\int \Psi_m^{(0)*} e^{\frac{iH_0 t}{\hbar}} \Psi(t) \, d^3 r = \int \left[e^{\frac{iH_0 t}{\hbar}} \Psi_m^{(0)} \right]^* \Psi(t) \, d^3 r. \tag{4.3.20}$$

The effect of the operator in the exponent can be evaluated by expansion of the exponential term in a Taylor series as

$$e^{\frac{iH_0 t}{\hbar}} = 1 + \frac{iH_0 t}{\hbar} + \frac{1}{2} \left(\frac{iH_0 t}{\hbar} \right)^2 + \cdots. \tag{4.3.21}$$

Applying this operator onto the definite eigenstate $\Psi_m^{(0)}$ yields

$$\left[1 + \frac{i E_m^{(0)} t}{\hbar} + \frac{1}{2}\left(\frac{i E_m^{(0)} t}{\hbar}\right)^2 + \cdots\right]\Psi_m^{(0)} = e^{\frac{i E_m^{(0)} t}{\hbar}}\Psi_m^{(0)}. \tag{4.3.22}$$

Therefore the left-hand side of Eq. (4.3.20) becomes

$$\int \Psi_m^{(0)*} e^{\frac{i E_m^{(0)} t}{\hbar}} \Psi(t)\, \mathrm{d}^3 r \tag{4.3.23}$$

and the coefficient $C_m(t)$ is

$$C_m(t) = \int \Psi_m^{(0)*} e^{\frac{i E_m^{(0)} t}{\hbar}} \Psi(t)\, \mathrm{d}^3 r. \tag{4.3.24}$$

In the presence of a perturbation V, the Schroedinger equation is

$$(H_0 + V)\Psi = i\hbar\frac{\mathrm{d}\Psi}{\mathrm{d}t}. \tag{4.3.25}$$

We can evaluate the right-hand side of Eq. (4.3.25) by substituting in for Ψ the expansion in terms of $\Psi_n^{(0)}$ given by Eq. (4.3.16) and recognizing that the exponential operator acting on $\Psi_n^{(0)}$ gives simply

$$e^{\frac{-i H_0 t}{\hbar}}\Psi_n^{(0)} = e^{\frac{-i E_0^{(0)} t}{\hbar}}\Psi_n^{(0)}. \tag{4.3.26}$$

The result is

$$i\hbar\frac{\mathrm{d}}{\mathrm{d}t}\sum_n C_n(t)e^{\frac{-i E_n^{(0)} t}{\hbar}}\Psi_n^{(0)}$$

$$= \sum_n -\frac{i}{\hbar}E_n^{(0)} i\hbar C_n(t)e^{\frac{-i E_n^{(0)} t}{\hbar}}\Psi_n^{(0)} + \sum_n i\hbar\, e^{\frac{-i E_n^{(0)} t}{\hbar}}\Psi_n^{(0)}\frac{\mathrm{d}C_n(t)}{\mathrm{d}t}$$

$$= \sum_n E_n^{(0)} C_n(t)e^{\frac{-i E_n^{(0)} t}{\hbar}}\Psi_n^{(0)} + \sum_n i\hbar\Psi_n^{(0)} e^{\frac{-i E_n^{(0)} t}{\hbar}}\frac{\mathrm{d}C_n(t)}{\mathrm{d}t}. \tag{4.3.27}$$

We can also evaluate the left-hand side of the time-dependent Schroedinger equation by substituting in the expansion for $\Psi(t)$ as

$$(H_0 + V)\sum_n C_n(t)e^{\frac{-i E_n^{(0)} t}{\hbar}}\Psi_n^{(0)}$$

$$= \sum_n E_n^{(0)} C_n(t)e^{\frac{-i E_n^{(0)} t}{\hbar}}\Psi_n^{(0)} + \sum_n V C_n(t)e^{\frac{-i E_n^{(0)} t}{\hbar}}\Psi_n^{(0)}. \tag{4.3.28}$$

When the left- and the right-hand sides are combined, the terms involving $E_n^{(0)}$ vanish, leaving

$$i\hbar\sum_n \Psi_n^{(0)} e^{\frac{-i E_n^{(0)} t}{\hbar}}\frac{\mathrm{d}C_n(t)}{\mathrm{d}t} = \sum_n V C_n(t)e^{\frac{-i E_n^{(0)} t}{\hbar}}\Psi_n^{(0)}. \tag{4.3.29}$$

Finally, we can determine the time rate of change of the coefficient $C_n(t)$ from Eq. (4.3.29) by multiplying by $\Psi_k^{(0)*}$ and integrating to give

$$
i\hbar \int \sum_n \Psi_k^{(0)*} \Psi_n^{(0)} e^{\frac{-i E_n^{(0)} t}{\hbar}} \frac{dC_n(t)}{dt} d^3r
$$

$$
= \int \sum_n \Psi_k^{(0)*} V \Psi_n^{(0)} e^{\frac{-i E_n^{(0)} t}{\hbar}} C_n(t) d^3r. \tag{4.3.30}
$$

Again use is made of the orthogonality condition of the time-independent eigenstates, and Eq. (4.3.30) simplifies to

$$
i\hbar e^{\frac{-i E_k^{(0)} t}{\hbar}} \frac{dC_k(t)}{dt} = \sum_n V_{kn} e^{\frac{-i E_n^{(0)} t}{\hbar}} C_n(t), \tag{4.3.31}
$$

where V_{kn} is defined as

$$
V_{kn} = \int \Psi_k^{(0)*} V \Psi_n^{(0)} d^3r. \tag{4.3.32}
$$

V_{kn} is the matrix element of the perturbation between the unperturbed states. We divide the exponential factor on the left-hand side, and Eq. (4.3.31) becomes

$$
i\hbar \frac{dC_k(t)}{dt} = \sum_n V_{kn} e^{\frac{-i \left[E_n^{(0)} - E_k^{(0)} \right] t}{\hbar}} C_n(t),
$$

$$
i\hbar \frac{dC_k(t)}{dt} = \sum_n V_{kn} C_n(t) e^{i \omega_{kn} t}, \tag{4.3.33}
$$

where ω_{kn} is defined as

$$
\omega_{kn} = \frac{E_k^{(0)} - E_n^{(0)}}{\hbar}. \tag{4.3.34}
$$

Physically, the above result can be understood as follows. Let the system at the initial time t_0 be in the definite state s. The state s is a definite state of the unperturbed Hamiltonian H_0. Therefore, at the initial time t_0, the coefficient C_s in the expansion over the time-independent wave functions must be one. This is due to the fact that, by design, the wave function is initially in the state s and hence the probability of finding it in the state s must be 1 at $t = t_0$. Therefore the square of the coefficient must be 1 from the expansion postulate and, since the coefficient is real, the coefficient itself must be one:

$$
C_s(t = t_0) = 1. \tag{4.3.35}
$$

Additionally, all the other coefficients at $t = t_0$ must be zero. Therefore

$$
C_k(t = t_0) = 0 \tag{4.3.36}
$$

for all coefficients $k \neq s$. The question we want to answer is what the coefficient C_k is at some later time t, $t > t_0$, given that a perturbation V has been applied to the system at $t = t_0$. The expression derived above for the time rate of change of the coefficient enables us to determine the probability that the system will be found in a different state at a later time than it was in initially. Hence, from the knowledge of how the probability of finding the system in a different state from its original state varies with time, we can determine the transition rate out of an initial state into a different final state.

To determine the probability of a state's undergoing a transition, what we need to know is what the probability is that the system is in a state k at some time $t > t_0$, given that it is initially in the state s. Equations (4.3.33) show how the coefficient $C_k(t)$ varies in time. From Eqs. (4.3.33) it is clear that $C_k(t)$ need not be zero at $t > t_0$. Now at t close to t_0, Eq. (4.3.33) is

$$i\hbar \frac{dC_k(t \sim t_0)}{dt} = V_{ks} C_s(t \sim t_0) e^{i\omega_{ks}t} = V_{ks} e^{i\omega_{ks}t} \qquad (4.3.37)$$

since we can approximate $C_s(t \sim t_0)$ by $C_s(t = t_0)$. $C_s(t = t_0)$ is one by design. Equation (4.3.37) can be integrated over time by use of the initial conditions that $C_k(t = t_0) = 0$ and $C_s(t = t_0) = 1$. The time rate of change of $C_k(t)$ given by Eq. (4.3.37) involves V_{ks}, the perturbation between the states k and s. If the perturbation V_{ks} is zero, C_k is zero for all time. This implies that the system cannot make a transition from the initial state s to the final state k through the perturbation V. However, if V_{ks} is not zero, then there is a chance that the system will make a transition from s to k through the action of the perturbation. In other words, the perturbation can act to change the system from its initial state s to some other final state k. To find what the probability is that a transition will occur, it is necessary to calculate $C_k(t)$ by integrating Eq. (4.3.37). This becomes

$$C_k(t) = -\frac{i}{\hbar} \int_{t_0}^{t} V_{ks} e^{i\omega_{ks}t'} dt'. \qquad (4.3.38)$$

As t approaches positive infinity the system settles down. The probability that the system will be found in the kth state at $t = +\infty$ is given as the square of the coefficient $C_k(t)$ then as

$$|C_k(+\infty)|^2 = \left| -\frac{i}{\hbar} \int_{t_0}^{\infty} V_{ks} e^{i\omega_{ks}t'} dt' \right|^2. \qquad (4.3.39)$$

Hence once the matrix element of the perturbation between the initial and the final states is determined, the probability that the system will make a transition to the final state is determined by integration of the product of the matrix element and the exponential factor $e^{i\omega_{ks}t'}$ over time, as given in Eq. (4.3.39).

4.4 **Fermi's Golden Rule**

Next we consider the rate at which transitions from one quantum-mechanical state to another occur. The quantum-mechanical transition rate is governed by what is referred to as Fermi's golden rule. In Sections 10.7 and 9.2 we will see that Fermi's golden rule has extensive application to describing electromagnetic transitions as well as phonon scatterings. In this section, we derive the mathematical statement of the transition rate.

Fermi's golden rule can be applied to a system if the potential itself is time independent but the time development of the system can still be described in terms of transitions between eigenstates of the unperturbed Hamiltonian H_0. An example of such a physical system is the action of a fixed potential on an incident beam of particles that acts to produce transitions from an initial momentum state into one of various possible final momentum states.

In Section 4.3 we presented a result for how the probability amplitude of a state changes with time under the action of a perturbation. The system is initially in a definite state s. On the application of a perturbation, the system can possibly undergo a transition into a new definite state k. To determine the probability that the transition from s to k occurs, it is necessary to determine the probability of finding the particle in the state k at a later time. To do so, we must know how the probability amplitude for the state k, C_k changes with time. In Section 4.3, we determined an expression for the time rate of change of C_k, given by Eq. (4.3.37) as

$$\frac{dC_k}{dt} = \frac{V_{ks}}{i\hbar}e^{i\omega_{ks}t} \quad (k \neq s). \tag{4.4.1}$$

Equation (4.4.1) can be readily solved for C_k once the initial conditions are specified. Let the initial time t_0 be zero; then $C_k = 0$. Therefore the particle is initially in state s and definitely not in state k. If the solution for $C_k(t)$ is assumed to be

$$C_k(t) = Ae^{i\omega_{ks}t} + B, \tag{4.4.2}$$

then A and B can be determined by substitution of Eq. (4.4.2) for C_k in Eq. (4.4.1) above. This yields

$$Ai\omega_{ks}e^{i\omega_{ks}t} + \frac{iV_{ks}}{\hbar}e^{i\omega_{ks}t} = 0. \tag{4.4.3}$$

Solving for A, we obtain

$$A = -\frac{V_{ks}}{\hbar\omega_{ks}}. \tag{4.4.4}$$

Therefore $C_k(t)$ becomes

$$C_k(t) = -\frac{V_{ks}}{\hbar\omega_{ks}}e^{i\omega_{ks}t} + B. \tag{4.4.5}$$

Recall the definitions of V_{ks} and ω_{ks}:

$$V_{ks} = \langle k| V |s \rangle, \qquad \hbar\omega_{ks} = E_k^{(0)} - E_s^{(0)}; \qquad (4.4.6)$$

the time-dependent coefficient $C_k(t)$ is

$$C_k(t) = -\frac{\langle k| V |s \rangle}{E_k^{(0)} - E_s^{(0)}} e^{i\omega_{ks}t} + B. \qquad (4.4.7)$$

At $t = 0$, $C_k = 0$. Therefore B can be determined readily as

$$0 = -\frac{\langle k| V |s \rangle}{E_k^{(0)} - E_0^{(0)}} e^{i\omega_{ks}0} + B,$$

$$B = \frac{\langle k| V |s \rangle}{E_k^{(0)} - E_s^{(0)}}. \qquad (4.4.8)$$

Therefore $C_k(t)$ is

$$C_k(t) = \frac{\langle k| V |s \rangle}{\left[E_k^{(0)} - E_s^{(0)}\right]} \left(1 - e^{i\omega_{ks}t}\right), \qquad (4.4.9)$$

where it is assumed that at $t = 0$, $C_k = 0$. The probability that the system will be in the state k at the time $t \neq 0$ when it is initially in the state s is then

$$|C_k(t)|^2 = \frac{|\langle k| V |s \rangle|^2}{\left[E_k^{(0)} - E_s^{(0)}\right]^2} |1 - e^{i\omega_{ks}t}|^2. \qquad (4.4.10)$$

Note that the above relationship is obtained essentially by the integration of Eq. (4.3.39) from $t = 0$ to $t = t$. We can simplify Eq. (4.4.10) by noting that

$$|1 - e^{i\theta}|^2 = |1 - \cos\theta - i\sin\theta|^2 = (1 - \cos\theta)^2 + \sin^2\theta$$
$$= 1 + \cos^2\theta - 2\cos\theta + \sin^2\theta$$
$$= 2 - 2\cos\theta$$
$$= 2(1 - \cos\theta). \qquad (4.4.11)$$

Finally, the expression for $|C_k(t)|^2$ becomes

$$|C_k(t)|^2 = \frac{2|\langle k| V |s \rangle|^2}{\left[E_k^{(0)} - E_s^{(0)}\right]^2} \left(1 - \cos\omega_{ks}t\right). \qquad (4.4.12)$$

The denominator of Eq. (4.4.12) is simply $(\hbar\omega_{ks})^2$. When $\omega_{ks} = 0$, the denominator dominates such that the transition rate is very large; in other words, the amplitude peaks prominently for $\omega_{ks} = 0$. This implies that for transitions to final states close in energy to that of the initial state, the transition rate is large.

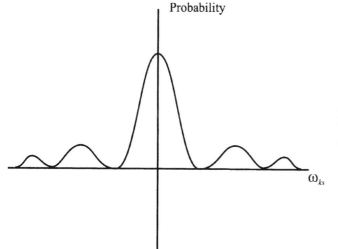

Figure 4.4.1. Sketch of the probability versus frequency for a transition.

The relative value of $|C_k(t)|^2$ is plotted in Figure 4.4.1. Inspection of Figure 4.4.1 shows that the peak occurs at $k = s$.

The total transition probability to all final states is given by summation over all the possible final states. Therefore the total transition probability is determined by summation of Eq. (4.4.12) over all the final states k. If the final states are all closely spaced in energy such that the density of final states is a quasi-continuum, then the final states can be expressed by a continuous function $\rho_f(E)$. The number of final states in the interval dE then is simply given by the product of the density-of-states function $\rho_f(E)$, and dE. The total transition probability from an initial state s to any one of the final states k in the quasi-continuum is found by integration of Eq. (4.4.12) over the final density of states as

$$\sum_k |C_k(t)|^2 = 2 \int \frac{|\langle k| V |s\rangle|^2}{\left[E_k^{(0)} - E_s^{(0)}\right]^2} (1 - \cos\omega_{ks} t)\rho_f\left[E_k^{(0)}\right] dE_k, \qquad (4.4.13)$$

where the integration is now over all possible values of the final energy. The transition rate then is determined from the time rate of change of Eq. (4.4.13). Denote the transition rate out of the initial state s into a final state k as W; W is then

$$
\begin{aligned}
W &= \frac{d}{dt} \sum_k |C_k(t)|^2 \\
&= \frac{d}{dt} 2 \int \frac{|\langle k| V |s\rangle|^2}{(\hbar\omega_{ks})^2} (1 - \cos\omega_{ks} t)\rho_f\left[E_k^{(0)}\right] dE_k \\
&= \frac{2}{\hbar^2} \int \frac{|\langle k| V |s\rangle|^2}{\omega_{ks}^2} \omega_{ks} \sin\omega_{ks} t\, \rho_f\left[E_k^{(0)}\right] dE_k \\
&= \frac{2}{\hbar^2} \int |\langle k| V |s\rangle|^2 \frac{\sin\omega_{ks} t}{\omega_{ks}} \rho_f\left[E_k^{(0)}\right] dE_k.
\end{aligned}
\qquad (4.4.14)
$$

The transition-rate expression derived above can be simplified if it is assumed that the matrix element and the density of states are roughly constant within the peak of $(\sin \omega_{ks} t)/\omega_{ks}$. The peak in the function occurs at $\omega_k = \omega_s$. We can approximate the density-of-states function for the states k by that at s. Therefore $\rho_f(E_k) \sim \rho_f(E_s)$. The transition rate is then

$$W = \frac{2}{\hbar^2} |\langle k| V |s\rangle|^2 \rho_f(E_s) \int \frac{\sin \omega_{ks} t}{\omega_{ks}} dE_k. \tag{4.4.15}$$

Since E_s is fixed, the integration can be written in terms of ω_{ks}. The bounds are then extended to range from $-\infty$ to $+\infty$, leaving

$$W = \frac{2}{\hbar} \langle k| V |s\rangle|^2 \rho_f[E_s^{(0)}] \int_{-\infty}^{+\infty} \frac{\sin \omega_{ks} t}{\omega_{ks}} d\omega_{ks}. \tag{4.4.16}$$

The integral can be evaluated as

$$\int_{-\infty}^{+\infty} \frac{\sin \omega_{ks} t}{\omega_{ks}} d\omega_{ks} = 2 \int_0^{+\infty} \frac{\sin \omega_{ks} t}{\omega_{ks}} d\omega_{ks} = 2 \frac{\pi}{2}$$
$$= \pi. \tag{4.4.17}$$

Therefore the transition rate is given as

$$W = \frac{2\pi}{\hbar} |\langle k| V |s\rangle|^2 \rho_f(E_s). \tag{4.4.18}$$

This result is known as Fermi's golden rule of time-dependent perturbation theory.

An alternative form of Fermi's golden rule can be derived in which a delta function is used in place of the final density-of-states function. This alternative form is valid over long time durations and applies to single-state transitions. The form derived above applies when there is a band of energies into which the electron can make a transition. The alternative form of Fermi's golden rule is

$$W = \frac{2\pi}{\hbar} |\langle k| V |s\rangle|^2 \delta(E_k - E_s). \tag{4.4.19}$$

The delta function implies that the energy is conserved in the transition.

4.5 **Second-Order Transitions**

In Section 4.4 we derived an expression, known as Fermi's golden rule, that describes the rate of transitions between two different quantum states. From the analysis, it was found that the transition rate depends on the coupling of the initial and the final states through the perturbation V as well as the availability of a final state. If the matrix element of the perturbation between the initial and the final states vanishes, then, according to Fermi's golden rule, the transition cannot occur. For example, some transitions between excited atomic states do not occur to first order since the matrix element vanishes between the initial and the final states. These transitions are called forbidden transitions. Additionally, in certain semiconductors, optical transitions also do not occur to first order because the matrix element coupling the initial and the final states vanishes.

Even though a transition is classified as forbidden, it is not necessarily true that the transition cannot occur altogether. Forbidden transitions cannot occur to first order; however, they may occur to second and higher order. Second- and higher-order transitions occur through some intermediate path. A transition occurs first to an intermediate state followed by another decay to the final state. In other words, the total transition from the initial to the final state occurs in two or more steps. The net result is the same as in a direct transition but the path is different.

To understand how second-order transitions occur, it is necessary to generalize the approach taken in the previous sections. The easiest way to understand the physics of second- and higher-order transitions is to introduce a different formalism, called the interaction picture. We start our discussion with the Schroedinger equation in the presence of some perturbation V:

$$(H_0 + V)\Psi(t) = i\hbar \frac{d\Psi(t)}{dt}. \tag{4.5.1}$$

Let us define a state vector $\Psi'(t)$ that is the time-evolved form of $\Psi(t)$ as

$$\Psi'(t) \equiv e^{\frac{iH_0 t}{\hbar}} \Psi(t). \tag{4.5.2}$$

It is interesting to construct the equation of motion for $\Psi'(t)$. This is obtained through differentiation of Eq. (4.5.2) with respect to t:

$$\frac{d\Psi'}{dt} = \frac{iH_0}{\hbar} e^{\frac{iH_0 t}{\hbar}} \Psi(t) + e^{\frac{iH_0 t}{\hbar}} \frac{d\Psi(t)}{dt}. \tag{4.5.3}$$

When both sides are multiplied by $i\hbar$, Eq. (4.5.3) becomes

$$i\hbar \frac{d\Psi'}{dt} = -H_0 e^{\frac{iH_0 t}{\hbar}} \Psi(t) + e^{\frac{iH_0 t}{\hbar}} i\hbar \frac{d\Psi(t)}{dt}. \tag{4.5.4}$$

But from the Schroedinger equation, the term involving the time rate of change of $\Psi(t)$ can be written as

$$i\hbar \frac{d\Psi(t)}{dt} = (H_0 + V)\Psi(t). \tag{4.5.5}$$

Therefore

$$
\begin{aligned}
i\hbar \frac{d\Psi'}{dt} &= -H_0 e^{\frac{iH_0 t}{\hbar}} \Psi(t) + e^{\frac{iH_0 t}{\hbar}} (H_0 + V)\Psi(t) \\
&= e^{\frac{iH_0 t}{\hbar}} V\Psi(t).
\end{aligned}
\tag{4.5.6}$$

Inverting Eq. (4.5.2),

$$\Psi(t) = e^{\frac{-iH_0t}{\hbar}}\Psi'(t) \qquad (4.5.7)$$

and substituting Eq. (4.5.7) for $\Psi(t)$ in Eq. (4.5.6) gives

$$i\hbar\frac{d\Psi'}{dt} = e^{\frac{iH_0t}{\hbar}}Ve^{\frac{-iH_0t}{\hbar}}\Psi'(t). \qquad (4.5.8)$$

If we define V' as

$$V' \equiv e^{\frac{iH_0t}{\hbar}}Ve^{\frac{-iH_0t}{\hbar}}, \qquad (4.5.9)$$

Eq. (4.5.8) becomes

$$i\hbar\frac{d\Psi'(t)}{dt} = V'\Psi'(t). \qquad (4.5.10)$$

The operator V' can be thought of as an effective Hamiltonian for the transformed state vector since the time dependence of Ψ is determined with V'. The above result implies that Ψ' changes in time only if there is a perturbation. If no perturbation exists then V' vanishes and by Eq. (4.5.10), the time rate of change of $\Psi'(t)$ is zero. Ψ' describes the same state as Ψ, but in what is called a different picture, the interaction picture.

Quantum mechanics can be formulated in terms of different but equivalent pictures. The two most noted pictures are the Schroedinger and the Heisenberg pictures. The Schroedinger picture is a formulation in which the time dependence of the system is built into the wave functions or states of the particles. The Heisenberg picture has the time dependence of the system built into the operators. In other words, the wave functions are time dependent while the operators are time independent in the Schroedinger picture and the operators are time dependent and the wave functions time independent in the Heisenberg picture.

In the interaction picture both the state vectors and the operators exhibit time dependence. The time evolution of the state vectors, though, is determined by the perturbation while that of the operators is determined by the unperturbed Hamiltonian H_0. Thus the interaction picture is intermediate between the Schroedinger and the Heisenberg pictures. Table 4.5.1 outlines these three different pictures.

Table 4.5.1
Different Pictures in Quantum Mechanics

	Schroedinger Picture	Heisenberg Picture	Interaction Picture
State vectors	Time dependent	Time independent	Time dependent
Operators	Time independent	Time dependent	Time dependent

Let us assume that the time development of a wave function within the interaction picture is given as

$$\Psi'(t) = U(t, t_0)\Psi'(t_0). \tag{4.5.11}$$

The time-development operators within the Schroedinger and the interaction pictures are related through use of Eq. (4.5.2) as

$$e^{\frac{iH_0 t}{\hbar}}\Psi(t) = U(t, t_0)e^{\frac{iH_0 t_0}{\hbar}}\Psi(t_0),$$
$$\Psi(t) = e^{\frac{-iH_0 t}{\hbar}}U(t, t_0)e^{\frac{iH_0 t_0}{\hbar}}\Psi(t_0). \tag{4.5.12}$$

But within the Schroedinger picture the time-evolution operator $T(t, t_0)$ is defined as

$$\Psi(t) = T(t, t_0)\Psi(t_0). \tag{4.5.13}$$

Therefore the time-development operators between the interaction and the Schroedinger pictures are related as

$$U(t, t_0) = e^{\frac{iH_0 t}{\hbar}}T(t, t_0)e^{\frac{-iH_0 t_0}{\hbar}}. \tag{4.5.14}$$

In Section 4.3, for small time increments, we found that the time-evolution operator $T(t, t_0)$ can be written as [see Eq. (4.3.13)]

$$T(t + \epsilon, t) = 1 - \frac{iH\epsilon}{\hbar}. \tag{4.5.15}$$

But

$$T(t + \epsilon, t_0) = T(t + \epsilon, t)T(t, t_0). \tag{4.5.16}$$

The derivative of $T(t, t_0)$ can be evaluated then as

$$\frac{dT(t, t_0)}{dt} = \lim_{\epsilon \to 0} \frac{T(t + \epsilon, t_0) - T(t, t_0)}{\epsilon}$$
$$= \lim_{\epsilon \to 0} \frac{[T(t + \epsilon, t) - 1]T(t, t_0)}{\epsilon}. \tag{4.5.17}$$

Substitution of the above expression for $T(t + \epsilon, t_0)$ yields

$$= -\frac{i}{\hbar}HT(t, t_0). \tag{4.5.18}$$

Therefore we obtain

$$i\hbar\frac{dT(t, t_0)}{dt} = HT(t, t_0). \tag{4.5.19}$$

Similarly, within the interaction picture, the time operator $U(t, t_0)$ obeys the relation

$$i\hbar\frac{dU(t, t_0)}{dt} = V'U(t, t_0),$$ (4.5.20)

where the operator V' is the effective Hamiltonian described above. Integrating Eq. (4.5.20) with respect to time yields

$$U(t, t_0) = U(t_0, t_0) - \frac{i}{\hbar}\int_{t_0}^{t} V'(t')U(t', t_0)\,dt'.$$ (4.5.21)

In Section 4.3, it was determined that there is an identity time-translation operator within the Schroedinger picture $T(t_0, t_0)$. Similarly, $U(t_0, t_0)$ is an identity operator within the interaction picture. With this observation, Eq. (4.5.21) becomes

$$U(t, t_0) = 1 - \frac{i}{\hbar}\int_{t_0}^{t} V'(t')U(t', t_0)\,dt'.$$ (4.5.22)

Integral equation (4.5.22) is in general difficult to solve. However, it can be approximated if the perturbation is small by use of an iterative procedure. First the value of $U(t', t_0)$ under the integral sign is approximated by $U = 1$. The integral is then evaluated, giving a zeroth-order value for $U(t, t_0)$. The zeroth-order value of U [given by the left-hand side of Eq. (4.5.22)] is next substituted into the integral on the right-hand side and the next order approximation is determined. This procedure continues until the level of desired accuracy is obtained. The iterative solution can be written then as

$$U(t, t_0) = 1 - \frac{i}{\hbar}\int_{t_0}^{t} V'(t')\,dt' + \left(-\frac{i}{\hbar}\right)^2 \int_{t_0}^{t} V'(t')\,dt' \int_{t_0}^{t'} V'(t'')\,dt'' + \cdots.$$
(4.5.23)

The general expression for the transition rate from an initial eigenstate s to a final eigenstate k can now be determined through use of the above formulas. Assume that the system is in state s of the unperturbed Hamiltonian initially, $t = t_0$. The probability amplitude that the electron will be found in a different state k at a later time t is given by the kth coefficient in the expansion of the state $\Psi'(t)$ in terms of the eigenstates of the unperturbed Hamiltonian. The kth coefficient is determined from the overlap of the state Ψ_k on the state $\Psi'(t)$ as given by Eq. (1.6.41). Therefore the kth coefficient is determined as

$$C_k = \int \Psi_k^* \Psi'(t)\,d^3r.$$ (4.5.24)

However, $\Psi'(t)$ can be written as $U(t, t_0)\Psi_s$, assuming that the system is in state s initially. Therefore C_k can be expressed in Dirac notation as

$$C_k = \langle k|U(t, t_0)|s\rangle.$$ (4.5.25)

Substituting the expansion for $U(t, t_0)$ given by Eq. (4.5.23) into Eq. (4.5.25) yields

$$C_k = -\frac{i}{\hbar} \int_{t_0}^{t} \langle k| V'|s\rangle dt' + \left(-\frac{i}{\hbar}\right)^2 \int_{t_0}^{t} dt' \int_{t_0}^{t'} \langle k| V'(t') V'(t'')|s\rangle\, dt'' + \cdots.$$

(4.5.26)

The first term above reduces to Eq. (4.3.38) when the potential within the interaction picture V' is replaced by Eq. (4.5.9). Clearly, the first term represents the first-order transition probability amplitude within the interaction picture. The second term above, then, represents the second-order transition probability within the interaction picture. Note that the second-order transition probability amplitude depends on two terms, $V'(t')$ and $V'(t'')$. The transition proceeds to second order through the product of these two terms.

In many cases, a transition will not occur to first order since the matrix element $\langle k| V'|s\rangle$ vanishes. The perturbation fails to couple the initial and the final states. Such a situation leads to forbidden transitions as explained above. However, a transition may still occur to second order. If it is assumed that the first-order matrix element vanishes, then $|C_k(t)|^2$ depends on the second-order term in Eq. (4.5.26). Recalling the definition of V' within the interaction picture given by Eq. (4.5.9), we can rewrite Eq. (4.5.26) as

$$C_k(t) = \left(-\frac{i}{\hbar}\right)^2 \int_{t_0}^{t} dt' \int_{t_0}^{t'} dt'' \sum_m \langle k|e^{\frac{iH_0 t}{\hbar}} V e^{\frac{-iH_0 t}{\hbar}}|m\rangle \langle m|e^{\frac{iH_0 t}{\hbar}} V e^{\frac{-iH_0 t}{\hbar}}|s\rangle,$$

(4.5.27)

in which a sum over a complete set of eigenstates, $\sum |m\rangle \langle m|$, has been inserted. We note that the exponential operators, when applied to the eigenstates, operate in the manner described by Eqs. (4.3.21) and (4.3.22). Applying the exponential operators gives then

$$C_k(t) = \left(-\frac{i}{\hbar}\right)^2 \int_{t_0}^{t} dt' \int_{t_0}^{t'} dt'' \sum_m \langle k| V |m\rangle \langle m| V |s\rangle e^{\frac{i(E_k - E_m)t'}{\hbar}} e^{\frac{i(E_m - E_s)t''}{\hbar}} e^{\eta(t' + t'')},$$

(4.5.28)

where we have added an exponential term to model a slow turn on of the potential. The factor η approaches zero in the limit. If we take t_0 to approach $-\infty$, let η approach zero in the limit, and perform the integrations over t' and t'', square, and take the time derivative, we can write the second-order transition rate as

$$W_{s \to k} = \frac{2\pi}{\hbar} \left| \sum_m \frac{\langle k| V |m\rangle \langle m| V |s\rangle}{E_s - E_m} \right|^2 \delta(E_k - E_s).$$

(4.5.29)

From the above result, we see that the system undergoes a second-order transition in two stages. First it goes from state s to an intermediate state m and then from m to the final state k. The transition probability is proportional to the product of both of these separate transition probabilities or matrix elements. Therefore, in general, the probability that a second-order transition will occur is far less than that for a first-order transition since it involves the product of two probabilities, not just one. To determine the overall transition rate we must sum over the complete set of all intermediate states m before squaring. The intermediate states need not be real. In other words, the system can make a transition to an intermediate virtual state and then to a final real state. We will discuss this point in more detail in Chapter 10 when we consider direct and indirect semiconductors.

PROBLEMS

1. A one-dimensional harmonic oscillator of charge q is perturbed by the application of an electric field E in the positive x direction so that the potential energy becomes

    $$V(X) = \frac{1}{2}m\omega^2 X^2 - qEX.$$

    a. Show that to first order in perturbation theory the shift in the energy levels is zero as a result of this perturbation.

    b. Calculate the shift in every level to second order.

    c. Show that the Schroedinger equation can be solved exactly in this system and find the lowest energy level.

2. An isotropic harmonic oscillator has a Hamiltonian given as

    $$H = H_0 + H'$$

    where

    $$H_0 = \frac{1}{2}\left(p_x^2 + p_y^2\right) + \frac{1}{2}(x^2 + y^2)$$

    and H', the perturbation Hamiltonian, is

    $$H' = \lambda xy.$$

    λ is a scalar constant. For simplicity assume that $\hbar = m = 1$ (natural units).

    Find the correct zeroth-order wave functions based on the linear combination of the zeroth-order states and the correct first-order energies for the degenerate states.

3. Consider an infinite one-dimensional square-well potential described as

    $$V(X) = 0, \qquad -\frac{a}{2} \le x \le \frac{a}{2},$$

    and

    $$V(X) = \infty \quad \text{elsewhere}.$$

    a. If a uniform electric field F_0 is applied to the system, determine to first order the shift in the energy of the ground state.

    b. If a linear electric field $F_0 x$ is applied, determine to first order the shift in energy of the ground state.

    c. If a uniform electric field F_0 is applied, determine the second-order shift in energy of the ground state.

4. a. Use first-order perturbation theory (nondegenerate) to find the lowest energy level in a V bottom whose potential is specified as

$$V(x) = \frac{\delta|x|}{a} \quad (|x| < a), \quad V(x) = \infty \quad \text{elsewhere.}$$

Assume that the unperturbed eigenstates are

$$\psi = \frac{1}{\sqrt{a}} \cos \frac{n\pi x}{2a} \quad \text{(even)},$$

$$\psi = \frac{1}{\sqrt{a}} \sin \frac{n\pi x}{2a} \quad \text{(odd)},$$

$$E = \frac{n^2 \pi^2 \hbar^2}{8ma^2}.$$

b. Expand out the corrected eigenfunction using only two terms involving ψ_1 and ψ_3 as

$$\psi' = a_1 \psi_1^{(0)} + a_3 \psi_3^{(0)},$$

where $\psi_1^{(0)}$ and $\psi_3^{(0)}$ are the unperturbed eigenstates for the infinite square well.

5

Equilibrium Statistical Mechanics

We next consider the dynamics of a collection or ensemble of particles. In the previous sections, we have discussed the dynamics of one particle or systems that can be reformulated as involving only one particle. In general, the macroscopic world consists of many particles simultaneously interacting with one another. For example, the description of the air molecules in a closed room requires tracking at least 10^{27} particles. Clearly, to attempt to describe the motions of each individual particle would exhaust the computational power of any conceivable machine, not to mention that of the investigator. Therefore it is necessary to construct a means by which the collective behavior of an entire system can be assessed without relying on the complete description of the underlying histories of each individual particle in the ensemble. A statistical approach, called statistical mechanics, has been invented to treat the behavior of a large collection of particles. The basic principles and applications of equilibrium statistical mechanics are the subject of this chapter. In Chapter 6 we will consider treating systems that depart from equilibrium.

5.1 Density of States

A large collection of particles is often referred to as an ensemble. To describe the collective behavior of a large ensemble it is necessary to adopt a statistical approach. The best that we can expect to do is to predict, by using statistics, the average values of the macroscopic observables of interest, particularly the energy, the momentum, etc., and the probability that any given particle has a specific value of one of these observables, for example whether a particle has energy $E + dE$.

Specifically, we are interested in determining the number of particles in a given ensemble in an energy range of E to $E + dE$. The total number of particles that lie within this range depends first on the total number of particles within the system N. Less obviously, the number of particles in a given energy range depends on two general factors; the number of available states in which the particles can be put (which, as we will see in Section 5.7, depends on the identity of the particle) and also on the probability that a particle can occupy a given state. The first factor, the number of states available, is called the density-of-states function $D(E)$. The second factor, the occupation probability function $f(E)$, is called the probability distribution function. The product of these two functions gives the probability that a quantum state of energy E is occupied, $N(E)$.

The functions defined above, $D(E)$, $f(E)$, and $N(E)$, can be understood from a situation familiar to anyone trying to park a car in a parking lot. In order to park one's car, two conditions must first be met; there must be parking spaces into which the car can be parked, and there must be at least one unoccupied space. The total number of parking spaces is described by the function $D(E)$, the density-of-states function [in this case it may be more appropriate to call $D(E)$ the density of parking spaces function]. If $D(E)$ is zero, then there are no parking spaces and obviously one cannot park his/her car. But if $D(E)$ is nonzero, then parking spaces exist. However, it is necessary that there be an unoccupied space, otherwise again one cannot park his/her car. The probability that a space is occupied is given by a separate function called $f(E)$, the occupation probability function. Therefore, in order to park one's car, there must be parking spaces that are unoccupied. A similar situation exists in arranging quantum-mechanical particles into macrostates: States into which the particle can be placed must exist and secondly there must be a vacancy into which the particle can be inserted.

The analogy between parking a car and arranging particles into different states can be stretched further. As we will see in Section 5.7, there are different types of particles in the universe, which can be subdivided in accordance to how they can be sorted or arranged in different states. Specifically, there are particles, called fermions, for which no two particles can exist within the same state simultaneously. There are other particles, called bosons, for which this restriction does not apply; any number of bosons can be put into any quantum state simultaneously. Parking cars is like arranging fermions; one cannot park two or more cars in the same parking space. Once a parking space is filled, it is impossible to add more cars. Likewise, once a quantum state is filled it is impossible to add more fermions. Alternatively, any number of bosons can be added to a state, provided (as we will see below) that the total energy and, in certain cases, particle number are conserved.

Now let us turn our attention to determining the form of the functions $D(E)$ and $f(E)$. If there are no accessible states of energy E then clearly there cannot be any particles present of energy E in the system. Therefore it is clear that the probability density function must depend on the number of available states in the system. In turn, $f(E)$ depends on the probability that an accessible state is occupied. For example, if the state lies at an extremely high energy, the probability of finding a particle within that state may be practically zero. Clearly, the product of the number of available states and the probability that each state is occupied yields the probability function that describes the number of particles in each state $N(E)$. $N(E)$ can be interpreted in one of two ways: either as the number of particles of energy E, or, when normalized, as the probability of finding a particle with energy E. Let us now find an expression for the three-dimensional density of states function $D(E)$. The determination of $f(E)$ is far more complicated and will be considered in the remainder of the chapter.

In general the density-of-states function $D(E)$ can be defined as

$$D(E) = \frac{1}{L^d} \frac{\partial N}{\partial E}, \tag{5.1.1}$$

where d is the dimensionality and N is the number of states. In words, $D(E)$ can be defined as the number of states at energy E in the volume L^d. The derivative $\partial N/\partial E$ can be rewritten in terms of k as

$$\frac{\partial N}{\partial E} = \frac{\partial N}{\partial k}\frac{\partial k}{\partial E}. \tag{5.1.2}$$

But

$$\frac{\partial N}{\partial k} = 2\frac{L^d}{(2\pi)^d}, \tag{5.1.3}$$

where we interpret $\partial N/\partial k$ to mean the number of states at a given k value. The factor of 2 arises from the spin degeneracy. The number of quantum states per dimension is given as $(L/2\pi)^d$ so that the volume occupied by one state is $(2\pi/L)^d$. Proof of this assertion will be given in Chapter 7. Therefore Eq. (5.1.2) can be written as

$$\frac{\partial N}{\partial E} = 2\frac{L^d}{(2\pi)^d}\frac{d^d k}{dE}. \tag{5.1.4}$$

The derivative of k with respect to E can be determined if it is assumed that the electrons are free particles. In that case, the energy of the electrons is fully kinetic, and it is given as

$$E = \frac{p^2}{2m}, \qquad E = \frac{\hbar^2 k^2}{2m}. \tag{5.1.5}$$

Then,

$$\frac{d^d k}{dE} = \frac{d^3 k}{dE} = 4\pi k^2 \frac{dk}{dE}. \tag{5.1.6}$$

The derivative with respect to E, assuming a free-particle spectrum, is

$$\frac{dk}{dE} = \frac{m}{\hbar^2 k}. \tag{5.1.7}$$

The density-of-states function $D(E)$ is found then by substitution of the above results into Eq. (5.1.4) as

$$D(E) = \frac{1}{L^d}\frac{L^d 2}{(2\pi)^3}\frac{m}{\hbar^2 k}4\pi k^2. \tag{5.1.8}$$

Simplifying, we obtain

$$D(E) = \frac{mk}{\pi^2 \hbar^2}. \tag{5.1.9}$$

When k is expressed in terms of E, $D(E)$ becomes

$$D(E) = \frac{m^{\frac{3}{2}}\sqrt{2E}}{\hbar^3 \pi^2},\qquad(5.1.10)$$

or, in a more convenient form,

$$D(E) = \frac{1}{2\pi^2}\left(\frac{2m}{\hbar^2}\right)^{\frac{3}{2}}\sqrt{E}.\qquad(5.1.11)$$

The above formula for the density-of-states function $D(E)$ gives the total number of quantum states available of energy E for a free-electron system. As can be seen from Eq. (5.1.11), the density of states increases with the square root of E for a free-electron system. Similar formulas for one- and two-dimensional systems can be derived with the above procedure. The derivation of the one- and the two-dimensional density-of-states formulas are left as an exercise.

5.2 The Fundamental Postulate of Statistical Mechanics

In Chapter 1 of this book, we discussed probability theory as it relates to classical and quantum-mechanical uncertainty. We often do not think of classical mechanics as embodying any uncertainty. However, there is some degree of uncertainty in classical physics owing to the fact that one cannot in practice determine the motion of every particle at all times in any large system. Uncertainty arises then in macrostates for which the calculations needed to characterize the system fully become impossible to perform in closed form. For example, even the three-body problem in classical mechanics cannot be solved exactly; only an iterative solution can be obtained. Systems that involve numerous particles, much larger than Avogadro's number of air molecules for example, clearly cannot be solved exactly such that the motions of each individual particle can be known precisely at all times. Nevertheless, in a philosophical sense, in classical mechanics the uncertainty arises from only our computational limitations, not from an inherent principle of nature. As we discussed in Chapter 1, quantum-mechanical uncertainty is somewhat different. Uncertainty in quantum mechanics implies that there is an inherent uncertainty in the description of any state of a system (single particle or many particles) arising from the fundamental wave–particle duality of nature. Therefore, in a quantum-mechanical ensemble, not only is there uncertainty in the classical sense, that arising from computational limitations in describing the motions of each of the constituent particles in the ensemble, but there is additional uncertainty in a quantum-mechanical sense, an inherent uncertainty of each individual particle's quantum state. In the remaining sections of this chapter, we will discuss the way in which macrostates of either classical or quantum-mechanical particles can be characterized and their macroscopic properties determined. The science by which the macroscopic behavior of ensembles of particles is described is called statistical mechanics.

We begin our study by considering the Fundamental Postulate of Statistical Mechanics. The Fundamental Postulate can be phrased as (see Kittel and Kroemer 1980, p. 29):

A closed system is equally probable to be in any of the quantum states accessible to it. All accessible quantum states of the system are equally likely.

In interpreting the meaning of the Fundamental Postulate it is helpful to first make some definitions. We define a configuration as the set of all possible microstates that yield the same set of macroscopic observables. In other words, a single configuration can have many different arrangements, with each arrangement yielding precisely the same macroscopic variables. For example, let us consider the air molecules in a closed room. Let the room be isolated, forming a closed system. There are a multitude of ways in which the air molecules can be arranged spatially and energetically such that the temperature of the air remains the same. The set of all such arrangements of the air molecules corresponding to one temperature in the room comprises a single configuration. The number of arrangements that comprise a single configuration is different between configurations. Obviously, the number of arrangements that will yield a configuration in which all the molecules are compacted into the same volume and have precisely the same energy is very much smaller than one that will yield a configuration corresponding to a highly random arrangement in space and energy. Clearly then, from the Fundamental Postulate, a system will most likely be found in the configuration that has the greatest number of ways in which it can be formed. As we will see below, the most likely configuration of a system is the equilibrium configuration. Therefore, any closed system, if left to itself over a sufficiently long time, will always be found in equilibrium because the equilibrium configuration is simply the most probable one; it contains the largest number of equivalent arrangements. In fact, the multiplicity of the equilibrium configuration is many orders of magnitude greater than that of any other configuration. Subsequently, it is obvious then why a closed system is always found in equilibrium; it is simply the most likely configuration in which to find the system because it has the greatest multiplicity. Therefore the key to understanding equilibrium is first to determine how many ways a system can be arranged to yield the same macrostate: one in which the macroscopic observables remain invariant even though the arrangements on a microscopic scale are different.

From the above discussion, it is clear that we need to construct a multiplicity function that enumerates the number of ways in which a system can be assembled. We can then determine the equilibrium configuration by finding the maximum of the multiplicity function. Let us consider a simple example of a system and its corresponding multiplicity function. Let an arrow represent the state of each cell of a system. For simplicity, assume that each arrow can be in one of only two states, up or down. Each arrow is said to have only two possible orientations or degrees of freedom, but the probability that it is up or down is assumed to be completely independent of all the other arrows; there are no correlation effects present. Such a system is said to be uncorrelated. The total number of arrangements of N arrows in N cells is given by

$$2 \times 2 \times 2 \times 2 \cdots = 2^N, \tag{5.2.1}$$

the product of the possible number of orientations of the arrows in all the cells. We define a state of the system then by giving the orientation of the arrow in each cell. Therefore there are 2^N possible microstates in which the system can be arranged.

If there are only two arrows, $N = 2$, then only two cells exist. Each arrow can be either up or down. How many microstates are possible? In other words, how many possible states can the system be arranged in? The answer is found in two ways. First, simply enumerate the possible states and count. Given that there are only two arrows and each arrow can be only up or down, there are only the following states,

$$\uparrow_1\uparrow_2 \quad \uparrow_1\downarrow_2 \quad \downarrow_1\uparrow_2 \quad \downarrow_1\downarrow_2, \tag{5.2.2}$$

which gives four possible states. Alternatively, we can determine the number of states from the simple formula $2^N = 2^2 = 4$.

As a further example, consider $N = 3$. In this case, three cells exist. The number of possible states is found again from $2^N = 2^3 = 8$. It is also easy to simply enumerate the number of possible states as

$$
\begin{aligned}
&\uparrow_1\uparrow_2\uparrow_3 \quad \downarrow_1\uparrow_2\uparrow_3 \quad \downarrow_1\downarrow_2\uparrow_3 \\
&\uparrow_1\uparrow_2\downarrow_3 \quad \uparrow_1\downarrow_2\downarrow_3 \quad \downarrow_1\downarrow_2\downarrow_3 \\
&\uparrow_1\downarrow_2\uparrow_3 \quad \downarrow_1\uparrow_2\downarrow_3 \, .
\end{aligned} \tag{5.2.3}
$$

These states can be arranged with a shorthand notation as

$$(\uparrow + \downarrow)^3 = \, \uparrow\uparrow\uparrow \, + 3 \uparrow\uparrow\downarrow \, + 3 \uparrow\downarrow\downarrow \, + \, \downarrow\downarrow\downarrow, \tag{5.2.4}$$

where $3 \uparrow\uparrow\downarrow$ represents three possible states that comprise of two arrows up and one arrow down with this general configuration. In other words, there are three microstates forming the configuration of two arrows up and one arrow down. Note that there is only one microstate corresponding to the configuration in which all the arrows are up or all the arrows are down.

All the possible states of a two-level system can be generated from the following expansion,

$$(x + y)^N = x^N + Nx^{N-1}y + \frac{1}{2}N(N-1)x^{N-2}y^2 + \cdots + y^N, \tag{5.2.5}$$

or, expressed as a sum,

$$(x + y)^N = \sum_{t=0}^{N} \frac{N!}{(N-t)!t!} x^{N-t} y^t. \tag{5.2.6}$$

In the above, use has been made of the binomial theorem, where t is defined as the number of arrows in one state and $(N-t)$ is the number of arrows in the second state. The function $(x+y)^N$ is called the generating function since it generates all the possible microstates of the two-state system. The generating function can be written in symmetric form if t is rewritten as $1/2N - S$. The sum then goes from $-N/2$ to $+N/2$. Equation (5.2.6) becomes $\uparrow$ substituted in place of x and $\downarrow$ in place of y:

$$(x + y)^N = \sum_{S=-\frac{N}{2}}^{S=+\frac{N}{2}} \frac{N!}{(\frac{N}{2}+S)!(\frac{N}{2}-S)!} x^{\frac{N}{2}+S} y^{\frac{N}{2}-S},$$

$$(\uparrow + \downarrow) = \sum_{S=-\frac{N}{2}}^{S=\frac{N}{2}} \frac{N!}{(\frac{N}{2}+S)!(\frac{N}{2}-S)!} \uparrow^{\frac{N}{2}+S} \downarrow^{\frac{N}{2}-S}. \tag{5.2.7}$$

The coefficient of each term in the expansion given in Eqs. (5.2.7) gives the number of microstates having $N_\uparrow$ arrows up $(N_\uparrow = N/2 + S)$ and $N_\downarrow$ arrows down $(N_\downarrow = N/2 - S)$. Let $g(N, S)$ be the multiplicity function that is simply the coefficient in Eqs. (5.2.7). $g(N, S)$ is defined then for a two-level system as

$$g(N, S) = \frac{N!}{(\frac{N}{2}+S)!(\frac{N}{2}-S)!} = \frac{N!}{N_\uparrow! N_\downarrow!}. \tag{5.2.8}$$

We can now write the generating function for the two-level system by using the multiplicity function as

$$(\uparrow + \downarrow)^N = \sum_{S=-\frac{N}{2}}^{\frac{N}{2}} g(N, S) \uparrow^{\frac{N}{2}+S} \downarrow^{\frac{N}{2}-S}. \tag{5.2.9}$$

The generating function produces all the possible microstates of the system.

Let us illustrate the use of the generating function by expanding out the $N = 2$ case again. The generating function given by Eq. (5.2.8) can be expanded as

$$(\uparrow + \downarrow)^2 = (\uparrow_1 + \downarrow_1)(\uparrow_2 + \downarrow_2)$$
$$= \uparrow_1\uparrow_2 + \uparrow_1\downarrow_2 + \downarrow_1\uparrow_2 + \downarrow_1\downarrow_2, \tag{5.2.10}$$

which agrees with that obtained before in Eq. (5.2.1). For N cells with 2 degrees of freedom, the generating function can be written as

$$(\uparrow_1 + \downarrow_1)(\uparrow_2 + \downarrow_2)(\uparrow_3 + \downarrow_3) \cdots (\uparrow_N + \downarrow_N). \tag{5.2.11}$$

The total number of microstates of each type, same number of arrows up and down, is given by the multiplicity function $g(N, S)$ [Eq. (5.2.7)]. For example, when $N = 2$, the number of microstates corresponding to the configuration that has both arrows up, $N_\uparrow = 2$, and no arrows down, $N_\downarrow = 0$, is

$$\frac{N!}{N_\uparrow! N_\downarrow!} = \frac{2!}{2!\,0!} = 1. \tag{5.2.12}$$

The number of states with one arrow up and the other arrow down, $N_\uparrow = 1$ and $N_\downarrow = 1$, is

$$\frac{N!}{N_\uparrow! N_\downarrow!} = \frac{2!}{1!1!} = 2, \tag{5.2.13}$$

and the number of states with both arrows down, $N_\downarrow = 2$, is

$$\frac{N!}{N_\uparrow! N_\downarrow!} = \frac{2!}{0!2!} = 1. \tag{5.2.14}$$

Comparison of these results with those determined directly from the generating function shows exact agreement.

As a further example consider the case for $N = 3$. From above, we know that there are eight different states. If all we are interested in is the number of cells that have an arrow up while the other arrows are down, not caring which specific arrows are pointed up and which specific arrows are pointed down, then the labels on each arrow can be disregarded. In this case, the number of equivalent microstates for each configuration is found from the multiplicity function $g(N, S)$. The number of states consisting of a single arrow pointing up for $N = 3$ is given by

$$g(N, S) = \frac{N!}{N_\uparrow! N_\downarrow!} = \frac{3!}{1!2!} = \frac{3\,2\,1}{1\,2\,1} = 3. \tag{5.2.15}$$

Therefore the configuration having two arrows down and one arrow up can be formed from three different microstates. This configuration is said to have a multiplicity of 3. Again, this agrees with the result we found from simply counting the number of possible microstates.

Let us now generalize these results to N items arranged in n containers. Let us take the case in which rearrangement of the objects within a container does not count. Subsequently, one must divide out all the possible rearrangements within each bin. The multiplicity function is given in this case as

$$g(N_1, N_2, N_3, \ldots, N_n) = \frac{N!}{N_1! N_2! \cdots N_n!}$$

$$= \frac{N!}{\prod\limits_{i=1}^{n} N_i!}, \tag{5.2.16}$$

where N_1 is the number of items in container 1, N_2 is the number of items in container 2, etc. The multiplicity function g gives the number of ways in which the system can be arranged and still have the same macroscopic value. Full justification for Eq. (5.2.16) is given in Section 5.4, Box 5.4.1.

The above multiplicity function is of great interest to us since it is applicable to classical systems. In classical systems, there are no restrictions on the number of particles that can be placed in any subcontainer and the order of the arrangment of the particles is unimportant. As we will see in Section 5.4, the extremalization of Eq. (5.2.17) will give the Maxwell–Boltzmann distribution which governs classical systems.

Before we end this section, it is useful to point out that some particles can be considered to be distinguishable while others must be considered to be indistinguishable. Particles that are considered to be distinguishable are such that one can determine each particle at all times from the other particles. For example, consider a collection of pocket billiard balls. Each ball is easily identified from one another; each is painted a different color and is numerically labeled 1, 2, 3, ..., etc. Clearly, for such a system the particles are distinguishable for all time. What about a collection of gas molecules though? Are these particles also distinguishable? In this case, the molecules cannot be painted or easily tagged. Nevertheless, we can film the motion of the molecules, label each one in each frame of the film, and then watch how they move in the following movie sequence. In principle then, each particle can be distinguished from one another. This is what is meant by distinguishable particles. In a classical system, it is always assumed that the particles are distinguishable and the counting statistics are chosen to reflect this. In Section 5.7, we will consider this point again when we talk about quantum-mechanical particles.

5.3 **Thermal Equilibrium**

In Section 5.2 we have learned that a closed system can be in any of the possible microstates accessible to it with equal probability. This is the underlying assumption of the Fundamental Postulate. We have also found that there are many microstates that have the same macroscopic values. The set of all such microstates with the same set of macroscopic observables is called a configuration. The multiplicity function enumerates the number of microstates within any given configuration. The most likely configuration in which the system will be is simply the configuration that comprises of the largest number of microstates. Therefore it is important to determine the multiplicity function for a given system since its maximum gives the most likely configuration of the system. The configuration with the greatest multiplicity, as we will see below, is defined as the equilibrium configuration of the system.

To best understand equilibrium, consider two subsystems that are placed into thermal contact. Thermal contact is defined as the condition in which energy can be freely exchanged between the two subsystems. The total system S is formed by the sum of the two subsystems S_1 and S_2. S is assumed to be closed, which

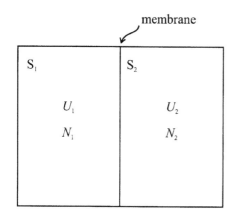

Figure 5.3.1. Schematic drawing of a container partitioned into two separate parts S_1 and S_2. The subcontainers are separated by a thin membrane that allows for energy transfer but not particle transfer.

means that energy cannot flow out of the total system or flow in from the external environment. Therefore the total energy of the system $U = U_1 + U_2$ must be conserved. For the time being, particle exchange between the two subsystems is assumed to be prohibited. Only energy can be exchanged between S_1 and S_2. An example of such a situation is that of a container of gas molecules partitioned by an impermeable membrane. Energy can be exchanged across the membrane, but the total number of particles on each side of the container remains the same. The question then is, what determines whether there is a net flow of energy from one system into the other?

Let the two subsystems S_1 and S_2 have energies U_1 and U_2, as shown in Figure 5.3.1. From the discussion in Section 5.2, the most likely state of the total system S is that which has the largest multiplicity. In other words, the most probable macroscopic configuration is the one that has the largest number of ways by which it can be configured. Let $g_1(N_1, U_1)$ be defined as the multiplicity function for subsystem 1 and $g_2(N_2, U_2)$ be defined as the multiplicity function for subsystem 2. If the two subsystems are placed into thermal contact, the total number of accessible states of the system is now much larger. This is due to the fact that the combined system $S_1 + S_2$ has many more states than either of the two separate subsystems. The number of accessible states of the combined system is given by the product of the separate multiplicities $g_1(N_1, U_1)$ and $g_2(N_2, U_2)$. The sum over all the possible configurations of the resulting system gives the total multiplicity of the combined system, $g(N, U)$. $g(N, U)$ is found as

$$g(n, U) = \sum_{U_1 \leq U} g_1(N_1, U_1) g_2(N_2, U_2)$$

$$= \sum_{U_1 \leq U} g_1(N_1, U_1) g_2(N_2, U - U_1). \tag{5.3.1}$$

What the above means in words is the following. Initially system 1 has energy U_1 and particle number N_1 with a configurational multiplicity of $g_1(N_1, U_1)$. System 2 has energy U_2 and particle number N_2 with a configurational multiplicity of $g_2(N_2, U_2)$. The total number of states in one configuration of the resulting combined system is equal to the product of the multiplicities of each of the component subsystems $g_1(N_1, U_1)$ and $g_2(N_2, U_2)$. However, there are many more configurations of the combined system S such that the total energy U is constant $U = U_1 + U_2$. For example, all the energy could be in subsystem 1. In this case, $U_1 = U$ and $U_2 = 0$. Alternatively, all the energy could be transferred into container 2. Then $U_2 = U$ and $U_1 = 0$. All partitions of the energy between the two containers are possible. The total number of new accessible states in the combined system is equal to the sum over all possible configurations (different values of U_1 in the sum above) multiplied by the multiplicity of each configuration. Therefore we find the total multiplicity of the combined system by summing the product of the multiplicities over all the possible ways in which the energy can be partitioned between the subsystems as described by Eq. (5.3.1).

What, though, does $g(N, U)$ look like for large N, a system that contains many particles? For purposes of illustration, let us first consider a simple two-state system consisting of N particles. As before, each particle can be in one of only two possible states. We found in Section 5.2 that the binomial distribution enumerates all the possible states of a two-state system. The multiplicity function $g(N, S)$ is equal to the coefficient in Eq. (5.2.7). This is

$$g(N, S) = \frac{N!}{(\frac{N}{2} + S)!(\frac{N}{2} - S)!},$$
(5.3.2)

where $N_\uparrow = N/2 + S$ and $N_\downarrow = N/2 - S$. We will see that as N becomes very large, $g(N, S)$ is very greatly peaked near $S = 0$. Taking the logarithm of $g(N, S)$ yields

$$\log g(N, S) = \log N! - \log(N/2 + S)! - \log(N/2 - S)!.$$
(5.3.3)

Substituting in Eq. (5.3.3) for $N/2 + S$, $N_\uparrow$ and $N/2 - S$, $N_\downarrow$,

$$\log g(N, S) = \log N! - \log N_\uparrow! - \log N_\downarrow!.$$
(5.3.4)

We can simplify Eq. (5.3.4) by using Stirling's approximation, which is given as

$$\log N! \sim \frac{1}{2} \log 2\pi + \left(N + \frac{1}{2}\right) \log N - N,$$

$$\log N_\uparrow! \sim \frac{1}{2} \log 2\pi + \left(N_\uparrow + \frac{1}{2}\right) \log N_\uparrow - N_\uparrow,$$

$$\log N_\downarrow! \sim \frac{1}{2} \log 2\pi + \left(N_\downarrow + \frac{1}{2}\right) \log N_\downarrow - N_\downarrow.$$
(5.3.5)

Another version of Stirling's approximation, which will be used extensively throughout this chapter, is

$$\log N! \sim N \log N - N,$$
(5.3.6)

where the terms $1/2 \log 2\pi$ and $1/2 \log N$ are neglected. This, of course, holds for very large N, which is typically the case.

We can simplify the first of approximations (5.3.5) by adding $1/2 \log N$ and subtracting $1/2 \log N$ to the right-hand side. We obtain

$$\log N! \sim \frac{1}{2} \log 2\pi + \frac{1}{2} \log N - \frac{1}{2} \log N + \left(N + \frac{1}{2}\right) \log N - N. \qquad (5.3.7)$$

The total number of particles N is simply equal to the sum of the number pointing up, $N_\uparrow$, and the number pointing down, $N_\downarrow$. Substituting in Eq. (5.3.7) for N, $N_\uparrow + N_\downarrow$, we find that approximation (5.3.7) becomes

$$\log N! \sim \frac{1}{2} \log \frac{2\pi}{N} + \frac{1}{2} \log(N_\uparrow + N_\downarrow)$$
$$+ \left(N_\uparrow + N_\downarrow + \frac{1}{2}\right) \log(N_\uparrow + N_\downarrow) - (N_\uparrow + N_\downarrow). \qquad (5.3.8)$$

When terms are combined, approximation (5.3.8) reduces to

$$\log N! \sim \frac{1}{2} \log \frac{2\pi}{N} + \left(\frac{1}{2} + \frac{1}{2} + N_\uparrow + N_\downarrow\right) \log(N_\uparrow + N_\downarrow) - (N_\uparrow + N_\downarrow). \qquad (5.3.9)$$

The multiplicity function g can now be simplified by substituting in for each logarithmic term in the expression for $\log g$,

$$\log g = \log N! - \log N_\uparrow! - \log N_\downarrow! \qquad (5.3.10)$$

as

$$\log g = \frac{1}{2} \log \frac{2\pi}{N} + \left(N_\uparrow + \frac{1}{2} + N_\downarrow + \frac{1}{2}\right) \log(N_\uparrow + N_\downarrow) - (N_\uparrow + N_\downarrow)$$
$$- \frac{1}{2} \log 2\pi - \left(N_\uparrow + \frac{1}{2}\right) \log N_\uparrow + N_\uparrow - \frac{1}{2} \log 2\pi$$
$$- \left(N_\downarrow + \frac{1}{2}\right) \log N_\downarrow + N_\downarrow. \qquad (5.3.11)$$

The above can be simplified and the terms involving 2π can be combined to yield

$$\log g = \frac{1}{2} \log \frac{1}{2\pi N} - \left(N_\uparrow + \frac{1}{2}\right) \log \frac{N_\uparrow}{N} - \left(N_\downarrow + \frac{1}{2}\right) \log \frac{N_\downarrow}{N}. \qquad (5.3.12)$$

Therefore $\log g$ becomes

$$\log g \sim \frac{1}{2} \log \frac{1}{2\pi N} - \left(N_\uparrow + \frac{1}{2}\right) \log \frac{N_\uparrow}{N} - \left(N_\downarrow + \frac{1}{2}\right) \log \frac{N_\downarrow}{N}. \qquad (5.3.13)$$

But $\log(N_\uparrow/N) = \log(1/2 + S/N)$ from the definition of $N_\uparrow$, since $N_\uparrow$ is equal to the sum of $N/2$ and S. Therefore

$$\log\left(\frac{N_\uparrow}{N}\right) = \log\frac{1}{2}\left(1 + \frac{2S}{N}\right)$$

$$= -\log 2 + \log\left(1 + \frac{2S}{N}\right) \sim -\log 2 + \frac{2S}{N} - \frac{2S^2}{N^2}, \tag{5.3.14}$$

where $\log(1 + 2\,S/N)$ has been expanded by use of the approximation

$$\log(1 + x) \sim x - \frac{x^2}{2} + \cdots . \tag{5.3.15}$$

Similarly

$$\log\left(\frac{N_\downarrow}{N}\right) \sim \log\frac{1}{2}\left(1 - \frac{2S}{N}\right) \sim -\log 2 - \frac{2S}{N} - \frac{2S^2}{N^2}. \tag{5.3.16}$$

We finally obtain

$$\log g \sim \frac{1}{2}\log\frac{2}{\pi N} + N\log 2 - \frac{2S^2}{N}, \tag{5.3.17}$$

which can be rewritten as

$$g(N, S) \sim \sqrt{\frac{2}{\pi N}}\, 2^N e^{\frac{-2S^2}{N}}, \tag{5.3.18}$$

which is the well-known Gaussian distribution. Note that when N, the number of items, is very large the distribution is sharply peaked about $S = 0$. Remember that $g(N, S)$ is the multiplicity function for a binary system wherein N is the total number of particles and S is defined as

$$S = \frac{1}{2}N - t, \tag{5.3.19}$$

where t is the number of arrows pointing down and $N - t$ is the number of arrows pointing up. If one-half of the arrows are up and the other half are down, then $S = 0$. From approximation (5.3.18) it is clear that the Gaussian distribution is peaked (attains its maximum value) when the system is such that half of the arrows are up while the remainder are down, that is, when $S = 0$. The Gaussian distribution given by approximation (5.3.18) is sketched in Figure 5.3.2.

To see how sharply peaked the Gaussian distribution is, let us compare two different configurations, one in which the number of arrows pointing up and down are the same, $S = 0$, and the other in which a relatively small amount more points up than down. What we are interested in determining is the relative value of $g(N, S)$ under both of these conditions. Recall that $g(N, S)$ is the multiplicity function and gives the total number of ways in which the system can be arranged

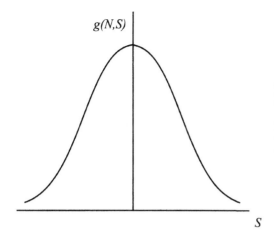

$g(N,S)$

S

Figure 5.3.2. Sketch of the Gaussian distribution $g(N, S)$ as a function of S. The distribution is peaked at $S = 0$, which corresponds to one-half of the two-level system's being in one state and the other half's being in the second state.

in each configuration. The dominant term in the expression for the distribution, approximation (5.3.18), is the exponential term. Therefore $g(N, S)$ goes as

$$g(N, S) \sim e^{-\frac{2S^2}{N}}. \tag{5.3.20}$$

When $S = 0$, this term is simply equal to 1. Consider its value when the system has a value S different from 0 by an amount δ. Let N be taken to be of the same order as Avogadro's number, 10^{23}. Let δ be 10^{13}, a large number, but relatively small compared with N. Then the fraction δ/N is exceedingly small. It is

$$\frac{10^{13}}{10^{23}} = 10^{-10}, \tag{5.3.21}$$

which implies a small fractional deviation from the condition $S = 0$. The question is, what is the probability, given by the Gaussian distribution, of finding the system in the configuration in which $S = 10^{13}$? Again, we recognize that the exponential term dominates in the expression for the Gaussian distribution. Hence, substituting $S = \delta$ in the exponent of approximation (5.3.20), the magnitude of the exponent becomes

$$\frac{2\delta^2}{N} = 2 \times 10^3, \tag{5.3.22}$$

which yields for $g(N, S)$

$$g(N, S) \sim e^{-2 \times 10^3} = e^{-2000} \sim 0. \tag{5.3.23}$$

Therefore the probability of finding the system in a state in which 10^{13} more arrows point up than down when the total system contains 10^{23} arrows is e^{-2000} times less likely than when the same number of arrows points up as down. Clearly, it is unlikely that the system will be observed to be in the configuration in which there is an appreciable difference in the number of arrows pointing up or down. Therefore there is essentially no probability of finding the system by chance in the configuration in which the fractional deviation from equal numbers of arrows pointing up and down is only 10^{-10}. This means that the probability of the system's having a deviation of only one in 10^{10} is essentially zero!

The most likely configuration for the system is then the one in which the same number of arrows points up as points down. This configuration is called the equilibrium configuration of the system. We define equilibrium then as the following.

> **The equilibrium configuration is the most likely configuration of the system. It is the configuration that has the greatest multiplicity.**

Therefore the reason why a closed system is always found to be in equilibrium is rather obvious. This is because the equilibrium configuration is simply the most likely configuration, and all other configurations that are measurably different from the equilibrium configuration are so highly unlikely that they are never observed. This point is discussed more in Box 5.3.1.

Box 5.3.1   Large Numbers and Thermodynamic Irreversibility

Let us consider the same two-level system as discussed in the text as the number of arrows increases toward a very large number. When only a few arrows are present, the probability of finding the system in a highly ordered state is relatively high. For example, if only three arrows are present, then the probability of finding all three arrows up is 1/8, as can be seen from Figure 5.3.3.a, since there is only one microstate in the configuration in which all the arrows are up and there are eight total microstates within the system. The probability of finding the system with two arrows up and one down is 3/8, somewhat higher than that for which all the arrows are up.

As the number of arrows increases, the probability of finding the system in the most ordered state becomes progressively less. To see this, consider part b in Figure 5.3.3. In this case, there are now five arrows present. There is still only one state in the configuration for which all arrows point up. However, there are now several configurations present, each of which comprises several microstates. For example, there are ten microstates in each of the configurations for three or two arrows up. The total number of possible microstates for this system can be counted up to thirty two or is given by the formula $2^N = 2^5 = 32$. Since every microstate is equally likely, then the probability of finding all the arrows up is 1/32. Comparing this with the case for $N = 3$ and all the arrows are up, we find that there is a significant reduction in probability.

Finally, consider the case in which the number of arrows is increased to 10^{23}. Again there is only one microstate comprising the configuration of all arrows up. Note that there are $N = 10^{23}$ microstates in the configuration of only one arrow down or $N - 1$ up. The probability then of finding all up as opposed to only one down is $1/10^{23}$, a remarkably small probability. The probability of finding the system in the configuration of all up overall is just $1/2^{(10^{23})}$. This is a virtually incomprehensibly small number.

The important issue to note from this example is that as the number of elements in an ensemble increases, the probability of finding the system in a highly ordered state (assuming of course that the system is closed) is incomprehensibly small. Only when the number of particles is very small is there a reasonably high probability of finding it in an ordered

continued

Box 5.3.1, cont.

state. However, even for only five particles, the probability of finding the system in the most ordered state is 1/32. Clearly, as the number of particles increases, the probability of finding the system in a highly ordered state is ridiculously small.

The above example provides the basis for understanding thermodynamic irreversibility. Consider the following thought experiment. Consider a closed system consisting of a box with some visible particulates in it. Let there be only three particulates in the box; the dimensions of the box are initially scaled such that they are of the order of, say, ten times the particulates' dimensions. The actual dimensions of the box are immaterial as long as the box is not so large that the particulates would be highly unlikely to sample any particular spatial region within it. Next we take a video image of the motions of the particulates in the box. If the video is replayed later at either forward or reverse speed, it is virtually impossible to tell which direction in time the video is shown. For example, let the particulates be initially together near the center of the box. They may later move outward away from one another toward the edges of the box. If the box is not very large (which it is not by design in our example) then there is a good chance that the particulates will be reflected by the walls back toward the center of the box. Therefore all three particulates can come together again at a later time with a reasonable probability, provided the box dimensions are relatively small. Therefore, if the video is shown forward or backward in time, there are no clues to the observer that enables him/her to determine the temporal sense of the process.

However, consider the case in which instead of only three particulates present there are now, say, 10,000, or 1 million. Let the dimensions of the box be rescaled to ensure the same average spacing between the particulates as before. Again let us take a video image of the system and start it with all the particulates clustered together in the center of the box. We accomplish this initial condition perhaps by first compressing the particulates and then releasing them. Now if the video is shown either forward or backward in time, it is obvious which replay direction is consistent with the original temporal direction. Anyone would recognize that a large number of particles do not go from a highly disordered state initially into a highly ordered state in a closed system. Whereas the converse is commonly observed, a closed system will move from a highly ordered state (when an internal constraint is removed, in this example after the particles are initially confined to the center and then released) to a highly random state in time. Consequently there is a clear temporal direction to the process when there are many particles present. If there are only a few particles, the system is temporally reversible. However, as the number of particles increases it ultimately becomes temporally irreversible. This irreversiblity we call thermodynamic irreversibility.

Thermodynamic irreversibility arises from the fact that a closed system will always be found when no internal constraints are applied or well after the release of an internal constraint in its most probable configuration. The most probable configuration, which we call the equilibrium configuration, in turn is simply the configuration with the greatest number of microstates. The configuration with the greatest randomness has the greatest number of possible microstates. Therefore the most probable configuration in which we will find a closed system is its most random configuration. Consequently, we do not expect to find a closed system in a highly ordered state if it has been left alone, particularly if the number of constituent particles in the system is very large. As the number of particles increases in the system, significant departures from the equilibrium configuration become extremely unlikely.

Configurations	**Type**	**# of Microstates**
a N=3		
↑↑↑	all up	1
↑↑↓ ↑↓↑ ↓↑↑	2 up	3
↑↓↓ ↓↑↓ ↓↓↑	1 up	3
↓↓↓	zero up	1
b N=5		
↑↑↑↑↑	all up	1
↑↑↑↑↓ ↑↑↑↓↑ ↑↑↓↑↑ ↑↓↑↑↑ ↓↑↑↑↑	4 up	5
↑↑↑↓↓ ↑↑↑↓↓ ↓↓↑↑↑ ↑↓↓↑↑ ↑↓↑↑↓ ↓↑↑↑↓ ↓↑↓↑↑ ↑↓↑↓↑ ↑↓↑↑↓ ↑↑↓↓↑	3 up	10
↑↑↓↓↓ ↓↑↑↓↓ ↓↓↑↑↓ ↓↓↓↑↑ ↑↓↓↓↑ ↑↓↓↑↓ ↓↓↑↓↑ ↓↑↓↑↓ ↑↓↑↓↓ ↓↑↓↓↑	2 up	10
↑↓↓↓↓ ↓↑↓↓↓ ↓↓↑↓↓ ↓↓↓↑↓ ↓↓↓↓↑	1 up	5
↓↓↓↓↓	0 up	1
c N = 10^{23}		
↑↑ ... ↑	all up	1
↑↑...↓ ↑↑...↓↑ etc.	N-1 up	10^{23}
	N/2 up	$N!/[N/2)! \ (N/2)!]$
↑↓...↓ ↓↑↓...↓ etc.	1 up	10^{23}
↓↓...↓	0 up	1

Figure 5.3.3. Sketch showing various configurations and number of microstates for a two-level system.

Next let us find the condition for thermal equilibrium to be achieved between two subsystems brought into contact. Above it was stated that the equilibrium configuration is defined as the most probable configuration of the system. To determine the nature of the equilibrium configuration we must maximize the multiplicity function. The configuration with the greatest multiplicity is then, by definition, the equilibrium configuration.

To maximize the multiplicity function of the combined systems $g(N, U)$ to find the equilibrium configuration, let us first recall that $g(N, U)$ is formed from two separate subsystems $g_1(N_1, U_1)$ (the multiplicity of subsystem 1) and $g_2(N_2, U_2)$ (the multiplicity of subsystem 2), as shown in Figure 5.3.1. Placed into thermal contact, the total multiplicity is given by Eq. (5.3.1) as

$$g(N, U) = \sum_{U_1 \leq U} g_1(N_1, U_1) g_2(N_2, U_2). \qquad (5.3.24)$$

If energy is freely exchanged between the two subsystems 1 and 2, many new configurations of the total system become possible. Each new configuration has a different partitioning of the total energy between the two subsystems 1 and 2. In other words, one possible configuration has all the energy within subsystem 1 while another has all the energy within subsystem 2. Of course all other manners of partitioning the energy between the two subsystems, as indicated by the sum over the energy in Eq. (5.3.24), are allowed. To appreciate fully the fact that the multiplicity function for the entire system is far greater than if the two subsystems did not interact, it is useful to reflect on what the multiplicity function would be in the absence of any interaction. If the two subsystems were not allowed to interact (no energy or particles are exchanged between the two subsystems) then the total multiplicity of the systems, 1 and 2, is simply equal to the product of the individual multiplicities $g_1(N_1, U_1)$ and $g_2(N_2, U_2)$, where the energy of each subsystem remains the same. The total multiplicity of the total system when the subsystems cannot interact differs from that given by Eq. (5.3.1) since there is no longer any sum involved. However, if the two subsystems interact through an exchange of energy, then, as we mentioned above, the energy can be partitioned between the two subsystems in any manner, from all of it in 1 to all of it in 2, including all possible arrangements in between. Therefore the total multiplicity of the interacting system is far greater than that of the noninteracting system.

To find the most probable configuration, the multiplicity function describing the combined system must be maximized. One may at this point ask whether one configuration or state is more favorable than another. In fact one may initially have the impression that the equilibrium configuration is a special configuration in some way since our experience is that a closed system will always return to equilibrium, given sufficient time. A naive view of equilibrium is that a closed system is always found in equilibrium because equilibrium is some mysterious condition. The system "likes it," so to speak!

It is important to realize that there is nothing inherently special or mysterious about the equilibrium configuration. From the Fundamental Postulate of Statistical Mechanics, every accessible state of a system is equally likely. This implies that no matter how unusual a state is (for example, the case discussed above in which all of the energy moves to container 1), it has the same probability of occurring as any other state. The important point is, however, that most configurations can only be formed in relatively very few ways. For example, there are far fewer ways in which the particles can be arranged such that all the energy is within container 1 and none is in container 2 corresponding to the configuration of $U_1 = U$, compared with the case in which the energy is equally distributed between the two subcontainers. The configuration corresponding to equal partitioning of the energy can be formed in many orders of magnitude more ways. Therefore as we saw in the case of the two-level system (Gaussian distribution), it is highly unlikely that the system will be observed in any state or configuration that departs measurably from the equilibrium configuration. Remember that as the number of items becomes large (greater than Avogadro's number) the Gaussian distribution becomes sharply peaked about $S = 0$. Therefore the most probable configuration of the two-level system, defined by approximation (5.3.18), is the one in which $S = 0$, equal numbers of arrows point up and down. This configuration is the equilibrium configuration of the system.

From the above discussion, it should now be obvious why a closed system that has been isolated for a sufficiently long time is always found in equilibrium. It is simply because the equilibrium configuration has the greatest multiplicity by very many orders of magnitude than any other configuration of the system. Therefore, for all practical purposes, an isolated system is always observed to be in any one state of the equilibrium configuration or one unmeasurably close to the equilibrium configuration. Departures from the equilibrium configuration or those close to it, though not forbidden, are extremely unlikely. Highly unusual states are relatively so unlikely that a system, for all practical purposes, can never be found in any one of them. This is of course true for only large N.

Before we derive mathematically the condition of thermal equilibrium, let us consider what is known as Maxwell's demon. Maxwell's demon is essentially defined by the following. From classical mechanics, there is no reason why the atoms in a closed room cannot move simultaneously to one corner, leaving you asphyxiated. Why this never seems to occur puzzled many physicists in the nineteenth century. To explain the fact that extreme configurations of a closed system are never observed, they postulated that these states can never occur because to do so would violate the Second Law of Thermodynamics: that all systems tend toward increasing disorder or entropy. Therefore all the atoms in a closed room cannot move to one corner because such a condition is far more ordered than one in which the atoms are randomly arranged throughout the room, thus violating the Second Law. This is of course true, but the explanation is akin to putting the cart in front of the horse. Any system (or system plus surroundings if the system is not isolated) will always experience a net increase in its entropy when a constraint is removed from the system and it is left alone for a sufficiently long time, owing to the fact that it will move into its most probable configuration, which has the greatest multiplicity and as such must have the greatest randomness or entropy. We see then that the increase in entropy, as required by the Second Law of Thermodynamics, arises quite naturally from the idea that an isolated system will always be found in its most likely configuration, which is also the most disordered or random arrangement possible. Therefore, if an isolated system starts in an ordered state and a constraint is removed, the final state of the system will naturally have a higher entropy than the initial state.

Let us now mathematically determine the conditions for equilibrium. As stated above, the equilibrium configuration is given by the extremum of the multiplicity function $g(N, U)$. Hence

$$dg(N, U) = 0. \tag{5.3.25}$$

The total differential of g is determined from

$$dg = 0 = \left(\frac{\partial g_1}{\partial U_1}\right)_{N_1} g_2 \, dU_1 + \left(\frac{\partial g_2}{\partial U_2}\right)_{N_2} g_1 \, dU_2, \tag{5.3.26}$$

where it is assumed that the particle numbers in each subsystem N_1 and N_2 remain fixed. This implies that there is no exchange of particles between the two subsystems.

The multiplicity function must be maximized under the constraint that the total energy of the system U remains invariant. This can be stated as

$$U = U_1 + U_2, \tag{5.3.27}$$

and for a closed system $dU = 0$. Hence

$$dU = dU_1 + dU_2 = 0. \tag{5.3.28}$$

Therefore $dU_1 = -dU_2$. Using this condition in Eq. (5.3.28) for dg, we obtain for equilibrium

$$\left(\frac{\partial g_1}{\partial U_1} \right)_{N_1} g_2 \, dU_1 = \left(\frac{\partial g_2}{\partial U_2} \right)_{N_2} g_1 \, dU_1,$$

$$\frac{1}{g_1} \left(\frac{\partial g_1}{\partial U_1} \right)_{N_1} = \frac{1}{g_2} \left(\frac{\partial g_2}{\partial U_2} \right)_{N_2},$$

$$\left[\frac{\partial (\log g_1)}{\partial U_1} \right]_{N_1} = \left[\frac{\partial (\log g_2)}{\partial U_2} \right]_{N_2}. \tag{5.3.29}$$

Let $\sigma_1 = \log g_1$ and $\sigma_2 = \log g_2$. The equilibrium condition becomes then

$$\left(\frac{\partial \sigma_1}{\partial U_1} \right)_{N_1} = \left(\frac{\partial \sigma_2}{\partial U_2} \right)_{N_2}. \tag{5.3.30}$$

We call σ the entropy of the system. σ is equal to the logarithm of the multiplicity of states, or more generally, the entropy is a measure of the number of ways in which a system can be arranged. A system will always move to the condition of maximum entropy simply because in that configuration (called the equilibrium configuration) there is the greatest multiplicity.

The common measure of the internal energy of a system is the temperature. If a system is in thermal equilibrium, its internal energy is everywhere the same. Hence the temperature of the two subsystems 1 and 2 must be equal:

$$T_1 = T_2. \tag{5.3.31}$$

We can then define the temperature in terms of the entropy as

$$\frac{1}{T_1} = k_B \left(\frac{\partial \sigma_1}{\partial U_1} \right)_{N_1}, \qquad \frac{1}{T_2} = k_B \left(\frac{\partial \sigma_2}{\partial U_2} \right)_{N_2}. \tag{5.3.32}$$

σ can be converted into S, which is called the conventional entropy and is commonly used in thermodynamics as $S = k_B \sigma$. k_B is called Boltzmann's constant and has the value 8.62×10^{-5} eV/K. Therefore the temperature can be written as

$$\frac{1}{T} = \frac{dS}{dU}, \tag{5.3.33}$$

or, in more familiar form,

$$T\,dS = dU,\tag{5.3.34}$$

which is simply the mathematical statement of the Second Law of Thermodynamics.

Let us next show mathematically that the total entropy change when two systems are brought into thermal contact is either unchanged or increased. Consider the total entropy change $\Delta\sigma$, given as the total differential of σ at constant N:

$$\Delta\sigma = \left(\frac{\partial\sigma_1}{\partial U_1}\right)_{N_1}\Delta U_1 + \left(\frac{\partial\sigma_2}{\partial U_2}\right)_{N_2}\Delta U_2.\tag{5.3.35}$$

Again the total energy of the system is invariant, implying that

$$\Delta U = \Delta U_1 + \Delta U_2 = 0.\tag{5.3.36}$$

Therefore the energy change of subsystem 1 must equal the negative of the energy change of subsystem 2. Defining Δ to be equal to the magnitude of the energy change, we have

$$\Delta U_1 = -\Delta U_2 = -\Delta.\tag{5.3.37}$$

The above definition implies that subsystem 1 has lost energy Δ while subsystem 2 has gained energy Δ. Given the above definition, there is a net transfer of energy from subsystem 1 to 2. With this substitution, the entropy change becomes

$$\Delta\sigma = \left[-\left(\frac{\partial\sigma_1}{\partial U_1}\right)_{N_1} + \left(\frac{\partial\sigma_2}{\partial U_2}\right)_{N_2}\right]\Delta.\tag{5.3.38}$$

But we defined

$$\left(\frac{d\sigma_1}{dU_1}\right)_{N_1} = \frac{1}{k_B T_1}\tag{5.3.39}$$

and

$$\left(\frac{d\sigma_2}{dU_2}\right)_{N_2} = \frac{1}{k_B T_2}.\tag{5.3.40}$$

The change in entropy then becomes

$$\Delta\sigma = \frac{\Delta}{k_B}\left[\frac{1}{T_2} - \frac{1}{T_1}\right].\tag{5.3.41}$$

Recall that energy is transferred from subsystem 1 to subsystem 2 by our choice of definition of Δ, that is, $\Delta U_1 = -\Delta$ and $\Delta U_2 = +\Delta$. From this convention, heat flows from subsystem 1 to subsystem 2. If heat flows from 1 to 2, then the temperature of subsystem 1 must be greater than the temperature of subsystem 2 (heat always flows from the hotter to the colder body). If $T_1 > T_2$ then, from

Eq. (5.3.41), $\Delta\sigma > 0$. Therefore the entropy change must be greater than or equal to zero. Hence we recover the full implications of the Second Law of Thermodynamics.

Before ending this section let us briefly list the four basic laws of thermodynamics:

Zeroth Law: If two systems are each in thermal equilibrium with a third system, they must be in thermal equilibrium with each other.

First Law: Conservation of energy as applied to heat:

$$dQ = dU + dW, \tag{5.3.42}$$

where dW is the work done by a system, dU is the internal energy, and dQ is the heat flow.

Second Law: The entropy of a closed system tends to remain constant or increases when an internal constraint of the system is removed.

Third Law: The entropy of a system approaches a constant value as the temperature approaches absolute zero. As T approaches zero, S goes to zero.

5.4 The Maxwell–Boltzmann Distribution

In Section 5.3 we determined that thermal equilibrium is established when the temperatures of the subsystems are equal. The question now becomes, what is the nature of the equilibrium distribution for a system of N noninteracting gas particles? The distribution function gives the probability of finding a particle at position r, with momentum $\hbar k$, at the time t. We first consider the simplest case, a system of N noninteracting classical gas particles.

The multiplicity function for N objects arranged in n containers is given as (see Box 5.4.1 for justification)

$$Q(N_1, N_2, \ldots, N_n) = \frac{N!}{N_1! N_2! \ldots, N_n!} = \frac{N!}{\prod\limits_{i=1}^{n} N_i!}, \tag{5.4.1}$$

Box 5.4.1   Multiplicity Function for a Classical System

To determine the number of ways in which n objects can be arranged into k distinct groups, consider the following example. Consider the first twelve letters of the alphabet, abcdefghijkl. Let there be three containers into which these letters can be placed. Let us place the further restriction that three and only three letters can be put into the first container, four letters into the second container, and five letters into the third container. The situation is sketched in Figure 5.4.1. Note that there are twelve different possibilities, since all twelve letters are available, for filling the first slot in the first bin. There are then eleven possibilities for filling the second slot in the first bin, since only eleven letters remain. Similarly, there are ten possibilities for filling the third slot since only ten letters remain. Therefore the number of arrangements is given as twelve! However, this gives the total number of arrangements, ignoring the order of the letters in any one container. Note that rearrangements of the letters within the same container do not constitute a different arrangement. In other words, if the letters a, b, and c are put into the first container, the arrangement abc is equivalent to acb, which is equivalent to cba, etc. Therefore it is necessary to divide the total number of arrangements by the number of distinct arrangements of the n_i letters in each group i. Hence the number of ways of partitioning twelve letters into the three groups becomes

$$\frac{12!}{3!4!5!}.$$

In general, the number of distinct arrangements of n particles into k groups containing $n_1, n_2, \ldots, n_k$ objects becomes

$$\frac{n!}{n_1! n_2! n_k!}.$$

The above result is the multiplicity function for n distinguishable particles into k groups. All the particles are distinguishable, and only one particle at a time can be placed into a slot. This is precisely the condition needed for modeling a classical system.

Figure 5.4.1. Determination of the number of arrangements of n particles into k groups with $n_1, n_2, \ldots$, groupings.

a Set of all 12 letters   a b c d e f g h i j k l

b Partitioning into 3 groups   a b c | d e f g | h i j k l

c Add the first letter   c _ _ | _ _ _ _ | _ _ _ _ _
12 possibilities exist

d Add the second letter   c h _ | _ _ _ _ _ | _ _ _ _ _
11 possibilities exist

e Add the third letter   c h j | _ _ _ _ | _ _ _ _ _
10 possibilities exist

Clearly, 12! possible arrangements exist of the 12 letters ignoring the ordering of the letters in any one container. If the orderings of the same elements are considered equal within any container, then the number of possible arrangements of the 12 letters, removing equivalent arrangements of the same letters within each container, is found by dividing by the number of arrangements in each container. This yields,

$$\frac{12!}{3!\ 4!\ 5!}$$

where N_i is the number of objects in container i. Physically, each container corresponds to a state in which each particle can be put. In the classical case there are no restrictions on how many particles can be put into any one container or state. In Section 5.7, when we discuss quantum mechanical particles, we will see that there are restrictions on the number of particles that can be put into any one state for some particles.

The probability of finding the system in a certain configuration is given by the ratio of the multiplicity of the specific configuration divided by the total multiplicity of the system. This follows immediately from the Fundamental Postulate since every state is equally likely. Generally, the microstates into which the particles are arranged are classified in terms of their energy E_i and they each have degeneracy g_i. Consequently each level has g_i degenerate states into which N_i particles can be arranged. The system is assumed to have n independent levels. The multiplicity of the system can be determined then as follows.

The system can be thought of as a set of n containers in which there are g_i separate subcontainers. Therefore the problem becomes, how can we arrange N_i particles in these g_i subcontainers subject to the counting conditions specified for each particle? In the case of a classical system, the particles are assumed to all be distinguishable and that any number of particles can be put into any state or subcontainer simultaneously. A distinguishable particle means that it can be tagged such that each particle in the system is labeled for all time. To discover how many arrangements of the system are possible, let us consider a simple example. Consider the case in which $g_i = 2$ and $N_i = 3$. In other words, there are three particles that are to be arranged into a twofold degenerate state. The number of distinguishable arrangements that can be made of $N_i = 3$ particles in $g_i = 2$ subcontainers is shown in Figure 5.4.2. As can be seen from the figure, three distinguishable particles can be arranged in eight possible arrangements in two subcontainers. This result can be generalized to the case of g_i subcontainers and N_i particles. This gives $g_i^{N_i}$ arrangements. In the case considered above, there are 2^3, or 8, arrangements. Therefore if we have a system that has a particular state containing

Figure 5.4.2. Possible arrangements of three distinguishable particles in two subcontainers. Note that there are eight possibilities or 2^3.

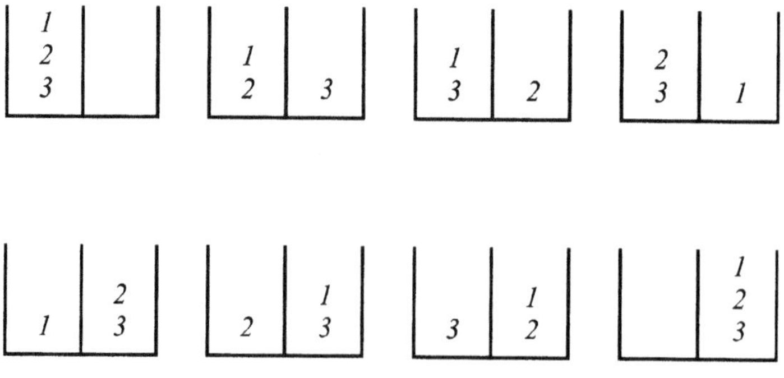

a degeneracy of two into which three particles can be placed, there is an additional multiplicity of eight for this particular state. The total multiplicity function given by Eq. (5.4.1) then has an additional factor of $g_i^{N_i}$ for each of the n states present. Hence the total multiplicity function for a collection of classical particles is

$$Q(N_1, N_2, \ldots, N_n) = \left[\frac{N!}{\prod\limits_{i=1}^{n} N_i!} \right] \prod_{i=1}^{n} g_i^{N_i}. \tag{5.4.2}$$

In the case of a system of classical particles, there are two physical constraints on the system: The total number of particles must be conserved, and the total energy of the system must also be conserved. The first constraint implies that the sum over the particles arranged in each of the n states must be constant and equal to N for all time. This yields

$$\sum_{i=1}^{n} N_i = N. \tag{5.4.3}$$

The second constraint implies that

$$\sum_{i=1}^{n} E_i N_i = U, \tag{5.4.4}$$

where U is the total energy of the system. As above, the equilibrium configuration corresponds to the most probable configuration of the system. Mathematically, this entails maximizing the multiplicity function Q subject to the above constraints. This requires the use of Lagrange multipliers.

Before we maximize the above multiplicity function, it is helpful to pause and review the extremization of a multivariable function subject to constraints. Consider a function of n variables defined as

$$f(x_1, x_2, \ldots, x_n), \tag{5.4.5}$$

where $\phi(x_1, x_2, \ldots, x_n)$ is held constant. The condition on ϕ is the system constraint. An extremum of f is found when $df = 0$. If ϕ is constant then $d\phi = 0$ as well. Therefore one can write

$$df + \alpha \, d\phi = 0, \tag{5.4.6}$$

where α is called a Lagrange multiplier and is simply a multiplicative factor.

The complete differentials in Eq. (5.4.6) can be expanded out in terms of partial derivatives of the coordinates of the system (denoted generally here as x_1, x_2, etc.) as

$$0 = \left(\frac{\partial f}{\partial x_1} + \alpha \frac{\partial \phi}{\partial x_1} \right) dx_1 + \cdots + \left(\frac{\partial f}{\partial x_n} + \alpha \frac{\partial \phi}{\partial x_n} \right) dx_n, \tag{5.4.7}$$

where $i = 1, 2, 3, \ldots n$. The multiplicity function Q now must be maximized. It is easier to use $\ln Q$ rather than Q. Taking the logarithm of Eq. (5.4.2), we obtain

$$\ln Q(N_1, N_2, \ldots, N_n) = \ln N! + \sum_{i=1}^{n} N_i \ln g_i - \sum_{i=1}^{n} \ln N_i!. \tag{5.4.8}$$

Use is made next of Stirling's approximation in the form $\ln x! \sim x \ln x - x$. With this approximation, Eq. (5.4.8) becomes

$$\ln Q = \ln N! + \sum_{i=1}^{n} N_i \ln g_i - \sum_{i=1}^{n} N_i \ln N_i + \sum_{i=1}^{n} \ln N_i!. \tag{5.4.9}$$

N, the total number of particles in the system, is constant. Therefore $\ln N!$ is a constant. We can maximize $\ln Q$ subject to the conditions that $\sum N_i = N$ and $\sum E_i N_i = U$ as

$$\frac{\partial}{\partial N_j} \ln Q + \alpha \frac{\partial \phi}{\partial N_j} - \beta \frac{\partial \psi}{\partial N_j} = 0. \tag{5.4.10}$$

Substituting the expression given by Eq. (5.4.9) in for $\ln Q$ yields

$$\frac{\partial}{\partial N_j} \left(\sum_i N_i \ln g_i - \sum_i N_i \ln N_i + \sum_i N_i \right)$$

$$+ \alpha \frac{\partial}{\partial N_j} \left(\sum_i N_i \right) - \beta \frac{\partial}{\partial N_j} \left(\sum_i E_i N_i \right) = 0. \tag{5.4.11}$$

Note that only those terms for which i is equal to j are nonzero in the derivatives taken above. This follows readily from the definition of a partial derivative; the derivative with respect to the jth variable acts only on the jth terms, leaving the ith terms alone. Equation (5.4.11) becomes

$$\ln g_j - \ln N_j - 1 + 1 + \alpha - \beta E_j = 0,$$

$$\ln N_j = \ln g_j + \alpha - \beta E_j,$$

$$\ln \left(\frac{N_j}{g_j} \right) = \alpha - \beta E_j,$$

$$\frac{N_j}{g_j} = e^{\alpha - \beta E},$$

$$N_j = g_j e^{\alpha} e^{-\beta E}. \tag{5.4.12}$$

The Lagrange multiplier β, as shown below, is equal to $1/k_B T$, where k_B is Boltzmann's constant. k_B is equal to 8.62×10^{-5} eV/K.

The value of β can be determined as follows. If the states are closely spaced in energy, forming a quasi-continuum, the total number of particles present N is given by the integral of $f(E)D(E)\,dE$, as we found in Section 5.1. Using the expression for the density of states $D(E)$ from Eq. (5.1.11) and the distribution function $f(E)$ determined above, we find that the total number of particles N is

$$N = \frac{8\pi m^{3/2}\sqrt{2}}{h^3}e^\alpha \int_0^\infty \sqrt{E}e^{-\beta E}\,dE. \tag{5.4.13}$$

The integral is standard and can be evaluated with the well-known result

$$\int_0^\infty x^n e^{-ax}\,dx = \frac{\Gamma(n+1)}{a^{n+1}}. \tag{5.4.14}$$

Thus

$$\int_0^\infty E^{1/2}e^{-\beta E}\,dE = \frac{\Gamma(3/2)}{\beta^{3/2}}. \tag{5.4.15}$$

The total energy of the N particles can be found from integration of the energy multiplied by the probability distribution and the density-of-states function as

$$E = \int_0^\infty Ef(E)D(E)\,dE. \tag{5.4.16}$$

Substituting in Eq. (5.4.16) for $D(E)$ Eq. (5.1.11) and $f(E)$ Eq. (5.4.12), yields

$$E = \frac{8\pi m^{3/2}\sqrt{2}}{h^3}e^\alpha \int_0^\infty E^{3/2}e^{-\beta E}\,dE, \tag{5.4.17}$$

which becomes

$$E = \frac{8\pi m^{3/2}\sqrt{2}}{h^3}e^\alpha \frac{\Gamma(5/2)}{\beta^{5/2}}. \tag{5.4.18}$$

Taking the ratio of N to E yields

$$\frac{N}{E} = \frac{\Gamma(3/2)}{\Gamma(5/2)}\beta = \beta\frac{\Gamma(3/2)}{3/2\Gamma(3/2)} = 2/3\beta. \tag{5.4.19}$$

The average energy per particle, though, is given as

$$\frac{\bar{E}}{N} = \frac{3}{2}k_B T. \tag{5.4.20}$$

Therefore β is equal to

$$\frac{3}{2}\frac{1}{\beta} = \frac{3}{2}k_B T, \qquad \beta = \frac{1}{k_B T}. \tag{5.4.21}$$

Equations (5.4.12) are called the Maxwell–Boltzmann distribution. They represent the number of particles in the jth state, given that the state has degeneracy g_j. For the distribution to reflect probabilities, it must be properly normalized. We normalize the distribution described by Eqs. (5.4.12) by dividing the number of particles in the jth state N_j by the total number of particles present N. The total number of particles N can be found by summation of Eqs. (5.4.12) over all possible states j. This yields

$$N = \sum_j N_j = e^\alpha \sum_j g_j e^{-\frac{E_j}{k_B T}}. \tag{5.4.22}$$

Solving for e^α, we obtain

$$e^\alpha = \frac{N}{\sum_j g_j e^{-\frac{E_j}{k_B T}}}. \tag{5.4.23}$$

Expression (5.4.23) for e^α can next be substituted into Eqs. (5.4.12) to obtain

$$N_j = \frac{N g_j e^{-\frac{E_j}{k_B T}}}{\sum_j g_j e^{-\frac{E_j}{k_B T}}}. \tag{5.4.24}$$

The normalized distribution is then given as the ratio of N_j to N as

$$\frac{N_j}{N} = P_j = \frac{g_j e^{-\frac{E_j}{k_B T}}}{\sum_j g_j e^{-\frac{E_j}{k_B T}}}, \tag{5.4.25}$$

where P_j is the probability of finding the particle in the jth state with energy E_j. If the degeneracy factor for each state is simply 1, then P_j is equal to

$$P_j = \frac{e^{-\frac{E_j}{k_B T}}}{\sum_j e^{-\frac{E_j}{k_B T}}}. \tag{5.4.26}$$

Recognizing that the index j in the denominator is a dummy index, we find that the probability of a particle having energy E_r is then

$$P_r = \frac{e^{-\frac{E_r}{k_B T}}}{\sum_j e^{-\frac{E_j}{k_B T}}}. \tag{5.4.27}$$

As we determined in Chapter 1, the mean value of a physical observable can be calculated from the probability distribution function as

$$\bar{y} = \frac{\sum_r P_r y_r}{\sum_r P_r}. \tag{5.4.28}$$

The mean value of a physical observable y in a system described by the Maxwell–Boltzmann distribution is then given as

$$\bar{y} = \frac{\sum_r y_r e^{-\beta E_r}}{\sum_r e^{-\beta E_r}}, \tag{5.4.29}$$

where β is $1/k_B T$.

As an example, the mean energy of a system can be determined from

$$\bar{E} = \frac{\sum_r E_r e^{-\beta E_r}}{\sum_r e^{-\beta E_r}}. \tag{5.4.30}$$

The denominator in Eq. (5.4.30) occurs in the evaluation of every mean value in the Maxwell–Boltzmann distribution. Since it occurs so often, the sum over the Boltzmann factor is given a special name, the classical partition function Z. The partition function is defined then as

$$Z = \sum_r e^{-\beta E_r}. \tag{5.4.31}$$

We can easily calculate the mean value of the energy from the partition function Z by noting the following relationships:

$$\sum_r E_r e^{-\beta E_r} = -\sum_r \frac{\partial}{\partial \beta} e^{-\beta E_r} = \frac{\partial}{\partial \beta} - \sum_r e^{-\beta E_r} = -\frac{\partial}{\partial \beta} Z. \tag{5.4.32}$$

Therefore

$$\bar{E} = -\frac{1}{Z} \frac{\partial}{\partial \beta} Z = -\frac{\partial}{\partial \beta} (\ln Z). \tag{5.4.33}$$

Finally,

$$\bar{E} = -\frac{\partial}{\partial \beta} (\ln Z). \tag{5.4.34}$$

Therefore the mean energy of a system can be found from differentiation of the logarithm of the partition function with respect to β.

Let us work a sample problem by using the classical partition function. A simple paramagnetic system can be understood from statistical mechanics. A paramagnet has permanent magnetic dipole moments but no long-range order until an external magnetic field is applied. These materials do not show any magnetic properties except when an external field is applied to the system.

Let H be an external magnetic field. Assume that each magnetic-moment vector μ has only 2 degrees of freedom, up or down. The energy of a magnetic moment in an external field H is given as (see Chapter 3)

$$E = -\vec{\mu} \cdot \vec{H}. \tag{5.4.35}$$

When μ is parallel to H, the energy of the system is given as $-\mu \cdot H$ while when μ is antiparallel to H the energy is equal to $+\mu \cdot H$. The mean magnetic moment of the system can be found with Eq. (5.4.28) as

$$\bar{\mu} = \frac{\sum\limits_r P_r \mu_r}{\sum\limits_r P_r}. \qquad (5.4.36)$$

But

$$P_r = e^{-\frac{E_r}{k_B T}}, \qquad E_r = \pm \mu H. \qquad (5.4.37)$$

Therefore

$$\bar{\mu} = \frac{\sum\limits_r \mu_r e^{-\beta E_r}}{\sum\limits_r e^{-\beta E_r}}. \qquad (5.4.38)$$

Performing the sum in the numerator yields

$$\bar{\mu} = \frac{\mu e^{-\beta(-\mu H)} + (-\mu)e^{-\beta(\mu H)}}{e^{-\beta(-\mu H)} + e^{-\beta(\mu H)}},$$

$$\bar{\mu} = \mu \frac{e^{\beta \mu H} - e^{-\beta \mu H}}{e^{\beta \mu H} + e^{-\beta \mu H}}. \qquad (5.4.39)$$

We further simplify the above result by recognizing that the hyperbolic tangent is defined as

$$\tanh y \equiv \frac{e^y - e^{-y}}{e^y + e^{-y}}. \qquad (5.4.40)$$

With this substitution, the mean value of the magnetic moment of a simple paramagnet can be written as

$$\bar{\mu} = \mu \tanh\left(\frac{\mu H}{k_B T}\right). \qquad (5.4.41)$$

The magnetization is defined as the product of the mean magnetic moment and the number of moments present in the system N_0. The magnetization is then

$$M = N_0 \bar{\mu}. \qquad (5.4.42)$$

Substituting into Eq. (5.4.42) for the mean value of μ Eq. (5.4.41) we find that the magnetization of a simple paramagnet is

$$M = N_0 \mu \tanh\left(\frac{\mu H}{k_B T}\right). \qquad (5.4.43)$$

If $(\mu H / k_B T) \ll 1$, which corresponds to the high-temperature limit, the hyperbolic tangent can be simplified as

$$\tanh\left(\frac{\mu H}{k_T}\right) \sim \frac{\mu H}{k_B T}. \qquad (5.4.44)$$

The mean magnetic moment becomes

$$\bar{\mu} \sim \frac{\mu^2 H}{k_B T} \qquad (5.4.45)$$

and the magnetization is

$$M = N_0 \frac{\mu^2 H}{k_B T}. \qquad (5.4.46)$$

The magnetization is commonly rewritten in terms of the magnetic susceptibility χ. The susceptibility is defined as the ratio of the magnetization to the applied magnetic field H. Hence χ is given as

$$M = \chi H, \qquad \chi = \frac{N_0 \mu^2}{k_B T}. \qquad (5.4.47)$$

Clearly, the susceptibility is inversely proportional to the temperature,

$$\chi \sim \frac{1}{T}, \qquad (5.4.48)$$

which is known as the Curie Law of Magnetism.

In the low-temperature limit, the term $(\mu H / k_B T) \gg 1$. The argument of the hyperbolic tangent then approaches ∞. Recall that $\tanh x$ approaches 1 as x approaches ∞. In this case the mean value of the magnetic moment becomes equal to μ. This is as expected since at low temperatures all the individual magnetic moments will become aligned. Hence the magnetization of the system is

$$M = N_0 \mu. \qquad (5.4.49)$$

The normalized magnetization for this system is plotted as a function of $(\mu H / k_B T)$ in Figure 5.4.3.

As a further example, let us consider the situation in which the magnetic moment of the particles can be oriented at any angle with respect to the applied magnetic field H. In this case, the energy of the interaction is still given by the dot vector product of μ and H. However, the energy of the interaction has a continuum of values, not just two, as was considered above. The interaction energy in this case is given as

$$E_i = -\vec{\mu}_i \cdot \vec{H} = -\mu_z H = -\mu H \cos\theta, \qquad (5.4.50)$$

where it is assumed that the applied field is along the z direction. The mean magnetic moment of the system can be determined in an analogous manner as was done for the two-level system. The major difference in this situation is that the sum over the partition function needs to be replaced by an integral since a continuum of states is now available. In other words, since a dipole can assume

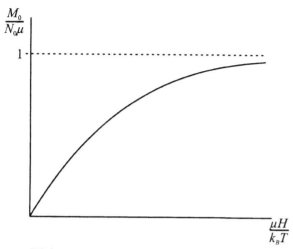

$\dfrac{M_0}{N_0\mu}$

1

$\dfrac{\mu H}{k_B T}$

High
Temperature

Low
Temperature

Figure 5.4.3. Normalized magnetization of a simple paramagnet plotted as
a function of $\mu H / kT$.

any orientation with respect to the applied magnetic field, the energy is a continuous function rather than a discrete function. Hence the sum in Eq. (5.4.29)
for the partition function Z needs to be replaced by an integral. The partition
function is then given as

$$Z = \int_0^\pi \mathrm{d}\theta \, \sin\theta \, e^{-\frac{E_j}{k_B T}} = \int_0^\pi \mathrm{d}\theta \, \sin\theta \, e^{\frac{\mu H\cos\theta}{k_B T}}. \tag{5.4.51}$$

We can evaluate the integral by making the following assignments:

$$y \equiv \frac{\mu H\cos\theta}{k_B T}, \qquad \mathrm{d}y = -\frac{\mu H}{k_B T}\sin\theta\,\mathrm{d}\theta. \tag{5.4.52}$$

The partition function Z can be evaluated as

$$Z = \frac{-k_B T}{\mu H}\int_{\frac{\mu H}{k_B T}}^{\frac{-\mu H}{k_B T}} e^y \, \mathrm{d}y,$$

$$Z = \frac{k_B T}{\mu H}\left(e^{\frac{\mu H}{k_B T}} - e^{\frac{-\mu H}{k_B T}}\right),$$

$$Z = \frac{2k_B T}{\mu H}\sinh\frac{\mu H}{k_B T}. \tag{5.4.53}$$

The mean magnetic moment along the z direction is obtained then by the integration of μ_z multiplied by the Boltzmann factor and all divided by the partition
function. This is given as

$$\bar{\mu}_z = \frac{\int_0^\pi \mu_z e^{\frac{\mu H\cos\theta}{k_B T}}\sin\theta\,\mathrm{d}\theta}{Z}. \tag{5.4.54}$$

We can reexpress the numerator in terms of Z by recognizing that

$$k_B T \left(\frac{\partial Z}{\partial H} \right)_T = \int \mu \cos \theta \, e^{\frac{\mu H \cos \theta}{k_B T}} \sin \theta \, d\theta. \qquad (5.4.55)$$

Therefore the mean value of the z component of the magnetic moment is

$$\overline{\mu_z} = \frac{k_B T}{Z} \left(\frac{\partial Z}{\partial H} \right)_T. \qquad (5.4.56)$$

From here, the mean magnetic moment can be determined by substituting the expression derived above in for Z.

Box 5.4.2 Example Problem

Consider a classical two-state system of energies 0 and ϵ. Determine the heat capacity of the system. The heat capacity is defined as

$$C_V = \frac{\partial \bar{E}}{\partial T},$$

where E is the average energy, T is the temperature, and the volume is fixed.
Solution:
From Eq. (5.4.34), the average energy is found by differentiation of the logarithm of the partition function:

$$\bar{E} = -\frac{\partial}{\partial \beta} \ln Z.$$

The partition function can be found from Eq. (5.4.31) as

$$Z = \sum_{s=0}^{1} e^{-\beta E_s} = e^{-\beta 0} + e^{-\beta \epsilon} = 1 + e^{-\beta \epsilon}.$$

Taking the logarithm yields

$$\ln Z = \ln[1 + e^{-\beta \epsilon}].$$

The average energy becomes then

$$\bar{E} = -\frac{\partial}{\partial \beta} \ln[1 + e^{-\beta \epsilon}] = \frac{\epsilon}{1 + e^{\beta \epsilon}}.$$

We can determine the heat capacity from the above by taking the derivative with respect to the temperature:

$$C_V = \frac{\partial \bar{E}}{\partial T} = \frac{\partial}{\partial T} \left[\frac{\epsilon}{1 + e^{\beta \epsilon}} \right].$$

Taking the derivative yields for the heat capacity

$$C_V = \frac{\frac{\epsilon^2}{kT^2} e^{\frac{\epsilon}{kT}}}{[1 + e^{\frac{\epsilon}{kT}}]^2},$$

which is the final result.

5.5 **Connection to Classical Thermodynamics**

It is interesting and useful to show how the results obtained from statistical mechanics are related to classical thermodynamics. In the study of classical thermodynamics, the most important quantities that are determined are the thermodynamic potentials. The most commonly studied potentials are the enthalpy H, the free energy (or the Helmholtz function) F, the free enthalpy (or Gibb's function) G, and the internal energy U. These potentials are related to one another through the Legendre transforms given as

$$
\begin{aligned}
H &= U + PV, \\
F &= U - TS, \\
G &= H - TS,
\end{aligned}
\tag{5.5.1}
$$

where P is the pressure, V is the volume, S is the entropy, and T is the absolute temperature. From these definitions and the First Law of Thermodynamics, $dU = T\,dS - P\,dV$ (where $P\,dV$ is the work done by the system), differential forms of the thermodynamic potentials can be obtained in terms of one another.

From the First Law of Thermodynamics the temperature can be related to the derivative of the internal energy with respect to the entropy as

$$
\begin{aligned}
dU &= T\,dS - P\,dV, \\
T &= \frac{dU}{dS} + P\frac{dV}{dS}, \\
T &= \left(\frac{\partial U}{\partial S}\right)_V.
\end{aligned}
\tag{5.5.2}
$$

The pressure can also be related to the internal energy as

$$
-P = \left(\frac{\partial U}{\partial V}\right)_S.
\tag{5.5.3}
$$

Finally, we can relate the derivatives by noting that the mixed partial derivatives must be equal, independent of the order in which they are taken. In other words,

$$
\frac{\partial^2 U}{\partial S\,\partial V} = \frac{\partial^2 U}{\partial V\,\partial S}.
\tag{5.5.4}
$$

The derivatives of the temperature T and the pressure P can be related as

$$
\left(\frac{\partial T}{\partial V}\right)_S = -\left(\frac{\partial P}{\partial S}\right)_V.
\tag{5.5.5}
$$

Many such relationships among the thermal potentials and the macroscopic thermodynamic variables can be constructed. These relationships are called the

Table 5.5.1
Maxwell Relations

Exact Differential	Maxwell Relations		
$dU = T\,dS - P\,dV$	$T = \left(\frac{\partial U}{\partial S}\right)_V$	$-P = \left(\frac{\partial U}{\partial V}\right)_S$	$\left(\frac{\partial T}{\partial V}\right)_S = \left(\frac{-\partial P}{\partial S}\right)_V$
$dH = T\,dS + V\,dP$	$T = \left(\frac{\partial H}{\partial S}\right)_P$	$V = \left(\frac{\partial H}{\partial P}\right)_S$	$\left(\frac{\partial T}{\partial P}\right)_S = \left(\frac{\partial V}{\partial S}\right)_P$
$dF = -S\,dT - P\,dV$	$-S = \left(\frac{\partial F}{\partial T}\right)_V$	$-P = \left(\frac{\partial F}{\partial V}\right)_T$	$\left(\frac{\partial S}{\partial V}\right)_T = \left(\frac{\partial P}{\partial T}\right)_V$
$dG = -S\,dT + V\,dP$	$-S = \left(\frac{\partial G}{\partial T}\right)_P$	$V = \left(\frac{\partial G}{\partial P}\right)_T$	$\left(\frac{-\partial S}{\partial P}\right)_T = \left(\frac{\partial V}{\partial T}\right)_P$

Maxwell relations. For reference, these relations along with the defining exact differential and the corresponding relations between the physical quantities, T, P, V, and S are collected in Table 5.5.1. From inspection of Table 5.5.1 it is clear that once one of the thermodynamic potentials is known, all the remaining thermodynamic potentials can be determined from it and knowledge of the macroscopic observables T, V, S, and P. The thermodynamic potential F, the free energy, can be determined from the microscopic physics of statistical mechanics. Let us see how F and the results of classical thermodynamics can be recovered from statistical mechanics.

Consider a system that is characterized by a parameter x. Let x change quasi-statically, meaning slowly in the sense that the system is restored to equilibrium at each instant during the change from x to $x + dx$. The change in energy in the state r arising from a change in x is

$$\Delta E_r = \frac{\partial E_r}{\partial x}\,dx. \qquad (5.5.6)$$

The macroscopic work done by the system dW, as a result of the displacement is found from use of Eq. (5.4.29) as

$$dW = \frac{\sum_r -\Delta E_r e^{-\beta E_r}}{\sum_r e^{-\beta E_r}}. \qquad (5.5.7)$$

Substituting in the expression for ΔE_r into Eq. (5.5.7) yields

$$dW = \frac{\sum_r e^{-\beta E_r}\left(-\frac{\partial E_r}{\partial x}\,dx\right)}{\sum_r e^{-\beta E_r}}. \qquad (5.5.8)$$

We can reexpress the expression for dW in terms of the partition function by noting that the derivative of E_r with respect to x can be written as

$$\sum_r e^{-\beta E_r} \frac{\partial E_r}{\partial x}\, dx = -\frac{1}{\beta} \frac{\partial}{\partial x} \left(\sum_r e^{-\beta E_r} \right) dx,$$

$$= -\frac{1}{\beta} \frac{\partial}{\partial x}(Z)\, dx,$$

$$dW = \frac{\frac{1}{\beta} \frac{\partial}{\partial x}(Z)}{Z}\, dx,$$

$$dW = \frac{1}{\beta} \frac{\partial}{\partial x}(\ln Z)\, dx. \tag{5.5.9}$$

Expand $\ln Z$ by using a complete differential in the variables x and β. This becomes

$$d(\ln Z) = \frac{\partial}{\partial x}(\ln Z)\, dx + \frac{\partial}{\partial \beta}(\ln Z)\, d\beta. \tag{5.5.10}$$

Equations (5.5.9) can be rewritten as

$$\frac{\partial}{\partial x}(\ln Z)\, dx = \beta\, dW. \tag{5.5.11}$$

The average energy can also be written in terms of Z from use of Eq. (5.4.34). This is given as

$$\frac{\partial}{\partial \beta}(\ln Z)\, d\beta = -\bar{E}\, d\beta. \tag{5.5.12}$$

When these substitutions are made, $d(\ln Z)$ can be written as

$$d(\ln Z) = \beta\, dW - \bar{E}\, d\beta. \tag{5.5.13}$$

To reduce $d(\ln Z)$ further, it is useful to note that

$$d(\bar{E}\beta) = \bar{E}\, d\beta + \beta\, d\bar{E}. \tag{5.5.14}$$

Substituting in Eq. (5.5.13) for $\bar{E}\, d\beta$ from Eq. (5.5.14) yields

$$d(\ln Z) = \beta\, dW - d(\bar{E}\beta) + \beta\, d\bar{E},$$
$$d(\ln Z) = \beta(dW + d\bar{E}) - d(\bar{E}\beta). \tag{5.5.15}$$

With the First Law of Thermodynamics,

$$dW + d\bar{E} = dQ. \tag{5.5.16}$$

$d(\ln Z)$ can be further reduced as

$$d(\ln Z) = \beta\, dQ - d(\bar{E}\beta),$$
$$d(\ln Z + \bar{E}\beta) = \beta\, dQ. \tag{5.5.17}$$

With the Second Law of Thermodynamics, $dS = dQ/T$ and the definition of $\beta = 1/k_B T$, the right-hand side of the above equation can be rewritten in terms of the entropy as

$$k_B \, d(\ln Z + \bar{E}\beta) = dS. \tag{5.5.18}$$

Therefore

$$S = k_B \left(\ln Z + \frac{\bar{E}}{k_B T} \right),$$
$$TS = k_B T \ln Z + \bar{E},$$
$$\bar{E} - TS = -k_B T \ln Z. \tag{5.5.19}$$

But the free energy F is defined as

$$F = \bar{E} - TS. \tag{5.5.20}$$

Therefore

$$F = -k_B T \ln Z. \tag{5.5.21}$$

Hence from knowledge of the partition function Z, the free energy F can be readily determined. As a result, all the thermodynamic potentials can then be determined, provided that the macroscopic observables $V, P, S,$ and T are known as well. Subsequently the results of classical thermodynamics are recovered from statistical mechanics.

5.6 **The Grand Partition Function**

In the previous sections we have discussed a closed system that is in thermal equilibrium. It was found that the temperature of the component subsystems is the same when the total system is in thermal equilibrium. This means that no gradient of the internal energy exists. However, in the system discussed in Section 5.3, no exchange of particles between the two subsystems was allowed. In this section, we consider the general case in which both particles and energy are exchanged across the boundary, one in which the system is in both thermal and diffusive contact.

Let us consider a subsystem, denoted in Figure 5.6.1 as S, of a much larger heat bath or reservoir R. The total number of particles present in the system, subsystem plus reservoir, is N while the total energy is U. For simplicity, it is assumed that the reservoir is very much larger than the subsystem. In the diagram, the system is shown in two different configurations. The first configuration has N_1 particles with energy E_1 in the subsystem, leaving $N - N_1$ particles and $U - E_1$ energy in the reservoir, while the second configuration has N_2 particles with energy E_2 in the subsystem, leaving $N - N_2$ particles and $U - E_2$ energy in the reservoir. Since S_1 is precisely specified, the multiplicity of the entire system S_1 and R is given by the multiplicity of just R. A similar situation holds for S_2.

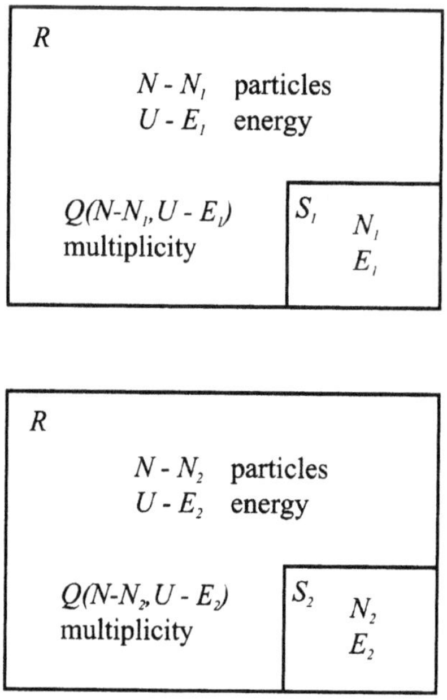

Figure 5.6.1. Two configurations of a system that comprises a reservoir R with a subsystem S. It is assumed that the reservoir is very much larger than the subsystem in either case. Therefore changes to the reservoir can be neglected with respect to those for the subsystem.

The probability of finding the system in the first configuration relative to that of finding it in the second configuration is given then by the ratio of the multiplicities of each configuration of R. Hence the ratio of the probabilities can be written as

$$\frac{P(N_1, E_1)}{P(N_2, E_2)} = \frac{Q_1(N - N_1, U - E_1)}{Q_2(N - N_2, U - E_2)}. \tag{5.6.1}$$

The entropy is defined as $S = k_B \ln Q$, so the multiplicity can be expressed in terms of S as

$$Q = e^{\frac{S}{k_B}}. \tag{5.6.2}$$

Then the ratio of the multiplicities becomes

$$\frac{Q_1}{Q_2} = \frac{e^{\frac{S_1}{k_B}}}{e^{\frac{S_2}{k_B}}} = e^{\frac{(S_1 - S_2)}{k_B}} = e^{\frac{\Delta S}{k_B}}. \tag{5.6.3}$$

But

$$\Delta S = S(N - N_1, U - E_1) - S(N - N_2, U - E_2). \tag{5.6.4}$$

Since the reservoir is assumed to be very large, the change in entropy can be accurately approximated through use of the linear terms in a Taylor series expansion. The expansion about the points U and N following the definition of a

Taylor series yields

$$f(x_0 + a) = f(x_0) + a\left(\frac{df}{dx}\right)_{x=x_0} + \frac{1}{2!}a^2\left(\frac{d^2 f}{dx^2}\right)_{x=x_0}$$

$$S(N - N_1, U - E_1) = S(N, U) - N_1\frac{\partial S}{\partial N} - E_1\frac{\partial S}{\partial U}$$

$$S(N - N_2, U - E_2) = S(N, U) - N_2\frac{\partial S}{\partial N} - E_2\frac{\partial S}{\partial U}. \tag{5.6.5}$$

Therefore the change in entropy becomes

$$\Delta S = -(N_1 - N_2)\frac{\partial S}{\partial N} - (E_1 - E_2)\frac{\partial S}{\partial U}. \tag{5.6.6}$$

The last term above can be reexpressed with the Second Law of Thermodynamics, $1/T = dS/dU$. If we define the chemical potential, μ as

$$-\frac{\mu}{T} = \frac{dS}{dN}, \tag{5.6.7}$$

then the change in entropy ΔS is

$$\Delta S = (N_1 - N_2)\frac{\mu}{T} - \frac{(E_1 - E_2)}{T}. \tag{5.6.8}$$

The chemical potential is a measure of diffusive equilibrium, much like temperature is a measure of thermal equilibrium. When the chemical potential is constant everywhere, there is no net diffusion in the system; there is no net exchange of particles between the subsystem and the reservoir. The chemical potential, like the temperature, describes the equilibrium condition. When the temperature and the chemical potential are the same everywhere within the system, equilibrium is established everywhere throughout the system.

When the expression for the change in entropy is used, the relative probabilities of each configuration's occurring can be related as

$$\frac{P(N_1, E_1)}{P(N_2, E_2)} = \frac{e^{\frac{S_1}{k_B}}}{e^{\frac{S_2}{k_B}}} = e^{\frac{\Delta S}{k_B}} = e^{\frac{(N_1 - N_2)\mu}{k_B T} - \frac{(E_1 - E_2)}{k_B T}}. \tag{5.6.9}$$

The ratio of the probabilities given above is called the Gibb's factor. If a sum is constructed over all the possible states of the system, a new partition function is formed. This partition function is called the grand partition function and has the following form:

$$Z \equiv \sum_n \sum_{E_s} e^{\frac{(\mu n - E_s)}{k_B T}}. \tag{5.6.10}$$

The grand partition function, like the partition function, is useful in determining the average value of the macroscopic variables of the system. Once the grand partition function has been constructed, physically useful parameters, such as the average energy of the system, the average number of particles of a certain energy, etc., can be calculated. To realize this, it is useful to do an example problem determining the grand partition function for a simple system.

Consider a system that comprises a particle that has three allowed occupation numbers, 0, 1, and 2. The particle is assumed to have no spin and obeys the following relationships:

$$E = \frac{p^2}{2m}, \qquad E_1 = E, \qquad E_2 = 2E. \tag{5.6.11}$$

The grand partition function for the system can be derived as follows. From the above definition of Z, it is necessary to construct a sum over all the possible energy states and particle occupation numbers. In other words, Z is formed by summation over all the possible ways in which the particles can be grouped and the corresponding energy of each group. Hence Z in this example is given as

$$Z = \sum_{\substack{n = 0, 1, 2 \\ E_s = 0, E, 2E}} e^{\frac{(\mu n - E_s)}{k_B T}}, \tag{5.6.12}$$

where the sum is formed over the three possible occupation values, $n = 0, 1,$ and 2. Corresponding to each of these possible number of particles in each state is a different energy. When $n = 0$, the energy is zero since the state is unoccupied. Therefore the first term in the sum is 1 since the exponent vanishes. Hence

$$e^{\frac{\mu 0 - 0}{k_B T}} = e^0 = 1. \tag{5.6.13}$$

When the state is occupied by one particle, its energy is E. Hence the second term in the sum is

$$e^{\frac{(\mu - E)}{k_B T}}. \tag{5.6.14}$$

Finally, for $n = 2$, $E = 2E$. Therefore the third and last term in the sum is simply

$$e^{\frac{2(\mu - E)}{k_B T}}. \tag{5.6.15}$$

The grand partition function then for this example is given by the sum of each of the above terms. Z is then

$$Z = 1 + e^{\frac{\mu - E}{k_B T}} + e^{\frac{2(\mu - E)}{k_B T}}. \tag{5.6.16}$$

From knowledge of the grand partition function for the system, the average energy, average particle number, etc., can be determined. In the present example, the average number of particles is found from use of Eq. (5.4.28):

$$\bar{N} = \frac{\sum_r P_r N_r}{\sum_r P_r},$$ (5.6.17)

where P_r is the probability distribution function. The probabilities in this case are given by the Gibb's factors. Therefore the denominator formed by the complete sum over all the possible probabilities is simply the grand partition function Z. We then obtain the average number of particles by expanding out the above sum over the probabilities and the occupation numbers as

$$\bar{N} = \frac{1e^{\frac{(\mu-E)}{k_BT}} + 2e^{\frac{2(\mu-E)}{k_BT}}}{Z}.$$ (5.6.18)

The average energy of the example system is obtained similarly from

$$\bar{E} = \frac{\sum_s E_s e^{\frac{(\mu n_s - E_s)}{k_BT}}}{Z},$$ (5.6.19)

which becomes, on expansion,

$$\bar{E} = \frac{Ee^{\frac{(\mu-E)}{k_BT}} + 2Ee^{\frac{2(\mu-E)}{k_BT}}}{1 + e^{\frac{(\mu-E)}{k_BT}} + e^{\frac{2(\mu-E)}{k_BT}}}.$$ (5.6.20)

As a further example of the grand partition function consider a collection of impurities intentionally introduced into a host semiconductor, such as phosphorous P, atoms in bulk silicon Si. Generally, the electrical properties of a semiconductor can be altered by the intentional addition of impurities, called dopants, that act to change the free-carrier concentration. Special types of impurities called donors form shallow energy levels below the normal free-electron energy band, called the conduction band, in the host semiconductor crystal. The impurity atoms are therefore readily ionized. The outermost electrons in the impurity atoms are so weakly held that they can be easily promoted to the conduction band. As a result, the free-electron carrier concentration within the material is increased. The host semiconductor material, when doped with donor impurities, is said to be *n* type, since it has an excess electron carrier concentration within the conduction band.

We seek to determine how many donor atoms are ionized in the host semiconductor. The system, in this case, consists of the host atoms and the impurity atoms. It can be visualized as a subsystem that comprises the impurity atoms in thermal and diffusive contact with a reservoir consisting of the host semiconductor atoms. The donor atom is assumed to have three states, ionized, un-ionized but filled with an electron of spin up, and un-ionized but filled with an electron of spin down. The ionized state can be treated as empty with energy defined as zero. Therefore both the occupation number N and the energy vanish

Table 5.6.1
Occupation Statistics for Donors in a Semiconductor

State	n	Energy
Donor ionized	0	0
Donor un-ionized, spin up	1	$-\epsilon$
Donor un-ionized, spin down	1	$-\epsilon$

in the unoccupied state. The un-ionized states are each assumed to have energy $-\epsilon$. To find the probability that an atom is ionized it is first necessary to construct the grand partition function. By using Eq. (5.6.9) and the grand partition function, we can determine the occupation probability in equilibrium. The possible states N and energy ϵ are listed in Table 5.6.1.

The grand partition function can be constructed from its definition as

$$Z = \sum_n \sum_{\epsilon_s} e^{\frac{(\mu n - E_s)}{k_B T}}. \tag{5.6.21}$$

The first term in the sum is given by substituting into the exponent of Eq. (5.6.21) $n = 0$ and $\epsilon = 0$. This yields

$$e^{\frac{(0-0)}{k_B T}} = 1. \tag{5.6.22}$$

The second and the third terms in the sum are characterized by $n = 1$ and $\epsilon = -\epsilon$ in each case. This yields

$$e^{\frac{(\mu+\epsilon)}{k_B T}}. \tag{5.6.23}$$

Hence the grand partition function is given by summation of each term, which gives

$$Z = 1 + 2e^{\frac{(\mu+\epsilon)}{k_B T}}. \tag{5.6.24}$$

Therefore the probability that a donor is ionized is simply given by the ratio of the first term above to the grand partition function as

$$P(n = 0, \epsilon = 0) = \frac{1}{1 + 2e^{\frac{(\mu+\epsilon)}{k_B T}}}. \tag{5.6.25}$$

The probability that the donor is not ionized is of course simply equal to 1 minus the probability that it is ionized. This is given then as

$$P(n = 1, \epsilon = -\epsilon) = 1 - P(n = 0, \epsilon = 0)$$

$$= 1 - \frac{1}{1 + 2e^{\frac{(\mu+\epsilon)}{k_B T}}}$$

$$= \frac{1}{\frac{1}{2}e^{\frac{-(\epsilon+\mu)}{k_B T}} + 1}. \tag{5.6.26}$$

5.7 **Quantum Distribution Functions**

In the previous sections, we have considered the statistics of a system of distinguishable particles that have no restrictions on the way in which each particle can be sorted. A system obeying such a relation is called a classical system. We found that the multiplicity function was given by that for N items arranged in n subcontainers, Eq. (5.2.17). Extremalization of this multiplicity function led to the Maxwell–Boltzmann distribution, characterized by the Boltzmann factor

$$e^{-\frac{E}{k_B T}}. \tag{5.7.1}$$

However, from our discussion of quantum mechanics, we found that the number of quantum-mechanical particles that can be placed into any one state simultaneously depends on the spin of the particle. If the particle has half-integral spin or integral spin, the number of particles that can be put into the same state is different. In Chapter 3, we found, based on empirical evidence, i.e., the Periodic Table of the Elements, that no two electrons can exist simultaneously within the same state. At the time, we postulated the Pauli exclusion principle in order to explain why all the electrons did not exist within the ground state of the atom. The requirements placed on the electrons by the exclusion principle explain why multielectron atoms have different chemistries. A more general statement of the Pauli exclusion principle is that no two quantum-mechanical particles of half-integral spin can occupy the same quantum state simultaneously.

The case for integral spin particles is quite different from that of half-integral spin particles. For integral spin particles, there are no restrictions on the number of particles that can be placed into any quantum state simultaneously. Integral spin particles do not obey the Pauli exclusion principle. Hence, all the particles in a system can be placed within the ground state or any other state. Subsequently, the ways in which an ensemble of quantum-mechanical particles can be arranged depends on the spin of the particle. Different equilibrium distributions will be obtained depending on the nature of the quantum-mechanical particle that is considered.

The difference between integral and half-integral spin particles arises from the restriction on the symmetry of the total wave function. As we will see in Chapter 7, the total wave function of half-integral spin particles must be antisymmetric while that for integral spin particles must be symmetric. With these restrictions, the consistency of half-integral and integral spin particles with the Pauli principle is preserved.

There is a further difference between quantum-mechanical and classical particles aside from counting restrictions imposed by the spin of the particle. Classical particles are all assumed to be distinguishable, as we discussed in Section 5.2. However, what is the situation for quantum-mechanical particles? Can two different quantum-mechanical particles be distinguished from one another? If two electrons come close enough together that their wave functions overlap, there is no way of distinguishing after they separate which one was incident from one side or the other. This is due to the fact that quantum-mechanical particles are not particles in the classical sense but are wave packets. Once the wave packets interfere the identity of each separate particle is lost. An interesting way of viewing this situation is to consider two water drops. If two water drops coalesce and later separate, it makes no sense to ask which one of the final drops was originally on one side or the other since the identity of the separate drops is lost when they combine (see McGervey 1971, p. 262). Similarly, identical quantum-mechanical particles cannot be distinguished from one another. The fact that the particles are no longer distinguishable leads to a significant difference in the counting distributions between classical and quantum-mechanical particles, as we will see below.

Let us consider how quantum-mechanical particles can be arranged in a system. There are three general cases we must consider. The particles can be sorted into separate categories according to their spin and restriction on total particle number. As discussed above, if the particle has integral or half-integral spin, the number of particles placed into any one state differs. In addition, there are some quantum-mechanical particles for which the particle number is not conserved even within a closed system. In other words, the particle number changes with time; some particles are removed or added to the system subject to the condition that the total energy remains conserved. A physical example of such a particle is a photon, the quantum of electromagnetic vibrations. For example, consider the case of a closed room with no windows. If an overhead light is extinguished, the room gets dark; the photons within the room disappear through absorption events. As a result, the total number of photons changes within the closed system as a function of time.

The three categories of quantum-mechanical particles of interest to us are:

1. fermions: particles with half-integral spin whose number is strictly conserved,
2. bosons: particles with integral spin whose number is strictly conserved,
3. planckians: particles with integral spin but whose number is not conserved.

Let us now consider the counting statistics of each class of particle.

Case 1: Fermions

Fermions are defined as quantum-mechanical particles with half-integral spin. As such, these particles obey the Pauli exclusion principle. Examples of fermions are electrons, protons, and neutrons. The number of fermions is always conserved; no fermions are added or lost to a closed system. In addition, the particles are indistinguishable and no two particles can occupy the same quantum state simultaneously. Therefore the ways in which a collection of fermions can be arranged are different from those of classical particles. The problem becomes, how do we arrange n particles into s bins, each of which has a degeneracy of states g_s? We start with the number of ways of adding particles to the first bin.

If the particles are distinguishable, we find the number of arrangements of n_s particles (where the subscript labels the bin) simply by putting the first particle into any of the g_s states within the first bin. Since no two particles can be put into the same state, the second particle can be put into only one of the $g_s - 1$ other states. Therefore the number of configurations of n_s particles arranged into a bin with g_s states, assuming that the number of states exceeds the number of particles, is simply

$$g_s(g_s - 1)(g_s - 2)\cdots(g_s - n_s + 1). \tag{5.7.2}$$

However, expression (5.7.2) includes all possible permutations of the n_s particles among themselves, ignoring the fact that the particles are indistinguishable. The total number of permutations of the n_s particles is simply $n_s!$. We find the total number of distinct arrangements then of indistinguishable particles within one bin of degeneracy g_s by dividing the number of arrangements given above in expression (5.7.2) by $n_s!$. This becomes

$$Q_s = \frac{g_s(g_s - 1)(g_s - 2)\cdots(g_s - n_s + 1)}{n_s!}. \tag{5.7.3}$$

Note that the numerator can be reexpressed as

$$\frac{g_s!}{(g_s - n_s)!}, \tag{5.7.4}$$

which enables rewriting the multiplicity of the sth bin as

$$Q_s = \frac{g_s!}{(g_s - n_s)!n_s!}. \tag{5.7.5}$$

The total number of configurations of the system with n_1 particles in bin 1, n_2 particles in bin 2, etc., is then obtained from the product of the multiplicities of each bin as

$$Q = Q(n_s) = \prod_{s=1}^{\infty} Q_s = \prod_{s=1}^{\infty} \frac{g_s!}{(g_s - n_s)!n_s!}. \tag{5.7.6}$$

We proceed exactly as before in Section 5.4 to find the equilibrium configuration by extremalizing the multiplicity function Q or, more conveniently, $\ln Q$. The logarithm of the multiplicity function can be written as

$$\ln Q = \sum_{s=1}^{\infty} [\ln g_s! - \ln(g_s - n_s)! - \ln n_s!]. \tag{5.7.7}$$

When Stirling's approximation is applied to each of the factorial expressions involving n_s, $\ln Q$ becomes

$$\ln Q = \sum_{s=1}^{\infty} [\ln g_s! - n_s \ln n_s + n_s - (g_s - n_s) \ln(g_s - n_s) + (g_s - n_s)], \tag{5.7.8}$$

which simplifies to

$$\ln Q = \sum_{s=1}^{\infty} [\ln g_s! - n_s \ln n_s - (g_s - n_s) \ln(g_s - n_s) + g_s]. \tag{5.7.9}$$

To determine the equilibrium configuration and hence the equilibrium distribution function, the multiplicity function for $\ln Q$ must be extremalized subject to the constraints that the energy and the particle number are conserved. To find the most probable configuration, we need to use the method of Lagrange multipliers defined in Section 5.4. Mathematically, the maximization is given as

$$\frac{\partial}{\partial n_j} \ln Q + \alpha \frac{\partial \phi}{\partial n_j} - \beta \frac{\partial \psi}{\partial n_j} = 0, \tag{5.7.10}$$

where α and β are Lagrange multipliers and ϕ and ψ are defined as

$$\phi \equiv \sum_s n_s = N, \qquad \psi \equiv \sum_s E_s n_s, \tag{5.7.11}$$

and n_j is the number of particles in the specific sth bin. Substituting in Eq. (5.7.10) for $\ln Q$ Eq. (5.7.9), ϕ and ψ Eqs. (5.7.11) we obtain

$$\frac{\partial}{\partial n_j} \left[\sum_s \ln g_s! - \sum_s n_s \ln n_s - \sum_s (g_s - n_s) \ln(g_s - n_s) + \sum_s g_s \right]$$

$$+ \alpha \frac{\partial}{\partial n_j} \sum_s n_s - \beta \frac{\partial}{\partial n_j} \sum_s E_s n_s = 0. \tag{5.7.12}$$

Performing the differentiations gives

$$-\ln n_j - 1 + \ln(g_j - n_j) + 1 + \alpha - \beta E_j = 0. \tag{5.7.13}$$

This simplifies to

$$-\ln n_j + \ln(g_j - n_j) + \alpha - \beta E_j = 0. \tag{5.7.14}$$

Solving for n_j yields

$$\ln \frac{(g_j - n_j)}{n_j} = -\alpha + \beta E_j,$$

$$n_j = (g_j - n_j) e^{\alpha - \beta E_j},$$

$$n_j = \frac{g_j e^{\alpha - \beta E_j}}{1 + e^{(\alpha - \beta E_j)}},$$

$$\frac{n_j}{g_j} = \frac{1}{1 + e^{-(\alpha - \beta E_j)}} = f(E_j), \tag{5.7.15}$$

where g_j is the degeneracy of the jth bin, n_j/g_j is the fraction of carriers in the jth bin, and α and β are the Lagrange multipliers. β has the same value as found earlier for the Maxwell–Boltzmann distribution, $1/k_B T$. The question is then, what is α? To find α, consider the following approach. From Section 5.5, the Helmholtz free energy F is defined in terms of the total internal energy U, the entropy S, and the temperature T:

$$F = U - TS. \tag{5.7.16}$$

The free energy F, though, is related to the partition function Z by

$$F = -k_B T \ln Z. \tag{5.7.17}$$

Substituting into Eq. (5.7.16), Eq. (5.7.17) for F yields

$$TS = k_B T \ln Z + U, \tag{5.7.18}$$

which reduces to

$$\ln Z = \frac{S}{k_B} - \beta U. \tag{5.7.19}$$

But U is

$$U = \sum_i n_i E_i \tag{5.7.20}$$

and $S = k \ln Q$. Using these definitions and the results of Box 5.7.1, we obtain

$$\ln Z = -\alpha N + \sum_s g_s \ln\left[1 + e^{+(\alpha - \beta E_s)}\right]. \tag{5.7.21}$$

Box 5.7.1   Derivation of the Relationship between ln Z and α

The log Z can be related to S and U as

$$\ln Z = \frac{S}{k_B} - \beta U = \ln Q - \beta U.$$

But $\ln Q$, the log of the multiplicity function is given as

$$\ln Q = \sum_s [\ln g_s! - \ln(g_s - n_s)! - \ln n_s!]$$

$$= \sum_s [g_s \ln g_s - g_s - (g_s - n_s) \ln(g_s - n_s) + (g_s - n_s) - n_s \ln n_s + n_s].$$

The system, though, is in equilibrium. Therefore the expression for n_s is simply

$$n_s = \frac{g_s}{1 + e^{-(\alpha - \beta E_s)}}.$$

$g_s - n_s$ is then

$$g_s - n_s = g_s \left[1 - \frac{1}{1 + e^{-(\alpha - \beta E_s)}}\right] = \frac{g_s e^{-(\alpha - \beta E_s)}}{1 + e^{-(\alpha - \beta E_s)}}.$$

Substituting in the above relationships for $g_s - n_s$ and n_s into the equation for $\ln Q$ yields

$$\ln Q = \sum_s g_s \ln g_s - \sum_s (g_s - n_s) \ln \frac{g_s e^{-(\alpha - \beta E_s)}}{1 + e^{-(\alpha - \beta E_s)}} - \sum_s n_s \ln \left[\frac{g_s}{1 + e^{-(\alpha - \beta E_s)}}\right].$$

continued

Box 5.7.1, cont.

Rewriting the logarithms, we obtain

$$\ln Q = \sum_s g_s \ln g_s - \sum_s (g_s - n_s)\{\ln g_s + \ln e^{-(\alpha - \beta E_s)} - \ln[1 + e^{-(\alpha - \beta E_s)}]\}$$
$$- \sum_s n_s\{-\ln[1 + e^{-(\alpha - \beta E_s)}]\} - \sum_s n_s \ln g_s.$$

Therefore $\ln Q$ is

$$\ln Q = \sum_s g_s \ln\left[\frac{1 + e^{-(\alpha - \beta E_s)}}{e^{-(\alpha - \beta E_s)}}\right] + \sum_s n_s \beta E_s - \sum_s n_s \alpha.$$

Simplifying yields

$$\ln Q = \sum_s g_s \ln[1 + e^{(\alpha - \beta E_s)}] + \beta U - \alpha N,$$

where U is the internal energy and N is the total number of particles. The logarithm of the partition function is expressed above in the first equation as

$$\ln Z = \ln Q - \beta U.$$

But $\ln Q - \beta U$ is

$$-\alpha N + \sum_s g_s \ln[1 + e^{(\alpha - \beta E_s)}].$$

Therefore $\ln Z$ becomes

$$\ln Z = -\alpha N + \sum_s g_s \ln[1 + e^{(\alpha - \beta E_s)}].$$

Since $\ln Z = -F/k_B T$, α is given by

$$\alpha = \frac{1}{k_B T}\frac{dF}{dN}. \tag{5.7.22}$$

We can relate α to the chemical potential μ by noting that μ was defined in Section 5.6 as

$$\mu \equiv -T\frac{dS}{dN}. \tag{5.7.23}$$

The entropy S can then be reexpressed in terms of U and F as

$$S = \frac{U - F}{T}. \tag{5.7.24}$$

Therefore the chemical potential is found by substitution of $(U - F)/T$ in for S. Hence μ is

$$\mu = -T\frac{d}{dN}\left(\frac{U - F}{T}\right) = -\frac{d}{dN}(U - F),$$

$$\mu = \left(\frac{dF}{dN}\right)_{U,T}. \tag{5.7.25}$$

Finally, α is given by

$$\alpha = \frac{1}{k_B T}\mu. \tag{5.7.26}$$

With this substitution, and assigning the degeneracy to be equal to 1, we find that the equilibrium distribution function for fermions becomes

$$f(E) = \overline{n_s} = \frac{1}{e^{\frac{E_s - \mu}{k_B T}} + 1}. \tag{5.7.27}$$

n_s physically gives the mean number of particles in the state s.

The Fermi distribution, commonly called the Fermi–Dirac distribution, governs the equilibrium behavior of fermions. Its properties can be understood as follows. At absolute zero degrees, 0 K, we can determine the mean particle

---

Box 5.7.2 Alternative Derivation of the Fermi–Dirac Distribution Function by use of the Grand Partition Function

Consider a system that comprises a single cell into which it is desired to add fermions. The mean occupation number of the cell, given that the particles are fermions, can be determined with the grand partition function. Let the energy of the system be zero, 0, if the cell is unoccupied, and equal to ϵ, if the cell is occupied. Since the particles are fermions, no more than one particle can be added to the cell at any given time. Therefore the occupation number for the system is simply equal to 0 or 1. The grand partition function for the system can be determined from use of the chart below. Let n_i represent the occupation number for the cell and E_i the energy of the state. Then

n_i	E_i	*Sum*
0	0	1
1	ϵ	$1 + e^{-\beta(\epsilon - \mu)}$

The grand partition function for the system is then

$$Z = 1 + e^{-\beta(\epsilon - \mu)}.$$

The mean number of particles in the cell $\langle n_i \rangle$ can be determined from

$$\langle n_i \rangle = \frac{\sum \sum n_i e^{-\beta(E_i - \mu n_i)}}{Z}.$$

Performing the sums yields

$$\langle n_i \rangle = \frac{0 + e^{-\beta(\epsilon - \mu)}}{Z},$$

$$\langle n_i \rangle = \frac{1}{1 + e^{\beta(\epsilon - \mu)}}.$$

The above result is equal to the mean occupation number for the ith cell, given that the particles to be arranged are fermions. This is precisely what is meant by the probability distribution function for fermions. Subsequently, the above result is simply the Fermi–Dirac distribution.

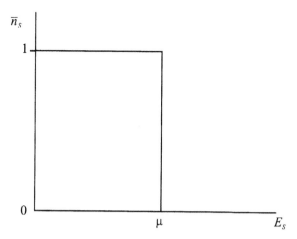

Figure 5.7.1. Sketch of the Fermi–Dirac distribution function at $T = 0$ K.

number in the state s from the above expression by noting the following: if $E_s > \mu$ then the exponential function is e^∞ and $n_s = 0$; if $E_s < \mu$ then the exponential function is $e^{-\infty}$ and $n_s = 1$. Therefore, at $T = 0$ K, the occupation probability is 100% for all states with energy below μ and is zero for all states with energy above μ. The distribution function at $T = 0$ K is plotted in Figure 5.7.1. As can be seen from the figure, all the states are occupied that have energy less than the chemical potential while all the states are empty that have energy greater than the chemical potential. From the Pauli principle, no two fermions can occupy the same state simultaneously. Therefore, at absolute zero, each energy state starting from the lowest energy state is progressively filled, leaving no vacancies, until the total number of particles in the system is exhausted. The energy of the state into which the last particle is put is equal to the chemical potential μ. Note that even at absolute zero degrees, there are many high-energy fermions, particles with energies well above the lowest possible energy state. This is of course due to the fact that multiple particles cannot condense into the same energy state. The lowest energy state of the ensemble then consists of many particles at energies well above the minimum possible for one individual particle.

The distribution at temperatures greater than 0 K is sketched in Figure 5.7.2. If the energy E_s is much greater than the chemical potential μ, the mean occupation probability

$$\overline{n_s} = f(E) = \frac{1}{e^{\frac{(E_s - \mu)}{k_B T}} + 1} \tag{5.7.28}$$

reduces since the exponential term is very much greater than 1. Therefore the 1 in the denominator in Eq. (5.7.28) can be neglected with respect to the exponential term, and n_s becomes

$$\overline{n_s} = e^{-\frac{E_s - \mu}{k_B T}}, \tag{5.7.29}$$

which is essentially the Maxwell–Boltzmann distribution. The high-energy tail of the Fermi–Dirac distribution resembles the Maxwell–Boltzmann distribution

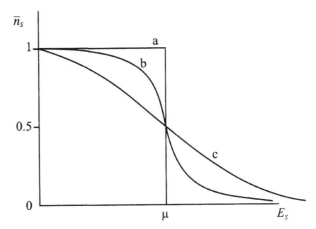

Figure 5.7.2. Sketch of the Fermi-Dirac distribution function at various temperatures: a. $T_1 = 0$ K. b. $T_2 > T_1$. c. $T_3 > T_2 > T_1$.

function. Note that as the temperature increases the distribution departs greatly from its rectangular shape at 0 K.

The effect of the restrictions on the number of arrangements of a group of fermions can be further understood through examination of Figure 5.7.3. Possible arrangements of five fermions into different quantum states are shown in the figure. The ground state corresponding to absolute zero degrees is created when one particle is precisely placed into each state until all the particles are used up. Note that the last electron has energy equal to the Fermi energy. The Fermi

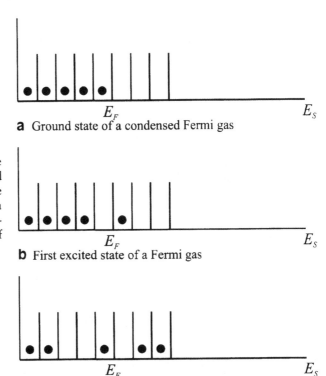

Figure 5.7.3. Arrangements of five fermions in quantum states: a. Ground state of a condensed Fermi gas of five particles. b. First excited state of a Fermi gas of five particles. c. A possible excited state of a Fermi gas of five particles.

a Ground state of a condensed Fermi gas

b First excited state of a Fermi gas

c A possible excited state of a Fermi gas

energy then is a function of the number of particles present in the system. The first excited state corresponding to a temperature greater than 0 K can be constructed when one electron is promoted into the first available state of energy greater than the Fermi energy, as is shown in the figure. Finally, we show another possible arrangement of the fermions for a state with temperature greater than 0 K.

Box 5.7.3   Example Problem by use of Fermi–Dirac Statistics

Find an expression for the internal energy per unit volume of a three-dimensional Fermi gas of free particles at 0 K in terms of the particle number N and the Fermi level E_f.

   Solution: The problem asks us to determine the average energy per unit volume of a free Fermi gas at 0 K. We recognize that the Fermi–Dirac distribution at 0 K is given by Fig. 5.7.1. Therefore the distribution is equal to 1 for $E < E_f$ and 0 for $E > E_f$. The average energy is given as

$$\bar{E} = \int E f(E) D(E) \, dE.$$

At absolute zero degrees, the Fermi–Dirac distribution simplifies as stated above. Substituting into the integral for $\bar{E}$ the expression for the density of states, we find that the average energy becomes

$$\bar{E} = \left(\frac{2m}{\hbar^2} \right)^{3/2} \frac{1}{(2\pi)^2} \int_0^{E_f} E^{3/2} \, dE.$$

Integrating out the energy, we find the average energy:

$$\bar{E} = \left(\frac{2m}{\hbar^2} \right)^{3/2} \frac{1}{(2\pi)^2} \frac{E_f^{5/2}}{5/2}.$$

The number of particles N in the system is determined from the integral over the energy of the product of the distribution function and the density of states as

$$N = \int_0^{E_f} f(E) D(E) \, dE.$$

Substituting into the integral for N Eq. (5.1.11) for $D(E)$ and 1 for $f(E)$ yields

$$N = \left(\frac{2m}{\hbar^2} \right)^{3/2} \frac{1}{(2\pi)^2} \int_0^{E_f} E^{1/2} \, dE.$$

On integration, N becomes

$$N = \left(\frac{2m}{\hbar^2} \right)^{3/2} \frac{1}{(2\pi)^2} \frac{E_f^{3/2}}{3/2}.$$

The average energy per particle $\bar{E}/N$ is given then as

$$\frac{\bar{E}}{N} = \frac{3}{5} E_f.$$

Finally, the average energy per unit volume V is

$$\bar{E} = \frac{3}{5} E_f N, \qquad \frac{\bar{E}}{V} = \frac{3 E_f N}{5 V}.$$

which is the desired result.

Case 2: The Bose–Einstein Distribution

The next system we consider is that of a collection of indistinguishable particles of integral spin. These particles do not obey the Pauli principle; any number of integral spin particles can be placed in a state simultaneously. Two different types of particles exist that have integral spin, those in which the particle number is not conserved and those for which the particle number is conserved. Both types of particles are commonly called bosons. However, to make a distinction between these two species, we define a boson as an indistinguishable particle of integral spin whose number is conserved. Indistinguishable, integral spin particles whose number is not conserved are called planckians. Let us first consider the case of bosons.

The first question one might ask when considering the counting distribution for bosons is, why is the distribution different from the classical distribution? In both cases, any number of particles can be placed into a state simultaneously. Remember, though, that quantum-mechanical particles are indistinguishable and thus the multiplicity function will be different from that derived for classical particles. Therefore the equilibrium distribution functions governing classical particles and bosons will be necessarily different.

The derivation of the equilibrium distribution function for bosons, called the Bose–Einstein distribution, follows that for fermions. As before, it is necessary to construct the multiplicity function Q for the system first and then maximize it subject to the constraints that the total energy and the particle number are conserved. We find the multiplicity function by determining the number of ways in which n_s particles can be arranged in a bin that contains g_s degenerate states under the condition that any number of particles can exist in any state simultaneously. Since any number of particles can exist in any state, the actual number of permutations of the n_s particles in the g_s states depends on the number of ways in which both the particles and the partitions can be arranged. It is helpful to use the graphical device shown in Figure 5.7.4 in order to understand how these permutations come about (see Yariv 1982, p. 182). The particles are represented by circles, and the states are formed by arrangement of the partitions. Note first that the number of states in each bin is given by the number of partitions plus one. In the figure, the total number of particles present in the bin n_s is twelve and the number of subbins or states is seven since there are six partitions forming seven subcontainers. The total number of ways of permuting the partitions and the particles is simply given by the number of ways in which

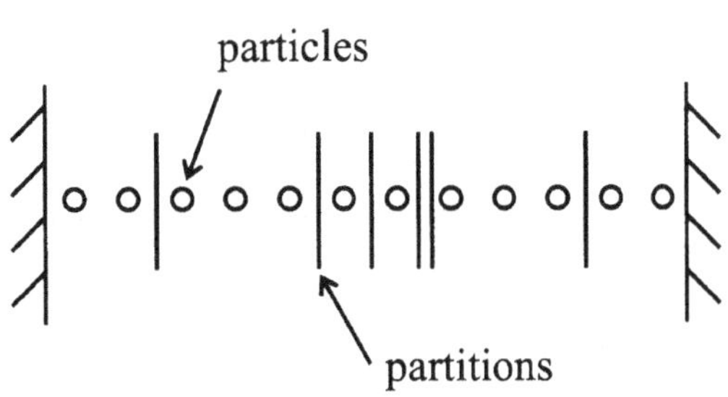

Figure 5.7.4. Graphical device for determining the number of distinct ways of arranging n_s particles into g_s states following the approach of Yariv (1982).

both the circles and the lines can be arranged in Figure 5.7.4. If both the circles and the lines are all distinguishable, the number of permutations is simply equal to the sum of the number of circles n_s and the number of lines $(g_s - 1)$ factorial:

$$(n_s + g_s - 1)!. \tag{5.7.30}$$

However, neither the circles nor the lines (particles or states) are distinguishable. Therefore the $n_s!$ permutations of the particles and the $(g_s - 1)!$ permutations of the partitions among themselves do not lead to indistinguishable arrangements. Therefore the total number of distinct arrangements of the particles and the states in a single bin s is given by

$$Q_s = \frac{(n_s + g_s - 1)!}{n_s!(g_s - 1)!}. \tag{5.7.31}$$

Consequently, the multiplicity of the entire system is given by the product of the separate multiplicities of each bin as

$$Q = \prod_{s=1}^{\infty} \frac{(n_s + g_s - 1)!}{n_s!(g_s - 1)!}. \tag{5.7.32}$$

To find the equilibrium distribution function for bosons, we follow the same procedure as we did for fermions. The logarithm of Q must again be maximized subject to the constraints that the particle number and the energy are conserved. The $\ln Q$ is simply

$$\ln Q = \sum_{s=1}^{\infty} [\ln(n_s + g_s - 1)! - \ln n_s! - \ln(g_s - 1)!], \tag{5.7.33}$$

which, when Stirling's approximation is used, becomes

$$\ln Q = \sum_{s=1}^{\infty} [(n_s + g_s - 1) \ln(n_s + g_s - 1) - (n_s + g_s - 1)$$
$$- n_s \ln n_s + n_s - (g_s - 1) \ln(g_s - 1) + (g_s - 1)]. \tag{5.7.34}$$

Simplifying Eq. (5.7.34) yields

$$\ln Q = \sum_{s=1}^{\infty} [(n_s + g_s - 1) \ln(n_s + g_s - 1) - n_s \ln n_s - (g_s - 1) \ln(g_s - 1)]. \tag{5.7.35}$$

The equation to be maximized is given in a way similar to that for fermions as

$$\frac{\partial}{\partial n_j} \ln Q + \alpha \frac{\partial}{\partial n_j} \sum_s n_s - \beta \frac{\partial}{\partial n_j} \sum_s E_s n_s = 0. \tag{5.7.36}$$

Substituting Eq. (5.7.35) for $\ln Q$ in Eq. (5.7.36) gives

$$\frac{\partial}{\partial n_j} \sum_{s=1}^{\infty} [(n_s + g_s - 1) \ln(n_s + g_s - 1) - n_s \ln n_s - (g_s - 1) \ln(g_s - 1)]$$

$$+ \alpha \frac{\partial}{\partial n_j} \sum_s n_s - \beta \frac{\partial}{\partial n_j} \sum_s E_s n_s = 0. \tag{5.7.37}$$

Taking the derivative with respect to n_j throughout Eq. (5.7.37) leaves

$$\ln(n_j + g_j - 1) + \frac{(n_j + g_j - 1)}{(n_j + g_j - 1)} - \ln n_j - \frac{n_j}{n_j} + \alpha - \beta E_j = 0,$$

$$\ln(n_j + g_j - 1) - \ln n_j + \alpha - \beta E_j = 0. \tag{5.7.38}$$

Therefore the distribution function can be found as before by solving for n_j. To simplify the result further, we assume that $n_j + g_j \gg 1$. Then the -1 in the first term can be neglected. Assuming that the sum of the number of particles and subcontainers is significantly greater than 1 is reasonable since any number of particles can be placed into any state. The number of particles within a quantum state n_j is generally quite large. The result for the distribution function becomes then

$$\ln \left(\frac{n_j + g_j}{n_j} \right) = -(\alpha - \beta E_j),$$

$$n_j [1 - e^{-(\alpha - \beta E_j)}] = -g_j. \tag{5.7.39}$$

Therefore n_j/g_j simplifies to

$$\frac{n_j}{g_j} = \frac{1}{e^{-(\alpha - \beta E_j)} - 1}, \tag{5.7.40}$$

which is called the Bose–Einstein distribution function. The Bose–Einstein distribution governs the equilibrium behavior of bosons.

The Lagrange multipliers α and β are precisely the same as those found earlier for the Fermi distribution. Hence

$$\alpha = \frac{\mu}{k_B T}, \qquad \beta = \frac{1}{k_B T}. \tag{5.7.41}$$

When these substitutions are made, the Bose–Einstein distribution function becomes

$$f(E) = \frac{1}{e^{\frac{E - \mu}{k_B T}} - 1}, \tag{5.7.42}$$

where μ is the chemical potential defined above.

Box 5.7.4 Alternative Derivation of the Bose–Einstein Distribution from the use of the Grand Partition Function

---

Let us consider a system that comprises a single cell into which it is desired to add bosons. From the discussion in the text, any number of bosons can be added to the cell from zero to infinity since there is no restriction on the number of particles allowed within a single state. The mean occupation number of bosons in a single cell can be determined with the grand partition function. Let us assume that the energy of the cell is zero when no particles are present and E_1 when one particle is present. The total energy when n_i particles are added to the cell is then $n_i E_1$. The grand partition function for the system can be determined by summation over the number of particles n_i from zero to infinity:

$$Z = \sum_{n_i=0}^{\infty} e^{\beta(\mu n_i - E_1 n_i)} = \sum_{n_i=0}^{\infty} e^{\beta(\mu - E_1)n_i}.$$

The above sum can be summed in closed form by use of the result

$$\sum_{s=0}^{\infty} x^s = \frac{1}{1-x}.$$

Therefore the grand partition function for the system is given as

$$Z = \sum_{n_i=0}^{\infty} e^{\beta(\mu - E_1)n_i} = \frac{1}{1 - e^{\beta(\mu - E_1)}}.$$

The mean occupation number of the cell is given by $\langle n_i \rangle$. This can be readily determined from the partition function as

$$\langle n_i \rangle = \frac{1}{\beta} \frac{\partial}{\partial \mu} \ln Z.$$

Substituting $\ln(\frac{1}{1 - e^{\beta(\mu - E_1)}})$ in for $\ln Z$ and performing the differentiation yields

$$\langle n_i \rangle = \frac{e^{\beta(\mu - E_1)}}{1 - e^{\beta(\mu - E_1)}},$$

which simplifies to

$$\langle n_i \rangle = \frac{1}{e^{-\beta(\mu - E_1)} - 1}.$$

This becomes

$$\langle n_i \rangle = \frac{1}{e^{\beta(E_1 - \mu)} - 1},$$

which is the Bose–Einstein distribution function.

Case 3: Statistics of Bosons for which the Particle Number Is Not Conserved, Planckians

The last assortment of quantum-mechanical particles that we consider is that of Planckians or, more generally, those particles that have integral spin but for which the total particle number is not conserved. The two most important examples of Planckians that concern us are photons and phonons. Photons were defined in Chapter 1 as the quantum of the electromagnetic field. Phonons will be discussed in Chapter 9. It suffices at present to mention that they are the quantum of lattice vibrations. As we found in Chapter 1, all forms of matter and energy exhibit both wavelike and particlelike behaviors. The particles associated with the vibrations of a crystal are phonons. For planckians, there are no restrictions on the number of particles in any state or mode or on the total number of particles present. Any number of photons can be created within the system (or destroyed in the system), provided that energy conservation is maintained. The particle number itself is free to change. This is obvious from the everyday experience that a closed room becomes dark if we turn off the lights. The photons are absorbed when they are removed from the system, in this case the room.

Richard Feynman offered an interesting way of looking at planckians. He suggested that photons are like words. One creates as many as are necessary to fulfill a task but one can never use them up (sometimes professors create an excessive barrage of verbiage in fulfilling a task!). For example, one cannot wear out one's name by saying it too often. We create a word as many times as we wish without running out of supplies of them. The same is true of photons. As many photons as are consistent with energy conservation are created or destroyed but there are in essence an unlimited supply of photons. Of course, if there is only a finite amount of energy in the universe there are then only a finite number of photons available, much like there are only a finite number of words I can say in a lifetime. This comes as a relief to my students, I am sure!

The Planckian equilibrium distribution is determined in precisely the same manner as that for bosons, excepting the fact that there is no constraint on the particle number. The multiplicity function is exactly the same as that for bosons since Planckians have integral spin and are indistinguishable particles. Therefore the number of ways in which photons or phonons can be sorted into various quantum states is exactly the same as for bosons. The multiplicity function Q then is given as

$$\ln Q = \sum_s (n_s + g_s - 1) \ln(n_s + g_s - 1) - n_s \ln n_s - (g_s - 1) \ln(g_s - 1). \quad (5.7.43)$$

To find the most likely configuration, the equilibrium configuration of the system, $\ln Q$ needs to be maximized subject to the constraint that the total energy of the system is constant. The resulting equation for the extremum of $\ln Q$ is

$$\frac{\partial}{\partial n_j} \ln Q - \beta \frac{\partial}{\partial n_j} \sum_s E_s n_s = 0, \quad (5.7.44)$$

where only one Lagrange multiplier β is present. There is no restriction on the particle number in this case. Subsequently, the Lagrange parameter α does not appear. Substituting into Eq. (5.7.44), Eq. (5.7.43) for $\ln Q$ and differentiating yields

$$\ln(n_j + g_j - 1) - \ln n_j - \beta E_j = 0. \quad (5.7.45)$$

Again, if the -1 is neglected with respect to $n_j + g_j$, we obtain

$$\ln\frac{n_j + g_j}{n_j} = \beta E_j, \tag{5.7.46}$$

which can be solved for n_j/g_j as

$$n_j(1 - e^{\beta E_j}) = -g_j,$$
$$\frac{n_j}{g_j} = -\frac{1}{1 - e^{\beta E_j}} = \frac{1}{e^{\beta E_j} - 1}. \tag{5.7.47}$$

β is equal to $1/k_B T$ as before. n_j/g_j is equal to $f(E_j)$, so

$$f(E) = \frac{1}{e^{\frac{E}{k_B T}} - 1}, \tag{5.7.48}$$

which is called the Planckian distribution function.

Before we discuss examples of each of the three quantal distribution functions, Fermi–Dirac, Bose–Einstein, and Planckian, it is useful to summarize their properties. It is important to realize that the equilibrium distribution functions, classical and quantum mechanical, are all derived from the same basic principle, that is, every microstate is equally likely. Given this postulate, it follows that the equilibrium configuration is simply the most likely configuration of the system. The equilibrium condition corresponds then to the configuration with the largest number of ways in which it can be arranged. The procedure for finding any one of the equilibrium configurations can be summarized as follows. First, it is necessary to determine the multiplicity function for the system. Once the multiplicity function Q is known, the equilibrium configuration is given by maximization of $\ln Q$, subject to any internal constraints on the system. In all the cases, the total energy of the system must be conserved. In some cases, the total particle number must also be conserved. The extremum of the multiplicity function thus found is, by definition, the equilibrium configuration. In Section 5.8 the physical nature of the equilibrium distributions determined here is discussed.

5.8 **Examples of Quantal Gas Systems**

There exist numerous macroscopic systems whose properties arise from the counting statistics implicit in the quantal gas distribution functions developed in Section 5.7. For example, systems as diverse as white dwarf stars, neutron stars, and metals obtain their distinguishing characteristics from the physics of the Fermi–Dirac distribution function. Let us first consider the heat capacity of an electron gas in a metal as an illustration of the physics of condensed Fermi systems.

It is instructive first to review the classical theory of the heat capacity of a free-electron gas to understand the significance of the inclusion of the Fermi–Dirac distribution. In classical statistical mechanics, a free particle has a heat capacity of $3/2\, k_B$ owing to the 3 independent degrees of freedom of its motion. Recall that each degree of freedom contributes $1/2\, k_B$ to the heat capacity of an individual particle. If there are N particles present, the total heat capacity of the system is $3/2\, Nk_B$. Therefore, if a system comprises free particles, its heat capacity must be linearly proportional to the number of particles present.

At the turn of the century, it was believed that a metal was comprised of essentially free electrons within a host atomic lattice. This model of a metal is consistent with the relatively high electrical and thermal conductivities observed in metals. The heat capacity of the metal depends on the heat capacity of the electrons and that of the lattice. From the above argument, it would seem that the electronic contribution to the heat capacity of a metal should be great. However, experimental measurements of the electronic heat capacity show that the heat capacity is very much less than that predicted from the simple formula, $3/2\, Nk_B$. The question then arose, how can a metal exhibit high electrical conductivity as if all the outermost valence electrons are free to move throughout the crystal, but not have a comparably high specific heat due to these electrons?

The answer to this question is based on the fact that the electrons obey the Fermi–Dirac distribution and not the classical Boltzmann distribution. To see this, let us consider a system of free electrons in three dimensions. The energy of a free-electron gas is of course simply

$$E = \frac{\hbar^2 k^2}{2m},\tag{5.8.1}$$

which is essentially the equation of a sphere in k space. The three-dimensional distribution function can be represented as a sphere in k space of radius k_f, called the Fermi wave vector. The energy corresponding to the Fermi wave vector is called E_f, the Fermi energy. The physical meaning of the Fermi energy can be understood as follows. From the discussion in Section 5.7, we found that at absolute zero degrees, all the states below the chemical potential μ are occupied while all those above μ are unoccupied. The chemical potential corresponds to the highest energy state of the completely condensed Fermi gas. From the above definition of the Fermi energy and the Fermi wave vector, the Fermi energy also corresponds to the highest energy of the condensed Fermi gas. Therefore the Fermi level and the chemical potential are precisely the same quantity. Historically, in honor of Enrico Fermi, the chemical potential in a solid is always referred to as the Fermi level and is represented as E_f. Therefore all the states within the sphere (called the Fermi sphere) of radius k_f are filled at $T = 0$ K, as shown in Figure 5.8.1.

What, though, is the effect on the heat capacity of treating the electron gas as a condensed Fermi gas as opposed to a free classical gas? From the Pauli

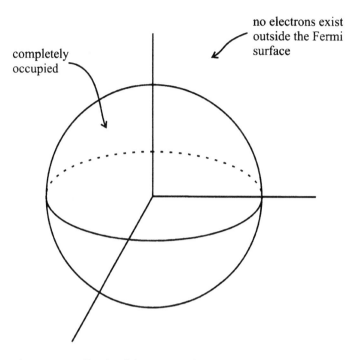

no electrons exist
outside the Fermi
surface

completely
occupied

Figure 5.8.1. Sketch of the Fermi sphere at $T = 0$ K.

principle, no two electrons can occupy the same quantum state simultaneously. An electron cannot make a transition to an occupied state; it can make transitions only to unoccupied states. The electrons near the Fermi surface can easily make transitions to unoccupied states since most of the surrounding states are unoccupied. Therefore the electrons near the Fermi surface can absorb small amounts of energy and contribute to the heat capacity of the metal. However, electrons deep within the filled Fermi sphere cannot make transitions to neighboring states since the vast majority of these states are occupied. Subsequently, the low-energy electrons in a metal, those deep within the Fermi sphere, do not contribute to the heat capacity since they cannot absorb relatively small amounts of energy because, on average, they cannot undergo a transition to a nearby state. Only the electrons near the Fermi surface contribute to the heat capacity of a metal since it is only these electrons that can absorb small amounts of energy. The heat capacity depends then on, in some way or other, the number of states within the surface area of the Fermi sphere and not its volume, as would be the case if the electrons were free classical particles.

How, though, can the high electrical conductivity of a metal be explained in light of the inclusion of Fermi–Dirac statistics? As is well known, the electrical conductivity of a metal is relatively large. The electrons in a metal act electrically as if all the conduction electrons are free and contribute to the conductivity of the metal. Yet, as we argued above, only those electrons near the Fermi surface can be promoted to vacant states and subsequently contribute to the heat capacity of the metal. How then can virtually all the conduction electrons contribute to the electrical current while only a small portion, those electrons near the Fermi surface, contribute to the heat capacity? To understand the answer to this question it is necessary to first ask what constitutes a net current.

A macroscopic current flow is due to a net flux of carriers across an interface at any given time. If the time-averaged number of carriers that flow forward and the number that flow backward across an interface are equal then there is no net current flow across that interface. If more carriers on average cross a plane moving in one direction than the other, then there is a net current flow in that direction. The carrier distribution must then have a net momentum in the forward direction. From Newton's Second Law, the time rate of change of the momentum must be equal to the applied force:

$$q\vec{F} = \frac{d\vec{p}}{dt} = \frac{d(\hbar\vec{k})}{dt},$$

$$\hbar\frac{d\vec{k}}{dt} = q\vec{F}. \tag{5.8.2}$$

Solving for k yields

$$\vec{k}_f = \vec{k}_i + \frac{q\vec{F}t}{\hbar}, \tag{5.8.3}$$

where k_f and k_i are the final and the initial k vectors of the electrons, respectively, and F is the applied electric field. Note that the k vectors are uniformly translated in k space by the amount $qFt/\hbar$. The electron distributions are sketched in Figure 5.8.2 for both $F = 0$ and $F \neq 0$. When the field is nonzero, all the carriers are displaced in k space in the direction of the field by the same amount. The net current can be determined by summation over all the electron wave vectors multiplied by the charge carried by each electron, q. If there is no applied field, the system must be in equilibrium. The most random configuration of a collection of free electrons would be one in which the momenta are completely randomized. Thus, for every electron with a positive momentum $+\vec{k}$, there is, on average, an electron with a compensating negative momentum $-\vec{k}$. Therefore the net momentum of the system is zero, since the sum of all the electrons with positively directed k vectors is canceled on average by the sum of all the electrons with negatively directed k vectors. This is the reason why in equilibrium no current flows, that is, the equilibrium configuration is the most random configuration and as such the net momentum and hence velocity of the electrons are zero.

If, on the other hand, a field is applied, there are uncompensated electrons with momentum in the field direction. The entire system of electrons suffer an increase in their momentum along the field direction. Subsequently, the net sum over the entire distribution no longer gives zero, since there are now electrons whose momentum is not compensated for on average. Therefore the distribution of electrons has a net forward momentum. The current is proportional to the field since, as the field increases, the Fermi sphere is displaced further from

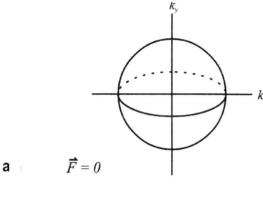

a $\vec{F} = 0$

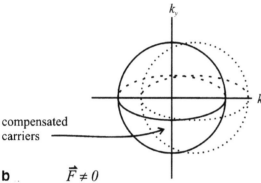

compensated
carriers

b $\vec{F} \neq 0$

Figure 5.8.2. a. Sketch of the Fermi sphere at zero applied electric field. Note that the sphere is symmetric about the point $k = 0, 0, 0$. Therefore every negative k vector is compensated for by a positive k vector, resulting in a net zero momentum for the system. b. Sketch of the Fermi sphere under an applied electric field. Note that the sphere shifts in k space such that there are now a net number of filled states with more positive k vectors than negative k vectors, resulting in a net value of the momentum for the system.

equilibrium, yielding more uncompensated electrons. It is important to note that all the electrons contribute to the electrical current while only those near the Fermi surface contribute to the heat capacity. All the electrons contribute to the current since the entire distribution is displaced in k space simultaneously by the same amount, producing vacancies around each carrier from the action of the applied field. Therefore each electron can make a transition to a neighboring state. Thermal excitations produce random vacancies in k space; therefore a vacancy is not necessarily produced in the near vicinity of an electron. As such the number of electrons that can absorb some small amount of thermal energy is greatly reduced.

The electrical conductivity is a volume effect while the heat capacity is a surface effect of the Fermi sphere. Hence, owing to the very many more electrons within the volume as opposed to only the surface, the number of electrons contributing to the heat capacity is very much less than the number of electrons contributing to the electrical current. This is the reason why the heat capacity of metals is relatively small while the electrical conductivity is very high.

Let us next derive a quantitative expression for the heat capacity. We follow the approach of Ashcroft and Mermin (1976). The heat capacity is defined as the change of the internal energy with respect to the temperature. Therefore it is necessary to determine an expression for the net increase in the internal energy of the electron gas. We give the increase in the total energy of an N electron

Fermi gas by finding the mean energy of the gas at nonzero temperature and subtracting off the mean energy of the electrons below the Fermi surface. The internal energy is then

$$U = \int_0^\infty dE\, E D(E) f(E) - \int_0^{E_f} dE\, E D(E), \qquad (5.8.4)$$

where it is assumed that $f(E) = 1$ for $E < E_f$ and $D(E)$ is the density of states. The number of particles N is given by

$$N = \int_0^\infty D(E) f(E)\, dE. \qquad (5.8.5)$$

Multiplying Eq. (5.8.5) by E_f yields

$$E_f N = E_f \int_0^\infty D(E) f(E)\, dE. \qquad (5.8.6)$$

Next differentiate Eq. (5.8.4) with respect to T to obtain the heat capacity C:

$$C = \frac{dU}{dT} = \frac{d}{dT}\left[\int_0^\infty E D(E) f(E)\, dE - \int_0^{E_f} E D(E)\, dE\right]. \qquad (5.8.7)$$

But E and $D(E)$ are temperature independent. Therefore Eq. (5.8.7) becomes

$$C = \int_0^\infty dE\, E D(E) \frac{df(E)}{dT}. \qquad (5.8.8)$$

Differentiating Eq. (5.8.6) with respect to T gives

$$\frac{d}{dT}(E_f N) = 0 = E_f \int_0^\infty D(E) \frac{df(E)}{dT}\, dE. \qquad (5.8.9)$$

Therefore the heat capacity is

$$C = \int_0^\infty dE(E - E_f) D(E) \frac{df}{dT}, \qquad (5.8.10)$$

but $D(E)\, df/dT$ is large only at energies near E_f. As E increases above E_f, df/dT approaches zero exponentially. For $E \ll E_f$ the density of states $D(E)$ approaches zero. Therefore the product of the two functions is appreciable only near the Fermi energy. The density-of-states function can then be approximated by the density of states at E_f, $D(E_f)$. If $D(E_f)$ is assumed to be constant, it can be removed outside of the integral. The heat capacity then becomes

$$C = D(E_f) \int_0^\infty dE(E - E_f) \frac{df}{dT}. \qquad (5.8.11)$$

But the derivative of f with respect to T is

$$
\frac{df}{dT} = \frac{d}{dT}\left(\frac{1}{e^{\frac{E-E_f}{k_B T}} + 1}\right)
$$
$$
= \frac{\frac{E-E_f}{k_B T^2} e^{\frac{E-E_f}{k_B T}}}{\left(e^{\frac{E-E_f}{k_B T}} + 1\right)^2}.
\tag{5.8.12}
$$

Therefore the expression for C becomes

$$
C = \frac{D(E_f)}{k_B T^2} \int_0^\infty \frac{(E-E_f)^2 e^{\frac{E-E_f}{k_B T}}\, dE}{\left(e^{\frac{E-E_f}{k_B T}} + 1\right)^2}.
\tag{5.8.13}
$$

Equation (5.8.13) can be integrated out by the assignments of

$$
u = \frac{E - E_f}{k_B T}, \qquad du = \frac{dE}{k_B T}.
\tag{5.8.14}
$$

With these assignments, the integral over E can be evaluated as

$$
C = D(E_f) k_B^2 T \int_0^\infty \frac{u^2 e^u\, du}{(e^u + 1)^2}.
\tag{5.8.15}
$$

The integral is equal to $\pi^2/3$. Therefore the electronic heat capacity is simply

$$
C = \frac{\pi^2}{3} D(E_f) k_B^2 T.
\tag{5.8.16}
$$

The heat capacity of a metal is found to depend on the density of states at the Fermi surface $D(E_f)$ and is linear in T. This is precisely what we expect based on our qualitative argument, that is, the heat capacity depends on the surface features of the Fermi sphere.

As a next example of the use of Fermi–Dirac statistics, consider the statistics of free carriers in semiconductors. What we seek to determine is how many carriers participate in electronic conduction processess. It is useful first to define a semiconductor. As we will see in Chapter 7, when many atoms are brought together to form a solid, the allowed energy states spread out to form a quasi-continuum of states called a band. There are two general types of bands, conduction and valence bands. The conduction band corresponds to electronic states that are not localized; electrons within these states are free to move throughout the crystal. The valence band consists of localized electronic states; electrons within these states are bound to atoms within the crystal lattice. In a semiconductor, a forbidden energy gap lies between the first conduction band and the valence band. No propagating electronic states exist within this forbidden band. Therefore the density-of-states function $D(E)$ for a perfect-host semiconductor vanishes for all energies within the energy gap.

The presence of carriers in the conduction band and vacancies within the valence band, which, as we will see later, are called holes and can also contribute to the conduction process in a semiconductor, provide the current-carrying species in semiconductors. As we discussed at the end of Section 5.6, the number of carriers within the conduction and the valence bands can be altered by the addition of impurity atoms into the host semiconductor material. Let us consider here the conduction properties of semiconductors that are undoped; there are no intentional impurities introduced into the material. These materials are called intrinsic semiconductors.

In an intrinsic semiconductor for every electron promoted to the conduction band a vacancy is left behind within the valence band. Therefore the concentration of free electrons and holes is the same and is called the intrinsic carrier concentration n_i.

How can n_i be determined? In Section 5.1 an expression for the number of carriers of energy E, $N(E)$, was determined from the density-of-states function $D(E)$ and the occupation probability function $f(E)$. The density-of-states function $D(E)$ was derived for a three-dimensional free-electron system in Section 5.1 as

$$D(E) = \frac{8\pi m^{\frac{3}{2}} V \sqrt{2E}}{h^3}. \tag{5.8.17}$$

For a collection of electrons, the occupation probability function $f(E)$ is simply given by the Fermi–Dirac distribution function

$$f(E) = \frac{1}{1 + e^{\frac{E-E_f}{k_B T}}}. \tag{5.8.18}$$

Therefore the concentration of free electrons in the conduction band n is found by integration of the product of the density-of-states function $D(E)$ by the equilibrium probability distribution function $f(E)$ over the full energy range of the conduction band. The electron concentration is given then as

$$n = \int_{E_{\text{bottom}}}^{E_{\text{top}}} D(E) f(E) \, dE, \tag{5.8.19}$$

where the bounds of the integral are defined as the energy at the top and the bottom of the conduction band, respectively.

In general, the density-of-states function in a real semiconductor system is not equal to the simple free-electron density-of-states function given above, but is far more complicated. For the moment, however, let us simply use Eq. (5.8.19) for purposes of illustration.

The expression for n given above cannot be evaluated in closed form when $f(E)$ is replaced by the Fermi distribution. The general solution is found in terms of a function called the Fermi–Dirac integral $F_{1/2}(\xi_f)$ as

$$n = N_c \frac{2}{\sqrt{\pi}} F_{1/2}(\xi_f), \tag{5.8.20}$$

where N_c is called the effective density of states,

$$N_c = 2 \left(\frac{2\pi m k_B T}{h^2} \right)^{\frac{3}{2}}, \tag{5.8.21}$$

and ξ_f is given by

$$\xi_f = \frac{E_f - E_c}{k_B T}. \tag{5.8.22}$$

The function $F_{1/2}(\xi_f)$ is defined as

$$F_{1/2}(\xi_f) = \int \frac{\xi^{\frac{1}{2}} \, d\xi}{1 + e^{\xi - \xi_f}}, \tag{5.8.23}$$

which can be evaluated numerically. The value of the Fermi–Dirac integral is presented graphically in Figure 5.8.3.

The above result for the electron concentration, though precise, is unwieldy to use since it involves the numerical evaluation of $F_{1/2}(\xi_f)$. The Fermi–Dirac integral can be greatly simplified if the Fermi–Dirac distribution is replaced by the Maxwell–Boltzmann distribution. The Fermi–Dirac distribution reduces to the Boltzmann distribution when $E - E_f \gg k_B T$. With this assumption, $f(E)$ becomes

$$f(E) = \frac{1}{1 + e^{\frac{E - E_f}{k_B T}}} \sim e^{\frac{-(E - E_f)}{k_B T}}. \tag{5.8.24}$$

The expression for the electron carrier concentration n can be reformulated with this approximation as

$$n = \int D(E) e^{\frac{-(E - E_f)}{k_B T}} \, dE. \tag{5.8.25}$$

If the density-of-states function $D(E)$ is substituted into Eq. (5.8.25) the above integral can be evaluated in closed form for the carrier concentration n if the upper bound is extended to infinity and the energy of the bottom of the conduction band is assumed to be zero. The upper bound can be extended to infinity with little error even though the band ends at some finite energy since the probability function decreases exponentially as the energy increases. Therefore the error introduced from counting to infinity is exceedingly small. Under these assumptions, the electron concentration is determined to be

$$n = 2 \frac{(2\pi m k_B T)^{\frac{3}{2}}}{h^3} e^{\frac{-(E - E_f)}{k_B T}} = N_c e^{\frac{-(E - E_f)}{k_B T}}, \tag{5.8.26}$$

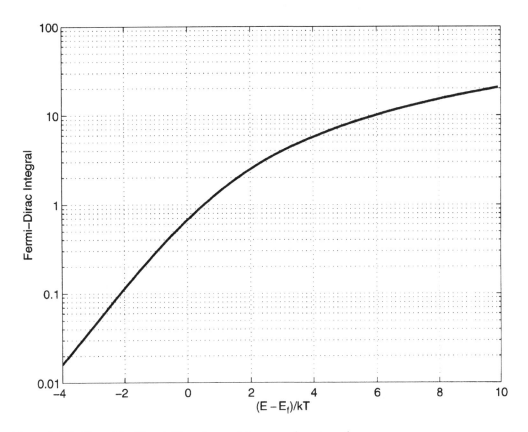

Figure 5.8.3. Plot of the Fermi–Dirac integral $F_{1/2}$ as a function of energy.

where N_c is the effective density of states defined above. Equation (5.8.26) is commonly used to determine the free-carrier concentration in an intrinsic semi-conductor.

As an example of a system that obeys Bose–Einstein statistics, we consider a Bose condensate. The most vivid example of Bose condensation is that of superfluid He^4. In a Bose system all the particles can occupy the same state simultaneously; there are no restrictions on orbital occupancy. For example, consider the possible arrangements of five bosons in different quantum states, as shown in Figure 5.8.4. As is shown in the figure, all the particles can be placed into the first state. This arrangement corresponds to the zero-temperature condition. As the temperature of the system is raised, a particle can be excited into any one of the higher-energy states. The first excited state of the system is shown in part b of the figure. Note that one boson is placed within the second state. Finally, we show an arbitrary excited state of the system.

From inspection of Figure 5.8.4, it is clear that as the system approaches absolute zero degrees, more and more of the particles condense into the lowest energy state. At absolute zero all the particles condense into the ground state of the system. Since the system is quantum mechanical, each state is separated from the next by a finite amount of energy. Therefore a boson must absorb a

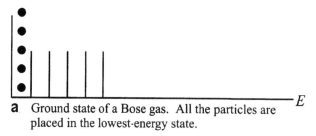

a Ground state of a Bose gas. All the particles are placed in the lowest-energy state.

Figure 5.8.4. Arrangements of five bosons in quantum states. a. Ground state of a Bose gas. All the five particles are placed within the lowest energy state. b. First excited state of a Bose gas. One particle is promoted to the first state above the ground state. c. Another possible excited state of a five-particle Bose gas.

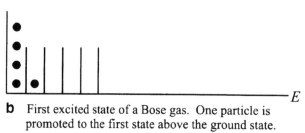

b First excited state of a Bose gas. One particle is promoted to the first state above the ground state.

c Another possible excited state of a Bose gas.

well-defined quantum of energy in order to be promoted from the ground state into the first excited state of the system. This is similar to the cases of the finite and the infinite square wells considered in Chapter 2. For a particle to make a transition from one quantum state to the next, it must absorb or emit a quantum of energy equal to the difference in energy between the two states. If the minimum energy needed for a transition to occur is not available, the particle cannot be promoted out of its original state.

He^4 nuclei are nuclei consisting of four nucleons, two protons, and two neutrons. The net spin of the nucleus is zero. Since a helium atom has two electrons in the ground state, both within the 1s orbital, the net spin of the electrons within a helium atom is also zero. Subsequently, the spin of the entire helium atom, sum of the electrons and nuclei, must be zero. Hence a collection of He^4 atoms form a Bose gas.

As the temperature of a collection of He^4 atoms approaches 0°K, most of the atoms condense into the ground state of the system. If the system reaches absolute zero, all the atoms would reside within the ground state. As we discussed above, in order to promote a He^4 atom out of the ground state into a higher-energy state, it must absorb a finite amount of energy equal to the difference in the energies of the states. If this amount of energy is not available, the He^4 atom cannot make a transition. At very low temperatures, < 2.2°K, the interactions between the helium atoms and the vessel in which they are contained are insufficiently large to promote excitations between the quantum levels. No excitations of the helium liquid can occur. Therefore the atoms cannot be excited out of the ground state by collisions with the vessel lining or by collisions with other atoms since these collisions cannot provide the quantum of energy necessary to excite a He^4 atom. As a result, the fluid of He^4 atoms flows without any resistance or viscosity. This property is called superfluidity.

As a final example of a quantal gas, let us consider an application of the Planckian distribution through the derivation of the blackbody radiation density. Physically, the blackbody radiation density corresponds to the amount of energy in an electromagnetic wave confined within a three-dimensional volume at frequency v. This result is of great importance in the derivation of the Einstein A and B coefficients that govern radiative emission and absorption processes.

The spectral density $P(v)\,dv$, or blackbody radiation density, is found in a way similar to finding $N(E)$. $P(v)\,dv$ is simply equal to the product of the photon density of states, often called the density of modes, and the mode occupation probability function $f(E)$, the Planckian distribution derived in the Section 5.7.

The density of modes $dN(v)$ can be determined as follows. It is easiest to first work in terms of k, the wave vector, and then convert back into frequency v. We assume that the photons are confined within a three-dimensional box whose sides are all equal to L. The photon modes are plane waves characterized as

$$e^{\pm ikr},\tag{5.8.27}$$

where k is the propagation constant or wave vector. The modes must vanish at the boundaries of the box. As in the case for the electron wave function in a three-dimensional infinite square box (see Chapter 3.1), the electromagnetic modes must vanish at the surfaces defined by

$$x = 0, \quad y = 0, \quad z = 0; \qquad x = L, \quad y = L, \quad z = L.\tag{5.8.28}$$

Since the modes are described by circular functions, the nontrivial solution requires that the k vectors satisfy

$$k_x = \frac{2\pi n_x}{L}, \qquad k_y = \frac{2\pi n_y}{L}, \qquad k_z = \frac{2\pi n_z}{L},\tag{5.8.29}$$

as we found for electrons confined within an infinite three-dimensional square well. The parameters n_x, n_y, and n_z are all integers. For each mode, we can associate a volume in k space. This volume is given as

$$dk_x\,dk_y\,dk_z = \left(\frac{2\pi}{L}\right)^3.\tag{5.8.30}$$

The number of modes available is given by the division of the total volume in k space by the volume per mode. The total volume in k space is simply given by the volume of a sphere of radius k:

$$\frac{4}{3}\pi k^3. \tag{5.8.31}$$

Therefore the number of modes is given by

$$N(k) = \frac{Vk^3}{3\pi^2}. \tag{5.8.32}$$

Neglecting dispersion effects and dividing by the volume, we find that the differential number of photon states per unit volume is given as

$$dN(k) = \frac{Vk^2 \, dk}{\pi^2} \frac{1}{V}. \tag{5.8.33}$$

The k vector can be reexpressed in terms of the frequency as

$$k = \frac{2\pi}{\lambda} = \frac{2\pi}{c}\bar{n}\upsilon, \tag{5.8.34}$$

where n is the index of refraction. Rewriting everything in terms of the frequency υ yields

$$dN(\upsilon) = \frac{8\pi\bar{n}^3\upsilon^2}{c^3} \, d\upsilon. \tag{5.8.35}$$

The blackbody radiation density is given then by the product of the density of modes function and the mode occupation probability as

$$P(\upsilon) \, d\upsilon = \frac{8\pi\bar{n}^3\upsilon^2}{c^3} \frac{d\upsilon}{e^{\frac{h\upsilon}{k_B T}} - 1}. \tag{5.8.36}$$

The above is called the spectral density of states.

Box 5.8.1 Example Problem: Photon Gas in Equilibrium

As an example of a photon gas in equilibrium, consider the following problem. A transmission line can be modeled as an electromagnetic system in one dimension. If the line has only two modes, one each for propagation in each direction, determine the heat capacity of the photons on the line when in equilibrium.

Solution: The heat capacity is defined as

$$C_\upsilon = \frac{\partial \bar{E}}{\partial T}.$$

The problem statement says that there are only two modes present in the system, one for propagation to the left and one for propagation to the right. Therefore the effective density of modes in this case is simply 2. The average energy of the system is given as

$$\bar{E} = \int E f(E) D(E) \, dE.$$

Substitution in for the equilibrium distribution function $f(E)$, the Planckian distribution yields

$$\bar{E} = \int \frac{2\hbar\omega}{e^{\frac{\hbar\omega}{kT}} - 1} \, dE.$$

Recognizing that $E = \hbar\omega$ and then setting $x = E/kT$ and $dx = dE/kT$, we find that the average energy becomes

$$\bar{E} = 2k^2 T^2 \int_0^\infty \frac{x \, dx}{e^x - 1},$$

but the integral

$$\int_0^\infty \frac{x \, dx}{e^x - 1} = \frac{\pi^2}{6}.$$

With this result, the average energy, $\bar{E}$, becomes,

$$\bar{E} = \frac{\pi^2 k^2 T^2}{3}.$$

The heat capacity C_v is then

$$C_v = \frac{\partial \bar{E}}{\partial T} = \frac{2\pi^2 k^2 T}{3},$$

which is the desired result. Note that the heat capacity is linear with the temperature.

PROBLEMS

1. Suppose that a system of n atoms of type A is placed in diffusive contact with a system of n atoms of type B at the same temperature and volume.

   a. Show that after diffusive equilibrium is established the entropy increase is equal to $2n \ln 2$.

   b. If the two atoms are identical, $A = B$, show that there is no entropy increase when diffusive equilibrium is established.

2. Calculate the internal energy of a photon gas assuming that the discrete sum can be replaced by an integral over the mode number.

3. Neglect the zero-point energy of a one-dimensional harmonic oscillator of quantum $\hbar\omega$. Determine

   a. the free energy of the system,

   b. the entropy of the system at high temperature.

4. At low temperatures the rotational partition function per molecule is

$$Z = 1 + 3e^{\frac{-\hbar^2}{Ik_B T}},$$

   where I is the molecular moment of inertia. Calculate the rotational specific heat for N molecules.

5. Determine the entropy of a two-level system whose energies are given as 0 and E. Let the system be at temperature T.

6. A simple one-dimensional harmonic oscillator has energy levels given by

$$E_n = \left(n + \frac{1}{2}\right)\hbar\omega, \quad (n = 0, 1, 2, 3, \ldots).$$

Suppose that the oscillator is in thermal contact with a reservoir at temperature T.

   a. Find the ratio of the probability of the oscillator's being in the first excited state to the probability of its being in the ground state.

   b. Assuming that only the ground state and the first excited state are appreciably occupied, find the mean energy of the oscillator.

7. Consider a simple model of the unwinding of a DNA molecule. Let the molecule have N nucleic acids, each one either being bound with energy 0 or open with energy ϵ. We require, though, that the DNA can unzip from only one end and that the nucleic acid can be open only if all the nucleic acids $1, 2, \ldots, s - 1$ in front of it are open.

   a. Determine the partition function Z of the system.

   b. Find the average number of open links.

8. Consider a system of N bosons of spin zero with orbitals at the single-particle energies 0 and ϵ. The chemical potential is μ and the temperature is assumed to be T. Find T such that the thermal average population of the lowest orbital is twice the population of the orbital at ϵ.

9. A quantum-mechanical particle moves in a circle of radius R with allowed energies

$$E = \frac{n^2 \hbar^2}{2I},$$

where n is an integer and I is the rotational moment of inertia. Find the heat capacity of the system if only the first two states, $n = 0$ and $n = 1$, are occupied.

10. Some Column IV elements (Si, Ge, C) can act as either a donor or acceptor impurity in a semiconductor. Assume that the impurity has energy -2ϵ when it accepts an electron, 0 when it donates an electron, and $-\epsilon$ when it is neutral; determine the fraction of donors present at temperature T.

11. Consider the following model of adsorption of atoms onto a surface. At the surface let there be N sites that are either vacant or occupied by a single atom; two or more atoms cannot occupy a site. The surrounding gas and surface are assumed to be in thermal equilibrium. Let the binding energy of a site be ϵ and the chemical potential be μ. Determine an expression for the average number of occupied sites.

12. A collection of N arrows, each of which can assume any orientation in space, is acted on by an external field $\vec{G}$. If the orientation energy of each arrow in the field $\vec{G}$ can be expressed as

$$E_i = -\vec{L}_i \cdot \vec{G},$$

find the average value of the orientation of $\vec{L}$ in the field $\vec{G}$.

13. a. Consider the competing processes of O_2 and CO adsorption by hemoglobin molecules in blood. If it is assumed that a hemoglobin molecule can either have

a vacancy or adsorb either O_2 or CO with binding energy ϵ_A or ϵ_B, respectively, what is the fraction of hemoglobin molecules bound with O_2? Assume equal amounts of CO and O_2 are present.

b. What is the probability that a CO molecule will be adsorbed compared with that of an O_2 molecule?

14. Consider a binary alloy system that comprises two atomic types, A and B. What is the entropy of the system due to the randomness in the arrangements of A and B alone? (Use Stirling's approximation, $\log N! \sim N \log N - N$, to simplify if necessary.)

15. What is the temperature of a hollow sphere of radius R if its moment of inertia is $I = 2/3MR^2$ and it rotates about a diameter with angular velocity Ω?

16. Let the upward and the downward transition rates for fermions between quantum states of energy E_i and E_j be described as

$$p_j S_{ji} q_i, \qquad p_i S_{ij} q_j,$$

where p_j is the occupation probability of the jth state (assume these lie lower than the ith states in energy), q_i is the probability that the ith state is empty, etc., and S_{ij} and S_{ji} are the transition probabilities. Determine the ratio of the transition probabilities

$$\frac{S_{ji}}{S_{ij}}.$$

17. Consider a classical gas at temperature T in an infinitely high vertical column at rest in a gravitational field of acceleration g.

a. If m is the mass of a molecule of the gas, determine the one-molecule partition function.

b. Determine the internal energy of the gas.

18. **a.** Determine the partition function of N noninteracting distinguishable two-dimensional harmonic oscillators of frequency v at temperature T. Reduce the expression into a compact form involving only the hyperbolic sine.

b. Determine the free energy of the system.

19. From use of the Maxwell relations and the definition of the thermodynamic potentials, show that in general the product of the pressure and volume are proportional to the temperature:

$$pV = AT.$$

Find an expression for the proportionality constant A in terms of $\ln P$.

20. Consider a simple one-dimensional polymer composed of N links each of length d with each link equally likely to be directed to the right or to the left. Find the force at extension l given that the force F is

$$F = -k_B T \left(\frac{d\sigma}{dl} \right).$$

Hint: Find $g(N, s)$, where s is the deviation from 1/2 to the right and 1/2 to the left. Assume that $s \ll N$. Also assume that the length of the polmer l is given by $l = 2|s|d$.

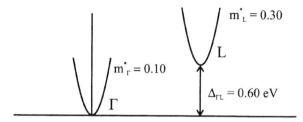

Figure Pr.5.21. Picture of the first conduction band in InP.

21. Consider the simplified picture of the first conduction band in InP, as shown in Figure Pr.5.21. If the electrons are assumed to have free-energy spectrums corresponding to

$$E = \frac{\hbar^2 k^2}{2m}$$

in each valley, Γ and L as marked, at what temperature is the system at if the concentration of electrons in Γ is five times that in L? Assume that $(E - E_f) \gg k_B T$.

6

Nonequilibrium Statistical Mechanics

The basics of equilibrium statistical mechanics were presented in Chapter 5. In general, most systems of interest in engineering and science are not in equilibrium but are in some nonequilibrium condition. A familiar example of a nonequilibrium state is that of a metal under the application of an applied electric field. An electrical current flows in the metal, resulting in a net transport of charge from one place to another. Such a state is highly unusual; there are relatively few ways in which the system can be arranged so as to provide a specific current flow. Other examples of systems that are in nonequilibrium are systems with a temperature or particle gradient. In these systems, there is a net transport of particles from one part of the system to another in order to establish equilibrium. Hence, in general, nonequilibrium statistical mechanics is concerned with the description of transport phenomena.

How are the states described above different from equilibrium? To answer this question, let us recall the definition of equilibrium from Chapter 5. In the discussion in Chapter 5, it was argued that the most random configuration of a system corresponds to the equilibrium configuration. If we consider a free isolated electron gas, the most random configuration of that gas would be one in which the momentum of the electrons is totally randomized in direction. When the individual values of the momentum are summed over, the net momentum would then be zero, since on average for every electron with a forward-directed momentum, there exists an electron with a compensating negative momentum equal in value. Consequently, the net or average momentum of the isolated system is zero. The average momentum of the system can be expressed mathematically as

$$\bar{p} = \frac{\sum\limits_{i=1}^{N} p_i}{N} = \frac{\sum\limits_{i=1}^{N} \hbar k_i}{N} = \hbar \bar{k}, \tag{6.0.1}$$

where it is assumed that there are N electrons present in the system. In equilibrium, the net momentum is equal to zero since the sum over the momentum is zero. It is important to recognize that the results developed for equilibrium statistical mechanics described in Chapter 5 apply to only isolated systems that have been left alone for some time. The importance of these two conditions will become evident in the ensuing discussion.

It is highly improbable that an isolated free-electron gas will spontaneously assemble itself into a state that has a large mean value of momentum in one direction for any appreciable length of time. This is much like the case of the arrows considered in Chapter 5. It is highly unlikely to find a large isolated system of arrows with an appreciable difference in the number of arrows pointing up or down. Therefore, in general, the mean value of the k vector and momentum for an isolated electron gas is zero, just like the mean direction of a collection of arrows is zero in a large ensemble of noninteracting arrows. These results are true unless some outside agent acts on the system. From common everyday experience, we know that a lightbulb does not emit light spontaneously without our first turning on the switch.

How can current transport be understood from the above arguments? The current density is defined as the net flux of charge across an interface at any fixed time. This can be written in terms of the mean carrier velocity of the electron gas as

$$j = qn\bar{v}, \tag{6.0.2}$$

where q is the electron charge, n is the carrier concentration, and v is the average carrier velocity. The average velocity of the electrons is related to the mean momentum by

$$m\bar{v} = \bar{p} = \hbar\bar{k}. \tag{6.0.3}$$

Therefore the mean velocity is

$$\bar{v} = \frac{\hbar\bar{k}}{m}. \tag{6.0.4}$$

The resulting expression for the current density is

$$j = qn\frac{\hbar\bar{k}}{m}. \tag{6.0.5}$$

An isolated system then exhibits no net current flow since the mean value of k is zero. This is a general result: **Any system in equilibrium experiences no net current flow.**

What happens, though, if an external field is applied to the system? In Chapter 5, it was determined that the Fermi sphere is displaced when a field is applied. The amount of the displacement of each k vector within the sphere is found to be $qF/\hbar\,\Delta t$, where F is the field and Δt is the time over which the field acts. Each electron in the system is displaced an equal amount in k space, which means pictorially that the Fermi sphere is displaced from the origin by the same amount. Clearly, summing over the k vectors of all the electrons within the system no longer yields zero. There are now uncompensated k vectors within the gas, giving rise to a net momentum in the field direction. Subsequently, from Eq. (6.0.5) above, a net current flows in the system.

6.1 **The Boltzmann Equation**

What we want to do now is to try to determine the form of the distribution function in the presence of an external driving force when the system is not in equilibrium. It will again be necessary to construct a probability distribution function much like the probability distribution functions determined for equilibrium conditions.

Let us define a distribution function $f(x, k, t)$ that represents the probability of finding a particle at position x, with k vector k at the time t. Clearly, this is a classical distribution function since it simultaneously specifies both the position x and the momentum k. (Remember that k is simply related to p for free particles.)

The question we seek to answer is, how does $f(x, k, t)$ evolve in space, time, and momentum space? The distribution $f(x, k, t)$ governs the behavior of the system in what is called phase space. Phase space is defined as the collective space consisting of three dimensions in real space and three dimensions in k or momentum space. Therefore six coordinates and time are needed to describe the system.

We can derive the basic equation that governs the behavior of a nonequilibrium system by recognizing that the probability distribution function $f(x, k, t)$ must be conserved. In other words, $f(x, k, t)$ must obey a continuity equation. In the special case in which there are no scatterings, the total derivative of $f(x, k, t)$ must be zero (see Box 6.1.1). Hence,

$$\frac{df}{dt} = 0. \tag{6.1.1}$$

Box 6.1.1 Justification of the Boltzmann Equation

The result $df/dt = 0$, which holds in the absence of scatterings, can be understood as follows. Let us begin by defining phase space as consisting of three real-space dimensions, d^3r, and three k-space dimensions, d^3k. A particle is located initially in the phase-space volume $d^3r \, d^3k$. As a result of diffusive and applied forces, the particle can move out of this phase-space volume into another phase-space volume at an infinitesimally later time. Calling the latter volume $d^3r'd^3k'$, we can relate the distribution functions

$$f(r, k, t) \, d^3r d^3k = f(r', k', t') \, d^3r'd^3k',$$

since the probability must remain the same during the time interval dt. In other words, since there are no scatterings present, if the particle is initially located within one definite phase-space volume at time t, at an infinitesimally later time, $t + dt$ it must exist in some other phase-space volume whose coordinates are displaced as

$$r' = r + v \, dt, \qquad k' = k + \frac{(qF)}{\hbar} \, dt.$$

Therefore, in the absence of scatterings, the probability of finding it in the new phase-space volume must be equal to that of finding it in the initial phase-space volume. It can be further shown that the Jacobian relating the two phase-space volumes above is simply equal to one (see Reif 1965, p. 499). Therefore the distribution functions can be related as

$$f(r', k', t') = f(r, k, t).$$

continued

Box 6.1.1, cont.

It should be further noted that r' and k' can be written as

$$r' = r + \frac{dr}{dt}dt, \qquad k' = k + \frac{dk}{dt}dt.$$

With these substitutions, the relation between the distributions is

$$f\left(r + \frac{dr}{dt}dt, k + \frac{dk}{dt}dt, t + dt\right) - f(r, k, t) = 0,$$

which is simply equal to

$$\frac{df}{dt} = 0.$$

Equation (6.1.1) states that the total time rate of change of the probability distribution function must be zero: There is no loss of probability and hence no loss of particles; the probability flux is conserved. However, in semiconductors, carriers undergo scatterings between themselves and their surroundings. As we will discuss in Chapter 9, carriers can be scattered by the ions within the lattice forming the host crystal or by impurities. In addition, carriers can transfer between bands through generation and recombination events. Therefore, in general, the distribution function changes because of scatterings and generation and recombination events. When all these effects are lumped together into one term, the general equation for the time rate of change of the nonequilibrium distribution function becomes

$$\frac{df}{dt} = \left(\frac{\partial f}{\partial t}\right)_{\text{scatterings}}. \tag{6.1.2}$$

Equation (6.1.2) is called the Boltzmann equation. We can recast it into a more convenient form by expanding out the total time derivative on the left-hand side. Since the nonequilibrium distribution function f is a function of x, k, and t, the total time derivative can be expanded as

$$\frac{df}{dt} = \frac{\partial f}{\partial t} + \frac{\partial f}{\partial x}\frac{dx}{dt} + \frac{\partial f}{\partial y}\frac{dy}{dt} + \frac{\partial f}{\partial z}\frac{dz}{dt} + \frac{\partial f}{\partial k_x}\frac{dk_x}{dt} + \frac{\partial f}{\partial k_y}\frac{dk_y}{dt} + \frac{\partial f}{\partial k_z}\frac{dk_z}{dt}. \tag{6.1.3}$$

We can simplify Eq. (6.1.3) by recognizing the following relationships:

$$\frac{dx}{dt}\hat{i} + \frac{dy}{dt}\hat{j} + \frac{dz}{dt}\hat{k} = \vec{v}; \qquad \frac{dk_x}{dt}\hat{i} + \frac{dk_y}{dt}\hat{j} + \frac{dk_z}{dt}\hat{k} = \frac{e\vec{F}}{\hbar}, \tag{6.1.4}$$

where $\vec{F}$ represents all the external forces acting on the system. Recognizing that,

$$\frac{\partial f}{\partial x}\hat{i} + \frac{\partial f}{\partial y}\hat{j} + \frac{\partial f}{\partial z}\hat{k} = \vec{\nabla}_x f, \qquad \frac{\partial f}{\partial k_x}\hat{i} + \frac{\partial f}{\partial k_y}\hat{j} + \frac{\partial f}{\partial k_z}\hat{k} = \vec{\nabla}_k f. \tag{6.1.5}$$

We find that the Boltzmann equation becomes

$$0 = \frac{\partial f}{\partial t} + \vec{\nabla}_x f \cdot \vec{v} + \vec{\nabla}_k f \cdot \frac{\vec{F}}{\hbar} - \left(\frac{\partial f}{\partial t}\right)_{\text{scatterings}}, \qquad (6.1.6)$$

where the last term on the right-hand side above includes the change in the distribution function due to scatterings between particles and their surroundings.

The scattering term is defined in the following way:

$$\left(\frac{\partial f}{\partial t}\right)_{\text{scatterings}} \equiv -\int \{ f(x, k, t)[1 - f(x, k', t)]S(k, k')$$
$$- f(x, k', t)[1 - f(x, k, t)]S(k', k)\} \, dk', \qquad (6.1.7)$$

where $[1 - f(x, k, t)]$ is the probability of a vacancy existing in the state denoted by (x, k) at the time t. $S(k, k')$ is the rate at which an electron makes a transition from the initial state k to the final state k'. Equivalently, $S(k', k)$ is the rate at which an electron makes a transition from the initial state k' to the final state k. The meaning of the terms in Eq. (6.1.7) can then be understood as follows. For a transition to occur from an initial state k to a final state k', a particle must first exist within the initial state. The probability that the initial state is occupied is given by $f(x, k, t)$. In addition, the final state k' must be unoccupied. This probability is simply $1 - f(x, k', t)$. Finally, there is some transition probability rate itself that is expressed by $S(k, k')$. The sum of the product of all these terms together gives the rate at which a transition occurs from k to k'. Hence the scattering term is related to the scattering rate of the particles from k into k' and from k' back to k.

Substituting for each term in Eq. (6.1.2) gives

$$\frac{\partial f}{\partial t} + \frac{\vec{F}_{\text{ext}}}{\hbar} \cdot \vec{\nabla}_k f + \vec{v} \cdot \vec{\nabla}_x f$$
$$= -\int \{ f(x, k, t)[1 - f(x, k', t)]S(k, k')$$
$$- f(x, k', t)[1 - f(x, k, t)]S(k', k)\} \, dk', \qquad (6.1.8)$$

which is known as the Boltzmann transport equation, or BTE. We will often refer to the BTE as simply the Boltzmann equation as well.

Before we consider some approximate methods of solving the Boltzmann equation, it is important to consider the assumptions made in its derivation. The principal assumption is that the distribution function can be specified in terms of position and momentum simultaneously. In other words, the distribution function, by design, is a strictly classical distribution function since it is specified as $f(x, k, t)$. The simultaneous description of the real-space position and momentum of the particle implicit in the definition of the distribution function is in direct contradiction to the Heisenberg Uncertainty Principle. Therefore the Boltzmann equation is not a valid description of transport when quantum-mechanical effects are evident: when the position and the momentum cannot be specified simultaneously. Less obvious, but nonetheless significant, assumptions made in the derivation of the Boltzmann equation are that the scattering probability is assumed to be independent of the external forces (those arising from external fields) and that all collisions are instantaneous. The latter assumption implies that there is no intracollisional field effect. The carriers are not accelerated by the field during the course of the collision. Although the Boltzmann equation suffers from these limitations, it nevertheless provides a useful description of many transport processes and can be used extensively in solid-state problems.

6.2 Approximate Solutions of the Boltzmann Equation

The Boltzmann equation is typically difficult to solve in all its generality since it is an integrodifferential equation. To obtain a result through direct solution of the Boltzmann equation, one must make drastic approximations that either are invalid for the system under investigation or obscure the relevant physics for which one is trying to solve. In this section, a useful means of solving the Boltzmann equation based on the relaxation time approximation is presented. We will see that this approximation, although often inappropriate, greatly reduces the complexity of the Boltzmann equation, allowing for a closed-form solution. Cases in which the relaxation-time approximation breaks down require more sophisticated solutions, particularly numerical ones. Among these techniques are the drift-diffusion and the Monte Carlo methods, which are discussed briefly in Section 6.3.

How can a solution of the Boltzmann equation be obtained? The basis of the relaxation-time approximation is that it is assumed that the scattering term, which involves an integral, can be replaced by a constant relaxation term, thus reducing the original integral–differential equation into only a differential equation. The partial derivative with respect to time due to collisions is assumed to be inversely proportional to a lifetime that characterizes the mean free time between collisions. To see how this works in practice, let us first make a few definitions. Let f_0 represent the equilibrium distribution function and f represent the nonequilibrium distribution function for which the Boltzmann equation is solved. The left-hand side of the Boltzmann equation can then be rewritten as

$$\left(\frac{\partial f}{\partial t}\right)_{collisions} = -\frac{f - f_0}{\tau}. \tag{6.2.1}$$

Equation (6.2.1) implies that the system will relax to the equilibrium distribution f_0 after a characteristic time τ, which is called the relaxation time. The time τ physically represents some average time in which, through the action of collisions, the system relaxes from an initial nonequilibrium state back into a final equilibrium state.

The meaning of τ can be further understood from the following. If an electric field F_0 is applied to the system at $t < 0$ and no diffusion gradients exist, $\nabla_x f = 0$, then the Boltzmann equation reduces to

$$\frac{\partial f}{\partial t} = -\frac{q\vec{F}_0}{m} \cdot \vec{\nabla}_v f(x, v, t) + \left(\frac{\partial f}{\partial t}\right)_{\text{collisions}}. \tag{6.2.2}$$

If at $t = 0$ the field is turned off, then the Boltzmann equation at $t > 0$ further reduces to

$$\frac{\partial f}{\partial t} = \left(\frac{\partial f}{\partial t}\right)_{\text{collisions}}. \tag{6.2.3}$$

With the relaxation-time approximation, the time rate of change of f due to collisions is given by Eq. (6.2.1). Subsequently, Eq. (6.2.3) becomes

$$\frac{\partial f}{\partial t} = -\frac{f - f_0}{\tau}. \tag{6.2.4}$$

The time rate of change of the equilibrium distribution function f_0 is zero, so it can be added to the left-hand side of Eq. (6.2.4) to give

$$\frac{\partial}{\partial t}(f - f_0) = -\frac{f - f_0}{\tau}, \tag{6.2.5}$$

which can be simply solved as

$$y \equiv (f - f_0), \qquad \frac{dy}{dt} = -\frac{y}{\tau},$$
$$y = y_0\, e^{-\frac{t}{\tau}}. \tag{6.2.6}$$

Therefore resubstituting for $y(f - f_0)$ in the last of Eqs. (6.2.6) yields

$$(f - f_0) = (f - f_0)e^{-\frac{t}{\tau}}. \tag{6.2.7}$$

The nonequilibrium distribution f is equal to

$$f(t) = f_0 + [f(t = 0) - f_0]\, e^{-\frac{t}{\tau}}. \tag{6.2.8}$$

From the above result it is clear that as t approaches infinity, the nonequilibrium distribution function f relaxes to f_0, the equilibrium distribution function. The system relaxes to equilibrium at a time τ, called the relaxation time. It is important to realize that the system relaxes through scattering events. In the absence of scatterings, the time rate of change of the distribution due to scatterings would be zero. Therefore Eq. (6.2.2) would reduce to

$$\frac{\partial f}{\partial t} = 0, \tag{6.2.9}$$

which clearly implies that the system never relaxes. Once driven out of equilibrium by the action of an external field, the system never returns to equilbrium. Such a condition exists in a superconducting system, which is discussed in Section 6.4.

Let us now seek a solution of the Boltzmann equation by using the relaxation-time approximation in the presence of a uniform applied electric field. For simplicity, we assume that the distribution is spatially uniform such that the spatial gradient of f is zero. The system is also assumed to be in steady state. These assumptions imply that the gradient of f with respect to x is zero and that the time rate of change of f is also zero. Hence

$$\vec{\nabla}_x f = 0, \qquad \frac{\partial f}{\partial t} = 0. \tag{6.2.10}$$

The Boltzmann equation without approximation is

$$\frac{\partial f}{\partial t} + \frac{\vec{F}_{\text{ext}}}{m} \cdot \vec{\nabla}_v f + \vec{v} \cdot \vec{\nabla}_x f = \left(\frac{\partial f}{\partial t} \right)_{\text{collisions}}, \tag{6.2.11}$$

which, under the above assumptions, reduces to

$$\frac{\vec{F}_{\text{ext}}}{m} \cdot \vec{\nabla}_v f = \left(\frac{\partial f}{\partial t} \right)_{\text{collisions}} = -\frac{f - f_0}{\tau}. \tag{6.2.12}$$

The external force can be written in terms of the field F_0 as

$$\vec{F}_{\text{ext}} = -q\vec{F}_0, \tag{6.2.13}$$

giving

$$-\frac{q\vec{F}_0 \cdot \vec{\nabla}_v f}{m} = -\frac{f - f_0}{\tau}. \tag{6.2.14}$$

If the field is assumed to act along only the z direction, Eq. (6.2.14) simplifies to

$$\frac{qF_0\tau}{m} \frac{\partial f}{\partial v_z} = (f - f_0). \tag{6.2.15}$$

Finally, if the nonequilibrium distribution function f is assumed to be not far from equilibrium, then the derivative of f can be approximated by the derivative of f_0 as

$$\frac{\partial f}{\partial v_z} \sim \frac{\partial f_0}{\partial v_z}. \tag{6.2.16}$$

The equilibrium distribution function is taken to be Maxwellian and is of the form

$$f_0 = e^{-\frac{E}{k_B T}}, \tag{6.2.17}$$

where the energy E is equal to $1/2\, mv^2$ for free particles. Hence f_0 is

$$f_0 = e^{\frac{-mv^2}{2k_B T}}.$$

(6.2.18)

Therefore the derivative of f_0 with respect to v_z becomes

$$\frac{\partial f_0}{\partial v_z} = \frac{\partial}{\partial v_z} e^{\frac{-mv^2}{2k_B T}} = -\frac{mv_z}{k_B T} e^{\frac{-mv^2}{2k_B T}}.$$

(6.2.19)

So

$$\frac{\partial f}{\partial v_z} = -\frac{mv_z}{k_B T} e^{-\frac{mv^2}{2k_B T}}.$$

(6.2.20)

Substituting Eq. (6.2.20) into Eq. (6.2.15) yields

$$-\frac{q F_0 \tau}{m} \frac{mv_z}{k_B T} e^{-\frac{mv^2}{2k_B T}} = f - f_0,$$

$$-\frac{q F_0 \tau}{k_B T} v_z f_0 = (f - f_0).$$

(6.2.21)

Therefore the nonequilbrium distribution function is found as

$$f = f_0 \left[1 - \frac{q F_0 \tau v_z}{k_B T} \right],$$

(6.2.22)

which gives an expression for the nonequilibrium distribution function f in terms of the equilibrium distribution function f_0. We can simplify this further by realizing that the z component of v is simply equal to $v \cos\theta$. Hence the distribution function can be written as

$$f = f_0 \left[1 - \frac{q F_0 \tau v \cos\theta}{k_B T} \right].$$

(6.2.23)

The nonequilibrium distribution function is compared with the equilibrium distribution function in Figure 6.2.1. As can be seen from the figure, the nonequilibrium distribution f is slightly shifted with respect to the equilibrium distribution function f_0. As expected, the equilibrium distribution function is centered about $v = 0$.

The current density j_z can be determined from the nonequilibrium distribution function. j_z is defined as

$$j_z = -nq \overline{v_z},$$

(6.2.24)

where v_z is the average velocity in the z direction. How is the average velocity determined? In Chapter 5, we determined the average value of a macroscopic variable by averaging that variable over the equilibrium distribution function. In nonequilibrium statistical mechanics, the equilibrium distribution is no longer an

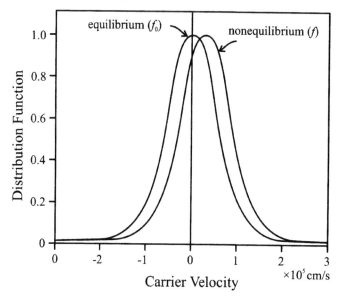

Figure 6.2.1. Plot of the equilibrium f_0 and nonequilibrium f distribution functions as functions of velocity. Note that f is slightly shifted with respect to f_0, which is centered about $v = 0.0$.

appropriate description of the system. The nonequilibrium distribution function f is used instead. The average value is found from

$$\overline{v_z} = \frac{\int v_z f(v) D(v) \, d^3v}{\int f(v) D(v) \, d^3v}, \tag{6.2.25}$$

where $D(v)$ is the density-of-states function per unit volume in velocity space given as $2m^3 V/h^3$ (where V is the volume), as shown below, $f(v)$ is the nonequilibrium distribution function, and v_z is the average value over the nonequilibrium distribution of the velocity along the z direction. The density of states in velocity space can be found from the expression for the density of states in k space. In Chapter 7 the explanation for the density-of-states result in k space is presented. For now, let us simply quote the result in both momentum and velocity space. The density of states in velocity space is simply a multiplicative factor that is determined as

$$D(v) \, d^3v = \frac{2m^3 \, \text{Vol}}{h^3} \, dv_x \, dv_y \, dv_z = \frac{2 \, \text{Vol}}{h^3} dp_x \, dp_y \, dp_z. \tag{6.2.26}$$

In the present problem, since the density of states is simply a multiplicative factor, it will divide out between the numerator and the denominator. Substituting in the expression for $f(v)$ from Eq. (6.2.23) and switching to polar coordinates, we can calculate v_z

$$\overline{v_z} = \frac{\int_0^\infty \int_0^\pi \int_0^{2\pi} d\phi f_0(v) \left[1 - \frac{qF_0\tau v\cos\theta}{k_B T}\right] v\cos\theta \, v^2 \, dv \sin\theta \, d\theta}{\int_0^\infty \int_0^\pi \int_0^{2\pi} d\phi f_0(v) \left[1 - \frac{qF_0\tau v\cos\theta}{k_B T}\right] v^2 \, dv \sin\theta \, d\theta}. \tag{6.2.27}$$

We can simplify the above integrals by noting that the integral

$$\int_0^\pi \cos\theta \sin\theta \, d\theta = 0. \tag{6.2.28}$$

Therefore the first term in the numerator and the second term in the denominator both vanish. The resulting expression for v_z is

$$\overline{v_z} = \frac{-\int_0^\infty \int_0^\pi \int_0^{2\pi} d\phi \, \frac{f_0(v) q F_0 \tau v^4 \cos^2\theta \sin\theta}{k_B T} \, d\theta \, dv}{\int_0^\infty \int_0^\pi \int_0^{2\pi} d\phi \, f_0(v) v^2 \sin\theta \, dv \, d\theta}. \tag{6.2.29}$$

The integrals over Φ divide out, leaving

$$\overline{v_z} = \frac{-\int_0^\infty \int_0^\pi f_0 \frac{q F_0 \tau v^4}{k_B T} \cos^2\theta \sin\theta \, d\theta \, dv}{\int_0^\infty \int_0^\pi f_0 v^2 \sin\theta \, d\theta \, dv}. \tag{6.2.30}$$

The θ dependence can be integrated out of the above expression if it is assumed that τ is independent of θ. Note that

$$\int_0^\pi \cos^2\theta \sin\theta \, d\theta = \frac{2}{3},$$

$$\int_0^\pi \sin\theta \, d\theta = 2. \tag{6.2.31}$$

With these substitutions, v_z becomes

$$\overline{v_z} = \frac{-\frac{1}{3} \frac{q F_0}{k_B T} \int_0^\infty v^2 \tau(v) [v^2 f_0(v)] \, dv}{\int_0^\infty v^2 f_0(v) \, dv}, \tag{6.2.32}$$

which is simply an averaging of $v^2 \tau(v)$ with respect to $v^2 f_0(v)$. Therefore

$$\overline{v_z} = -\frac{q F_0}{3 k_B T} \overline{v^2 \tau(v)}. \tag{6.2.33}$$

We can simplify Eq. (6.2.33) further by recognizing that the mean kinetic energy of a free particle is

$$\overline{\frac{1}{2} m v^2} = \frac{3}{2} k_B T,$$

$$k_B T = \frac{1}{3} \overline{m v^2}. \tag{6.2.34}$$

Therefore

$$\overline{v_z} = -\frac{q F_0}{3 k_B T} \overline{v^2 \tau} \frac{3 k_B T}{\overline{m v^2}}, \tag{6.2.35}$$

since

$$\frac{3k_B T}{m\overline{v^2}} = 1.$$

(6.2.36)

v_z becomes then

$$\overline{v_z} = -\frac{q F_0}{m} \frac{\overline{v^2 \tau}}{\overline{v^2}},$$

(6.2.37)

or

$$\overline{v_z} = -\frac{q F_0}{m} \bar{\tau},$$

(6.2.38)

where

$$\bar{\tau} \equiv \frac{\overline{v^2 \tau}}{\overline{v^2}}.$$

(6.2.39)

The mobility of an electron often represented as μ (not to be confused with the chemical potential whose standard representation is also μ) is defined as the ratio of the carrier velocity in the field direction to the magnitude of the electric field F_0. Hence the mobility is given as

$$\mu = \left| \frac{\overline{v_z}}{F_0} \right|,$$

(6.2.40)

so substituting in for v_z yields

$$\mu = \frac{q\bar{\tau}}{m}.$$

(6.2.41)

The conductivity of a free-electron gas can also be determined from the knowledge of v_z. The conductivity is defined as the ratio of the current density j_z to the electric field F_0. The current density is

$$j_z = -nq\overline{v_z},$$
$$j_z = (-nq)\left(-\frac{q F_0 \bar{\tau}}{m}\right) = \frac{nq^2 F_0 \bar{\tau}}{m} = \sigma F_0.$$

(6.2.42)

The conductivity of a free-electron gas is then

$$\sigma = \frac{nq^2 \bar{\tau}}{m} = nq\mu.$$

(6.2.43)

6.3 **Derivation of the Drift-Diffusion Equation**

In this section we again solve the Boltzmann equation by using the relaxation-time approximation to determine the basic equations governing transport in semiconductors and semiconductor devices, drift-diffusion and Poisson's equations. The drift-diffusion equations cannot always be applied to describe transport in semiconductor devices. During transients, such as accelerations under rapidly changing electric fields, the drift-diffusion equations no longer retain their validity. In these situations, the Boltzmann equation or a modified formulation of the drift-diffusion equations including energy balance must be used instead. Nevertheless, the drift-diffusion equations can be used in many semiconductor device-modeling applications, specifically in long-channel field-effect transistors, conventional, large-basewidth bipolar transistors, and diodes.

The extent to which the drift-diffusion equations can be used to study semiconductor devices is, at one level, quite surprising. Device structures have presently matured to have feature lengths of less than 0.5 μm. At these dimensions, one would argue that the drift-diffusion equations no longer hold. However, advanced drift-diffusion simulators can be used to reliably calculate the static current-voltage characteristics of these devices. The principal reason why the drift-diffusion solvers can be applicable in some situations is that they reflect the behavior of the mean of the distribution. In other words, quantities such as the current and the average carrier velocity depend principally on the first moment of the distribution function, that is, the mean of an observable. The drift-diffusion equations, as we will see below, are calculated by taking the first two moments of the Boltzmann equation, and, as such, they reasonably reflect the behavior of the mean of the distribution. In situations in which more information is desired beyond the mean, the drift-diffusion equations provide little information. For example, if one would like to calculate the carrier impact ionization rates or hot-electron charging effects, etc., knowledge of the high-energy tail of the distribution is required. As we will see in Chapter 10, to describe the impact ionization rate properly, one must determine the nature of the high-energy tail of the distribution function, which is not accurately described by knowledge of the mean alone. For these reasons, the drift-diffusion equations are of little help in studying systems whose behaviors are dictated principally by the details of the tail of the distribution function.

In several relatively new semiconductor device structures, such as ultrasubmicron gate length field-effect transistors, high-electron-mobility transistors, heterostructure bipolar transistors, and multiple-quantum-well devices, important nonstationary transport effects dominate the behavior of the device. These effects are due physically to the departure of the system from quasi-steady-state conditions. In other words, the distribution function is typically different from a Maxwellian or even a drifted Maxwellian distribution. Therefore it is necessary to solve for more than just the mean of the distribution function and associated quantities based on the mean. As a result, in devices in which nonstationary effects become important, the transport physics cannot be treated realistically by only the drift-diffusion models. As stated above, the Boltzmann equation must then be solved either exactly or with a higher-order approximation than that provided by the drift-diffusion method.

An exact solution of the Boltzmann equation can be made by direct simulation of the flights of the carriers subject to the physics of the system. Chief among these techniques is the Monte Carlo simulation, in which the microscopic trajectories of the charge carriers are simulated. The Monte Carlo method consists of tracing the trajectory of one or more carriers in a crystal or device structure subject to the action of applied electric fields, device geometry, and phonon- or carrier-scattering mechanisms. The carrier free-flight time and the phonon-scattering agents are selected stochastically in accordance with some given probabilities describing the microscopic process. Therefore the Monte Carlo method, which draws its name from the stochastic aspect of the technique, provides a rigorous solution to the Boltzmann equation limited only by the extent to which the full underlying physics of the problem is included. The Monte Carlo method, although it affords a complete solution of the Boltzmann equation and therefore provides a complete description of the microscopic transport processes present in a real device, is nontrivial to formulate and execute. In particular, the Monte Carlo method is highly computationally demanding. Therefore it is beneficial to choose a simpler, although far less detailed, approach to solving the Boltzmann equation, which still preserves much of the essential physics governing the device. The drift-diffusion equations enable a simpler, yet accurate device-modeling analysis in many situations.

The drift-diffusion equation can be derived from the Boltzmann equation as follows. We start with Eq. (6.2.11), substituting the relaxation-time approximation given by Eq. (6.2.1) for the scattering term. This yields

$$\frac{\partial f}{\partial t} + \frac{\vec{F}_{\text{ext}}}{\hbar} \cdot \vec{\nabla}_k f + \vec{v} \cdot \vec{\nabla}_x f = -\frac{f - f_0}{\tau}. \tag{6.3.1}$$

An approximate solution of the Boltzmann equation can now be obtained through what is known as the method of moments. In the method of moments, both sides of the Boltzmann equation are multiplied by a function Θ, which is called a moment-generating function. Then each term in Eq. (6.3.1) above can be evaluated with the moment-generating function. Multiplying both sides of Eq. (6.3.1) by Θ and integrating over all k yields

$$\int \Theta \frac{\partial f}{\partial t} d^3k + \int \Theta \left(\frac{\vec{F}_{\text{ext}}}{\hbar} \cdot \vec{\nabla}_k f \right) d^3k + \int \Theta (\vec{v} \cdot \vec{\nabla}_x f) \, d^3k$$
$$= -\int \Theta \frac{f - f_0}{\tau} d^3k. \tag{6.3.2}$$

Each of the terms in Eq. (6.3.2) can be evaluated with some important results from solid-state physics for the case in which the moment-generating function is equal to the velocity. For the present we simply assume the validity of the following expressions, delaying to subsequent chapters their derivation. The velocity of an electron in a crystal is given as

$$v = \frac{1}{\hbar}\frac{dE}{dk},$$ (6.3.3)

where E is the energy of the carrier and k is its k vector. The effective mass of a particle moving in a solid, which accounts for the effects of the action of the potential of the lattice ions on the electron, is given by

$$\frac{1}{m} = \frac{1}{\hbar^2}\frac{d^2 E}{dk^2}.$$ (6.3.4)

Using these results, we can evaluate each of the terms of Eq. (6.3.2).

Let us consider the result for the more general moment-generating function first. We can rewrite the first term,

$$\int \Theta \frac{\partial f}{\partial t} d^3 k,$$ (6.3.5)

by noting that the function Θ depends on the k vector only. The function Θ is therefore independent of t but dependent on k. Therefore the derivative with respect to t can be taken outside the integral as

$$\int \Theta \frac{\partial f}{\partial t} d^3 k = \frac{\partial}{\partial t} \int \Theta f \, d^3 k.$$ (6.3.6)

We can evaluate the integral on the right-hand side of Eq. (6.3.6)

$$\int \Theta f \, d^3 k$$ (6.3.7)

further by noting that the carrier concentration n is given as the integral of the probability density function $f(E)$ multiplied by the density-of-states function $D(E)$ over energy. However, the above integral is given in terms of k. How can it be evaluated? It is useful to consider how an integral over k can be transformed into an integral over E. Consider the integral over E of the density-of-states function multiplied by the distribution function:

$$\int D(E) f(E) \, dE.$$ (6.3.8)

We would like to rewrite expression (6.3.8) into an integral over k; we can do this by realizing that $D(E)$ can be written in terms of the number of states by making use of the relation

$$\frac{dN}{dk} = 2\frac{1}{(2\pi)^3},$$ (6.3.9)

where the factor of 2 is due to the spin degeneracy. The number of states per k is

$$\frac{dN}{dk} = \frac{1}{4\pi^3}.$$

(6.3.10)

But dN/dk is related to the density of states $D(E)$. It is essentially the number of states at a given k, much like $D(E)$ is the number of states at a given energy. Consequently, the expression for the carrier concentration n can be written in terms of an integral over k as

$$n = \int \frac{1}{4\pi^3} f(k)\, d^3k.$$

(6.3.11)

Hence,

$$\int f(k)\, d^3k = 4\pi^3 n.$$

(6.3.12)

Typically, $f(k)$ is defined as to include the term $4\pi^3$. Therefore the integral over k of $f(k)$ is simply n:

$$\int f(k)\, d^3k = n.$$

(6.3.13)

With this definition, the first term in Eq. (6.3.2) can then be written as

$$\frac{\partial \Theta}{\partial t}.$$

(6.3.14)

We can rewrite the second term in Eq. (6.3.2),

$$\int \Theta \vec{v} \cdot \vec{\nabla}_x f\, d^3k,$$

(6.3.15)

by making use of the following identity

$$\Theta \vec{v} \cdot \vec{\nabla}_x f = \vec{\nabla}_x \cdot (f\Theta\vec{v}) - f\vec{\nabla}_x \cdot (\Theta\vec{v}).$$

(6.3.16)

Note that Θ has been placed inside the differential operator. This is valid since Θ is a function of only k, not of x. The second term becomes then

$$\int \Theta\vec{v} \cdot \vec{\nabla}_x f\, d^3k = \int [\vec{\nabla}_x \cdot (f\Theta\vec{v}) - f\vec{\nabla}_x \cdot \Theta\vec{v}]\, d^3k$$

$$= \nabla_x \cdot \overline{\Theta v} - \overline{\nabla_x \cdot \Theta v}.$$

(6.3.17)

The third term,

$$\int \Theta \frac{\vec{F}}{\hbar} \cdot \vec{\nabla}_k f\, d^3k,$$

(6.3.18)

can be simplified through use of the same identity given by Eq. (6.3.16) to

$$\int \Theta \frac{\vec{F}}{\hbar} \cdot \vec{\nabla}_k f \, d^3 k = \int \frac{1}{\hbar} [\vec{\nabla}_k \cdot (f \Theta \vec{F}) - f \vec{\nabla}_k \cdot (\Theta \vec{F})] \, d^3 k. \qquad (6.3.19)$$

The integral on the right-hand side of Eq. (6.3.19) can be broken up into two separate integrals. The first integral,

$$\int \frac{1}{\hbar} \vec{\nabla}_k \cdot (f \Theta \vec{F}) \, d^3 k, \qquad (6.3.20)$$

can be transformed into a surface integral through use of the divergence theorem. The first integral becomes then

$$\frac{1}{\hbar} \int \vec{\nabla}_k \cdot (f \Theta \vec{F}) \, d^3 k = \frac{1}{\hbar} \oint (f \Theta F)_n \, d\sigma, \qquad (6.3.21)$$

where, in the term on the right-hand side, $d\sigma$ is the differential surface element and the subscript n represents the normal component. The surface extends out to infinity. Since the distribution function f must be finite, f vanishes on the boundary. Subsequently, the first term integrates to zero. We can simplify the second term in Eq. (6.3.19) further by applying the vector identity as

$$\vec{\nabla}_k \cdot (\Theta \vec{F}) = \Theta \vec{\nabla}_k \cdot \vec{F} + \vec{F} \cdot \vec{\nabla}_k \Theta. \qquad (6.3.22)$$

Equation (6.3.19) becomes then

$$\int \Theta \frac{\vec{F}}{\hbar} \cdot \vec{\nabla}_k f \, d^3 k = -\frac{1}{\hbar} \left[\int f (\Theta \vec{\nabla}_k \cdot \vec{F} + \vec{F} \cdot \vec{\nabla}_k \Theta) \, d^3 k \right]. \qquad (6.3.23)$$

The entire integrand is averaged over the distribution function. Therefore the third term in Eq. (6.3.2) is

$$-\frac{1}{\hbar} [\overline{\Theta \nabla_k \cdot F} + \overline{F \cdot \nabla_k \Theta}]. \qquad (6.3.24)$$

Finally, we can easily integrate the fourth term in Eq. (6.3.2),

$$\int \Theta \frac{(f - f_0)}{\tau} \, d^3 k, \qquad (6.3.25)$$

assuming that the relaxation time τ is independent of k. With this assumption, the fourth term becomes

$$\frac{1}{\tau} \int \Theta (f - f_0) \, d^3 k = \frac{(\bar{\Theta} - \overline{\Theta_0})}{\tau}, \qquad (6.3.26)$$

where $\bar{\Theta}$ and $\bar{\Theta}_0$ are the nonequilibrium and the equilibrium averages of the parameter Θ, respectively. Substituting into Eq. (6.3.2) for each term, we can write Eq. (6.3.2) finally in terms of the general moment function Θ as

$$\frac{\partial}{\partial t}\bar{\Theta} + \nabla_x \cdot \overline{\Theta v} - \overline{\nabla_x \cdot \Theta v} - \frac{1}{\hbar}[\overline{\Theta \nabla_k \cdot F} + \overline{F \cdot \nabla_k \Theta}] = -\frac{\bar{\Theta} - \bar{\Theta}_0}{\tau}. \qquad (6.3.27)$$

Let us first attempt a solution of Eq. (6.3.27) by choosing $\Theta = 1$. It is necessary to examine each term in the equation. The first term, when $\Theta = 1$, is simply

$$\frac{\partial}{\partial t}\bar{\Theta} = \frac{\partial}{\partial t}\int \Theta f(k)\,\mathrm{d}^3 k = \frac{\partial n}{\partial t}. \qquad (6.3.28)$$

The second term becomes

$$\nabla_x \cdot \overline{\Theta v} = \nabla_x \cdot (nv). \qquad (6.3.29)$$

The third and the fourth terms can be combined as follows. Consider first the fourth term. The force F is equal to the negative gradient with respect to x of the potential energy E. Therefore the fourth term is equal to with $\Theta = 1$,

$$\overline{\Theta \nabla_k \cdot F} = \overline{\nabla_k \cdot -\nabla_x E}. \qquad (6.3.30)$$

But k and x are independent so the order of the differentiation can be interchanged. This gives

$$\overline{\nabla_k \cdot -\nabla_x E} = \overline{\nabla_x \cdot \nabla_k(-E)}. \qquad (6.3.31)$$

But from Eq. (6.3.3), the gradient of the energy with respect to k is equal to $\hbar v$. Therefore the above can be simplified as

$$\overline{\nabla_x \cdot \nabla_k(-E)} = \int \vec{\nabla}_x \cdot (-\vec{\nabla}_k E)f\,\mathrm{d}^3 k = \int \vec{\nabla}_x \cdot (-\hbar\vec{v})f\,\mathrm{d}^3 k. \qquad (6.3.32)$$

The fourth term is then

$$-\frac{1}{\hbar}\overline{\nabla_k \cdot F} = -\frac{1}{\hbar}\int \vec{\nabla}_x \cdot (-\hbar\vec{v})f\,\mathrm{d}^3 k = \int \vec{\nabla}_x \cdot (\vec{v})f\,\mathrm{d}^3 k. \qquad (6.3.33)$$

However the third term of Eq. (6.3.27) is simply the negative of the above. Hence the third and the fourth terms of Eq. (6.3.27) subtract out.

The fifth term of Eq. (6.3.27) with $\theta = 1$,

$$-\frac{1}{\hbar}\overline{F \cdot \nabla_k 1}, \qquad (6.3.34)$$

is clearly zero since the gradient of 1 is zero. Finally, the right-hand side of Eq. (6.3.27) is simply

$$-\frac{\bar{\Theta} - \bar{\Theta}_0}{\tau} = -\frac{n - n_0}{\tau}, \qquad (6.3.35)$$

where n is the nonequilibrium carrier concentration and n_0 is the equilibrium carrier concentration. If there are no sources or sinks of particles, the carrier concentration remains invariant with time and $n = n_0$. Under these conditions, the right-hand side is equal to zero. When the above results are combined, Eq. (6.3.27) becomes

$$\frac{\partial n}{\partial t} + \vec{\nabla}_x \cdot (n\vec{v}) = 0, \tag{6.3.36}$$

which is the carrier continuity equation. In words Equation (6.3.36) states that the time rate of change of the carrier concentration is equal to the net flux of the carriers across some Gaussian surface. The standard current continuity equation is obtained if n is defined as a charge-carrier concentration; it has q multiplied by it.

To obtain the drift-diffusion equation for electrons, we must solve Eq. (6.3.2) by substituting $\vec{v}$ the velocity in for Θ. Again each term must be evaluated. In this case, Eq. (6.3.2) becomes

$$\int \vec{v} \frac{\partial f}{\partial t} d^3k + \frac{1}{\hbar} \int \vec{v}(\vec{F} \cdot \vec{\nabla}_k f) d^3k + \int \vec{v}(\vec{v} \cdot \vec{\nabla}_x f) d^3k$$
$$= - \int \vec{v} \frac{f - f_0}{\tau} d^3k. \tag{6.3.37}$$

We can rewrite the first term by noting that the velocity is an explicit function only of k. Therefore the partial derivative of v with respect to t is zero. Hence the derivative with respect to t can be taken outside of the integral, giving

$$\int \vec{v} \frac{\partial f}{\partial t} d^3k = \frac{\partial}{\partial t} \int \vec{v} f \, d^3k. \tag{6.3.38}$$

We can evaluate the integral on the right-hand side further by noting that the average of v over the distribution function is simply equal to nv. The first term is then

$$\int \vec{v} \frac{\partial f}{\partial t} d^3k = \frac{\partial}{\partial t}(n\vec{v}). \tag{6.3.39}$$

The second term in Eq. (6.3.37) can be written with aid of the vector and tensor identities in Box 6.3.1 as

$$\frac{1}{\hbar} \int \vec{v}(\vec{F} \cdot \vec{\nabla}_k f) d^3k$$
$$= \frac{1}{\hbar} \left\{ \int [\vec{v}(\vec{\nabla}_k \cdot f\vec{F}) \right.$$
$$\left. - f\vec{v} \vec{\nabla}_k \cdot \vec{F}] d^3k \right\}. \tag{6.3.40}$$

The first term on the right-hand side of

Box 6.3.1 Standard Vector and Tensor Identities

Vector:

$$\vec{\nabla} \cdot (g\vec{F}) = g\vec{\nabla} \cdot \vec{F} + \vec{F} \cdot \vec{\nabla}g.$$

Tensor:

$$\vec{\nabla} \cdot (g\vec{F}\vec{G}) = \vec{G}\vec{\nabla} \cdot (g\vec{F}) + (g\vec{F} \cdot \vec{\nabla})\vec{G}.$$

Eq. (6.3.40), though can be simplified as

$$\frac{1}{\hbar}\int \vec{v}\,(\vec{\nabla}_k \cdot f\vec{F})\,\mathrm{d}^3 k = \frac{1}{\hbar}\left\{ \int [(\vec{\nabla}_k \cdot (f\vec{F}\vec{v}) - (f\vec{F}\cdot\vec{\nabla}_k)\vec{v}]\,\mathrm{d}^3 k \right\}. \qquad (6.3.41)$$

But the integral of the divergence of $f F v$ vanishes since it is equal to a surface integral at infinity. Because f must be finite, the surface integral vanishes identically. The second term in Eq. (6.3.37) is then

$$\frac{1}{\hbar}\int \vec{v}\,(\vec{F}\cdot\vec{\nabla}_k f)\,\mathrm{d}^3 k = \frac{1}{\hbar}\int f(\vec{\nabla}_x E \cdot\vec{\nabla}_k)\vec{v}\,\mathrm{d}^3 k + \frac{1}{\hbar}\int f\vec{v}\,(\vec{\nabla}_k \cdot\vec{\nabla}_x E)\,\mathrm{d}^3 k. \qquad (6.3.42)$$

The order of differentiation can be exchanged in the last term above. Again, by using Eq. (6.3.3), we can rewrite the derivative with respect to k of E in terms of the velocity of the particle as

$$\vec{\nabla}_k \cdot \vec{\nabla}_x E = \vec{\nabla}_x \cdot \vec{\nabla}_k E = \hbar\vec{\nabla}_x \cdot \vec{v}. \qquad (6.3.43)$$

Therefore the second term becomes

$$= \frac{1}{\hbar}\int f(\vec{\nabla}_x E \cdot \vec{\nabla}_k)\vec{v}\,\mathrm{d}^3 k + \int f\vec{v}\,(\vec{\nabla}_x \cdot \vec{v})\,\mathrm{d}^3 k. \qquad (6.3.44)$$

The third term in Eq. (6.3.37) can be simplified as

$$\int \vec{v}\,(\vec{v}\cdot\vec{\nabla}_x f)\,\mathrm{d}^3 k = \int \vec{v}\,(\vec{\nabla}_x \cdot f\vec{v} - f\vec{\nabla}_x \cdot \vec{v})\,\mathrm{d}^3 k. \qquad (6.3.45)$$

When the tensor identity in Box 6.3.1 is used, Eq. (6.3.45) simplifies to

$$= \int [\vec{\nabla}_x \cdot (f\vec{v}\vec{v}) - f(\vec{v}\cdot\vec{\nabla}_x)\vec{v} - f\vec{v}\,(\vec{\nabla}_x \cdot \vec{v})]\,\mathrm{d}^3 k. \qquad (6.3.46)$$

Finally, we can easily integrate the fourth term, assuming that the relaxation time is independent of k. With this assumption, the fourth term becomes

$$-\frac{1}{\tau}\int v(f - f_0)\,\mathrm{d}^3 k = -n\frac{\bar{v} - \bar{v_0}}{\tau}, \qquad (6.3.47)$$

where v and v_0 are the average nonequilibrium and equilibrium velocities, respectively. The average equilibrium velocity by definition is zero, since in equilibrium the ensemble has no net momentum. Therefore the fourth term simplifies to

$$-\frac{n\bar{v}}{\tau}. \qquad (6.3.48)$$

Collecting all the terms yields

$$\frac{\partial}{\partial t}(n\vec{v}) + \frac{1}{\hbar}\int f(\vec{\nabla}_x E \cdot \vec{\nabla}_k)\vec{v}\, d^3k + \int f\vec{v}(\vec{\nabla}_x \cdot \vec{v})\, d^3k$$
$$+ \int [\vec{\nabla}_x \cdot (f\vec{v}\,\vec{v}) - f(\vec{v} \cdot \vec{\nabla}_x)\vec{v} - f\vec{v}(\vec{\nabla}_x \cdot \vec{v})]\, d^3k = -\frac{n\vec{v}}{\tau}. \tag{6.3.49}$$

Note that the third and the sixth terms subtract out, leaving

$$\frac{\partial}{\partial t}(n\vec{v}) + \frac{1}{\hbar}\int f(\vec{\nabla}_x E \cdot \vec{\nabla}_k)\vec{v}\, d^3k + \int \vec{\nabla}_x \cdot (f\vec{v}\,\vec{v})\, d^3k - \int f(\vec{v} \cdot \vec{\nabla}_x)\vec{v}\, d^3k$$
$$= -\frac{n\vec{v}}{\tau}. \tag{6.3.50}$$

Equation (6.3.50) can be reduced to the standard drift-diffusion equation for electrons or holes with several assumptions. The first assumption is that the energy of the carriers is simply given as

$$E = \frac{\hbar^2 k^2}{2m}, \tag{6.3.51}$$

where the mass is assumed to be isotropic. In general, the mass in a crystal differs in different directions. The full details of how the results differ with an anisotropic mass are discussed in the reference by Smith (1992).

We can simplify the third term in Eq. (6.3.50) by recognizing that if the mass is constant and isotropic, the product of $\vec{v}\,\vec{v}$ is a diagonal tensor and the energy is simply given as the sum over the energy in each direction:

$$\bar{E} = \sum_{i=1}^{3} E_i = \sum_{i=1}^{3} \frac{1}{2}mv_i^2. \tag{6.3.52}$$

It is further assumed that

$$\bar{E} = 3E_i: \tag{6.3.53}$$

then

$$\int \vec{\nabla}_x \cdot (f\vec{v}\,\vec{v})\, d^3k = \frac{2}{m}\left[\sum \hat{a}_i \hat{a}_i \cdot \left(\frac{1}{3}\bar{E}\vec{\nabla}_x n + n\vec{\nabla}_x E_i\right)\right], \tag{6.3.54}$$

where E_i is the average energy along each direction and the vectors $\hat{a}_i$ are used to denote the fact that a tensor operation is being performed. Since E represents the average energy of the carriers, it can be reexpressed in terms of the temperature as

$$\bar{E} = \frac{3}{2}k_B T. \tag{6.3.55}$$

If it is assumed that there is no spatial gradient of the energy, which is physically equivalent to stating that the temperature is the same everywhere in the system, then the third term simplifies to

$$\frac{2}{3}\frac{1}{m}\bar{E}\nabla_x n,\tag{6.3.56}$$

where the system is assumed to be both isotropic and constant. Therefore the spatial gradient of the velocity is zero since the velocity is equal to $\hbar k/m$, which are all independent of x.

Finally, we can simplify the second term in Eq. (6.3.50) by noting that $\vec{F}_{ext}(\text{force}) = -\vec{\nabla}_x E$. Therefore the second term is

$$\frac{1}{\hbar}\int f\vec{\nabla}_x E \cdot \vec{\nabla}_k v\, d^3k = -\frac{1}{\hbar}\int f\vec{F}_{ext} \cdot \vec{\nabla}_k v\, d^3k.\tag{6.3.57}$$

Making use of the fact that the velocity is simply equal to $1/\hbar\vec{\nabla}_k E$ and taking the derivative of v with respect to k yields a diagonal tensor with elements

$$\frac{1}{\hbar}\frac{d^2E}{dk^2}.\tag{6.3.58}$$

The effective mass, though, was defined in Eq. (6.3.4) as

$$\frac{1}{m} = \frac{1}{\hbar^2}\frac{d^2E}{dk^2}.\tag{6.3.59}$$

With these substitutions and recognizing that the vector product of $\vec{F}_{ext}$ with the diagonal tensor gives a vector parallel to $\vec{F}_{ext}$, we reduce the second term to

$$-\frac{1}{\hbar}\int f\vec{F}_{ext} \cdot \vec{\nabla}_k v\, d^3k = -\frac{\vec{F}_{ext}}{m}\int f\, d^3k,\tag{6.3.60}$$

which is simply

$$-\frac{\vec{F}_{ext}}{m}\int f\, d^3k = -\frac{\vec{F}_{ext}}{m}n.\tag{6.3.61}$$

When all the terms are collected, Eq. (6.3.50) becomes

$$\frac{\partial}{\partial t}(n\vec{v}) + \frac{2}{3m}\bar{E}\vec{\nabla}_x n - \frac{\vec{F}_{ext}(\text{force})n}{m} = -\frac{n\vec{v}}{\tau}.\tag{6.3.62}$$

The force $\vec{F}_{ext}$ is simply equal to the product of $-q$ the electron charge and the field $\vec{F}$. Therefore

$$\frac{\partial}{\partial t}(n\vec{v}) + \frac{k_B T}{m}\vec{\nabla}_x n + \frac{q\vec{F}n}{m} = -\frac{n\vec{v}}{\tau},\tag{6.3.63}$$

where the mean energy has been replaced with $3/2\ k_B T$. When each side is multiplied by $-q\tau$, the above becomes

$$\tau \frac{\partial}{\partial t}(-qn\vec{v}) - \frac{q\tau}{m}k_B T \vec{\nabla}_x n - \frac{q\tau}{m}q\vec{F}n = qn\vec{v}. \tag{6.3.64}$$

Note that Eq. (6.3.64) involves expressions for qnv and its first derivative, which is simply the electron current density and its first derivative. Therefore Eq. (6.3.64) is a function of J_n and dJ_n/dt. If Eq. (6.3.64) is considered to only zero order, the first derivative can be neglected. This is a reasonable assumption if the average collision time is very small, $\sim ps$, which is typically the case. Therefore the first term above involving the time derivative vanishes. When the definitions of the mobility and the current density

$$\mu = \frac{q\tau}{m}, \qquad J_n = -qn\bar{v}, \tag{6.3.65}$$

are used, Eq. (6.3.64) becomes

$$\vec{J}_n = \mu k_B T \vec{\nabla}_x n + qn\mu \vec{F}. \tag{6.3.66}$$

The diffusion constant D_n is defined as

$$D_n = \mu_n \frac{k_B T}{q}, \tag{6.3.67}$$

where the subscript n has been added to differentiate between electrons and holes. With the above definition, the electron current density becomes

$$\vec{J}_n = q\mu_n n\vec{F} + q D_n \vec{\nabla}_x n, \tag{6.3.68}$$

which is the electron drift-diffusion equation.

A similar equation holds for holes. The only difference is that the holes move in the opposite direction from the electrons. The hole drift-diffusion equation is

$$\vec{J}_p = q\mu_p p\vec{F} - q D_p \vec{\nabla}_x p. \tag{6.3.69}$$

The above drift-diffusion equations are important in semiconductor device analysis and are used to model most semiconductor devices.

Before we end this section, it is useful to review some of the implications of the approximations made in deriving the drift-diffusion equations above. In the derivation, it was assumed accurate to only zero order in the current density. As a consequence, all time-dependent conductivity phenomena, like velocity overshoot effects, etc., are neglected. In many cases, this is reasonable to assume because of the very fast relaxation processes present in semiconductors. Typically, electron and hole relaxation times are of the order of 10^{-13} or 10^{-14} s. Only in few semiconductor devices in which the transit times or the carrier dynamics change at rates comparable with the relaxation processes do the drift-diffusion equations lose their validity. Therefore, for the majority of cases, the drift-diffusion models can be used. Nevertheless, in many new semi-conductor devices such as submicrometer gate length field-effect transistors and narrow-width bipolar transistors, the electric field shows large spatial fluctuations leading to a strong departure from the quasi-steady-state. In these cases, the Boltzmann equation must be solved instead or energy balance terms must be added to the drift-diffusion equations.

6.4 **Superconductivity**

Superconductivity has recently undergone a renaissance owing to the discovery of superconducting materials at relatively high temperatures. These materials, commonly referred to as high-temperature superconductors, are typically made from the La—Ba—Cu—O family or related material systems. Superconductive behavior has been observed at temperatures well above that of liquid nitrogen, 77 K, but still below room temperature, 300 K. Nevertheless, there is reason to hope that room-temperature superconductors may one day be feasible. The realization of room-temperature superconductivity would almost certainly produce a new revolution in technology, having an impact on all aspects of electrical engineering from power distribution systems to microdevices. In this section, a brief sketch of the physics of superconductivity as currently understood is presented. At the time of this writing, the complete picture of the physics of high-temperature superconductivity remains unknown.

Let us begin our discussion with a review of the basic empirical aspects of superconductivity. A superconductor has the following macroscopic properties:

1. absence of all measurable dc electrical resistance;
2. exhibits perfect diamagnetic behavior; a magnetic field cannot penetrate into the interior of a superconductor (provided that the field is not too strong);
3. exhibits a sharp transition temperature above which no superconductive effects are observed where the system behaves normally;
4. behaves as if there were an energy gap centered about the Fermi energy.

Let us first consider a macroscopic theory of superconductivity originally developed to explain the diamagnetism of a superconductor. The effect that an external magnetic field will be excluded from within a superconducting material is called the Meisner effect. Mathematically, if the superconductor cannot support a magnetic field within it, the magnetic field obviously must vanish within the material. Subsequently, $\vec{B}$ must equal zero inside the superconductor. To account for the Meisner effect, it was originally postulated that the current density within the superconductor is related to the external magnetic field $\vec{B}$ as

$$\vec{\nabla} \times \vec{j} = -\frac{n_s q^2}{mc} \vec{B}, \tag{6.4.1}$$

where n_s is the concentration of superconducting electrons and cgs units are used. The applied magnetic field is assumed to be uniform. The magnetic field is related to $\vec{j}$ through one of Maxwell's equations as

$$\vec{\nabla} \times \vec{B} = \frac{4\pi}{c} \vec{j} \quad \text{(cgs)}.$$ (6.4.2)

Applying the curl operator to both sides of Eq. (6.4.1) yields

$$\vec{\nabla} \times \vec{\nabla} \times \vec{j} = -\frac{n_s q^2}{mc} \vec{\nabla} \times \vec{B},$$ (6.4.3)

and applying the curl operator to both sides of Eq. (6.4.2) yields

$$\vec{\nabla} \times \vec{\nabla} \times \vec{B} = \frac{4\pi}{c} \vec{\nabla} \times \vec{j}.$$ (6.4.4)

Using the vector identity

$$\vec{\nabla} \times \vec{\nabla} \times \vec{B} = \vec{\nabla}(\vec{\nabla} \cdot \vec{B}) - \nabla^2 \vec{B},$$ (6.4.5)

we can rewrite both Eqs. (6.4.3) and (6.4.4). The divergence of $\vec{B}$ is zero from Maxwell's equations. Therefore Eq. (6.4.4) becomes

$$\nabla^2 \vec{B} = \frac{4\pi n_s q^2}{m c^2} \vec{B}.$$ (6.4.6)

Equation (6.4.6) implies that the magnetic field within the superconductor must be zero. This is because $\vec{B} = $ constant is not a solution to Eq. (6.4.6) since $\nabla^2 \vec{B} = 0$ for constant $\vec{B}$. Therefore $\vec{B}$ cannot be equal to an arbitrary constant within the superconductor but must equal zero. Subsequently, when a weak, uniform magnetic field is applied to a superconductor, the magnetic field within the superconductor is zero; the field is excluded from within the superconductor.

The above formulation accounts for the Meisner effect by postulating the relationship given by Eq. (6.4.1). For a material to obey the Meisner effect there must be some concentration of superconducting particles n_s within the material that obey Eq. (6.4.1). The question is then, what is the origin of this concentration of superconducting particles and how do these particles behave? Before we answer this question, it is useful to first discuss the materials that are empirically observed to be superconducting.

Up until recently, superconductivity was observed to occur only in metallic systems, both elements and alloys. The range in which these materials become superconducting is from near-absolute-zero degrees kelvin to 23 K. Niobium germanium, Nb_3Ge, has one of the highest transition temperatures, defined as the temperature below which the material exhibits superconducting properties, of 23 K.

Table 6.4.1
Some High-Temperature
Superconducting Materials

Material	Transition Temp (K)
LaBaCuO	30–40
$La_{2-x}Sr_xCuO_{4-y}$	~70
$YBa_2Cu_3O_7$	~95
$Bi_2Sr_2CaCu_2O_{8+y}$	~120
$Tl_2Ba_2Ca_2Cu_3O_{10}$	~125

The more recently discovered high-temperature superconductors all seem to be oxides containing various metallic elements such as copper. A short list of high-temperature superconducting materials and their transition temperatures is provided in Table 6.4.1. As can readily be seen from the table, these materials all superconduct at temperatures near or above liquid-nitrogen temperature, 77 K. The relative inexpense of liquid nitrogen compared with liquid helium accounts for the great excitement as to the practical utility of these materials in engineering.

Qualitative Aspects of the Theory of Superconductivity

The macroscopic theory of superconductivity discussed above accounts for the Meisner effect but does not explain the origin of the superconducting carrier concentration or its meaning. Before 1957, there was no consistent microscopic theory of superconductivity, even though superconductivity had been known for almost fifty years. The first successful microscopic theory of superconductivity was developed by Bardeen, Cooper, and Schrieffer and is known as the BCS theory. The BCS theory is one of the most successful theories of modern physics. It has successfully predicted the transition temperatures and behaviors of the low-temperature superconducting materials flawlessly. Only the recently discovered high-temperature superconductors seem to lie outside its domain. Yet it is still possible that some variation of the original theory may explain the behavior of high-temperature superconductors. The nature of the high-temperature superconductors remains open.

In this section, we sketch a qualitative picture of the origin of superconductivity, delaying a quantitative approach to Section 6.5. To understand the BCS theory it is first necessary to review conduction and to understand how a superconducting state can arise. Electronic current flow arises when the ensemble of electrons has a net momentum, as discussed in Section 5.8. As we found above, after the driving field has been removed, the electron distribution relaxes back to equilibrium by scattering processes. The nonequilibrium distribution relaxes to the equilibrium distribution in some characteristic time called the relaxation time. Scatterings occur between the electrons and the lattice, transferring energy from the more energetic electrons to the lattice ions. Ultimately, the electrons regain thermal equilibrium, and no net transfer of energy between the electrons and the lattice then occurs.

The above argument can be restated in the following way. A solid can be viewed as two separate subsystems; a collection of free or nearly free electrons and a lattice. As we will see in a later chapter (see Section 7.1), because of the different masses of the electrons and the host atoms, the response of these two different species to an applied electric field is different. For this reason, the dynamics of the solid can be separated into two different dynamical subsystems, the electrons and the ions. The dynamics of each of these subsystems can be treated independently to zeroth order. Coupling between the electrons and the ions occurs through electron–lattice scatterings. For example, an applied electric field acts to increase the energy of the electron system, called field heating, without significantly changing the internal energy of the ions within the lattice. Remember the electrons can easily respond to an applied electric field since they are basically free to move throughout the solid while the atomic centers are held in place by bonding forces. Hence the electronic system is driven out of equilibrium with the lattice system. The electrons attain a quasi-steady-state, nonequilibrium condition sustained by the competing influences of field heating and relaxation through scatterings with the lattice. If the driving force is removed there is no longer any means of sustaining nonequilibrium; the electrons relax by means of lattice collisions (see Chapter 9) to equilibrium. In other words, an equilibrium is ultimately established between the two subsystems from the coupling between them in the form of lattice scatterings once the driving electric field is removed.

The question becomes, what happens if no scatterings are present? Scatterings form the only coupling between the electron and the lattice subsystems. Again if the electrons are heated by an applied electric field they will become significantly hotter than the lattice ions; the electrons have a much higher energy. If no scatterings can occur, then there no longer is any means of transferring energy between the two subsystems; the two subsystems become decoupled. This implies that the electron gas will remain at a higher energy than the lattice for an indefinite time. Additionally, if the electrons suffer no scatterings, if they have an initial net momentum or equivalently net current, then they will retain the same net momentum or current for an indefinite time. Physically, the two subsystems do not interact and each can have a different internal energy from the other for all time. This is similar to the two subsystems, S_1 and S_2 shown in Figure 5.3.1, that are in neither thermal nor diffusive contact. If the partition between S_1 and S_2 is not removed or changed to a semipermeable membrane, then the two subsystems will never interact and no exchange of energy or particles will occur between them. Subsequently, subsystem S_1, if initially hotter than S_2, will remain hotter than S_2 for an indefinite time.

Clearly, if two subsystems become decoupled they will remain at unequal temperatures. Therefore, if we decompose a solid into electron and lattice ion subsystems and if these two subsystems become decoupled somehow, then they can each have different temperatures for an indefinite period of time.

Can such a situation, in which no scatterings are possible between the electrons and the lattice, occur in a solid? The answer is yes. The superconducting state is just such a state; no energy can be exchanged between the conduction system and the lattice. In the superconducting state, the current is not carried by single electrons but by paired electrons called Cooper pairs. A Cooper pair arises from a very weak coupling between different electrons due to a complex electron–lattice interaction. One electron attracts the positively charged lattice ions, slightly distorting the lattice and altering its local potential. This creates a region of enhanced positive charge that then attracts another electron. The two electrons become coupled by this weak interaction, forming a new bound state that lies lower in energy than the free-electron states. This state is called the BCS ground state. The BCS state is the lowest possible energy state of the system and is separated by a finite energy from the free single-particle electron states of the Fermi gas. A finite amount of energy must be supplied to the system in order to break the Cooper pair and return the electrons to the normal single-particle states. The energy separation is related to the binding energy of a Cooper pair, which is denoted here as Δ. An energy of Δ then must be supplied to dissociate a Cooper pair and thereby excite the system from the superconducting state to the normal resistive state.

On the basis of the above discussion, it is possible to understand how no carrier scatterings can arise in a superconductor. There are no available states between the BCS superconducting state and the normal resistive state. As we know from our study of quantum mechanics, a system can exist in only well-defined, discrete-energy states. If insufficient energy is supplied to the system to raise its energy from one quantum state to another, then a transition cannot occur. For example, in the infinite quantum well discussed in Chapter 2, we found that an electron cannot be promoted from the ground state to the first excited state unless it absorbs energy equal to the difference in these two states. In this example that is

$$E = \frac{n^2 \pi^2 \hbar^2}{2ma^2}, \qquad \Delta E = (4 - 1)\frac{\pi^2 \hbar^2}{2ma^2}. \tag{6.4.7}$$

Therefore, if the energy ΔE is not available, then the system will remain in its original state. The same physical situation exists for a Cooper pair. Clearly, if insufficient energy is supplied to break a Cooper pair, the system remains in the superconducting state. Therefore, if the existing scattering agents cannot provide an energy of at least Δ, then no transition can occur. As a result, the pair cannot suffer any scatterings. The pair is not broken, and the single electrons are not promoted to the usual conduction states. As a result, the system remains superconducting. Hence the system does not relax since no scatterings can occur. The current flows without resistance.

Why does the material have a finite transition temperature however? As long as one Cooper pair exists within the material, it will exhibit superconducting behavior since the superconducting current short circuits the normal current. Of course, if a large current is forced through the superconductor, electron–electron collisions will break the Cooper pair, reverting the system to its normal state. This is why phenomena such as magnetic fields and current densities affect the superconducting state, since they also lead to Cooper pair dissociation.

6.5 **Quantitative Aspects of the Bardeen, Cooper, and Schrieffer Theory**

In this section, we discuss a simplified picture of the BCS theory of superconductivity. The full details of the BCS theory are well beyond the level of this textbook. The interested reader should consult the references for a more detailed description of the theory. To understand the basic physics of the pairing interaction and its relationship to superconductivity, let us consider first the general problem of the energy of a free-electron gas in the presence of an attractive electron–electron interaction. One would first expect that there is only a repulsive electron–electron interaction owing to Coulomb repulsion. However, we will see later that there is an attractive interaction between two electrons in the presence of a filled Fermi sea of electrons. For the moment, however, it is not necessary to identify this mechanism only to assume its existence. Let us consider the behavior of two electrons described by coordinates r_1 and r_2 in the presence of a filled Fermi gas of electrons. The condition that the Fermi sphere is filled acts to prohibit the two electrons from occupying states whose k vectors are less than that of the Fermi wave vector k_f. The collective wave function of the two electrons can be written in terms of the difference of r_1 and r_2 if it is assumed that the center of mass of the two electrons is at rest. Therefore the two-particle wave function is given as

$$\psi = \psi(r_1 - r_2). \tag{6.5.1}$$

The two-particle wave function can be written as a linear superposition of eigenstates of a Hermitian operator based on the fundamental expansion postulate from Chapter 1. Plane waves form a basis set and are eigenfunctions of the momentum operator. Therefore the two-particle wavefunction can be expanded in terms of a complete Fourier series (or equivalently on a plane-wave basis). This expansion is

$$\Psi(\vec{r}_1 - \vec{r}_2) = \sum_k g(k) e^{i\vec{k} \cdot (\vec{r}_1 - \vec{r}_2)}, \tag{6.5.2}$$

where k is defined as

$$\vec{k} = \frac{1}{2}(\vec{k}_1 - \vec{k}_2), \tag{6.5.3}$$

and k_1 and k_2 are the k vectors of the single-particle electron states. The Schroedinger equation for the two electrons is given then as

$$-\frac{\hbar^2}{2m}(\nabla_1^2 + \nabla_2^2)\Psi(\vec{r}_1 - \vec{r}_2) + V(\vec{r}_1 - \vec{r}_2)\Psi = E\Psi, \tag{6.5.4}$$

where E is the energy of the two electrons, called a pair, relative to the state in which the two electrons are at the Fermi level. In other words, if the two electrons are not paired their energy would each be E_f, which we set to zero for convenience. Therefore, on pairing, the total energy of the pair is simply the pairing energy E. The interesting result that occurs is that the energy of the pair is less than the sum of the separate single-particle energies. In other words, pairing reduces the energy of the system. It is just this energy reduction that leads to superconductive behavior, as we will see below.

Substituting the expansion given above in for $\Psi(r_1 - r_2)$ yields

$$
-\frac{\hbar^2}{2m}(\nabla_1^2 + \nabla_2^2) \sum_k g(k) e^{i\vec{k} \cdot (\vec{r}_1 - \vec{r}_2)} + V(\vec{r}_1 - \vec{r}_2) \sum_k g(k) e^{i\vec{k} \cdot (\vec{r}_1 - \vec{r}_2)}
$$
$$
= E \sum_k g(k) e^{i\vec{k} \cdot (\vec{r}_1 - \vec{r}_2)}. \tag{6.5.5}
$$

Applying the kinetic-energy operators to the first term yields

$$
\sum_k -\frac{\hbar^2}{2m}(-2k^2) g(k) e^{i\vec{k} \cdot (\vec{r}_1 - \vec{r}_2)} + V(\vec{r}_1 - \vec{r}_2) \sum_k g(k) e^{i\vec{k} \cdot (\vec{r}_1 - \vec{r}_2)}
$$
$$
= E \sum_k g(k) e^{i\vec{k} \cdot (\vec{r}_1 - \vec{r}_2)}. \tag{6.5.6}
$$

To arrive at an expression for the energy of the pair, we follow the usual procedure used throughout Chapters 2 and 3 of multiplying each side by $e^{ik' \cdot r}$ and integrating over all space. For convenience, we define r to be equal to the difference in r_1 and r_2. Performing this operation gives

$$
\sum_k \frac{\hbar^2}{m} k^2 g(k) \int e^{ikr} e^{-ik'r} \mathrm{d}^3 r + \int \sum_k g(k) e^{-ik'r} V(r) e^{ikr} \mathrm{d}^3 r
$$
$$
= \int E \sum_k g(k) e^{-ik'r} e^{ikr} \mathrm{d}^3 r. \tag{6.5.7}
$$

Note that the integral

$$
\int e^{i(\vec{k} - \vec{k}') \cdot \vec{r}} \mathrm{d}^3 r = \delta(k - k') \tag{6.5.8}
$$

is simply the Dirac delta function. When this result is used, Eq. (6.5.7) becomes

$$
\frac{\hbar^2}{m} k^2 g(k) + \sum_{k'} g(k') V_{kk'} = E g(k), \tag{6.5.9}
$$

where $V_{kk'}$ is the matrix element of the potential between the two plane-wave states defined as ($V_{kk'}$ is the Fourier transform of the potential V)

$$
V_{kk'} \equiv \int V(r) e^{i(k - k') \cdot r} \mathrm{d}^3 r. \tag{6.5.10}
$$

Let us next assume that the potential V is attractive and can be approximated as constant over a range of energy values near the Fermi surface. This range is defined as $E_f + \delta$. When $E_f = 0$ is set, as before, the range is simply from 0 to δ. Let us further assume that the potential has the following form:

$$V_{kk'} = -V \quad \frac{\hbar^2 k^2}{2m} < \delta,$$
$$V_{kk'} = 0 \quad \text{otherwise.} \tag{6.5.11}$$

Substituting in Eq. (6.5.9) for the potential of Eq. (6.5.11) and rearranging terms yields

$$\left[\frac{-\hbar^2 k^2}{m} + E \right] g(k) = -V \sum_{k'} g(k'). \tag{6.5.12}$$

If we define the sum over all the Fourier coefficients as equal to K, then

$$K = \sum_{k'} g(k'), \tag{6.5.13}$$

and then $g(k)$ can be solved for as

$$g(k) = \frac{VK}{-E + \frac{\hbar^2 k^2}{m}}. \tag{6.5.14}$$

Summing Eq. (6.5.14) also over k yields

$$\sum_k g(k) = \sum_k \frac{VK}{-E + \frac{\hbar^2 k^2}{m}}. \tag{6.5.15}$$

But the sum over k of $g(k)$ is again just equal to K, so Eq. (6.5.15) reduces to

$$1 = V \sum_k \frac{1}{-E + \frac{\hbar^2 k^2}{m}}. \tag{6.5.16}$$

Define ζ as $\hbar^2 k^2 / 2m$. The sum can be transformed into a sum over $\zeta(k)$ instead of k. With these changes, Eq. (6.5.16) becomes

$$1 = V \sum_\zeta \frac{1}{2\zeta(k) - E}. \tag{6.5.17}$$

However, the sum can be replaced by an integral over ζ for the case of a quasi-continuum of states. Hence Eq. (6.5.17) is

$$1 = V \int_0^\delta \frac{N(\zeta) \, d\zeta}{2\zeta - E}, \tag{6.5.18}$$

where $N(\zeta)$ is the density-of-states function.

Equation (6.5.18) can be evaluated if it is assumed that the largest contribution to the integral occurs near the Fermi surface. Then the density-of-states function $N(\zeta)$ can be approximated by its value at the Fermi surface $N(E_f)$, which is independent of ζ. Consequently, $N(\zeta)$ can be taken outside of the integral and the expression can be easily evaluated as

$$1 = -VN(E_f) \int_0^\delta \frac{d\zeta}{E - 2\zeta},$$

(6.5.19)

which integrates to

$$1 = \frac{1}{2} VN(E_f) \ln(E - 2\zeta)\big|_0^\zeta.$$

(6.5.20)

Substituting in the bounds and rearranging gives

$$\frac{2}{VN(E_f)} = \ln[(E - 2\delta)] - \ln[E],$$

(6.5.21)

which becomes

$$\frac{E - 2\delta}{E} = e^{\frac{2}{VN(E_f)}}.$$

(6.5.22)

Solving for the energy E, we finally obtain

$$E = \frac{2\delta}{1 - e^{\frac{2}{VN(E_f)}}},$$

(6.5.23)

which simplifies for small values of the potential V to

$$E = -2\delta e^{-\frac{2}{VN(E_f)}},$$

(6.5.24)

where $N(E_f)$ is the density of states at the Fermi surface and δ is the energy range for which the attractive potential V is nonzero. Note that the energy of the pair of electrons E is negative, signifying the presence of a bound state. Recall our assignment of zero energy to the Fermi surface. The above result shows that the paired electrons have energy less than the minimum energy of the two electrons if they were not paired, which in this case would be zero if both electrons are on the Fermi surface. In other words, the energy of the pair of electrons is less than that if the two electrons were totally free. Hence the presence of the attractive potential V serves to lower the energy of the electron pair to a value below the free-electron energy of the individual electrons. The attractive potential therefore leads to the formation of a new two-electron bound state. This new two-electron bound state is called a Cooper pair. It is important to note that, in the presence of some attractive potential coupling two electrons together, the excitation of two electrons out of states below the Fermi energy E_f to states above it leads to a lower energy of the system. This implies that the filled Fermi sphere is unstable to Cooper pair formation. Hence the system will, under appropriate conditions, form Cooper pairs.

Before we discuss the origin of the attractive potential, it is useful to consider the effect of the attractive potential in light of our discussion on Cooper pairs and their connection to superconductivity. The result obtained above suggests that in the presence of an attractive potential, two electrons condense into a new bound state that lies lower in energy than if the electrons were completely free. The energy of this new state, a Cooper pair, is defined as Δ. From the calculation above, Δ is clearly equal to just that amount of energy subtracted from the energy of the two free electrons. This is

$$\Delta = 2\delta e^{-\frac{2}{N(E_f)V}}.$$ (6.5.25)

The energy Δ is the energy of formation of the Cooper pair. In order to dissociate a pair then, it is necessary to add an energy Δ to the system.

As discussed above, if the system is at low enough energy such that the binding energy of the Cooper pair exceeds that available from collisions with the lattice, then the Cooper pair cannot be dissociated by a lattice collision. Subsequently, the pair moves through the lattice unaltered in its motion by lattice scatterings. Since there is insufficient energy for a scattering event to induce a transition of the pair to two free-particle states, no transitions can occur. This is the physical origin of the superconducting state.

It remains to understand the origin of the attractive potential V. In the above we have assumed that some coupling occurs among the electrons in a collective state. However, we have not stated how this coupling arises physically. It at first may seem surprising that any coupling can occur that reduces the energy of the electrons since the electrons, having the same charge, undergo Coulomb repulsion. The Coulomb repulsion between two electrons, though, is greatly reduced in the presence of a free-electron gas. The origin of the reduction in the Coulomb repulsion is somewhat complicated mathematically and is omitted here, but we can understand it qualitatively as follows. The strong long-range interaction between two electrons due to the Coulomb force embedded in an electron gas is essentially a screened interaction. In other words, two electrons do not repel one another with a force simply equal to the Coulomb force between them. Instead, the repulsion is far less, owing to the fact that the electrons act to avoid one another and hence are always at some distance apart. There are two reasons for this. First, according to the Pauli principle, electrons of parallel spin repel one another, leading to a natural separation of the two electrons. Second, even if their spins are opposite, the electrons will arrange themselves to minimize the energy of the entire system. Since the Coulomb force increases the energy of the system, the electrons naturally keep their distance. Hence the electrons behave as if they are surrounded by a region in space completely deficient of other electrons. The collective system adjusts itself to reach a minimum energy configuration that greatly reduces the electron–electron interaction between any two given electrons.

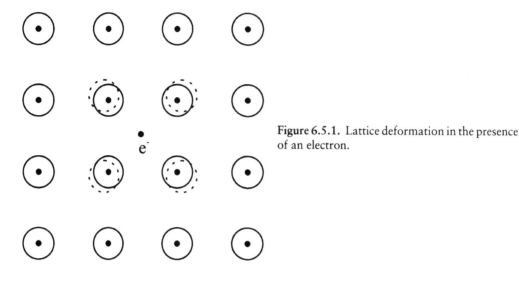

Figure 6.5.1. Lattice deformation in the presence of an electron.

Now let us consider what happens when an electron moves through a crystalline lattice. As an electron moves through the lattice, it attracts the positive ions toward itself. The electron becomes surrounded by a region where the lattice is slightly denser than usual. The lattice is effectively deformed because of the motion of the ions toward the electron. This effect is illustrated in Figure 6.5.1. As a result, a region of net positive charge is formed that serves to attract another neighboring electron. Although the attraction is necessarily weak, because of the slightness with which the lattice ions are deformed, the attraction can still be enough at low temperatures to lead to an effective lowering of the electron energies. As such, the two electrons effectively form a pair. This is the origin of the Cooper pair.

The actual calculation of the ground-state energy with the BCS theory requires some rather sophisticated aspects of quantum and quantum field theory that would be inappropriate to this book. The interested reader is referred to the many excellent books on superconductivity, some of which are listed in the bibliography at the end of this book.

PROBLEMS

1. **a.** Assume the relaxation time approximation in using the Boltzmann equation and determine the form of the diffusivity D. Neglect the electric-field term, assume one-dimensional motion, and assume the Boltzmann form for the equilibrium distribution.

   b. Assume that the relaxation time τ_c is a constant and derive an expression for the electrical conductivity σ.

2. Using the Boltzmann equation, derive an expression for the time rate of change of the momentum of the ensemble. To do this, use the method of moments, substituting in $\hbar k_z$, and evaluate each term. Assume that the scattering rate is independent of

k, that the relaxation time approximation holds, and that the relaxation time is k independent. Also assume that diffusion can be neglected.

3. Derive an expression for the time rate of change of the average energy using the same approach and approximations as in Problem 2.

4. Consider a uniform isotropic substance at constant temperature in the presence of a constant applied electric field F. Assume that steady-state conditions are achieved. If the nonequilibrium distribution function f can be written as

$$f = f_0(1 + \lambda),$$

determine the parameter λ, if it is further assumed that f_0 is given by the Fermi–Dirac distribution function and that the energy relaxation-time approximation holds.

5. Derive the London equation for superconductivity,

$$\vec{\nabla} \times \vec{j} = -\frac{nq^2}{mc} \vec{B} \quad \text{(cgs)},$$

where q is the charge of a Cooper pair. Assume that the velocity of a particle can be expressed as

$$\vec{v} = \frac{1}{m}\left(\vec{p} - \frac{q}{c}\vec{A}\right)$$

and that the pair amplitudes are given as

$$\Psi = \sqrt{n}\, e^{i\theta(r)}.$$

6. Calculate the coherence length for a Cooper pair by assuming that the kinetic energy of the pair cannot exceed the energy gap of the superconducting system Δ. With this assumption, the material will revert back to the normal state. Assume that the pair amplitude is given as

$$\phi(x) = \frac{1}{\sqrt{2}}\left[e^{i(k+p)x} + e^{ikx}\right].$$

Find an expression for $1/p$, the coherence length, by determining the kinetic energy of the pair and then comparing it with Δ. Assume that p^2 is small.

7 Multielectron Systems and Crystalline Symmetries

In this chapter, we begin our study of crystalline solids by concentrating on multielectron systems. The first topic we consider is called the Born–Oppenheimer or adiabatic approximation. This approximation enables us to treat the electronic and the ionic components of a solid as two separate, distinct subsystems coupled only through the electron–lattice scatterings. The next topic is that of multielectron systems. The nature of the exchange interaction is examined, and the energies or multielectron systems are studied. Finally, we examine crystalline symmetries and how these properties facilitate our understanding of crystalline solids.

7.1 The Born–Oppenheimer Approximation

The basic assumption implicit to the theory of crystalline solids is called the Born–Oppenheimer or the adiabatic approximation. Simply put, the Born–Oppenheimer approximation assumes that a solid can be treated as being composed of two separate subsystems, the electronic and the lattice systems. The electrons are considered as moving in a stationary lattice, and the lattice system is treated as being embedded within a uniform electron gas. The dynamics of each subsystem can then be treated independently of the dynamics of the other system, which holds to zeroth order. Coupling between the two subsystems occurs through electron–lattice scatterings. Therefore the Born–Oppenheimer approximation enables us to treat a solid symbolically in the same way as the two separate subsystems shown in Figure 5.3.1 when energy exchange is allowed between them.

The use of the adiabatic approximation is justified in the following way. The masses of the electrons and the ions differ by several orders of magnitude. Hence the ions respond much more slowly to external perturbations as well as changes in the electron system while the electrons respond almost instantaneously to external perturbations as well as motions of the lattice ions. Therefore the Hamiltonian describing the entire solid can be decomposed into separate parts: one that describes the electron motions H_{electron}, one that describes the ion motions H_{ion}, and an interaction term that describes the coupling of the ions and the electrons $H_{\text{el–ion}}$. The Hamiltonian then for the complete system, electrons plus ions, can be written as

$$H_{\text{total}} = H_{\text{electron}} + H_{\text{ion}} + H_{\text{el–ion}}. \tag{7.1.1}$$

The part of the Hamiltonian describing the electrons has the form

$$H_{\text{electron}} = H_{\text{el,kin}} + H_{\text{el–el}} + H_{\text{exchange}} + H_{\text{el–ion}}, \tag{7.1.2}$$

358

where each term is defined as follows. The first term on the right-hand side of Eq. (7.1.1), $H_{el,kin}$, is the kinetic-energy term of a free-electron gas whose solution we have already discussed in Chapters 5 and 6; H_{el-el} is the electron–electron interaction term; $H_{exchange}$ is the exchange interaction, which we discuss below; and H_{el-ion} is the interaction term between the lattice and the electrons.

The Hamiltonian describing the lattice system is

$$H_{ion} = H_{ion,kinetic} + H_{ion-ion} + H_{el-ion}, \tag{7.1.3}$$

where the first term is the ion kinetic energy, the second is the ion–ion interaction, and the third term again represents the coupling between the ions and the electrons. $H_{ion-ion}$ is the ion–ion interaction that determines the crystal's structural properties. Because of the bonding strengths and angles, various crystalline materials can be formed.

Let us now consider the nature of the electron–ion interaction term. In general, H_{el-ion} consists of two parts, H_{el-ion}^0 and $H_{el-phonon}$. The first term represents the interaction of the electrons with the ions when the lattice is in its equilibrium configuration, while the second term represents the interaction of the electrons with the ions in nonequilibrium. The first term gives rise to what is known as the electronic band structure of the solid. The second term, $H_{el-phonon}$, describes the electron–lattice scatterings. As we will see, the quantum of the lattice vibrations is called a phonon and the electron–lattice interaction is typically treated as an electron–phonon interaction.

How, though, do we solve the Schroedinger equation for the complete crystalline system, including both the electrons and the ions? As we saw in Chapters 5 and 6, the free-electron gas system is treated statistically. What we have to do now in the complete problem is to determine what effect the electron–ion interaction, the exchange interaction, and the electron–electron interaction have on the behavior of the free-electron gas system. Additionally, we must analyze the ion system, determining the crystalline lattice structure and the behavior of the ions when they are displaced from equilibrium.

The first topic we consider below is the exchange interaction and how it affects the behavior of the electron gas.

7.2 **Multielectron Atoms and the Exchange Interaction**

The simplest multielectron system that we can consider is the helium atom. The helium atom consists of two electrons. If the atom is in its ground state, then the two electrons are both within the 1s state, one with spin up and the other with spin down. The Hamiltonian of the system must describe the separate kinetic energies of each electron, the separate potential energies of each electron with respect to the central-force potential of the ion, and the electron–electron potential. The electron–ion potentials for the two electrons can be written as

$$-\frac{Zq^2}{r_1}, \qquad -\frac{Zq^2}{r_2}, \tag{7.2.1}$$

where it is assumed that the ions are stationary; they maintain their equilibrium positions. Note that cgs units have been used in expressions (7.2.1). If one chooses to use mks units, which we will sometimes want to do, an additional factor of $1/(4\pi\epsilon_0)$ appears in each term. The electron–electron interaction is (in cgs units as well)

$$\frac{q^2}{|r_1 - r_2|}, \tag{7.2.2}$$

where r_1 and r_2 describe the positions of electrons 1 and 2, respectively. Substituting into the Hamiltonian the expressions for electron–ion potentials Eqs. (7.2.1), and the electron–electron potentials Eq. (7.2.2), the two-particle Hamiltonian becomes

$$H = -\frac{\hbar^2}{2m}\nabla_1^2 - \frac{Zq^2}{r_1} - \frac{\hbar^2}{2m}\nabla_2^2 - \frac{Zq^2}{r_2} + \frac{q^2}{|r_1 - r_2|}. \tag{7.2.3}$$

With this form of the Hamiltonian, the Schroedinger equation can be written as

$$\left[-\frac{\hbar^2}{2m}\nabla_1^2 - \frac{Zq^2}{r_1} - \frac{\hbar^2}{2m}\nabla_2^2 - \frac{Zq^2}{r_2} + \frac{q^2}{|r_1 - r_2|}\right]\Psi = E\Psi. \tag{7.2.4}$$

Under the assumption that both nuclear motion and the electron interaction can be neglected, the solution is given immediately since the resulting equation is separable as

$$\left[\left(-\frac{\hbar^2}{2m}\nabla_1^2 - \frac{Zq^2}{r_1}\right) + \left(-\frac{\hbar^2}{2m}\nabla_2^2 - \frac{Zq^2}{r_2}\right)\right]\Psi = E\Psi. \tag{7.2.5}$$

Because Eq. (7.2.5) is separable in terms of particles 1 and 2, it is reasonable to expect that the total wave function that describes the motion of the two particles can be written as the product of the single-particle wave functions. Hence it is useful to let Ψ, the two-particle wave function, be

$$\Psi = \Psi_1\Psi_2. \tag{7.2.6}$$

With this assumption for Ψ and multiplying the resulting equation by $1/\Psi$, we obtain

$$\frac{1}{\Psi_1}\left(-\frac{\hbar^2}{2m}\nabla_1^2 - \frac{Zq^2}{r_1}\right)\Psi_1 + \frac{1}{\Psi_2}\left(-\frac{\hbar^2}{2m}\nabla_2^2 - \frac{Zq^2}{r_2}\right)\Psi_2 = E. \tag{7.2.7}$$

Note that the two terms on the left-hand side are each independent of the other: the first term depends on only particle 1 and the second term depends on only particle 2. Since the sum of the two terms equals a constant, each term must then be separately equal to a constant. We choose term 1 to be equal to E_1 and term 2 to equal E_2. The sum of these two constants is of course equal to E. Hence,

Eq. (7.2.7) can be broken into two separate equations as

$$\frac{1}{\Psi_1}\left(-\frac{\hbar^2}{2m}\nabla_1^2 - \frac{Zq^2}{r_1}\right)\Psi_1 = E_1, \qquad \frac{1}{\Psi_2}\left(-\frac{\hbar^2}{2m}\nabla_2^2 - \frac{Zq^2}{r_2}\right)\Psi_2 = E_2,$$
$$(7.2.8)$$

where $E_1 + E_2 = E$. Both equations are one-particle hydrogenlike Schroedinger equations. The solution for these equations was determined in Chapter 3. Therefore Eqs. (7.2.8) must have the same solution as that obtained earlier for the hydrogen atom. The energies E_1 and E_2 are then

$$E_1 = -\frac{Z^2 q^4 m}{2n_1^2 \hbar^2}, \qquad E_2 = -\frac{Z^2 q^4 m}{2n_2^2 \hbar^2}, \qquad (7.2.9)$$

where n_1 and n_2 are the quantum numbers describing the allowed energies of particles 1 and 2, respectively.

E_1 and E_2 must sum to E. Therefore E, the energy eigenvalue of the two-electron state, must be

$$E = -\frac{Z^2 q^4 m}{2n_1^2 \hbar^2} - \frac{Z^2 q^4 m}{2n_2^2 \hbar^2}. \qquad (7.2.10)$$

Additionally, the single-particle wave functions that solve Eqs. (7.2.8) must be hydrogenlike. These wave functions are then

$$\Psi_1 \sim u_{n_1 l_1 m_1}(r_1), \qquad \Psi_2 \sim u_{n_2 l_2 m_2}(r_2). \qquad (7.2.11)$$

The total wave function for the system then is found from the product of the single-electron wave functions as

$$\Psi = \Psi_1 \Psi_2. \qquad (7.2.12)$$

Ψ is clearly a function of the two variables r_1 and r_2. To denote this, we write the two-particle wave function as

$$\Psi(r_1 - r_2). \qquad (7.2.13)$$

The two-particle wave function is often abbreviated as

$$\Psi(r_1, r_2). \qquad (7.2.14)$$

Note that Ψ and E are found above assuming no interaction between the two electrons. In this case the solution is simple; the net energy is equal to the sum of the separate single-particle energies and the net wave function is the product of the two separate single-particle wave functions. However, if the electron–electron interaction is considered, then the total wave function will no longer be equal to the product of the single-particle wave functions. The electron–electron interaction will lead to a coupling of the electrons.

Let us now consider what happens when the electron–electron interaction is added to the Hamiltonian. The product of the simple single-particle wave functions is no longer the correct solution since the equation can no longer be simply separated as before. The Hamiltonian of the system is given by Eq. (7.2.3). The Hamiltonian can be written in two different ways by an exchange of the particle labels on the operators. The two forms of the Hamiltonian are then

$$H(1, 2) = -\frac{\hbar^2}{2m}\nabla_1^2 - \frac{\hbar^2}{2m}\nabla_2^2 - \frac{Zq^2}{r_1} - \frac{Zq^2}{r_2} + \frac{q^2}{|r_1 - r_2|} \tag{7.2.15}$$

or

$$H(2, 1) = -\frac{\hbar^2}{2m}\nabla_2^2 - \frac{\hbar^2}{2m}\nabla_1^2 - \frac{Zq^2}{r_2} - \frac{Zq^2}{r_1} + \frac{q^2}{|r_2 - r_1|}. \tag{7.2.16}$$

It is important to note that the two expressions above are equivalent, $H(1, 2) = H(2, 1)$. The Hamiltonians are identical under the exchange of particle labels. This means that the system is invariant under an exchange of particle labels: label 1 permuted with label 2 leads to an equivalent description of the system. This seems quite obvious but it is due to the fact that the particles are indistinguishable. Particles 1 and 2 cannot be considered as separate distinguishable particles for all time.

Since the Hamiltonians $H(1, 2)$ and $H(2, 1)$ are equivalent, their zeroth-order energy eigenvalues must also be equivalent. Additionally, the two-particle wave functions

$$\Psi(r_1, r_2), \qquad \Psi(r_2, r_1) \tag{7.2.17}$$

must be related. Either possibility, $\Psi(r_1, r_2)$ and $\Psi(r_2, r_1)$, is a feasible solution to the Schroedinger equation with identical energies. Since these two solutions are degenerate they must be related by a multiplicative constant as

$$\Psi(r_1, r_2) = c\Psi(r_2, r_1), \tag{7.2.18}$$
$$\Psi(r_2, r_1) = c\Psi(r_1, r_2). \tag{7.2.19}$$

Therefore, in order for both of the above relationships to hold, we must have

$$\Psi(r_1, r_2) = c\Psi(r_2, r_1) = cc\Psi(r_1, r_2),$$
$$\Psi(r_1, r_2) = c^2\Psi(r_1, r_2). \tag{7.2.20}$$

Subsequently c must equal 1 or negative 1. The two-particle wave function must then obey

$$\Psi(r_1, r_2) = +\Psi(r_2, r_1) \quad \text{symmetric} \tag{7.2.21}$$

or

$$\Psi(r_2, r_1) = -\Psi(r_1, r_2) \quad \text{antisymmetric}. \tag{7.2.22}$$

The first relation is called the symmetric state, and it physically represents a state that is symmetric under particle exchange; the wave functions are equal to one another when the labels are interchanged. The second relation is called the antisymmetric state, and it physically represents a state that is antisymmetric under particle exchange; the first wave function is equal to the negative of the other wave function. The two-particle state is doubly degenerate and is either a spatially symmetric or antisymmetric wave function. The wave functions can be symbolically represented as

$$\Psi_+ = \frac{1}{\sqrt{2}}[\Psi_{n_1}(1)\Psi_{n_2}(2) + \Psi_{n_1}(2)\Psi_{n_2}(1)],$$

$$\Psi_- = \frac{1}{\sqrt{2}}[\Psi_{n_1}(1)\Psi_{n_2}(2) - \Psi_{n_1}(2)\Psi_{n_2}(1)], \tag{7.2.23}$$

where the subscript represents the state and the argument labels the particle. The prefactor is simply a normalization constant.

What, though, is the effect of the spin on the system? The wave functions developed above are only the spatially dependent parts. In addition to the spatial dependence, the spin of the electron must also be specified. As we learned in Chapter 4, the dynamics of the electron must be generally specified in not only three-dimensional space but in space–time. The spin arises from consideration of space–time in the description of an electron. The spin is an additional independent degree of freedom of the electron. The total wave function of the electron then must reflect the spin as well as the spatial states of the electron. Since the spin is independent of the spatial variables (neglecting any spin–orbit coupling), the total wave function can be formed by the product of the spin and the spatial components of the wave function.

Let us symbolically represent the spin-up state as α and the spin-down state as β. The state corresponding to electron 1 spinning up is represented as $\alpha(1)$ and electron 2 spinning down as $\beta(2)$. Therefore, for a two-particle state with electron 1 spinning up and electron 2 spinning down, the spin state is given as

$$\alpha(1)\beta(2). \tag{7.2.24}$$

In Chapter 3 we introduced the Pauli principle to account for the fact that no two electrons are observed to ever be in the same quantum state simultaneously. The Pauli principle can be more generally stated as the requirement that the total wave function of any fermion including, of course, electrons must be antisymmetric. As we see below, requiring that the total wave function of a collection of electrons be antisymmetric will guarantee that no two electrons occupy the same quantum state simultaneously. To guarantee that the total wave function of the electrons is antisymmetric, if the spatial part is symmetric, then the spin part must be antisymmetric and vice versa. From the above discussion, it was found that the spatial part of a two-particle wave function can be either symmetric or antisymmetric. Similarly, the spin part of the wave function can also be either symmetric or antisymmetric.

Let us consider how to form a completely antisymmetric wave function for two fermions. If the spatial part of the wave function Ψ is symmetric, then the spin part χ must be antisymmetric. The antisymmetric spin state for two electrons is given as

$$\chi_- = \frac{1}{\sqrt{2}}[\alpha(1)\beta(2) - \alpha(2)\beta(1)]. \tag{7.2.25}$$

Note that if the particle labels are exchanged, $1 \rightarrow 2$ and $2 \rightarrow 1$, then the wave function goes into the negative of itself. Given that χ is antisymmetric, Ψ must then be symmetric. If Ψ is antisymmetric then χ must be symmetric. There are three ways in which a symmetric spin state for two particles can be constructed. These are

$\alpha(1)\alpha(2)$ both spins up,

$\frac{1}{\sqrt{2}}[\alpha(1)\beta(2) + \alpha(2)\beta(1)],$

$\beta(1)\beta(2)$ both spins down. $\tag{7.2.26}$

Therefore when Ψ is spatially symmetric there is only one two-particle state. This state is called a singlet state. However, when Ψ is spatially antisymmetric, there are three possible states corresponding to the different symmetric spin states. These states form a triplet state.

With this information, we can now proceed to construct the wave function for the ground state of the helium atom including spin. The total wave function must of course be antisymmetric. In the ground state, both electrons have identical spatial wave-function quantum numbers. The spatial states of the two electrons in the ground state are

$$\Psi_{100}(r_1), \qquad \Psi_{100}(r_2), \tag{7.2.27}$$

where the subscripts denote the spatial quantum numbers n, l, and m_l. Therefore the antisymmetric spatial wave function vanishes in the ground state. This is obvious from inspection of the antisymmetric spatial wave function, in this case,

$$\Psi = \frac{1}{\sqrt{2}}[\Psi_{100}(r_1)\Psi_{100}(r_2) - \Psi_{100}(r_2)\Psi_{100}(r_1)], \tag{7.2.28}$$

which is clearly equal to zero. Hence the only possible ground-state wave function for the helium atom is composed of a spatially symmetric part. The total wave function can be formed from the product of the spatially symmetric part and the antisymmetric spin part. Neglecting the normalization constants and delaying proper normalization of the total wave function, we find that the product of the spatial and the spin parts gives

$$[\Psi_{100}(1)\Psi_{100}(2) + \Psi_{100}(2)\Psi_{100}(1)] \cdot [\alpha(1)\beta(2) - \alpha(2)\beta(1)]. \tag{7.2.29}$$

Multiplying out yields

$$[\Psi_{100}(1)\Psi_{100}(2)\alpha(1)\beta(2) - \Psi_{100}(1)\Psi_{100}(2)\beta(1)\alpha(2)$$
$$+ \Psi_{100}(2)\Psi_{100}(1)\alpha(1)\beta(2) - \Psi_{100}(2)\Psi_{100}(1)\beta(1)\alpha(2)]. \tag{7.2.30}$$

Combining like terms and normalizing the above simplifies to

$$\frac{1}{\sqrt{2}}[\Psi_{100}(1)\Psi_{100}(2)\alpha(1)\beta(2) - \Psi_{100}(1)\Psi_{100}(2)\beta(1)\alpha(2)], \tag{7.2.31}$$

which is the resultant antisymmetric wave function of the system.

The total antisymmetric wave function can be formed in an alternative way by use of what is known as the Slater determinant. The Slater determinant provides a far more convenient way in which an antisymmetric wave function for a many-particle system can be constructed as well as compactly written. To find the resultant antisymmetric wave function for a two-particle state, a determinant is formed as

$$\Psi = \frac{1}{\sqrt{2}} \begin{vmatrix} \Psi_{100}(1)\alpha(1) & \Psi_{100}(1)\beta(1) \\ \Psi_{100}(2)\alpha(2) & \Psi_{100}(2)\beta(2) \end{vmatrix}. \tag{7.2.32}$$

Evaluation of this determinant, following the usual rules, yields

$$\Psi = \frac{1}{\sqrt{2}}[\Psi_{100}(1)\alpha(1)\Psi_{100}(2)\beta(2) - \Psi_{100}(2)\alpha(2)\Psi_{100}(1)\beta(1)], \tag{7.2.33}$$

which is precisely the same result we obtained above through direct multiplication of the spatial and the spin parts of the two-particle wave function.

Generally, the wave function for an N-particle system can be expressed with a Slater determinant. The form of the Slater determinant for N particles is

$$\Psi = \frac{1}{\sqrt{N!}} \begin{vmatrix} \Psi_1(r_1, s_1) & \Psi_1(r_2, s_2) & \Psi_1(r_3, s_3) & \cdots & \Psi_1(r_n, s_n) \\ \Psi_2(r_1, s_1) & \Psi_2(r_2, s_2) & \Psi_2(r_3, s_3) & \cdots & \Psi_2(r_n, s_n) \\ & \cdot & & \cdot & \\ & \cdot & & \cdot & \\ \Psi_n(r_1, s_1) & \Psi_n(r_2, s_2) & \Psi_n(r_3, s_3) & \cdots & \Psi_n(r_n, s_n) \end{vmatrix}, \tag{7.2.34}$$

where Ψ_1 represents the first particle wave function, r_1 and s_1 are the spatial and the spin parts of the wave function in the first state, r_2 and s_2 are the spatial and the spin parts of the wave function in the second state, etc. The expansion of the Slater determinant gives the correct normalized, antisymmetrized wave function for an N particle system.

Note that for the case in which the particles are put into the same state, the determinant vanishes. For example, in the two-particle case, when the two electrons are placed into exactly the same state, the determinant becomes

$$\Psi = \frac{1}{\sqrt{2}} \begin{vmatrix} \Psi_{100}(1)\alpha(1) & \Psi_{100}(1)\alpha(1) \\ \Psi_{100}(2)\alpha(2) & \Psi_{100}(2)\alpha(2) \end{vmatrix},$$
(7.2.35)

which is clearly equal to zero since all the elements within the rows are the same. Use of the Slater determinant then guarantees that a properly antisymmetrized wave function will be obtained.

Now we are ready to examine the physics of a multielectron system in the presence of an electron–electron interaction. The electron Hamiltonian is written as

$$H_{\text{el}} = -\sum_k \frac{\hbar^2}{2m} \nabla_k^2 + \frac{1}{2} \sum_{k \neq k', k, k'} \frac{q^2}{|r_k - r'_k|} + \sum_{k,i} V(r_k - R_i),$$
(7.2.36)

where the first term is the kinetic-energy part for each electron, the second term accounts for the electron–electron interaction, and the third term is the electron-ion interaction. The sum in the second term is over both k and k'. This sum is taken over all the electrons in the ensemble, excluding the term $k = k'$, since one electron cannot interact with itself through the Coulomb potential. The extra factor of $1/2$ in the second term is due to the double counting in the sum. Since the sum is over both k and k', we count the interaction between the kth and the k'th electrons twice by doing the double sum. Therefore it is necessary to divide by 2 to evaluate the term correctly. The sum in the third term is over both the ions and the electrons.

The Schroedinger equation corresponding to the above Hamiltonian would be easy to solve if it could be reduced into a single-particle problem yet still

---

Box 7.2.1 Two-body Operators and Sums

The two body operator for the Coulomb potential,

$$\sum_k \sum_{k'} \frac{q^2}{|r_k - r'_k|}$$

excluding the term $k = k'$, involves a double sum over both indices, k and k'. Each sum is over all N electrons in the system. This sum consequently describes all of the two body interactions present. Take $k = 1$ fixed. The sum over all $k', k \neq k'$, gives the potential on the kth electron from all of the other electrons in the system. This gives only the total potential on one electron. To find all of the possible interactions, an additional sum over the kth electron must be performed. Notice however, that the double sum results in double counting. Each Coulomb interaction between any two electrons is counted twice through the double sum. Therefore, to eliminate the double counting, the prefactor is divided by 2.

retain the essential physics of the many-particle problem. If the electron–electron term can be neglected, the Schroedinger equation for Ψ becomes separable since all the other remaining operators are one-body sums. Consequently, the total wave function would be given as the product of single-particle wave functions, $\Psi_1, \Psi_2, \ldots, \Psi_n$, as before. Therefore, when the interaction term is neglected, the total wave function for the N-electron system is

$$\Psi(r_1, r_2, \ldots, r_n) = \Psi_1(r_1)\Psi_2(r_2), \ldots, \Psi_n(r_n), \tag{7.2.37}$$

with the corresponding energy of the system E of

$$E = \sum_k E_k. \tag{7.2.38}$$

The presence of the electron–electron interaction term complicates the solution from that corresponding to noninteracting particles. Nevertheless, the zeroth-order solution to the Schroedinger equation with the Hamiltonian of Eq. (7.2.36) is given by Eq. (7.2.37).

Substituting in the expression for Ψ into the Schroedinger equation yields

$$H\Psi = E\Psi, \tag{7.2.39}$$

where the energy E is found from

$$E = \langle \Psi | H | \Psi \rangle. \tag{7.2.40}$$

The Hamiltonian given by Eq. (7.2.36) consists of single-particle sums and double-particle sums. The kinetic-energy and the electron–ion terms are both single-particle sums; they involve summing over a single electron at a time. The electron–electron interaction term is a double-particle sum since it involves a sum over two electrons at a time. If the single-particle sums are combined into a new single-particle operator H_k as

$$H_k = -\frac{\hbar^2}{2m}\nabla_k^2 + V(r_k - R_i), \tag{7.2.41}$$

then the energy is given as

$$E = \langle \Psi | H | \Psi \rangle = \sum_k \langle \Psi_k | H_k | \Psi_k \rangle + \frac{q^2}{2} \sum_{kk', k \neq k'} \langle \Psi_k \Psi_{k'}' | \frac{1}{|r_k - r_k'|} | \Psi_k \Psi_{k'}' \rangle. \tag{7.2.42}$$

The Ψ_k that minimize E represent the best set of functions for the ground state within the framework of the assumption of Eq. (7.2.37). Equation (7.2.42) is minimized with respect to variations in Ψ_k (see box 7.2.2). To do this, the

---

Box 7.2.2   Variational Methods in Quantum Mechanics

The variational method is used to find the solution to the Schroedinger equation based on the variational theorem. The variational theorem states that the functions, Ψ, for which the variation of the mean value, $\langle a \rangle$, as defined by,

$$\langle a \rangle = \frac{\int \Psi^* A \Psi \, d^3 r}{\int \Psi^* \Psi \, d^3 r}$$

is zero, satisfy eigenvalue equations. The eigenvalue equations are given as

$$A\Psi = a\Psi.$$

To show the validity of this theorem, assume that the variation of the mean value, $\langle a \rangle$, is zero. This implies

$$\delta \langle a \rangle = \delta \frac{\int \Psi^* A \Psi \, d^3 r}{\int \Psi^* \Psi \, d^3 r} = 0.$$

Let C and D be defined as

$$C \equiv \int \Psi^* A \Psi \, d^3 r; \quad D \equiv \int \Psi^* \Psi \, d^3 r.$$

Then the variation of $\langle a \rangle$ can be written as

$$\delta \langle a \rangle = \delta \left(\frac{C}{D} \right) = \frac{D \delta C - C \delta D}{D^2} = 0,$$

which can be simplified to

$$\langle a \rangle \delta D - \delta C = 0.$$

The values of δC and δD are

$$\delta C = \int \delta \Psi^* A \Psi \, d^3 r + \int \Psi^* A \delta \Psi \, d^3 r,$$

$$\delta D = \int \delta \Psi^* \Psi \, d^3 r + \int \Psi^* \delta \Psi \, d^3 r.$$

Substituting in for δC and δD above obtains

$$\langle a \rangle \left[\int \delta \Psi^* \Psi \, d^3 r + \int \Psi^* \delta \Psi \, d^3 r \right] - \left[\int \delta \Psi^* A \Psi \, d^3 r + \int \Psi^* A \delta \Psi \, d^3 r \right] = 0,$$

which is equal to

$$\int \delta \Psi^* (\langle a \rangle - A) \Psi \, d^3 r + \int \delta \Psi (\langle a \rangle - A) \Psi^* \, d^3 r = 0.$$

Since the variations in Ψ and Ψ^* are arbitrary, the real and imaginary parts satisfy the above equation only if

$$(\langle a \rangle - A)\Psi_r = 0; \quad (\langle a \rangle - A)\Psi_i = 0.$$

Therefore, the condition is that the wave functions must satisfy an eigenvalue relationship for A. So when we perform a variational calculation, an eigenvalue equation is obtained for the wave functions.

---

variation with respect to Ψ_k is taken of Eq. (7.2.42) and set equal to zero. Mathematically, this is equivalent to

$$\delta[E - \langle \Psi | H | \Psi \rangle] = 0. \tag{7.2.43}$$

Substituting Eq. (7.2.42) for the matrix element in Eq. (7.2.43) results then in

$$0 = \langle \delta\Psi_j | H_j | \Psi_j \rangle + q^2 \sum_{k \neq j} \langle \delta\Psi_j, \Psi_k | \frac{1}{|r_k - r_j|} | \Psi_j, \delta\Psi_k \rangle - E_j \langle \delta\Psi_j | \Psi_j \rangle,$$

(7.2.44)

where E_j is the energy associated with the jth state. Note that the two-particle sum has been replaced by a single-particle sum since the variation is taken with respect to a fixed state j. Equation (7.2.42) can be rewritten as

$$0 = \langle \delta\Psi_j | \left[H_j + q^2 \sum_{k \neq j} \langle \Psi_k | \frac{1}{|r_k - r_j|} | \Psi_k \rangle - E_j \right] | \Psi_j \rangle,$$

(7.2.45)

which is true regardless of the variation of Ψ_j. H_j is the single-particle part of the Hamiltonian given by Eq. (7.2.41). Therefore Eq. (7.2.45) becomes

$$\left[-\frac{\hbar^2}{2m} \nabla^2 + V(r) + q^2 \sum_{k \neq j} \langle \Psi_k | \frac{1}{|r_k - r_j|} | \Psi_k \rangle \right] \Psi_j = E_j \Psi_j.$$

(7.2.46)

But the second term above can be written in integral form as

$$\langle \Psi_k | \frac{1}{|r_k - r_j|} | \Psi_k \rangle = \int \frac{|\Psi_k(r')|^2}{|r - r'|} \, d^3r'.$$

(7.2.47)

With the above substitution, the Schroedinger equation becomes

$$\left[-\frac{\hbar^2}{2m} \nabla^2 + V(r) + q^2 \sum_{k \neq j} \int \frac{|\Psi_k(r')|^2}{|r - r'|} \, d^3r' \right] \Psi_j(r) = E_j \Psi_j(r).$$

(7.2.48)

Equation (7.2.48) is a single-particle equation known as the Hartree equation. It describes the dynamics of a specific electron within the ensemble, in this case denoted as j, at location r in the potential field of the ions $V(r)$ and in the Coulomb potential of an average distribution of electrons given by the sum over $k \neq j$. The parameters E_j are the single-particle energies.

The Hartree equation is derived under the assumption that the multiparticle wave function can be constructed as the product of single-particle wave functions, as given by Eq. (7.2.37). However, the wave function given by Eq. (7.2.37) is not properly symmetrized. It is not antisymmetric under the exchange of particle label and as such cannot properly describe a multielectron system. Subsequently, the Hartree approach does not provide a reasonable approximation to the solution of the multielectron system problem.

To overcome the limitations of the Hartree model, it is possible to go one further step and to develop an equation for the system by use of a properly symmetrized wave function. The multiparticle wave function can be constructed with a Slater determinant of the form

$$
\Psi = \frac{1}{\sqrt{N!}}
\begin{vmatrix}
\Psi_1(r_1, s_1) & \cdots & \Psi_1(r_n, s_n) \\
\Psi_2(r_1, s_1) & \cdots & \Psi_2(r_n, s_n) \\
 & \cdot & \\
 & \cdot & \\
 & \cdot & \\
\Psi_n(r_1, s_1) & \cdots & \Psi_n(r_n, s_n)
\end{vmatrix},
\tag{7.2.49}
$$

where r represents the spatial coordinates and s represents the spin coordinates. The expectation value of the energy of the electronic system can again be determined from the Schroedinger equation with the above form for the wave function rather than that given by Eq. (7.2.37). The spatial coordinates need to be integrated over, while the spin coordinates need to be summed. Solving for the energy E yields

$$
\begin{aligned}
E &= \sum_k \sum_s \int \Psi_k^* H_k \Psi_k \, d^3 r_k \\
&\quad + \frac{q^2}{2} \sum_{kk', k \neq k'} \sum_s \int \frac{|\Psi_k(r_1, s_1)|^2 |\Psi_{k'}(r_2, s_2)|^2}{|r_1 - r_2|} \, d^3 r_1 \, d^3 r_2 \\
&\quad - \frac{q^2}{2} \sum_{kk', k \neq k'} \sum_s \int \frac{\Psi_k^*(r_1, s_1) \Psi_k(r_2, s_2) \Psi_{k'}^*(r_2, s_2) \Psi_{k'}(r_1, s_1)}{|r_1 - r_2|} \, d^3 r_1 \, d^3 r_2,
\end{aligned}
\tag{7.2.50}
$$

where the two last terms arise from the original Coulomb term. The integrations are over the spatial coordinates.

What is the nature of the last term in Eq. (7.2.50)? The first term is a single-particle term as before. The second and the third terms arise from the following. The two-particle matrix element between the Coulomb potential can be rewritten in terms of a two-particle wave function Ψ as

$$
\langle \Psi | \frac{q^2}{|r_1 - r_2|} | \Psi \rangle = \sum_s \int \frac{q^2}{|r_1 - r_2|} \Psi^* \Psi \, d^3 r_1 \, d^3 r_2.
\tag{7.2.51}
$$

Let us consider the value of the term $\Psi^* \Psi$. Generally, the wave function will involve n terms for the n electrons present. To simplify writing, let us consider only a two-particle wave function. Substituting in the antisymmetrized form of the wave function and disregarding normalization (since that will be implicitly included in the final result), we find that the product is given as

$$
\begin{aligned}
&[\Psi_k(r_1, s_1) \Psi_{k'}(r_2, s_2) - \Psi_k(r_2, s_2) \Psi_{k'}(r_1, s_1)]^* \\
&\cdot [\Psi_k(r_1, s_1) \Psi_{k'}(r_2, s_2) - \Psi_k(r_2, s_2) \Psi_{k'}(r_1, s_1)].
\end{aligned}
\tag{7.2.52}
$$

Multiplying expression (7.2.52) out term by term yields

$$
\begin{aligned}
&= [|\Psi_k(r_1, s_1)|^2 |\Psi_{k'}(r_2, s_2)|^2 + |\Psi_k(r_2, s_2)|^2 |\Psi_{k'}(r_1, s_1)|^2 \\
&\quad - \Psi_k^*(r_1, s_1) \Psi_{k'}^*(r_2, s_2) \Psi_k(r_2, s_2) \Psi_{k'}(r_1, s_1) \\
&\quad - \Psi_k^*(r_2, s_2) \Psi_{k'}^*(r_1, s_1) \Psi_k(r_1, s_1) \Psi_{k'}(r_2, s_2)].
\end{aligned}
\tag{7.2.53}
$$

The first two terms in Eq. (7.2.53) can be combined to give

$$\frac{1}{2} \sum_{kk', k \neq k'} |\Psi_k(r_1, s_1)|^2 |\Psi_{k'}(r_2, s_2)|^2. \tag{7.2.54}$$

The sum in expression (7.2.54) can be readily expanded to show its equivalence to the first two terms of Eq. (7.2.53). Similarly, the third and the fourth terms can be combined into a sum as

$$-\frac{1}{2} \sum_{kk', k \neq k'} \Psi_k^*(r_1, s_1) \Psi_{k'}^*(r_2, s_2) \Psi_k(r_2, s_2) \Psi_{k'}(r_1, s_1). \tag{7.2.55}$$

Therefore the expectation value of the energy can be written as

$$\begin{aligned} E = &\sum_k \sum_s \int \Psi_k^* H_k \Psi_k \, d^3 r_k \\ &+ q^2 \frac{1}{2} \sum_{kk', k \neq k'} \sum_s \int \frac{|\Psi_k(r_1, s_1)|^2 |\Psi_{k'}(r_2, s_2)|^2}{|r_1 - r_2|} \, d^3 r_1 \, d^3 r_2 \\ &- \frac{q^2}{2} \sum_{kk', k \neq k'} \sum_s \int \frac{\Psi_k^*(r_1, s_1) \Psi_{k'}^*(r_2, s_2) \Psi_k(r_2, s_2) \Psi_{k'}(r_1, s_1)}{|r_1 - r_2|} \, d^3 r_1 \, d^3 r_2. \end{aligned} \tag{7.2.56}$$

Use is again made of the variational method to minimize the energy through variation of the wave function. Equation (7.2.56) becomes

$$\begin{aligned} 0 = &\langle \delta \Psi_k | H_k | \Psi_k \rangle + \frac{q^2}{2} \sum_{kk', k \neq k'} \sum_s \langle \delta \Psi_k | \int \frac{|\Psi_{k'}(r_2, s_2)|^2}{|r_1 - r_2|} \, d^3 r_2 | \Psi_k \rangle \\ &- \frac{q^2}{2} \sum_{kk', k \neq k'} \sum_s \langle \delta \Psi_k | \int \frac{\Psi_{k'}^*(r_2, s_2) \Psi_k(r_2, s_2)}{|r_1 - r_2|} \, d^3 r_2 | \Psi_{k'} \rangle - \langle \delta \Psi_k | E_k | \Psi_k \rangle. \end{aligned} \tag{7.2.57}$$

The variation of Ψ can be anything, so Eq. (7.2.57) must be generally true for all Ψ_k. Therefore, substituting in for the single-particle operator H_k, we find that Eq. (7.2.57) becomes

$$\begin{aligned} &\left[-\frac{\hbar^2}{2m} \nabla_1^2 + V(r_1) \right] \Psi_k(r_1, s_1) + q^2 \sum_{k', k \neq k'} \sum_s \int \frac{|\Psi_{k'}(r_2, s_2)|^2}{|r_1 - r_2|} \, d^3 r_2 \Psi_k(r_1, s_1) \\ &- q^2 \sum_{k', k \neq k'} \sum_s \int \frac{\Psi_{k'}(r_2, s_2) \Psi_k(r_2, s_2)}{|r_1 - r_2|} \, d^3 r_2 \Psi_{k'}(r_1, s_1) = E_k \Psi_k(r_1, s_1). \end{aligned} \tag{7.2.58}$$

The integrals are taken over r and the spin is summed over. If no spin–orbit coupling is present, then the wave functions can be written as the product of the spatial and the spin parts. The sum over the spin variable in the last term vanishes when the spins are opposite because of the orthogonality of the spin states. Therefore only states of parallel spin give a nonzero contribution to the integral in the last term since the spatial terms are the same. The spin affects only the last term. The expressions reduce to integrations over only the real-space variables within Eq. (7.2.58). The resulting expression is

$$
\left[-\frac{\hbar^2}{2m}\nabla^2 + V(r) \right]\Psi_j(r) + q^2 \sum_{k \neq j} \int \frac{|\Psi_k(r')|^2}{|r - r'|}\, d^3r'\Psi_j(r)
$$

$$
- q^2 \sum_{k \neq j, \text{spin}\|} \int \frac{\Psi_k^*(r')\Psi_j(r')}{|r - r'|}\, d^3r'\Psi_k(r) = E_j\Psi_j(r), \tag{7.2.59}
$$

which is known as the Hartree–Fock equation. The presence of the last term is the only difference to the Hartree equation derived above.

Let us explore the meaning of the Hartree–Fock equation. The first two terms are single-particle energy terms arising from the kinetic energy of the electrons and their potential energy in the field of the ions. The third term is the same two-particle term as appeared in the Hartree equation. It represents the energy of the electron in the presence of an electronic charge cloud of density ρ defined as

$$
\rho = q \int |\Psi_k(r')|^2\, d^3r'. \tag{7.2.60}
$$

In other words, the first three terms follow directly from classical arguments. The energy of an electron is given simply by the sum of its kinetic energy, its potential energy in the ion field, and its potential energy due to the charge distribution of the remaining $N - 1$ electrons. What, though, does the fourth term represent physically?

The presence of the new term in Eq. (7.2.59) has an important physical significance. This term arises from the exchange interaction, the fact that the particles are indistinguishable under the exchange of particle label. The exchange interaction influences the energy of the system. It implies that the jth electron interacts with electrons of the same spin. Hence the motion of electrons of the same spin is correlated. The exchange interaction acts to lower the energy of the system, which compensates somewhat for the classical electron–electron Coulomb repulsion. Note that the sign of the exchange term is different from the third term, often called the direct Coulomb term, in Eq. (7.2.59).

The difference between the Hartree and the Hartree–Fock equations is that in the Hartree model an electron interacts with all the remaining electrons independently of their position while in the Hartree–Fock model the electron interaction is not location independent. The Pauli principle leads to a reduction in the strength of the direct Coulomb interaction through the exchange term. The electrons of parallel spin are excluded from the near vicinity of the electron in question, thereby reducing the effect of the direct Coulomb interaction. As a result, the electron energy in the free-electron gas is greater than if the exchange term is omitted.

Another way to view the effect of the exchange term is to recognize that for the case in which the spin part of the wave function is symmetric (parallel spins), the spatial part of the wave function is antisymmetric. Inspection of the spatially antisymmetric wave function, though, shows that the probability density that the electrons are at the same position is zero. Conversely, for the state of antiparallel spins, in which the spin state is antisymmetric, the spatial component of the wave function must be symmetric. In this case, there is a sizable probability that the two electrons can approach one another spatially. Therefore, in the case in which the electrons have parallel spins, they remain farther apart on average than in the case in which the electrons have antiparallel spins. Since the Coulomb interaction is a function of the spatial separation of the electrons, it is clearly smaller for electrons of parallel spin than for those of antiparallel spin. This is the physical origin of the exchange term.

It is important to note that both the Hartree and the Hartree–Fock equations reduce the many-electron problem into one-electron equations. The effects of all other electrons on the single-particle energy are included through both the Coulomb and the exchange terms present in the Hartree–Fock equation. Therefore, we have succeeded in developing a means of treating an interacting many-electron system such as that in a solid. In Section 7.3 we consider multielectron states as calculated with the computer code developed in Chapter 2.

7.3   **Molecular and Solid-State Formation from Atomic Levels**

In this section, let us consider what happens when two atoms are brought close together. In Section 7.2 we found that when two or more electrons interact, they form new collective states that must be fully antisymmetric. Since the spatial and the spin parts are independent, either a spatially symmetric state with a corresponding antisymmetric spin part or a spatially antisymmetric state with a symmetric spin part is a possible state. Depending on the spatial symmetry, either a singlet or a triplet state can occur.

Figure 7.3.1 illustrates the potential diagram of two separate atoms initially separated by a distance R. For simplicity, let us assume that the atoms have only a single electron in the outermost energy level. If R is large such that the atoms

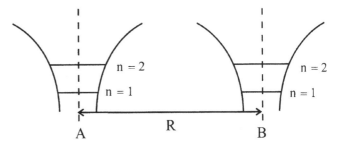

Figure 7.3.1. Potential-energy diagram of two isolated atoms A and B separated from their ion centers by a distance R, which is sufficiently large that the two atoms do not interact. For illustration two quantized energy states are shown, $n = 1$ and $n = 2$.

are essentially isolated from one another, then the valence electrons in each separate atom A and B can be represented as single-particle wave functions $\Psi_A(r_1)$ and $\Psi_B(r_2)$. The electronic states are the simple atomic states similar to those determined in Chapter 3. Of course, they are not simply the hydrogen atom wave functions in general, but they are similarly determined. As the distance of separation between the two atoms decreases, the electronic wave functions localized originally on each separate isolated atom core begin to overlap. As a result, the two electrons must now be treated as interacting particles, as discussed in Section 7.2. A multielectron wave function must be constructed to describe the dynamics of the two electrons since the treatment of the electrons as two noninteracting particles is no longer valid.

The collective wave function must be either spatially symmetric or antisymmetric and can be written as

$$\Psi_{AB}(1,2) = \frac{1}{\sqrt{2}}[\Psi_A(r_1)\Psi_B(r_2) + \Psi_A(r_2)\Psi_B(r_1)] \tag{7.3.1}$$

or

$$\Psi_{AB}(1,2) = \frac{1}{\sqrt{2}}[\Psi_A(r_1)\Psi_B(r_2) - \Psi_A(r_2)\Psi_B(r_1)]. \tag{7.3.2}$$

The spatially symmetric wave function is even under particle exchange while the antisymmetric wave function is odd under exchange. The subscripts A and B denote the atomic core states from which the collective states are derived.

The spatially symmetric and antisymmetric collective states are sketched in Figure 7.3.2. We determine the form of the states by realizing that the state must be either symmetric or antisymmetric spatially. Hence, under reflection about

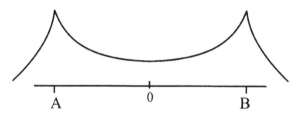

a symmetric combination

Figure 7.3.2. a. Spatially symmetric, b, spatially antisymmetric wave functions for a two-atom system once they have been brought close enough to interact.

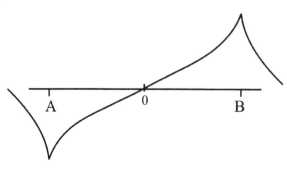

b antisymmetric combination

$R = 0$, the wave function must go back into either itself or the negative of itself. Note that the antisymmetric state vanishes midway between the two ion cores while the symmetric state does not. Since the square of the wave function is equal to the probability density, for the spatially antisymmetric state there is no chance of finding the electrons at $R/2$, midway between the two ions. In fact, for the spatially antisymmetric state the greatest probability density occurs near each ion core and decreases strongly away from the ion center.

What is the origin of this behavior? From the Pauli principle it is impossible to confine two electrons to the same spatial point when their spins are the same. Hence the two electrons repel one another, as we have seen from the discussion in Section 7.2. The electrons when in the symmetric spin state maintain a sizeable real-space separation between them. They stay in the near vicinity of their respective ion cores. This state is called an antibonding state since it does not lead to molecular formation.

Conversely, for the spatially symmetric state, the wave function clearly does not vanish midway between the two ions; there is a nonzero probability of finding the electrons at $R/2$. In this case, the corresponding spin state is antisymmetric in order to render the total state antisymmetric. Hence, when the electrons are spatially brought together, owing to their opposite spins, no repulsion occurs. In fact the resulting energy of the state is less than the corresponding isolated atomic states. The spatially symmetric state is called a bonding state since it leads to molecular formation.

The electronic energies as functions of interatomic separation are plotted in Figure 7.3.3. The most striking aspect of this curve is that the electronic energy reaches a minimum below that corresponding to the separate, noninteracting atomic levels at an ion separation distance of R_0. As mentioned above, this arises from the spatially symmetric combination of the atomic valence

Figure 7.3.3. Electronic energy as a function of interatom separation distance R for the two interaction possibilities, spatially symmetric or spatially antisymmetric states. The quantity R_0 is the equilibrium separation distance between the atoms or the bond length. Note that as the separation increases, both states converge to the atomic orbital energy of the separate atoms.

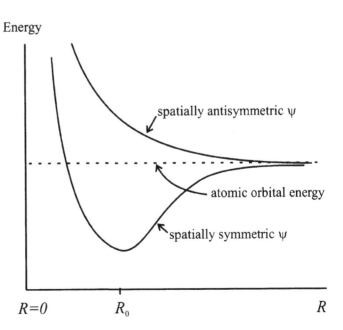

electrons. Clearly, when the two electrons combine in the spatially symmetric state, molecular formation occurs owing to the energetically favorable condition of this state; physical systems always move to a minimum energy configuration. It is also interesting to note that for the spatially antisymmetric state, the energy increases everywhere as R decreases. Hence no bonding can occur under these conditions.

What is the origin of the sharp rise in energy as R approaches zero? As the interatomic separation distance R decreases, the nuclear–nuclear Coulomb repulsion dominates and the molecule tries to restore itself to equilibrium. Inspection of the diagram shows that equilibrium is established at an interatomic separation distance of R_0, since movement in either direction, further compression, or expansion of the molecule leads to a higher energy state. As R approaches infinity, the electron energies approach the values corresponding to the isolated atomic levels, as expected.

To understand why the bonding state has a lower energy than the independent atomic states and the antibonding state, it is helpful to consider a different, but somewhat analogous system. We consider a system that comprises two quantum wells, as shown in Figure 7.3.4. If the barrier layer separating the two wells is very long or very high, then the wave functions of the electrons within each well do not overlap; there is no penetration of the wave function from one well to the other. Under these circumstances, the two wells behave as if they are completely decoupled and independent. However, if the separating barrier is shortened and reduced in height, then it is possible that the wave functions will leak from one well into the other. The electronic states within the two wells interact, leading to some degree of coupling of the system.

To see the effect of the coupling of the electronic states on the allowed energies of the system, it is useful to perform a series of computer experiments. From the discussion in Chapter 2, we found that when an electron is confined within a potential well of dimensions approximately equal to its de Broglie wavelength, spatial quantization effects occur; the electron can then have only certain discrete-energy values. To determine the allowed energies of the confined electronic states, we solved the Schroedinger equation in each region and matched the boundary conditions at all the interfaces. Although this technique will properly determine the eigenenergies of the system, another approach is often more useful computationally. As discussed in Section 2.5, the confined state energies correspond to the transmission resonances of the well. The peaks of the transmissivity of the well–barrier system occur precisely at the eigenenergies of the well.

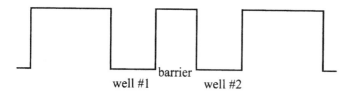

well #1   barrier   well #2

Figure 7.3.4. Sketch of a two-well, single-barrier quantum-well device. We call this a single-barrier device because only one barrier separates the quantum wells.

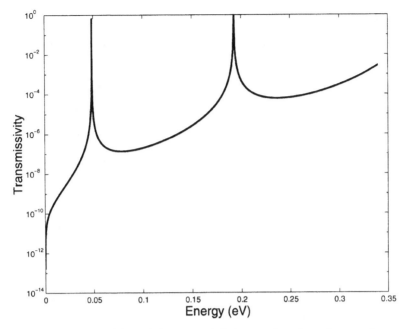

Figure 7.3.5. Logarithm of the transmissivity as a function of energy for a single GaAs quantum well with a barrier height of 0.347 eV, width of 75 Å, and barrier width of 50 Å. Note that two resonances appear at energies of 0.048 and 0.193 eV, corresponding to the confined quantum states within the potential well.

As an example, consider a one-well, two-barrier GaAs/AlGaAs structure similar to the wells examined in Section 2.3. In this case, the well width is chosen to be 75 Å, the barrier widths are each selected to be 50 Å, and the barrier heights are 0.347 eV. The calculated logarithm of the transmissivity as a function of incident electron energy for this structure is plotted in Figure 7.3.5. As can be seen from the figure, two peaks in the transmissivity curve are present that occur at energies of 0.048 and 0.193 eV. These energies correspond to the spatial quantization levels within the structure, $n = 1$ at $E = 0.048$ eV and $n = 2$ at $E = 0.193$ eV. The absolute values of the corresponding wave functions for the two states are plotted in Figures 7.3.6 and 7.3.7, respectively. Note that the $n = 1$ wave function has even parity while the $n = 2$ level has odd parity, as expected.

It is interesting to see what happens when additional quantum wells are added. We first consider adding one additional quantum well to the original single-quantum-well structure considered above. The structure comprises two identical wells of width 75 Å with a barrier sandwiched between them of 50 Å in width. Again, the materials are assumed to be GaAs and AlGaAs with a barrier height of 0.347 eV. In this way, through comparison with the results presented in Figures 7.3.5–7.3.7, we can examine what happens to the allowed energy states as a system transforms from one well to two interacting wells. The log of the transmissivity as a function of incident electron energy for the two-well structure is plotted in Figure 7.3.8. As can be seen from the figure, there are

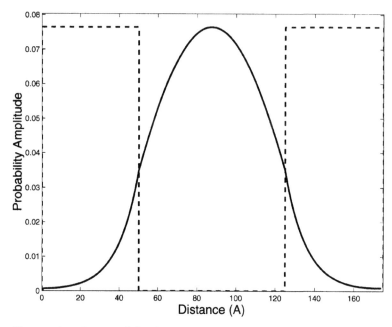

Figure 7.3.6. Square of the electronic wave function for the $n = 1$ state of the single quantum well of Figure 7.3.5. Note that the wave function has no nodes, as is expected from the discussion in Chapter 2. This wave function has even parity.

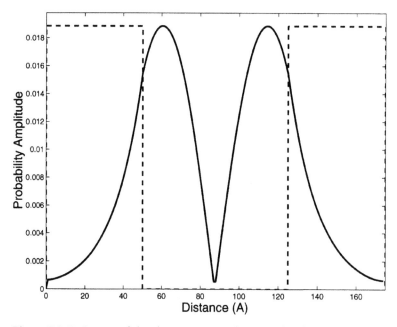

Figure 7.3.7. Square of the electronic wave function for the $n = 2$ state of the single quantum well of Figure 7.3.5. Note that the wave function has a single node in the center of the well. This wave function has odd parity.

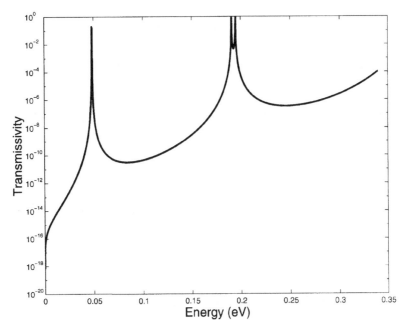

Figure 7.3.8. Logarithm of the transmissivity versus energy for a two-well GaAs/AlGaAs quantum-well system. Note that there is a single peak at 0.048 eV while there is a double peak at 0.1908 and 0.195 eV. The doublet appears because of the interaction between the quantum states between the two wells.

three peaks in the transmissivity. At low energy, $E = 0.048$ eV, there is a single peak while at higher energy there are two closely spaced peaks, $E = 0.1908$ and 0.195 eV, respectively. The low-energy peak matches the $n = 1$ state within the single-well structure. The high-energy peaks are roughly centered about the $n = 2$ peak within the single-well structure. It is important to note that one of the two peaks lies at a higher energy while the other peak lies at a lower energy with respect to the $n = 2$ peak in the single well. The behavior of these peaks can be understood as follows.

The $n = 1$ state within the two wells is clearly degenerate since there is only one peak in the transmissivity curve. The energy is precisely the same in each well for this state. The degeneracy is due to the fact that the electronic wave functions within each well do not overlap for these states. Therefore electrons within these states act independently of one another and hence have the same energy eigenvalues. The electronic wave functions of the particles in the $n = 1$ state are plotted in Figure 7.3.9. As can be seen from the figure, the wave functions within each well reach zero midway within the barrier separating the two wells, indicating that no interaction occurs.

Interestingly, at higher energy, the one peak corresponding to the $n = 2$ state in the single well splits into two peaks in the two-well system. One peak lies lower while the other one lies higher in energy than the $n = 2$ peak in the single well. This implies that there are two different states. The degeneracy of the $n = 2$

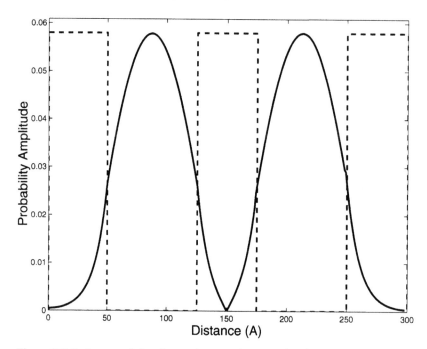

Figure 7.3.9. Square of the electronic wave function for the $n = 1$ state of the two-well system of Figure 7.3.4. Note that the probability amplitude is zero within the barrier between the wells. Subsequently, there is no leakage of probability amplitude between the two wells and hence no interaction between them.

state has been broken by the interaction of the electronic wave functions between the two wells. As discussed above, when the electronic wave functions overlap, the particles can no longer be treated as two separate electrons in well-defined localized states. Instead a collective state of the system is necessary to characterize the structure fully. Either symmetric or antisymmetric combinations of the spatial wave functions are possible, leading to a splitting of the degeneracy of the two states. The electronic wave functions within the wells for the $n = 2$ case are plotted in Figure 7.3.10. As can be seen from the figure, the wave functions overlap significantly within the barrier region. It is this interaction that leads to the splitting of the degeneracy of the $n = 2$ levels. In contrast, for the $n = 1$ case, the electrons are at sufficiently low energy that their respective wave functions do not overlap significantly and, as such, the $n = 1$ state remains degenerate.

It is interesting to note the similarities between the situation considered above for electrons in two different quantum wells and that of electrons in atomic states brought together to form a molecule. As we stated above, when two atoms are brought close together, the wave functions of the outermost valence electrons within each atom overlap, leading to antibonding and bonding states. The bonding state lies lower in energy than the noninteracting atomic-energy levels while the antibonding state lies higher in energy than the noninteracting atomic-energy levels. Because of the energetically favorable configuration of the bonding state, the two atoms are held together to form a molecule. Similarly, in the

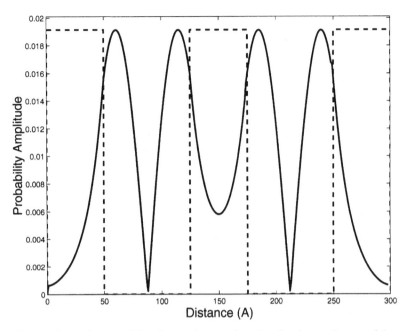

Figure 7.3.10. Square of the electronic wave function for the $n = 2$ state of the two-well system shown in Figure 7.3.4. In this case, the probability amplitude of the wave function is not zero within the barrier between the wells. As a result, there is a leakage of probability between the two wells, resulting in an interaction between their energy levels. As a result, spatially symmetric and antisymmetric quantum states are formed that result in lower and higher energies, respectively.

quantum-well system, when two quantum wells are brought into close proximity, the wave functions of the highest confined states overlap between the two wells, producing two new states. One state lies lower in energy while the other state lies at a higher energy than the confined states within the noninteracting wells. The lower-energy state is like the bonding state while the higher-energy state resembles the antibonding state in molecules.

Usually it is in only the highest-energy confined states of the quantum-well system that the wave functions overlap. Overlap of the lower-energy confined states occurs if the wells are brought even closer together. However, overlap occurs between electrons in the higher-energy confined states sooner because the high-energy wave functions penetrate more into the potential barrier. Similarly, the wave functions of the inner-shell electrons within an atom do not extend far from the atomic cores. Only the outermost valence electrons within an atom contribute to the chemical bond. There is no bonding from the inner valence electrons. This is why chemical reactions generally involve only the outer-shell electrons and not the inner-shell electrons.

The formation of a solid can be understood in a similar manner as the formation of molecules. Instead of only two atoms interacting within a solid, a much larger number of atoms, closer to Avogadro's number, interact. Just as in the case of two atoms, for a large number of atoms the wave functions of the out-ermost valence electrons interact. As a result, the atomic levels split into many levels, one generally for each atom present. Subsequently, if 10^{23} are present, 10^{23} different levels form about the original isolated atomic energy level. Sub-sequently, the energy spacing between these levels is very small and the levels form a quasi-continuum or band of allowed-energy states.

Similarly, a band of allowed energies forms in a multiquantum-well system when the wells are brought into close proximity such that the wave functions of the highest-energy states overlap. As in the case of atoms, for each well in the system another state forms. Ten coupled wells lead to ten states.

Multiquantum-well structures currently exist and are grown rather routinely from alternating layers of narrow- and wide-bandgap materials. These structures are presently used in a wide range of new devices, called multiquantum-well and superlattice structures. Multiquantum-well structures, in which the electrons interact to form bands, called minibands, are given a special name, superlattices. Through the use of superlattices, the band structures of materials can be directly engineered. The modern semiconductor device physicist no longer has to settle for the the band structures inherent in a material; he/she can artificially create one to suit his/her needs. Let us consider how the study of the electronic states in these systems can illuminate the physics of solids.

The logarithm of the electronic transmissivity as a function of energy is plotted for an eight-well/nine-barrier GaAs/AlGaAs multiquantum-well system in Figures 7.3.11 and 7.3.12. In Figure 7.3.11 the barrier widths are all equal to

Figure 7.3.11. Logarithm of the transmissivity versus energy for an eight-potential-well system. In this case, the well widths are 20 Å and the barrier widths are 30 Å wide.

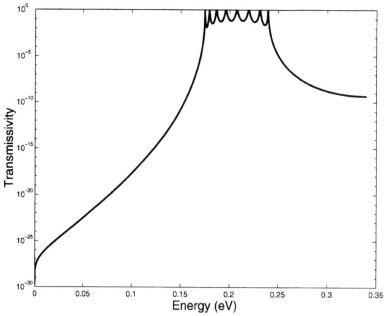

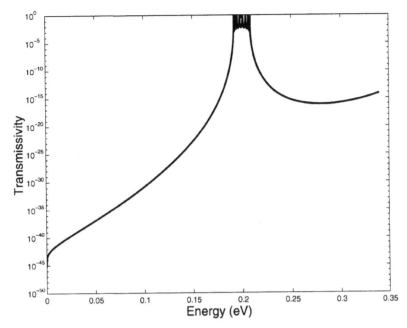

Figure 7.3.12. Logarithm of the transmissivity versus energy for an eight-potential-well system. In this case the barrier widths are widened from those in Figure 7.3.11 to 50 Å. As a result, there is less interaction between the levels.

30 Å , while in Figure 7.3.12 the barriers are widened to 50 Å each. Inspection of Figures 7.3.11 and 7.3.12 shows that eight peaks occur, corresponding to the collective states arising from the interaction within the eight-well system. As the barrier widths increase, the interactions between the states decrease, leading to a sharpening of the resonance. In other words, the system approaches the noninteracting case wherein the electronic levels are all degenerate. As the number of wells increases, the number of peaks in the main resonance peak increases proportionately. Hence, as the number of wells approaches a large number, the number of allowed states increases, also giving rise to a quasi-continuum of allowed energies called a band.

The formation of bands in a solid arises from the overlap of localized atomic orbitals in exactly the same way as minibands form in a superlattice from overlapping confined quantum-well wave functions. At very large separation distances between the atoms, each atom behaves as an isolated system. As such, the atomic levels all have the same energies. As the atoms come closer together, the outermost valence-electron levels interact, leading to collective state formation and the subsequent lifting of the degeneracy, as shown in Figure 7.3.13. Both bonding and antibonding states form. The bonding states correspond to the levels that split to lower energy, while the antibonding states are those splitting to higher energy. The bonding states are formed from spatially symmetric and antisymmetric spin combinations of the constituent wave functions, while the antibonding states correspond to the spatially antisymmetric states. As can be seen from Figure 7.3.13, when the atoms are within a distance

Energy

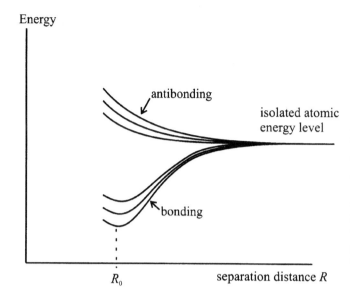

antibonding

isolated atomic
energy level

bonding

R_0

separation distance R

Figure 7.3.13. Energy-level scheme for an N-atom system, plotted as a function of separation distance between the atoms. Note that states form to both lower and higher energy than the isolated atomic-energy level. The lower-energy states are the bonding orbitals while the higher-energy states are the anti-bonding orbitals. The bonding orbitals are formed by spatially symmetric wave functions, while the antibonding orbitals are formed by spatially antisymmetric wave functions.

R_0 of one another, they attain the lowest energy configuration of the system. This results in the formation of a solid. If the atoms are brought closer together than R_0, the nuclear repulsion becomes appreciable, leading to an increase in energy of the system. In Chapter 8 we will consider the implications of the energy bands in solids and how these bands affect the electronic properties of the material.

7.4 Crystalline Lattices and Symmetries

From the discussion in the previous sections, it is apparent that the formation of a solid depends to some degree on the chemical bond formed between the constituent atoms. If the nature of the chemical bond is known, the detailed structure of a solid can often be determined. A thorough discussion of chemical bonds would require a book in itself and is thus beyond the scope of this book. However, our goal here is to understand the effect the potential of the ions has on the dynamics of the electrons within the material. However, to understand how the ionic potential affects the electronic properties of the solid, it is important to know in what ways a crystalline solid can form.

The first question one might raise is, what is a crystal? A crystal is defined as a material that exhibits perfect or nearly perfect periodicity in atomic structure. Any crystalline solid can be generated by appropriate translations along three independent directions of a basic unit cell, which is the smallest such unit that can fully generate the lattice. Therefore the basic unit cell contains, by definition, all the symmetries present within the crystalline solid as a whole. The smallest such unit cell is also called the primitive unit cell.

It is of use to define a set of basis vectors that fully span the space and are all linearly independent. Every lattice point can be described by an appropriate linear combination of the basis vectors. The basis vectors of a crystal are denoted as $\hat{a}$, $\hat{b}$, and $\hat{c}$. Hence any lattice point at position r can be represented as

$$\vec{r} = l\hat{a} + m\hat{b} + n\hat{c}, \tag{7.4.1}$$

where l, m, and n are all integers.

Since a crystal has perfect periodicity and can be fully generated from appropriate translations of a primitive unit cell, it has inherent symmetry properties. The most obvious such symmetry is translational symmetry; the crystal goes back into itself under a uniform rectilinear translation. Physically, if the crystal is translated by an appropriate amount in a symmetry direction, then the crystal looks exactly the same from any point as before. The physical properties of the crystal remain invariant. Other such symmetries may exist in a particular crystal. For example, the crystal may be invariant under a rotation of 60°. Therefore the potential as well as all the electronic properties must remain the same after a 60° rotation. Hence it is important to know the symmetry properties of a crystal since its physical properties must have the same symmetries.

All crystalline solids have translational symmetry by the definition of a crystal. Aside from translational symmetry, three additional symmetries of the unit cell may exist. These are rotation, reflection, and inversion symmetry. A reflection symmetry is present if one half of the unit cell is a mirror image of the other half through a particular plane drawn through the center of the unit cell. An inversion symmetry occurs when the crystal goes into the negative of itself; in other words, on inversion of the coordinates, $r \rightarrow -r$, the crystal is invariant. An inversion symmetry is equivalent to a rotation plus a reflection. Finally, rotational symmetry implies that the unit cell goes back into itself under a rotation about a fixed axis.

The most obvious question then is, how many three-dimensional lattices can be formed that have crystalline symmetry, meaning the lattice appears the same from any lattice point? There are only seven unique three-dimensional crystal systems that can be generated from integer translations of the unit vectors such that the system has crystalline symmetry. Of these seven crystal systems, each is formed from primitive unit cells for which lattice points exist at only the corners of the cell.

The crystal systems can be generalized to fourteen different lattice types. These are called the fourteen Bravais lattices. The Bravais lattices are defined as the set of lattices formed by primitive and nonprimitive unit cells such that from each lattice point the lattice everywhere appears the same.

The seven crystal systems are characterized by the lengths of their primitive unit vectors, $\vec{a}$, $\vec{b}$, and $\vec{c}$, and the angles formed between these vectors, α, β, and γ, as defined in Figure 7.4.1. Table 7.4.1 lists the conditions on the primitive unit vectors and angles that describe each unique crystalline system. The different Bravais lattices

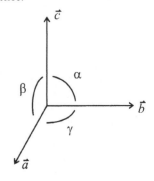

Figure 7.4.1. Primitive unit vectors and their corresponding angles used in describing a Bravais lattice.

Table 7.4.1
Seven Crystal Systems and the Relationship
Between the Primitive Unit Vectors and Angles

Crystal System	Unit Vectors	Angles
Triclinic	$a \neq b \neq c$	$\alpha \neq \beta \neq \gamma$
Monoclinic	$a \neq b \neq c$	$\alpha = \gamma = 90 \neq \beta$
Orthorhombic	$a \neq b \neq c$	$\alpha = \gamma = \beta = 90$
Tetragonal	$a = b \neq c$	$\alpha = \beta = \gamma = 90$
Rhombohedral	$a = b = c$	$\alpha = \beta = \gamma \neq 90$
Hexagonal	$a = b \neq c$	$\alpha = \beta = 90 \; \gamma = 120$
Cubic	$a = b = c$	$\alpha = \beta = \gamma = 90$

and their names are listed in Table 7.4.2. There are fourteen Bravais lattices within the seven crystal groups. The fourteen Bravais lattices are sketched in Figure 7.4.2. The Bravais lattices are grouped by crystal type in Figure 7.4.2.

The set of symmetries present in a crystal can be classified in two general ways, as either point-group or space-group symmetries. A point group is the collection of all symmetry operations that, when applied about a lattice point, leaves the lattice invariant. There are only thirty-two point groups. A space group is the set of all combined point-group and translational symmetry operations that leaves the lattice invariant. Knowledge of the point-group symmetries is quite helpful in determining the electronic properties of a crystalline solid. The full details of point group symmetries are discussed in the book by Bassani and Pastori Parravicini (1977).

Table 7.4.2
Bravais Lattices and Crystal Types

Crystal Type	Bravais Lattices
Triclinic	Simple triclinic
Monoclinic	Simple monoclinic
	Base-centered monoclinic
Orthorhombic	Simple orthorhombic
	Base-centered orthorhombic
	Body-centered orthorhombic
	Face-centered orthorhombic
Rhombohedral	Rhombohedral
Tetragonal	Simple tetragonal
	Body-centered tetragonal
Hexagonal	Hexagonal
Cubic	Simple cubic
	Body-centered cubic
	Face-centered cubic

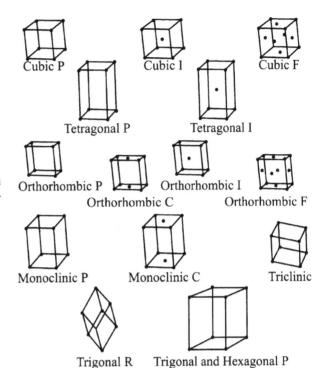

Figure 7.4.2. Sketches of the fourteen Bravais lattices formed with the conventional cells.

Cubic P   Cubic I   Cubic F

Tetragonal P   Tetragonal I

Orthorhombic P   Orthorhombic I

Orthorhombic C   Orthorhombic F

Monoclinic P   Monoclinic C   Triclinic

Trigonal R   Trigonal and Hexagonal P

There are only five rotation angles that restore a Bravais lattice into itself. These rotations are 360°, 180°, 120°, 90°, and 60°. This assertion can be proved as follows. Consider four lattice points, as shown in Figure 7.4.3. The lattice points marked 1 and 2 in the drawing are assumed to be separated by one lattice constant a. The points 1′ and 2′, which are also lattice points, must be separated by an integral multiple of a, which we call m. Hence the product of m and a,

Figure 7.4.3. Construction used in proving that only five angles exist for which a Bravais lattice is fully restored into itself under rotation.

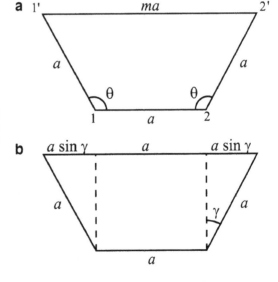

ma, can be determined from the geometry of Figure 7.4.3.b as

$$ma = a + 2a \sin \gamma, \qquad (7.4.2)$$

where

$$\gamma = \theta - 90. \qquad (7.4.3)$$

The $\cos \theta$ is simply equal to $-\cos(90 - \gamma)$ that is in turn equal to $-\sin \gamma$ from use of the double-angle formula for $\cos(90 - \gamma)$:

$$\cos(90 - \gamma) = \cos 90 \cos \gamma + \sin 90 \sin \gamma = \sin \gamma. \qquad (7.4.4)$$

Therefore Eq. (7.4.2) becomes

$$ma = a - 2a \cos \theta, \qquad (7.4.5)$$

where *m* is an integer. Solving for $\cos \theta$ yields

$$\cos \theta = \frac{(1 - m)}{2}. \qquad (7.4.6)$$

Inspection of Eq. (7.4.6) clearly shows that $m < -1$ is impossible since it would otherwise lead to an imaginary argument for the cosine. Similarly, *m* cannot be greater than 3. The only possible values of *m* then are the integers between -1 and 3. These are

$$
\begin{aligned}
m &= -1, & \theta &= 0, 360, \\
m &= 0, & \theta &= 60, \\
m &= 1, & \theta &= 90, \\
m &= 2, & \theta &= 120, \\
m &= 3, & \theta &= 180.
\end{aligned} \qquad (7.4.7)
$$

Therefore there are five rotation elements that restore a Bravais lattice into itself.

There is a standard notation used by crystallographers to describe the directions and the planes of a crystal. We will find it useful to refer to different crystallographic directions and planes, and it is best to use the standard notation to do so. This notation is as follows:

[100] describes the specific crystallographic direction 100 (which lies along the *x* axis, as we will see below).
⟨100⟩ describes the family of equivalent 100 directions.
(100) describes the specific 100 planes.
{100} describes the family of equivalent 100 planes.

Equivalent directions and planes mean the following: Any plane or direction that is equivalent by symmetry is a member within the family of equivalent planes or directions. For example, the plane (100) is equivalent to the plane (010) and (001) in a cubic crystal. Therefore all three of these planes are in the family {100} of equivalent planes.

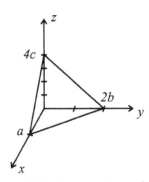

Figure 7.4.4. Construction used for determining the Miller indices.

To denote a plane or direction that cuts through the axis along a negative direction, the negative index is represented by a small bar over the top of the index. An example would be the plane intersecting at -1 along the x axis and 0 along both the y and the z axes. This is represented as

$$(\bar{1}00). \tag{7.4.8}$$

How, though, do we find the appropriate indices used in the above notation? The indices used to describe a plane or family of planes are called the Miller indices. The Miller indices can be determined in a systematic manner by use of the following prescription. First, construct the xyz axes and draw the plane of interest such that it intersects at the (abc) multiple, as shown in Figure 7.4.4. Second, form the triad $(x/a, y/b, z/c)$. For example, when the intercepts given in Figure 7.4.4 are used, the triad is given as

$$\left[\frac{1a}{a}, \frac{2b}{b}, \frac{4c}{c}\right] = [1, 2, 4]. \tag{7.4.9}$$

Invert the triad. In the above example this becomes

$$[1, 2, 4] = \left[\frac{1}{1}, \frac{1}{2}, \frac{1}{4}\right]. \tag{7.4.10}$$

Next, multiply by the least common denominator such that only integers are obtained. In the example the least common denominator is 4, so the triad becomes [4, 2, 1]. The desired plane is then written as (421), which are the Miller indices of this plane.

It is important to determine the symmetries of a crystal since they can be exploited in the evaluation of the electronic properties of the material. The basic problem that confronts us is the solution of the Schroedinger equation, including the exchange interaction as well as the electron–lattice interaction. The electrons interact with the ions through a Coulomb potential that must have the full symmetry of the lattice since each lattice site is occupied by an ion. Consequently, the solutions of the Schroedinger equation, the electronic energies and the wave functions, must then exhibit the same symmetries. In this way, the complicated electron–ion interaction can be simplified.

The translational symmetry present in any crystalline solid places constraints on the allowed values of the electron wave functions and their corresponding energy eigenvalues. In other words, since the ion potential is periodic then the electronic wave functions and energies must be periodic as well. Let us first assume that the electrons behave as free particles whose wave functions are

$$\Psi \sim e^{ikx}, \tag{7.4.11}$$

where k is given as

$$k = \sqrt{\frac{2mE}{\hbar^2}}. \tag{7.4.12}$$

The electrons are, of course, in sharp momentum eigenstates of the eigenvalue $\hbar k$. If a potential arising from the presence of the ions is superimposed on the electrons, the resulting one-particle Hamiltonian is

$$-\frac{\hbar^2}{2m}\nabla^2 + V(r), \tag{7.4.13}$$

where the potential $V(r)$ is periodic with periodicity R. Therefore

$$V(r) = V(r + R). \tag{7.4.14}$$

Since the potential is periodic, the wave functions must also be periodic. At the same time, though, we would like to retain the free-particle-like behavior of the wave functions. Therefore the wave functions are chosen as plane-wave states modulated by a periodic function $u_{nk}(r)$, which has the full periodicity of the lattice

$$u_{nk}(r) = u_{nk}(r + R). \tag{7.4.15}$$

The wave functions are then defined as

$$\Psi_{nk}(r) = e^{ikr}u_{nk}(r), \tag{7.4.16}$$

which are called Bloch functions.

Bloch functions have the following properties. If the function is translated in real space by a distance R, which is equal to a Bravais lattice vector (in other words it is an integer multiple of the Bravais lattice unit vectors), then the Bloch function is

$$\Psi(r + R) = u_{nk}(r + R)e^{ik(r+R)}. \tag{7.4.17}$$

When Eq. (7.4.15) is used, the Bloch function reduces to

$$\Psi(r + R) = u_{nk}(r)e^{ikr}e^{ikR}. \tag{7.4.18}$$

Substituting into Eq. (7.4.18) the expression given by Eq. (7.4.16) yields

$$\Psi(r + R) = \Psi_{nk}(r)e^{ikR}. \tag{7.4.19}$$

Therefore the Bloch function $\Psi(r)$ is periodic, differing from itself only under a translation of R (the periodicity of the lattice) by a phase factor. Since the translation leads to a change in phase only, the magnitude and the probability density remain unchanged.

Consider the effect of periodic boundary conditions on the Bloch states. In one dimension, periodic boundary conditions require that the wave function, when translated by a certain amount, return into itself. Mathematically, if the boundary is given as Na, then after a translation of Na, the wave function must be invariant. This is stated as

$$\Psi(x + Na) = \Psi(x) = e^{ikNa}\Psi(x). \tag{7.4.20}$$

Clearly,

$$e^{ikNa} = 1. \tag{7.4.21}$$

Taking the Nth root of each side of Eq. (7.4.21) yields

$$e^{ika} = 1^{1/N}. \tag{7.4.22}$$

But 1 can be rewritten as $e^{in2\pi}$. Hence

$$\begin{aligned} e^{ika} &= e^{i2\pi n/N}, \\ ka &= \frac{2\pi n}{N}. \end{aligned} \tag{7.4.23}$$

Therefore the allowed values of k are

$$k = \frac{2\pi n}{Na}. \tag{7.4.24}$$

If N is very large then the allowed k values are very closely spaced and the system behaves like a quasi-continuum. This is the basis of the electronic theory of solids in that the electronic states can be treated as a quasi-continuum in energy.

In summary, in this section we have determined that a crystal is a material characterized by nearly perfect regularity. This perfect periodicity requires that the electronic states and energies also be periodic. To satisfy the periodicity of the potential, the electronic states must be Bloch functions. Imposition of periodic or circular boundary conditions leads to the conclusion that the possible wave vectors of the electrons are all closely spaced, forming a quasi-continuum or band of allowed values. In Chapter 8 we will discuss in detail how these energy bands form in crystals.

PROBLEMS

1. Write the Schroedinger equation for heliumlike atoms (any atomic number but containing only two electrons), neglecting spin–orbit effects and nuclear motion.

a

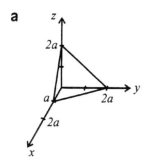

b

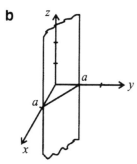

Figure Pr.7.4. a, Plane intersecting along the xyz axes at a, $2a$, and $2a$ respectively. b, Plane intersecting at points $x=a$ and $y=a$. Note the plane does not intersect the z-axis.

Introduce scaled variables for a perturbative treatment of the electron–electron interaction,

$$\frac{e^2}{r_{12}},$$

and show that in the resulting dimensionless equation the perturbation is strictly proportional to $1/Z$, where Z is the atomic number.

2. Show the equivalence of

$$\Psi(r + R) = e^{ikR}\Psi(r), \qquad \Psi(r) = e^{ikr}u(r),$$

by using Bloch's Theorem, where $u(r)$ is periodic in r with period R.

3. Four identical particles are described by the Hamiltonian

$$H = \sum_{i=1}^{4} H_{sp}(x_i),$$

where H_{sp} is given as

$$H_{sp} = \frac{p^2}{2m} + \frac{1}{2}m\omega^2 x^2.$$

a. Find the eigenfunctions and the energies of the system, disregarding proper symmetrization.

b. Give the degeneracy of the lowest three states ($n = 0, 1, 2$), and list the quantum numbers for each state.

c. Consider spinless particles and now list the degeneracies for the lowest three states $n = 0, 1, 2$.

4. Determine the Miller indices for the planes of Figures Pr.7.4.a and Pr.7.4.b.

5. Sodium crystallizes into a body-centered cubic lattice with a lattice constant of 5.868 Å . Let the basis vectors defining the lattice be

$$\bar{a} = a\hat{i}, \qquad \bar{b} = a\hat{j}, \qquad \bar{c} = a\hat{k}.$$

Compute the distance in amgstroms from the origin to the nearest atom in the [111] direction.

8

Motion of Electrons in a Periodic Potential

In this chapter, we consider the dynamics of electrons in a periodic potential. We begin by discussing some aspects of the properties of solids. Specifically, we discuss the Brillouin zone and Bragg reflection. Next we focus on the solution of the Schroedinger equation for electrons in a periodic potential, assuming that the electrons have plane-wave characteristics. We can generalize this to the study of electron propagation in a crystal, assuming that the electronic states are Bloch states. These results can be extended to the study of realistic systems. Only the very fundamentals of calculations for realistic band structures are presented. We also discuss holes and their properties in a semiconductor.

8.1 Effective Mass Theory and the Brillouin Zone

As was found in Chapter 1, electrons propagating in plane-wave states, corresponding to free particles, have sharp values of the momentum operator with momentum eigenvalues equal to $\hbar k$. However, in a crystal, because of the periodicity of the ionic potential, the electronic wave functions must also be periodic. The electronic states in a crystal are Bloch states, as discussed in Chapter 7, and have the form

$$\Psi_k(x) = u_{nk}(x)\, e^{ikx}. \tag{8.1.1}$$

In a rigorous sense, the Bloch functions are not definite eigenstates of the momentum operator, and therefore electrons within a crystal do not have sharp values of the momentum p. Nevertheless, as we will see in the following sections, the electrons in a crystal can often be treated as nearly free particles. Under such conditions the k vector of the electron can still be associated with the momentum of the electron. The k vector is referred to as the crystalline momentum and bears essentially the same relationship to p as it did for the free-particle case.

Let us now consider the motion of an electron in a crystalline lattice under the influence of an applied electric field. The question we seek to answer is, how can the effect of the applied electric field on the electron's state be accounted for? The answer is immediate for a free electron; its behavior is determined entirely from Newton's Second Law of Motion; the time rate of change of the linear momentum is equal to the applied force $dp/dt = F_{ext}$. What, though, is the effect of including the ionic potential? Equivalently, how does the state of the electron change if the electron is in a Bloch state as opposed to a free-particle state?

To understand the influence of the periodic ionic potential on the dynamics of an electron in a crystal, it is useful to determine first the velocity of an electron in a crystal. The quantum-mechanical instantaneous velocity of a moving electron at any point is equal to

$$\vec{v} = \frac{\hbar}{2mi} \frac{\Psi^* \vec{\nabla}\Psi - \Psi \vec{\nabla}\Psi^*}{|\Psi|^2}. \tag{8.1.2}$$

The average value of the velocity is obtained through use of Eq. (1.4.10). Therefore

$$\bar{v} = \int v |\Psi|^2 \, d^3r, \tag{8.1.3}$$

which becomes, on substitution of Eq. (8.1.2) for v,

$$\bar{v} = \frac{\hbar}{2mi} \int (\Psi^* \nabla \Psi - \Psi \nabla \Psi^*) \, d^3r. \tag{8.1.4}$$

The wave function for an electron in a crystal, though, is a Bloch state given by Eq. (7.4.16). The Bloch state satisfies the Schroedinger equation for the electron in a periodic potential $V(r)$:

$$\nabla^2 \Psi + \frac{2m}{\hbar^2}[E - V(r)]\Psi = 0. \tag{8.1.5}$$

Differentiating the Schroedinger equation with respect to k_x yields

$$\nabla^2 \frac{\partial \Psi}{\partial k_x} + \frac{2m}{\hbar^2} \left\{ \Psi \frac{\partial E}{\partial k_x} + [E - V(r)]\frac{\partial \Psi}{\partial k_x} \right\} = 0, \tag{8.1.6}$$

while differentiating the Bloch function with respect to k_x yields

$$\frac{\partial \Psi}{\partial k_x} = ix\Psi + e^{ikr}\frac{\partial u}{\partial k_x}. \tag{8.1.7}$$

The kinetic-energy term in the Schroedinger equation for the Bloch state in Eq. (8.1.6) can be determined by application of the Laplacian operator onto Eq. (8.1.7) as

$$\nabla^2 \frac{\partial \Psi}{\partial k_x} = 2i\frac{\partial \Psi}{\partial x} + ix\nabla^2 \Psi + \nabla^2 \left(e^{ikr}\frac{\partial u}{\partial k_x} \right). \tag{8.1.8}$$

Substituting the expressions given by Eqs. (8.1.8) and (8.1.7) into Eq. (8.1.6) gives

$$2i\frac{\partial \Psi}{\partial x} + \frac{2m}{\hbar^2}\Psi\frac{\partial E}{\partial k_x} + \left\{ \nabla^2 + \frac{2m}{\hbar^2}[E - V(r)] \right\}\left(e^{ikr}\frac{\partial u}{\partial k_x} \right)$$
$$+ ix\left\{ \nabla^2 \Psi + \frac{2m}{\hbar^2}[E - V(r)]\Psi \right\} = 0. \tag{8.1.9}$$

The last term of Eq. (8.1.9) is zero from the Schroedinger equation. Multiplying both sides by Ψ^* and integrating over all space yields

$$\frac{2m}{\hbar^2}\frac{\partial E}{\partial k_x}\int \Psi^*\Psi\, d^3r = -2i\int \Psi^*\frac{\partial \Psi}{\partial x}\, d^3r - \int \left[\Psi^*\nabla^2\left(e^{ikr}\frac{\partial u}{\partial k_x}\right)\right.$$

$$\left. +\frac{2m}{\hbar^2}\Psi^*[E-V(r)]\left(e^{ikr}\frac{\partial u}{\partial k_x}\right)\right]d^3r. \tag{8.1.10}$$

The integral over the volume in the first term of the last part of Eq. (8.1.10) can be simplified through use of Eq. (1.5.33), Green's theorem. This becomes

$$\int \left[\Psi^*\nabla^2\left(e^{ikr}\frac{\partial u}{\partial k_x}\right) - \left(e^{ikr}\frac{\partial u}{\partial k_x}\right)\nabla^2\Psi^*\right]d^3r$$

$$= \int \left[\Psi^*\frac{\partial}{\partial n}\left(e^{ikr}\frac{\partial u}{\partial k_x}\right) - \left(e^{ikr}\frac{\partial u}{\partial k_x}\right)\frac{\partial}{\partial n}\Psi^*\right]dS, \tag{8.1.11}$$

where n represents the normal direction. The integral over S is over the surface while the integral over r is over the volume. The integral over the surface vanishes in the same manner as that found in Section 1.5, only the integrals over the volume then survive and they are related as

$$\int \Psi^*\nabla^2\left(e^{ikr}\frac{\partial u}{\partial k_x}\right)d^3r = \int \left(e^{ikr}\frac{\partial u}{\partial k_x}\right)\nabla^2\Psi^*\, d^3r. \tag{8.1.12}$$

Therefore

$$\int \Psi^*\left\{\nabla^2 + \frac{2m}{\hbar^2}[E-V(r)]\right\}\left(e^{ikr}\frac{\partial u}{\partial k_x}\right)d^3r$$

$$= \int e^{ikr}\frac{\partial u}{\partial k_x}\left\{\nabla^2 + \frac{2m}{\hbar^2}[E-V(r)]\right\}\Psi^*\, d^3r. \tag{8.1.13}$$

The right-hand side of Eq. (8.1.13), though, is zero since Ψ^* is a solution of the Schroedinger equation. Thus Eq. (8.1.13) simplifies to

$$\int \Psi^*\left\{\nabla^2 + \frac{2m}{\hbar^2}[E-V(r)]\right\}\left(e^{ikr}\frac{\partial u}{\partial k_x}\right)d^3r = 0. \tag{8.1.14}$$

With this substitution, Eq. (8.1.10) becomes

$$i\int \Psi^*\frac{\partial \Psi}{\partial x}d^3r = -\frac{m}{\hbar^2}\frac{\partial E}{\partial k_x}. \tag{8.1.15}$$

Generalizing to three dimensions yields

$$i\int \Psi^*\vec{\nabla}\Psi\, d^3r = -\frac{m}{\hbar^2}\vec{\nabla}_k E. \tag{8.1.16}$$

Similarly,

$$-i\int \Psi\vec{\nabla}\Psi^*\, d^3r = -\frac{m}{\hbar^2}\vec{\nabla}_k E. \tag{8.1.17}$$

The average velocity can then be reexpressed as

$$\bar{v} = \frac{\hbar}{2mi} \left[-\frac{m}{i\hbar^2} \nabla_k E - \frac{m}{i\hbar^2} \nabla_k E \right], \tag{8.1.18}$$

which becomes

$$\bar{v} = \frac{1}{\hbar} \nabla_k E. \tag{8.1.19}$$

The velocity of an electron in a crystal is simply proportional to the gradient of the energy as a function of k. The above result is quite general and holds for all crystalline solids in which the electronic wave function can be written as a Bloch function.

From the above expression for the velocity of an electron in a crystal, it is common to define an effective mass of the electron. The incremental change in the energy of a particle dE can be determined from the applied field $\vec{F}$ as

$$dE = \frac{dE}{dk} dk = -q\vec{F} \cdot d\vec{x} = -q\vec{F} \cdot \vec{v} \, dt. \tag{8.1.20}$$

When the expression derived above for the velocity of an electron is used, the magnitude of the differential work done on an electron by an applied field $\vec{F}$ is

$$dE = -\frac{qF}{\hbar} \frac{dE}{dk} \, dt,$$

$$\frac{dE}{dk} dk = -\frac{qF}{\hbar} \frac{dE}{dk} \, dt,$$

$$dk = -\frac{qF}{\hbar} \, dt,$$

$$\hbar \, dk = -qF \, dt,$$

$$\hbar \frac{dk}{dt} = -qF = \frac{dp}{dt}, \tag{8.1.21}$$

which is simply Newton's Second Law of Motion. If p is the crystalline momentum discussed above, then the force is simply equal to the time rate of change of p. The crystalline momentum of an electron changes under the influence of an applied electric field in much the same way as the true momentum of a free particle does.

Consider the time rate of change of the magnitude of the group velocity v:

$$\frac{dv}{dt} = \frac{1}{\hbar} \frac{d}{dt} \left(\frac{dE}{dk} \right) = \frac{1}{\hbar} \frac{d^2 E}{dk^2} \frac{dk}{dt},$$

$$\frac{dv}{dt} = \frac{1}{\hbar} \frac{d^2 E}{dk^2} \frac{dk}{dt}. \tag{8.1.22}$$

The force, though, is equal to $\hbar\,dk/dt$. Multiplying the right-hand side of Eqs. (8.1.22) by $\hbar/\hbar$ yields

$$\frac{dv}{dt} = \frac{1}{\hbar^2}\frac{d^2E}{dk^2}\hbar\frac{dk}{dt}. \tag{8.1.23}$$

Equation (8.1.23) is analogous to Newton's Second Law if the left-hand side is associated with the acceleration and the right-hand side is associated with the ratio of the force to the mass. From the definition of the force and the acceleration, an effective mass of the electron can be defined as

$$\frac{1}{m^*} = \frac{1}{\hbar^2}\frac{d^2E}{dk^2}. \tag{8.1.24}$$

With this definition, Eq. (8.1.23) becomes a simple statement of Newton's Second Law. The effective mass can be envisioned as arising from the physical effect of the crystalline potential on the dynamics of an electron. In other words, the motion of an electron in a crystal can be treated identically to that of a free particle but with the effective mass m^* replacing the free-electron mass.

The effective mass can be obtained in a different way from the above. If it is assumed that there is a minimum in the energy relation $E(k)$, then near the minimum a Taylor series expansion can be performed. The linear term in the Taylor series expansion vanishes since the first derivative is equal to zero at a minimum. Ignoring terms greater than the quadratic term, which is valid near the minimum point, we find that the expansion is given as

$$E(k) = E(k_0) + \frac{1}{2}\sum_{a,b}\frac{\partial^2 E(k)}{\partial k_a \partial k_b}(k_a - k_{a0})(k_b - k_{b0}). \tag{8.1.25}$$

Since the energy is a scalar and k is a vector, the energy varies generally along different directions in k. Therefore the second-derivative term is generally a tensor. This term reduces to a scalar for the specific case in which the tensor is reduced along the principal directions of the crystal and if the system is isotropic. If it is further assumed that the energy of the minimum is zero, then the expansion of the energy becomes

$$E(k) = \frac{1}{2}\frac{\partial^2 E}{\partial k^2}k^2. \tag{8.1.26}$$

Multiplying and dividing Eq. (8.1.26) by $\hbar^2$ yields

$$E(k) = \frac{\hbar^2 k^2}{2}\frac{1}{\hbar^2}\frac{\partial^2 E}{\partial k^2}. \tag{8.1.27}$$

The energy, though, of a free electron is given as

$$E = \frac{p^2}{2m} = \frac{\hbar^2 k^2}{2m}. \tag{8.1.28}$$

A free-energy-like spectrum is recovered then if an effective mass m^* is defined as

$$\frac{1}{m^*} = \frac{1}{\hbar^2} \frac{\partial^2 E}{\partial k^2}.$$

(8.1.29)

The energy of the electron in the crystal, near the band minimum, is given then as

$$E = \frac{\hbar^2 k^2}{2m^*},$$

(8.1.30)

where m^* is the effective mass of the electron in the crystal.

The use of an effective mass is an especially useful technique for describing the motion of an electron in a crystal. Physically, the effective mass model enables us to account for the effect of the ionic potential on the electron dynamics by simply changing the mass of the electron in all the equations describing its behavior. Subsequently, the complicated problem of an electron moving within a crystal under the interaction of a periodic but complicated potential can be reduced to that of a simple free particle but with a modified mass. The effective-mass model is, of course, not always valid, since it is derived only near the band edge. However, its use often provides a reasonably accurate model of electronic transport in many situations of interest.

From the above discussion, we have found that electrons in a crystal can, for the most part, be treated as free or nearly free particles with a well-defined crystalline momentum characterized by a wave vector k. According to the Uncertainty Principle, if a particle has a well-defined momentum, then its real-space position is poorly defined. Thus it makes no sense to try and describe an electron in terms of its real-space position since it is not localized. Instead, it is far more effective to use the electron's k vector to describe the particle since it has well-defined momentum. We choose to work then in k space or reciprocal lattice space. The k space is often referred to as reciprocal lattice space since the k vector has units of inverse length.

The use of reciprocal lattice space is no stranger than the use of the frequency domain in signal analysis. Both position and momentum and time and frequency are conjugate variables connected by means of simple Fourier transformations, as was discussed in Chapter 1. In signal analysis, the frequency domain is often of greater use to formulate a problem than the time domain. The same is true in the present situation. It is far more convenient to formulate the electron states in a crystalline solid in terms of k rather than x.

Let us first define a reciprocal lattice vector, abbreviated as RLV, as a vector such that the dot product formed between it and a direct lattice vector is an integer multiple of 2π. Let G be a RLV and R be a direct lattice vector. Then G and R are related as

$$\vec{G} \cdot \vec{R} = 2\pi n,$$

(8.1.31)

where n is an integer. The real-space vector $\vec{R}$ is easily described in terms of the real basis vectors $\hat{a}$, $\hat{b}$, and $\hat{c}$ as

$$\vec{R} = l_1\hat{a} + l_2\hat{b} + l_3\hat{c}. \tag{8.1.32}$$

The reciprocal lattice has the same symmetry properties that the direct or the real-space lattice has. As in the case for the direct lattice, the reciprocal lattice is generated by three basis vectors, $\hat{a}^*$, $\hat{b}^*$, and $\hat{c}^*$. These basis vectors are determined from the direct lattice vectors as

$$\hat{a}^* = 2\pi\frac{(\hat{b} \times \hat{c})}{V}, \qquad \hat{b}^* = 2\pi\frac{(\hat{c} \times \hat{a})}{V}, \qquad \hat{c}^* = 2\pi\frac{(\hat{a} \times \hat{b})}{V}, \tag{8.1.33}$$

where V is the volume of the unit cell in real space.

Consequently, any RLV can be written as a linear combination of the RLV vectors. For example, the RLV G can be written as a linear combination of the reciprocal lattice basis vectors as

$$\vec{G} = n_1\hat{a}^* + n_2\hat{b}^* + n_3\hat{c}^*. \tag{8.1.34}$$

The unit cell of the reciprocal lattice fully reconstructs the entire reciprocal lattice under appropriate translations similar to the basic unit cell in direct lattice space. The unit cell in reciprocal lattice space is given a special name, the first Brillouin zone, because of its immense importance. Before we discuss the physical significance of the first Brillouin zone, let us first determine how it is constructed.

As an example, let us construct the first Brillouin zone for a two-dimensional lattice. Consider the two-dimensional direct lattice shown in Figure 8.1.1.a.

a Two-dimensional direct lattice

b First Brillouin zone in the reciprocal lattice

Figure 8.1.1. Sketch showing the construction of a two-dimensional Brillouin zone for a square lattice. a. Two-dimensional direct lattice; b. first Brillouin zone in the reciprocal lattice; c. second Brillouin zone; d. third Brillouin zone.

c ———— Second Brillouin zone

d ———— Third Brillouin zone

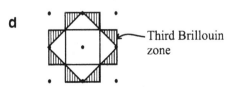

The reciprocal lattice corresponding to the square direct lattice is also square, although the dimensions are different. This follows directly from the fact that the reciprocal lattice must have the same symmetries as the direct lattice. We find the first Brillouin zone by connecting an arbitrary reciprocal lattice point to all its nearest-neighbor lattice points by using straight lines. Next we construct the perpendicular bisectors to all these lines. Once the perpendicular bisectors are constructed, the smallest area enclosed is the unit cell in reciprocal lattice space or the first Brillouin zone. The first Brillouin zone within the square two-dimensional lattice is sketched in Figure 8.1.1.b.

The second and the third Brillouin zones, although not as important as the first Brillouin zone, are also of interest. We can construct these from the first Brillouin zone by connecting the reciprocal lattice point in question to all its second-nearest neighbors. For the two-dimensional lattice of Figure 8.1.1, the central point is connected to points along the diagonals of the central square. Again these lines are bisected, and the enclosed region between the bisecting lines and the first Brillouin zone boundaries forms the second Brillouin zone. In Figure 8.1.1c the lined region is the second Brillouin zone. Finally, we form the third Brillouin zone in the same manner by connecting the original point to its third-nearest neighbors and bisecting the resulting lines. The area enclosed by these lines and the boundaries of the second and the first Brillouin zones form the third Brillouin zone (see Fig. 8.1.1.d).

In three dimensions, the same procedure is followed as in two dimensions except that the lines are now planes and the enclosed area becomes an enclosed volume. The smallest enclosed volume formed by the bisecting planes is the corresponding first Brillouin zone in three dimensions. Sketches of the Brillouin zone for the body-centered and the face-centered cubic lattices are given in Figure 8.1.2.

Why, though, is the Brillouin zone important? To understand the physical significance of the Brillouin zone we must consider it in terms of the Bragg law of scattering. Let us first briefly review the origin of the Bragg law.

As we discussed in Chapter 1, an electron can behave as either a particle or a wave. An electron beam impinging on a crystal acts essentially as a wave front. Depending on the nature of the crystal and the electron energy, the beam will be scattered at some angle ϕ with respect to the surface of the material. The situation is diagrammatically represented in Figure 8.1.3. The electron wave is assumed reflected from two different planes separated by the lattice constant a. For constructive interference to occur, the path difference between the two reflected rays must be equal to an integral multiple of 2π rad in phase difference. In general, the ratio of the path difference of the two rays to the wavelength of the incident beam is equal to the ratio of the phase difference to 2π. From the construction in Figure 8.1.3, the path difference is readily seen to be equal to $2a \cos(90-\phi)$. Therefore

$$\frac{\text{path difference}}{\lambda} = \frac{\text{phase difference}}{2\pi}. \tag{8.1.35}$$

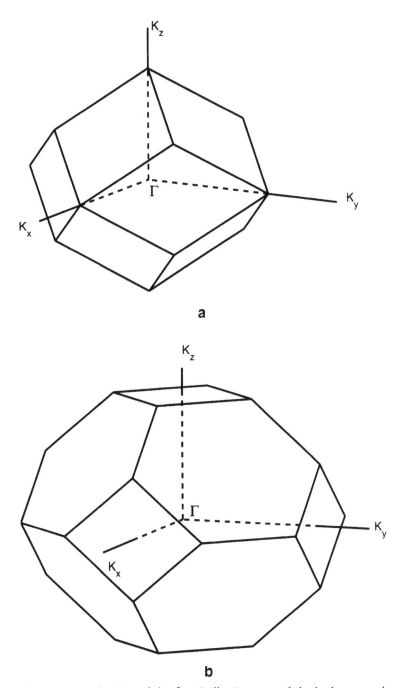

Figure 8.1.2. Sketches of the first Brillouin zones of the body-centered and the face-centered cubic lattices.

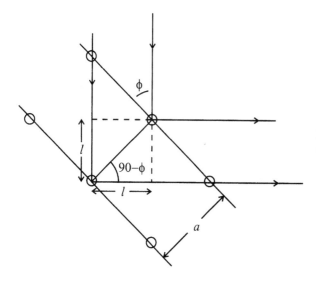

Figure 8.1.3. Bragg reflection condition.

Substituting into Eq. (8.1.35) the expression for the path difference yields

$$\frac{2a \sin \phi}{\lambda} = \frac{\Delta \gamma}{2\pi}. \tag{8.1.36}$$

Constructive interference occurs when the phase difference $\Delta \gamma$ is equal to an integer multiple of 2π. Therefore

$$2a \sin \phi = n\lambda, \tag{8.1.37}$$

which is the Bragg law of scattering.

The Bragg law can be derived in an alternative way that reveals the importance of the first Brillouin zone. It is assumed that constructive interference occurs between the two reflected rays when

$$\Delta \vec{k} \cdot \vec{a} = 2\pi n, \tag{8.1.38}$$

where $\Delta \vec{k}$ is the difference between the incident and the reflected wave vectors, as shown in Figure 8.1.4, and $\vec{a}$ is a direct lattice vector equal in magnitude to the lattice constant. A proof of this assertion, which is equivalent to stating that $\Delta \vec{k}$ is a RLV, is given in Box 8.1.1. k_0 is taken as the incident-wave vector and k is the reflected-wave vector. For an elastic collision, one in which the kinetic energy of the electrons is conserved, the magnitudes of the incident and the reflected k vectors are equal. The law of cosines can be used to determine a relationship between the magnitude of Δk and the incident- and the reflected-wave vectors as

Figure 8.1.4. Condition used to evaluate the change in the k vector after it suffers a Bragg reflection.

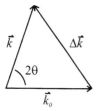

$$\Delta k^2 = k_0^2 + k^2 - 2kk_0 \cos 2\theta,$$

$$\Delta k^2 = 2k_0^2 - 2k_0^2 \cos 2\theta,$$

$$\Delta k^2 = 2k_0^2 (1 - \cos 2\theta). \tag{8.1.39}$$

But the term involving $\cos 2\theta$ can be reexpressed with a double-angle formula to give

$$\Delta k^2 = 2k_0^2(2\sin^2\theta),$$
$$\Delta k = 2k_0\sin\theta. \tag{8.1.40}$$

From the definition of a reciprocal lattice vector, given by Eq. (8.1.31), the vector Δk is clearly a RLV. Subsequently the magnitude of Δk must be equal to an integer multiple of the shortest RLV parallel to Δk. For any family of lattice planes in a Bravais lattice, the RLVs are perpendicular to the direct lattice planes. If the direct lattice planes are spaced a distance a apart, then the smallest RLV is equal to $2\pi/a$. Therefore, since Δk is a RLV, its magnitude must be

$$\Delta k = \frac{2\pi n}{a}. \tag{8.1.41}$$

Equating Eqs. (8.1.40) and (8.1.41) yields then

$$2k_0\sin\theta a = 2\pi n. \tag{8.1.42}$$

But the incident k vector k_0 is

$$k_0 = \frac{2\pi}{\lambda}. \tag{8.1.43}$$

Making this substitution into Eq. (8.1.42) yields

$$2a\sin\theta = n\lambda. \tag{8.1.44}$$

The angle θ is simply equal to the angle ϕ, as can be seen from the construction shown in Figure 8.1.5. Therefore the Bragg law of scattering is recovered.

The above derivation of the Bragg law hinges on the assumption of Eq. (8.1.38). As stated above, this statement implies that $\Delta \vec{k}$ is a RLV. In Box 8.1.1

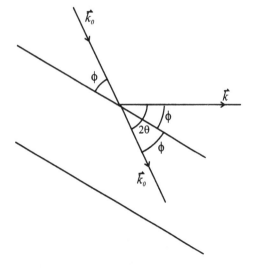

Figure 8.1.5. Construction showing the equivalence of angles ϕ and θ used in deriving the Bragg condition.

Box 8.1.1 Demonstration that Δk is a Reciprocal Lattice Vector

In this box, we demonstrate that the change in the k vector during a Bragg reflection is a reciprocal lattice vector. To see this, consider the event sketched in Figure Box 8.1.1. Let the incident electron beam be represented as e^{ikr} and the outgoing electron beam be $e^{ik'r}$. The phase difference between the two beams is then simply $e^{i(k-k')r}$. From the diagram above, the path difference is simply $r \sin\phi$. Using Eq. (8.1.35),

$$\frac{\text{path difference}}{\lambda} = \frac{\text{phase difference}}{2\pi},$$

yields

$$\frac{2\pi}{\lambda} r \sin\phi = \text{phase difference}.$$

Note that the vector dot product of Δk and r is

$$\Delta \vec{k} \cdot \vec{r} = \frac{2\pi}{\lambda} r \sin\phi,$$

which is precisely the phase difference obtained above. Therefore the phase difference between the incident and the reflected waves is simply given as $\Delta \vec{k} \cdot \vec{r}$. The phase factor difference is $e^{i(k-k')r}$. The amplitude of the scattered wave A can be written in terms of an expansion over a real-space periodic function $n(r)$ as

$$A = \int d^3r\, n(r) e^{-i\Delta \vec{k} \cdot \vec{r}}.$$

But $n(r)$ is periodic and as such can be expanded in terms of a Fourier series. Note first that $n(r + a) = n(r)$, given that the periodicity of the lattice is equal to a. Thus we can expand either $n(r)$ or $n(r + a)$ in a Fourier series and obtain equivalent results. $n(r)$ can be expressed in terms of an expansion over the reciprocal lattice vectors G as

$$n(r) = \sum_G n_G e^{i\vec{G} \cdot \vec{r}}.$$

With this substitution, the expression for the scattering amplitude is

$$A = \sum_G \int d^3r\, n_G e^{i(\vec{G} - \Delta \vec{k}) \cdot \vec{r}}.$$

Note that the scattering amplitude is very small if G is not equal to Δk. Appreciable values of A occur then only when $G = \Delta k$. Hence Δk must be a reciprocal lattice vector.

Figure Box.8.1.1. Sketch of the incident- and the reflected-wave fronts on a crystalline sample.

a proof is given that demonstrates that the change of the k vector $\Delta \vec{k}$ is indeed a RLV. From the above argument, it is clear that Bragg reflection occurs if the change in the wave vector $\Delta \vec{k}$ is equal to a RLV $\vec{G}$; $\Delta \vec{k} = \vec{G}$.

How, though, does the above condition relate to the Brillouin zone and Bragg reflection? The incident k vector k_0 can be written in terms of $\Delta \vec{k}$ and $\vec{k}$ as

$$\vec{k}_0 = \vec{k} - \Delta \vec{k}. \tag{8.1.45}$$

Taking the dot vector product of k_0 on itself gives

$$(\vec{k} - \Delta \vec{k}) \cdot (\vec{k} - \Delta \vec{k}) = (\vec{k}_0 \cdot \vec{k}_0). \tag{8.1.46}$$

When $\vec{G}$ is substituted in for $\Delta \vec{k}$, Eq. (8.1.46) becomes

$$(\vec{k} - \vec{G}) \cdot (\vec{k} - \vec{G}) = k_0^2,$$
$$k^2 - 2\vec{k} \cdot \vec{G} + G^2 = k_0^2. \tag{8.1.47}$$

For an elastic collision, $k^2 = k_0^2$. Therefore

$$G^2 = 2\vec{k} \cdot \vec{G}, \tag{8.1.48}$$

which can be rewritten as

$$\left(\frac{1}{4} G^2 \right) = \vec{k} \cdot \frac{1}{2} \vec{G}, \tag{8.1.49}$$

since

$$\frac{1}{4} G^2 = \frac{1}{4} (2\vec{k} \cdot \vec{G}). \tag{8.1.50}$$

The meaning of the above result can be understood through the construction shown in Figure 8.1.6. We construct a plane perpendicular to the vector $\vec{G}$ at its midpoint. $\vec{G}$ is a RLV since it connects two points A and B in the reciprocal lattice. The line connecting point A to the bisector of G has length

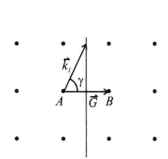

$1/2\,G$. Any vector k_1 drawn from A to the bisector of G satisfies

$$\vec{k}_1 \cdot \frac{1}{2}\vec{G} = |k_1| \left|\frac{1}{2}G\right| \cos\gamma. \qquad (8.1.51)$$

From the geometry of Figure 8.1.6,

$$k_1 \cos\gamma = \frac{1}{2}G. \qquad (8.1.52)$$

Therefore Eq. (8.1.51) becomes

$$\vec{k}_1 \cdot \frac{1}{2}\vec{G} = \left|\frac{1}{2}G\right|^2, \qquad (8.1.53)$$

Figure 8.1.6. Construction used in interpreting the relation of Bragg reflection to the first Brillouin zone.

which is the Bragg condition given by Eq. (8.1.50). Hence any vector originating from the reciprocal lattice point A connecting to the plane that bisects G will satisfy the Bragg condition.

Physically, the above implies the following. We construct the first Brillouin zone by forming the bisectors of the lines or planes connecting a reciprocal lattice point to its nearest-neighbor points. The bisector of the RLV $\vec{G}$ clearly forms part of the surface of the first Brillouin zone associated with the reciprocal lattice point A. From the construction shown in Figure 8.1.6, the vector $\vec{k}_1$ connects a reciprocal lattice point to any point lying on the surface of the first Brillouin zone. Consequently, since all such vectors $\vec{k}_1$ suffer Bragg reflection, then the surface of the Brillouin zone consists of just those k vectors for which Bragg reflection occurs.

What, though, is so important about Bragg reflection? Why is it important to the theory of solids? To answer these questions, let us consider the following argument. The surface of the first Brillouin zone is formed by just those vectors whose magnitude is equal to exactly half of a unit RLV, $1/2\,G$. The unit RLV has magnitude equal to $2\pi/a$, where a is the lattice constant. Hence half of this is simply π/a. If the electron k vector is equal to π/a, then the electronic wave function has the form

$$\Psi \sim e^{-\frac{i\pi x}{a}} \quad \text{or} \quad \Psi \sim e^{+\frac{i\pi x}{a}}. \qquad (8.1.54)$$

The first expression above corresponds to an incident wave while the second corresponds to a reflected wave. The electronic wave function is in general equal to a linear combination of the incident and the reflected waves. Consider the situation in which the linear combination is of equal amplitude. The wave function becomes then

$$\Psi(+) = e^{\frac{i\pi x}{a}} + e^{\frac{-i\pi x}{a}} = 2\cos\left(\frac{\pi x}{a}\right) \qquad (8.1.55)$$

or

$$\Psi(-) = e^{\frac{i\pi x}{a}} - e^{\frac{-i\pi x}{a}} = 2i\sin\left(\frac{\pi x}{a}\right). \qquad (8.1.56)$$

The above expressions are simply standing waves. Hence when the electron k vector lies on the surface of the Brillouin zone, standing waves are produced because of Bragg reflections.

The square of the electronic wave function represents the probability density $\rho(x)$. For the first case $\Psi(+)$, the probability density is simply

$$\rho(+) = \Psi^*(+)\Psi(+) \sim \cos^2\left(\frac{\pi x}{a}\right). \tag{8.1.57}$$

When $x = a$, the probability density attains its maximum value:

$$\rho(+) \sim \cos^2 \pi = +1. \tag{8.1.58}$$

Physically, this means that the electron, when in the $\Psi(+)$ state, is most likely to be found at $x = a$, near the ion centers. The charge spends most of its time in the near vicinity of the ion potential, where the overall potential energy of the electron is low. This is, of course, due to the attractive nature of the ion potential.

The probability density of electrons in the $\Psi(-)$ state is given as

$$\rho(-) = \Psi^*(-)\Psi(-) \sim \sin^2\left(\frac{\pi x}{x}\right). \tag{8.1.59}$$

At $x = a$, near the ion centers, the electronic charge density vanishes in this case: $\rho(-) = 0$. Therefore the electronic charge density is zero near the lattice sites and reaches its maximum midway between the ions where the electron potential energy is large. Subsequently, the $\Psi(-)$ state has a higher energy than the $\Psi(+)$ state.

It is important to note that the presence of Bragg reflection, which occurs for those k vectors that lie on the surface of the first Brillouin zone, leads to standing electron waves. There are two possible nondegenerate electron standing-wave states, $\Psi(+)$ or $\Psi(-)$. The $\Psi(+)$ state has lower energy than the $\Psi(-)$ state. Hence for the same k vector, there are two different states of different energy that form a discontinuity in the energy versus k spectrum. In other words, an energy gap is produced in the $E(k)$ spectrum for just those k states for which Bragg reflection occurs. The situation is sketched in Figure 8.1.7.

From inspection of Figure 8.1.7, it is clear that near the zone edge, $k = \pi/a$, the band structure undergoes a discontinuity in energy. An energy gap forms. The physical origin of this discontinuity is due to Bragg reflection of electrons whose k vectors lie on the surface of the first Brillouin zone. Interestingly, at the zone edge, the curvature of the $E(k)$ relation is negative; the second derivative has a negative sign. The effective mass is defined in terms of the second derivative in accordance with Eq. (8.1.29). Therefore the effective mass at the zone edge is negative. What does this mean physically? A negative mass implies that when an electron is accelerated from an initial k state into a different k state, equal to $k + \Delta k$, the momentum transfer from the electron to the lattice is larger than the applied force supplied to the electron. Although k is increased by the amount Δk through the action of the applied field, an overall decrease in the

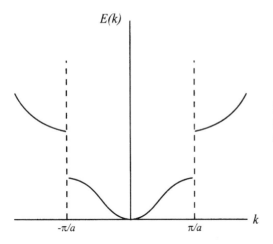

Figure 8.1.7. $E(k)$ relation showing the band structure in the first Brillouin zone and in the extended zone.

forward momentum occurs because of Bragg reflection; the electron increases its k vector by means of an acceleration, yet its forward momentum does not increase. In fact, at the zone edge, for those electron states whose k vectors are equal to $\pm\pi/a$, Bragg reflection occurs and the electron is specularly reflected, leading to an abrupt change in its forward momentum. Consequently, the electron behaves as a particle with negative mass; a positive change in k from the action of an applied field leads to a decrease in the forward momentum of the particle.

In summary, the Brillouin zone plays an important role in solid-state physics since it defines the set of all k vectors for which Bragg reflection occurs. Bragg reflection at the zone edge leads to energy-gap formation in the $E(k)$ spectrum. In the next sections we will examine how the formation of energy gaps occurs in systems with periodic potentials.

8.2 The Kronig–Penney Model

In this section, we examine the solution of the Schroedinger equation in a periodic potential. Interestingly, for any periodic potential, only certain discrete energies are allowed and these energies spread out to form a band. In addition, a band of forbidden energies appears. The forbidden band represents the set of states for which no propagation of the electron wave function can occur. Subsequently, any system that has a translationally invariant potential always gives rise to energy-band formation. In Section 8.3 we extend this result to the nearly-free-electron problem.

Let us examine the solution of the Schroedinger equation for an arbitrary periodic potential, as shown in Figure 8.2.1. For simplicity we will restrict ourselves to only one real space dimension. Nevertheless, the results obtained are general and apply to three dimensions as well. For simplicity, the potential will be assumed to be that of a periodic rectangular barrier as shown in Figure 8.2.1. Any periodic potential can be assumed. However, to obtain a closed-form analytical solution, it is necessary that the potential be regular, for which a closed-form

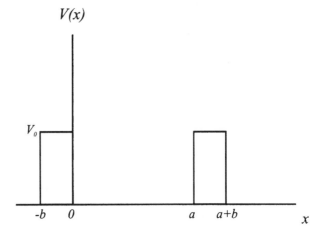

Figure 8.2.1. One dimensional periodic potential of height V_0 and width b. Note that the potential is periodic with a unit cell width of $a + b$.

analytical solution of the Schroedinger equation is available. The potential chosen has magnitude V_0 and width b. The basic periodic unit cell has length equal to $a + b$. The Schroedinger equation for the system is

$$H\Psi = E\Psi, \tag{8.2.1}$$

which, for a one-dimensional system, is

$$\frac{d^2\Psi}{dx^2} + \frac{2m}{\hbar^2}[E - V(x)]\Psi = 0. \tag{8.2.2}$$

The potential $V(x)$ is periodic with period $a + b$. Therefore

$$V(x) = V[x + (a + b)]. \tag{8.2.3}$$

Given that the potential is periodic, then the wave functions must also be periodic and be of Bloch form. Hence Ψ can be taken as

$$\Psi(x) = e^{ikx}u(x). \tag{8.2.4}$$

Substituting Ψ into Eq. (8.2.2) yields

$$\frac{d^2u}{dx^2} + 2ik\frac{du}{dx} - k^2u - \frac{2mV(x)}{\hbar^2}u + \frac{2mE}{\hbar^2}u = 0. \tag{8.2.5}$$

To simplify notation, it is convenient to define

$$\alpha \equiv \sqrt{\frac{2mE}{\hbar^2}}, \qquad \beta \equiv \sqrt{\frac{2m(E - V_0)}{\hbar^2}}. \tag{8.2.6}$$

With these substitutions, Eq. (8.2.5) becomes

$$\frac{d^2u}{dx^2} + 2ik\frac{du}{dx} - \left[k^2 - \alpha^2 + \frac{2mV(x)}{\hbar^2}\right]u = 0. \tag{8.2.7}$$

The solution of the problem now is found in a manner similar to that used in Chapter 2. The general form of the solution is obtained in each region of the structure, between 0 and a and a and b in the sketch shown in Figure 8.2.1. Once the solution is known in each region, the boundary conditions must be applied. From the discussion in Chapter 2, the boundary conditions that apply are the continuity of the wave function and the probability current density. If it is further assumed that the effective mass is the same within each region, then the boundary condition of the conservation of probability current density can be replaced by the continuity of the first derivative of the wave function.

From inspection of Figure 8.1.1, it is clear that the structure can be subdivided into a unit cell that comprises only two unique regions. The entire structure is obtained through repeated translations of the unit cell. The first region is defined as $0 < x < a$, in which the potential $V(x)$ is zero. The second region is defined over the range in x between a and $a + b$ or equivalently between $-b$ and 0. In the second region, the potential is defined as V_0. Schroedinger's equation in regions 1 and 2 becomes then

region 1: $(0 < x < a)$,

$$\frac{d^2 u_1}{dx^2} + 2ik\frac{du_1}{dx} - (k^2 - \alpha^2)u_1 = 0; \tag{8.2.8}$$

region 2: $(-b < x < 0)$,

$$\frac{d^2 u_2}{dx^2} + 2ik\frac{du_2}{dx} - (k^2 - \beta^2)u_2 = 0, \tag{8.2.9}$$

where α and β are defined from Eqs. (8.2.6). The solutions of these two equations are simply plane-wave-like and have the form

$$u_1 = A\,e^{i(\alpha-k)x} + B\,e^{-i(\alpha+k)x} \quad (0 < x < a),$$
$$u_2 = C\,e^{i(\beta-k)x} + D\,e^{-i(\beta+k)x} \quad (-b < x < 0). \tag{8.2.10}$$

The coefficients A, B, C, and D for u_1 and u_2 can be determined from application of the boundary conditions. Assuming that the effective mass of the electron is the same in both regions, the boundary conditions are that the wave functions and their first derivatives must be continuous across each boundary. Since the system is periodic, it is necessary to consider the application of the boundary conditions at only two dissimilar interfaces. The boundary conditions are applied at $x = 0$ and at $x = a$, which is equivalent to $x = -b$. Therefore the boundary conditions are

$$u_1(a) = u_2(-b),$$
$$\frac{du_1(a)}{dx} = \frac{du_2(-b)}{dx},$$
$$u_1(0) = u_2(0),$$
$$\frac{du_1(0)}{dx} = \frac{du_2(0)}{dx}. \tag{8.2.11}$$

Substituting in the expressions for the wave functions u_1 and u_2 into the above four conditions yields

$$A e^{i(\alpha-k)a} + B e^{-i(\alpha+k)a} = C e^{-i(\beta-k)b} + D e^{i(\beta+k)b},$$

$$i(\alpha - k) A e^{i(\alpha-k)a} - i(\alpha + k) B e^{-i(\alpha+k)a} = i(\beta - k) C e^{-i(\beta-k)b}$$
$$-i(\beta + k) D e^{i(\beta+k)b},$$

$$A + B = C + D,$$

$$i(\alpha - k)A - i(\alpha + k)B = i(\beta - k)C - i(\beta + k)D. \qquad (8.2.12)$$

Excluding the trivial solution, we determine the coefficients $A, B, C,$ and D by requiring that the determinant of the system vanish. Recall that the trivial solution is recovered if the matrix of the coefficients has an inverse. The inverse of the matrix does not exist if its determinant vanishes. This leads to

$$0 = \begin{vmatrix} 1 & 1 & 1 & 1 \\ (\alpha - k) & -(\alpha + k) & (\beta - k) & -(\beta + k) \\ e^{i(\alpha-k)a} & e^{-i(\alpha+k)a} & e^{-i(\beta-k)b} & e^{i(\beta+k)b} \\ (\alpha - k) e^{i(\alpha-k)a} & -(\alpha + k) e^{-i(\alpha+k)a} & (\beta - k) e^{-i(\beta-k)b} & -(\beta + k) e^{i(\beta+k)b} \end{vmatrix}.$$

$$(8.2.13)$$

The determinant can be expanded out and simplified to yield

$$-\frac{\alpha^2 + \beta^2}{2\alpha\beta} \sin \alpha a \, \sin \beta b + \cos \alpha a \, \cos \beta b = \cos k(a + b), \quad (E > V_0).$$

$$(8.2.14)$$

Equation (8.2.14) is valid provided that the energy of the electron is greater than the potential energy V_0. If the electron energy is less than V_0, a different equation is obtained. From the definition of β, when the energy is less than V_0, β is imaginary. The sine and the cosine functions in Eq. (8.2.14) then have imaginary arguments. Sine and cosine functions of purely imaginary arguments can be expressed in terms of hyperbolic functions as

$$\sin ix = i \sinh x, \qquad \cos ix = \cosh x. \qquad (8.2.15)$$

Subsequently, if $i\gamma$ is substituted for β in Eq. (8.2.14) and the expressions given in Eqs. (8.2.15) are then used, Eq. (8.2.14) becomes

$$\frac{\gamma^2 - \alpha^2}{2\alpha\gamma} \sinh \gamma b \sin \alpha a + \cosh \gamma b \cos \alpha a = \cos k(a + b), \quad (E < V_0).$$

$$(8.2.16)$$

The two equations, (8.2.14) and (8.2.16), apply under different conditions. Equation (8.2.14) holds for $E > V_0$ while Eq. (8.2.16) holds for $E < V_0$. The latter case corresponds to the electrons in bound states while the former applies to propagating states.

What is the nature of the solutions to Eqs. (8.2.14) and (8.2.16)? In either case, the right-hand side of the equation remains the same, $\cos k(a + b)$. The allowed values of k can be determined from the boundary conditions at infinity of the electronic states. Since the wave functions are Bloch functions, they must retain their plane-wave character at either positive or negative infinity. Therefore propagating states result only if k is real. Otherwise, if k is imaginary, the exponent of the term e^{ikx} becomes real. As such, exponentially increasing or decreasing functions of result that are clearly nonpropagating. Hence, for propagating solutions, the cosine term on the right-hand side of Eqs. (8.2.14) and (8.2.16) must always be between $+1$ and -1.

Consider the first case above, $V_0 < E < \infty$. In this case, only propagating solutions are expected. For simplicity, let us define

$$K_2 = -\frac{(\alpha^2 + \beta^2) \sin \beta b}{2\alpha\beta}, \tag{8.2.17}$$

$$K_1 = \cos \beta b. \tag{8.2.18}$$

With these definitions, Eq. (8.2.14) becomes

$$K_2 \sin \alpha a + K_1 \cos \alpha a = \cos k(a + b). \tag{8.2.19}$$

Equation (8.2.19) can be rewritten with some trigonometric relations. Note that

$$A\cos(\alpha a - \delta) = A\cos \alpha a \cos \delta + A \sin \alpha a \sin \delta. \tag{8.2.20}$$

Therefore Eq. (8.2.19) can be further simplified if additional definitions are made:

$$K_1 = A\cos \delta, \qquad K_2 = A \sin \delta. \tag{8.2.21}$$

Then,

$$\begin{aligned} K_1^2 &= A^2 \cos^2 \delta, \\ K_2^2 &= A^2 \sin^2 \delta, \\ A &= \sqrt{K_1^2 + K_2^2}, \\ \delta &= \tan^{-1}(K_2/K_1). \end{aligned} \tag{8.2.22}$$

The sum of the squares of K_1 and K_2 can be rewritten in terms of a new quantity K_3. Defining K_3 as

$$K_3 = \sqrt{K_1^2 + K_2^2} \tag{8.2.23}$$

and using Eq. (8.2.20), we find that the left-hand side of Eq. (8.2.19) is equal then to

$$K_2 \sin \alpha a + K_1 \cos \alpha a = K_3 \cos(\alpha a - \delta). \tag{8.2.24}$$

Therefore, Eq. (8.2.14) finally can be rewritten as

$$K_3 \cos(\alpha a - \delta) = \cos k(a + b). \tag{8.2.25}$$

Substituting into Eq. (8.2.23) for K_1 and K_2 Eqs. (8.2.18) and (8.2.17) respectively, we find that K_3 is

$$K_3 = \left[\cos^2 \beta b + \sin^2 \beta b \frac{(\alpha^2 + \beta^2)^2}{4\alpha^2 \beta^2} \right]^{\frac{1}{2}},$$

$$K_3 = \left[\frac{4\alpha^2 \beta^2 \cos^2 \beta b + \alpha^4 \sin^2 \beta b + 2\alpha^2 \beta^2 \sin^2 \beta b + \beta^4 \sin^2 \beta b}{4\alpha^2 \beta^2} \right]^{\frac{1}{2}},$$

$$K_3 = \left[1 + \frac{(\alpha^2 - \beta^2)^2 \sin^2 \beta b}{4\alpha^2 \beta^2} \right]^{\frac{1}{2}}. \tag{8.2.26}$$

δ can be found from

$$\delta = \tan^{-1}(K_2/K_1), \tag{8.2.27}$$

which is equal to

$$\tan \delta = -\frac{(\alpha^2 + \beta^2) \tan \beta b}{2\alpha\beta}. \tag{8.2.28}$$

Therefore Eq. (8.2.14) can be written in a convenient form as

$$\left[1 + \frac{(\alpha^2 - \beta^2)^2}{4\alpha^2 \beta^2} \sin^2 \beta b \right]^{\frac{1}{2}} \cos(\alpha a - \delta) = \cos k(a + b). \tag{8.2.29}$$

Once δ and K_3 are known in terms of α and β, which are in turn functions of the energy and the potential, the energy as a function of k can be plotted. Before we examine the solution of Eq. (8.2.29), let us first determine a simplified version of Eq. (8.2.16). Then we will plot both solutions.

In the case in which the energy is less than the potential barrier, $E < V_0$, the circular functions become hyperbolic functions. K_1' and K_2' can be defined in a similar manner as

$$K_1' = \frac{(\gamma^2 - \alpha^2) \sinh \gamma b}{2\alpha\gamma},$$

$$K_2' = \cosh \gamma b. \tag{8.2.30}$$

A similar analysis to the above leads again to expressions for a magnitude K_3' and phase angle δ' as

$$K_3' = \left[1 + \frac{(\alpha^2 + \gamma^2)^2}{4\alpha^2 \gamma^2} \sinh^2 \gamma b \right]^{\frac{1}{2}},$$

$$\tan \delta' = \frac{\alpha^2 + \gamma^2}{2\alpha\gamma} \tanh \gamma b, \tag{8.2.31}$$

which becomes

$$\left[1 + \frac{(\alpha^2 + \gamma^2)^2}{4\alpha^2\gamma^2} \sinh^2 \gamma b\right]^{\frac{1}{2}} \cos(\alpha a - \delta') = \cos k(a + b). \qquad (8.2.32)$$

It is interesting to examine the solutions of Eqs. (8.2.29) and (8.2.32). In both cases the left-hand side has an amplitude greater than 1 under all conditions. From the condition on k, the right-hand side can never exceed $+1$ or be less than -1 for propagating solutions to exist. Therefore, there are real values of k for only certain values of the argument of the cosine function on the left-hand side of these equations. Clearly, there will also be imaginary solutions for k for different values of the argument of the cosine function. The solutions to Eqs. (8.2.29) and (8.2.32) then involve both real and imaginary values of k. This is always true provided that the coefficient on the left-hand side is greater than 1. Inspection of both equations, though, reveals that the coefficient must always be greater than 1. Therefore it is required that both real and imaginary solutions for k always occur in a system consisting of a periodic potential.

Note that $\alpha^2 - \beta^2$ is given as

$$\alpha^2 - \beta^2 = \frac{2mE}{\hbar^2} - \frac{2m(E - V_0)}{\hbar^2} = \frac{2mV_0}{\hbar^2} = \text{constant}. \qquad (8.2.33)$$

$z(E)$ is defined as

$$z(E) = \left[1 + \frac{(\alpha^2 - \beta^2)^2}{4\alpha^2\beta^2} \sin^2 \beta b\right]^{\frac{1}{2}} \cos(\alpha a - \delta). \qquad (8.2.34)$$

When the expressions for α and β are substituted into Eq. (8.2.34), $z(E)$ becomes

$$z(E) = \left[1 + \frac{V_0^2}{4E(E - V_0)} \sin^2 \sqrt{\frac{2m(E - V_0)}{\hbar^2}} b\right]^{\frac{1}{2}} \cos\left(\sqrt{\frac{2mE}{\hbar^2}} a - \delta\right). \qquad (8.2.35)$$

The function $z(E)$ is equal, though, to $\cos k(a + b)$. When k is

$$k = \frac{n\pi}{a + b}, \qquad (8.2.36)$$

where n is an integer, the cosine function $\cos k(a + b)$ equals ± 1. Therefore $z(E)$ must also equal ± 1. There are real values of k in the range for which $z(E)$ lies between $+1$ and -1. When $z(E) > +1$ or when $z(E) < -1$, then $\cos k(a + b)$ is either greater than $+1$ or less than -1, respectively. Consequently, k must then be imaginary. Under these conditions, there are no allowed values of the energy since imaginary k violates the boundary conditions at infinity, as discussed above.

The function $z(E)$ versus E is plotted in Figure 8.2.2. The shaded regions represent the range of energies for which the corresponding k vector is imaginary:

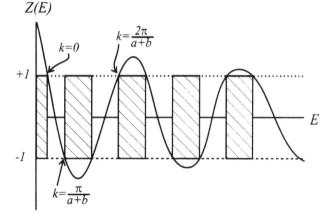

Figure 8.2.2. Sketch of the function $z(E)$, as defined in the text, versus E. The shaded areas correspond to regions for which no real solutions for k exist.

regions for which $z(E) > +1$ or $z(E) < -1$. This range of energies comprises the set of forbidden energies, in which there are no allowed propagating electron states. This range of forbidden energies is called the forbidden band. The unshaded regions in Figure 8.2.2 correspond to the allowed energies: that range of energies for which there are propagating electron states.

It is important to realize that the above derivation is quite general in that no restrictions on the magnitude of V_0, a, or b exist. The only requirement is that the system be translationally invariant. Hence, for a periodic potential, energy bands, both forbidden and propagating, will always occur. This result is quite general. As we will see in Section 8.3, any periodic potential of any form will always lead to energy-band formation.

It is interesting to examine the allowed energies as a function of k. The allowed energy spectrum as a function of k is plotted in Figure 8.2.3. As can be seen from this figure, it is clear that energy gaps are introduced in the $E(k)$ relation. Discontinuities in the allowed energies arise in going along the k-vector spectrum. Remember, all real k-vector values lead to propagating electron states in the crystal. Subsequently, these states are allowed and all of them have corresponding real energies, as shown in the figure. However, the spectrum is not continuous since, for some energies, only imaginary k vectors occur. As discussed above, the electron states corresponding to imaginary values of k are evanescent; they cannot propagate through the crystal. Such states are forbidden and represent a gap in the allowed energies of the system.

In summary, in this section, we have found that a periodic potential profile always gives rise to an energy-band spectrum that contains both forbidden (those for which only imaginary k vectors exist) and allowed (those corresponding to real k vectors) energy bands. What, though, does a state with an imaginary k vector represent physically? This can be easily understood from consideration of a plane-wave state with a pure imaginary propagation constant k. Such a state has the dependence e^{ikx}. If k is purely imaginary, then the function becomes either an increasing or decreasing exponential. Clearly, the resulting electronic state described by a Bloch function does not propagate. The electron wave

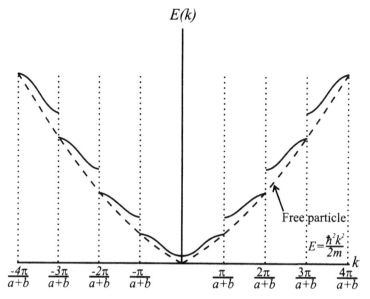

Figure 8.2.3. Sketch of $E(k)$ versus k for both a free particle (dashed curve) and a particle in a periodic potential. Note the presence of gaps at multiples of $\pi/(a+b)$ that correspond to the imaginary values of k in Fig. 8.2.2.

function must decay exponentially (the exponentially increasing solution is not physically acceptable for the reasons discussed in Chapter 2), leading to electron localization.

8.3 The Nearly-Free-Electron Model

In Section 8.2 we examined the solution of the Schroedinger equation for a periodic potential formed by a repeating rectangular potential barrier of arbitrary height and width. It was found that the solution of the Schroedinger equation in such a system always leads to a set of allowed and forbidden energies of the electron called the allowed and the forbidden energy bands of the system. Although the system examined in Section 8.2 was only one dimensional, its extension to three dimensions is apparent. However, one might wonder if the results obtained in Section 8.2 apply to any arbitrary periodic potential. In this section we examine the solution of the Schroedinger equation for any arbitrary periodic potential. It will again be seen that energy-band formation is a necessary consequence of a periodic system.

To solve the Schroedinger equation for an arbitrary, yet periodic potential, it is useful to expand the potential in a Fourier series. Since the potential is periodic, the wave functions must also be periodic. Therefore the wave functions can also be expanded in terms of a Fourier series. Once again, we restrict ourselves to solving the problem in one dimension for simplicity. No generality is lost since the problem can be readily extended to three dimensions and a similar, although slightly more complicated, result is obtained.

The Schroedinger equation in one dimension is then

$$H\Psi = E\Psi$$

$$-\frac{\hbar^2}{2m}\frac{d^2}{dx^2}\Psi + [V(x) - E]\Psi = 0,$$

$$\frac{d^2\Psi}{dx^2} + \frac{2m}{\hbar^2}[E - V(x)]\Psi = 0. \tag{8.3.1}$$

The potential $V(x)$ is periodic, although otherwise completely arbitrary. Since the potential is periodic, it can be expanded in terms of a Fourier series. For simplicity, let us define a periodic function $f(x)$ that has the same periodicity as the potential and thus the lattice. The potential $V(x)$ can be written in terms of $f(x)$ as

$$-\frac{2m}{\hbar^2}V(x) = \gamma f(x). \tag{8.3.2}$$

To simplify the notation further, let

$$k_0^2 = \frac{2mE}{\hbar^2}. \tag{8.3.3}$$

With these substitutions, the Schroedinger equation becomes

$$\frac{d^2\Psi}{dx^2} + [k_0^2 + \gamma f(x)]\Psi = 0. \tag{8.3.4}$$

Since the function $f(x)$ is periodic, it can be expanded in terms of a Fourier series. Let us assume that $f(x)$ can be expanded as

$$f(x) = \sum_{n=-\infty}^{+\infty} C_n e^{\frac{-i2\pi nx}{a}}, \tag{8.3.5}$$

where a is assumed to be the lattice constant or the spatial period of the potential. Equation (8.3.5) follows readily from the definition of a complex Fourier series. From the orthonormality properties of a Fourier series, the coefficients C_n in the expansion can be readily determined as

$$C_n = \frac{1}{a}\int_0^a f(x) e^{\frac{i2\pi nx}{a}} dx. \tag{8.3.6}$$

The electronic wave functions must also be periodic with periodicity a as well as satisfy the Bloch theorem. The Bloch functions developed in Section 7.4 are

$$\Psi_{nk}(r) = e^{ikr} u_{nk}(r), \tag{8.3.7}$$

where the function $u_{nk}(r)$ must be periodic. Reducing these functions to only one dimension and restricting to only one band (the subscript n then will be neglected) gives

$$\Psi_k(x) = e^{ikx} u_k(x). \tag{8.3.8}$$

$u_k(x)$ is the periodic part of the wave function and can thus be expanded in a Fourier series. This expansion is given as

$$u_k(x) = \sum_{n=-\infty}^{+\infty} b_n e^{\frac{-i2\pi nx}{a}}. \tag{8.3.9}$$

Therefore the electronic wave functions are

$$\Psi(x) = u_k(x) e^{ikx} = e^{ikx} \sum_n b_n e^{\frac{-i2\pi nx}{a}}. \tag{8.3.10}$$

If the electrons are completely free, then the potential $V(x)$ is zero and $\gamma = 0$. The solutions of the Schroedinger equation given by Eq. (8.3.4) are simply plane waves of the form

$$\Psi(x) = b_0 e^{ik_0 x}. \tag{8.3.11}$$

When γ is not zero, the Schroedinger equation is simply given by Eq. (8.3.4). If γ is small, which corresponds to the weak potential assumption, then the Bloch function can be expanded as

$$\Psi(x) = b_0 e^{ikx} + \gamma e^{ikx} \sum_{n \neq 0} b_n e^{\frac{-i2\pi nx}{a}}, \tag{8.3.12}$$

where the part corresponding to the function $u_k(x)$ is simply

$$u_k(x) = b_0 + \gamma \sum_{n \neq 0} b_n e^{\frac{-i2\pi nx}{a}}. \tag{8.3.13}$$

The wave function given by Eq. (8.3.12) is simply equal to the sum of a plane-wave part and a periodic part of the order of the potential γ. Hence the wave function can be thought of as being a plane wave with a small periodic correction. This is the origin of the name, nearly-free-electron model, since the wave functions correspond to nearly free electrons. Substituting the expression for the wave function into the Schroedinger equation, Eq. (8.3.4), yields

$$0 = \frac{d^2}{dx^2} \left[b_0 e^{ikx} + \gamma e^{ikx} \sum_{n \neq 0} b_n e^{\frac{-i2\pi nx}{a}} \right] + \left(k_0^2 + \gamma \sum_{n \neq 0} C_n e^{\frac{-i2\pi nx}{a}} \right)$$

$$\times \left[b_0 e^{ikx} + \gamma e^{ikx} \sum_{n' \neq 0} b_{n'} e^{\frac{-i2\pi n'x}{a}} \right]. \tag{8.3.14}$$

Expanding out the second-derivative term yields

$$-k^2 b_0 \, e^{ikx} - k^2 \gamma \, e^{ikx} \sum_{n \neq 0} b_n \, e^{\frac{-i2\pi nx}{a}} - \frac{4\pi^2 n^2}{a^2} \gamma \, e^{ikx} \sum_{n \neq 0} b_n \, e^{\frac{-i2\pi nx}{a}}$$

$$+ 2\gamma k \frac{2\pi n}{a} \, e^{ikx} \sum_{n \neq 0} b_n \, e^{\frac{-i2\pi nx}{a}}. \tag{8.3.15}$$

Substituting the above result into Eq. (8.3.14) and combining like terms yields

$$0 = (k_0^2 - k^2) b_0 \, e^{ikx} + \gamma \sum_{n \neq 0} (k_0^2 - k^2) b_n \, e^{i\left(k - \frac{2\pi n}{a}\right)x}$$

$$+ b_0 \gamma \sum_{n \neq 0} C_n \, e^{i\left(k - \frac{2\pi n}{a}\right)x} + \gamma^2 \sum_{n \neq 0} \sum_{n' \neq 0} C_n b_n' \, e^{i\left(k - \frac{2\pi n}{a} - \frac{2\pi n'}{a}\right)x}$$

$$+ \gamma \sum_{n \neq 0} \left(\frac{4\pi n}{a} k - \frac{4\pi^2 n^2}{a^2}\right) b_n \, e^{i\left(k - \frac{2\pi n}{a}\right)x}. \tag{8.3.16}$$

The term of the order of γ^2 can be neglected when the potential is weak, as is our assumption. As in the cases we discussed in Chapter 4, we will calculate the wave function to only first order in the perturbation but determine the energy to both first and second order. So in calculating Ψ, we neglect the terms in γ^2. We can simplify Eq. (8.3.15) further by combining the last and the second terms by using the following argument. The second and the last terms above are

$$\gamma \sum_{n \neq 0} (k_0^2 - k^2) b_n \, e^{i\left(k - \frac{2\pi n}{a}\right)x} + \gamma \sum_{n \neq 0} \left(\frac{4\pi n}{a} k - \frac{4\pi^2 n^2}{a^2}\right) b_n \, e^{i\left(k - \frac{2\pi n}{a}\right)x}. \tag{8.3.17}$$

Define the following relation:

$$k = k_n + \frac{2\pi n}{a}; \tag{8.3.18}$$

then,

$$k^2 = k_n^2 + \frac{4 k_n \pi n}{a} + \frac{4\pi^2 n^2}{a^2}. \tag{8.3.19}$$

The second and the last terms can then be combined as

$$\gamma \sum_{n \neq 0} b_n \, e^{i\left(k - \frac{2\pi n}{a}\right)x} \left[k_0^2 - k^2 + \frac{4\pi n}{a} k - \frac{4\pi^2 n^2}{a^2} \right]. \tag{8.3.20}$$

We can write the above substituting $k_n + 2\pi n/a$ in for k:

$$= \gamma \sum_{n \neq 0} b_n \, e^{i\left(k - \frac{2\pi n}{a}\right)x} \left[k_0^2 - k_n^2 - \frac{4 k_n \pi n}{a} - \frac{4\pi^2 n^2}{a^2} + \frac{4\pi nk}{a} - \frac{4\pi^2 n^2}{a^2} \right].$$

$$\tag{8.3.21}$$

Further simplification finally gives for the second and the last terms,

$$= \gamma \sum_{n \neq 0} b_n e^{i\left(k - \frac{2\pi n}{a}\right)x} [k_0^2 - k_n^2]. \tag{8.3.22}$$

With these substitutions, Eq. (8.3.16) reduces to

$$b_0(k_0^2 - k^2) e^{ikx} + \gamma \sum_{n \neq 0} [(k_0^2 - k_n^2)b_n + b_0 c_n] e^{i\left(k - \frac{2\pi n}{a}\right)x} = 0. \tag{8.3.23}$$

Next multiply Eq. (8.3.23) by

$$e^{-ik_m x}, \tag{8.3.24}$$

where k_m is defined as

$$k_m = k - \frac{2\pi m}{a}, \tag{8.3.25}$$

and integrate over a full period. Equation (8.3.23) becomes

$$b_0(k_0^2 - k^2) \int_0^a e^{ikx} e^{-ik_m x} \, dx + \gamma \sum_{n \neq 0} [(k_0^2 - k_n^2)b_n + b_0 C_n] \int_0^a e^{\frac{2\pi i(m-n)x}{a}} \, dx = 0, \tag{8.3.26}$$

since

$$e^{-ik_m x} = e^{-i\left(k - \frac{2\pi m}{a}\right)x}. \tag{8.3.27}$$

Therefore Eq. (8.3.26) becomes

$$b_0(k_0^2 - k^2) \int_0^a e^{\frac{i2\pi mx}{a}} \, dx + \gamma \sum_{n \neq 0} [(k_0^2 - k_n^2)b_n + b_0 C_n] \int_0^a e^{\frac{i2\pi(m-n)x}{a}} \, dx = 0. \tag{8.3.28}$$

If $m = 0$, then the last integral in Eq. (8.3.28) vanishes for all n since n does not equal zero in the sum. We can simplify Eq. (8.3.28) by noting the following relationships. In this case

$$\int_0^a e^{-\frac{i2\pi nx}{a}} \, dx = 0, \quad n \neq 0; \qquad \int_0^a \, dx = a, \quad n = 0;$$
$$\int_0^a e^{\frac{i2\pi(m-n)x}{a}} \, dx = \begin{cases} 0 & \text{if } m \neq n \\ a & \text{otherwise} \end{cases}. \tag{8.3.29}$$

Equation (8.3.28) reduces to

$$b_0(k_0^2 - k^2)a = 0, \tag{8.3.30}$$

since the integrand in the first term is simply 1 when $m = 0$. In general, then, k must equal k_0.

When $m \neq 0$, then the first integral vanishes and the second integral is zero unless $m = n$. In this case, Eq. (8.3.28) becomes

$$\gamma\left[(k_0^2 - k_m^2)b_m + b_0 C_m\right]a = 0, \tag{8.3.31}$$

or, when we solve for the coefficients b_m,

$$b_m = \frac{b_0 C_m}{(k_m^2 - k^2)}, \tag{8.3.32}$$

where $k = k_0$ has been used. The wave function $\Psi(x)$ can now be written with the results for the coefficients in the expansion and the definition given by Eq. (8.3.12) as

$$\Psi(x) = b_0 \, e^{ikx}\left[1 - \gamma \sum_{n \neq 0} \frac{C_n}{(k^2 - k_n^2)} e^{\frac{-i2\pi nx}{a}}\right]. \tag{8.3.33}$$

It remains to find the first- and the second-order corrections to the energy. The first-order correction is zero since $k = k_0$. Hence the energy is given as to first order:

$$E = \frac{\hbar^2 k^2}{2m} = \frac{\hbar^2 k_0^2}{2m}. \tag{8.3.34}$$

The second-order correction to the energy is obtained from the term of second order in γ in Eq. (8.3.16). Again noting that the last term can be combined with the second term in Eq. (8.3.16), we find that Eq. (8.3.16) becomes

$$b_0(k_0^2 - k^2)e^{ikx} + \gamma \sum_{n \neq 0}\left[(k_0^2 - k_n^2)b_n + C_n b_0\right]e^{ik_n x}$$

$$+ \gamma^2 \sum_{n \neq 0}\sum_{n' \neq 0} b_{n'} C_n \, e^{\frac{-i2\pi(n+n')x}{a}} e^{ikx} = 0, \tag{8.3.35}$$

where k_n is defined as

$$k_n = k - \frac{2\pi n}{a}. \tag{8.3.36}$$

We can simplify Eq. (8.3.35) by multiplying by e^{-ikx} and integrating over a full period 0 to a to give

$$b_0(k_0^2 - k^2)a + \int_0^a \gamma \sum_{n \neq 0}\left[(k_0^2 - k_n^2)b_n + C_n b_0\right]e^{\frac{-i2\pi nx}{a}} \, dx$$

$$+ \gamma^2 \int_0^a \sum_{n \neq 0}\sum_{n' \neq 0} b_{n'} C_n \, e^{\frac{-i2\pi(n+n')x}{a}} \, dx = 0. \tag{8.3.37}$$

We can simplify Eq. (8.3.36) by using the results given by Eqs. (8.3.29) to

$$b_0(k_0^2 - k^2)a + \gamma^2 \sum_{n' \neq 0} b_{n'} C_{-n'} a = 0. \tag{8.3.38}$$

The sum over n' can be replaced by a sum over $-n$. The coefficient $C_{-n'}$ though, is simply equal to $C_{n'}^*$. This follows from the definition of the function $f(x)$:

$$f(x) = \sum_{n=-\infty}^{\infty} C_n e^{\frac{-i2\pi n x}{a}} \quad n \to -n \quad \sum_{n=-\infty}^{\infty} C_{-n} e^{\frac{i2\pi n x}{a}}, \tag{8.3.39}$$

where the substitution of $-n$ for n has been made. $f(x)$ is a real function since the potential is a real function. Therefore $f(x) = f^*(x)$. Subsequently,

$$f^*(x) = \sum_{n=-\infty}^{+\infty} C_n^* e^{\frac{i2\pi n x}{a}}. \tag{8.3.40}$$

Comparing Eq. (8.3.40) with Eq. (8.3.39) shows that $C_{-n} = C_n^*$. Hence $C_{n'}^*$ can be substituted in for $C_{-n'}$ in Eq. (8.3.38). Equation (8.3.38) becomes then

$$b_0(k_0^2 - k^2) + \gamma^2 \sum_{n' \neq 0} b_{n'} C_{n'}^* = 0. \tag{8.3.41}$$

But the coefficients $b_{n'}$ are given by Eq. (8.3.32). (Note n' and m are dummy indices.) Substituting them into Eq. (8.3.41) yields

$$(k_0^2 - k^2) - \gamma^2 \sum_n \frac{C_n C_n^*}{(k^2 - k_n^2)} = 0. \tag{8.3.42}$$

Therefore k_0^2 can be solved to obtain

$$k_0^2 = k^2 + \gamma^2 \sum_{n \neq 0} \frac{C_n^* C_n}{(k^2 - k_n^2)}. \tag{8.3.43}$$

But $k_n = k - 2\pi n/a$ and $E = \hbar^2 k_0^2 / 2m$. Multiplying Eq. (8.3.43) by $\hbar^2 / 2m$ gives

$$\frac{\hbar^2 k_0^2}{2m} = \frac{\hbar^2 k^2}{2m} + \frac{\hbar^2 \gamma^2}{2m} \sum_{n \neq 0} \frac{C_n^* C_n}{\left[k^2 - \left(k - \frac{2\pi n}{a}\right)^2\right]}. \tag{8.3.44}$$

V_n is defined as

$$V_n = -\sqrt{\frac{\hbar^2 \gamma^2}{2m}} C_n. \tag{8.3.45}$$

With this definition, the energy E becomes

$$E = \frac{\hbar^2 k^2}{2m} + \sum_{n \neq 0} \frac{|V_n|^2}{\left[\frac{\hbar^2 k^2}{2m} - \frac{\hbar^2}{2m}\left(k - \frac{2\pi n}{a}\right)^2\right]}. \tag{8.3.46}$$

Equation (8.3.46) for the energy includes a free-electron component, that given by $\hbar^2 k^2/2m$, and a correction term that involves V_n^2. k is related to k_n through Eq. (8.3.36):

$$k_n = \left(k - \frac{2\pi n}{a}\right). \tag{8.3.47}$$

The denominator in the second term vanishes in Eq. (8.3.46) if $k^2 = k_n^2$. Under this condition, the second term diverges and the energy becomes indeterminant. Additionally, the wave function also diverges under these conditions, as can be seen from Eq. (8.3.33). The coefficients b_n in the expansion of the wave function diverge, and subsequently the product of γ and b_n is no longer small.

The presence of a singularity in the solution for the energy and the wave function has significant physical implications. The energy and the wave function diverge when $k = \pm k_n$. If the negative sign is chosen, then

$$k = -k_n, \tag{8.3.48}$$

which can be reexpressed with the definition of k_n as

$$k = -k + \frac{2\pi n}{a},$$

$$2k = \left(\frac{2\pi n}{a}\right). \tag{8.3.49}$$

Therefore the value of k at which the energy diverges is

$$k = \frac{\pi n}{a}. \tag{8.3.50}$$

If the positive sign is chosen instead, then

$$k = k - \frac{2\pi n}{a}, \tag{8.3.51}$$

which requires that $n = 0$. The value of the energy then diverges whenever the k vector is equal to some integer multiple of π/a. From the discussion in Section 8.1, however, k vectors equal to an integer multiple of π/a lie on the surface of the first Brillouin zone. Subsequently, there is a discontinuity in the energy versus k relation for those k vectors that lie on the surface of the Brillouin zone, implying the existence of an energy gap at these points. As in the Kronig–Penney model, energy gaps are found to form. In the above situation, we have made no assumptions about the nature of the potential except that it is periodic. Energy gaps are found as a natural consequence of this periodicity.

Let us further determine the nature of the solution at the points $k = n\pi/a$. These points lie at the edge of the first Brillouin zone. Therefore we expect then that the energy will be multivalued when k lies on the surface of the Brillouin zone. To obtain an expression for the energy near the zone edge, it is useful to examine the behavior of the wave function near the singularity. We can approximate the expression for the wave function $\Psi(x)$ by recognizing that the coefficient b_n for which $k_n = k$ dominates the expansion when k lies on the surface of the Brillouin zone. (The surface of the Brillouin zone is often referred to as the zone edge.) Under these conditions, the sum over n can be approximated by a single term, that of b_n alone. $\Psi(x)$ becomes

$$\Psi(x) = b_0\, e^{ikx} + \gamma b_n\, e^{ikx} e^{\frac{-i2\pi nx}{a}}. \tag{8.3.52}$$

The exponential factors can be combined as

$$\Psi(x) = b_0\, e^{ikx} + \gamma b_n\, e^{ik_n x}. \tag{8.3.53}$$

We can obtain an expression for the energy of the electron at the zone edge by solving the Schroedinger equation by using Eq. (8.3.53) for the wave function. On substitution of Eq. (8.3.53) for $\Psi(x)$ into the Schroedinger equation, Eq. (8.3.4) becomes

$$-b_0 k^2\, e^{ikx} - \gamma b_n k_n^2\, e^{ik_n x} + \gamma b_0 \sum_{n' \neq 0} C_{n'}\, e^{i\left(k - \frac{2\pi n'}{a}\right)x}$$

$$+ \gamma^2 b_n \sum_{n' \neq 0} C_{n'}\, e^{i\left(k - \frac{2\pi n}{a} - \frac{2\pi n'}{a}\right)x} + k_0^2 b_0\, e^{ikx} + k_0^2 \gamma b_n\, e^{ik_n x} = 0. \tag{8.3.54}$$

Collecting terms and simplifying obtains

$$b_0\left(k_0^2 - k^2\right) e^{ikx} + \gamma b_n\left(k_0^2 - k_n^2\right) e^{i\left(k - \frac{2\pi n}{a}\right)x}$$

$$+ \gamma b_0 \sum_{n' \neq 0} C_{n'}\, e^{i\left(k - \frac{2\pi n'}{a}\right)x} + \gamma^2 b_0 \sum_{n' \neq 0} C_{n'}\, e^{i\left(k - \frac{2\pi n}{a} - \frac{2\pi n'}{a}\right)x} = 0. \tag{8.3.55}$$

We can simplify Eq. (8.3.55) further by multiplying through by e^{-ikx} and integrating from 0 to a. This becomes

$$\int_0^a b_0\left(k_0^2 - k^2\right) dx + \int_0^a \gamma^2 b_n \sum_{n' \neq 0} C_{n'}\, e^{-i\left(\frac{2\pi n}{a} + \frac{2\pi n'}{a}\right)x} dx = 0, \tag{8.3.56}$$

where the second and the third terms in Eq. (8.3.55) evaluate to zero since the integral in these terms is over the exponential of $i2n\pi x/a$ in both cases. Evaluating the remaining integrals in Eq. (8.3.56) by using the results of Eqs. (8.3.29) yields

$$b_0\left(k_0^2 - k^2\right) + \gamma^2 b_n C_n^* = 0. \tag{8.3.57}$$

Alternatively, Eq. (8.3.55) can be reduced if it is multiplied by

$$e^{-ik_n x} \tag{8.3.58}$$

and is integrated over a full period to obtain

$$C_n b_0 + (k_0^2 - k_n^2) b_n = 0. \tag{8.3.59}$$

Equations (8.3.57) and (8.3.59) are two linear equations in two unknowns, b_0 and b_n. The factors b_0 and b_n can be determined then by the requirement that the determinant of the coefficients vanish. This becomes

$$\begin{vmatrix} (k_0^2 - k^2) & \gamma^2 C_n^* \\ C_n & (k_0^2 - k_n^2) \end{vmatrix} = 0, \tag{8.3.60}$$

which evaluates to

$$(k_0^2 - k^2)(k_0^2 - k_n^2) - \gamma^2 C_n^* C_n = 0. \tag{8.3.61}$$

Equation (8.3.61) can be solved for k_0^2 through use of the quadratic formula. By assigning $k_0^2 = x$, we find that Eq. (8.3.61) becomes

$$x^2 - 2k^2 x + k^2 k_n^2 - \gamma^2 C_n^* C_n = 0. \tag{8.3.62}$$

Solving for x yields

$$x = \frac{2k^2 \pm \sqrt{4k^4 - 4k^2 k_n^2 + 4\gamma^2 C_n^* C_n}}{2}. \tag{8.3.63}$$

We can simplify Eq. (8.3.63) by adding zero to the right-hand side in the form $k_n^2 - k_n^2$. This becomes

$$k_0^2 = \frac{1}{2} \left[(k^2 + k_n^2) \pm \sqrt{(k^2 - k_n^2)^2 + 4\gamma^2 C_n^* C_n} \right]. \tag{8.3.64}$$

Finally, the energy spectrum is determined to be

$$E(k) = \frac{\hbar^2}{4m} \left\{ k^2 + \left(k - \frac{2\pi n}{a} \right)^2 \pm \sqrt{\left[k^2 - \left(k - \frac{2\pi n}{a} \right)^2 \right]^2 + \left(\frac{4m|V_n|}{\hbar^2} \right)^2} \right\}. \tag{8.3.65}$$

The solution for the energy at the zone edge, $k = k_n = n\pi/a$, can now be obtained from the above result. The first term under the square root in Eq. (8.3.65) becomes zero. The first two terms are simply $2k^2$. With these substitutions, Eq. (8.3.65) becomes

$$E = \frac{\hbar^2}{2m} \left(\frac{n\pi}{a} \right)^2 \pm V_n. \tag{8.3.66}$$

The resulting $E(k)$ relation shows a clear gap at the edge of the first Brillouin zone equal in magnitude to $2V_n$.

In conclusion, then, we have found that energy gaps appear at the edge of the first Brillouin zone for any arbitrary periodic potential. The only requirement made in determining Eq. (8.3.66) is that the potential be periodic with period a. No other restriction was made in its derivation. The presence of an energy gap at the zone edge is a natural consequence of any periodic potential and arises physically from Bragg reflection at the Brillouin zone boundary. In Section 8.4 we will discuss how the presence of bandgaps influences the physical properties of solids.

8.4 Energy Bandgaps and the Classification of Solids

The presence of energy bandgaps drastically affects the physical properties of a material. In the previous sections we have found that energy gaps arise at the Brillouin zone boundaries because of Bragg reflection. Within these bandgaps no allowed energy states are available. Hence these regions are often called forbidden bands. The allowed energy bands can be either conduction or valence bands. The conduction bands correspond to bands formed from overlapping atomic levels that are either totally empty or partially empty. The valence bands are those bands formed from overlapping filled atomic levels. As the name implies, electrons within the conduction band are free to propagate throughout the crystal. It is often convenient to classify a solid in terms of its electrical conductivity, which is related to the properties of the conduction and the valence bands. In this section, we discuss how the electrical conductivity is influenced by the presence of bandgaps and the properties of the conduction and the valence bands.

In Chapter 6, we learned that an electrical current arises when the electronic distribution has a net momentum in the field direction; there are electrons whose momentum is uncompensated. In a free-electron system, there is no limitation on the number of available states outside of the Fermi sphere. This is because the free-electron energy spectrum is simply

$$E = \frac{\hbar^2 k^2}{2m},$$
(8.4.1)

which is continuous; for all energies there is a corresponding real k vector. In the free-electron system, if the electrons are accelerated by an applied field, the carriers can be accelerated into new k states. This is because there is a continuum of energy states into which the electrons can be placed. As a result, the distribution is shifted in k space.

What would happen, though, if an energy gap existed in the system? Can a current always flow? As we will see below, the answer to this question will help our understanding of the differences between metals, semimetals, semiconductors, and insulators. Let us now consider current flow in the context of band theory.

In equilibrium, no current flows since the sum over all the k vectors of the electrons in the system vanishes:

$$\sum_i \vec{k}_i = 0.$$
(8.4.2)

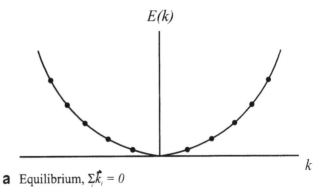

a Equilibrium, $\sum_i \vec{k}_i = 0$

Figure 8.4.1. Conduction processes in a single band of a solid. In case a. the net momentum of the system is zero; no current flows. In case b. the distribution is shifted in k space, leading to current flow.

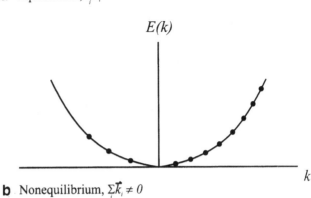

b Nonequilibrium, $\sum_i \vec{k}_i \neq 0$

For every electron directed in the positive direction with a k vector $+\vec{k}$, there is another electron, on average, with a compensating k vector equal to $-\vec{k}$. Subsequently, the net current flow is zero. If the electrons are considered to be confined to a finite set of states within an energy band, then if the system is in equilibrium, there are equal numbers of electrons with positive $\vec{k}$ as with negative $\vec{k}$. The situation is sketched in Figure 8.4.1.a. On the application of an applied external electric field, the electrons are accelerated by the action of the field, resulting in a change in the k vectors of each of the electrons within the ensemble. From the results of Eqs. (8.1.21), the k vector of each electron changes by an amount $qF\Delta t/\hbar$, as

$$\vec{k}_f = \vec{k}_i + \frac{q\vec{F}\Delta t}{\hbar}. \tag{8.4.3}$$

Provided that the final state k_f is not filled and that it exists, an electron can be accelerated into that state. As a result, a current can flow. For a free-electron gas, every electron can undergo a transition since the displaced Fermi sphere intersects available, previously unfilled states. The electrons are displaced in k space such that the sum over all k vectors is no longer zero, as shown in Figure 8.4.1.b.

However, if forbidden energy bands exist then empty states are not necessarily available. For instance, if the highest occupied energy band is completely filled and separated from all other bands by a nonzero energy gap, then no current can flow. This is a general result; a completely filled energy band cannot conduct any current since there are no vacancies that the electrons can move into. This can be understood as follows. The highest-energy electron cannot be accelerated into another state since only forbidden states exist at energies greater than its energy. Since the band is completely filled, all the states lying at energies less than the highest-energy state are also filled. Therefore there are no vacancies anywhere within the band and no transitions can occur because of the Pauli principle. The sum over all the k vectors in the ensemble is zero both before and after the application of the field. No current flows in the system. Such a system is called an insulator.

For a current to flow in a solid, there must be vacancies as well as electrons within the band itself. The band must be only partially filled. A highly conductive material then must contain at least one partially empty band. On the application of an external field, the electrons can be accelerated into vacancies within the band. Hence the net momentum of the system changes under the application of the field, resulting in a current flow. An example of such a system is a metal. A metal contains at least one partially empty band.

What causes a material to behave then as a metal or an insulator? From the above discussion, a metal has a partially empty band, even at absolute zero degrees. The lowest-lying collective energy state of a metal is such that not all the vacancies in the highest-energy band are filled. Consequently, there are more electronic states than electrons in the highest-energy band even when the electrons are assembled into their lowest-energy configuration. Conversely, an insulator is a material in which the number of electronic states in each band matches precisely the number of electrons in the band. Therefore the lowest-energy configuration of the system is one in which all the states within bands that contain any electrons are completely filled. Additionally, in an insulator the highest-energy band that contains electrons and is thus completely filled is separated by a large amount of energy from the next-nearest higher-energy band, which is completely empty.

Semiconductors and semimetals are intermediate cases of the two extremes of metals and insulators. A semimetal is formed when the first conduction band, defined as the lowest-lying unfilled band, overlaps a completely filled valence band. Under the application of an applied electric field, electrons within the valence band can be accelerated into the conduction band, giving rise to a current. A semiconductor is similar to an insulator in that it has a completely filled valence band and a completely empty conduction band at zero degrees. Subsequently, a semiconductor behaves as an insulator at zero degrees. The primary difference between a semiconductor and an insulator is that the energy gap between the valence and the first conduction band is sufficiently small that electrons can be thermally or optically excited into the first conduction band, leaving vacancies behind in the valence band. The number of electrons present in the conduction band depends then on the temperature of the material as well as on whether any optical excitation is present.

A semiconductor can conduct a current at temperatures above absolute zero or under optical excitation. The origin of electrical conduction in semiconductors is due to the excitation of electrons out of the valence band into the conduction band. Carriers are introduced into the conduction band, leaving behind vacancies in the valence band. As a result, both bands are partially empty and both bands can then conduct a current. More is discussed in Section 8.5 about conduction in an unfilled valence band.

Is there a means of determining whether a material forms a metal, insulator, semiconductor, or semimetal from its fundamental properties? To some extent, the answer is yes. However, there are many materials that defy easy classification, and a detailed analysis of their chemistry is required for ascertaining their electrical properties. Nevertheless, there are some simple rules that can be helpful in classifying solids into one of the above four categories. It should be stressed that these rules are most useful, not as a steadfast means of predicting the electrical properties of materials, but more as a means of illustrating the effects the band structure has on electrical conduction.

The key issue that determines whether a material is a metal or an insulator is whether the highest-energy band that contains electrons at zero temperature is completely filled or only partially filled. If the band is completely filled, it cannot conduct. If no carriers populate the next-highest-energy band, no current can flow in this system under application of an external field and the material behaves as an insulator. As mentioned above, as the temperature increases, if electrons can be excited into the next energy band, then the material is a semiconductor. If the energy gap is too large so that no carriers can be excited across the gap, then the material behaves as an insulator. On the other hand, in a metal the highest-energy band that contains electrons at zero temperature is only partially filled. To determine then whether a material is a metal or an insulator it is necessary to count the number of available states and the number of electrons present in the unit cell. If the number of available states is greater than the number of electrons, then the material is a metal. Otherwise it may be an insulator, semiconductor, or semimetal, depending on the bandgap energy separation with the next-nearest band.

If the tight-binding approximation is valid, then the allowed energy bands in the crystal arise from the overlap of the atomic levels. For every electronic level of the isolated atom, there are $2N$ states per band for the collective crystal, where N is the number of atoms present. The factor of 2 arises from the Pauli spin degeneracy of every electronic state. Any solid with only one valence electron per atom contributes only N electrons to the collective crystal. Since there are $2N$ electronic states per band, the highest-energy band is only partially filled, even at absolute zero. As a result, such a material is a metal. If, on the other hand, two valence electrons are contributed per atom, then the resulting band is filled and the material is not metallic. The primary exception to this rule is when another band overlaps the highest filled band. If the overlap is substantial, that is, it occurs throughout most of the Brillouin zone, then the material behaves as a metal. If the overlap is limited to only a small portion of the band, then the material is semimetallic.

The structure into which a material crystallizes influences its electronic properties as well. For example, some materials such as the tetravalent elements, those having a valence of four, can be either semiconductors, semimetals, insulators, or even metals, depending on the crystal structure and the resulting band structure of the material. For example, silicon and germanium are always semiconductors but tin can be either a semiconductor or a metal, depending on its crystal structure. Carbon, on the other hand, forms a nearly perfect insulator when it crystallizes as a diamond. For a more elaborate discussion of these and other materials the reader is referred to the references.

Before we end our discussion of the various material types, it is important to examine them in light of the bond picture of solids. In particular, it is of interest to examine the behavior of semiconductors in light of the bond model. One can envision a solid from either the electronic band approach, in which the electronic states are labeled in accordance with their k vector, or from the atomic bonding approach, in which the real-space picture of the bond orbitals is examined. The energy gap between the first conduction and the highest valence band in a semiconductor can be thought of as that amount of energy needed for an electron in a bound valence orbital to escape. In other words, the energy bandgap can be thought of as the amount of energy necessary to break a bond in the crystal. The stronger the bond, the greater the bandgap, resulting in fewer carriers promoted to the conduction band. Therefore, for every electron promoted to the conduction band from the valence band, there is a vacancy left behind in the valence band. This can alternatively be examined from the atomic bond model. When an interatomic bond is broken, the electron is no longer localized and is free to move about the crystal, leaving behind an unfilled bond into which another electron can enter. We will see in Section 8.5 that the existence of vacant bonds within the crystal, called holes, greatly affects the physical properties of semiconductors.

8.5 **Holes**

One of the more interesting yet confusing issues in solid-state physics is that of holes. A hole is defined as a vacant orbital in an otherwise filled band. Holes are of great importance since they provide an additional current-carrying species in a solid aside from electrons. One must be careful, though, to not overcount the amount of current delivered to the outside world when considering the current flow arising from the simultaneous transit of both electrons and holes. This point is developed below. First, we must define a hole and specify its properties.

The most important property of a hole is that it acts in an applied electric field as if it has unit positive charge $+q$. This behavior arises from the fact that a partially filled band can conduct a current. In Section 8.4, we found that a completely filled band cannot conduct an electrical current since there are no vacancies into which the electrons can be accelerated. Hence the sum over all the k vectors of the system is zero, and the system has no net momentum or current.

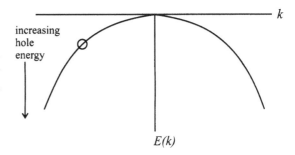

Figure 8.5.1. Hole energy plotted as a function of wave vector. Note that hole energy increases as the hole moves downward in the diagram.

If one electron is missing from an orbital of wave vector $\vec{k}_e$ in an otherwise filled band containing N orbitals, then the total wave vector of the band is simply equal to $-\vec{k}_e$. Rather than consider the band as filled by $N-1$ electrons we say that it is filled instead by one hole of wave vector $\vec{k}_h = -\vec{k}_e$. By definition, the wave vector of a hole is equal to the negative of the missing electron's wave vector.

The energy of the hole increases the lower in the band the hole moves, as shown in Figure 8.5.1. Physically, this arises from the fact that it takes more work to remove an electron from a deep orbital than from a shallow one. Let us pause here and elaborate on this point. As discussed in Section 8.4, in both semiconductors and insulators there is an energy gap between the topmost valence band and the lowest conduction band. At absolute zero degrees, the valence band is completely filled with electrons and the conduction band is completely empty in these materials. Depending on the magnitude of the energy gap at higher temperatures, electrons can be thermally promoted to the conduction band, leaving behind vacancies or holes in the valence band. For an electron to be promoted to the conduction band it must have at least an energy equal to the bandgap difference between the conduction and the valence bands. If both bands are assumed to be free-electron-like, in other words the energy versus k relation is parabolic, then the smallest energy gap occurs at $k = 0$, as shown in Figure 8.5.2. As the electron moves away from the $k = 0$ point in either direction, the energy difference between the electron and the lowest-lying energy state in the conduction band increases. Therefore more energy is required for promoting the electron from the valence band into the conduction band as the magnitude of the electron k vector increases in the valence band. Hence holes have greater energy at larger k or, equivalently, at energies deeper in the valence band. A nice phyisical analogy to this situation is that of bubbles in water. The deeper the water is, the greater the amount of energy needed to form a bubble.

The valence band sketched in Figure 8.5.2 has a negative electron mass since its second derivative with respect to k has negative curvature. The mass of the hole, though, is defined as a positive quantity. This is because the hole energy is essentially equal to the negative of the electron energy, that is,

$$E_h(k_h) = -E_e(k_e), \tag{8.5.1}$$

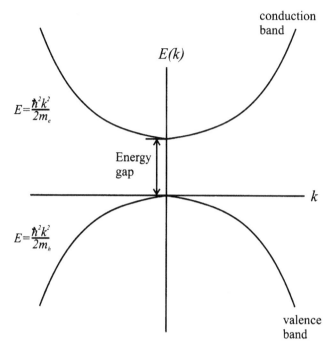

Figure 8.5.2. Sketch of the conduction and the valence bands in a semiconductor in which it is assumed that both the conduction and the valence bands are free-electron-like. In the present example, the energy gap between the conduction and the valence bands is formed at $k = 0$.

as discussed above. (It takes more energy to remove an electron from a low-energy orbital to form a hole than from a high-energy orbital. Hence the energies of the electron and the hole bands are negatives of one another.)

Consider next what happens when an electric field is applied to a partially filled band. The field acts to accelerate the electrons in one direction and in effect accelerates the hole in the opposite direction, as can be seen from Figure 8.5.3. Let an electric field point along the negative k direction. The resulting electric force on the electrons in the valence band is then directed in the positive k direction. As a result, the electrons move to states of more positive k value. When an electron accelerates from its initial state into its final state, it leaves behind a vacancy in its initial state and fills a vacancy in its final state. Hence the vacancy in effect moves in the direction opposite to that of the electrons.

What then is the current carried by the hole? If the band is completely filled, then no current flows. The current density arising from the motions of all the electrons within the band is given by summation over the entire band as

$$\vec{j} = -q \int \vec{v}(k) \frac{2}{(2\pi)^3} \, d^3k. \tag{8.5.2}$$

The current density for a filled band is clearly zero since the integral of the vector velocity over the entire ensemble must be zero. We can relate the current density corresponding to the motions of the electrons within a partially filled band to that due to the motion of the vacancies by recognizing that

$$\vec{j}_{\text{filled}} - \vec{j}_{\text{occupied}} = \vec{j}_{\text{vacancies}}. \tag{8.5.3}$$

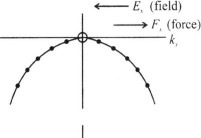

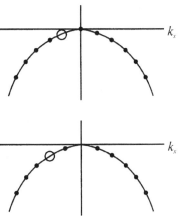

Figure 8.5.3. Time progression of the motion of electrons and a hole in the valence band under the influence of an electric field aligned in the negative k_x direction.

But the current density due to the filled band is zero. Consequently the current density due to the motions of the vacancies must be negative to that due to the motions of the electrons. The current density due to the motion of the holes is then

$$\vec{j}_{\text{holes}} = +q \int \vec{v}(k) \frac{2}{(2\pi)^3} \, d^3k. \qquad (8.5.4)$$

A hole behaves like a positively charged particle of positive mass. Therefore the current flows in the same direction as the holes while the current flows in the opposite direction from the electron motion.

Holes are of course not real particles in the sense that one can hold one in one's hand. They of course do not exist outside of a solid. Nevertheless, one can model the dynamics of a nearly filled band by assuming that it is essentially empty except for some positively charged particles called holes. In other words, holes are a fictitious means of representing the current-carrying dynamics of a nearly filled band.

Finally, one might ask about the nature of the current when both electrons and holes contribute to the conduction process. What is the amount of current delivered to the outside circuit when both electrons and holes are present in the semiconductor? If an electron and a hole are generated within the material, the resulting electron–hole pair delivers a complete charge q to the external circuit. Therefore each individual electron and hole created as a pair induce only a portion of the total charge q in the external circuit such that the total charge arising from the transit of both carriers taken together is q. Another way of understanding this is to recall that current flow is defined as the passage of a charge through a complete circuit. When an electron–hole pair is generated within a semiconductor, the total path within the semiconductor is traversed partly by the electron and partly by the hole (remember that the electron and the hole move in opposite directions) such that the total distance traveled is equal to the total semiconductor length.

Alternatively, if an electron enters the semiconductor from the outside world at one terminal and leaves it at the other terminal, it effectively delivers a full charge q to the circuit. Only when an electron–hole pair is generated within the semiconductor between the two external terminals through which the current flows do we encounter the situation discussed above. In Chapter 10, we will return to this point when we discuss generation and recombination processes in semiconductors.

8.6 Cellular Methods of Calculating Band Structure

Let us turn now to the subject of determining the energy-band structures of realistic crystalline materials. In the previous sections, we have discussed energy-band formation from both the tight-binding approach and the nearly-free-electron method. Both formulations have been used to demonstrate qualitatively the formation of energy bands in crystals. We have succeeded in showing that for any periodic one-dimensional potential, allowed and forbidden energy bands are produced. Generalization to three dimensions is, of course, straightforward. It should now be apparent to the reader that energy bands occur in any crystalline solid, since by definition it has a periodic potential. The question is, though, how does one calculate the nature of the energy bands in a realistic crystalline material? The methods examined up to now are greatly simplified and are not readily applicable to realistic crystalline systems. How then can the electronic energy bands in real systems be determined? To answer this question, we must extend our results to three dimensions and consider realistic atomic core potentials.

We can simplify the problem of calculating realistic energy-band structures by exploiting the translational symmetry of the lattice. The form of the Bloch functions implies that it is necessary to construct the solution within only the primitive unit cell of the reciprocal lattice or the first Brillouin zone. The solution for all other k vectors outside of the Brillouin zone differs by only a phase factor e^{ikR}. In other words, every k vector can be reduced to an equivalent k vector within the first Brillouin zone; the first Brillouin zone contains all the unique k vectors in the system.

Similarly, the solutions for $\Psi(x)$ in real space need be specified within only the primitive unit cell of the direct lattice. From translational symmetry, once the solution within the primitive unit cell of the direct lattice is known everywhere, it is known throughout the crystal as well. However, not every solution of the Schroedinger equation within the primitive unit cell leads to an acceptable solution for the crystal. The solution must also satisfy the correct boundary conditions at the cell edge. As we found in Chapter 2, the wave function and its first derivative must be continuous across the interface to yield an acceptable solution. In terms of the Bloch function, the continuity of the wave function and its first derivative yield

$$\Psi(r) = e^{-ikR} \Psi(r + R),$$

$$\hat{n}(r) \cdot \vec{\nabla}\Psi(r) = e^{-ikR}\hat{n}(r + R) \cdot \vec{\nabla}\Psi(r + R), \tag{8.6.1}$$

where $\hat{n}$ is an outward normal vector and $\vec{r}$ and $\vec{r} + \vec{R}$ are vectors that connect points on the surface of the unit cell in real space.

The problem we are faced with is to solve the Schroedinger equation in three dimensions within the primitive unit cell subject to the boundary conditions on the surface of the cell. This technique is commonly referred to as a cellular method, since it requires finding the solution within only one cell. The problem is similar to that of the nearly-free-electron model. In solving for the energy in the nearly-free-electron model, the solution of the Schroedinger equation was found throughout the unit cell in real space and the boundary conditions were applied. Similarly, in the cellular methods, the solution of the Schroedinger equation must be found throughout the entire unit cell and the boundary conditions at all points along the surface must be applied. The easiest way to construct the solution for a three-dimensional unit cell is simply to take linear combinations of plane-wave states since these readily satisfy the boundary conditions at the edges of the unit cell. However, to represent the solution near the ion cores, where the potential changes strongly, the superposition of many plane waves is needed. Alternatively, one could choose as the basis set atomic core wave functions, which are guaranteed to satisfy the solution near the ion cores. Unfortunately, it is difficult to satisfy the boundary conditions at the surface of the primitive unit cell by using these functions. Clearly, a trade-off exists. To satisfy the boundary conditions along the surface of the cell, simple plane waves are the most suitable solutions. However, in the near vicinity of the ionic cores, where the potential changes greatly, the choice of plane-wave solutions is not fully suitable. We will see below that hybrid wave functions, combining both plane-wave and atomic core state aspects, provide the best choice of basis states in constructing the solution. First, as a means of illustrating cellular methods, we seek a solution by using plane waves only.

From the discussion in Chapter 1, it was found that plane-wave states form a basis set of wave functions. Any arbitrary periodic function can be expanded in terms of a series of plane-wave states (this is of course just the Fourier theorem). The electronic wave functions within a crystal can then be represented as a linear combination of normalized plane waves as

$$|k\rangle = \frac{1}{\sqrt{\Omega}} e^{ikr}, \tag{8.6.2}$$

where Ω represents the volume over which the state is defined. The electronic wave function $\Psi(r)$ can be expanded in terms of the set of $|k\rangle$'s as

$$\Psi(r) = \sum_k a_k |k\rangle. \tag{8.6.3}$$

With the above substitution for $\Psi(r)$ the Schroedinger equation becomes

$$-\frac{\hbar^2}{2m}\nabla^2\Psi + V(r)\Psi = E\Psi,$$

$$\sum_k -\frac{\hbar^2}{2m}\langle k'|\nabla^2 a_k|k\rangle + \sum_k \langle k'|V(r)a_k|k\rangle = \sum_k Ea_k\langle k'|k\rangle, \tag{8.6.4}$$

but $\langle k'|k\rangle = \delta_{k,k'}$ and $\nabla^2|k\rangle = -k^2|k\rangle$, which follows readily from the definition of plane waves and Dirac notation (see Chapter 1). With these substitutions, Eqs. (8.6.4) become

$$\frac{\hbar^2}{2m}a_{k'}k'^2 + \sum_k a_k\langle k'|V(r)|k\rangle = Ea_{k'}. \tag{8.6.5}$$

The potential $V(r)$ is periodic and can be written as

$$V(r) = \sum_j v(r - r_j), \tag{8.6.6}$$

where the sum is over the total number of atoms present and r_j labels each lattice site. Therefore the potential at the point r is found from the sum of all the ion core potentials at each of their respective points r_j, with separation distance $r - r_j$. Substituting Eq. (8.6.6) in for $V(r)$ and writing the wave functions in the coordinate representation, we find that Eq. (8.6.5) becomes

$$\frac{\hbar^2}{2m}a_{k'}k'^2 + \frac{\sum_k a_k}{\Omega}\int e^{-i\vec{k'}\cdot\vec{r}}\sum_j v(r - r_j)e^{i\vec{k}\cdot\vec{r}}\,\mathrm{d}^3r = Ea_{k'}. \tag{8.6.7}$$

We can simplify Eq. (8.6.7) by multiplying and dividing the term involving the potential $v(r - r_j)$ by $e^{i(\vec{k'}-\vec{k})\cdot\vec{r}}$ and by interchanging the integral and the sum to give

$$\frac{\hbar^2}{2m}a_{k'}k'^2 + \frac{\sum_k a_k}{\Omega}\sum_j e^{-i(\vec{k'}-\vec{k})\cdot\vec{r_j}}\int e^{-i(\vec{k'}-\vec{k})\cdot(\vec{r}-\vec{r_j})}v(r - r_j)\,\mathrm{d}^3r = Ea_{k'},$$

$$\tag{8.6.8}$$

where Ω is the total volume of the crystal over which the wave functions are normalized. The total volume can be reexpressed as the product of the total number of unit cells in the crystal multiplied by the volume of one unit cell:

$$\Omega = N\Omega_0, \tag{8.6.9}$$

where Ω_0 is the volume of the unit cell and N is the total number of unit cells present in the crystal. The term

$$\frac{1}{\Omega} \sum_j e^{-i(\vec{k}'-\vec{k})\cdot\vec{r}_j} \tag{8.6.10}$$

is called the structure factor. The structure factor is a sum over all j, the atom positions. The vector $\vec{r}_j$, which describes each lattice site, can be written as the sum of two different vectors, $\vec{T}_j$ and $\vec{\delta}_j$. $\vec{T}_j$ is defined as a vector that describes the position of the unit cell in question relative to a coordinate axis while $\vec{\delta}_j$ is a vector that describes each lattice site within the unit cell. Therefore the position of the jth lattice site can be written as

$$\vec{r}_j = \vec{T}_j + \vec{\delta}_j. \tag{8.6.11}$$

The location of every ion core can be represented then as the sum of two vectors; the first vector connects the primitive unit cell of interest to a fixed point while the second connects the potential center to a reference point within the unit cell. The sum over δ_j is of course the same for each unit cell since all unit cells are equivalent. Subsequently, this sum can be taken outside of the sum over the lattice translations. The structure factor can be broken into two separate sums. The first sum is over the unit cell itself while the second sum is over all the unit cells within the crystal. The structure factor is

$$\sum_j e^{-i(\vec{k}'-\vec{k})\cdot\vec{\delta}_j} \sum_{n_1,n_2,n_3} e^{-i(\vec{k}'-\vec{k})\cdot(n_1\hat{a}+n_2\hat{b}+n_3\hat{c})}, \tag{8.6.12}$$

where the sum over j is taken within one cell. The vector $(n_1\hat{a}+n_2\hat{b}+n_3\hat{c})$ is the general expression for $\vec{T}$. It is important to note that the second sum vanishes unless $(k - k')$ is a reciprocal lattice vector, RLV. From the definition of a RLV, the dot product of a RLV and a direct lattice vector such as T is always equal to some integer multiple of 2π. This will give a nonzero value of the second sum. When $(k - k')$ is not a RLV, the structure factor becomes the product of three geometric series, one or more of which must vanish. The matrix element of the potential $\langle k'|V|k\rangle$ is zero then unless $k' - k$ is a RLV. For a given k' the only nonvanishing coefficients a_k in the Schroedinger equation are those for which k differs from k' by a RLV. A different set of equations is then obtained for each k' in the Brillouin zone as

$$\frac{\hbar^2}{2m}a_{k'}k'^2 + \sum_k a_k\langle k'|V(r)|k\rangle = Ea_{k'}. \tag{8.6.13}$$

Equation (8.6.13) has in principle many energy eigenvalues. The energy versus k spectrum $E(k)$ is found by the solution of Eq. (8.6.13) for each wave vector k'. This ultimately yields the complete band structure in the Brillouin zone. Of course, the Brillouin zone can be subdivided into an infinite number of points.

The solution is determined in practice at sufficient points such that the $E(k)$ relation is known to the level of precision desired.

Although the above approach is conceptually simple, it is highly computationally intensive since many k' values and many terms in the expansion of the wave function are required for correctly obtaining $E(k)$. This is of course due to the fact that the electronic wave functions are different from plane waves near the potential centers. In the vicinity of the ion cores we find the solution by taking linear combinations of the plane-wave states. For complex potentials, this requires many terms in the expansion for achieving reasonable accuracy. Of course, the advantage of the plane-wave method is that it readily satisfies the boundary conditions at the cell boundary, but in doing so it sacrifices both elegance and accuracy near the ion cores.

The opposite approach would be to expand the electronic wave functions in terms of the atomic core states. The atomic core states also form a basis set, as we found in Chapter 3. In Chapter 3, it was determined that the solutions of the Schroedinger equation for a spherically symmetric potential are given by the spherical harmonics $\Psi_{l,m}(\theta, \phi)$ as

$$\Psi_{l,m}(r) = R(r)Y_{l,m}(\theta, \phi), \qquad (8.6.14)$$

where $R(r)$ is the radial part of the wave function. This choice of basis set has an overall spherical symmetry, as shown in Figure 8.6.1. The electronic wave functions for the crystal $\Psi(r)$ are well behaved near the ion cores and thus represent the solution well in that region. However, these solutions must also satisfy the boundary conditions at the zone edge: continuity of $\Psi(r)$ and its first derivative.

From the overall symmetry of the wave functions it is clear that the boundary conditions cannot be readily satisfied since the cell edge is not spherical in shape. Computationally, the boundary conditions can be readily applied at only a finite number of points on the surface of the unit cell. Again, many terms are needed in the expansion of $\Psi(r)$ to satisfy the boundary conditions at the cell boundary. In addition, the use of a spherical potential is not fully justified, since it neglects

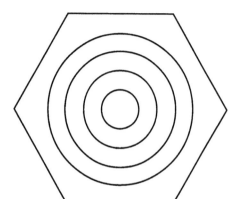

Figure 8.6.1. Spherically symmetric wave functions within the first Brillouin zone.

the effects of other neighboring ions on the overall potential. Away from the immediate vicinity of each ion core, the potential becomes far less spherical in shape. Subsequently, the approximation of a spherical potential everywhere within the primitive unit cell is not always useful.

From the above discussion, it is clear that neither plane waves nor atomic core states provide an adequate basis set for constructing the form of the solution within the primitive unit cell of the lattice. Plane waves readily satisfy the boundary conditions at the cell edge and represent the solution well away from the potential center. However, near the ion cores, the wave functions are different from plane-wave states and are more spherical in nature. In the vicinity of the potential centers, the solution is more appropriately represented in terms of atomic core states. This choice of basis, though, has the disadvantage of not readily matching the boundary conditions. Clearly, the best approach then is to choose a basis set that combines the distinct advantages of the plane waves and atomic core states.

The form of the solution is chosen to be plane-wave-like far away from the potential centers and similar to atomic core states in the vicinity of the ions. This choice of solution is consistent with a potential that resembles a muffin tin. The potential is assumed to be similar to that of an atomic core within a certain region, say $r < r_0$, and to be constant everywhere else, $r > r_0$. The potential can be pictured as looking something like a muffin tin, as shown in Figure 8.6.2.

Mathematically, the muffin-tin potential is of the form

$$v(r) = V(r - R), \quad |r - R| < r_0 \text{ (core region)},$$
$$v(r) = 0, \qquad\qquad |r - R| > r_0 \text{ (interstitial region)}. \tag{8.6.15}$$

Within the interstitial region, far from the ion cores, the potential is essentially flat and is reasonably well modeled by a constant potential. In this region, the solutions are best approximated by plane waves. Within the core regions, the

Figure 8.6.2. Schematic representation of a muffin-tin potential.

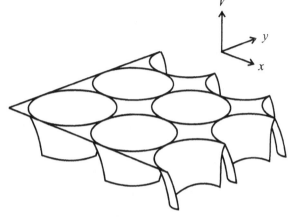

potential is essentially spherical and is better modeled by the ionic potentials. The solution here is best approximated by the atomic core state functions. A hybrid solution is thus formed. We construct the total wave function by linking the plane-wave solution within the interstitial region to the atomic core state solution within the core region. At the spherical boundary $r = r_0$, the plane-wave part and the ion core states must be matched, which requires some expansion. Nevertheless, it is easier to match the boundary conditions on the wave function and its first derivative on the surface of a sphere rather than on the surface of the primitive unit cell, which is typically a polyhedron.

The wave functions are generally some hybrid of plane waves and atomic core states. To show how they are constructed, let us define the plane-wave and the core states as follows. Let the plane-wave states be represented as kets of the form $|k\rangle$, and the core states as $|t, j\rangle$, where the index t labels each particular core state and j labels each atom position. We choose the wave functions such that they are orthogonal to the core states. The wave functions, called orthogonal plane-wave (OPW) states, can be constructed as

$$\text{OPW} = |k\rangle - \sum_{t,j} |t, j\rangle\langle t, j \mid k\rangle, \tag{8.6.16}$$

where

$$\langle t, j \mid k\rangle = \frac{1}{\sqrt{\Omega}} \int \Psi_t^*(r - r_j)e^{ikr} \, \mathrm{d}^3 r. \tag{8.6.17}$$

The OPW states are orthogonal to the core states by design, which can be demonstrated as follows. Multiply the OPW state on the left by $|t, j\rangle$ and integrate over all space. The OPW state becomes

$$\int [\langle t, j \mid k\rangle - \sum_{t,j}\langle t, j|t, j\rangle\langle t, j \mid k\rangle] \, \mathrm{d}^3 r. \tag{8.6.18}$$

The sum over t and j of the second term can be simplified. Recall that the core states are orthogonal. Hence the matrix element of $|t, j\rangle$ on itself must simply be 1 when $|t, j\rangle$ and $\langle t, j|$ are the same state and zero otherwise. The sum must vanish for all j except $j = j$. With this simplification, expression (8.1.18) becomes

$$\int [\langle t, j \mid k\rangle - \langle t, j \mid k\rangle] \, \mathrm{d}^3 r = 0. \tag{8.6.19}$$

Clearly, the states $|t, j\rangle$ are orthogonal to all other core states.

In general, the potential $V(r)$ within the region $r < r_0$ is not exactly equal to that of a single free ion. This is because the potentials arising from neighboring ions overlap each core region. Consequently, the net potential is somewhat different from that of a single isolated ion. The contributions from all the neighboring ions is such that their collective effect provides a nearly constant shift in the potential. Therefore the energies are all shifted uniformly from the single-free-atom values. A uniform shift in the energy has no real effect on the solution since the reference value can always be redefined. Therefore the wave functions are essentially those given by the atomic core states.

The energy of the collective system can be obtained in the following way. Let us first define an operator P, called the projection operator, as

$$P = \sum |t, j\rangle\langle t, j|. \tag{8.6.20}$$

The OPW state can then be rewritten with P as

$$\text{OPW} = |k\rangle - \sum_{t,j} |t, j\rangle\langle t, j||k\rangle$$

$$= (1 - P)|k\rangle. \tag{8.6.21}$$

The electronic states of the system Ψ_k can then be constructed from an expansion over the OPW states as

$$\Psi_k = \sum_k a_k(1 - P)|k\rangle$$

$$= (1 - P)\sum_k a_k|k\rangle, \tag{8.6.22}$$

where the a_k's are the expansion coefficients. When the above expansion is used, the electronic states of the system can be determined, and the expansion is found to rapidly converge.

Let us now seek the solution of the Schroedinger equation by using the OPW wave functions Ψ_k. For convenience, let us define ϕ to be

$$\phi = \sum_k a_k|k\rangle, \tag{8.6.23}$$

which is called a pseudo-wave-function. Outside of the core regions, the projection operator is simply equal to zero. In this region,

$$\Psi_k = \phi, \qquad r > r_0. \tag{8.6.24}$$

The wave function Ψ_k can be written then in terms of ϕ as

$$\Psi_k = (1 - P)\phi, \tag{8.6.25}$$

where use has been made of Eq. (8.6.22). Substituting this expression for Ψ_k into the Schroedinger equation gives

$$-\frac{\hbar^2}{2m}\nabla^2(1 - P)\phi + V(r)(1 - P)\phi = E(1 - P)\phi$$

$$-\frac{\hbar^2}{2m}\nabla^2\phi + V(r)\phi - \left[-\frac{\hbar^2}{2m}\nabla^2 + V(r)\right]P\phi + EP\phi = E\phi. \tag{8.6.26}$$

Let

$$\left\{ V(r) + \left[\frac{\hbar^2 \nabla^2}{2m} - V(r) \right] P \right\} \phi + E P \phi \equiv W \phi. \tag{8.6.27}$$

The function W is called the pseudopotential. It contains both the self-consistent potential $V(r)$ and a term involving the core states. The definition of W reduces the Schroedinger equation for ϕ into a simple form,

$$-\frac{\hbar^2}{2m} \nabla^2 \phi + W \phi = E \phi, \tag{8.6.28}$$

which can be solved in terms of the pseudopotential W.

What, though, is W? W is given through Eq. (8.6.27) above as

$$W = V(r) + \left[\frac{\hbar^2 \nabla^2}{2m} - V(r) \right] P + E P, \tag{8.6.29}$$

which can be evaluated with the following argument. Consider the action of the kinetic-energy and the potential-energy operators on the projection operator P:

$$\left[-\frac{\hbar^2 \nabla^2}{2m} + V(r) \right] P = \left[-\frac{\hbar^2 \nabla^2}{2m} + V(r) \right] \sum_{t,j} |t, j\rangle \langle t, j|. \tag{8.6.30}$$

But the states $|t, j\rangle$ are eigenstates of the kinetic- and the potential-energy operators with eigenvalues equal to the core state energies $E_{t,j}$:

$$\left[-\frac{\hbar^2 \nabla^2}{2m} + V(r) \right] |t, j\rangle = E_{t,j} |t, j\rangle. \tag{8.6.31}$$

Therefore Eq. (8.6.30) can be simplified as

$$\left[-\frac{\hbar^2 \nabla^2}{2m} + V(r) \right] \sum_{t,j} |t, j\rangle \langle t, j| = \sum_{t,j} E_{t,j} |t, j\rangle \langle t, j|. \tag{8.6.32}$$

With the above substitutions, W becomes

$$W = V(r) - \sum_{t,j} E_{t,j} |t, j\rangle \langle t, j| + E \sum_{t,j} |t, j\rangle \langle t, j|. \tag{8.6.33}$$

Note that the term involving $E_{t,j}$ is negative. The negative sign can be traced back to Eq. (8.6.29). The kinetic- and the potential-energy operators in this equation are the negative of the Hamiltonian operators. Subsequently, a negative sign appears. Finally, the expression for the pseudopotential is

$$W = V(r) + \sum_{t,j}(E - E_{tj})|t, j\rangle\langle t, j|. \tag{8.6.34}$$

It is instructive at this point to examine the qualitative meaning of the pseudopotential. Note that W is smaller than the true potential $V(r)$ because of the ion cores. This follows mathematically since $V(r)$ is always attractive, and hence has a negative sign, while the difference $E - E_{tj}$ is always positive: the energy E is always greater than the core state energies. Therefore W must be less than $V(r)$ in magnitude. Hence the projection operator term acts to cancel some of the effects of the attractive potential of the ion cores, leading to a weaker effective crystalline potential. This is the underlying reason why the nearly-free-electron model seems to work; the effective potential that the electrons experience in the crystal is actually far less than that expected from the ions alone.

The question remains, though, what is the physical origin of this reduced potential? Why is the effective potential less than $V(r)$, that arising from the ion cores? The answer is due to the fact that the electrons are accelerated near the ion centers and hence spend less time in the near vicinity of the attractive potential. Therefore the electrons do not experience the action of the attractive ion potential as strongly as one would expect for the simple reason that on average the electrons are not near the ion cores but are located within the interstitial region where the potential is relatively constant. The electrons spend most of their time then within the interstitial regions and subsequently are not as strongly influenced by the ion potentials. Because of this, a crystal can behave as a collection of nearly free electrons that interact only weakly with the lattice potential.

From use of the cellular methods, the band structure of different semiconductor materials can be derived. The band structure of bulk GaAs, InP, and silicon calculated from the pseudopotential method of Cohen and Bergstresser (1966) are shown in Figures 8.6.3, 8.6.4, and 8.6.5, respectively. The points marked by a Γ represent the (000) points in k space. The points marked by an L are at the (111) point in the Brillouin zone while the points marked by an X are at the (100) point in the first Brillouin zone. In all these calculations, the hole energy minimum corresponds to 0.0 eV in the figures and occurs at the Γ point in the Brillouin zone. The figures show the energy-band structure calculated along the three principal directions in the first Brillouin zone. Recall that for a cubic symmetry crystal the three principal directions are along [100], [110], and [111]. In the figures shown, these directions correspond to moving along k as Γ–X, Γ–K, and Γ–L, respectively.

Note that three valence bands are present in the plots for GaAs and InP. These bands are the heavy, light, and split-off hole bands, respectively. The light hole band is the band splitting from the Γ point in each figure with the greater curvature and hence lighter mass. The split-off valence band arises from the

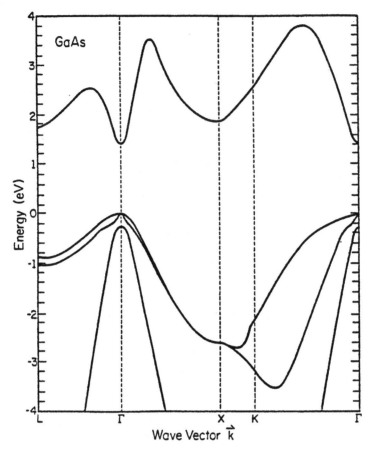

Figure 8.6.3. Calculated energy-band structure of bulk GaAs based on the pseudopotential approach.

spin–orbit effect discussed in Chapter 4. This band is termed the split-off band since it is typically no longer degenerate with the heavy and the light hole bands at the Γ point, as can be seen from Figures 8.6.3 and 8.6.4.

It is interesting to examine the first conduction bands of the calculated band structures of GaAs, InP, and silicon, as shown in Figures 8.6.3, 8.6.4, and 8.6.5. As can be seen from the figures, the minimum of the first conduction band in silicon occurs at the X point in the first Brillouin zone while in GaAs and InP the conduction-band minimum occurs at the Γ point. Since the minimum hole energy in all the materials occurs at the Γ point, the difference in location in k space of the minimum energy of the conduction band has important ramifications in the electrical and the optical properties of these materials. As we will see in Chapter 10, a material such as GaAs or InP, in which the conduction-band minimum and the valence-band minimum (minimum hole energy) occur at the same point in k space, is called a direct-gap semiconductor. A material such as silicon, in which the conduction-band minimum appears at a different point to the valence-band minimum, is called an indirect-gap

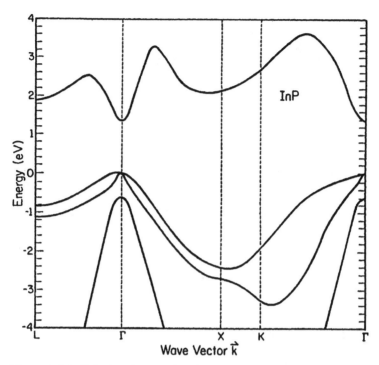

Figure 8.6.4. Calculated energy-band structure of bulk InP based on the pseudopotential approach.

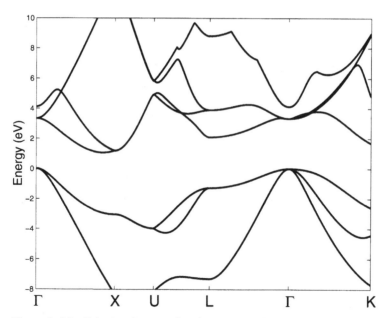

Figure 8.6.5. Calculated energy-band structure of bulk silicon based on the pseudopotential approach.

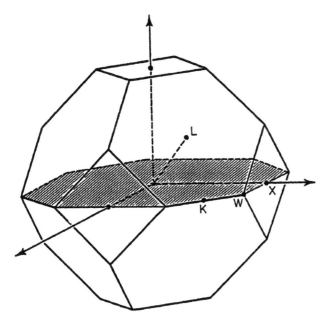

Figure 8.6.6. Sketch of a cross-sectional cut within the first Brillouin zone, which includes the X point.

semiconductor. Since the minima appear at the same point in k space in direct-gap semiconductors, an electron can be promoted from the valence band across the energy gap into the conduction band without any change in momentum. As a result, optical transitions can occur to first order in direct-gap semiconductors. In indirect-gap semiconductors, optical transitions require an intermediary step before they can occur. As we will see in Chapter 10, in indirect-gap semiconductors a two-step process must occur in order for an optical transition to occur.

It is of further interest to examine the isoenergy surfaces in the conduction bands of these materials. For purposes of illustration, isoenergy surfaces of the first conduction band of InP are presented for two different cuts in the first Brillouin zone. Figure 8.6.6 shows the first Brillouin zone of InP. The shaded region of the figure shows the cut of interest. Note that the shaded plane includes the Γ, X, and K points. The isoenergy surface for the first conduction band of InP corresponding to the shaded plane of Figure 8.6.6 is illustrated in Figure 8.6.7. As can be seen from Figure 8.6.7, near the Γ point, the energy surface is essentially spherical. Therefore the approximation of a parabolic energy band modeled with an effective mass is reasonable. However, as the energy increases to 1.5 eV and greater, note that the energy bands depart strongly from a spherical shape. In fact, the energy bands are highly distorted at energies equal to and greater than 1.5 eV. For these energies, usage of a parabolic approximation is inappropriate to describe the energy-band structure.

Finally, Figures 8.6.8 and 8.6.9 present a different cut in the first Brilloin zone and the corresponding isoenergy surface of the first conduction band of InP. In this case, the L and the K points are included. Inspection of Figure 8.6.9 shows that the energy bands are again highly distorted at high energies. Again, there is a strong departure from parabolic energy bands.

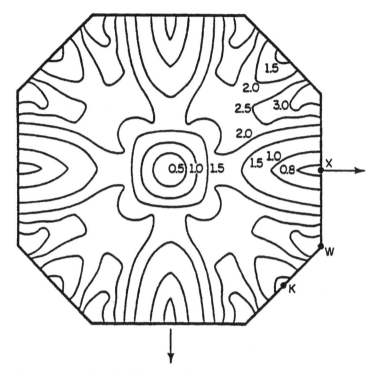

Figure 8.6.7. Sketch of the isoenergy surfaces of the first conduction band of bulk InP within the cross-sectional cut of the first Brillouin zone shown in Figure 8.6.6.

Figure 8.6.8. Sketch of a cross-sectional cut within the first Brillouin zone, which includes the L point.

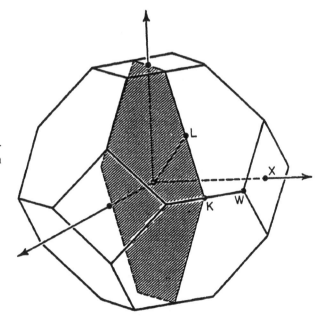

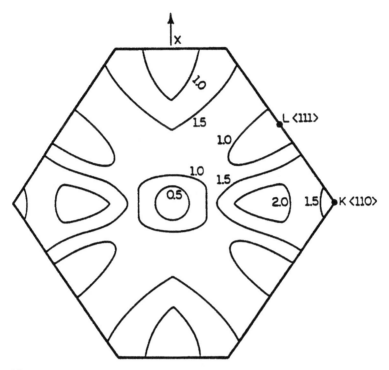

Figure 8.6.9. Sketch of the isoenergy surfaces of the first conduction band of bulk InP within the cross-sectional cut of the first Brillouin zone shown in Figure 8.6.8.

8.7 $\boldsymbol{k \cdot p}$ Calculation of Band Structure in Semiconductors

An alternative method of calculating the band structure in semiconductors is to use a technique known as the $k \cdot p$ method. This method is particularly useful in calculating the valence energy bands since it can readily account for the spin–orbit interaction. To see how to calculate the band structure by using this technique, we first start with the Schroedinger equation with Bloch wave functions:

$$H\Psi = E\Psi, \qquad \Psi = e^{i\vec{k} \cdot \vec{r}} u_{nk}(r). \tag{8.7.1}$$

Substituting in for ψ into the Schroedinger equation yields

$$\left(-\frac{\hbar^2}{2m}\nabla^2 + V\right)u_{nk}(r)e^{i\vec{k} \cdot \vec{r}} = E u_{nk}(r)e^{i\vec{k} \cdot \vec{r}}. \tag{8.7.2}$$

To evaluate Eq. (8.7.2), first apply the Laplacian operator onto the wave function as

$$\nabla^2\left[e^{i\vec{k} \cdot \vec{r}} u_{nk}(r)\right] = -k^2 e^{i\vec{k} \cdot \vec{r}} u + e^{i\vec{k} \cdot \vec{r}}\nabla^2 u + 2i\, e^{i\vec{k} \cdot \vec{r}}\vec{k} \cdot \vec{\nabla}u. \tag{8.7.3}$$

With this substitution, the Schroedinger equation becomes

$$-\frac{\hbar^2}{2m}[-k^2 + \nabla^2 + 2i\vec{k}\cdot\vec{\nabla}]u + Vu = Eu. \tag{8.7.4}$$

Recall the definition of the linear-momentum operator:

$$\vec{p} = \frac{\hbar}{i}\vec{\nabla} = -i\hbar\vec{\nabla}. \tag{8.7.5}$$

Therefore the Schroedinger equation becomes

$$\left[\frac{\hbar^2 k^2}{2m} + \frac{p^2}{2m} + \frac{\hbar}{m}\vec{k}\cdot\vec{p} + V\right]u = Eu. \tag{8.7.6}$$

Note that in addition to the usual kinetic-energy and potential-energy terms on the left-hand side, two new terms appear. The first is simply an energy term while the other term involves the dot vector product of the k vector and the momentum operator, or $\vec{k}\cdot\vec{p}$. The presence of this term gives rise to the origin of the name, $\vec{k}\cdot\vec{p}$ method.

Let us next choose a specific point in k space, k_0. The functions $u_{nk}(r)$, where n is a band index running over the complete set of bands, form a complete set for any given k. Hence the wave function for any given k can be written as an expansion in terms of k_0 as

$$u_{nk}(r) = \sum_{n'} c_{n'}(k-k_0)u_{n'k_0}(r). \tag{8.7.7}$$

To proceed further, it is useful to make the following definition. Let

$$H_{k_0} = \frac{p^2}{2m} + \frac{\hbar}{m}\vec{k}_0\cdot\vec{p} + \frac{\hbar^2 k_0^2}{2m} + V(r). \tag{8.7.8}$$

Given the definition of the Bloch functions, the following condition must be satisfied:

$$H_{k_0}u_{nk_0} = E_n(k_0)u_{nk_0}. \tag{8.7.9}$$

With these definitions, the Schroedinger equation becomes

$$\left[H_{k_0} + \frac{\hbar}{m}(\vec{k}-\vec{k}_0)\cdot\vec{p} + \frac{\hbar^2}{2m}(k^2-k_0^2)\right]u_{nk} = E_n(k)u_{nk}. \tag{8.7.10}$$

Substitute in the expansion for u_{nk}:

$$\left[H_{k_0} + \frac{\hbar}{m}(\vec{k}-\vec{k}_0)\cdot\vec{p} + \frac{\hbar^2}{2m}(k^2-k_0^2)\right]\sum_{n'} c_{n'}u_{n'k_0} = \sum_{n'} c_{n'} E_n(k)u_{n'k_0}. \tag{8.7.11}$$

Multiply Eq. (8.7.11) by $u^*_{nk_0}$ and integrate over all space. Each of the four terms becomes the following.

First term:

$$\sum_{n'} \int u^*_{nk_0} E_{n'}(k_0) u_{n'k_0} c_{n'} \, d^3r = \sum_{n'} \delta_{nn'} E_{n'} c_{n'} = E_n(k_0) c_n. \tag{8.7.12}$$

Second term:

$$\sum_{n'} \frac{\hbar}{m} \int u^*_{nk_0} (\vec{k} - \vec{k}_0) \cdot \vec{p} u_{n'k_0} c_{n'} \, d^3r = \sum_{n'} \frac{\hbar}{m} (\vec{k} - \vec{k}_0) \cdot \vec{p}_{nn'} c_{n'}, \tag{8.7.13}$$

where $\vec{p}_{nn'}$ is defined as

$$\vec{p}_{nn'} \equiv \int u^*_{nk_0} \vec{p} u_{n'k_0} \, d^3r. \tag{8.7.14}$$

Third term:

$$\sum_{n'} \int \frac{\hbar^2}{2m} (k^2 - k_0^2) u^*_{nk_0} c_{n'} u_{n'k_0} \, d^3r = \sum_{n'} \frac{\hbar^2}{2m} c_{n'} (k^2 - k_0^2) \delta_{nn'}$$

$$= \frac{\hbar^2}{2m} c_n (k^2 - k_0^2). \tag{8.7.15}$$

Fourth term:

$$\sum_{n'} c_{n'} \int E_n(k) u^*_{nk_0} u_{n'k} \, d^3r = E_n(k) c_n. \tag{8.7.16}$$

Combining all four terms yields

$$E_n(k_0) c_n + \frac{\hbar^2}{2m} (k^2 - k_0^2) c_n + \sum_{n'} \frac{\hbar}{m} (\vec{k} - \vec{k}_0) \cdot \vec{p}_{nn'} c_{n'} = E_n(k) c_n. \tag{8.7.17}$$

Note that the only nondiagonal term within Eq. (8.7.17) is the last term on the left-hand side involving the dot product.

To find the solutions of the Schroedinger equation under the above assumptions, standard perturbation theory can be applied. The solutions can be found about the point k_0. To zero order in correction, the off-diagonal term can be neglected and the energy eigenvalue at a point k near k_0 is simply given as

$$E_n(k) = E_n(k_0) + \frac{\hbar^2}{2m} (k^2 - k_0^2). \tag{8.7.18}$$

The first-order energy shift can be determined from perturbation theory. The perturbation is assumed to be given by

$$\frac{\hbar}{m} (\vec{k} - \vec{k}_0) \cdot \vec{p}. \tag{8.7.19}$$

When perturbation theory is used, the first-order shift in the energy at the point k near k_0 is

$$\frac{\hbar}{m}(\vec{k} - \vec{k}_0) \cdot \vec{p}_{nn},\tag{8.7.20}$$

where p_{nn} is given as

$$\vec{p}_{nn} = \int u^*_{nk_0} \vec{p} \, u_{nk_0} \, \mathrm{d}^3 r.\tag{8.7.21}$$

Note that if the crystal has a center of symmetry about the point k_0, then p_{nn} will vanish about that point and the first-order correction to the energy is zero. Band extremum points have just such a symmetry. Subsequently, it is necessary to take the perturbation to at least second order. The energy eigenvalue for the system to second order in the perturbation is given then as

$$\begin{aligned}
E_n(k) = E_n(k_0) &+ \frac{\hbar}{m}(\vec{k} - \vec{k}_0) \cdot \vec{p}_{nn} + \frac{\hbar^2}{2m}(k^2 - k_0^2) \\
&+ \frac{\hbar^2}{m^2} \sum_{n'} \frac{|(\vec{k} - \vec{k}_0) \cdot \vec{p}_{nn'}|^2}{E_n(k_0) - E_{n'}(k_0)}.
\end{aligned}\tag{8.7.22}$$

If the point k_0 is an extremum, then p_{nn} vanishes, as discussed above. Note that the sum is taken over all the bands in the system. In practice, only the first few bands are of importance in evaluating the sum.

Consider the case in which k_0 is zero and is also an extremum of the system. Then p_{nn} again vanishes, and the energy is given as

$$E_n(k) = E_{0n} + \frac{\hbar^2 k^2}{2m} + \frac{\hbar^2}{m^2} \sum_{n'} \frac{|\vec{k} \cdot \langle u_{0n'} | \vec{p} | u_{0n} \rangle|^2}{E_{0n'} - E_{0n}}.\tag{8.7.23}$$

Note that a more general form of the effective mass is obtained in Eq. (8.7.23) from that determined in Section 8.1. The energy $E_n(k)$ can be written with an effective-mass tensor $\bar{m}/m^*$ as

$$E_n(k) = E_{0n} + \frac{\hbar^2}{2m} \vec{k} \cdot \frac{\bar{m}}{m^*} \cdot \vec{k}.\tag{8.7.24}$$

The elements of the effective-mass tensor can be defined by a comparison of Eq. (8.7.23) with Eq. (8.7.24), yielding

$$\left.\frac{\bar{m}}{m^*}\right|_{ij} = \delta_{ij} + \frac{2}{m} \sum_{n'} \frac{\langle u_{0n} | p_i | u_{0n'} \rangle \langle u_{0n'} | p_j | u_{0n} \rangle}{E_{0n} - E_{0n'}}.\tag{8.7.25}$$

Therefore, near the band edge we once again recover a simple effective-mass method for describing the band structure. In the more general case derived above, the effective mass is a tensor quantity and not a simple scalar quantity. This implies that the mass can be different in different directions. This is often the case in many semiconductors at different extremum points. The effective mass can be determined along three principal directions, enabling a description of the mass at any extremum in terms of these directions. In most semiconductors, many of the most important extremum points resemble ellipsoids such that only a transverse and longitudinal effective mass must be specified. For example, in silicon the minimum of the first conduction band occurs along the (100) direction near the edge of the first Brillouin zone. In this case, the energy surface is an ellipsoid with the symmetry axis as the axis of revolution. The effective mass at the conduction-band minimum in silicon must be specified in terms of two effective masses, the transverse and the longitudinal masses.

PROBLEMS

1. A semiconductor is characterized as having a bandgap typically less than 3 eV. Assume that the Fermi–Dirac distribution can be approximated by the Boltzmann factor and calculate the effective electron carrier concentration in a semiconductor of electron effective mass m^* and temperature T. Use the three-dimensional density-of-states function determined in Chapter 5.

2. Determine the carrier velocity at the Brillouin zone edge if the energy versus k vector relation is given by the one-dimensional nearly-free-electron model as

$$E = \frac{\hbar^2}{4m} \left\{ k^2 + \left(k - \frac{2\pi n}{a} \right)^2 \pm \sqrt{\left[k^2 - \left(k - \frac{2\pi n}{a} \right)^2 \right]^2 + \left(\frac{4m|V_n|}{\hbar^2} \right)^2} \right\}.$$

3. Consider the equation for the Kronig–Penney model derived in Section 8.2, Eq. (8.2.14). Determine what happens when the barriers become delta functions such that the product of the potential and the barrier width $V_0 b$ remains constant.

4. Derive an expression for the electronic density-of-states function for one-, two-, and three-dimensional systems. The two-dimensional system is of interest in the physical description of the inversion layer in a metal-oxide semiconductor field-effect transistor. Assume parabolic bands.

5. See the paper by Allen, Phys. Rev. **91**, 531 (1953), and show that one can obtain a relationship between α and β defined in the text as

$$\tan u = -\frac{(\pi - \omega)V}{u\omega} \tan V, \qquad \cot u = -\frac{(\pi - \omega)V}{u\omega} \cot V,$$

where

$$u = (\pi - \omega)\sqrt{E - V}, \qquad V = \omega\sqrt{E}.$$

6. Consider a system quantized in the z direction but having free particlelike solutions in the x and the y directions. Determine the mean number of particles in the system if the energy in each quantized level is $E_{1,2}$ and the energy in the xy directions is simply

$$E = \frac{\hbar^2 k^2}{2m^*},$$

where k^2 is

$$k^2 = k_x^2 + k_y^2.$$

Use Fermi statistics.

9

Phonons and Scattering Mechanisms in Solids

In this chapter, we consider what happens when the lattice is no longer in equilibrium. In Chapter 8 we considered the motion of electrons in a crystal assuming that the ions were at rest for all times about their equilibrium positions. The presence of the ionic potential on the electrons results in energy-band formation: allowed ranges of energies for the electrons. However, this is not a physically realistic situation, since at nonzero lattice temperatures, the ions undergo some oscillation about their equilibrium positions. This additional motion, called lattice vibration, has important consequences on the behavior of the electrons. In Section 9.1 we consider lattice vibrations and the quanta of lattice vibrations, phonons. In the subsequent sections, we then examine how phonons influence the electronic properties of a crystal.

9.1 Lattice Vibrations and Phonons

From the discussion in Chapter 7, we found that the equilibrium configuration of the ions results in the minimum potential energy of the crystal. As the ions become more compressed, a repulsive force due to the nuclear–nuclear interaction acts to restore the system back to equilibrium. Similarly, as the lattice is stretched, the attractive force from the molecular bonds formed between adjacent ions acts again to restore equilibrium. Hence a lattice can be thought of as a collection of mass centers situated at definite positions and held together by the action of an elastic restoring force. The system has a definite minimum potential energy. As a consequence, the system will oscillate as a harmonic oscillator about its equilibrium position if it is slightly perturbed. This can be easily seen from the following argument.

Consider the potential-energy diagram sketched in Figure 9.1.1. As can be seen from this figure, the potential energy undergoes a minimum near $r = r_0$. Expanding the potential energy in a Taylor series about $r = r_0$ yields

$$U(r) = U(r_0) + (r - r_0)\frac{dU}{dr}\bigg|_{r_0} + \frac{1}{2}(r - r_0)^2\frac{d^2U}{dr^2}\bigg|_{r_0} + \cdots. \qquad (9.1.1)$$

Since the potential is a minimum at $r = r_0$, the first derivative is zero. When only the second-order term is retained, the potential energy is

$$U(r) = U(r_0) + \frac{1}{2}(r - r_0)^2\frac{d^2U}{dr^2}\bigg|_{r_0}, \qquad (9.1.2)$$

which is essentially the potential energy for a harmonic oscillator, given that the

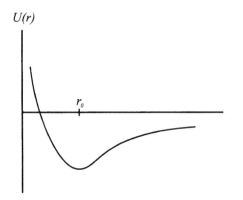

U(r)

Figure 9.1.1. Potential-energy diagram of a bound system, akin to a crystalline solid.

spring constant can be defined as

$$k = \frac{d^2 U}{dr^2}\bigg|_{r=r_0}. \qquad (9.1.3)$$

Subsequently, if the ions of a crystal are excited to energies greater than absolute zero, they will undergo oscillations about their equilibrium positions. As such, the lattice can be modeled as a collection of masses connected together by a series of springs, as illustrated in Figure 9.1.2. The masses oscillate about their equilibrium positions in response to a perturbation. In a solid, thermal excitations induce lattice vibrations of varying degrees until the crystal melts. For the purposes of our discussion, it is assumed that all the lattice vibrations considered here are linear and obey Hooke's Law for springs:

$$F = -\beta x, \qquad (9.1.4)$$

where β is the linear-force constant.

The ions in a crystal are free to vibrate in many ways. Nevertheless, the vibrations can always be decomposed into motions along the three principal directions of the crystal. These motions are independent of one another. To study the behavior of the lattice vibrations, it is helpful to model the crystal as a collection of masses on springs, as shown in Figure 9.1.2. Although the system in Figure 9.1.2 is one dimensional, no loss of generality occurs since the motion of a real system can always be recovered by analysis of the motion along three independent directions. Therefore the motion along one direction will still be instructive.

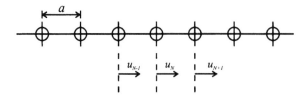

Figure 9.1.2. Schematic representation of lattice vibrations of a one-dimensional monatomic lattice of lattice constant a. The u_n are the amplitudes of vibration.

The equilibrium separation of the ions is simply equal to the lattice constant of the system a. The force on the nth atom due to its nearest neighbors $n - 1$ and $n + 1$ is

$$F_n = \beta(u_{n+1} - u_n) - \beta(u_n - u_{n-1}), \tag{9.1.5}$$

where Hooke's Law has been assumed. Simplifying Eq. (9.1.4) by combining like terms and using Newton's Second Law of Motion gives

$$F_n = \beta(u_{n+1} + u_{n-1} - 2u_n) = m\frac{d^2u_n}{dt^2}. \tag{9.1.6}$$

Since the force is harmonic, the amplitudes u_n must also be harmonic. Therefore u_n is assumed to be of the form

$$u_n \sim A e^{i(\omega t - nka)}, \tag{9.1.7}$$

where the position of the nth atom can be represented as na, the product of the lattice position and the lattice constant. Expressions for u_{n+1} and u_{n-1} follow readily from Eq. (9.1.6) as

$$u_{n+1} = A e^{i[\omega t - (n+1)ka]}, \qquad u_{n-1} = A e^{i[\omega t - (n-1)ka]}. \tag{9.1.8}$$

Substituting expressions (9.1.8) for the amplitudes into Eq. (9.1.5) yields

$$-m\omega^2 e^{i(\omega t - nka)} = \beta\{ e^{i[\omega t - (n+1)ka]} + e^{i[\omega t - (n-1)ka]} - 2 e^{i(\omega t - nka)}\},$$
$$-m\omega^2 e^{-inka} = \beta[e^{-i(n+1)ka} + e^{-i(n-1)ka} - 2 e^{-inka}],$$
$$-\frac{m\omega^2}{\beta} = [e^{-ika} + e^{ika} - 2]. \tag{9.1.9}$$

But we can simplify the above by noting that

$$2\cos ka = e^{-ika} + e^{ika}, \tag{9.1.10}$$
$$1 - \cos 2\theta = 2\sin^2\theta. \tag{9.1.11}$$

Therefore Eqs. (9.1.8) reduce to

$$m\omega^2 = 4\beta \sin^2\frac{ka}{2},$$
$$\omega^2 = \frac{4\beta}{m}\sin^2\frac{ka}{2}. \tag{9.1.12}$$

The angular frequency of vibration of the system is given then as

$$\omega = \sqrt{\frac{4\beta}{m}}\left| \sin\frac{ka}{2}\right|, \tag{9.1.13}$$

which is called the lattice vibration dispersion relation or simply the dispersion relation. Note that the angular frequency ω is not a linear function of k, as is the case in electromagnetic vibrations.

The dispersion relation can be simplified at long wavelengths, small k, by approximation of the sine by its argument. With this approximation, the dispersion relation becomes

$$\omega = \sqrt{\frac{4\beta}{m}}\left(\frac{ka}{2}\right) = ka\sqrt{\frac{\beta}{m}}. \tag{9.1.14}$$

In this limit, ω is linearly related to k. In Chapter 1, we found that the phase velocity of a wave is given by the ratio of ω to k. Therefore, in the long-wavelength limit, the phase velocity of the lattice wave is a constant given by

$$v_p = \frac{\omega}{k} = a\sqrt{\frac{\beta}{m}}. \tag{9.1.15}$$

The group velocity is found from the derivative of the dispersion relation with respect to k as

$$v_g = \frac{d\omega}{dk}. \tag{9.1.16}$$

Clearly, in the long-wavelength limit, the group velocity and the phase velocity of the lattice wave are equal. As a consequence, the lattice waves behave as elastic linear vibrations of constant velocity. Since the angular frequency and the wave vector are linearly related for these lattice modes, they are called the acoustic modes of vibration.

The dispersion relation is plotted in Figure 9.1.3. As can be seen from the figure, the angular frequency attains its maximum value at $k = \pm\pi/a$. This follows readily from Eq. (9.1.13) since the sine has its maximum value at these limits. As can be seen from Figure 9.1.3, the dispersion relation can be approximated as a straight line at small k. Owing to the linear relation at long wavelengths

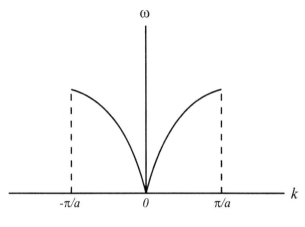

Figure 9.1.3. Dispersion relation for a one-dimensional monatomic lattice of lattice constant a.

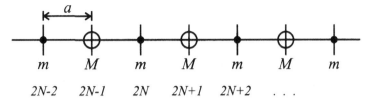

Figure 9.1.4. Schematic representation of a one-dimensional diatomic lattice consisting of two centers of mass m and M connected by springs of force constant β.

between the angular frequency and the wave vector, as k approaches zero the angular frequency also approaches zero. This mode of vibration is called the acoustic mode since the phase velocity obeys a linear relation between the frequency and the wave vector, $v_p = \omega/k$. In the acoustic mode of vibration, the ion cores all oscillate in phase with one another.

It is interesting to consider a more complicated system than that of the single monatomic lattice in one dimension. A diatomic one-dimensional lattice is sketched in Figure 9.1.4. For simplicity, we assume that the force constant β is the same between each mass and that the lattice constant a is also the same. With these assumptions, the force on the $2n$ and the $2n+1$ atoms is

$$F_{2n} = m\frac{d^2u_{2n}}{dt^2} = \beta(u_{2n+1} + u_{2n-1} - 2u_{2n}),$$

$$F_{2n+1} = M\frac{d^2u_{2n+1}}{dt^2} = \beta(u_{2n+2} + u_{2n} - 2u_{2n+1}). \qquad (9.1.17)$$

Since the masses of the two atoms are different, the amplitudes of vibration of each center will not necessarily be the same. To distinguish between the two possibilities, the amplitude of vibration of the smaller mass particle m is defined as A and that of the larger mass particle M is defined as B. The solutions must again be harmonic and are assumed to be of the form

$$u_{2n} = A e^{i(\omega_1 t - 2nka)},$$

$$u_{2n+1} = B e^{i[\omega_2 t - (2n+1)ka]},$$

$$u_{2n+2} = A e^{i[\omega_1 t - (2n+2)ka]} = u_{2n} e^{-2ika},$$

$$u_{2n-1} = B e^{i[\omega_2 t - (2n-1)ka]} = u_{2n+1} e^{2ika}. \qquad (9.1.18)$$

Substituting these expressions for the amplitudes into Eqs. (9.1.17) yields

$$-m\omega_1^2 u_{2n} = \beta\big[(1 + e^{2ika})u_{2n+1} - 2u_{2n}\big],$$

$$-M\omega_2^2 u_{2n+1} = \beta\big[(1 + e^{2ika})u_{2n} - 2u_{2n+1}\big]. \qquad (9.1.19)$$

If Eqs. (9.1.19) hold for all time, then ω_1 must equal ω_2, which we now call ω. u_{2n+1} can be expressed in terms of u_{2n} from use of the relationships given by

Eqs. (9.1.17) as

$$u_{2n+1} = \beta \frac{(1 + e^{-2ika})}{(2\beta - M\omega^2)} u_{2n}. \tag{9.1.20}$$

Substituting the expression for u_{2n+1} given in Eq. (9.1.20) into the first of Eqs. (9.1.19) yields

$$(2\beta - m\omega^2)u_{2n} = \frac{\beta^2(1 + e^{2ika})(1 + e^{-2ika})}{(2\beta - M\omega^2)} u_{2n},$$

$$(2\beta - m\omega^2)(2\beta - M\omega^2) = \beta^2(1 + e^{2ika})(1 + e^{-2ika}),$$

$$(2\beta - m\omega^2)(2\beta - M\omega^2) = 2\beta^2(1 + \cos 2ka). \tag{9.1.21}$$

The cosine term can be simplified with a double-angle formula to give

$$(2\beta - m\omega^2)(2\beta - M\omega^2) = 4\beta^2 \cos^2 ka. \tag{9.1.22}$$

Solving for ω^2 through use of the quadratic formula yields

$$\omega_\pm^2 = \frac{\beta(m+M)}{mM} \left[1 \pm \sqrt{1 - \frac{4mM\sin^2 ka}{(m+M)^2}} \right]. \tag{9.1.23}$$

The dispersion relation defined above is plotted in Figure 9.1.5. It is interesting to note that the dispersion relation is double valued. Note that two possibilities exist, ω_- and ω_+. The lower branch in the figure corresponds to the ω_- solution while the upper branch is the ω_+ solution. It readily follows that ω_- is the lower branch since, at $k = 0$, ω_- becomes

$$\omega_- = \frac{\beta(m+M)}{mM}(1 - 1) = 0. \tag{9.1.24}$$

Let us examine the solution for ω_- in the long-wavelength limit. The sine function can then be represented by its argument

$$\sin^2 ka \sim (ka)^2. \tag{9.1.25}$$

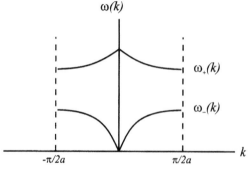

Figure 9.1.5. Dispersion relation for a diatomic one-dimensional lattice.

With this substitution, Eq. (9.1.23) becomes

$$\omega_-^2 = \frac{\beta(m+M)}{mM}\left[1 - \sqrt{1 - \frac{4mMk^2a^2}{(m+M)^2}}\right].$$

(9.1.26)

Expanding out the square root by using a Taylor series expansion, we can approximate Eq. (9.1.26) as

$$\omega_-^2 = \frac{\beta(m+M)}{mM}\left[1 - 1 + \frac{1}{2}\left(\frac{4mMk^2a^2}{(m+M)^2}\right)\right],$$

(9.1.27)

where only the first two terms have been retained in the expansion. Simplifying Eq. (9.1.27) yields

$$\omega_-^2 = \frac{2\beta k^2 a^2}{(m+M)}.$$

(9.1.28)

The angular frequency ω_- is then

$$\omega_- = ka\sqrt{\frac{2\beta}{(m+M)}}.$$

(9.1.29)

Clearly the ω_- root corresponds to the acoustic mode of vibration of the crystal similar to what was found above for the monatomic one-dimensional lattice.

What is the nature of the second part of the solution, ω_+? ω_+^2 is given as

$$\omega_+^2 = \frac{\beta(m+M)}{mM}\left[1 + \sqrt{1 - \frac{4mM\sin^2 ka}{(m+M)^2}}\right].$$

(9.1.30)

Again in the long-wavelength limit, small k, the sine function can be approximated by its argument. Using this approximation and expanding the square root in a Taylor series, we find that Eq. (9.1.30) reduces to

$$\omega_+^2 = \frac{2\beta(m+M)}{mM}\left[1 - \frac{mMk^2a^2}{(m+M)^2}\right].$$

(9.1.31)

From examination of Eq. (9.1.31), it readily follows that the ω_+ root at $k = 0$ is not zero. In fact, the frequency of vibration at $k = 0$ is

$$\omega_+ = \sqrt{\frac{2\beta(m+M)}{mM}}.$$

(9.1.32)

Before we consider the physics behind this result, let us examine the solutions for the frequencies ω_- and ω_+ at large k, near the edge of the Brillouin zone. Let us first consider the behavior of ω_- at $k = \pi/2a$. In this limit, ω_-^2 becomes

$$\omega_-^2 = \frac{\beta(m+M)}{mM}\left[1 - \sqrt{1 - \frac{4mM}{(m+M)^2}}\right].$$

(9.1.33)

Multiplying through by $\beta(m+M)/mM$ yields

$$\omega_-^2 = \frac{\beta(m+M)}{mM} - \sqrt{\left[\frac{\beta(m+M)}{mM}\right]^2 - \frac{4\beta^2}{mM}}. \tag{9.1.34}$$

Combining the terms under the square root in Eq. (9.1.34) by using the least common denominator m^2M^2 yields

$$\omega_-^2 = \frac{\beta(m+M)}{mM} - \sqrt{\frac{m^2\beta^2 + M^2\beta^2 - 2\beta^2mM}{m^2M^2}}. \tag{9.1.35}$$

Simplifying the square-root term above by factoring the numerator into the square of $\beta(m - M)$ yields

$$\omega_-^2 = \frac{\beta(m+M)}{mM} - \frac{\beta(M - m)}{mM},$$

$$\omega_- = \sqrt{\frac{2\beta}{M}}. \tag{9.1.36}$$

Similarly, at $k = \pi/2a$, ω_+ is found to be

$$\omega_+ = \sqrt{\frac{2\beta}{m}}. \tag{9.1.37}$$

Now consider the behavior of the amplitudes u_{2n+1} and u_{2n}; their ratio is given from Eq. (9.1.20) as

$$\frac{u_{2n+1}}{u_{2n}} = \frac{\beta(1 + e^{-2ika})}{2\beta - M\omega_\pm^2} = \frac{B}{A}. \tag{9.1.38}$$

With the small k limit, the ratio becomes

$$\frac{B}{A} = \frac{2\beta}{(2\beta - M\omega_\pm^2)}. \tag{9.1.39}$$

For the acoustic mode of vibration, ω_- approaches zero as k approaches zero. Hence, the ratio of the amplitudes becomes

$$\frac{B}{A} = \frac{2\beta}{2\beta} = 1. \tag{9.1.40}$$

For the ω_+ root, called the optical mode of vibration, as k approaches zero, ω_+ approaches

$$\omega_+ = \sqrt{\frac{2\beta(m+M)}{mM}}. \tag{9.1.41}$$

Hence the ratio of the amplitudes B/A becomes

$$\frac{B}{A} = \frac{2\beta}{2\beta - \frac{M2\beta(m+M)}{mM}},$$

$$\frac{B}{A} = \frac{1}{1 - \frac{(m+M)}{m}} = -\frac{m}{M}, \qquad (9.1.42)$$

which holds as k approaches zero. Note that for the acoustic mode of vibration, the amplitudes of the oscillation of each type of atom are the same as k goes to zero. More importantly, the motion of the atoms is in the same direction; the atoms all oscillate in phase with one another. For the optical branch, the vibrations of the atoms are in opposite directions, and as k approaches zero their amplitudes are inversely related as the ratio of their masses. In the optical mode of vibration, the center of mass of the unit cell remains fixed during the motion. In the acoustic mode of vibration, the center of mass does not remain fixed since all the atoms vibrate in phase. The ω_+ branch is called the optical branch or the optical mode since, if the ions are oppositely charged, an electric field will tend to move them in opposite directions. The electric field of an electromagnetic wave can induce the ω_+ mode of vibration.

Let us reconsider Eq. (9.1.22), which describes the one-dimensional diatomic lattice dispersion relation,

$$(2\beta - M\omega^2)(2\beta - m\omega^2) - 4\beta^2 \cos^2 ka = 0. \qquad (9.1.43)$$

Solving for $\cos ka$ yields

$$\cos ka = \sqrt{\left(1 - \frac{M\omega^2}{2\beta}\right)\left(1 - \frac{m\omega^2}{2\beta}\right)}. \qquad (9.1.44)$$

Note that an imaginary solution for k can occur depending on the value of each term in the square root. When

$$\frac{2\beta}{M} < \omega^2 < \frac{2\beta}{m}, \qquad (9.1.45)$$

then the first term in Eq. (9.1.43) is negative. Therefore Eq. (9.1.44) becomes

$$\cos ka = i\sqrt{\left(\frac{M\omega^2}{2\beta} - 1\right)\left(1 - \frac{m\omega^2}{2\beta}\right)}. \qquad (9.1.46)$$

Let δ be defined as the square-root term in Eq. (9.1.46). Then the expression for $\cos ka$ becomes

$$\cos ka = i\delta. \qquad (9.1.47)$$

If we further define $(x + iy) = ka$, then the cosine can be expanded as

$$\cos(x + iy) = \cos x \cosh y - i \sin x \sinh y = i\delta. \qquad (9.1.48)$$

Comparison of the real and the imaginary parts gives

$$\cos x \cosh y = 0, \tag{9.1.49}$$
$$\sin x \sinh y = -\delta. \tag{9.1.50}$$

For both of the above relationships to hold, x must be equal to $\pm \pi/2$. With this assignment, y is simply

$$y = \mp \sinh^{-1} \delta. \tag{9.1.51}$$

Therefore

$$ka = \pm \frac{\pi}{2} \mp i \sinh^{-1} \delta. \tag{9.1.52}$$

Substituting for ka into the expression for the amplitude u_{2n} as given by Eqs. (9.1.18) yields

$$
\begin{aligned}
u_{2n} &= A e^{i\omega t} e^{-2ikna} \\
&= A e^{i\omega t} e^{-2in\left(\pm\frac{\pi}{2} \mp i \sinh^{-1}\delta\right)} \\
&= A e^{i\omega t} e^{\mp in\pi} e^{\mp 2n \sinh^{-1}\delta}.
\end{aligned}
\tag{9.1.53}
$$

Recognizing that $e^{-in\pi}$ is simply equal to ± 1, we find that u_{2n} becomes

$$u_{2n} = \pm A e^{i\omega t} e^{\mp 2n \sinh^{-1}\delta}. \tag{9.1.54}$$

If we define K as

$$K = \frac{\sinh^{-1} \delta}{a}, \tag{9.1.55}$$

u_{2n} can finally be written as

$$u_{2n} = A e^{i\omega t} e^{-2nKa}, \tag{9.1.56}$$

where we neglect the growing exponential term. Clearly, any mode with the above amplitude will not propagate through the crystal because of the presence of the decaying exponential function e^{-2nKa}. Consequently, there is a range of angular frequencies ω for which no lattice waves can propagate. These frequencies are determined from the condition

$$\frac{2\beta}{M} < \omega^2 < \frac{2\beta}{m}. \tag{9.1.57}$$

Therefore lattice waves cannot propagate at frequencies within the above range in a diatomic crystal.

Before we end our discussion of lattice waves, it is important to introduce the concept of phonons. Phonons are the quanta of lattice vibrations, much like photons are the quanta of electromagnetic vibrations. In fact, the name phonons is derived from this analogy. As we discussed in Chapter 1, quantum mechanics ascribes a duality to all forms of energy and matter. Particles have wavelike aspects and waves have particlelike aspects. This duality underlies all physical phenomena. Therefore a quantum of lattice vibrations must correspond. These quanta are called phonons.

Similar to what we learned about photons, the energy of a phonon is given as the product of Planck's constant and the frequency of vibration of the mode $h\upsilon$ or $\hbar\omega$. The collective energy of a mode is

$$E = \left(n + \frac{1}{2}\right)\hbar\omega, \tag{9.1.58}$$

where n is the number of phonons excited in each mode.

The factor of 1/2 arises from the zero-point motion of the lattice. In the discussion in Chapter 2 about the harmonic oscillator, it was determined that a harmonic oscillator always exhibits a zero-point motion so as not to violate the Uncertainty Principle.

Phonons play an important role in the treatment of scattering processes in crystals. In Section 9.2 we consider the interactions of phonons and electrons. Carrier–phonon interactions serve as the coupling between the electron and the lattice gases. Energy is transferred from one system to the other by means of electron–phonon or hole–phonon scattering events.

9.2 **Electron–Phonon Interactions**

In Section 9.1 we have found that at finite temperatures the atoms comprising a crystalline lattice undergo vibrations about their fixed equilibrium positions. These lattice vibrations are quantized, and the quantum of lattice vibrations are called phonons. As one might suspect, an electron moving within a crystal in which the atoms are vibrating experiences a slightly different potential than if it is traversing a crystal whose atoms are completely frozen in space. In this section, we discuss how lattice vibrations influence the motions of electrons within a crystal.

As an electron moves through a perfectly periodic crystal, the periodic potential of the lattice affects the dynamics of the electron. As we found in Chapter 8, the periodic potential of the lattice leads to forbidden and allowed ranges of energies for which the electron can propagate through the crystal. These allowed and forbidden energy ranges are called energy bands and forbidden bands, respectively. What happens, though, when the potential is no longer perfectly periodic? The perfect periodicity of the lattice can be altered by several different mechanisms. If the atoms undergo oscillations about their equilibrium positions, the potential that an electron experiences is no longer perfectly uniform at all times. The presence of a foreign atom either substitutionally or interstitially within the host lattice alters the perfect periodic potential of the lattice. In addition, the lattice itself can be deformed through dislocations or grain boundaries. The presence of a surface also acts to alter the periodic boundary conditions assumed in the simple bulk infinite crystal solution determined in Chapter 8. Each of the above departures from perfect periodicity has significant ramifications for the behavior of electrons in crystals.

For the most part, the effect of lattice vibrations and impurities on the motions of electrons in a crystalline solid can be treated perturbatively. It is typically assumed that the effect of lattice vibrations on the electron dynamics is sufficiently weak that a first-order perturbation treatment is adequate to describe it. In this section, we discuss how the effect of lattice vibrations on electron dynamics is treated perturbatively by using the results derived in Chapter 4. Our discussion is restricted to only electron–phonon interactions and only a brief discussion is presented at that.

As we found in Chapter 4, when a disturbance perturbs a system for a period of time, it can induce a transition of an electron from an initial state into a different final state. In Chapter 4, we determined how the transition rate from one state into another is specified through Fermi's golden rule. In the treatment of electron–phonon interactions it is typical to adopt Fermi's golden rule, which is what we will do here.

Of greatest interest to us are the description of electron–phonon interactions in both column IV and compound semiconductors. Column IV semiconductors are materials such as silicon and germanium in which the interatomic bonds are strictly covalent and the constituent atoms exhibit no permanent electric dipole moment due to the electronic bond formation. In a covalent bond between two identical atoms, the electron shared by the two atoms is just as likely to be around one atom as the other, so no permanent dipole moment occurs. However, if a bond is formed between two different atoms of different electronegativity, the bond will necessarily exhibit some polar behavior. This is the situation in a compound semiconductor.

Because of the different natures of the interatomic bonds in polar and non-polar semiconductors, different electron–phonon scattering mechanisms appear. The two principal means by which the electrons are coupled to the phonons are through the deformation potential and polarization waves.

Deformation-potential scattering arises from the deformation of the lattice by the motion of an electron. This is similar to the mechanism that couples two electrons in a superconductor, as was discussed in Chapter 6. The presence of an electron in the near vicinity of the positively charged ions leads to an attractive interaction that results in the local ions' moving toward the electron. As a result, the lattice is slightly deformed locally by the presence of the electron. The deformation leads then to a vibration of the ions about their equilibrium positions. Either the acoustic or the optical modes of vibration can be excited. Deformation-potential scattering arising from acoustic-mode excitation is typically called acoustic scattering while deformation-potential scattering arising from optical-mode excitation is called nonpolar optical scattering. Both acoustic and nonpolar optical scattering typically exhibit only small momentum transfer, often referred to as small $\vec{Q}$. Another type of deformation-potential scattering occurs for electrons called intervalley scattering. Intervalley scattering involves large momentum transfer or large $\vec{Q}$. Since intervalley scattering occurs for large $\vec{Q}$, the phonons governing the process must necessarily have a relatively large wave vector and must therefore be near the edge of the Brillouin zone. At the zone edge, the acoustic and the optical branches can overlap so that either branch can contribute to an intervalley scattering event.

Coupling of the electrons to the phonons by means of polarization waves can also occur but only in polar materials. The permanent electric dipole moment of the constituent ions in a polar material affects the motions of electrons moving through the crystal. The electrons interact with the electric dipole moment, leading to an additional coupling to the ions. This coupling can be mediated through either the acoustic or the optical branches. The interaction with the acoustic branch is called piezoelectric scattering while that with the optical branch is called polar optical scattering.

The state vector for the system, which comprises both electrons and ions, can be represented as

$$|k, c\rangle, \tag{9.2.1}$$

where k represents the state vector of the electrons and c represents the state vector of the ions. If the adiabatic approximation is made (see Section 7.1) then the state vector of the system can be written as the product of the independent electron and ion state vectors as

$$|k, c\rangle = |k\rangle|c\rangle. \tag{9.2.2}$$

The transition probability per unit time from an initial state defined as $|k, c\rangle$ to a final state defined as $|k', c'\rangle$ due to a perturbation Hamiltonian H' is determined from Fermi's golden rule as

$$P(k, c; k', c') = \frac{2\pi}{\hbar} |\langle k', c'|H'|k, c\rangle|^2 \delta[E(k', c') - E(k, c) \pm \hbar\omega]. \tag{9.2.3}$$

The perturbation is assumed to be harmonic. Therefore H' can be expanded in terms of a Fourier series as

$$H'(r, y, r') = \frac{(2\pi)^{3/2}}{V} \sum_Q H'(Q, y, r') e^{iQr}, \tag{9.2.4}$$

where r represents the coordinate of the electron of interest, y are the coordinates of the ion displacement from their equilibrium positions, and r' are the coordinates of the other electrons in the electron gas. If the adiabatic approximation is made and the Fourier series expansion of H' is substituted into Eq. (9.2.3) the matrix element for the transition rate becomes

$$\langle k', c'|H'|k, c\rangle = \frac{(2\pi)^{3/2}}{V} \int \sum_Q \langle c'|H'(Q)|c\rangle \Psi_k^*(r) e^{iQr} \Psi_{k'}(r) \, d^3r, \tag{9.2.5}$$

where the electronic states are taken to be Bloch states. The normalized Bloch states are given then as

$$\Psi_k(r) = \frac{1}{\sqrt{N}} u_k(r)\, e^{ikr}, \tag{9.2.6}$$

where N is the number of unit cells in the crystal. Following Jacoboni and Lugli (1989), we can rewrite r by using a direct lattice vector $\vec{R}$ and $\vec{r}''$ as

$$\vec{r} = \vec{r}'' - \vec{R}. \tag{9.2.7}$$

It is assumed that the vector $\vec{r}$ can be decomposed into a vector $\vec{r}''$ and a direct lattice vector $\vec{R}$. As we found in Chapter 8, any direct lattice vector can be decomposed into a direct lattice vector that connects to a unit cell along with an additional vector within the unit cell that connects to the point in question. The integration over r can then be reexpressed as a combined sum over all the unit cells multiplied by an integration over one unit cell of the vector r''. This can be expressed quantitatively as

$$\int \mathrm{d}^3 r\, \Psi_k^*(r)\, e^{iQr}\, \Psi_{k'}(r) = \sum_R e^{i(k-k'+Q)R} \frac{1}{N} \int_{\text{cell}} \mathrm{d}^3 r''\, u_{k'}(r'') u_k^*(r'')\, e^{i(k-k'+Q)r''}. \tag{9.2.8}$$

The sum over R requires that $\vec{k} - \vec{k}' + \vec{Q}$ be a reciprocal lattice vector $\vec{G}$ in order that the sum does not vanish. Given that $\vec{k} - \vec{k}' + \vec{Q}$ is a reciprocal lattice vector, the sum over R yields simply N, the total number of unit cells present. Collecting these results, we find the transition rate as

$$P(k, c; k', c') = \frac{(2\pi)^4}{\hbar V^2} \left| \sum_Q \langle c' | H'(Q) | c \rangle \right|^2 \zeta \delta[E(k', c') - E(k, c) \pm \hbar\omega], \tag{9.2.9}$$

where ζ is called the overlap integral and is given as

$$\zeta = \left| \int_{\text{cell}} \mathrm{d}^3 r\, u_k^*(r) u_{k'}(r)\, e^{iGr} \right|^2. \tag{9.2.10}$$

From the above analysis, it is clear that the electron–phonon transition rate can be calculated once the nature of the perturbation Hamiltonian is known. Below we consider the form of the perturbing Hamiltonian for the cases described above, that is, the deformation-potential and polarization-wave scattering.

The first case we consider is that of deformation-potential scattering. Although there are several different types of deformation-potential scattering mechanisms active in semiconductors, we consider only one example since the calculations are nearly identical for each mechanism. For a complete discussion the reader is referred to the book by Jacoboni and Lugli (1989). The deformation potential physically arises from a displacement of the ions about their equilibrium positions. When the lattice spacing is changed from its equilibrium value, the potential energy of the lattice is locally increased. This is obvious from inspection of Figure 9.1.1. As we have discussed in Section 9.1, the crystal attains its minimum potential energy when all the ions are at their equilibrium positions. If the lattice expands or contracts about its equilibrium spacing, the local potential energy increases.

The potential energy of the lattice varies as a function of strain within the lattice. As the strain increases, the potential energy increases. The potential-energy shift is essentially linear with the strain or dilatation of the lattice. Hence the relationship between the perturbation Hamiltonian H' arising from the deformation potential and the strain in the lattice can be modeled as

$$H' = \sum_{lj} \epsilon_{lj} S_{lj}, \tag{9.2.11}$$

where S_{lj} is the strain tensor and ϵ_{lj} is known as the deformation-potential tensor. The strain is generally written as the derivative of the ionic displacement y.

The displacement y can be written in terms of phonon creation and annihilation operators, $a_Q^\dagger$ and a_Q, respectively. As was discussed in Chapter 2, creation and annihilation operators can be used to formulate the harmonic oscillator problem. A creation operator $a\dagger$ was determined to raise the energy of a harmonic oscillator state by one quantum, which for a one-dimensional oscillator is equal to $\hbar\omega$. Similarly, an annihilation operator acts to lower the energy of a state by one quantum. Any harmonic field can be treated quantum mechanically in much the same manner as was done in Chapter 2 for a simple harmonic oscillator. As in the simple harmonic oscillator, it is common to introduce both creation and the annihilation operators in describing a harmonic field. In the case of interest here, that of the harmonic vibration of a crystal, the creation and the annihilation operators create or destroy a quantum of energy equal to that of a phonon in the crystal. Subsequently, the harmonic displacement of the crystal y can be written as

$$y = \sum_{Q} \sqrt{\frac{\hbar}{2\rho V \omega}} \xi \left(a_Q + a_{-Q}^\dagger \right) e^{iQr}, \tag{9.2.12}$$

where ρ is the crystal density, V is the volume of the unit cell, ω is the phonon angular frequency, and ξ is the phonon polarization. Differentiating y with respect to r and substituting dy/dr for the strain tensor, we can write the perturbing Hamiltonian as

$$H' = \sum_{Q} i \sqrt{\frac{\hbar}{2\rho V \omega}} \left(a_Q + a_{-Q}^\dagger \right) \epsilon_{lj} Q_l \xi_j \, e^{iQr}. \tag{9.2.13}$$

Let us consider the special case of acoustic phonons. In this situation, for most cubic semiconductors the deformation-potential tensor ϵ_{lj} has cubic symmetry and is diagonal in form. Since cubic symmetry ensures that all the (100) directions are equivalent, then the diagonal elements should all be equal and have the same value of the deformation potential. Two different phonon polarizations are possible, either transverse or longitudinal. For the transverse acoustic phonons, the polarization ξ is perpendicular to the displacement ϵ. In the absence of off-diagonal terms in the deformation-potential tensor, the dot product between the polarization and the displacement vectors is clearly zero. Therefore there is no electron–phonon scattering by transverse acoustic phonons in materials that exhibit cubic crystalline symmetry. The longitudinal acoustic-phonon scattering does not vanish, however, since the phonon polarization and displacement are collinear. When the deformation-potential tensor is reduced to a scalar value, the perturbation becomes

$$H' = \sum_Q i \sqrt{\frac{\hbar}{2\rho V\omega}} (a_Q + a^\dagger_{-Q}) \epsilon_l Q \xi \, e^{iQr}. \tag{9.2.14}$$

We can now determine the longitudinal acoustic-phonon scattering rate by substituting in the expression for H' given by Eq. (9.2.14) into the matrix element and applying Fermi's golden rule. From Eq. (9.2.9) the operator H' acts on only the states $|c\rangle$ and $|c'\rangle$ through the operators a_Q and $a^\dagger_Q$. The matrix element vanishes for all terms in the sum over Q except those for which $|c'\rangle$ is changed by either $+Q$ or $-Q$ from $|c\rangle$. Hence the matrix elements of interest are then

$$|\langle c'|a_Q|c\rangle|^2, \qquad |\langle c'|a^\dagger_Q|c\rangle|^2. \tag{9.2.15}$$

The first matrix element above is simply equal to N_Q, the number of phonons in the state $|c\rangle$. Similarly, the second matrix element is equal to $N_Q + 1$. The proof of these results is given in Box 9.2.1. Because phonons are bosons, the phonon occupation factor, N_Q, is given by the Bose–Einstein distribution (see Chapter 5).

Box 9.2.1 Second Quantization Operators for Bosons

From our discussion in Chapter 2, the operators a and $a\dagger$ are destruction and creation operators, respectively. The Schroedinger equation for the harmonic oscillator is written as [Eq. (2.6.73)]

$$H\Psi = (aa\dagger - 1/2)\hbar\omega\Psi = E_n\Psi$$

or as [Eq. (2.6.75)]

$$H\Psi = (a\dagger a + 1/2)\hbar\omega\Psi = E_n\Psi.$$

The energy eigenvalue is given as [Eq. (2.6.90)]

$$E_n = (n + 1/2)\hbar\omega.$$

Therefore, when the above two equations are compared, n can be written in terms of operators as

$$a\dagger a\Psi = n\Psi.$$

The operator $a\dagger a$ is essentially a number operator. When applied to a state Ψ, it measures or counts the number of quanta n, present in that state.

Let the state Ψ be a multiparticle state. We represent this multiparticle state as

$$|n_1, n_2, \ldots, n_k, \ldots\rangle,$$

where n_1 represents the number of quanta present in the first state, n_2 is the number of quanta in the second state, etc. With this definition of the multiparticle state, the operator $a\dagger a$ needs to be subscripted to count the number of particles in a specific state. For example,

$$a\dagger_k a_k|n_1, n_2, \ldots, n_k, \ldots,\rangle = n_k|n_1, n_2, \ldots, n_k, \ldots\rangle,$$

where n_k represents the number of particles in the kth state. With the results of Chapter 2, the following commutation relation holds for the number operator $a\dagger_k a_k$:

$$[a_k, a\dagger_{k'}] = \delta_{kk'}.$$

To realize the meaning of these operators, consider the following:

$$a\dagger_k a_k a\dagger_k|n_1, n_2, \ldots, n_k, \ldots\rangle = (n_k + 1)a_k\dagger|n_1, n_2, \ldots, n_k, \ldots\rangle,$$

which follows from use of the commutation relation. Similarly,

$$a\dagger_k a_k a_k|n_1, n_2, \ldots, n_k, \ldots\rangle = (n_k - 1)a_k|n_1, n_2, \ldots, n_k, \ldots\rangle.$$

In each of the above equations, there is an undetermined coefficient that is due to the application of the creation or the destruction operator onto the multiparticle state. For simplicity, define the coefficients as

$$a\dagger_k|n_1, n_2, \ldots, n_k, \ldots\rangle = A^+(n_k)|n_1, n_2, \ldots, n_k, \ldots\rangle,$$

$$a_k|n_1, n_2, \ldots, n_k, \ldots\rangle = A^-(n_k)|n_1, n_2, \ldots, n_k, \ldots\rangle.$$

To find the effect of the application of each operator on the multiparticle state, consider

$$\langle \ldots n_k, \ldots, n_2, n_1 | n_1, n_2, \ldots n_k, \ldots\rangle = 1,$$

$$\langle \ldots n_k, \ldots n_2, n_1 | a\dagger_k a_k | \ldots, n_k, \ldots, n_2, n_1 \rangle = n_k.$$

But the application of a_k and $a\dagger_k$ in succession on $|\ldots n_k, \ldots, n_1\rangle$ is equal to the product of the coefficients $A^+(n_k - 1)A^-(n_k)$, where the argument types follow from the definitions of the creation and the destruction operators. Similarly,

$$\langle \ldots n_k, \ldots n_2, n_1 | a_k a\dagger_k | \ldots, n_k, \ldots, n_2, n_1 \rangle = n_k + 1,$$

which again must equal $A^+(n_k)A^-(n_k + 1)$. The above equations are satisfied if $A^\dagger(n_k)$ and $A^-(n_k)$ are defined as

$$A^\dagger(n_k) = \sqrt{n_k + 1}, \qquad A^-(n_k) = \sqrt{n_k}.$$

Therefore we obtain

$$a\dagger_k|n_1, n_2, \ldots, n_k, \ldots\rangle = \sqrt{(n_k + 1)}|n_1, n_2, \ldots, n_k, \ldots\rangle,$$

$$a_k|n_1, n_2, \ldots, n_k, \ldots\rangle = \sqrt{n_k}|n_1, n_2, \ldots, n_k, \ldots\rangle.$$

A common approximation made in calculating the acoustic-phonon scattering rate is to assume that the phonon energy is small. This approximation, called the equipartition approximation, is typically acceptable if the scattering occurs within a single valley (intravalley scattering) and the lattice temperature is near room temperature, 300 K. In the equipartition approximation, the Bose–Einstein distribution for the phonons can be approximated by

$$N_Q = \frac{1}{e^{\frac{\hbar\omega}{k_B T}} - 1} \sim \frac{k_B T}{\hbar Q s}, \tag{9.2.16}$$

where s is the sound velocity. With this assumption, the resulting scattering probability as a function of energy assuming parabolic energy bands is then obtained by evaluation of the integral over the final density of states as

$$\frac{1}{\tau} = \frac{\sqrt{2m^3}k_B T}{\pi \hbar^4 s^2 \rho}\sqrt{E}, \tag{9.2.17}$$

where m is the effective mass of the electron and E is the electron energy. Note that the scattering rates are the same for either phonon absorption or emission in the quipartition approximation since, by definition, the phonon energy is small with respect to the lattice temperature and as such the absorption and emission rates are comparable. The acoustic-phonon scattering rate in bulk GaAs as a function of electron energy within the equipartition approximation is plotted in Figure 9.2.1.

The second case of interest is electron-phonon scattering through polarization waves. As mentioned above, these scatterings occur from the interaction of the electron with the permanent electric dipole moment of the lattice. Obviously the host lattice needs to have some degree of polarization. A host lattice formed completely of identical atoms, such as silicon or germanium, exhibits no permanent electric dipole moment and as such cannot have polarization-wave

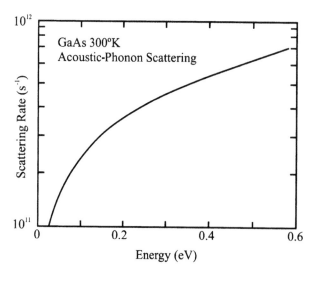

Figure 9.2.1. Acoustic-phonon scattering in the gamma valley in bulk GaAs.

scattering. As in the case of deformation-potential scattering, polarization-wave scattering can occur either with acoustic or optical branch phonons. In this section, we discuss the interaction with optical phonons only and again refer the reader to the references for a complete discussion.

The interaction Hamiltonian for the system is given as

$$H' = q \int \rho(r)\phi(r)\, \mathrm{d}^3 r, \tag{9.2.18}$$

where $\rho(r)$ is the charge density of the electron and is equal to the square of the electron wave function $|\Psi(r)|^2$. $\phi(r)$ is the electric potential associated with the polarization $P(r)$ of the crystal. As discussed above, polar optical phonon scattering arises from the interaction of the electrons with the polar phonons by means of the dipole moment induced by the polarization waves that accompany the phonons. The polarization $\vec{P}$ can be expressed by relation of the electric field $\vec{E}$ to the displacement $\vec{D}$ as

$$\vec{D} = \epsilon \vec{E} + \vec{P}. \tag{9.2.19}$$

If it is assumed that the material is neutral, then there is no free charge, and, from Maxwell's equations,

$$\vec{\nabla} \cdot \vec{D} = 0. \tag{9.2.20}$$

The electric field is then generally derived from a potential as

$$\vec{E} = -\vec{\nabla}\Phi. \tag{9.2.21}$$

From here the interaction Hamiltonian is typically derived with the help of quantum field theory. Since this is beyond the level of this book, the reader is referred to the books by Kittel (1963) and Callaway (1991) for a full discussion. From Shichijo (1980), the square of the interaction Hamiltonian is given as

$$\frac{2\pi q^2 \hbar \omega}{4\pi \epsilon_0 V Q^2} \left[\frac{1}{\epsilon_0} - \frac{1}{\epsilon_\infty} \right] G(k, k') \left[\frac{N_0}{N_0 + 1} \right], \tag{9.2.22}$$

where N_0 holds for absorption and $N_0 + 1$ holds for emission. In expression (9.2.22) the phonon energy $\hbar \omega$ is the optical phonon energy, ϵ_0 is the static dielectric constant, ϵ_∞ is the high-frequency dielectric constant, Q is the phonon wave vector, and $G(k, k')$ is the overlap integral between the initial and the final electronic states. In evaluating the total scattering rate, complications enter because of nonparabolic energy-band effects. If the energy bands are assumed to be parabolic, then the total scattering rate is calculated as

$$\frac{1}{\tau} = \frac{q^2 \sqrt{m}\, \omega_0}{4\pi \epsilon_0 \sqrt{2}\, \hbar} \left[\frac{1}{\epsilon_\infty} - \frac{1}{\epsilon_0} \right] \frac{1}{\sqrt{E}} \ln \left| \frac{\sqrt{E} + \sqrt{E'}}{\sqrt{E} - \sqrt{E'}} \right| \begin{Bmatrix} N_0 & \text{absorption} \\ N_0 + 1 & \text{emission} \end{Bmatrix}. \tag{9.2.23}$$

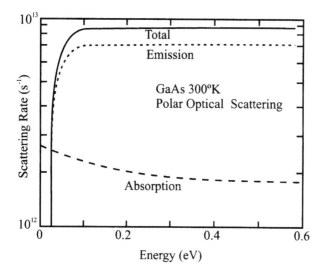

Figure 9.2.2. Polar optical scattering within the gamma valley of bulk GaAs.

The electron–polar optical scattering rate in bulk GaAs is plotted in Figure 9.2.2 as a function of electron energy. A comparison of Figures 9.2.1 and 9.2.2 shows that the polar optical scattering rate is substantially greater than the acoustic-phonon scattering rate in bulk GaAs. In addition, the final electron state is more likely directed near the initial electron state after a polar optical event than after an acoustic-phonon scattering event. For a more elaborate discussion, the reader is referred to the paper by Fawcett, Boardman, and Swain (1970) in the references.

9.3 **Electron–Electron Interactions**

In addition to scattering from phonons, there are several other scattering mechanisms of importance in semiconductors. Generally, there are two additional classes of scattering mechanisms: Coulombic scattering and disorder scattering. Coulombic scattering arises from the Coulomb interaction between a free electron and either a fixed impurity charge in the lattice, called ionized impurity scattering, or other free carriers, referred to as electron–electron or electron–hole scattering. Disorder scattering can arise from macroscopic crystalline imperfections, such as edge dislocations, grain boundaries, etc., from surface roughness and from the random fluctuations in the potential caused by the random arrangement of the atoms in an alloy. In this section we discuss scatterings arising from the interaction of a free carrier and other additional free carriers. Specifically, we consider electron–electron interactions, and in Section 9.4 we consider ionized impurity scattering. The reader is referred to the references, specifically to the book by Jacoboni and Lugli (1989) for a thorough presentation of disorder scattering.

The electron–electron interaction arises from the long-range nature of the Coulombic force. Bohm and Pines (1953) first recognized that the electron–electron interaction can be separated into two different components owing to the fact that the electron gas exhibits both collective and individual particle behaviors. The collective response arises from the collective oscillation of the electron gas due to the long-range nature of the Coulomb interaction. The resulting oscillations, called plasma oscillations, are manifest over distances greater than the characteristic screening length of the system. The quantum of the plasma oscillations is called a plasmon. On shorter-distance scales, the electron gas behaves more as a collection of individual charged particles.

The influence of the electron–electron interaction on the carrier dynamics is most pronounced in degenerate material. A semiconductor is said to be degenerate if the Fermi level lies near or above the conduction- or the valence-band edges. Perhaps a more precise definition of degeneracy is that in a degenerate material, the free-carrier statistics cannot be approximated by the Boltzmann distribution but require instead the use of the Fermi–Dirac distribution.

The short-range component of the electron–electron interaction is typically treated something like a billiard ball collision. One carrier scatters off another as if the carriers were hard spheres. During such a collision, the net momentum and the energy of the colliding particles are conserved. As a result, this mechanism cannot lead to any net loss of energy or momentum of the entire electron-gas system in bulk material under steady-state conditions.

The short-range electron–electron transition rate can be expressed through use of Fermi's golden rule as

$$S(k, k') = \frac{2\pi}{\hbar} |M|^2 f_{k_0} f_k (1 - f_{k_0})(1 - f_{k'}) \delta(E), \tag{9.3.1}$$

where $\delta(E)$ is a delta function that requires conservation of energy, f_k is the occupation distribution function, and $1 - f_{k'}$ is the probability that a vacancy exists in the state k'. The matrix element M is determined from a two-body screened Coulomb interaction given as

$$V(r - r') = \frac{q^2}{4\pi\epsilon} \frac{e^{-\beta|r-r'|}}{|r - r'|}, \tag{9.3.2}$$

where β is the inverse screening length. M is found by integration of the potential $V(r - r')$ over the two final electron states k'_0 and k' as

$$M = \frac{q^2}{4\pi\epsilon} \iint d^3r' d^3r \frac{e^{-\beta|r-r'|}}{|r - r'|} \Psi_{k_0}^*(r) \Psi_k^*(r') \Psi_{k'_0}(r) \Psi_{k'}(r'). \tag{9.3.3}$$

If the electronic states are assumed to be plane-wave states, then the above expression for M can be evaluated through use of a Fourier transform. When u and r are defined as

$$u = r - r', \quad r = u + r', \tag{9.3.4}$$

M becomes

$$M = \frac{q^2}{4\pi\epsilon} \frac{1}{V^2} \iint du \, e^{i(k'_0 - k_0)u} \frac{e^{-\beta u}}{u} \int e^{i(k'_0 - k_0 + k' - k)r'} d^3r'. \tag{9.3.5}$$

When the integrations are carried out, M can finally be evaluated to give

$$M = \frac{q^2}{\epsilon V} \frac{\delta(k'_0 + k', k_0 + k)}{\beta^2 + |k'_0 - k_0|^2}.$$ (9.3.6)

Typically the scattering rate is calculated then by the substitution of the above expression for M into Fermi's golden rule. It is further assumed that all the final electron states are unoccupied and that the initial state is occupied. The final expression for the short-range electron–electron scattering rate is given as

$$\frac{1}{\tau} = \frac{mq^4}{4\pi \hbar^3 \epsilon^2 \beta^2} \frac{k}{(\beta^2 + k^2)},$$ (9.3.7)

where k is defined as the magnitude of the difference between the two initial k vector states. In deriving Eq. (9.3.7) complications due to the electron spin have been neglected.

Far more interesting is the long-range component of the electron–electron interaction. The long-range component of the interaction can be pictured semiclassically as the excitation of a plasma wave from the disturbance of the free-carrier gas. Fundamentally, a semiclassical plasma wave arises from the response of the electron gas to the motion of a test charge or group of charges. The motion of the test charge disturbs the electron gas, leading to a collective excitation of the system. Put another way, a plasma wave originates from a displacement of the free charge from its equilibrium configuration. For example, if some of the electron charge is allowed to accumulate in one area, forming a depletion region in an adjacent region, a small dipole is created. Such a displacement will lead to a plasma oscillation as the charge seeks to move to restore equilibrium. If the electron gas is confined to a flat slab, the net polarization P will give rise to a field F of magnitude:

$$F = -4\pi n q^2 x,$$ (9.3.8)

where q is the electron charge and x is the displacement. The equation of motion of an electron in response to the dipole is

$$m \frac{d^2 x}{dt^2} = -4\pi n q^2 x.$$ (9.3.9)

Clearly, the system will respond with an oscillatory motion.

The frequency of oscillation of the collective electron system described above is typically denoted as ω_p, which is called the plasmon frequency. We can find the plasma frequency by finding the zeroes of the dielectric function for the system. As the reader may recall, the dielectric function relates the macroscopic to the microscopic electric fields in the solid. Owing to the presence of a free-carrier charge in the system, the effective potential experienced by an electron in a solid is not necessarily equal to the actual applied potential. Instead, the effective potential U is given by the quotient of the applied potential V and the dielectric constant of the system $\epsilon(Q, \omega)$, which is in general a function of the wavelength and the frequency of the applied potential. U can be expressed then as

$$U = \frac{V}{\epsilon(Q, \omega)}. \tag{9.3.10}$$

When $\epsilon(Q, \omega)$ vanishes, the effective potential that the electrons experience approaches infinity. This implies that a small external perturbation will lead to a large internal effective field; the electron gas becomes self-exciting.

To realize how a semiclassical plasma wave behaves in a highly degenerate electronic system, it is useful to solve numerically the Boltzmann equation self-consistently for a slab of degenerate GaAs material by use of the ensemble Monte Carlo method. The Monte Carlo method was described briefly in Section 6.3. As mentioned in Section 6.3, the Monte Carlo technique provides a microscopic solution of the Boltzmann equation through the direct simulation of the trajectories of the free carriers subject to the band structure, potential, and boundary conditions present. The simulation begins with some of the electrons accumulated in a small region, leaving behind a depleted region of uniform positive charge. At $t = 0$, a small dipole exists within the GaAs layer, as can be seen from Figure 9.3.1.a. The self-consistent Monte Carlo simulation is then used to study how the charge responds to the initial dipole excitation. No external applied electric field is assumed present. The only force acting on the electrons in the simulation is that due to the electron–electron interaction and the attraction from the uniform background positive charge. As can be seen from the successive frames of Figure 9.3.1, the dipole first begins to contract as charge moves into the depleted region. The charge movement then overshoots the equilibrium condition (charge balance everywhere throughout the device), giving rise to another dipole of inverted polarity, Fig. 9.3.1.d. The oscillation continues going through one complete cycle, reforming the original dipole, as shown in Figure 9.3.1.f.

From Figure 9.3.1 it is clear that the charge distribution undergoes oscillations on being disturbed from equilibrium. It is also clear that, because of inelastic scatterings, the plasma oscillation is damped. Physically, the coupling of the plasma to the lattice through phonon scatterings results in a net decrease in the energy of the electron gas, resulting in a decay of the plasma oscillation. The maximum amplitude of the dipole progressively decreases with time, as can be seen from the time evolution of Figure 9.3.1. The oscillation decay rate can be estimated from observation of how much the amplitude of the dipole decays after one cycle.

As we found in Chapter 1, all physical phenomena manifest themselves as both waves and particles but not simultaneously. Subsequently, the plasma waves in an electron gas can be pictured as quantum-mechanical particles called plasmons. As is the case for photons (the quanta of electromagnetic vibrations), plasmons (the quanta of plasma oscillations) have only a discrete energy equal to $\hbar\omega$, where ω, in this case, is the plasma frequency. In the quantum-mechanical picture of an electronic plasma, an electron can either absorb or emit a plasmon of energy $\hbar\omega$.

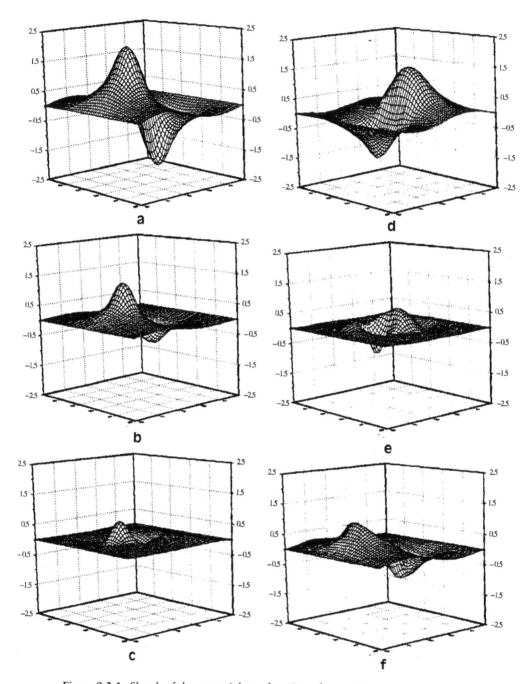

Figure 9.3.1. Sketch of the potential as a function of xy position in micrometers at various simulation times. The sequence of figures illustrates the oscillation of a plasma wave, within a degenerately doped semiconductor, in response to an initial disturbance. No applied field is present.

There are several different techniques commonly used to describe the collective behavior of electrons in a solid or a free-electron gas. As the reader can easily see, the quantum-mechanical description of the full nonequilibrium dynamics of an electron in the presence of many other electrons is complicated and almost hopeless to calculate directly. Nevertheless, there are two different models for treating collective excitations of an electron gas. These models are called elementary excitations and quasi-particles. Elementary excitations can be envisioned as arising from the excitation of several individual oscillators within the system. For example, each atom in a solid can undergo small oscillations about its equilibrium position. The collective oscillation of the entire lattice can then be divided into several independent normal modes. The quanta of these modes are the phonons.

The quasi-particle model of collective excitations in a many-particle system treats the interaction differently. Basically, instead of modeling the system as a collection of interacting particles, the system is modeled as a collection of noninteracting particles, whose characteristics are altered to reflect the influence of the surrounding particles on its behavior. Each quasi-particle is a particle that is accompanied by a compensating cloud of other charges. The interaction arising from the action of the other particles on the particular particle of interest is replaced by the inertia of the charge cloud itself. Such a particle is often called a dressed particle or a quasi-particle. The quasi-particle picture is often useful in describing the dynamics of electrons in solids since it replaces the complicated many-particle interactions by a suitably dressed particle.

The electron–plasmon interaction can be formulated with either the elementary excitation or the quasi-particle pictures. Let us first consider the elementary excitation model of the plasmon interaction. In this model, it is assumed that perturbations in the electron density generate fluctuations that can be characterized by a collective oscillation of the electron gas. These oscillations result in a polarization that generates an electric field described by a vector potential $\vec{A}$. The interaction of an individual electron and the field due to the vector potential $\vec{A}$ forms the basis of the electron–plasmon interaction. Following the approach of Madelung (1978), we find that the Hamiltonian of the system is given as

$$H^2 = \frac{q^2 \hbar^3}{8 V \epsilon_0 m^2} \sum_Q \frac{(\pm 2 \vec{Q} \cdot \vec{k} + Q^2)^2}{\omega_p(Q) Q^2} \left\{ \begin{matrix} N_Q \\ (N_Q + 1) \end{matrix} \right\}, \tag{9.3.11}$$

where the collective behavior is defined for momentum transfer of only less than a specified cutoff wave vector Q_c, N_Q is the Bose–Einstein occupation number for plasmons, $\omega_p(Q)$ is the plasma frequency, ϵ_0 is the static background dielectric constant, and k and Q are the electron and the plasmon momenta, respectively. V is the volume in which the electron gas is enclosed. There are two terms in Eq. (9.3.11) corresponding to absorption and emission. The top term $(+2\vec{Q} \cdot \vec{k})$ is for absorption while the bottom term $(-2\vec{Q} \cdot \vec{k})$ is for emission. The scattering rate can be computed with Fermi's golden rule with Eq. (9.3.11) for the matrix element within the Born approximation.

Applying Fermi's golden rule yields for absorption,

$$\frac{1}{\tau} = \frac{q^2\hbar^2}{32\epsilon m^2\pi^2} \int d^3Q \frac{4(\vec{Q}\cdot\vec{k})^2 + 4(\vec{Q}\cdot\vec{k})Q^2 + Q^4}{Q^2\omega_p(Q)} N_Q\delta(E_f - E_i - \hbar\omega_p).$$

$$(9.3.12)$$

To obtain the emission rate, the expression for the matrix element given by Eq. (9.3.11) must be used instead. Depending on the choice of the plasmon dispersion relation $\omega_p(Q)$, the integral in expression (9.3.12) can be evaluated. The most precise manner by which we can obtain the dispersion relation is by locating the zeroes of the dielectric function as described above. In practice only the real part of the dielectric function vanishes while the imaginary part, although typically very small, determines the lifetime of the excitation. Coupling between the plasmon and the phonon modes can be modeled through the dielectric function. The dielectric function includes lattice effects as well as the free-carrier motions. One means of expressing the dielectric function that includes lattice effects is given by Young and Kelly (1993) as

$$\epsilon(Q, \omega) = \epsilon_\infty - 4\pi[\mathcal{X}_{ij}(Q, \omega) + \mathcal{X}_L(\omega)],$$

$$(9.3.13)$$

where $\mathcal{X}_{ij}$ represents all the intraband and the interband contributions, that is, electron–electron, heavy hole–heavy hole, light hole–light hole, heavy–light hole, etc., and $\mathcal{X}_L$ represents the lattice contribution.

Often the plasmon dispersion relation is simplified through approximation. To a first-order approximation, the plasmon dispersion relation can be written as

$$\omega_p(Q) = \omega_0\left[1 + \frac{3}{10}\left(\frac{Qv_f}{\omega_0}\right)^2\right],$$

$$(9.3.14)$$

where ω_0 is defined as

$$\omega_0 = \sqrt{\frac{4\pi n_{\text{free}}q^2}{\epsilon_0 m}},$$

$$(9.3.15)$$

where n_{free} is the free-electron concentration. Equation (9.3.15) for the dispersion relation is useful since it greatly reduces the labors involved in determining the ω_p from the detailed dielectric function. However, it should be recognized that the above is only approximate and can lead to significant error in some situations.

The approximate form of the dispersion relation can be used to illustrate the calculation of the plasmon scattering rate as follows. Let us start with Eq. (9.3.12) with the dispersion relation given by Eq. (9.3.15). In deriving a formula for the plasmon interaction we will make several assumptions. In addition to the assumption of a constant value for the plasmon frequency, the band structure will be assumed to be parabolic and the effective-mass approximation is assumed to hold. By assuming a constant value for $\omega_p(Q)$ we can perform the calculation analytically. We can simplify the delta function in the energy by noting that the final and the initial energies are given as

$$E_f = E_{k+Q} = \frac{\hbar^2}{2m}(k + Q)^2 = \frac{\hbar^2 k^2}{2m} + \frac{\hbar^2 Q^2}{2m} + \frac{\hbar^2 \vec{k} \cdot \vec{Q}}{m},$$

$$E_i = \frac{\hbar^2 k^2}{2m}. \tag{9.3.16}$$

The argument of the delta function for absorption is then

$$E_f - E_i - \hbar\omega = \frac{\hbar^2 Q^2}{2m} - \hbar\omega + \frac{\hbar^2 \vec{k} \cdot \vec{Q}}{m}. \tag{9.3.17}$$

The expression for the scattering rate becomes

$$\frac{1}{\tau} = \frac{q^2 \hbar^2}{32\epsilon\pi^2 m^2} \int_{Q < Q_c} \frac{N_Q \sin\theta}{\omega_p} [4(\vec{k} \cdot \vec{Q})^2 + 4Q^2(\vec{k} \cdot \vec{Q}) + Q^4]$$

$$\times \delta\left[\frac{\hbar^2 Q^2}{2m} - \hbar\omega_p + \frac{\hbar^2 \vec{k} \cdot \vec{Q}}{m}\right] dQ \, d\theta \, d\phi. \tag{9.3.18}$$

If we define x as $\cos\theta$ and $dx = -\sin\theta \, d\theta$ and perform the integration over ϕ, the result becomes

$$\frac{1}{\tau} = \frac{q^2 \hbar^2}{16\epsilon\pi m^2} \int_{Q < Q_c} \frac{N_Q}{\omega_p} [4k^2 Q^2 x^2 + 4Q^3 kx + Q^4]$$

$$\times \delta\left[\frac{\hbar^2 Q^2}{2m} - \hbar\omega_p + \frac{\hbar^2 kQx}{m}\right] dQ \, dx. \tag{9.3.19}$$

Using the properties of delta functions, we can rewrite the delta function as

$$\delta\left[\frac{\hbar^2 Q^2}{2m} - \hbar\omega_p + \frac{\hbar^2 kQx}{m}\right] = \frac{m}{\hbar^2 kQ} \delta\left[x - \frac{m}{\hbar^2 kQ}\left(\hbar\omega_p - \frac{\hbar^2 Q^2}{2m}\right)\right]. \tag{9.3.20}$$

Make the following definition,

$$x_0 = \frac{m}{\hbar^2 kQ}\left(\hbar\omega_p - \frac{\hbar^2 Q^2}{2m}\right), \tag{9.3.21}$$

and integrate over x. Simplifying the resulting integral over q, taking into account the definition of x_0, we find that the integral reduces to

$$\frac{1}{\tau} = \frac{q^2 m}{4\pi\epsilon\hbar^2 k} \int_{Q < Q_c} \frac{N_Q \omega_p}{Q} \, dQ. \tag{9.3.22}$$

Performing the integration, denoting the lower and the upper bounds on Q by Q_{lower} and Q_{upper}, respectively, yields

$$\frac{1}{\tau} = \frac{q^2 m \omega}{4\pi \epsilon \hbar^2 k} N_Q \ln\left[\frac{Q_{upper}}{Q_{lower}}\right]. \tag{9.3.23}$$

Typically, the cutoff wave vector Q_c is less than the upper bound defined by the condition implied by the delta function. Therefore the integral over Q has an upper bound given by Q_c. By using the lower bound obtained from the delta function condition, we obtain the absorption rate by substituting these bounds into Eq. (9.3.23) to yield

$$\frac{1}{\tau} = \frac{q^2 m \omega}{4\pi \epsilon \hbar^2 k} \ln\left[\frac{Q_c/k}{\sqrt{1+\frac{\hbar\omega}{E}} - 1}\right] N_Q. \tag{9.3.24}$$

Similarly, the emission rate can be found as

$$\frac{1}{\tau} = \frac{q^2 m \omega}{4\pi \epsilon \hbar^2 k} \ln\left[\frac{Q_c/k}{1 - \sqrt{1-\frac{\hbar\omega}{E}}}\right] (N_Q + 1). \tag{9.3.25}$$

Another way of picturing the quantum-mechanical nature of the plasmon interaction is as a decay of a quasi-particle excitation into the Fermi sea of carriers. A quasi-particle is obtained by the addition to a filled Fermi sea of another electron that occupies an originally empty state. The presence of the Coulomb interaction acts as a mechanism by which the energy of the filled Fermi sea and quasi-particle can then be redistributed, causing the state of the quasi-particle to decay. The lifetime of the quasi-particle represents the mean time between emission of plasmons. The plasmon scattering rate, in this formulation, is obtained from the lifetime of the quasi-particle, which is in turn related to the imaginary part of the dielectric function. This can be seen from the following argument.

The wave function for a single excited electron has a time dependence given as

$$\Psi \sim e^{-i(E_r + iE_i)t}, \tag{9.3.26}$$

where E_r and E_i are the real and the imaginary parts of the electron energy, respectively. Clearly, the imaginary part E_i is associated with the lifetime of the particle, since it leads to an exponentially damped component of the wave function. Physically, after a time equal to one half the reciprocal of the imaginary part of the self-energy, the quasi-particle state will have transferred most of its excitation energy to the collective electron gas. This arises from the fact that the excitation itself is not a stationary state of the collective system. Quinn (1962) has shown that the imaginary part of the electron self-energy can be related to the dielectric function as

$$E_i = Im \frac{q^2}{2\pi^2} \int \frac{d^3 Q}{Q^2 \epsilon(Q, \omega)}. \tag{9.3.27}$$

Therefore, from a determination of the imaginary part of the dielectric function, the lifetime of the quasi-particle excitation can be found.

Either model of the electron–plasmon interaction, the elementary excitation or quasi-particle picture, can be invoked. Much work still remains on the subject of carrier–plasmon interactions. In this section, our attempt is only to broach the subject. The interested reader is referred to the current literature and to the references for more details.

9.4 Ionized Impurity Scattering

Aside from phonon and carrier–carrier scattering, a carrier is subject to scattering from an ionized impurity within the host semiconductor. Often impurities are unintentionally incorporated within a semiconductor and the control and avoidance of these impurities is one of the principal problems of semiconductor device engineering. In addition, it is often desirable to add impurities to a semiconductor to alter its electrical properties. Intentional introduction of impurities is called doping, and the dopants can be either n or p type. When a dopant is introduced into a semiconductor it will donate either an electron or a hole, leaving behind an ionized charged impurity center. These ionized impurity centers can act as scattering agents for free carriers as they propagate through the crystal.

The carrier-ionized impurity scattering rate can be evaluated in a similar way as was done in Section 9.3 for the short-range electron–electron interaction. In both situations, the interaction is Coulombic. As is well known, the scattering cross section for the Coulomb interaction between two bare charges diverges. This is due to the fact that the unscreened Coulomb interaction has infinite spatial extent. It is necessary to then cut off the interaction either through the use of screening, as was done in Section 9.3, or by simply cutting off the interaction after a certain distance. The latter method was adopted by Conwell-Weisskopf (Phys. Rev. 77, 388 [1950]). In this section, we adopt the screening technique in keeping with the method used in Section 9.3.

The screening length for charged impurity scattering can be treated as arising from static screening of the ionized charge. The effect of the electrostatic potential of a charged impurity on a carrier far removed from the impurity is not due to the action of the impurity alone. The charged impurity acts to attract electrons within the near vicinity of the center producing an induced charge density, $q\langle\rho(r)\rangle$. This induced charge density alters the potential in a manner that leads to the potential specified by Eq. (9.3.2). In nondegenerate semiconductors the screening length can be characterized by the inverse Debeye length, a characteristic length in semiconductors. The magnitude of the Debeye length and its physical origin can be sketched out as follows.

In general, the Coulomb potential due to an ionized impurity is given as

$$V(r) = \frac{q}{4\pi\epsilon(0)|r|},\tag{9.4.1}$$

where $\epsilon(0)$ is the static dielectric constant of the material. The Fourier transform of this potential yields (where the divergence at infinity has been cut off)

$$V(Q) = \frac{q}{\epsilon(0)Q^2}. \tag{9.4.2}$$

The dielectric constant is more generally written in terms of the dielectric function, which includes the frequency and the wave vector dependency of ϵ. Therefore $V(Q)$ becomes

$$V(Q) = \frac{q}{Q^2\epsilon(Q, \omega)}. \tag{9.4.3}$$

It can be shown (see Ferry 1991, p. 431) that the dielectric function for a non-degenerate semiconductor in the static limit approaches

$$\epsilon(Q \to 0, \omega = 0) = \epsilon(0) + \frac{nq^2}{Q^2 k_B T}, \tag{9.4.4}$$

where n is the free-electron concentration. In the case in which electrons and holes are both present, n should be replaced by $n + p$. The potential $V(Q)$ can be written as

$$V(Q) = \frac{q}{Q^2\left[\epsilon(0) + \frac{nq^2}{Q^2 k_B T}\right]} = \frac{q}{\epsilon(0)(Q^2 + Q_D^2)}, \tag{9.4.5}$$

which can be inverse transformed to

$$V(r) = \frac{q}{4\pi\epsilon(0)r} e^{-Q_D r}. \tag{9.4.6}$$

The quantity Q_D is defined as the inverse Debeye length. In Section 9.3, the screened potential was written in terms of a parameter β, which in the case of nondegenerate statistics becomes equal to Q_D, the inverse Debeye length described above.

When the system is no longer degenerate, the Coulomb potential can still be treated as screened but with a different inverse screening length than Q_D. In this case, the inverse screening length is described by the Thomas–Fermi screening wave vector, given as

$$Q_{TF} = \left(\frac{3\pi \hbar^2 \epsilon_0}{2m q^2}\right)^{1/2} \left(\frac{\pi}{3n}\right)^{1/6}, \tag{9.4.7}$$

where n is the electron concentration.

The scattering cross section from a screened Coulomb potential is a classic problem in quantum mechanics. It is solved in virtually every quantum-mechanics textbook. Consequently, not all the details are provided here, but the salient features of the calculation are summarized. The method we will adopt is often called the Born approximation. In the Born approximation, it is assumed that the scattered wave can be determined with only the first-order correction in a perturbation series expansion. The incident-wave function is assumed to propagate along the z direction and then has the form

$$\Psi_{incoming} = e^{ikz}. \tag{9.4.8}$$

The scattered wave is also assumed to be a plane wave. If the scattering is isotropic, meaning the scattering into all directions is equally probable, the outgoing wave should be spherically symmetric with the form

$$\Psi_{outgoing} \sim \frac{e^{ikr}}{r}. \tag{9.4.9}$$

However, in general, the scattering is not isotropic and has an angular dependence $f(\theta)$. In this case the outgoing wave has the form

$$\Psi_{outgoing} \sim \frac{f(\theta)\,e^{ikr}}{r}. \tag{9.4.10}$$

The quantity $f(\theta)$ is called the scattering amplitude and is a measure of the probability of the particle's being scattered at a certain angle θ. The form of the total wave function then for steady-state scattering is given as the sum of the incident and the scattered amplitudes as

$$\Psi_{total} = e^{ikz} + \frac{f(\theta)\,e^{ikr}}{r}. \tag{9.4.11}$$

The number of particles scattered into an element of the solid angle $d\Omega$ (see Figure 9.4.1) can be determined from the probability current density in the radial direction at this angle. The radial component of the probability current density is

$$J = \frac{\hbar}{2mi}\left[\Psi_{out}^{*}\frac{d}{dr}\Psi_{out} - \Psi_{out}\frac{d}{dr}\Psi_{out}^{*}\right]. \tag{9.4.12}$$

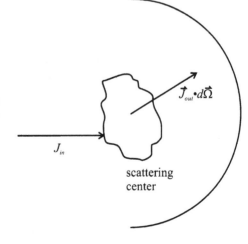

Figure 9.4.1. Diagram showing the incident and the scattered current densities around a scattering center.

Substituting Eq. (9.4.11) for the outgoing wave Eq. (9.4.12) becomes

$$J_{\text{out}} = \frac{\hbar k}{m r^2} |f(\theta)|^2. \tag{9.4.13}$$

The differential-scattering cross section is defined as $d\sigma$, which represents the effective area by which the incident carriers are scattered into some solid angle $d\Omega$. Therefore $d\sigma$ is given as the ratio of the scattered current to the incident current in the solid angle $d\Omega$. The probability density scattered at the output is simply given by $r^2 \, d\Omega \, J_{\text{out}}$. The incoming current is J_{in}. Therefore the differential-scattering cross section becomes

$$d\sigma = \frac{r^2 J_{\text{out}} \, d\Omega}{J_{\text{in}}}. \tag{9.4.14}$$

The incoming wave propagates only along z. Therefore the incoming current density is given as

$$J_{\text{in}} = \frac{\hbar k}{m}. \tag{9.4.15}$$

With these substitutions, the differential-scattering cross section becomes

$$d\sigma = |f(\theta)|^2 \, d\Omega. \tag{9.4.16}$$

The total scattering cross section σ is determined by integration of Eq. (9.4.16) over the solid angle.

Conservation of probability current density implies that the input current density must be equal to the output scattered probability density. The product of the input current density J_{in} and the differential-scattering cross section $d\sigma$ is equal to the amount of probability current density scattered in the direction $d\Omega$. The probability density scattered at the output is simply given by $r^2 \, d\Omega \, J_{\text{out}}$. The output can be determined in a different way, however, when use is made of the scattering transition rate.

In the physical situation considered here, the incident electron makes a transition from its initial state into a final state through the action of the scattering. The rate at which the incident electron undergoes a transition is given by the quantum-mechanical transition rate $1/\tau$. The probability of being scattered into the final solid element angle $d\Omega$ depends on the transition rate and the relative magnitude of $d\Omega$ with respect to all possible directions, that is, 4π. The scattered density into a differential solid-angle element $d\Omega$, called the scattering cross section, is then simply given by the product of the transition rate $W_{kk'}$ and $d\Omega$ as

$$r^2 J_{\text{out}} \, d\Omega = J_{\text{in}} \, d\sigma = \frac{1}{\tau} \frac{d\Omega}{4\pi}. \tag{9.4.17}$$

The quantum-mechanical transition rate from one state into another was found in Chapter 4 to be given by Fermi's golden rule. Making use of Eq. (4.4.18), we can write the transition rate as

$$\frac{1}{\tau} = W_{kk'} = \frac{2\pi}{\hbar} |\langle k'|V|k\rangle|^2 \rho(E_{k'}),$$
(9.4.18)

where k' and k are the final and the initial states, respectively, and $\rho(E_{k'})$ is the density of final states. For a three-dimensional system the density of states is given by Eq. (5.1.11). However, the form derived in Chapter 5 includes spin. In a scattering event, typically the spin of the particle is conserved. Therefore the final density of states available after the scattering is given by the number of states of the same spin as the initial case. This gives a density of states with a factor of 2 less than that specified by Eq. (5.1.11). Substituting the parabolic form in for the energy yields

$$\rho(E_{k'}) = \frac{mk'}{2\pi^2\hbar^2}.$$
(9.4.19)

When the above results are combined, the differential-scattering cross section per differential solid angle is

$$\frac{d\sigma}{d\Omega} = \frac{1}{4\pi} W_{kk'} \frac{1}{J_{in}}.$$
(9.4.20)

Substituting into Eq. (9.4.20) the expressions for $W_{kk'}$ and $1/J_{in}$ yields

$$\frac{d\sigma}{d\Omega} = \frac{1}{4\pi^2} \frac{m^2}{\hbar^4} \frac{k'}{k} |\langle k'|V|k\rangle|^2.$$
(9.4.21)

But for ionized impurity scattering, the interaction is elastic. This implies that $k' = k$, which gives

$$\frac{d\sigma}{d\Omega} = \frac{1}{4\pi^2} \frac{m^2}{\hbar^4} |\langle k'|V|k\rangle|^2.$$
(9.4.22)

The scattering amplitude $f(\theta)$ can now be determined with the above result and Eq. (9.4.16) $f(\theta)$ becomes

$$f(\theta) = \frac{1}{2\pi} \frac{m}{\hbar^2} |\langle k'|V|k\rangle|.$$
(9.4.23)

We can now calculate the scattering cross section for a screened Coulomb potential by substituting in the form of the screened Coulomb potential energy into Eq. (9.4.23). Substituting in and integrating over the angles yields

$$f(\theta) = \frac{mq^2}{2\pi\epsilon(0)\hbar^2} \frac{1}{K} \int e^{-Q_D r} \sin Kr \, dr,$$
(9.4.24)

where K is

$$K = |\vec{k} - \vec{k'}|. \tag{9.4.25}$$

Integrating over r yields a final expression for $f(\theta)$:

$$f(\theta) = \left[\frac{mq^2}{2\pi\epsilon(0)\hbar^2} \frac{1}{K^2 + Q_D^2} \right]. \tag{9.4.26}$$

The screened Coulomb scattering cross section is obtained from squaring $f(\theta)$ to yield

$$\frac{d\sigma}{d\Omega} = \left[\frac{mq^2}{2\pi\epsilon(0)\hbar^2} \frac{1}{K^2 + Q_D^2} \right]^2. \tag{9.4.27}$$

Generally, the Coulomb center can be multiply ionized. In other words, the impurity may be doubly or even triply ionized. For these cases, Eq. (9.4.27) should be multiplied by an additional factor Z as

$$\frac{d\sigma}{d\Omega} = \left[\frac{Zmq^2}{2\pi\epsilon(0)\hbar^2} \frac{1}{K^2 + Q_D^2} \right]^2. \tag{9.4.28}$$

The total scattering cross section is obtained from Eq. (9.4.28) by integration over all solid angles.

It is interesting to consider the above result for the scattering cross section in the limit of very highly incident energies. In this case, $K^2 \gg Q_D^2$. The cross section reduces then to

$$\frac{d\sigma}{d\Omega} = \left[\frac{Zq^2}{16\pi\epsilon(0)E} \right]^2 \frac{1}{\sin^4 \frac{\theta}{2}}, \tag{9.4.29}$$

which is precisely the classical cross section for Coulomb scattering. Hence in the limit of highly incident electron energies, the classical result is obtained.

In many applications the total scattering rate $1/\tau$ is of greater interest than the scattering cross section derived above. The total scattering rate can be determined directly with Fermi's golden rule. Here we derive the total scattering rate by using the second form of Fermi's golden rule given by Eq. (4.4.19). Since the scattering is elastic, the delta function between the energies is simply $\delta(E_f - E_i)$. The total scattering rate is obtained by summation of the transition rate over all the final states. If the final states are assumed to be in a quasi-continuum, which is the case for a solid, then the sum can be replaced by an integral over k'. The total scattering rate can be written then as

$$\frac{1}{\tau} = \frac{V}{(2\pi)^3} \int W_{kk'} \, d^3k'. \tag{9.4.30}$$

Eq. (4.4.19) for $W_{kk'}$ yields

$$W_{kk'} = \frac{2\pi}{\hbar} |\langle k' | V(r) | k \rangle|^2 \delta(E'_k - E_k). \tag{9.4.31}$$

Assuming normalized plane-wave states for the initial and the final states of the system, we can find the matrix element as

$$\frac{1}{V} \int e^{-ik'r} \frac{Zq^2}{4\pi \epsilon(0) r} e^{-Q_D r} e^{ikr} \, d^3r, \tag{9.4.32}$$

which is simply the Fourier transform of the screened Coulomb potential. The integration is similar to that performed in deriving Eq. (9.4.26). The result is

$$\frac{Zq^2}{\epsilon(0) V} \frac{1}{4k^2 \sin^2 \frac{\theta}{2} + Q_D^2}. \tag{9.4.33}$$

The transition rate $W_{kk'}$ is then

$$W_{kk'} = \frac{2\pi}{\hbar} \frac{Z^2 q^4}{\epsilon^2(0) V^2} \frac{1}{\left[4k^2 \sin^2 \frac{\theta}{2} + Q_D^2\right]^2} \delta(E_{k'} - E_k). \tag{9.4.34}$$

Integrating $W_{kk'}$ over all final states, assuming the states are those of free electrons, that is,

$$E = \frac{\hbar^2 k^2}{2m}, \tag{9.4.35}$$

we can ultimately calculate the total scattering rate per unit volume to be (see problem 5)

$$\frac{1}{\tau} = \frac{16\pi^2 Z^2 q^4 m}{\hbar^3 \epsilon^2(0)} \frac{k}{(2\pi)^3} \left[\frac{1}{2Q_D^2(4k^2 + Q_D^2)}\right]. \tag{9.4.36}$$

The above gives the scattering rate in inverse seconds for the scattering of an incident electron by a charged impurity.

PROBLEMS

1. The scattering probability per unit time for scattering by long-wavelength acoustic phonons in a two-dimensional system is given by

$$S(k, k') = \frac{2\pi^2 q^2 k_B T}{\rho A \hbar \omega^2} D \delta(E_{k'} - E_k),$$

where A is the area of the crystal, ρ is the mass density, D is the deformation potential constant, q is the electron charge, ω is the frequency, and T is the lattice temperature.

a. Determine the two-dimensional relaxation time defined by

$$\frac{1}{\tau} = \sum_{k'} S(k, k')(1 - \cos\theta).$$

b. Find the expression for the electron mobility in the system.

2. Consider the case of carrier–carrier scattering. The interaction between any two particles is a screened Coulomb interaction of the form

$$V(r - r') = \frac{q^2}{4\pi\epsilon} \frac{e^{-\beta|r-r'|}}{|r - r'|},$$

where β is the screening length and q is the electron charge. Determine the transition probability $S(k, k')$, if it is assumed that the system is nondegenerate. To do this, evaluate the matrix element of $V(r - r')$ between two particle states that are assumed to be plane waves and substitute into Fermi's golden rule.

3. Using the result of problem 2, determine the total scattering rate between two particles of equal mass. Assume that parabolic energy bands govern the system. Assume that the initial states are both occupied and that the final states are empty. To determine the total scattering rate, sum the transition rate determined in problem 2 over all the possible final and initial states. Convert the summations into integrals and evaluate.

4. Rework problem 3 assuming that the two particles have different masses. Let one particle be an electron of mass m_e and the other particle be a hole of mass m_h. Again, assume parabolic energy bands.

5. Derive the expression for the ionized impurity scattering rate given by Eq. (9.4.36) starting with the expression given by Eq. (9.4.34).

10

Generation and Recombination Processes in Semiconductors

In this chapter, we consider generation and recombination processes in semiconductors. In a semiconductor, electrons and holes can be either generated or recombined within a given volume, thereby changing the local carrier concentrations. In a sense, there are sources and sinks of particles within the semiconductor itself. Although generation and recombination events change the local carrier concentrations, the entire semiconductor must always remain space-charge neutral. This requirement leads to the injection or extraction of charge at the contacts. In this chapter, we examine the different types of generation and recombination processes and outline their behaviors.

10.1 Basic Generation–Recombination Mechanisms

There are in general three basic generation–recombination channels available in semiconductors. These are

1. Auger,
2. radiative,
3. thermal.

These mechanisms are defined as follows.

An Auger process is defined as an electron–hole pair (EHP) recombination followed by a transfer of energy from the recombined EHP to a free carrier, which is then excited to high energy within the band. The inverse Auger effect, in which an EHP is produced, is called impact ionization. In this case, a high-energy free carrier collides with the lattice and transfers its excess kinetic energy to an electron in the valence band, promoting it to the conduction band. Hence an EHP is produced after the event.

In a radiative recombination event, an EHP recombines with the emission of a photon. The electron recombines from the conduction band with a hole in the valence band. The energy lost by the electron is equal to the sum of its excess kinetic energy and the energy gap of the material. A photon of precisely this amount of energy is produced. Radiative recombination events can be either spontaneous or stimulated. Stimulated emission events arise from the perturbation of an excited system by an electromagnetic field. Spontaneous events occur without any external perturbation to the system. Radiative generation events of an EHP occur from the absorption of a photon of energy greater than or equal to the bandgap energy.

Thermal recombination and generation events arise from phonon emission or absorption respectively. Thermal processes are generally responsible for the presence of free carriers in the conduction and the valence bands in intrinsic semiconductors (those for which there is no intentional introduction of impurities) under equilibrium conditions, no photopumping or current injection. Of course, as the temperature decreases, the number of thermal generation events decreases and the free-carrier concentration is accordingly smaller.

There are, as well, two types of states from which a carrier can be generated or into which a carrier can be recombined, either band or bound states. Band states are defined simply as states in either the conduction or the valence band of the host semiconductor. Bound states, on the other hand, are defined as states arising from the presence of impurities or imperfections in the host semiconductor crystal. These states can be either intentionally or unintentionally introduced.

Intentional introduction of impurities is generally performed to change the free-carrier concentrations of electrons and holes. If an impurity is introduced that forms a shallow energy state for electrons near the valence-band edge, it is called an acceptor. These states act as traps, capturing thermally excited electrons from the valence band and leaving behind holes in their place. The material is said then to be p type since its equilibrium hole concentration exceeds its electron carrier concentration. Impurities that form shallow levels below the conduction-band edge act to increase the free-electron carrier concentration in the conduction band. Electrons are readily promoted from the shallow impurity levels to the conduction band through thermal processes. These shallow impurity levels are called donors. As a result, the material has an excess concentration of free electrons in the conduction band and is said to be n type.

Unintentionally added impurities typically form levels nearer the middle of the bandgap. As a result, unlike donor or acceptor states, these impurity states do not readily ionize. They behave more like traps, capturing and later reemitting electrons and holes. These states are typically referred to as deep-level traps. Electrons and holes can recombine or be generated from deep-level traps, thereby changing the free-carrier concentrations.

There are two types of transitions possible, band to band and band to bound. A band-to-band transition is one, for example, in which an electron in the conduction band directly recombines with a hole in the valence band. A bound-to-band transition, on the other hand, leads to EHP recombination through an intermediate impurity or defect state.

Let us consider band-to-band and band-to-bound transitions in detail. First, it is useful to define a notation to describe the action of each of these processes. Let g_n be the electron generation rate, g_h be the hole generation rate, r_n be the electron recombination rate, and r_h be the hole recombination rate. The above notation specifies the general class of generation–recombination events. Superscripts are used to specify the precise type of event, T for thermal, o for optical, and A for Auger. A recombination event removes a free carrier while a generation event produces a free carrier.

Let us consider each type of process, thermal, optical, and Auger, generation and recombination events qualitatively. Let us focus first on band-to-band

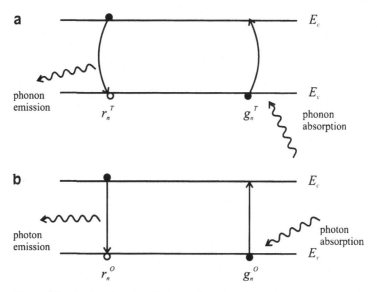

Figure 10.1.1. Band-to-band thermal and optical recombination and generation mechanisms. a. Electron thermal recombination r_n^T and generation g_n^T mechanisms. Note that these mechanisms are mediated by phonon emission and absorption, respectively. b. Electron optical recombination r_n^0 and generation g_n^0 mechanisms. In this case, these mechanisms are mediated by photon emission and absorption, respectively.

transitions. Each of these different mechanisms is sketched in Figures 10.1.1 and 10.1.2. Thermal generation and recombination events are mediated by phonon absorption and emission processes, respectively, as shown in Figure 10.1.1.a. Thermal processes depend strongly on the temperature of the lattice.

Band-to-band optical generation and recombination events proceed by means of photon absorption or emission, respectively. Typical radiative generation and recombination events are sketched in Figure 10.1.1.b. The degree to which a semiconductor either emits or absorbs a photon depends on the band structure. Materials such as silicon and germanium are called indirect semiconductors since the conduction-band minimum and the valence-band maximum (hole-energy minimum) occur at different points in k space, as shown in Figure 10.1.3. As we found in Chapter 8, the electron k vector is the crystalline momentum of the electron. Therefore, for indirect transitions, those that do not occur at the same place in k space, the electron momentum is different between the initial and the final states. In a radiative transition, a photon cannot either supply or remove the excess momentum necessary for an indirect transition to occur. Therefore indirect radiative transitions are forbidden to first order. These transitions are called then forbidden transitions.

Although forbidden transitions cannot occur to first order, these transitions can nevertheless occur to second order. Subsequently, an indirect-gap semiconductor can still have band-to-band radiative transitions. The second-order

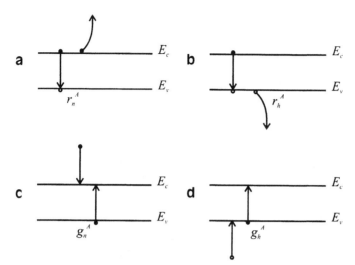

Figure 10.1.2. Band-to-band Auger generation and recombination mechanisms. a. Auger electron recombination. b. Auger hole recombination. c. Auger electron generation. d. Auger hole generation. c and d are also called electron and hole initiated impact ionization events, respectively.

transitions are mediated by a phonon. The phonon supplies the necessary momentum for which the generation or recombination event proceeds, as shown in Figure 10.1.3.

In direct-gap semiconductors, such as GaAs, InP, ZnS, etc., both momentum and energy are conserved directly in an optical generation or recombination

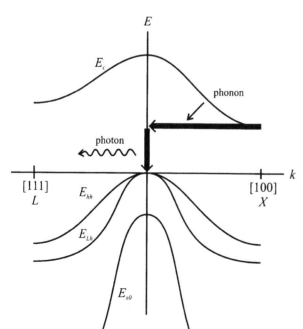

Figure 10.1.3. Energy-band structure of an indirect-energy-gap semiconductor illustrating indirect optical recombination.

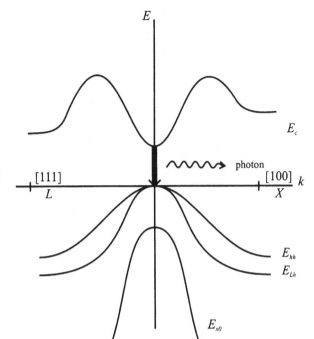

Figure 10.1.4. Energy-band structure of a direct-energy-gap semiconductor illustrating optical recombination.

event. This is due to the fact that the minima of the conduction and the valence bands occur at the same point in k space, as shown in Figure 10.1.4. A radiative transition can then occur to first order.

What, though, is the nature of first- and second-order radiative transitions? A first-order transition proceeds in accordance with Fermi's golden rule given in Chapter 4 as

$$W = \frac{2\pi}{\hbar} |\langle k|V|s\rangle|^2 \delta(E_k - E_s), \tag{10.1.1}$$

where V is the perturbing potential that induces the transition. In the case of optical transitions, V is the electric dipole moment. Summing over all the final states k in Eq. (10.1.1) above gives the total transition rate out of the initial state s due to the action of the perturbation V. Clearly, the probability of the transition occurring depends on the square of the matrix element $\langle k|V|s\rangle$ for the transition. Provided that the matrix element does not vanish, that is, V couples the initial state $|s\rangle$ to the final state $|k\rangle$, the transition can occur.

For indirect semiconductors, radiative transitions cannot occur to first order since the matrix element in Eq. (10.1.1) vanishes because the momentum is not conserved. The transition can occur, though, to second order when a transition is first made from the initial to an intermediate state followed by a transition from the intermediate to the final state. The second-order transition rate is given by Eq. (4.5.29):

$$W = \frac{2\pi}{\hbar} \left| \sum_m \frac{\langle k|V|m\rangle\langle m|V'|s\rangle}{(E_s - E_m)} \right|^2 \delta(E_k - E_s). \tag{10.1.2}$$

Physically, Eq. (10.1.2) implies that the transition from the initial state $|s\rangle$ to the final state $|k\rangle$ occurs through an intermediate state $|m\rangle$. The state $|m\rangle$ is not necessarily real but can be a virtual state instead. The intermediate transition is mediated by a phonon event. In other words, a phonon absorption or emission event takes the electron from the initial state $|s\rangle$ to the intermediate state $|m\rangle$, which lies at the same point in k space as the final state $|k\rangle$. The electron can then recombine radiatively from $|m\rangle$ to the final state $|k\rangle$. Hence an indirect recombination event is a two-step process; the first step is a thermal transition from the initial state to an intermediate state governed by the second matrix element in Eq. (10.1.2) followed by a transition from the intermediate state to the final state by a radiative process. The total probability of the transition occurring then depends on the product of the separate probabilities of each step occurring. Generally, the probability of each separate process occurring together is not highly likely. The probability that a second-order transition will occur is typically much less than that of a first-order transition. Therefore indirect semiconductors are not as efficient light emitters or detectors as direct semiconductors.

Auger generation and recombination processes involve the transfer of energy from one carrier to another through a collision. For example, in an electron Auger recombination event, as shown in Figure 10.1.2.a, when an EHP recombines, the energy of the recombination event is transferred to another electron, exciting it to high energy within the conduction band. Similarly, in an Auger hole recombination event, the recombination energy is transferred to a free hole that is promoted to high energy in the valence band.

Auger generation processes are also called impact ionization events. In an impact ionization event, an EHP is produced from the collision of an energetic carrier, either a hot electron or hole, with an electron in the valence band. The excess kinetic energy of the incident hot carrier is transferred to the valence-band electron, promoting it to states within the conduction band. Clearly, the impact ionization process, like all band-to-band generation mechanisms, is a threshold process; the initiating carrier must have sufficient energy to produce an EHP. The mimimum energy required for an impact ionization event to occur is called the threshold energy and must be at least equal to the bandgap energy. Typically, the threshold energy is somewhat higher than the bandgap energy. Impact ionization is an important process in many semiconductor devices since it can provide an internal current gain.

In addition to band-to-band generation–recombination processes, band-to-bound events can occur. Band-to-bound transitions occur between free-carrier states and deep-level trap or bound states located within the energy bandgap. These generation–recombination transitions occur at a fixed impurity center within the crystal. Physically, a deep-level trap state arises when an impurity atom has a significantly different electron binding energy or affinity from that of the host atoms or dopant atoms in the crystal. Therefore an electron is more tightly bound to the impurity than a dopant state and the level lies deeper in the band, often near midgap. Because of the large energy difference between deep-level trap states and the conduction band, these states are not readily ionized at room temperature. Therefore electrons falling into deep-level traps from the conduction band are trapped for a considerable time before they are reemitted. On being reemitted, these electrons generally recombine with holes in the valence band rather than return to the conduction band. Subsequently, an EHP is ultimately removed from the system.

Even a very small amount of deep-level traps in a semiconductor can greatly alter the electron–hole recombination rate. Sometimes deep-level impurities are intentionally added to a material to increase the electron–hole recombination rate. For example, in high-speed switching diodes, deep-level recombination centers are added to accelerate the rate with which the excess carriers will recombine after the diode is switched from forward to reverse bias.

Electron and hole band-to-bound generation–recombination events are sketched in Figure 10.1.5. Figures 10.1.5.a and 10.1.5.b illustrate electron capture and emission from a defect center. Hole capture and emission from a deep-level trap are shown in parts c and d of Figure 10.1.5. An empty deep-level trap state can capture a free electron within the conduction band. If a hole is then captured by the filled trap, or equivalently if the electron within the trap recombines with a free hole in the valence band (Figure 10.1.5.c), the net result is equivalent to a band-to-band EHP recombination event. Consequently, an EHP band-to-bound recombination event occurs in two steps, processes a and c as shown in Figure 10.1.5; a free electron in the conduction band is first captured by a deep-level trap state with its subsequent reemission and recombination with a free hole in the valence band.

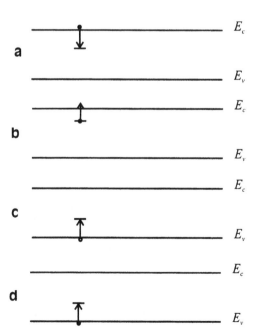

Figure 10.1.5. Shockley–Read–Hall recombination events. a. Electron capture by means of a defect center. b. Electron emission from a defect center. c. Hole capture by means of a defect center. d. Hole emission, electron captured from the valence band.

Similarly, an EHP can be generated through two band-to-bound transitions, d followed by b in Figure 10.1.5. An electron within the valence band is first captured by an empty trap state, leaving a free hole behind in the valence band. The electron can then be emitted from the trap state into the conduction band, giving rise to a free EHP in the semiconductor. The general class of processes involving transitions through a deep-level trap are called Shockley–Read–Hall recombination events.

10.2 Band-to-Band Thermal Generation and Recombination

Band-to-band generation and recombination processes occur in all semiconductor materials. Even in equilibrium, EHPs are generated and recombined by means of thermal processes. However, the net increase or decrease of EHPs from generation and recombination events does not change when the system is in equilibrium. However, for a system that is initially in nonequilibrium, generation–recombination processes alter the electron and the hole concentrations significantly. For example, consider a system that is originally illuminated by light of a frequency greater than the cutoff frequency for interband optical excitation. The incident light will generate EHPs in the material. As a result, the electron and hole carrier concentrations will become greater than the equilibrium carrier concentrations within the semiconductor. The resulting electron and hole carrier concentrations are said to be equal to the sum of the equilibrium and excess concentrations and are written as

$$n = n_0 + \delta n, \qquad p = p_0 + \delta p, \tag{10.2.1}$$

where p_0 and n_0 are the equilibrium hole and electron concentrations, respectively.

The continuity equation is helpful in understanding the physics governing generation–recombination in general and band-to-band generation–recombination specifically. In this section, we consider only band-to-band processes, delaying until Section 10.3 discussion of band-to-bound processes. The continuity equation is given by Eq. (6.3.36) as

$$\frac{1}{q}\vec{\nabla} \cdot \vec{J}_n = \frac{\partial n}{\partial t}. \tag{10.2.2}$$

In deriving the above relationship (see Section 6.3), it is assumed that there are no sources or sinks of particles within a phase-space volume. However, if generation and recombination events can occur, then electrons can be generated or removed within a particular phase-space volume. The above expression for the continuity equation must then be changed to take into account generation and recombination events. The resulting expression for the continuity equation is simply

$$\frac{1}{q}\vec{\nabla} \cdot \vec{J}_n + G - R = \frac{\partial n}{\partial t}. \tag{10.2.3}$$

If the current density J_n is zero, the continuity equation reduces to

$$G - R = \frac{\partial n}{\partial t}. \tag{10.2.4}$$

In words, Eq. (10.2.4) states that the time rate of change of the electron concentration is equal to the difference of the generation and recombination rates when the current density is zero. This is precisely what we would expect since in the absence of a current, the only way in which the electron concentration can change is through generation and recombination events. If the net generation rate exceeds the recombination rate, then the carrier concentration increases; the time rate of change of n is positive in Eq. (10.2.4). Conversely, if the recombination rate exceeds the generation rate, then the carrier concentration decreases.

If the system is in equilibrium, then the time rate of change of n is equal to the time rate of change of n_0:

$$\frac{\partial n}{\partial t} = \frac{\partial n_0}{\partial t} = 0, \tag{10.2.5}$$

which is clearly equal to zero. Therefore in equilibrium the net generation rate must be equal to the net recombination rate:

$$G = R \quad \text{(equilibrium)}. \tag{10.2.6}$$

From the discussion in Section 10.1, it is clear that the band-to-band recombination rate depends on the concentration of electrons in the conduction band and the concentration of holes in the valence band. Certainly no recombination events can occur if there are neither any electrons in the conduction band nor any holes in the valence band. The recombination rate then can be written as

$$R = K(np), \tag{10.2.7}$$

where K is a rate constant. In equilibrium, R is given as

$$R = R_0 = Knp = Kn_0 p_0 = Kn_i^2 = G_0, \tag{10.2.8}$$

where R_0 and G_0 are defined as the equilibrium recombination and generation rates, respectively. Outside of equilibrium, it is convenient to write the recombination and generation rates as

$$R = R_0 + R', \qquad G = G_0 + G'. \tag{10.2.9}$$

Therefore Eq. (10.2.4) becomes

$$(R' + R_0) - (G' + G_0) = -\frac{\partial(\delta n)}{\partial t}. \tag{10.2.10}$$

But $R_0 = G_0$, so

$$R' - G' = -\frac{\partial(\delta n)}{\partial t}. \tag{10.2.11}$$

If it is assumed that the nonequilibrium generation rate G' is zero, implying that there is no external optical pumping of the system, etc., then Eq. (10.2.11) reduces to

$$R' = -\frac{\partial(\delta n)}{\partial t}. \tag{10.2.12}$$

Equation (10.2.12) implies that the excess concentration decays through the recombination rate. As such, the recombination effectively acts to restore the system back to equilibrium. In other words, the excess electron concentration will decay, in the absence of a current flow and a nonequilibrium generation rate, back to its equilibrium concentration in some characteristic time by the action of the recombination mechanisms.

An expression for the excess electron concentration as a function of time in the presence of band-to-band recombination can be obtained as follows. The total recombination rate R can be written as

$$\begin{aligned}
R &= K(np) = K(n_0 + \delta n)(p_0 + \delta p) \\
&= K(n_0 p_0 + n_0 \delta p + p_0 \delta n + \delta n \delta p).
\end{aligned} \tag{10.2.13}$$

Therefore R' becomes

$$R' = Kn_0\delta p + Kp_0\delta n + K\delta n\delta p = -\frac{\partial(\delta n)}{\partial t}. \tag{10.2.14}$$

If it is assumed that $\delta n = \delta p$ everywhere, then

$$Kn_0\delta n + Kp_0\delta n + K\delta n^2 = -\frac{\partial(\delta n)}{\partial t}. \tag{10.2.15}$$

Solving for δn yields

$$K[n_0 + p_0 + \delta n]\delta n = -\frac{\partial(\delta n)}{\partial t},$$
$$\int_0^t \frac{\frac{\partial(\delta n)}{\partial t}\, dt}{(n_0 + p_0 + \delta n)\delta n} = \int_0^t -K\, dt'. \tag{10.2.16}$$

But we can integrate the left-hand side of the second of Eqs. (10.2.16) by noting that it is equal to

$$\int \frac{d(\delta n)}{(n_0 + p_0 + \delta n)\delta n}, \tag{10.2.17}$$

which can be integrated with the integral:

$$\int \frac{dx}{(A + Bx)x} = -\frac{1}{A} \log\left(\frac{A + Bx}{x}\right). \tag{10.2.18}$$

Therefore the expression integrates out to

$$-\frac{1}{n_0 + p_0}\left\{\log\left[\frac{n_0 + p_0 + \delta n(t)}{\delta n(t)}\right] - \log\left[\frac{n_0 + p_0 + \delta n(0)}{\delta n(0)}\right]\right\} = -Kt,$$

$$\log\left\{\frac{[n_0 + p_0 + \delta n(t)]\delta n(0)}{[n_0 + p_0 + \delta n(0)]\delta n(t)}\right\} = K(n_0 + p_0)t. \tag{10.2.19}$$

Solving for $\delta n(t)$ yields

$$\delta n(t) = \frac{\delta n(0)(n_0 + p_0)}{e^{K(n_0+p_0)t}[n_0 + p_0 + \delta n(0)] - \delta n(0)}. \tag{10.2.20}$$

Checking the bounds of the above result as t goes to zero, we find that the excess concentration becomes simply $\delta n(0)$. As t approaches infinity, the excess carrier concentration approaches zero; the electron concentration relaxes back to its equilibrium concentration n_0 through the action of the recombination.

The expression for the excess electron concentration can be simplified if the excess concentration is assumed to be small with respect to the sum of the equilibrium electron and hole concentrations. This implies that

$$\delta n(0) \ll (n_0 + p_0). \tag{10.2.21}$$

Under this assumption the excess concentration as a function of time becomes

$$\delta n(t) = \delta n(0)e^{-K(n_0+p_0)t}. \tag{10.2.22}$$

Therefore the excess electron concentration is found to decay exponentially at some characteristic time determined by the product of the recombination rate constant K and the sum of the equilibrium concentrations n_0 and p_0. Generally, the characteristic time constant of the decay is called the carrier lifetime for band-to-band recombination and is denoted as τ. The expression for the excess electron concentration becomes then

$$\delta n(t) = \delta n(0)e^{-\frac{t}{\tau}}. \tag{10.2.23}$$

10.3 **Shockley–Read–Hall Recombination Processes**

Let us now consider band-to-bound generation–recombination mechanisms quantitatively. It is helpful to first make some symbolic definitions:

R_{cn} is the recombination rate of electrons into traps,
C_n is the trap capture cross section (considered a constant),
E_n is the trap emission cross section (considered a constant),

R_{en} is the electron emission rate from traps,
N_t is the total number of trap states present,
$f(E_t)$ is the probability that a trap of energy E_t is occupied,
$1 - f(E_t)$ is the probability that a trap state is empty.

The trap capture rate is proportional to the number of empty traps, which is given as the product of the total number of traps and the probability that a trap is unoccupied:

$$N_t[1 - f(E_t)]. \tag{10.3.1}$$

The trap capture rate must also depend on the concentration of free carriers available n, since no capture events can occur if there are no electrons available to be trapped. The trap emission rate depends on the number of filled traps,

$$N_t f(E_t), \tag{10.3.2}$$

as well as on the number of vacancies present in the conduction band. Under nondegenerate conditions (defined as the condition in which the use of Maxwell–Boltzmann statistics is applicable in determining the electron carrier concentration in the conduction band) the conduction band is virtually empty and the Fermi–Dirac filling factor for electrons in the conduction band can be safely neglected in determining the trap emission rate.

The trap emission cross section E_n is assumed to be constant, a function of only the material identity. An expression for E_n can be obtained as follows. In equilibrium the trap emission and capture rates must be equal. Therefore

$$R_{en} = R_{cn}. \tag{10.3.3}$$

The recombination or capture rate is directly proportional to the product of the number of empty traps and the concentration of electrons in the conduction band, as discussed above. R_{cn} is given then as

$$R_{cn} = C_n N_t[1 - f(E_t)]n, \tag{10.3.4}$$

where C_n is the capture cross section. Similarly, the emission rate is proportional to the number of filled traps and can be written as

$$R_{en} = E_n N_t f(E_t), \tag{10.3.5}$$

where the concentration of vacancies in the conduction band has been neglected. Equating the two rates and solving for E_n yields

$$E_n = n_0 C_n \frac{[1 - f_0(E_t)]}{f_0(E_t)}, \tag{10.3.6}$$

where n_0 is defined as the equilibrium electron carrier concentration and $f_0(E_t)$ is the equilibrium distribution function. The equilibrium distribution function is simply the Fermi–Dirac function,

$$f_0(E_t) = \frac{1}{1 + e^{\frac{E_t - E_f}{k_B T}}}.$$ (10.3.7)

Therefore $1 - f_0(E_t)$ is simply

$$1 - f_0(E_t) = \frac{e^{\frac{E_t - E_f}{k_B T}}}{1 + e^{\frac{E_t - E_f}{k_B T}}}.$$ (10.3.8)

With these substitutions, the emission cross section becomes

$$E_n = n_0 C_n e^{\frac{E_t - E_f}{k_B T}}.$$ (10.3.9)

We can simplify expression (10.3.9) by noting that we can obtain the equilibrium electron concentration n_0 easily by integrating the product of the equilibrium distribution function by the density of states by using the approach in Section 5.8. The equilibrium electron concentration is given by Eq. (5.8.26):

$$n_0 = N_c e^{\frac{-(E_c - E_f)}{k_B T}},$$ (10.3.10)

where N_c is defined as the effective conduction-band density of states. N_c is defined by Eq. (5.8.26) as

$$N_c = 2 \left(\frac{2\pi m^* k_B T}{h^2} \right)^{\frac{3}{2}},$$ (10.3.11)

where m^* is the electron effective mass and only one equivalent conduction-band minimum is assumed present. The concentration n_0 can be rewritten in terms of the intrinsic concentration n_i as

$$n_0 = n_i e^{\frac{(E_f - E_i)}{k_B T}},$$ (10.3.12)

where E_i is called the intrinsic level and is equal to the Fermi level in pure, intrinsic material. The expression for E_n can be simplified further if it is assumed that the trap energy is aligned with the Fermi level, $E_t = E_f$. With this assumption, the expression for the emission cross section is

$$E_n = C_n n_i e^{\frac{(E_t - E_i)}{k_B T}} = C_n n_1,$$ (10.3.13)

where n_1 is defined as the equilibrium electron concentration corresponding to the special case of the trap energy's being coincident with the Fermi level. The emission cross section is found then to be equal to the product of the capture cross section and the equilibrium electron concentration.

The recombination rate outside of equilibrium is typically different from the generation rate. For example, if the excess carrier concentration exceeds the equilibrium carrier concentration and all external perturbations are then removed, recombination will act to restore the system to equilibrium. Similarly, if the system is initially constructed such that the carrier concentrations are less than the equilibrium concentrations and then the driving forces are removed, generation events will restore the system back to equilibrium. The net trap capture rate is given by the difference between the trap capture and trap emission rates as

$$R_n = R_{cn} - R_{en}. \tag{10.3.14}$$

Substituting the expressions for R_{cn} and R_{en} given by Eqs. (10.3.4) and (10.3.5) into Eq. (10.3.14) yields

$$R_n = C_n N_t [1 - f(E_t)] n - E_n N_t f(E_t). \tag{10.3.15}$$

When the expression for E_n given by Eq. (10.3.13) is used, the net recombination rate becomes

$$R_n = C_n N_t \{ n[1 - f(E_t)] - n_1 f(E_t) \}, \tag{10.3.16}$$

where n_1 is the equilibrium electron carrier concentration and n is the nonequilibrium electron carrier concentration. It is important to recognize that the distribution functions, $f(E_t)$ and $1 - f(E_t)$ used in Eq. (10.3.16), are nonequilibrium distribution functions. Below we will derive an expression for $f(E_t)$.

Similar capture and emission rates can be obtained for holes. The hole capture rate depends on the number of holes and the number of filled traps. A trap emits an electron that recombines with a free hole in the valence band during a hole capture event. A hole emission event depends on the number of empty traps. In a hole emission event, an electron within the valence band is captured by an empty trap, leaving a free hole behind in the valence band. Under nondegenerate conditions, the concentration of electrons in the valence band is sufficiently large that hole emission is not limited by it. The capture and the emission rates can then be written as

$$R_{cp} = C_p N_t f(E_t) p,$$
$$R_{ep} = E_p N_t [1 - f(E_t)]. \tag{10.3.17}$$

The hole emission cross section E_p can be found from equilibrium conditions in a manner similar to that for electrons. This yields

$$E_p = C_p p_1, \tag{10.3.18}$$

where p_1 is the equilibrium hole concentration; it is again assumed that the trap and the Fermi levels are coincident:

$$p_1 = n_i e^{\frac{-(E_t - E_i)}{k_B T}}. \tag{10.3.19}$$

The net hole recombination rate R_p is given by the difference of the hole capture and emission rates. R_p becomes

$$R_p = R_{cp} - R_{ep} = C_p N_t \{p f(E_t) - p_1 [1 - f(E_t)]\}. \tag{10.3.20}$$

Under steady-state conditions, the electron and hole recombination rates R_n and R_p are equal. Equating the expressions given by Eqs. (10.3.16) and (10.3.20) for R_n and R_p yields

$$C_n N_t \{n[1 - f(E_t)] - n_1 f(E_t)\} = C_p N_t \{p f(E_t) - p_1 [1 - f(E_t)]\},$$
$$C_n N_t n - C_n N_t n f(E_t) - C_n N_t n_1 f(E_t) = C_p N_t p f(E_t) - C_p N_t p_1$$
$$+ C_p N_t p_1 f(E_t). \tag{10.3.21}$$

Combining terms and simplifying yields

$$f(E_t)[C_p N_t (p + p_1) + C_n N_t (n + n_1)] = N_t (C_n n + C_p p_1). \tag{10.3.22}$$

Solving for the nonequilibrium distribution function $f(E_t)$ yields

$$f(E_t) = \frac{C_n n + C_p p_1}{C_n (n + n_1) + C_p (p + p_1)}, \tag{10.3.23}$$

where n_1 and p_1 are the equilibrium electron and hole carrier concentrations defined by Eqs. (10.3.13) and (10.3.19), respectively. The electron and hole concentrations n and p in Eq. (10.3.23) are the nonequilibrium carrier concentrations. The equilibrium carrier concentrations n_1 and p_1 are related through the law of mass action. The law of mass action states that the product of the equilibrium electron and hole concentrations is a constant. From Eq. (5.8.26) the electron concentration in the conduction band is

$$n = N_c e^{\frac{-(E - E_f)}{k_B T}}. \tag{10.3.24}$$

A similar relationship can be derived for the hole concentration in the valence band. The equilibrium hole concentration is given as

$$p = N_v e^{\frac{-(E_f - E_v)}{k_B T}}. \tag{10.3.25}$$

The product of the equilibrium concentrations n and p is then

$$np = N_c N_v e^{\frac{-(E_c - E_f)}{k_B T}} e^{\frac{-(E_f - E_v)}{k_B T}},$$
$$np = N_c N_v e^{\frac{-(E_c - E_v)}{k_B T}}. \tag{10.3.26}$$

But $(E_c - E_v)$ is simply equal to the bandgap energy E_g. Therefore the np product becomes

$$np = N_c N_v e^{-\frac{E_g}{k_B T}}. \tag{10.3.27}$$

The np product is constant at fixed temperature. This constant is equal to the square of the intrinsic carrier concentration n_i, defined as the electron and hole concentration in pure material at a fixed temperature. The np product is then

$$np = n_i^2. \tag{10.3.28}$$

Subsequently, the product of n_1 and p_1 must also be equal to n_i^2, since both n_1 and p_1 are equilibrium concentrations. Therefore

$$n_1 p_1 = n_i^2. \tag{10.3.29}$$

With this result and the expression given by Eq. (10.3.23) for the nonequilibrium distribution function, the electron and hole trap recombination rates become

$$R_n = R_p = C_n N_t \{n[1 - f(E_t)] - n_1 f(E_t)\},$$
$$R_n = R_p = C_n N_t \left\{ n \left[1 - \frac{C_n n + C_p p_1}{C_n(n + n_1) + C_p(p + p_1)} \right] \right.$$
$$\left. - n_1 \left[\frac{C_n n + C_p p_1}{C_n(n + n_1) + C_p(p + p_1)} \right] \right\}. \tag{10.3.30}$$

We can simplify the first term in Eqs. (10.3.30) by rewriting it in terms of its least common denominator as

$$C_n N_t \left[\frac{n C_n n_1 + n C_p p}{C_n(n + n_1) + C_p(p + p_1)} \right]. \tag{10.3.31}$$

With the above simplification, the electron and hole trap recombination rates are

$$R_n = R_p = C_n N_t \left[\frac{C_n n n_1 + C_p n p - C_n n n_1 - C_p n_1 p_1}{C_n(n + n_1) + C_p(p + p_1)} \right],$$
$$R_n = R_p = \frac{C_n N_t C_p (np - n_i^2)}{C_n(n + n_1) + C_p(p + p_1)}. \tag{10.3.32}$$

If it is assumed that the electron and hole trap capture cross sections C_p and C_n are the same, then the expressions for the trap recombination rates reduce to

$$R_n = R_p = \frac{C N_t (np - n_i^2)}{(n + n_1) + (p + p_1)}, \tag{10.3.33}$$

where C is the trap capture cross section. Note that the recombination rate depends on the np product. In equilibrium, the np product satisfies the law of mass action and is thus simply equal to n_i^2. As a result, the numerator of Eq. (10.3.33) vanishes and the net electron and hole recombination rates are zero. However, when the np product is greater than n_i^2, there is an excess concentration of carriers in the semiconductor. The system will then try to restore itself to equilibrium through recombination. The expression for R_n and R_p above is positive, implying a net recombination rate. If the np product is less than n_i^2, there is a net depletion of free carriers. The system will then try to restore itself to equilibrium through generation. The expression for R_n and R_p in this case is negative, implying a net generation rate.

The electron and hole recombination rates can be reformulated in terms of an expression involving only the excess minority carrier concentration and a characteristic lifetime. It is common to define electron and hole lifetimes τ_n and τ_p as

$$\tau_p = \tau_n \equiv \frac{1}{CN_t}. \tag{10.3.34}$$

With these definitions, R_n and R_p become

$$R_n = R_p = \frac{np - n_i^2}{\tau_p(n + n_1) + \tau_n(p + p_1)}. \tag{10.3.35}$$

The nonequilibrium electron and hole concentrations n and p can be written in terms of the equilibrium concentrations n_0 and p_0 and the excess concentrations δn and δp as

$$n = n_0 + \delta n, \qquad p = p_0 + \delta p. \tag{10.3.36}$$

A semiconductor is classified as n type if the equilibrium electron concentration n_0 exceeds the equilibrium hole concentration p_0. Conversely, a material is p type if the equilibrium hole concentration is greater than the corresponding electron concentration. In an n-type semiconductor the holes are said to be the minority carriers in the material. If the excess electron concentration is small with respect to the equilibrium electron concentration in an n-type semiconductor, then the excess electron concentration can be neglected with respect to the equilibrium electron concentration. The excess hole concentration, however, may be comparable with or greater than the equilibrium hole concentration. The generation rate is given then as

$$R = \frac{(n_0 + \delta n)(p_0 + \delta p) - n_i^2}{\tau_p(n_0 + \delta n + n_1) + \tau_n(p_0 + \delta p + p_1)}. \tag{10.3.37}$$

But n_1 and p_1 are defined as the equilibrium electron and hole concentrations, assuming the trap energy lies near the Fermi level, $E_t \sim E_F$. If it is further assumed that the trap energy is around midgap, then n_1 and p_1 are essentially equal to the intrinsic electron and hole concentrations and can be approximated as being equal to n_i. With this approximation and the assumptions that for n-type material,

$$n_0\delta p \gg p_0\delta p, \qquad n_0\delta p \gg \delta n\delta p, \tag{10.3.38}$$

R can be written as

$$R = \frac{n_0 \delta p}{\tau_p (n_0 + \delta n + n_i)}, \tag{10.3.39}$$

where the second term in the denominator of Eq. (10.3.37) has been neglected since all its terms are much less than n_0. We can simplify the expression for R in Eq. (10.3.39) further by neglecting n_i and δn in the denominator with respect to n_0. R finally can be written as

$$R = \frac{n_0 \delta p}{n_0 \tau_p} = \frac{\partial p}{\tau_p}, \tag{10.3.40}$$

which expresses the recombination rate in terms of the excess minority carrier concentration and its lifetime.

10.4 Impact Ionization Transition Rate

The inverse Auger effect is commonly referred to as impact ionization. During an Auger event, an EHP recombines, while after an impact ionization event an EHP is produced. As discussed in Section 10.1, an impact ionization event occurs when a high-energy carrier makes a collision with the lattice, transferring its excess kinetic energy to a bound electron in the valence band, promoting it into the conduction band, and leaving a hole behind in the valence band. The initiating carrier can be either an electron or a hole. To generate an EHP, the incident carrier must have a kinetic energy at least equal to the bandgap energy. As such, impact ionization is a threshold process with a threshold energy of at least the bandgap energy. Typically, the threshold energy for impact ionization, defined as the minimum energy for which an impact ionization event will occur, is greater than the bandgap energy due to momentum conservation.

We can derive an expression for the impact ionization rate assuming that the interaction arises from a screened Coulomb scattering event. Let us consider an incident high-energy electron that makes a collision with an electron in the valence band, producing two electrons in the conduction band. The momentum of the collision must be conserved, which requires that the k vectors of the carrier species must be conserved. Defining the initial k vector as k_1 and the k vectors of the final states as k_1' and k_2', the k-vector conservation requirement gives

$$\vec{k}_1 + \vec{k}_2 = \vec{k}_1' + \vec{k}_2'. \tag{10.4.1}$$

The electron–electron interaction is essentially a two-body collision. The matrix element of the interaction can be written as

$$\langle k_1', k_2' | \frac{q^2 e^{-Q_D |r_1 - r_2|}}{4\pi \epsilon |r_1 - r_2|} | k_1, k_2 \rangle, \tag{10.4.2}$$

where a screened Coulomb interaction is assumed. The screening constant is given by Q_D. The electronic states of the system are assumed to be Bloch states of the form

$$|k_1\rangle = \frac{1}{\sqrt{V}} u_{k_1}(r_1) e^{i \vec{k}_1 \cdot \vec{r}_1}, \qquad |k_2\rangle = \frac{1}{\sqrt{V}} u_{k_2}(r_2) e^{i \vec{k}_2 \cdot \vec{r}_2},$$

$$|k_1'\rangle^* = \frac{1}{\sqrt{V}} u_{k_1'}^* e^{-i \vec{k}_1' \cdot \vec{r}_1}, \qquad |k_2'\rangle^* = \frac{1}{\sqrt{V}} u_{k_2'}^* e^{-i \vec{k}_2' \cdot \vec{r}_2}. \tag{10.4.3}$$

When the above forms are substituted in for the wave functions, the matrix element becomes

$$\frac{1}{V^2} \iint d^3 r_1 \, d^3 r_2 \left(u_{k_1'}^* u_{k_2'}^* u_{k_1} u_{k_2} \right) e^{-i(\vec{k}_1' \cdot \vec{r}_1 + \vec{k}_2' \cdot \vec{r}_2)} \frac{q^2 e^{-Q_D |r_1 - r_2|}}{4\pi \epsilon |r_1 - r_2|}$$

$$\times e^{i(\vec{k}_1 \cdot \vec{r}_1 + \vec{k}_2 \cdot \vec{r}_2)}. \tag{10.4.4}$$

Let u be defined as the difference between r_1 and r_2 and v be defined as r_2; then expression (10.4.4) can be rewritten as

$$\frac{1}{V^2} \frac{q^2}{4\pi \epsilon} \iint e^{-i(\vec{k}_1' - \vec{k}_1) \cdot (\vec{u} + \vec{v})} e^{-i(\vec{k}_2' - \vec{k}_2) \cdot \vec{v}} \frac{e^{-Q_D |u|}}{u} u_{k_1'}^* u_{k_2'}^* u_{k_1} u_{k_2} \, du \, dv, \tag{10.4.5}$$

where the Jacobian mapping $r_1 r_2$ to $du\, dv$ can be shown to equal 1. The functions u_k are integrated over only one unit cell as compared with the exponential functions, which are integrated over all space. Since the functions u_k are periodic, the integrals of these functions over all space will not converge. Therefore the integrations must be performed over only a unit cell, while the remainder of the integrand must be integrated over all space. If it is assumed that the remainder of the integrand is slowly varying over a unit cell, then we can approximate expression (10.4.5) by breaking the integral into the product of two separate integrals. One double integral is over u and v while the others are over r_1 and r_2 involving the u_k. The integrals over the u_k are called overlap integrals and we abbreviate these as I_1 and I_2. This is a standard approximation that is commonly used in solid-state physics. Rewriting expression (10.4.5) yields

$$\frac{q^2}{4\pi \epsilon} \frac{1}{V^2} I_1 I_2 \int e^{-i(\vec{k}_1' - \vec{k}_1) \cdot \vec{u}} \frac{e^{-Q_D u}}{u} du \int e^{-i(\vec{k}_1' - \vec{k}_1 + \vec{k}_2' - \vec{k}_2) \cdot \vec{v}} \, dv. \tag{10.4.6}$$

But the integral over v is simply a delta function of the form

$$V \delta(\vec{k}_1' + \vec{k}_2' - \vec{k}_1 - \vec{k}_2). \tag{10.4.7}$$

Expression (10.4.6) then becomes

$$\frac{q^2}{4\pi\epsilon V} I_1 I_2 \int_0^\infty \int_0^\pi \int_0^{2\pi} du \frac{e^{-Q_D u}}{u} u^2 \sin\theta \, d\theta \, d\phi \, e^{-i|\vec{k}'_1 - \vec{k}_1| u \cos\theta}$$

$$\times \delta(\vec{k}'_1 + \vec{k}'_2 - \vec{k}_1 - \vec{k}_2). \tag{10.4.8}$$

Integrating out the θ and ϕ dependencies yields

$$\frac{q^2}{4\pi\epsilon V} I_1 I_2 \frac{2\pi}{i|k'_1 - k_1|} \int_0^\infty e^{-Q_D u} \, du [e^{i|k'_1 - k_1|u} - e^{-i|k'_1 - k_1|u}]$$

$$\times \delta(\vec{k}'_1 + \vec{k}'_2 - \vec{k}_1 - \vec{k}_2). \tag{10.4.9}$$

Finally, expression (10.4.6) reduces to

$$\frac{q^2}{\epsilon V} \frac{I(k_1, k'_1) I(k_2, k'_2)}{(Q^2 + Q_D^2)} \delta(\vec{k}_1 + \vec{k}_2 - \vec{k}'_1 - \vec{k}'_2), \tag{10.4.10}$$

where Q is defined as $|k'_1 - k_1|$.

The total impact ionization transition rate can be calculated through use of Fermi's golden rule. The total transition rate is then

$$W_{ii} = \frac{2\pi}{\hbar} \left[\frac{V}{(2\pi)^3} \right]^2 \int |M|^2 \delta(E_f - E_i) \, d^3k_2 \, d^3k'_1, \tag{10.4.11}$$

where $|M|^2$ is given as

$$2 \left(\frac{q^2}{\epsilon V} \right)^2 \left| \frac{I(k_1, k'_1) I(k_2, k'_2)}{(Q^2 + Q_D^2)} \right|^2, \tag{10.4.12}$$

where the additional factor of 2 appears by neglecting collisions between electrons of like spin. The details of this derivation can be found in the book by Ridley (1988) and arise from the fact that the particles have definite spin and that different cases, both distinguishable and indistinguishable, are possible. The impact ionization transition rate can then be written as

$$W_{ii} = \frac{4\pi}{\hbar} \left(\frac{q^2}{\epsilon V} \right)^2 \left[\frac{V}{(2\pi)^3} \right]^2 \iint \frac{|I(k_1, k'_1) I(k_2, k'_2)|^2}{(Q^2 + Q_D^2)^2} \delta(E_f - E_i) \, d^3k_2 \, d^3k'_1. \tag{10.4.13}$$

The above result for the impact ionization transition rate is quite general and holds, provided that Fermi's golden rule remains valid.

We can simplify Eq. (10.4.13) by making assumptions about the energy-band structure $E(k)$. The most common assumption made to simplify Eq. (10.4.13) is to assume that the bands are parabolic, $E(k) = \hbar^2 k^2 / 2m$, or are nonparabolic, $E(1 + \alpha E) = \hbar^2 k^2 / 2m$. In either case, the band structure is assumed to obey an analytic relation between E and k. Such an approximation is not fully justified for most semiconductor materials, however, since the band structure, especially at the high energy at which impact ionization occurs, is not parabolic or even closely parabolic. Inspection of the band structures of silicon and GaAs shown in Section 8.6 shows that these band structures are not parabolic at high energy. In many narrow-bandgap materials such as InAs and HgCdTe, the parabolic band or nonparabolic band approximation is more accurate. As an example of the extension of Eq. (10.4.13) in determining the impact ionization transition rate, let us apply Eq. (10.4.13) to study bulk InAs.

InAs is a relatively narrow-bandgap material, having a bandgap of 0.41 eV at 77 K. In addition to the fact that the energy bandgap is relatively small so that impact ionization will occur at relatively low carrier energies, the electrons stay almost completely within the gamma valley since the L and the X valleys lie at very high energies, 1.31 eV and 1.95 eV, respectively, from the gamma valley minimum. The gamma valley typically can be well represented by an analytic relationship between E and k. Therefore it is expected that use of an analytic relationship between E and k in simplifying Eq. (10.4.13) will not incur great error.

It is common in discussing impact ionization to refer to a threshold energy. The threshold energy is defined as the minimum energy on which an impact ionization event can occur. In addition to the obvious requirement that the energy of the incident carrier be sufficiently high so that an EHP can be created from the aftermath of its collision with the valence band, the momentum of the system must also be conserved in the manner prescribed by Eq. (10.4.1). The conservation of energy requires then that

$$E_1 = E_1' + E_2' - E_2. \tag{10.4.14}$$

The threshold energy is determined by minimization of Eq. (10.4.14) subject to the condition of momentum conservation specified by Eq. (10.4.1). Therefore the minimization can be performed with Lagrange multipliers. The minimum condition then must be

$$\frac{\partial E_1'}{\partial k_1'} = \frac{\partial E_2'}{\partial k_2'} = \frac{\partial E_2}{\partial k_2}, \tag{10.4.15}$$

which implies that the group velocities of the states must all be equal. If parabolic energy bands are assumed, then the magnitudes of the k vectors corresponding to each state can be simply related. The k vector of the electron initially in the valence band, k_2, is equal to the negative k vector of the corresponding hole that is produced from the collision. Call the k vector of the hole k_v; application of Eq. (10.4.15) leads to

$$|k_1'| = |k_2'|, \qquad |k_v| = \frac{m_h}{m_e}|k_1'|. \tag{10.4.16}$$

The conservation of energy gives

$$E_{\text{th}} = 2E_1' + E_g + \frac{\hbar^2 k_v^2}{2m_h}, \tag{10.4.17}$$

where we have used the result that $k_1' = k_2'$ at threshold and that the hole energy is equal to the negative of the electron energy. If γ is defined as the ratio of the electron to the hole masses, the threshold energy can be written as

$$E_{th} = E_g + \frac{\hbar^2 k_c^2}{2m_e}(2 + \gamma^{-1}),$$
(10.4.18)

where k_c represents the k-vector magnitude for all three species. We can simplify the expression for the threshold energy further by noting that the conservation of k vectors gives

$$K_{th} = k_1' + k_2' - k_2.$$
(10.4.19)

But k_2 is equal to $-k_v$. Therefore the threshold k vector becomes

$$K_{th} = k_1' + k_2' + k_v = 2k_c + \gamma^{-1}k_c = (2 + \gamma^{-1})k_c.$$
(10.4.20)

The threshold energy is then, assuming parabolic energy bands,

$$E_{th} = \frac{\hbar^2 K_f^2}{2m_e} = \frac{\hbar^2 (2 + \gamma^{-1})^2 k_c^2}{2m_e}.$$
(10.4.21)

Equating expression (10.4.21) to that given by Eq. (10.4.18) yields

$$E_{th} = \frac{\hbar^2 (2 + \gamma^{-1})^2 k_c^2}{2m_e} = E_g + \frac{\hbar^2 k_c^2}{2m_e}(2 + \gamma^{-1}).$$
(10.4.22)

Solving for E_{th} in terms of E_g yields

$$E_{th} = E_g \frac{(2\gamma + 1)}{(\gamma + 1)}.$$
(10.4.23)

If the electron and hole masses are assumed to be the same, then $\gamma = 1$ and the threshold is

$$E_{th} = \frac{3}{2} E_g \quad (m_e = m_h).$$
(10.4.24)

The expression for the impact ionization transition rate W_{ii} can be simplified with the above relationships.

We can simplify the denominator of Eq. (10.4.13) by neglecting the screening constant Q_D. Substituting in for Q from Eq. (10.4.8), we find that the denominator becomes

$$\frac{1}{(Q^2 + Q_D^2)^2} = \frac{1}{Q^4} = \frac{1}{|k_1' - k_1|^4}.$$
(10.4.25)

But k_1' can be related to k_1 as

$$k_1' = \frac{\gamma k_1}{(2\gamma + 1)}.$$
(10.4.26)

The denominator can then be simplified as

$$|k_1' - k_1|^4 = k_1^4 \left| \frac{(1+\gamma)}{(1+2\gamma)} \right|^4,$$

(10.4.27)

with the threshold energy defined as

$$E_{\text{th}} = \frac{\hbar^2 k_1^2}{2m_e}, \qquad k_1^4 = \frac{E_{\text{th}}^2 (2m_e)^2}{\hbar^4}.$$

(10.4.28)

Using the above two results and rewriting the overlap integrals as I_c and I_v, we find that the expression for the impact ionization transition rate becomes

$$W_{ii} = \frac{2\hbar^3}{(2\pi)^5} \left(\frac{q^2}{\epsilon} \right)^2 \frac{I_c^2 I_v^2}{E_{\text{th}}^2 (2m_e)^2} \left[\frac{1+2\gamma}{1+\gamma} \right]^4 \iint \delta(E_f - E_i) \, \mathrm{d}^3 k_1' \, \mathrm{d}^3 k_2.$$

(10.4.29)

We can evaluate the double integrals in Eq. (10.4.29), assuming parabolic energy bands and much algebra, to be

$$\iint \delta(E_f - E_i) \, \mathrm{d}^3 k_1' \, \mathrm{d}^3 k_2 = \left(\frac{2m_e}{\hbar^2} \right)^3 \frac{\pi^3}{2} \frac{(1+\gamma)^2}{(1+2\gamma)^{\frac{7}{2}}} (E - E_{\text{th}})^2.$$

(10.4.30)

The impact ionization transition rate reduces then to

$$W_{ii} = \left(\frac{q^2}{4\pi\epsilon} \right)^2 \frac{m_e}{\hbar^3} \frac{I_c^2 I_v^2}{(1+2\gamma)^{\frac{3}{2}}} \left(\frac{E - E_{\text{th}}}{E_g} \right)^2.$$

(10.4.31)

Note that the impact ionization rate depends quadratically on the energy for energies greater than the threshold energy. The quadratic relationship of the impact ionization rate with energy was first determined by Keldysh (1965).

The impact ionization transition rate depends on the overlap integrals I_c and I_v. There is a simple sum rule that is often used to determine the values of the overlap integrals. However, recent numerical studies (see Burt 1985) have shown that the overlap integrals are widely overestimated by the sum rule and are typically very much lower. Therefore the overlap integrals involved in Eq. (10.4.31) are generally difficult to determine unless a direct numerical evaluation is performed with the calculated band structure. As a result, Eq. (10.4.31) is essentially treated as a parameterized equation and is usually rewritten as

$$W_{ii} = p \frac{1}{\tau(E_{\text{th}})} \left(\frac{E - E_{\text{th}}}{E_{\text{th}}} \right)^2,$$

(10.4.32)

where $1/\tau(E_{\text{th}})$ is the scattering rate at the threshold energy and p is a parameter that is adjusted along with the threshold to agree with experimental results.

As mentioned above, the Keldysh formula is most properly applied to materials for which the parabolic approximation to the band structure is appropriate and for which the electrons are confined chiefly within the central valley near the gamma point. For these reasons, Eq. (10.4.31) is best applied to narrow-bandgap materials such as InAs and HgCdTe. Nevertheless, the Keldysh formula is typically applied to the study of impact ionization in wider-bandgap materials such as silicon and GaAs since it is relatively simple to use. More advanced theories are currently being examined but because of their complexity are still not commonly applied (see the reference by Bude, Hess, and Iafrate [1992b], for example).

10.5 **Theories of Impact Ionization**

In Section 10.4 we determined an expression for the impact ionization transition rate. This expression determines the rate at which an impact ionization event will occur as a function of the initiating carrier's energy once it has attained threshold. The threshold energy was found to be equal to at least the bandgap energy and is typically much higher, depending on the conservation-of-momentum condition. If the bands are all assumed to be parabolic and are described by an effective mass, then a simple expression for the threshold energy is obtained that depends on the bandgap energy and the effective masses of the electrons and holes. Therefore the initiating carrier must attain a relatively high energy before an impact ionization event can occur. In order then to know what the total impact ionization rate in a material is, it is necessary to determine not only the impact ionization transition rate but also the rate at which carriers attain the threshold energy. Consequently, the total impact ionization rate depends on the transition rate for impact ionization as well as on the survival rate of high-energy carriers.

Shockley first suggested that impact ionization is due mainly to "lucky electrons," those that suffer no collisions in attaining the threshold energy (see Shockley 1961). In other words, carriers starting at or near the band edge drift under the action of an applied electric field to energies at which they can impact ionize. In so doing, the carriers suffer no collisions with phonons. Therefore the carriers lose energy only through impact ionization events. Clearly, not all the carriers within the semiconductor impact ionize, only those "lucky enough" to avoid phonon interactions during their flight.

A quantitative expression for the impact ionization rate based on the lucky electron theory can be developed as follows. We need first to derive the probability of a carrier's escaping collisions. If $1/\tau(E)$ is the probability per unit time that a particle is scattered, then the probability that a particle is scattered in time dt is

$$dt/\tau(E). \tag{10.5.1}$$

Let $P(t)$ be the probability of a carrier's not being scattered in the time range $[0, t]$. The question we seek to answer then is, what is the probability of a carrier's not being scattered in time $t + dt$, $P(t + dt)$? $P(t + dt)$ is simply given as the product of the probability of a carrier's drifting for time t without collision $P(t)$

and the probability that a carrier is not scattered in the time dt, $[1 - dt/\tau(E)]$ as

$$P(t + dt) = P(t)[1 - dt/\tau(E)]. \tag{10.5.2}$$

$P(t + dt)$ can also be expanded by a power series as

$$P(t + dt) = P(t) + dP/dt\, dt. \tag{10.5.3}$$

Equating expressions (10.5.2) and (10.5.3) for $P(t + dt)$ yields

$$P(t + dt) = P(t) + dP/dt\, dt = P(t) - P(t)dt/\tau(E),$$
$$dP/P = -dt/\tau(E). \tag{10.5.4}$$

Integrating both sides yields

$$\ln P(t_0) = -\int_0^{t_0} \frac{dt'}{\tau(E)},$$
$$P(t_0) = e^{-\int_0^{t_0} \frac{dt'}{\tau(E)}}, \tag{10.5.5}$$

which is the resulting expression for the probability that a carrier will escape collisions. t_0 is defined as the mean free time between impact ionizations.

Again assuming parabolic energy bands $E = \hbar^2 k^2/2m$, we can readily determine t_0 in terms of the ionization threshold energy E_{th}. The final k state at which the ionization occurs, k_f, can be found from the equation of motion of the carrier as

$$\hbar\, dk/dt = qF,$$
$$\hbar(k_f - k_i) = qF t_0, \tag{10.5.6}$$

where t_0 is the mean time necessary for a carrier to drift to the impact ionization threshold energy. If the carrier is assumed to start its drift from the band edge, then $k_i = 0$ and k_f becomes

$$k_f = qF t_0/\hbar. \tag{10.5.7}$$

When the assumption of parabolic energy bands is used, the threshold energy E_{th} is simply

$$E_{th} = \hbar^2 k_f^2/2m. \tag{10.5.8}$$

Substituting the expression for k_f given by Eq. (10.5.7) into Eq. (10.5.8) yields

$$E_{th} = q^2 F^2 t_0^2/2m. \tag{10.5.9}$$

Therefore t_0 is found to be

$$t_0 = (2mE_{th})^{1/2}/qF. \tag{10.5.10}$$

Finally, the expression for the probability of a carrier's escaping collisions within the time t_0 is

$$P(t_0) = e^{-\int_0^{\frac{\sqrt{2mE_{th}}}{qF}} \frac{dt'}{\tau(E)}}. \tag{10.5.11}$$

To find an expression for the impact ionization rate, we need to make one further assumption: that the relaxation time $\tau(E)$ can be approximated as a constant. If we let $1/\tau(E)$ be defined as C_1, then the value of $P(t_0)$ is simply

$$P(t_0) = e^{-\frac{C_1 \sqrt{2mE_{th}}}{qF}}. \tag{10.5.12}$$

The impact ionization rate has dimensions of inverse distance and is typically defined as the reciprocal of the mean free path for impact ionization. The mean free path for impact ionization, $\langle d \rangle$, is defined as the ratio of the ballistic impact ionization distance L_b to the probability that a carrier will drift this distance without scattering $P(t_0)$ and is given as

$$\langle d \rangle \equiv L_b / P(t_0). \tag{10.5.13}$$

The ballistic ionization distance L_b is simply the minimum distance a carrier must travel to reach threshold:

$$L_b = E_{th}/qF. \tag{10.5.14}$$

With this definition, the mean free path for impact ionization, $\langle d \rangle$, becomes

$$\langle d \rangle = E_{th}/qF \; 1/P(t_0). \tag{10.5.15}$$

The impact ionization rate, denoted as α for electrons and β for holes, is defined as the inverse of the mean free path as

$$\alpha = 1/\langle d \rangle = P(t_0)qF/E_{th}. \tag{10.5.16}$$

Substituting the expression for $P(t_0)$ into Eq. (10.5.16) yields

$$\alpha = qF/E_{th} \exp -[C_1\sqrt{(2mE_{th})}/qF], \tag{10.5.17}$$

which is the impact ionization rate when it is assumed that only lucky electrons contribute to the process. Note that the ionization rate goes exponentially with the reciprocal of the electric field, $\alpha \sim \exp[-1/F]$.

An alternative theory was proposed by Wolff (1954). Wolff suggested an essentially opposite picture of impact ionization from that of Shockley. Rather than lucky electron events, impact ionization was predicted to arise from a diffusion of carriers upward in energy. According to this theory, the electrons undergo many collisions while gaining energy, yet the energy lost per collision is small. The result is that the electron distribution becomes nearly isotropic and can thus be characterized as a drifted Maxwellian with an effective electron temperature T_e. Impact ionization events occur from those carriers within the high-energy tail of the distribution whose energies exceed the threshold energy. As the electric field is increased, the distribution is shifted to higher energy, resulting in more carriers within the high-energy tail. As a result the impact ionization rate increases accordingly.

In steady state, the energy gain from the field is balanced on average by the energy loss to the phonons. This can be written as a rate equation for the average energy as

$$dE/dt = qFv_d - E_p v_\theta/\lambda = 0, \tag{10.5.18}$$

where v_θ is the random Maxwellian thermal velocity, v_d is the drift velocity in the field direction (hence the power input into the electron system is simply the product of the drift velocity and the field), and λ is the mean free path between collisions. E_p is the average phonon energy. v_θ can be related to the mean thermal energy as

$$3/2k_B T_e = 1/2mv_\theta^2. \tag{10.5.19}$$

Expression (10.5.19) implies that the average electron energy is simply equal to the average kinetic energy. The time rate of change of the linear momentum is found as (see Problem 2, Chapter 6)

$$dp/dt = qF - mv_d v_\theta/\lambda = 0, \tag{10.5.20}$$

which leads to

$$v_d = qF\lambda/mv_\theta. \tag{10.5.21}$$

The energy rate equation, Eq. (10.5.18), readily gives a relationship for v_θ as

$$v_\theta = qFv_d\lambda/E_p. \tag{10.5.22}$$

Substituting Eq. (10.5.22) into Eq. (10.5.21) yields

$$v_d = qF\lambda E_p/(mqFv_d\lambda). \tag{10.5.23}$$

When Eq. (10.5.23) is simplified, v_d becomes

$$v_d = (E_p/m)^{1/2}. \tag{10.5.24}$$

Equation (10.5.19) relates the electron temperature to the Maxwellian velocity as

$$3k_B T_e = mv_\theta^2. \tag{10.5.25}$$

Substituting Eqs. (10.5.22) and (10.5.24) into Eq. (10.5.25) yields

$$3k_B T_e = \left(mq^2 F^2 \lambda^2 / E_p^2\right)(E_p/m),$$
$$k_B T_e = (qF\lambda)^2 / 3E_p. \tag{10.5.26}$$

As discussed above, the electron distribution is considered to be Maxwellian and is characterized by an electron temperature T_e. Therefore the probability that a carrier reaches the threshold energy E_{th} is simply given by the Boltzmann factor:

$$e^{-\frac{E_{th}}{k_B T_e}}. \tag{10.5.27}$$

Consequently, the probability that a carrier attains the threshold energy is

$$P(E_{th}) = e^{\frac{-3E_{th} E_p}{(qF\lambda)^2}}. \tag{10.5.28}$$

The impact ionization rate α is found in a similar way as before as the inverse of the mean free path for impact ionization. Hence α is given as

$$\alpha = \frac{qF}{E_{th}} e^{\frac{-3E_{th} E_p}{(qF\lambda)^2}}. \tag{10.5.29}$$

Note that the impact ionization rate is proportional to the exponential of the inverse square of the electric field. This is in contrast to the Shockley result in which the ionization rate is proportional to the exponential of the inverse field. In summary, the Shockley and Wolff theories predict that

$$\alpha \sim \exp[-1/F] \quad \text{Shockley,}$$
$$\alpha \sim \exp[-1/F^2] \quad \text{Wolff.} \tag{10.5.30}$$

Baraff (1962) first presented an alternative theory to that of either Wolff or Shockley that contains elements of both theories. As mentioned above, Shockley's lucky electron theory and Wolff's thermalized distribution theory are essentially opposite approaches. Shockley's theory predicts that impact ionization events arise solely from carriers that undergo lucky flights, those in which collisions are completely avoided, from the band edge to the ionization threshold energy. Wolff's theory, on the other hand, assumes that the ionization events occur from carriers within the high-energy tail of the electron energy distribution. Baraff combines the two approaches by assuming that the carrier distribution function is composed of two separate parts as

$$f(v, \cos\theta) = A(v) + B(v)k(1 - \cos\theta), \tag{10.5.31}$$

where $A(v)$ is a spherical or nearly isotropic velocity distribution similar to that proposed by Wolff and $B(v) k(1 - \cos\theta)$ is peaked in the field direction corresponding to the lucky electron distribution of the Shockley theory. In other words, the velocity distribution is composed of two separate parts, one component corresponding to an isotropic, randomized velocity distribution and the other composed of carriers that accelerate along the field direction, suffering no collisions along the way.

If one substitutes in expression (10.5.31) for the velocity distribution function into the Boltzmann transport equation, the resulting energy distribution function obtained is (see Capasso 1985)

$$f_0(E) = E^{-(\frac{2}{3}+\frac{qF\lambda}{3E_p})} e^{-\frac{E_{th}}{\frac{(qF\lambda)^2}{3E_p}+\frac{2}{3}qF\lambda}}, \tag{10.5.32}$$

where the variables are the same as defined above. In the limit as $qF\lambda$ approaches the average phonon energy E_p, Eq. (10.5.32) reduces to

$$f(E) = \frac{1}{E} e^{-\frac{E_{th}}{qF\lambda}}, \tag{10.5.33}$$

which is essentially Shockley's result, Eq. (10.5.17). This corresponds to the low electric-field limit; the field heating is comparable with the phonon energy. In the limit of high electric fields, $qF\lambda \gg E_p$, where the field heating is much larger than the average phonon energy, the distribution becomes

$$f(E) \sim E^{-\frac{qF\lambda}{3E_p}} e^{-\frac{3E_p E_{th}}{(qF\lambda)^2}}, \tag{10.5.34}$$

which is roughly Wolff's result. Therefore, at low electric fields, it is expected that Shockley's lucky electron theory governs the impact ionization process while at high electric fields, Wolff's theory is most applicable.

Baraff further developed an expression for the ionization rate in terms of the mean number of phonons emitted between impact ionization events as

$$\alpha = \frac{qF}{E_{th} + \bar{n}E_p}, \tag{10.5.35}$$

where n is the mean number of phonons emitted and E_p is the mean phonon energy as before. The actual evaluation of Eq. (10.5.35) is usually performed with a polynomial approximation of the impact ionization rate. The ionization rate can then be formulated as a function of three different variables, the mean free path between phonon collisions, the average phonon energy E_p, and the ionization threshold energy E_{th} as

$$\alpha = \frac{1}{\lambda} e^{[C_0(r)+C_1(r)x+C_2(r)x^2+C_3(r)x^3]}, \tag{10.5.36}$$

where $C_0(r)$, $C_1(r)$, $C_2(r)$, and $C_3(r)$ are polynomial functions. These are defined as

$$C_0(r) = -0.07238 - 51.5r + 239.6r^2 + 3357r^3,$$
$$C_1(r) = -0.4844 + 12.45r + 363r^2 - 5836r^3,$$
$$C_2(r) = 0.02982 - 0.07571r - 148.1r^2 + 1627r^3,$$
$$C_3(r) = -1.841 \times 10^{-5} - 0.1851r + 10.41r^2 - 95.65r^3, \tag{10.5.37}$$

where $r \equiv E_p/E_{th}$ and $x \equiv E_{th}/F\lambda$.

The principal disadvantage of the Baraff theory is that it relies on the determination of three adjustable parameters, E_p, E_{th}, and λ, none of which can be known with any degree of certainty from first principles. Typically, we determine these parameters by comparing the calculated ionization rates with experimentally determined measurements. As a result, the Baraff theory clearly does not provide a first-principles theory of impact ionization. It cannot be used as a means of predicting the ionization rate in advance of experimental measurements. Even more limiting is the fact that it is not clear just what each variable represents physically since these parameters absorb such complicated effects on the ionization rate as the band structure and the impact ionization transition rate.

To progress beyond the limitations of Baraff's theory, Shichijo and Hess (1981) used a numerical approach to reduce the number of assumptions and parameterizations. The numerical solution of the Boltzmann equation avoids many of the simplifying assumptions typically used in analytical approaches. Many of these assumptions are invalid or obscure the physics for which one is trying to solve. The numerical methods, on the other hand, make few assumptions and thereby retain much of the critical physics of the carrier transport dynamics. Chief among the numerical techniques is the ensemble Monte Carlo simulation, which was adopted by Shichijo and Hess (1981). In the Monte Carlo method, the Boltzmann equation is solved directly by simulation of the carrier motions throughout the material subject to the effects of the band structure and phonon scattering agents present (Jacoboni and Lugli 1989).

In addition to the numerically based Monte Carlo studies, other advanced theories of interband impact ionization have been proposed (Ridley 1983; Burt 1985; and Marsland 1987). In these theories and the numerically based approaches, most interband impact ionization events are predicted to arise from carriers that "lucky drift," not from the band minimum, as in the Shockley theory, but from the average energy of the distribution. Ridley (1983) has identified three main sources of ionizing carriers and has determined an expression for the electron interband impact ionization rate α as

$$\alpha = \frac{qF}{E_i}\left\{ P_1(F, E_i) + P_2(F, E_i) + P_T\left[P_1^T(F, E_i) + P_2^T(F, E_i)\right]\right\}, \qquad (10.5.38)$$

where E_i is the ionization threshold energy, F is the electric field, $P_1(F, E_i)$ is the probability of ballistic ionization's occurring from a carrier launched at zero energy, $P_2(F, E_i)$ is the probability of lucky drift from zero energy to the threshold energy, P_T is the probability of hot-electron thermalization, $P_1^T(F, E_i)$ is the probability of ballistic ionization's occurring from a carrier initially at the average thermal energy, and $P_2^T(F, E_i)$ is the probability of lucky-drift-induced impact ionization from a carrier initially at the average thermal energy.

Ridley (1983) defines lucky drift in terms of a carrier that drifts in the field and relaxes only its momentum but not its energy. This is plausible since at high carrier energies, a single phonon-scattering event will relax the energy of the carrier by typically no more than 50 meV but the momentum can be completely randomized. Lucky drift can be pictured as the physical situation in which the carrier drifts for some time under the action of the applied electric field with a velocity determined principally from momentum-relaxing collisions without significant energy relaxation. Lucky drift is an intermediate state between Shockley's ballistic state and Wolff's thermalized, steady-state. Capasso (1985) neatly summarizes Ridley's theory by pointing out that in the lucky-drift model most carriers start initially from the average energy and ionize after suffering only a few phonon collisions. Therefore the manner by which a carrier reaches sufficient energy to impact ionize can be viewed as a modified Shockley theory.

The lucky-drift model of impact ionization is also in good agreement with the numerically based theories of impact ionization. The most commonly used numerical method for studying impact ionization is the Monte Carlo technique. In this method, the Boltzmann equation is solved directly by simulation of the carrier motions throughout the material subject to the applied external electric field, device geometries, and phonon- and carrier-scattering mechanisms. The carrier's free-flight time and the phonon-scattering agents are selected stochastically (hence the name Monte Carlo) in accordance with some given probabilities describing the microscopic process. From tracing the carrier trajectories through the device in both k and real space, the macroscopic quantities of interest, that is, average energy, drift velocity, etc., can be ascertained. The Monte Carlo models make few assumptions and thereby retain much of the critical aspects of the carrier transport dynamics.

What happens once a carrier reaches an energy above which it can impact ionize? Early theories, both lucky drift (Ridley 1983) and numerical (Shichijo and Hess 1981), assumed the existence of a hard threshold, meaning that once a carrier achieved this energy it would readily suffer an impact ionization event. More recent numerical theories, such as those of Bude et al. (1992b), as well as analytical theories, such as those of Marsland (1987), have called this assumption into question. In their work, these authors argue that the threshold energy is typically soft, meaning that once a carrier has achieved the threshold energy, it is not highly likely to ionize until it drifts to a substantially higher energy.

Until recently, the impact ionization transition rate incorporated most often within the Monte Carlo models was based on the Keldysh formula described earlier (Keldysh 1965). The use of the Keldysh formula, although universally accepted, is not totally satisfactory since it is derived with the assumption of parabolic energy bands, contains two parameters p and E_{th} that must be determined empirically, and fails to account for any possible wave-vector dependence (k dependence) of the rate. Owing to its reliance on empirically determined parameters, the use of the Keldysh formula is restricted to the study of materials for which the electron and hole impact ionization rates have been experimentally measured. Therefore a simulator that incorporates the Keldysh formula is of limited usefulness when one is investigating the behavior of materials and devices that comprise technologically immature materials systems for which no reliable experimental data of the ionization rates are available. Nevertheless, the Keldysh formula has found extensive use owing to the ease with which it can be incorporated into a Monte Carlo study as well as its prediction of a soft or hard threshold state in the material. By studying how the ionization rate depends on the choice of p, it is possible to predict whether a material exhibits a hard or soft threshold. A soft threshold is characterized by a small value of p in the Keldysh formula, while a hard threshold is characterized by a large value of p. There remains some ambiguity, though, in the use of the Keldysh formula since more than one set of parameters can often reproduce the same ionization rate in a given material.

The first attempt to improve the impact ionization formulation was provided by Kane (1967). In Kane's formulation the pair cross section for the impact ionization transition rate in bulk silicon is calculated with pseudopotential generated energy bands and wave functions. Kane compared the transition rate calculated assuming momentum conservation as well as a random-k approximation in which momentum conservation is effectively ignored. Interestingly, he found reasonable agreement in the calculated transition rates averaged over energy in either approach. In other words, if the transition rate is determined by averaging over a suitable set of wave vectors (k vectors) for a given energy, then the k dependence is effectively averaged out. Nevertheless, the transition rate at any one initial k state may be quite different from that at another k state of comparable energy. More recently, Sano and Yoshii (1994) and Bude et al. (1992a) have investigated the threshold energy in silicon and GaAs. Their work indicates that the threshold is highly k dependent and that the relative softness of the threshold originates from the k dependence of the impact ionization transition rate.

New theories of interband impact ionization have been suggested that treat the problem both semiclassically (Bude et al. 1992a and Wang et al. 1994) and quantum mechanically (Bude et al. 1992b). These works have demonstrated that the threshold energy is always relatively soft owing to energy-broadening effects from the electron–phonon interaction. Bude et al. (1992a) and Wang et al. (1994) have shown that the ionization rate does not necessarily truncate the high-energy tail until very high energies are attained. In other words, some of the electrons survive to relatively high energies before undergoing impact ionization, in contrast to that for a hard threshold model.

The Monte Carlo calculations further indicate that the impact ionization rates in bulk GaAs are nearly isotropic (independent of the field direction) at high applied electric-field strengths. At lower applied fields, at which impact ionization can just occur, some anisotropy does exist. This is consistent with Baraff's prediction that the impact ionization process is due mainly to lucky electrons at low applied fields and that Wolff's theory holds at high applied fields. An anisotropy is expected to exist if the threshold energy is different in different crystallographic directions and if lucky electrons dominate the impact ionization process. Therefore, if the major contribution to the ionization process is due to lucky electrons, these electrons will reach threshold in some directions more readily than in others. The ionization rate then will be greater in one direction than in another. On the other hand, if the majority of impact ionization events arise from a randomized distribution, as in Wolff's theory, the carriers are distributed everywhere throughout the Brillouin zone and no preferred direction for impact ionization occurs. This is what is observed at high electric fields. In summary, Monte Carlo calculations indicate that the impact ionization process is dominated by a Wolff-like mechanism at high electric fields while at low electric fields most events are due to lucky electrons, consistent with Shockley's theory.

10.6 Physics of Confined-State Impact Ionization

Aside from interband ionization processes, an impact ionization event can occur between a free carrier and a carrier confined within an impurity state, an impurity band, or a quantum-well state. Several different types of photodetectors are based on the principles of confined-state impact ionization. In this section, the physics of these processes, focusing primarily on ionization out of confined quantum states, is presented.

Photodetectors based on impact ionization of electrons confined within a quantum well were first proposed by Smith et al. (1983) and independently by Chuang and Hess (1986). In this device, called the confined-state photomultiplier, photogenerated electrons are injected into a multiplication region. The multiplication region consists of degenerately doped quantum or classical wells separated by nearly intrinsic wide-bandgap semiconducting material. As in impurity band impact ionization, the threshold energy for confined-state ionization is much lower than that for an interband ionization. In impurity band or state ionization devices, the impurity level lies at relatively low energy with respect to the band edge, much less than the bandgap. As a result, the amount of energy needed to impact excite an electron out of an impurity state is roughly equal to the depth of the level itself. The confined-state devices operate in principle similarly to devices based on impurity states or impurity band impact ionization.

The impact ionization process in a confined-state device involves the impact excitation of an electron confined within a quantum or classical heterostructure well by a high-energy electron injected into the well. The basic mechanism of confined-state ionization is sketched in Figure 10.6.1. An incident electron with an excess kinetic energy at least equal to the difference between the conduction-band edge of the trailing edge of the well and the confined-state energy can cause an impact excitation of a confined-state electron. As a result, two electrons with kinetic energies greater than the trailing edge of the well are produced. A vacancy is left behind in the quantum state.

It is important to recognize that confined-state impact ionization involves only one carrier species. Such a multiplication process will exhibit ultralow noise behavior. After an ionization event, only free electrons are produced within the conduction band and no holes are generated within the valence band. A vacancy

Initial State

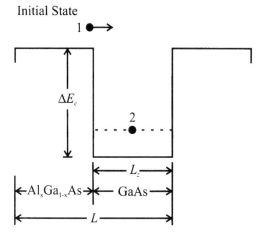

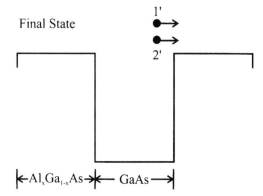

Figure 10.6.1. Schematic drawing of a confined state impact ionization event showing the initial and the final states of the system.

is produced instead within the confined quantum state. This vacancy, however, cannot propagate through the device since it is typically localized by the potential well formed within the valence band. As such, the vacancy cannot undergo an impact ionization event itself.

The buildup of vacancies within the quantum well following the impact excitation events ultimately results in space-charge neutrality's no longer being maintained. This alters the electric-field profile within the device and subsequently the device operation. To sustain proper dc operation of the confined-state device, the vacancies must be continuously replenished with electrons. Although carrier replenishment is a fundamental limitation of confined-state devices, it does not necessarily imply that these devices cannot be useful since replenishment can be accomplished through an optical reset or through thermal generation.

The physics of confined-state impact ionization was first developed by Chuang and Hess (1986). The impact excitation transition rate is determined from a screened Coulomb interaction between an incident-free electron and an electron confined spatially in one dimension. The final states of the electrons after the interaction are propagating states at energies greater than the trailing barrier height. The transition rate is given generally as

$$P_{\text{tr}} = \frac{1}{V} \sum_{k_1} \sum_{k_2} \sum_{k_1'} \sum_{k_2'} P_{k_1 k_2}^{k_1' k_2'} f(k_1) f(k_2) [1 - f(k_1')][1 - f(k_2')], \qquad (10.6.1)$$

where f is the Fermi distribution, k_1 is the initial state of the incident electron, k_2 is the initial state of the target electron within the well, and k_1' and k_2' are the final states of the electrons after the interaction. In expression (10.6.1) the final states are typically assumed always to be initially empty. Subsequently, the terms that involve $(1 - f)$ are simply assumed to be equal to 1. The quantum-mechanical transition rate per unit time P can be determined to first order by Fermi's golden rule as

$$P_{k_1 k_2}^{k_1' k_2'} = \frac{2\pi}{\hbar} |\langle 12 | H_s | 1'2' \rangle|^2 \delta(E_1 + E_2 - E_1' - E_2'), \qquad (10.6.2)$$

where the matrix element $\langle 12 | H_s | 1'2' \rangle$ is simply the two-body screened Coulomb interaction in which spin effects are neglected. If the incident-electron state is assumed to be given by a plane wave and the target electron wave function is approximated by that corresponding to an infinite square well, the total impact ionization rate out of confined quantum states was shown by Chuang and Hess (1986) and Wang, Park, and Brennan (1990b) to be

$$\frac{1}{\tau} = \frac{1}{n_0} \frac{1}{(2\pi)^3} \int f(k_0) \, d^3 k_0 \frac{1}{\tau(E_0)}, \qquad (10.6.3)$$

where n_0 is the incident-electron concentration, $f(k_0)$ is the incident-electron distribution function, and $1/\tau(E_0)$ is the impact ionization transition rate. $1/\tau(E_0)$ is given as

$$\frac{1}{\tau(E_0)} = \int \frac{4\pi q^4}{\hbar \epsilon^2} \frac{1}{L_{zn}} \left(\frac{L_{zn}}{L} \right)^2 \frac{f(k_{2t})}{(2\pi)^2} [S(k_1, k_{2n}^+) + S(k_1, k_{2n}^-)] \, dk_{2t}, \qquad (10.6.4)$$

where L_{zn} is the effective well width, L is the total width of the well and barrier regions, and $f(k_{2t})$ is the electron distribution function within the quantum well of the transverse component of the confined electron. The functions S are

$$S(k_1, k_{2n}^{\pm}) = \int_{k_1', k_2' > k_{\text{th}}'} \frac{d^3 k_1}{(2\pi)^3} \frac{\delta(E_1 + E_2 - E_1' - E_2')}{\left[Q_D^2 + |k_1' - k_1|^2 \right]^2} \bigg|_{k_2' = k_1 + k_{2n}^{\pm} - k_1'}, \qquad (10.6.5)$$

where Q_D is the inverse screening length, typically given by the inverse Debeye length.

From the above relationships, it is clear that the total impact ionization rate $1/\tau$ depends on several factors: the electron concentration within the well n_0, the ratio of the well width to the total width of the unit cell (sum of the well and barrier widths), the transition rate $1/\tau(E_0)$, and the incident-electron distribution $f(k_0)$. The geometric factor $(L_{zn}/L)^2$ arises from the overlap of the incident- and the confined-electron wave functions. In the normalization of the wave functions, the incident electron is assumed normalized over one full period while the confined electron is normalized over the effective width of the well. As a result, the overlap occurs within only the well region, leading to a reduced impact ionization cross section.

Wang et al. (1990a) also derived the physics of impact excitation of electrons that are spatially confined within a classical well, one whose dimensions are sufficiently large that spatial quantization effects do not occur. The primary difference between the classically confined case and the quantum confined case discussed above is the fact that the wave function of the target electron is assumed to be given by a three-dimensional plane-wave state normalized within the well. In addition, the screening parameter is defined now for a three-dimensional system rather than a two-dimensional system.

In either the classically confined system or the quantum confined system the overall transition rate depends on the threshold conditions for the process. The final states of the electrons after the interaction k'_1 and k'_2 must each lie at energies greater than the conduction-band-edge discontinuity on the trailing side of the well. If it is assumed that the incident electron k_1 has only a component in the direction of propagation z, then from momentum conservation in the z direction,

$$k_{1z} + k_{2z} = k'_{1z} + k'_{2z}. \qquad (10.6.6)$$

The threshold condition for impact ionization requires that both k'_{1z} and k'_{2z} be greater than k'_{th}, which is defined as the corresponding value of k_z necessary for the electrons to exit the well:

$$k'_{th} = \left(\frac{2m^* \Delta E'_c}{\hbar^2} \right)^{\frac{1}{2}}, \qquad (10.6.7)$$

where $\Delta E'_c$ is the conduction-band-edge discontinuity at the trailing side of the well. Therefore the condition on the initial k vectors is then

$$2k'_{th} \leq k_{1z} + k_{2z}, \qquad (10.6.8)$$

which holds independently of whether the electrons are classically or quantum mechanically confined.

The threshold condition illustrates an inherent advantage of the quantum confined device over the classically confined structure. If the confined-state wave vector k_{2z} is small, then k_{1z} needs to be large in order for an event to occur and vice versa. Quantization pushes electrons to higher-energy states, such that there are more electrons present within the well at higher k_{2z}, enabling ionization events from a greater range of incident electrons than in a classically confined device of the same barrier height and carrier concentration. Hence the inequality is more easily satisfied in the quantum case and the ionization rate should be greater than in the classically confined device.

Recent experimental work by Allam et al. (1987) and Levine et al. (1987) indicates that gain can be achieved through the impact excitation of confined-state carriers, although at some compromise in overall device speed and dark current. Levine et al. (1987) showed excellent agreement between theoretically predicted values and experimental measurements for the avalanche gain from confined-state multiplication. In their work they examined confined-state multiplication initiated by 10-μm intersubband absorption. In addition, Capasso et al. (1986) have experimentally observed confined-state impact ionization out of classically confined states as well as quantum confined states.

Much of the physics presented in this section applies to the physics of impurity state or impurity band impact ionization. In either process, a localized electron is impact ionized by a free carrier as in the quantum or classically confined-state devices. The only major difference in these processes is the nature of the target electron's wave function. In the calculations presented above, the electronic state of the target electron is assumed to either be that of a one-dimensional confined electron or a free three-dimensional electron. In impurity state ionization, the target electron state is best assumed to be an atomic level. Subsequently, we obtain the transition rate for these devices readily by following a similar analysis as done above but by using an atomic state wave function instead.

10.7 **Optical Absorption in Semiconductors**

As discussed in Section 10.2, the optical absorption and emission rates depend on whether a semiconductor is direct or indirect. It is usually possible to ignore the momentum of a photon with respect to that of the electrons and holes for most of the transitions of interest in semiconductors. Consequently, in order that momentum be conserved in an unaided optical transition, either emission or absorption, it is necessary that the system be direct. If a semiconductor is indirect, an optical transition can still occur, provided that an additional agent is present to enable momentum conservation. Typically, an optical transition can occur in an indirect semiconductor through the mediation of a phonon absorption or emission event. In this section, we develop the microscopic physics of interband optical transitions for both direct and indirect semiconductors.

The transition rate for a quantum-mechanical process is generally determined with time-dependent perturbation theory. As was discussed in Chapter 4, the transition rate to first order is given by Fermi's golden rule as

$$W = \frac{2\pi}{\hbar} |\langle k | H' | s \rangle|^2 \delta(E_k - E_s) \tag{10.7.1}$$

for a time-independent potential H'. To calculate the optical transition rate from Fermi's golden rule it is essential to evaluate the matrix element $|\langle k | H' | s \rangle|$ between the initial and the final states. The first step in the process then is to determine the form of the perturbation H'.

The Hamiltonian describing the motion of a carrier in a semiconductor in the presence of an electromagnetic field is

$$H = \frac{1}{2m}(\vec{p} - q\vec{A})^2 + V(r), \tag{10.7.2}$$

where $\vec{A}$ is the vector potential associated with the electromagnetic field and $V(r)$ is the periodic crystalline potential. When the squared term is expanded out, the Hamiltonian becomes

$$H = \frac{1}{2m}[p^2 - q(\vec{p} \cdot \vec{A} + \vec{A} \cdot \vec{p}) + q^2 A^2] + V(r). \tag{10.7.3}$$

H can be written in terms of an unperturbed Hamiltonian H_0 and a perturbation in the usual manner as

$$H = H_0 + H'. \tag{10.7.4}$$

The unperturbed Hamiltonian is, of course, just

$$H_0 = \frac{p_2}{2m} + V(r), \tag{10.7.5}$$

while the perturbation is

$$H' = -\frac{q}{2m}(\vec{p} \cdot \vec{A} + \vec{A} \cdot \vec{p}) + \frac{q^2 A^2}{2m}. \tag{10.7.6}$$

The term involving the square of A can usually be neglected with respect to the terms linear in A in H'. Additionally, $\vec{\nabla} \cdot \vec{A} = 0$ in the Coulomb gauge (that is, $\vec{\nabla} \cdot \vec{A} = 0$). We must be careful, however, with the term $\vec{p} \cdot \vec{A}$ in Eq. (10.7.6). Recall from Chapter 1 that a differential operator acts on everything to its right. Therefore the term $\vec{p} \cdot \vec{A}$ gives rise to two terms, one from the differentiation of $\vec{A}$ and the other from the differentiation of a wave function to the right of $\vec{A}$. So the term $\vec{p} \cdot \vec{A}$ becomes $\vec{p} \cdot \vec{A} = \hbar/i[\vec{\nabla} \cdot \vec{A} + \vec{A} \cdot \vec{\nabla}]$. When the Coulomb gauge is used, the term $\vec{\nabla} \cdot \vec{A}$ vanishes, which leaves $\vec{p} \cdot \vec{A} = \vec{A} \cdot \vec{p}$. With this result, H' becomes then

$$H' = -\frac{q}{m}\vec{A} \cdot \vec{p}, \tag{10.7.7}$$

where the vector potential for a monochromatic wave $\vec{A}$ is written as

$$\vec{A} = \frac{A_0}{2}\left[e^{i(\vec{Q} \cdot \vec{r} - \omega t)} + e^{-i(\vec{Q} \cdot \vec{r} - \omega t)}\right]\hat{a}, \tag{10.7.8}$$

and $\hat{a}$ is a unit vector in the direction of propagation. The first term in the above expression for $\vec{A}$ characterizes an absorption event while the second term characterizes an emission event. The interaction Hamiltonian can be written then as the sum of absorption and emission terms as

$$H' = H_{abs} + H_{em}, \tag{10.7.9}$$

where

$$H_{abs} = \frac{-q A_0}{2m} e^{iQr} \hat{a} \cdot \vec{p}, \tag{10.7.10}$$

$$H_{em} = \frac{-q A_0}{2m} e^{-iQr} \hat{a} \cdot \vec{p}, \tag{10.7.11}$$

where the time dependence is absorbed into the delta function term in Fermi's golden rule. The absorption rate can now be determined as

$$W_{abs} = \frac{2\pi}{\hbar} |\langle f | H_{abs} | i \rangle|^2 \delta(E_f - E_i - \hbar\omega) \tag{10.7.12}$$

and the emission rate as

$$W_{em} = \frac{2\pi}{\hbar} |\langle f | H_{em} | i \rangle|^2 \delta(E_f - E_i + \hbar\omega). \tag{10.7.13}$$

We can evaluate the expressions for the absorption and the emission rates by integrating the matrix elements over all the final states. In the case of transitions within a semiconductor, the final states are distributed within an energy band. Substituting Eq. (10.7.10) in for H_{abs} into Eq. (10.7.12), we find that the absorption rate becomes

$$W_{abs} = \frac{2\pi}{\hbar} \frac{A_0^2 q^2}{4m^2} |\langle f | e^{iQr} \hat{a} \cdot \vec{p} | i \rangle|^2 \delta(E_f - E_i - \hbar\omega). \tag{10.7.14}$$

Since only the magnitude of the matrix element is retained, the phase term e^{iQr} gives only unity. Therefore the absorption rate becomes

$$W_{abs} = \frac{2\pi}{\hbar} \frac{A_0^2 q^2}{4m^2} |\langle f | \hat{a} \cdot \vec{p} | i \rangle|^2 \delta(E_f - E_i - \hbar\omega). \tag{10.7.15}$$

An expression for the magnitude of the vector potential can be obtained directly from the definitions of the electric and the magnetic fields. The expression for the vector potential can be rewritten in terms of a cosine function as

$$\vec{A} = A_0 \cos(Qr - \omega t)\hat{a}. \tag{10.7.16}$$

The electric and the magnetic fields can be found from the vector potential by use of Eqs. (1.2.4) and (1.2.7):

$$\vec{E} = -\omega A_0 \sin(Qr - \omega t)\hat{a}, \tag{10.7.17}$$

$$\vec{H} = \frac{1}{\mu}\vec{\nabla} \times \vec{A} = -\frac{A_0}{\mu} \sin(Qr - \omega t)(\vec{Q} \times \hat{a}). \tag{10.7.18}$$

The Poynting vector $\vec{S}$ is obtained from the vector cross product of $\vec{E}$ and $\vec{H}$ as

$$\vec{S} = \frac{\omega A_0^2}{\mu} \sin^2(Qr - \omega t)(\hat{a} \times \vec{Q} \times \hat{a}). \tag{10.7.19}$$

But the vector cross product can be simplified with the standard identity

$$\hat{a} \times \hat{b} \times \hat{c} = (\hat{a} \cdot \hat{c})\hat{b} - (\hat{a} \cdot \hat{b})\hat{c}. \tag{10.7.20}$$

When applied to the expression for $\vec{S}$, the double cross product becomes equal to $\vec{Q}$. Therefore $\vec{S}$ becomes

$$\vec{S} = \frac{\omega A_0^2}{\mu} \sin^2(Qr - \omega t)\vec{Q}. \tag{10.7.21}$$

Time averaging yields

$$\langle S \rangle = \frac{A_0^2}{2\mu}\omega Q. \tag{10.7.22}$$

The phase velocity v is simply equal to ω/Q. With this substitution the average Poynting vector is

$$\langle S \rangle = \frac{A_0^2}{2\mu} Q^2 v. \tag{10.7.23}$$

The energy density is equal to the ratio of $\langle S \rangle$ to v. But the energy density ρ_E of a monochromatic wave is equal to the product of the number of photons and the energy of each photon divided by the volume:

$$\rho_E = \frac{N\hbar\omega}{V}. \tag{10.7.24}$$

Therefore the magnitude of the vector potential A_0 can be determined as

$$A_0^2 = \frac{2N\hbar\omega\mu}{Q^2 V}, \tag{10.7.25}$$

where V is the volume and μ is the magnetic permeability. With this expression for A_0 the absorption and the emission rates can be evaluated once the initial and the final states are described.

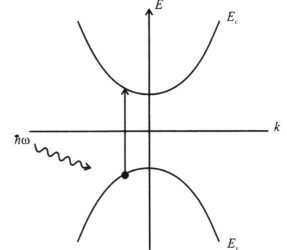

Figure 10.7.1. Sketch of a direct optical transition in a semiconductor.

As an example, let us consider the simple case of optical absorption in a direct-gap semiconductor. To calculate the optical absorption transition rate, the result for the magnitude of the vector potential can be substituted into the expression for the absorption rate as

$$W_{\text{abs}} = \frac{2\pi}{\hbar} \frac{2N\hbar\omega\mu}{Q^2V} \frac{q^2}{4m^2} |\langle f| \hat{a} \cdot \vec{p} |i\rangle|^2 \delta(E_f - E_i - \hbar\omega). \qquad (10.7.26)$$

Let us now evaluate the matrix element in Eq. (10.7.26) for a direct transition. A direct transition, we recall, occurs when the valence- and the conduction-band energy minimums occur at the same place in momentum space. An example of a direct transition is sketched in Figure 10.7.1.

To evaluate the matrix element, it is necessary to choose the initial and the final states. For a band-to-band transition, the initial and final states are assumed to be Bloch-like and of the form

$$\Psi_k(r) = e^{ikr} u_k(r). \qquad (10.7.27)$$

With Bloch-like wave functions chosen for the initial and the final states, the matrix element can be written in coordinate representation as

$$\int [u_{k'}^*(r)e^{-ik'r}]e^{iQr}\hat{a} \cdot \vec{p}\,[u_k(r)e^{ikr}]\,\mathrm{d}^3r. \qquad (10.7.28)$$

From Chapter 2, we recall that the momentum operator $\vec{p}$ is simply $\hbar/i\vec{\nabla}$ in three dimensions. Applying $\vec{p}$ yields

$$\frac{\hbar}{i} \int u_{k'}^*(r)e^{-ik'r}e^{iQr}[\hat{a} \cdot \vec{\nabla}u_k(r)e^{ikr} + i\hat{a} \cdot \vec{k}e^{ikr}u_k(r)]\,\mathrm{d}^3r, \qquad (10.7.29)$$

which simplifies to

$$\frac{\hbar}{i} \int u_{k'}^*(r) e^{i(k-k'+Q)r} [\hat{a} \cdot \vec{\nabla} u_k(r) + i\hat{a} \cdot \vec{k} u_k(r)] \, d^3r. \tag{10.7.30}$$

As in Section 10.4, the integration of the Bloch functions over all space is restricted to over one unit cell. The exponential terms are integrated over all space and yield a volume multiplied by a delta function in the momentum, as in Section 10.4. This yields

$$\frac{\hbar}{i} V\delta(k - k' + Q) \int_\Omega u_{k'}^*(r)[\hat{a} \cdot \vec{\nabla} u_k(r) + i\hat{a} \cdot \vec{k} u_k(r)] \, d^3r. \tag{10.7.31}$$

Note that the matrix element contains two terms, one involving $u_k(r)$ and the other $\vec{\nabla} u_k$. It is important to recognize that the Bloch functions are orthogonal to one another for different bands. Therefore the overlap integral between $u_{k'}(r)$ and $u_k(r)$ vanishes. The term in expression (10.7.31) involving the product of $u_{k'}(r)$ and $u_k(r)$ vanishes and is thus called a forbidden transition. The term involving $\vec{\nabla} u_k$ does not vanish and is called the first-order allowed transition.

Consider first the allowed transition. Making use of the definition of the momentum operator $\vec{p} = \hbar/i\vec{\nabla}$, we find that the allowed transition is proportional to

$$\sim \int u_{k'}^*(r)\hat{a} \cdot \vec{p} u_k(r) \, d^3r. \tag{10.7.32}$$

Substituting expression (10.7.32) back into Eq. (10.7.26) yields the general result for the absorption transition rate for allowed transitions in direct-gap semiconductors:

$$W_{abs} = \frac{2\pi}{\hbar} \frac{A_0^2 q^2}{4m^2} \delta(E_{k'} - E_k - \hbar\omega) \left[V\delta(k - k' + Q) \int u_{k'}^*(r)\hat{a} \cdot \vec{p} u_k(r) d^3r \right]^2, \tag{10.7.33}$$

where we have used expression (10.7.25) for the magnitude of A_0. At this point, we can determine the transition rate further by choosing the appropriate forms of the Bloch functions and the density of states. A simplified expression, assuming parabolic energy bands, has been derived by Wolfe, Holonyak, and Stillman (1989). The reader is referred to their work for further information.

The expression derived above holds for allowed transitions within direct-gap semiconductors. In expression (10.7.31) there are two terms. The first term corresponds to allowed transitions, while the second term corresponds to forbidden transitions. In either case, these hold for direct-gap semiconductors. Consider now the second term in the matrix element of expression (10.7.31). Recognizing that the term $\hat{a} \cdot \vec{k}$ can be taken outside of the integration over r, we find that the second term becomes

$$\hbar V \delta(k - k' + Q) \hat{a} \cdot \vec{k} \int u_{k'}^*(r) u_k(r) \, d^3r. \tag{10.7.34}$$

We can find the transition rate as before by substituting in expression (10.7.34) for the matrix element into Fermi's golden rule, Eq. (10.7.26), which becomes

$$W_{\text{abs}} = \frac{2\pi}{\hbar} \frac{A_0^2 \hbar^2 q^2}{4m^2} \delta(E_{k'} - E_k - \hbar\omega)$$

$$\times |\hat{a} \cdot \vec{k}|^2 \left[V \delta(k - k' + Q) \int u_{k'}^*(r) u_k(r) \, d^3r \right]^2. \tag{10.7.35}$$

Note that the forbidden transition rate depends on the overlap integral between the initial and the final Bloch functions, which always vanishes when $k = k'$ for different bands. In the situation here, k of course is not exactly equal to k' since a photon of wave vector Q is absorbed. However, for most optical transitions, the wave vector of the photon is very small compared with k or k', as we will see below. Therefore the difference between k and k' is extremely small and the overlap integral is then between states of very nearly equal wave vectors. As a result, the value of the overlap integral is extremely small in these cases and the transition has a low probability. Hence the name, forbidden transitions. Wolfe et al. (1989, pp. 211–14) provide expressions for the allowed and the forbidden transitions assuming parabolic energy bands. The reader is referred to this work for more details. Here it suffices for us to know that the allowed direct-gap transitions are much stronger than the forbidden transitions.

Alternatively, a material may exhibit only indirect transitions. Indirect transitions occur predominately in materials in which the conduction-band minimum occurs at a different point in k space than the valence-band minimum. An example indirect transition is sketched in Figure 10.7.2. As can be seen from the figure, the minimum energy point in the conduction band occurs at the X point while the minimum energy point in the valence band occurs at the Γ point. Silicon is an example of such a material. An optical transition between these two minimum points can occur but is far less likely than when the minimums occur at the same point in k space, as described above. These transitions occur to second order, as was discussed in Section 4.5. Let us consider an indirect transition qualitatively in some detail.

The k-vector difference between the center and the edge of the Brillouin zone was shown in Section 8.1 to be equal to π/a, where a is the lattice constant. For a typical semiconductor material such as silicon, this separation is $\sim 5.78 \times 10^7$ 1/cm. In order for momentum to be conserved for a transition such as the one shown in Figure 10.7.2, the photon would have to supply a momentum equal to $\hbar k$, which in this case is $\sim 3.8 \times 10^{-8}$ eV s/cm. Energy must also be conserved during the transition. Again for silicon, the indirect energy gap is ~ 1.12 eV. The wavelength of the photon of energy 1.12 eV can be determined

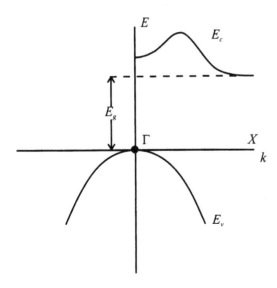

Figure 10.7.2. Sketch of the energy-band structure of an indirect-gap semiconductor.

from

$$E = \frac{hc}{\lambda} = \frac{1.24 \times 10^{-4}}{\lambda}, \tag{10.7.36}$$

which yields a wavelength of 1.1×10^{-4} cm and a corresponding momentum of 3.76×10^{-11} eV s/cm. Note that the momentum of the absorbed photon is 3 orders of magnitude smaller than what is necessary to preserve both momentum and energy conservation for this transition. Therefore the process cannot occur to first order since both energy and momentum are not conserved.

An indirect optical transition can occur only if another agent supplies the necessary momentum to satisfy the momentum conservation requirement. Phonons can have wave vectors as large as π/a. An indirect optical transition can occur if accompanied by the simultaneous absorption or emission of a phonon. A sketch of a phonon-assisted indirect transition is shown in Figure 10.7.3. As can be seen from the figure, the process occurs diagrammatically in two steps. First, the electron absorbs a photon and makes a transition to an intermediate state within the bandgap. Next, the electron absorbs or emits a phonon of large Q (wave vector) and completes the transition into the conduction band. As was discussed in Section 4.5, the process can be described physically by use of second-order time-dependent perturbation theory as

$$W_{s \to k} = \frac{2\pi}{\hbar} \left| \sum_m \frac{\langle k| V |m\rangle \langle m| V |s\rangle}{E_s - E_m} \right|^2 \delta(E_k - E_s + \hbar\omega). \tag{10.7.37}$$

The sum over m in Eq. (10.7.36) represents a sum over intermediate states between the initial and the final states, s and k, respectively. The intermediate states need not be real. As was discussed in Section 4.5, the intermediate states can be virtual; in other words, they do not have to be eigenstates of the Hamiltonian. An electron can make a transition into a virtual state, provided that it does not

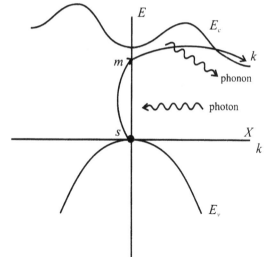

Figure 10.7.3. Phonon-assisted indirect-gap radiative transition.

remain there for a long time. Owing to the Uncertainty Principle for energy and time, an electron can occupy a virtual state, provided the duration for which it remains in is relatively short. As we found in Chapter 8, there are no eigenstates within the energy gap of a semiconductor. Therefore the intermediate states for the indirect transition sketched in Figure 10.7.3 are virtual.

Since the simultaneous absorption of a photon and a phonon is a higher-order process, an indirect transition is said to occur only to second or higher order. The probability of an indirect transition's occurring depends then on the probability of two different processes' simultaneously occurring, that is, the absorption of a photon and of a phonon. The combined probability is lower than the probability of either event's occurring separately. Hence the occurrence of a second-order indirect transition is far less likely than that of a first-order direct transition. It is for this reason that an indirect material, such as silicon, is not an efficient emitter of optical radiation. In addition, silicon is a much less efficient absorber of optical radiation, particularly for wavelengths near the cutoff wavelength of silicon. Direct-gap materials such as GaAs are far more efficient optical emitters and absorbers. Subsequently, GaAs and other direct-gap semiconductors will lase while bulk silicon does not. In Chapter 13 we will discuss lasing and laser structures.

In the results derived above, it was assumed that the energy gap remained fixed at its bulk value. On the application of a field, there is a shift in the cutoff wavelength toward longer wavelengths. In the presence of the field, the crystal will absorb light for wavelengths at which the crystal is usually transparent. Field-aided absorption of photons in semiconductors is called the Franz-Keldysh effect. This results in the absorption of photons with energies less than the energy bandgap of the semiconductor.

The Franz–Keldysh effect can be understood qualitatively as follows. If an electric field is applied to a semiconductor, the energy bands tilt, as shown in Figure 10.7.4. Let the lines sketched in Figure 10.7.4.b marked as L and U denote the valence- and the conduction-band states, respectively. Note that the

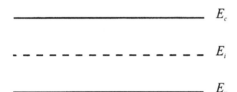

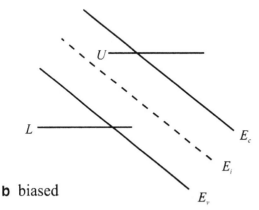

a unbiased

b biased

Figure 10.7.4. Sketch of the energy-band diagrams for a semiconductor under a, unbiased, b, biased conditions.

states marked as U and L extend into the forbidden gap. This extension arises from the evanescent tail of the carrier wave function; the carriers tunnel into the band gap. As the field increases, the distance separating these two states decreases. Essentially, the wave functions associated with the two different states in the conduction and the valence bands are brought physically closer together in real space. As we discussed in Section 7.3, when the wave functions of two or more electrons are brought into close proximity, there is a coupling of the two states, leading to increased interaction between them. Since the energy separation of U and L is less than the band gap energy, an absorption event involving a photon of energy less than E_g can occur between these two states. As such, the absorption rate is increased because of the increased interaction between the initial and the final band states in the presence of an external field.

10.8 **Stimulated and Spontaneous Emission**

In Section 10.7 we discussed radiative absorption in semiconductors. It was found that the absorption rate is significantly larger for direct-gap semiconductors than for indirect-gap semiconductors. In this section, we investigate spontaneous and stimulated emission processes. First, let us define spontaneous and stimulated emission. Spontaneous emission corresponds to the emission of radiation following a spontaneous decay of an electron from a high-energy state into a lower-energy state. Stimulated emission, on the other hand, is induced by the action of an external electromagnetic field. The external field perturbs an electron within a high-energy state, inducing it to decay to a lower-energy state with the subsequent emission of a photon. Absorption can obviously occur only under the action of an external field; a photon must be present and absorbed in order to promote an electron from a lower-energy state into a higher-energy state.

In Section 10.7 we derived the microscopic physics for the absorption transition rate. The total absorption rate is found by integration of Eq. (10.7.33) over all the possible final states in the crystal. The total absorption rate is found then as

$$r_{abs} = \frac{2V}{(2\pi)^3} \int W_{abs}\, f_1(1 - f_2)\, d^3k, \tag{10.8.1}$$

where the additional factor of 2 accounts for a possible spin change during an absorption event. The factors f_1 and $(1 - f_2)$ are the distribution functions describing the occupancy of the valence band and the vacancies in the conduction band, respectively. Obviously, an absorption event depends on the presence of an electron in the valence band and a vacancy in the conduction band. If the conduction band was completely full, no absorption event could occur since an additional electron cannot be added to a full band. Likewise, if the valence band is completely unoccupied, then there are no electrons available to absorb the photon. The distribution functions are typically assumed to be slowly varying with respect to k and can thus be taken outside of the integral in Eq. (10.8.1) above. This yields

$$r_{abs} = \frac{2V}{(2\pi)^3} f_1(1 - f_2) \int W_{abs}\, d^3k. \tag{10.8.2}$$

We can rewrite Eq. (10.8.2) by defining two new quantities $P(E_{21})$ and B_{12} as

$$r_{abs} = B_{12}\, f_1(1 - f_2) P(E_{21}). \tag{10.8.3}$$

B_{12} is known as the Einstein coefficient for absorption. $P(E_{21})$ is the density of incident photons of energy E_{21}, equal to the difference between the initial and the final states. The distribution functions f_1 and $1 - f_2$ are nonequilibrium distribution functions. The exact form of the nonequilibrium distribution function is in general difficult to determine. In semiconductors the nonequilibrium distribution function can be well approximated through use of quasi-Fermi levels (see Box 10.8.1 for the details of quasi-Fermi levels). As discussed in Box 10.8.1,

---

Box 10.8.1 Quasi-Fermi Levels

(See Hess 1988) Consider a semiconductor in nonequilibrium through either optical excitation or electrical injection. In either case, excess carriers are generated. The recombination time of an excess electron–hole pair is relatively long compared with the relaxation time of carriers in their separate bands. In other words, since the electron–electron, hole–hole,

continued

Box 10.8.1, cont.

and carrier–phonon scattering rates are much greater (by several orders of magnitude) than the interband recombination rate, the excess electrons and holes effectively thermalize well before they recombine. Therefore the electrons within the conduction band and the holes within the valence band will each separately be in equilibrium even though the entire combined system of the electrons and holes is not in equilibrium. In other words, even though the entire system, electron and hole concentrations taken together, is not in equilibrium, each component is. Under these conditions, the electrons and the holes each separately obey a quasi-equilibrium distribution that is Fermi–Dirac-like. The form of the distributions is given as

$$f_1(E) = \frac{1}{1 + e^{\frac{(E - q\phi_1)}{kT}}},$$

where ϕ_1 is called the quasi-Fermi level for system 1.

We can visualize the usage of quasi-Fermi levels by using the system shown in Figure 10.8.1. As shown by the figure, two tanks filled with gas are connected by a thin, narrow tube. It is assumed that the tank volumes are very much larger than the volume of the tube. The amount of gas flowing between the tanks at any given time then is very small with respect to the volume of gas in each tank. The tanks are only very weakly coupled. Hence the gases in each tank are relatively undisturbed by the gas flowing between them in the tube. Therefore each gas can be separately considered to be in equilibrium with itself, yet the entire system is clearly not in equilibrium since some transport between the two tanks still occurs.

The carrier concentrations can be written with the quasi-Fermi levels. The most common usage is for nondegenerate conditions. In this case, the quasi-Fermi distribution can be replaced by a quasi-Boltzmann distribution. The resulting electron and hole concentrations become

$$n = n_i e^{\frac{q(\psi - \phi_n)}{kT}},$$
$$p = n_i e^{\frac{q(\phi_p - \psi)}{kT}},$$

where n_i is the intrinsic concentration and ψ is the electrostatic potential defined as $-E_i/q$. Note that the above expressions are similar to the expressions for the equilibrium carrier concentrations given by Eq. (10.3.11).

The physical significance of the quasi-Fermi levels can be understood from the product of the nonequilibrium electron and hole distributions. With the above expressions, the np product becomes

$$np = n_i^2 e^{\frac{q(\psi - \phi_n)}{kT}} e^{\frac{q(\phi_p - \psi)}{kT}} = n_i^2 e^{\frac{q(\phi_p - \phi_n)}{kT}}.$$

In equilibrium, the law of mass action, derived in Section 10.3, holds. Inspection of the above result indicates that in equilibrium $\phi_p = \phi_n$; the quasi-Fermi levels are equal. The difference in the quasi-Fermi levels provides a measure of how far from equilibrium the system is. In fact, the difference in the quasi-Fermi levels is equal to the magnitude of the external applied bias or effective optical bias. Note that if the np product is greater than n_i^2, the system must be in forward bias, $V > 0$. Alternatively, under reverse bias the np product is less than n_i^2.

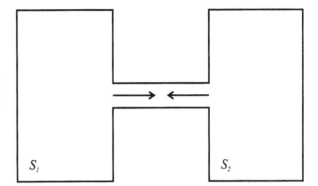

Figure 10.8.1. Two tanks of gas weakly coupled as an analogous system to that of carriers in a semiconductor described by quasi-Fermi levels.

the nonequilibrium distribution functions can be written as

$$f_1(E_1) = \frac{1}{1 + e^{\frac{(E_1 - q\phi_1)}{kT}}}, \qquad f_2(E_2) = \frac{1}{1 + e^{\frac{(E_2 - q\phi_2)}{kT}}}, \qquad (10.8.4)$$

where ϕ_1 is the quasi-Fermi level for the valence band and ϕ_2 is the quasi-Fermi level for the conduction band.

As mentioned above, there are two emission processes, stimulated and spontaneous. The stimulated emission process, like the absorption process, depends on the incident electromagnetic radiation field. The stimulated emission rate can thus be written in analogy to the absorption rate as

$$r_{21}(\text{stim}) = B_{21} f_2 (1 - f_1) P(E_{21}), \qquad (10.8.5)$$

where the rate depends now on the occupancy of the conduction band and vacancies within the valence band. B_{21} is called the Einstein B coefficient for stimulated emission.

Finally, we can determine the spontaneous emission rate by recognizing that it does not depend on any incident radiation field. In this case, the emission rate is given as

$$r_{21}(\text{spontaneous}) = A_{21} f_2 (1 - f_1), \qquad (10.8.6)$$

where A_{21} is the transition probability for a spontaneous emission event. A_{21} is different from B_{21}, as will be shown below.

The relationship between the Einstein A and B coefficients can be obtained with the principle of detailed balancing. Detailed balancing simply implies that the reverse reaction rate can be obtained once the forward reaction rate is known. In thermal equilibrium, the net upward transition rate must be equal to the net downward transition rate. Hence

$$r_{12} = r_{21}(\text{stim}) + r_{21}(\text{spont}). \qquad (10.8.7)$$

In equilibrium the quasi-Fermi levels are equal to one another, as discussed in

Box 10.8.1. Substituting in the expressions for each of the rates into Eq. (10.8.7) yields

$$B_{12} f_1(1 - f_2)P(E_{21}) = B_{21} f_2(1 - f_1)P(E_{21}) + A_{21} f_2(1 - f_1). \qquad (10.8.8)$$

Solving for $P(E_{21})$ yields

$$P(E_{21}) = \frac{A_{21} f_2(1 - f_1)}{[B_{12} f_1(1 - f_2) - B_{21} f_2(1 - f_1)]}. \qquad (10.8.9)$$

An alternative relationship can be obtained for $P(E_{21})$ directly from the radiation field. $P(E_{21})$ is defined as the spectral density at an energy E_{21}, which can be determined from the product of the photon density of states and the occupation probability of each photon mode. In Section 5.8 we derived the spectral density in Eq. (5.8.36). Rewriting in terms of E yields

$$P(E) = \frac{8\pi n^3 E^2}{(hc)^3} \frac{1}{\left(e^{\frac{E}{kT}} - 1\right)}, \qquad (10.8.10)$$

where n is the index of refraction.

The right-hand side of Eq. (10.8.9) can be simplified as follows. First divide both the numerator and the denominator by $f_2(1 - f_1)$. This yields

$$P(E) = \frac{A_{21}}{\left[B_{12} \frac{f_1(1 - f_2)}{f_2(1 - f_1)} - B_{21}\right]}. \qquad (10.8.11)$$

But the term $f_1(1 - f_2)/f_2(1 - f_1)$ is simply equal to $e^{(E/kT)}$, where E is the difference between E_2 and E_1 and the fact that the quasi-Fermi levels are equal has been used. Therefore the spectral density can be written in terms of the Einstein coefficients as

$$P(E) = \frac{8\pi n^3 E^2}{(hc)^3} \frac{1}{\left(e^{\frac{E}{kT}} - 1\right)} = \frac{A_{21}}{B_{12} e^{\frac{E}{kT}} - B_{21}}. \qquad (10.8.12)$$

The temperature-dependent terms in Eq. (10.8.12) must be equal as well as the temperature-independent terms. Equating the temperature-independent terms results in

$$A_{21} = \frac{8\pi n^3 E^2}{(hc)^3} B_{21}, \qquad (10.8.13)$$

which yields a relationship between the Einstein A and B coefficients. Typically, the Einstein A coefficient is significantly less than the B coefficient. Equating the temperature-dependent terms yields

$$\frac{8\pi n^3 E^2}{(hc)^3} B_{12} e^{\frac{E}{kT}} = A_{21} e^{\frac{E}{kT}}. \qquad (10.8.14)$$

Substituting Eq. (10.8.13) for A_{21}, we find that Eq. (10.8.14) becomes

$$\frac{8\pi n^3 E^2}{(hc)^3} B_{12} e^{\frac{E}{kT}} = \frac{8\pi n^3 E^2}{(hc)^3} B_{21} e^{\frac{E}{kT}}, \tag{10.8.15}$$

which yields the simple relation

$$B_{12} = B_{21}. \tag{10.8.16}$$

Subsequently, the Einstein B coefficients for absorption and stimulated emission are shown to be equal.

The spontaneous emission rate is very much lower than the stimulated emission rate. Empirical evidence makes this abundantly clear. It is generally not dangerous to look at a lit incandescent bulb with the naked eye. Light emission in this case is strictly spontaneous. However, it is extremely dangerous to look directly into an active laser. Unless the power is extremely low (as is used in eye surgery), such an action will result in almost instantaneous blindness. For this reason it is extremely important to take proper precautions while working around lasers.

PROBLEMS

1. An impurity center in silicon has the following Shockley–Read–Hall parameters at 300 K:

   $$E_n = 100 \text{ s}^{-1}, \qquad C_n = 1.0 \times 10^{-9} \text{ cm}^3/\text{s}, \qquad E_p = 1.0 \times 10^{-5} \text{ s}^{-1},$$
   $$n_i = 1.0 \times 10^{10} \text{ cm}^{-3}.$$

   a. What is the hole capture rate C_p?

   b. What is the thermal activation rate of holes?

   c. What is the thermal activation rate of electrons?

   The thermal activation rate is defined as the difference between the trap energy and the band-edge energy, conduction-band edge for electrons and valence-band edge for holes.

2. There are eight Auger generation and recombination nontunneling band-to-bound processes. Sketch all eight processes and write down rate equations for each one.

3. Consider a nondegenerate semiconductor with N_t traps at energy E_t. The sample is illuminated with low-intensity light of energy $\hbar\omega > E_g$, giving rise to a generation rate G, of electron–hole pairs/s. Let B be the interband recombination rate coefficient, C_n be the electron trap capture rate coefficient, C_p be the hole trap capture rate coefficient, g_n be the electron trap generation rate coefficient, and G be the interband generation rate coefficient. Neglect the hole generation rate out of the traps. The scheme is shown in Figure Pr.10.3.

   a. Write the rate equation for the electron concentration in the conduction band n in steady state. Write the rate equation for the number of traps filled with electrons n_t.

   b. Define η as the ratio of Bnp to G, the interband absorption rate to the interband generation rate. Determine η if all the photons are absorbed.

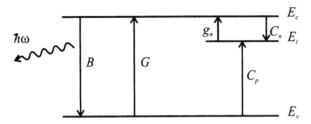

Figure Pr.10.3. Sketch of a nondegenerate semiconductor, showing interband and trap assisted generation/recombination processes.

4. From the expression for the electron and hole recombination rates through traps given by Eqs. (10.3.30), determine the maximum recombination rate if $C_n = C_p$.

5. Determine the excess minority carrier concentration if an n-type semiconductor is illuminated by light of intensity

$$G(t) = G_0|\sin(\omega t)|, \quad t > 0.$$

Find the minority carrier concentration for all time t. Assume that the net current density in the material is zero.

6. Impact ionization can also occur between free electrons and electrons confined within quantum wells. An incident electron can collide with another electron initially confined within a quantum well, as shown in Figure Pr.10.6.a.

   a. What is the threshold energy for the event if the confined electron is assumed to be in the $n = 1$ state and its energy is assumed to be given by the infinite square-well

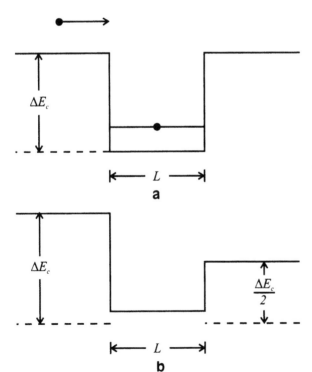

Figure Pr.10.6. a. Incident electron colliding with an electron confined within a quantum well. b. Same as a, but for an asymmetric well.

formula,

$$E = \frac{n^2\pi^2\hbar^2}{2mL^2}.$$

Assume that

$$L = 100 \text{ Å}, \qquad m = 0.067m_{\text{el}}, \qquad \Delta E_c = 0.347 \text{ eV}.$$

b. If the well is now made asymmetric, as shown in Figure Pr.10.6.b, what is the threshold energy? Assume that the energy of the state remains the same as in part a.

c. Draw a schematic representation of the process in part a. Also write an expression in terms of the carrier concentrations for the rate.

7. A bulk GaAs sample is optically excited such that an induced voltage of 0.259 V is produced. The relaxation times within the sample are given as

$$\tau_n = \tau_p = 1.0 \times 10^{-8} \text{ s}.$$

Determine the excess electron and hole concentrations if the equilibrium electron concentration present is $1.0 \times 10^{10} \text{ cm}^{-3}$. Assume that

$$n_i = 1.79 \times 10^6 \text{ cm}^{-3}, \qquad kT = 0.0259 \text{ eV}.$$

8. Consider a system consisting of two types of deep-level traps as shown in Figure Pr.10.8. Assume that the number of traps of type 1 are N_1 and of type 2 are

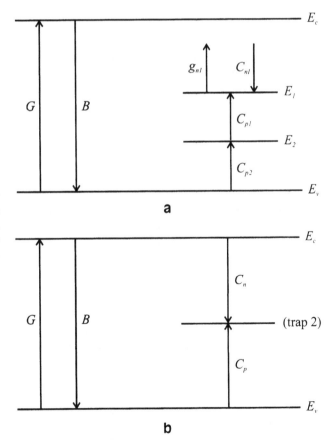

Figure Pr.10.8. a. System consisting of two types of deep-level traps and interband processes. b. System consisting of only one type of deep-level traps and interband processes.

N_2, the interband recombination rate is B, the interband generation rate is G, and the only processes involving the traps are as shown. Define E_1 as the energy of traps 1 and E_2 as the energy of traps 2. Assume the system is in steady state through both parts a and b.

a. Write the rate equations for electrons in the conduction band and in the two types of traps.

b. If traps 1 are removed and only electron capture from the conduction band into traps 2 can occur, determine the fraction of interband recombination events. The corresponding diagram is shown in Figure Pr.10.8.b.

9. Explain the prominent features of the Shockley and Wolff theories of interband impact ionization. Specifically, discuss the following:

a. How does the ionization rate depend on the applied electric field in each model?

b. What is the physical mechanism in each model by which the carriers reach threshold?

c. Which theory best approximates the solution at low fields; at high fields?

d. What do the numerical theories based on the Monte Carlo method predict?

10. It is interesting to compare different materials systems for confined-state ionization devices. Consider confined-state photomultipliers made from GaN/AlGaN and HgTe/HgCdTe. In both cases assume that the binary compound forms the well and the ternary forms the barriers. Use only the energy conservation condition.

a. If the effective mass of HgTe is three times smaller than in GaN and the potential discontinuity is two times greater in HgTe/HgCdTe than in GaN, what is the difference in the threshold condition between the two devices?

b. How would the threshold state change in the GaN device alone if an asymmetric device was used instead? Assume that the trailing edge is 1/2 of the leading edge.

11. Determine the optical excitation g_{op} in units of EHP/cm^3 s for a semiconductor sample if the resistivity changes by 50% on illumination from the dark resistivity. Use the following information for InGaAs:
$\tau_n = \tau_p = 2.2 \times 10^{-4}$ s, $N_d = 7.0 \times 10^{14}$ cm^{-3}, $\mu_n = 12,000$ cm^2/V s, $\mu_p = 450$ cm^2/V s, area A $= 2.5 \times 10^{-5}$ cm^2. Assume that all the donors are ionized.

12. Consider an infinite, uniform semiconductor sample under constant illumination by a source that generates excess carriers within only the yz plane at $x = 0$. Assume that there is no applied electric field, steady-state conditions apply, and no excess carrier generation anywhere outside of the plane at $x = 0$. The sample is strongly n type such that the excess electron concentration is everywhere much smaller than the majority carrier concentration. Determine the flux strength of the source, defined as the total net hole current density flowing outward from the $x = 0$ plane. Recall that the diffusion length $L_p = \sqrt{D_p \tau_p}$.

13. Determine the form of the interband impact ionization rate in the Wolff theory, if the semiconductor is assumed to be a two-dimensional system only. Start with the rate equations for the average energy,

$$\frac{dE}{dt} = eFv_d - \frac{E_p v_\theta}{\lambda},$$

where v_d is the drift velocity, F is the field, E_p is the mean phonon energy, v_θ is the thermal velocity, and λ is the mean free path between collisions and the momentum:

$$\frac{dp}{dt} = eF - \frac{mv_d v_\theta}{\lambda}.$$

Consider the situation in steady state.

14. Determine the conditions on the electron and the hole concentrations in terms of the mobilities for which a semiconductor has its minimum electrical conductivity. Assume that the semiconductor is in equilibrium.

15. Consider an n-type semiconductor slab. Determine the excess hole concentration, $\delta p(x)$, under the following conditions:

    a. A steady-state light source of flux ϕ creates a generation rate $G = \alpha \phi e^{-\alpha x}$.

    b. There is no electric field present, $F = 0$.

    c. The semiconductor is in steady state.

    d. $J_p(0) = S\delta p(0)$, where S is a surface recombination velocity.

    e. Assume that the sample extends from 0 to infinity along the x direction.

11

Junctions

In this chapter, we discuss the physics of any two junctions formed between two crystalline solids both in equilibrium and nonequilibrium. In general there are many types of junctions that can be formed between two different crystalline materials. Specifically, the junctions we are most concerned with are *p–n* homojunctions, *p–n* or *n–n* heterojunctions, metal–semiconductor junctions, and metal-insulator–semiconductor (MIS) junctions. The *p–n* homojunction consists of *n-* and *p*-type layers made from the same material type, a common silicon *p–n* junction diode, for example. A heterojunction is formed from two dissimilar material types that are often doped differently as well. For example, a common heterojunction of great use in modern semiconductor devices is that formed from *n*-type AlGaAs on either intrinsic GaAs or *p*-type GaAs.

In addition to semiconductor–semiconductor junctions, metal–semiconductor and MIS junctions can be formed as well. The two most important types of metal-semiconductor junctions are Schottky barriers, which have diodelike, rectifying current-voltage characteristics, or ohmic contacts, which have linear current-voltage characteristics.

Knowledge of the equilibrium and the nonequilibrium properties of these junction types, along with the earlier topics covered in this book, will provide us with sufficient background to study advanced semiconductor devices in the next chapters. First we consider the equilibrium properties of each junction type. Next we consider the nonequilibrium current-flow processes in each junction. Our method is to treat these different junction types, when possible, by using a unified approach.

11.1 Homojunctions in Equilibrium

In Chapter 5 we found that for any two subsystems in thermal and diffusive contact, both the temperature and the chemical potential are uniform throughout when the collective system is in equilibrium. The temperature is defined as a measure of the internal energy of the system. Since there is no net exchange of energy between two subsystems in thermal equilibrium, the temperature then is everywhere the same. In Chapter 5, we defined the chemical potential as a measure of the particle concentration. In the absence of a concentration gradient, the chemical potential is everywhere uniform and there is no net exchange of particles between the component subsystems. Hence, when a junction is formed between any two types of materials in equilibrium, both the temperature and the chemical potential are uniform throughout.

The chemical potential is often called the Fermi level in solid-state physics. The chemical potential or Fermi level is typically a function of the temperature of the system. The value of the Fermi level at absolute zero degrees is equal to the Fermi energy. As we found in Chapter 5, the Fermi energy is defined as the energy of the topmost filled orbital at absolute zero degrees. In this book, we denote E_f as the Fermi level, the temperature-dependent value of the chemical potential.

In equilibrium then, the chemical potential or, equivalently, the Fermi level is uniform throughout the material. This is an extremely important result, that is, **in equilibrium the Fermi level is flat throughout the entire device structure.** We will repeatedly use this concept in understanding the current flow and the behavior of semiconductor junctions.

To aid us in our discussion of the p–n homojunction it is helpful to first define notation. In general, n and p represent the carrier type, electrons and holes, respectively. Subscripts to these variables describe which side of the junction the carriers are on. For example, the electron carrier concentration on the n side of the junction is labeled as n_n. Similarly, the hole carrier concentration on the n side is p_n.

Let us now consider the behavior of a p–n homojunction in equilibrium. Our aim is to describe what happens when p-type and n-type layers of the same material are placed into contact. Although no p–n junctions are made any longer by simply joining p- and n-type material by alloying, it is useful to treat the formation of a junction theoretically as if the two materials are physically brought into contact at a certain time.

It is interesting to note that most of the important information concerning a p–n homojunction can be extracted directly from the band diagram. As we will see below, the built-in voltage and other useful quantities can be found immediately once the band diagram is known. Therefore it is important first to determine the correct band diagram for the junction. Fortunately, it is a relatively simple task to construct the band diagram if one simply follows three simple to remember rules. These are outlined below.

The most useful concept for drawing the junction in equilibrium is to recognize that the Fermi level must be flat throughout the junction. To illustrate further that the Fermi level is flat everywhere in equilibrium, consider the following. The equilibrium electron concentration on the n side of the junction, n_n, is given by Eq. (10.3.11) as

$$n_n = n_i e^{\frac{(E_f - E_i)}{kT}}. \tag{11.1.1}$$

The Fermi level E_f can be readily found from the intrinsic level E_i and the intrinsic electron carrier concentration n_i as

$$E_f = E_i + kT \ln\left(\frac{n_n}{n_i}\right). \tag{11.1.2}$$

Differentiating Eq. (11.1.2) on both sides by x gives

$$\frac{dE_f}{dx} = \frac{dE_i}{dx} + kT\frac{1}{n_n}\frac{dn_n}{dx}, \tag{11.1.3}$$

where we recognize that the derivative of n_i with respect to x is zero since n_i is a constant. We can simplify Eq. (11.1.3) further by noting that the derivative of the intrinsic level with respect to x is related to the electric field. The electric field is in general defined as the gradient of the electrostatic potential $V(x)$ as

$$F(x) = -\frac{dV(x)}{dx}. \tag{11.1.4}$$

From the discussion in Chapter 8, it was found that the band diagram simply represents the electron's potential energy. Therefore the potential and the energy bands are simply related as

$$V(x) = \frac{E_c(x)}{(-q)} = \frac{E_i(x)}{(-q)}. \tag{11.1.5}$$

Consequently, the electric field can be written in terms of the intrinsic level as

$$F(x) = \frac{1}{q}\frac{dE_i}{dx}. \tag{11.1.6}$$

Substituting expression (11.1.6) that relates the field $F(x)$ to the gradient of the intrinsic level into Eq. (11.1.3) yields

$$\frac{dE_f}{dx} = qF(x) + \frac{kT}{n_n}\frac{dn_n}{dx}. \tag{11.1.7}$$

We can transform Eq. (11.1.7) into an expression for the current density by multiplying both sides by $\mu_n n_n$. Equation (11.1.7) becomes

$$\mu_n n_n\frac{dE_f}{dx} = q\mu_n n_n F(x) + \mu_n kT\frac{dn_n}{dx}. \tag{11.1.8}$$

But the diffusion coefficent D_n is defined by Eq. (6.3.67) as

$$D_n = \frac{kT}{q}\mu_n. \tag{11.1.9}$$

When the definition of the diffusion coefficient in Eq. (11.1.8) is used, it becomes

$$\mu_n n_n\frac{dE_f}{dx} = q\mu_n n_n F(x) + qD_n\frac{dn_n}{dx}. \tag{11.1.10}$$

The right-hand side of Eq. (11.1.10) is simply the electron current density in one dimension [Eq. (6.3.68)]. Therefore we finally obtain

$$\mu_n n_n\frac{dE_f}{dx} = J_n(x). \tag{11.1.11}$$

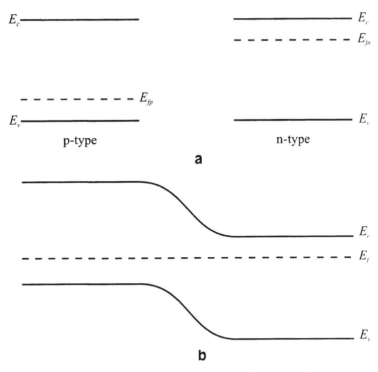

Figure 11.1.1. a. Energy-band diagrams for isolated *p*- and *n*-type material in equilibrium. b. Energy-band diagram for *p*- and *n*-type material in contact in equilibrium.

In equilibrium, the current density must of course be zero. Hence we see immediately that the gradient of the Fermi level must be zero in equilibrium in accordance with our argument based on statistical mechanics.

How can we use the result that the Fermi level remains flat throughout the junction in equilibrium to understand the behavior of the junction? The energy-band structures for two separate slabs of *p*- and *n*-type materials are sketched in Figure 11.1.1.a. If the *p*- and the *n*-type materials are placed into contact, when equilibrium is established between them the Fermi level must line up straight across the junction, as shown in Figure 11.1.1.b. The energy-band diagram for a homojunction in equilibrium can be constructed from three simple rules:

1. The Fermi level must be flat. One simply draws a straight horizontal line throughout the junction to represent the Fermi level.
2. Far from the junction the bulk properties of each layer must be recovered. In other words, far from the junction on the *p* side the energy-band diagram must be exactly like that of the isolated *p*-type material. Similarly, far from the junction on the *n* side the energy-band diagram must be exactly like that of isolated *n*-type material.
3. In the junction region itself, the energy bands are simply connected. No discontinuity appears in the energy bands for a homojunction.

So to draw the band diagram for a homojunction in equilibrium, first draw a straight horizontal line to represent the Fermi level. On either side of the junction away from the junction region, draw the energy band structures of both the *n*- and the *p*-type layers as if they were isolated. Finally, simply connect the energy bands in the junction region. Note that connecting the conduction and the valence bands on either side of the junction requires that the bands bend down in going from the *p* to the *n* side.

What, though, is the physical meaning of the band bending? Remember that the energy bands represent the electron energy in the material. The band bending implies then that the electron energies are greater on the *p* side than on the *n* side or, equivalently, that the electrostatic potential is greater on the *n* side than on the *p* side (note that the potential is found from the energy by dividing by $-q$). How does this arise? For the Fermi levels to align, electrons must transfer from the *n*-type to the *p*-type material and holes must move from the *p*-type to the *n*-type material. The electrons leave behind ionized donors with a net positive charge on the *n* side of the junction and the holes leave behind ionized acceptors with a net negative charge on the *p* side of the junction. The ionized acceptors and donors, which are frozen in the lattice and cannot move, produce a built-in electric field that is directed from the *n* side to the *p* side. A built-in voltage then appears across the junction given by the band bending in the band diagram. The *n* side lies at a higher electrostatic potential than the *p* side owing to the presence of the positively charged ionized donors.

The resulting space-charge arrangement in the junction after equilibrium has been established is sketched in Figure 11.1.2 along with the potential and the corresponding energy-band diagram. The metallurgical junction is located at the position marked x_j in the diagram. The region denoted by *W* is called the depletion region since it is fully depleted of free-charge carriers. The width of the depletion region, as we will see below, depends on the doping concentrations of the *p* and the *n* layers. It can be thought of as the amount of space charge uncovered in the process of equilibrating the Fermi levels across the junction. Free electrons and holes are exchanged between the two sides of the junction until the electrostatic potential difference increases to the point that the diffusion of charge is balanced from one side of the junction to the other.

The edges of the depletion region on the *n* and the *p* sides are labeled as x_n and $-x_p$, respectively. The potential difference established across the junction is called the built-in potential, V_{bi}. As is shown in Figures 11.1.2.b and 11.1.2.c, the built-in potential is equal to the full amount by which the bands bend in equilibrium. This is a general result, that is, **the built-in potential for a homojunction is equal to the full band bending in equilibrium.** The value of the built-in potential can be determined from inspection of Figure 11.1.3. As can be seen from the figure, the built-in potential is simply equal to the difference in the Fermi levels on the *p* and the *n* sides before contact. Therefore V_{bi} is given as

$$q V_{bi} = E_g - E_{fp} - E_{fn}, \tag{11.1.12}$$

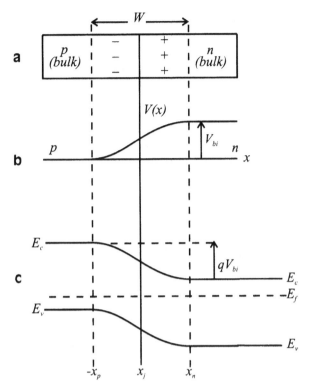

Figure 11.1.2. *p–n* homojunction in equilibrium. a. Doping schematic showing the *p* and *n* regions, depletion region *W*, and edges of the depletion region $-x_p$ and x_n. b. Potential diagram for the junction, showing the built-in potential V_{bi}. c. Energy-band diagram.

where E_{fp} and E_{fn} are the positions of the Fermi levels on the *p* and the *n* sides of the junction, respectively, as shown in Figure 11.1.3. The energy gap, E_{fn}, and E_{fp} can be written in terms of the effective densities of states N_c and N_v by use of Eqs. (10.3.27), (10.3.24), and (10.3.25) to give

$$q V_{bi} = kT \ln\left(\frac{N_c N_v}{n_i^2}\right) - kT \ln\left(\frac{N_c}{n_{n0}}\right) - kT \ln\left(\frac{N_v}{p_{p0}}\right), \qquad (11.1.13)$$

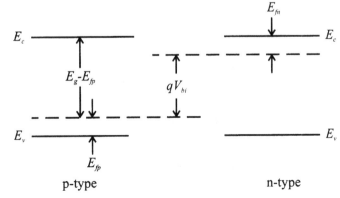

Figure 11.1.3. Energy-band diagrams used to illustrate the determination of the built-in voltage in a *p–n* homojunction.

where n_{n0} is the equilibrium electron concentration on the n side and p_{p0} is the equilibrium hole concentration on the p side. Combining the logarithms and simplifying, we find that the built-in potential becomes

$$V_{\text{bi}} = \frac{kT}{q} \ln\left(\frac{n_{n0}\, p_{p0}}{n_i^2}\right). \tag{11.1.14}$$

If it is further assumed that all the donors and acceptors are ionized, V_{bi} can be written in terms of the known acceptor N_a and donor N_d concentrations as

$$V_{\text{bi}} = \frac{kT}{q} \ln\left(\frac{N_a\, N_d}{n_i^2}\right). \tag{11.1.15}$$

Another form for the built-in potential can be also be obtained. Since the system is in equilibrium, the law of mass action holds on each side of the junction. This implies that

$$n_{n0}\, p_{n0} = n_i^2 = n_{p0}\, p_{p0}. \tag{11.1.16}$$

When the law of mass action is applied to Eq. (11.1.15), another sometimes useful formula for the built-in potential is obtained as

$$V_{\text{bi}} = \frac{kT}{q} \ln\left(\frac{n_{n0}}{n_{p0}}\right) = \frac{kT}{q} \ln\left(\frac{p_{p0}}{p_{n0}}\right). \tag{11.1.17}$$

It is important to note that much of the above information has been gleaned from inspection of the band diagram for the junction. It is far easier to remember or deduce the rules for forming a homojunction than it is to remember the formulas for the built-in potential. In only a few steps, one can recover the correct energy-band diagram and a quantitative expression for the built-in potential by following the above procedure.

The electric field can be readily determined from the solution of Poisson's equation. A one-dimensional solution is all that is generally necessary. Poisson's equation in one dimension is then

$$-\frac{d^2 V}{dx^2} = \frac{q}{\epsilon_s}\left[p(x) - n(x) + N_d^+(x) - N_a^-(x)\right], \tag{11.1.18}$$

where ϵ_s is the dielectric constant for the semiconductor and $p(x)$ and $n(x)$ are the free hole and electron concentrations, respectively. In Eq. (11.1.18) it is assumed that the dielectric function is constant, independent of x. The usual assumption made in solving the Poisson equation in a homojunction is that the free-carrier concentrations are assumed to be zero within the depletion region. This assumption is called the depletion approximation. The depletion approximation is based on the fact that the electric field is nonzero within W and hence any free carriers will be swept out of the depletion region by the action of the built-in field. Therefore both $n(x)$ and $p(x)$ are set equal to zero within W. If it is further assumed that only ionized acceptors are present on the p side and ionized donors on the n side, then on the n side of the junction within W, $N_d^-(x)$ is the only nonzero term left on the right-hand side of the Poisson equation. Therefore, Eq. (11.1.18) becomes

$$-\frac{d^2 V}{dx^2} = \frac{q N_d}{\epsilon_s}, \quad 0 \le x \le x_n. \tag{11.1.19}$$

Similarly, on the p side of the junction, the donor concentration is zero, which leads to

$$\frac{d^2 V}{dx^2} = \frac{q N_a}{\epsilon_s}, \quad -x_p \le x < 0. \tag{11.1.20}$$

Integrating the potential in Eqs. (11.1.19) and (11.1.20) yields expressions for the electric field as

$$F(x) = -\frac{q N_a (x + x_p)}{\epsilon_s}, \quad -x_p \le x < 0, \tag{11.1.21}$$

$$F(x) = \frac{q N_d (x - x_n)}{\epsilon_s}, \quad 0 \le x \le x_n. \tag{11.1.22}$$

Checking the limits of expressions (11.1.21) and (11.1.22) for the field reveals that at $x = -x_p$ and $x = x_n$ the field vanishes, as expected, since the built-in voltage is dropped across only the depletion region. At $x = 0$, the boundary condition on the fields, as specified by classical electromagnetics, is that the normal component of the D fields must be continuous if there is no surface charge present. In the case of a homojunction, there cannot be a buildup of surface charge at the metallurgical junction due to the presence of the built-in field at that point. Therefore the normal components of D must be continuous at x_j, the metallurgical junction. Equating the D fields yields

$$D_p(x = 0) = \epsilon_s \frac{-q N_a x_p}{\epsilon_s} = \epsilon_s \frac{-q N_d x_n}{\epsilon_s} = D_n(x = 0). \tag{11.1.23}$$

When simplified, Eq. (11.1.23) becomes

$$N_a x_p = N_d x_n, \tag{11.1.24}$$

which states that the amount of charge uncovered on the p side of the junction must be equal to the amount of charge uncovered on the n side. This is of course in accordance with space charge neutrality requirements, that is, the total amount of charge enclosed by a Gaussian surface surrounding the entire junction must be balanced. Clearly the maximum value of the electric field occurs at $x = 0$, and it is given by

$$F_m = \frac{-q N_d x_n}{\epsilon_s} = \frac{-q N_a x_p}{\epsilon_s}. \tag{11.1.25}$$

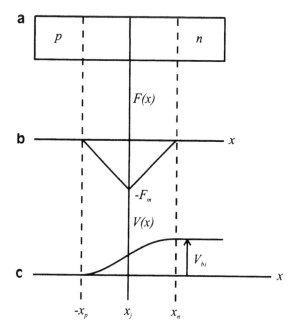

Figure 11.1.4. p–n homojunction device in equilibrium. a. Diagram showing the p and the n regions. b. Electric-field profile within the junction derived assuming the depletion approximation. c. Potential profile.

The electric field in the junction is plotted in Figure 11.1.4 along with the potential profile and junction scheme.

Finally, we conclude this section by observing that we can determine the depletion region width W by integrating the field over x. The area under the electric-field profile versus x is equal to the built-in potential V_{bi}. V_{bi} is given then as

$$V_{bi} = \frac{1}{2} F_m W. \tag{11.1.26}$$

Substituting Eq. (11.1.25) for F_m, we find that Eq. (11.1.26) for V_{bi} becomes

$$V_{bi} = \frac{1}{2} \frac{q N_d x_n W}{\epsilon_s}. \tag{11.1.27}$$

Recognizing that $W = x_n + x_p$ and using Eq. (11.1.24), we can write x_n in terms of W as

$$x_n = \frac{W N_a}{N_a + N_d}. \tag{11.1.28}$$

With this substitution, the built-in voltage becomes

$$V_{bi} = \frac{1}{2} \frac{q}{\epsilon_s} \frac{N_a N_d W^2}{N_a + N_d}. \tag{11.1.29}$$

Inverting Eq. (11.1.29) yields an expression for the depletion region width W as

$$W = \sqrt{\frac{2\epsilon_s V_{bi}}{q} \frac{(N_a + N_d)}{N_a N_d}}. \tag{11.1.30}$$

If the junction is formed between a highly doped layer and a much less doped layer, then the expression for the depletion region can be simplified to

$$W = \sqrt{\frac{2\epsilon_s V_{bi}}{q N_B}}, \tag{11.1.31}$$

where $N_B = N_d$ if $N_a \gg N_d$ and $N_B = N_a$ if $N_d \gg N_a$.

11.2 **Heterojunctions in Equilibrium**

Consider next the case of placing two dissimilar semiconductor materials together into contact. The two materials need not be doped differently, yet in most device applications they are. Such a junction is called a heterojunction. The most commonly used heterojunction at present is that formed between GaAs and an $Al_x Ga_{1-x}As$ alloy composition. The GaAs/AlGaAs heterojunction system is important since it is lattice matched: the lattice constants of GaAs and all compositions of $Al_x Ga_{1-x}As$ are the same to within a small percent. For purposes of illustration, in the following discussion we consider the GaAs/AlGaAs system only. However, the results presented are quite general and can be applied to any heterojunction system.

When a junction is formed between two different semiconductor materials, there is necessarily a discontinuity produced in the energy-band structure at the interface due to the difference in the energy bandgaps between the two semiconductors. For example, the energy bandgap of GaAs at 300 K is 1.42 eV while the gap in $Al_x Ga_{1-x}As$ varies with the aluminum composition as

$$E_g = 1.424 + 1.247x, \quad 0 < x < 0.45. \tag{11.2.1}$$

For aluminum compositions in excess of 45%, the alloy becomes indirect. At all compositions the energy bandgap of $Al_x Ga_{1-x}As$ is greater than that of GaAs. Heterojunctions formed from GaAs and AlGaAs contain an energy-bandgap discontinuity at the interface.

There are several different ways in which the energy-bandgap discontinuity is accounted for at the interface. Generally, there are three different classes of heterojunctions, which are referred to as type I, II, and III heterostructures. The three different types are sketched in Figure 11.2.1. The first type, type I, is the most common arrangement. The GaAs/AlGaAs material system is of type I. In this configuration the bandgap discontinuity is equal to the sum of the conduction-band-edge and valence-band-edge discontinuities:

$$\Delta E_g = \Delta E_c + \Delta E_v. \tag{11.2.2}$$

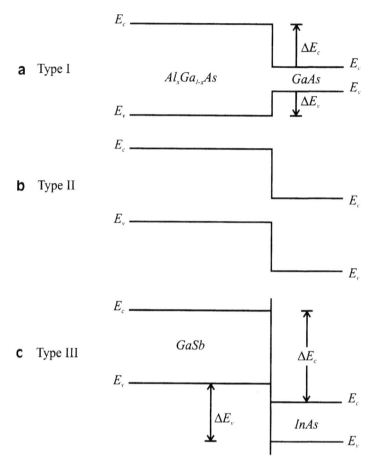

Figure 11.2.1. Sketches of the energy-band structure for three different heterojunction classes. a. Type I heterojunction. b. Type II heterojunction. c. Type III heterojunction.

The type II heterostructure, as shown in the figure, is such that the discontinuities have different signs. Note that the bandgap discontinuity is given as the difference between the conduction-band-edge and the valence-band-edge discontinuities in this case.

For the type III heterostructure, the band structure is such that the top of the valence band of one of the compounds lies above the bottom of the conduction band of the other compound. As for the type II case, the bandgap discontinuity is equal to the difference between the conduction-band-edge and the valence-band-edge discontinuities. An example of this type of heterostructure is the GaSb/InAs system.

Let us consider, for example, the formation of an n-type GaAs, p-type AlGaAs heterojunction in equilibrium. The energy-band diagrams for the constituent bulk materials are sketched in Figure 11.2.2. It is helpful first to define several physical quantities for characterizing the behavior of the junction. These are

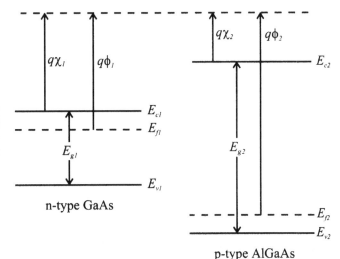

Figure 11.2.2. Equilibrium energy band diagrams of GaAs and AlGaAs in equilibrium and spatially separated showing the work functions ϕ and affinities χ.

1. ϕ, the work function. As discussed in Chapter 1, $q\phi$ is the energy needed to move an electron from the Fermi level to the vacuum level or, equivalently, free space.
2. χ, the electron affinity. $q\chi$ is the energy needed to move an electron from the conduction-band minimum to the vacuum level.
3. ΔE_c and ΔE_v, the conduction-band-edge and the valence-band-edge discontinuities, respectively.

We can determine the equilibrium energy-band diagram for the heterojunction system easily by again using the first two rules found for a homojunction; the Fermi level is flat throughout the junction, and far from the junction the bulk energy-band diagrams must be recovered. Figure 11.2.3 illustrates how the energy-band diagram can be determined from these rules. First, the Fermi level is drawn as a horizontal line extending throughout the junction. Far from the junction, the bulk energy-band diagrams of Figure 11.2.2 must be valid. Therefore the GaAs and AlGaAs bulk energy-band diagrams are next drawn on either side of the junction, as shown in Figure 11.2.3.a.

What happens, though, at the junction itself? Again we recognize that in order for the Fermi level to align, electrons must flow from the n side into the p side and holes must flow from the p side into the n side, producing a space-charge layer. The electrostatic potential is then higher on the n side than on the p side. Hence the electron energies are correspondingly lower on the n side and the bands must bend accordingly (see Figure 11.2.3.b). As mentioned above, there is a discontinuity in the conduction and the valence bands at the heterointerface owing to the different bandgaps in each constituent material. A possible representation of the energy bands is sketched in Figure 11.2.3.b. In this case, the valence-band discontinuity forms a notch at the interface. The actual way in which the bands bend, forming or not forming notches at the

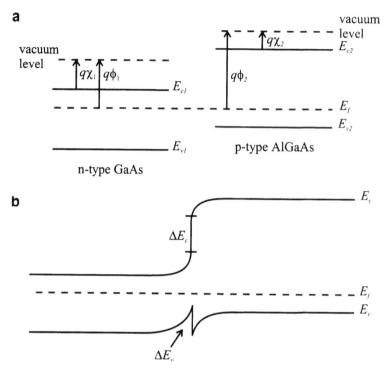

Figure 11.2.3. *n*-type GaAs – *p*-type AlGaAs heterojunction in equilibrium. a. Energy-band diagrams far from the junction. b. Junction region showing the conduction- and the valence-band discontinuities.

interface, depends on the doping concentration in each layer and on whether the Fermi level is pinned by a metal contact on one side of the junction (more will be stated about metal–semiconductor junctions shortly). To obtain the band diagram in general, it is necessary first to solve Poisson's equation for the band bending and then superimpose the band-edge discontinuities. The band diagram presented in Figure 11.2.3 is only meant as a possible representation of the resultant band structure. To obtain the band diagram correctly it is necessary to solve the Poisson equation for the system. However, with some practice, one can generally ascertain what the band diagram will look like, given the nature of the materials.

The built-in potential, as discussed above, is equal to the full band bending in equilibrium. From inspection of the band diagram in Figure 11.2.3, the full band bending is given by the difference in the work functions as

$$V_{bi} = \phi_2 - \phi_1. \tag{11.2.3}$$

Therefore the built-in potential energy is simply

$$q V_{bi} = q(\phi_2 - \phi_1), \tag{11.2.4}$$

which is equal to the difference in the work functions of the two constituent

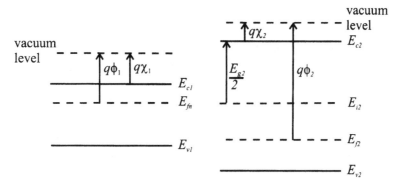

Figure 11.2.4. Sketches for determining the values of the work functions in terms of the affinities.

materials. The work functions can be written in terms of the affinities and the Fermi levels from inspection of Figure 11.2.4 as

$$q\phi_2 = q\chi_2 + \frac{E_g}{2} + (E_{i2} - E_{fp}), \qquad q\phi_1 = q\chi_1 + (E_c - E_{fn}). \qquad (11.2.5)$$

The Fermi levels can be expressed in terms of the doping concentrations through use of Eqs. (10.3.24) and (10.3.12) as

$$-(E - E_{fn}) = kT \ln\left(\frac{n_{10}}{N_{c1}}\right), \qquad E_{i2} - E_{fp} = kT \ln\left(\frac{p_{20}}{n_{i2}}\right). \qquad (11.2.6)$$

With these substitutions, the expressions for the work functions can be written as

$$q\phi_2 = q\chi_2 + \frac{E_{g2}}{2} + kT \ln\left(\frac{p_{20}}{n_{i2}}\right),$$

$$q\phi_1 = q\chi_1 - kT \ln\left(\frac{n_{10}}{N_{c1}}\right). \qquad (11.2.7)$$

The built-in voltage can now be determined from the difference in the work functions as

$$V_{bi} = \phi_2 - \phi_1 = (\chi_2 - \chi_1) + \frac{E_{g2}}{2q} + \frac{kT}{q} \ln\left(\frac{p_{20}n_{10}}{n_{i2}N_{c2}}\right). \qquad (11.2.8)$$

Expression (11.2.8) for the built-in voltage can be readily computed, provided that all the parameters are known. It is important to note that the built-in voltage for a heterojunction can be readily obtained from the band diagram along with knowledge of basic aspects of the junction itself, that is, doping concentrations, electron affinities, etc.

Up to now, we have avoided the question as to how the bandgap discontinuity is distributed between the conduction- and the valence-band discontinuities. Briefly, there is no complete theory as of yet that enables a prediction of the values of the discontinuities in all situations. The simplest model is called the Anderson model and assumes that the conduction-band-edge discontinuity is equal to the difference in the electron affinities. Therefore, if this situation is assumed,

$$\Delta E_c = q(\chi_2 - \chi_1). \tag{11.2.9}$$

Even though Expression (11.2.9) provides a simple estimate of the conduction-band-edge discontinuity, it is ambiguous when applied to direct- or indirect-gap semiconductor heterostructures. Another major drawback of Eq. (11.2.9) for estimating the discontinuity is that the affinities are a bulk property of the materials, yet they are determined from surface measurements. Affinity experiments measure the electron ionization energy from the surface of the material, which is not necessarily the same as the electron affinity defined for the bulk due to additional complications from surface states, etc. In addition, there is no general theory for determining the relative bulk electron affinities from first principles. Therefore the determination of the conduction- and the valence-band discontinuities is generally made empirically by the formation of heterostructures and optical or electrical measurements. For further discussion, the reader is referred to the book by Ferry (1991).

In a heterojunction, band bending will still occur even if the constituent materials are doped with the same type of dopants, both n type, for example. The formation of such a junction is sketched in Figure 11.2.5. When the two materials are put into contact, again the Fermi levels must align in equilibrium, and, as before, the bulklike properties of the materials are recovered far from the junction interface. For the example shown in Figure 11.2.5, $E_{f2} > E_{f1}$. When the junction is formed in this case, electrons transfer from material 2 to material 1, leaving behind ionized donors in material 2. The way in which the bands bend can be determined as follows. If one draws a Gaussian surface around the entire system, material 1 and 2, the net charge is zero. The system has to be space-charge neutral. Nevertheless, since there has been a net transfer of electrons from material 2 to 1, 1 has a net negative charge compared with 2. There is a positive charge in material 2 because of the ionized donors that precisely balance the negative charge due to the electron transfer. As an electron in material 2 approaches material 1, it sees a net negative charge that acts to raise its energy. Hence the bands bend upward from 2 to 1, as drawn in the diagram. From the other side, as an electron is moved from 1 to 2, it sees a net positive charge due to the ionized donors present in material 2. Thus the closer the electron comes to material 2, the more net positive charge it sees, which reduces its energy. Hence the bands bend downward.

The presence of the conduction-band-edge discontinuity forms a notch in the conduction band of the narrow-gap material, as shown in Figure 11.2.5. Electrons within this region of the semiconductor are then confined by a triangular well-like potential, as sketched in Figure 11.2.6. Typically, the dimensions of this triangular well are very small, of the order of ~10.0 nm or less, depending on

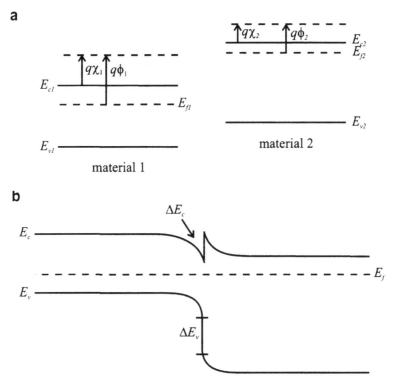

Figure 11.2.5. *n–n* heterojunction in equilibrium. a. Bulk systems: material 1 is a weaker *n*-type material than material 2. b. Energy-band diagram for the resulting *n–n* heterojunction.

the doping differential. Since the potential well has dimensions smaller or comparable with the electron's de Broglie wavelength, spatial quantization effects occur, as in the finite rectangular quantum well discussed in Chapter 2. Spatial quantization occurs in the direction perpendicular to the interface, forming quantized energy levels. These levels are called subbands since in the directions parallel to the interface the electronic energies are not quantized and the electrons retain their band properties. Therefore the electrons have 1 fewer spatial degree of freedom when confined to the triangular potential well formed at the heterointerface. The resulting system is called a two-dimensional electron gas.

The electron energies within the two-dimensional system are easily found from the sum of the confined-state energy E_i arising from the spatial quantization in the direction z normal to the interface and the free-electron energies in the directions x and y parallel to the interface as

$$E = E_i + \frac{\hbar^2}{2m^*}\left(k_x^2 + k_y^2\right). \tag{11.2.10}$$

We can find the subband energies E_i from solving the Schroedinger equation self-consistently with the Poisson equation. However, it is far easier to estimate the subband energies by approximating the notch by a triangular potential well,

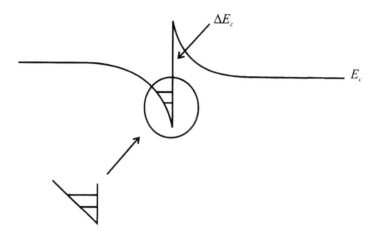

Figure 11.2.6. Conduction-band diagram showing the band bending at the *n–n* heterojunction. Note the presence of spatial quantization levels. The inset shows the resulting triangular potential well formed at the interface.

as shown by the inset in Figure 11.2.6. An analytic solution for the energy can be obtained if the triangular potential well is assumed to be infinite in extent. As a rough approximation, the well can be taken as infinite in extent and triangular in shape. The resulting energy levels are given as

$$E_i = \left(\frac{\hbar^2}{2m^*}\right)^{\frac{1}{3}} \left(\frac{3}{2}\pi q F\right)^{\frac{2}{3}} \left(i + \frac{3}{4}\right)^{\frac{2}{3}}, \tag{11.2.11}$$

where F is the electric-field strength corresponding to the slope of the energy band and i is an integer representing the band index. From Eqs. (11.2.10) and (11.2.11) the energies in the two-dimensional system can be crudely approximated.

Two-dimensional electron-gas systems occur in many semiconductor devices. Example devices are metal-oxide–semiconductor field-effect devices (MOSFETs) and other field-effect devices, such as high-electron-mobility transistors (HEMTs).

The built-in voltage of an *n–n* heterojunction can be calculated in a similar way as was done for a *p–n* heterojunction. Again, the built-in voltage is simply determined from the difference in the work functions of the two constituent materials forming the junction. From Figure 11.2.5 the work functions are given as

$$q\phi_2 = q\chi_2 + (E_{c2} - E_{f2}), \qquad q\phi_1 = q\chi_1 + (E_{c1} - E_{f1}). \tag{11.2.12}$$

The built-in potential can then be found from

$$q V_{\text{bi}} = q(\chi_2 - \chi_1) + (E_{c2} - E_{f2}) + (E_{c1} - E_{f1}). \tag{11.2.13}$$

The last two terms in Eq. (11.2.13) can be rewritten in terms of the carrier concentrations as

$$E_{c1} - E_{f1}) = kT \ln\left(\frac{n_{10}}{N_{c1}}\right),$$

$$(E_{c2} - E_{f2}) = kT \ln\left(\frac{n_{20}}{N_{c2}}\right). \tag{11.2.14}$$

With these substitutions, the built-in voltage becomes

$$q V_{bi} = q(\chi_2 - \chi_1) + kT \ln\left(\frac{n_{20} N_{c1}}{n_{10} N_{c2}}\right). \tag{11.2.15}$$

Note that, if the affinities are known or if the Anderson model is invoked and the conduction-band-edge discontinuities are known, then the built-in potential can be found from the doping concentrations and the effective density of states on either side of the junction.

It is important to note that when equilibrium is established within an n–n or an n–i heterojunction, electrons transfer from the highly doped side to the less highly doped side in order to equilibrate the Fermi levels. If the wide-gap material is highly doped while the narrow-gap material is intrinsic or nearly intrinsic, there will be a large transfer of electrons from the wide-gap layer to the narrow-gap layer. Because of the presence of the triangular quantum well formed at the heterointerface, the free electrons will be trapped in the corresponding spatial quantization levels. The transfer continues until the Fermi levels are aligned and the electrostatic field that is due to the positively charged ionized donors is balanced by the field that is due to the spatially localized electrons in the two-dimensional well. Hence free electrons are introduced into the narrow-gap layer, greatly increasing the electron concentration over that of the bulk material without introducing any dopants into that layer. The electrons within the two-dimensional well come from the ionized donors within the wide-gap semiconductor layer and are therefore spatially isolated from the parent ions. This is called modulation doping. The importance of modulation doping is that the free-electron concentration in a material can be greatly increased without sacrificing the carrier mobility by spatially separating the electrons and their corresponding parent donor ions. The mobility decreases rapidly with increasing scattering, as found in Chapter 9. One of the most important scattering mechanisms is ionized impurity scattering between the parent donor ions and the free electrons introduced into the semiconductor. Ionized impurity scattering is essentially a screened Coulomb interaction and hence depends on the distance of separation of the two charges, electron and ion. The modulation doping technique therefore greatly reduces the ionized impurity scattering between the donor ions and the free electrons by spatially separating the two species. Modulation doping can be used to greatly increase the current-carrying capability of a GaAs field-effect transistor without sacrificing the intrinsic mobility advantages of GaAs. This is the underlying principle of the HEMT. A brief discussion of the HEMT is presented in Chapter 14.5.

11.3 **Metal–Semiconductor Junctions in Equilibrium**

As mentioned at the beginning of Chapter 11, metal–semiconductor junctions can be either rectifying or ohmic, depending on the nature of the metal and the underlying semiconductor. In this section, we discuss the equilibrium properties of metal–semiconductor junctions. Rectifying metal–semiconductor junctions are called Schottky barriers while ohmic metal–semiconductor junctions are called ohmic contacts. We discuss Schottky barriers first and finish our discussion with a brief introduction to ohmic contacts.

Consider a junction formed between an *n*-type semiconductor and a metal, as shown in Figure 11.3.1. To obtain the equilibrium band diagram we again follow the two rules used for heterojunctions; the Fermi levels must align in equilibrium and far from the junction the bulklike properties of the material must be recovered. There is an important difference in this case because of the presence of the metal. To a good approximation, a metal can be considered as a perfect conductor. By definition, a perfect conductor cannot support any potential difference across it. Therefore no band bending can occur in the metal. The relative position of the Fermi level within the metal with respect to the vacuum remains unchanged then after contact. Therefore the band diagram of the metal appears the same at the interface as it does far from the junction. As a consequence, the Fermi level in the semiconductor is pinned at the interface.

The resulting band diagram on contact is presented in Figure 11.3.1.b. Note that in order for the Fermi levels to align in this case, electrons must transfer

Figure 11.3.1. Formation of a Schottky barrier in equilibrium. a. Energy-band diagrams of the metal and *n*-type semiconductor material spatially separated. b. Energy-band diagram of the metal and *n*-type semiconductor material in contact in equilibrium.

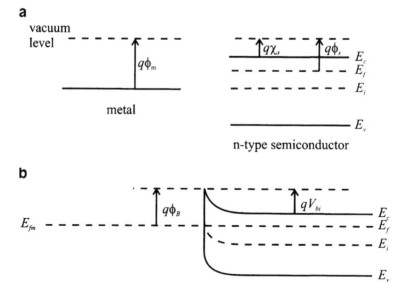

from the semiconductor to the metal, leaving behind ionized donors in the semi-conductor. The bands bend down away from the interface going into the semi-conductor.

From inspection of Figure 11.3.1, the effective potential barrier height be-tween the metal and the semiconductor as seen looking into the semiconductor from the metal is given as

$$\phi_B = \phi_m - \chi_s, \tag{11.3.1}$$

where ϕ_m is the metal work function and χ_s is the semiconductor affinity. ϕ_B is called the Schottky barrier height. Equation (11.3.1) is not entirely correct since it neglects the image force of the metal. A more complete expression for ϕ_B is found below.

The built-in potential is defined, as before, as the complete band bend-ing in equilibrium. Since there is no band bending in the metal, V_{bi} is found from the conduction-band bending in the semiconductor. From inspection of Figure 11.3.1.b the built-in voltage is easily determined as

$$V_{bi} = \phi_m - \phi_s, \tag{11.3.2}$$

which corresponds to the difference in the metal and the semiconductor work functions.

We find a more precise expression for the Schottky barrier height by consid-ering the image-force barrier lowering. In general, when an electron is moved away from a surface of a metal an equal but opposite charge is induced on the surface of the metal in order to satisfy the Gaussian law. The force between an electron and the induced surface charge on the metal is exactly equal to the force between the electron and a positive image charge located at the mirror-image point behind the plane defined by the position of the conductor. This follows readily from the fact that an infinite conducting sheet can always be inserted midway between a point-charge dipole without affecting the field and potential lines of the system. All the field lines must be perpendicular to the plane bisecting the dipole because of the geometry of the problem. Since an infinite conducting sheet cannot sustain any tangential electric-field component, the field lines in the metal can have only a perpendicular component. Hence the field due to an electron in the vicinity of a conducting sheet is precisely equal to that due to an opposite but equal image charge.

The force on an electron near the surface of a metal can be found through use of Coulomb's law for two point charges as

$$F = -\frac{q^2}{4\pi\epsilon_0(2x)^2}, \tag{11.3.3}$$

where x is the distance of the charge from the conducting sheet. The work done by this force on the electron as it moves from $+\infty$ to x is

$$U = \int_{\infty}^{x} \vec{F} \cdot d\vec{x}. \tag{11.3.4}$$

The dot vector product of F and dx is negative since dx points in the negative x direction. Hence the work done is simply

$$U = \int_{\infty}^{x} \frac{q^2}{16\pi\epsilon_0 x^2} dx, \tag{11.3.5}$$

which integrates out to

$$U = -\frac{q^2}{16\pi\epsilon_0 x}. \tag{11.3.6}$$

The potential energy is plotted as a function of x in Figure 11.3.2. Note that U approaches zero as x approaches infinity.

It can be readily shown (see problem 1) that in the depletion approximation, the electrostatic potential due to space charge within the semiconductor region in the Schottky barrier is

$$V(x) = \frac{qN_d}{\epsilon_s}\left(Wx - \frac{x^2}{2}\right). \tag{11.3.7}$$

The net potential energy within the semiconductor is given by the sum of the band bending from the space charge and effect of the image force. The resulting

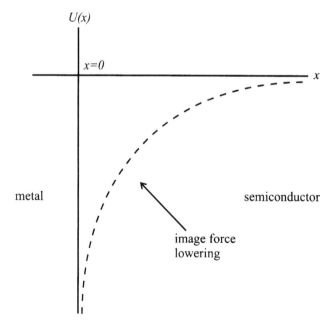

Figure 11.3.2. Potential energy as a function of position in a metal–semiconductor Schottky barrier showing only the image-force lowering.

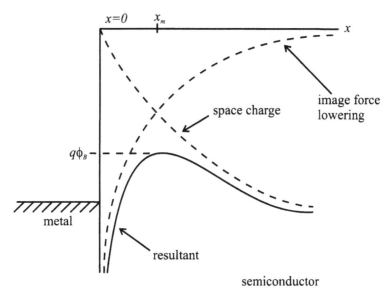

Figure 11.3.3. Potential-energy diagram showing the net potential energy due to the image-force lowering and the space charge in the semiconductor.

potential energy becomes

$$U(x) = \frac{-q^2 N_d}{\epsilon_s} \left(Wx - \frac{x^2}{2} + \frac{1}{16\pi N_d x} \right). \tag{11.3.8}$$

The resultant potential energy arising from the sum of the space-charge contribution and the image-force lowering is plotted in Figure 11.3.3. As can be seen from the diagram, the potential barrier height, marked as ϕ_B, is lower than what would be expected if the image force is not accounted for. Note that the maximum in the potential energy has moved from the interface to somewhere within the semiconductor. The position of the potential maximum is marked as x_m in the diagram. We can obtain the value of x_m by taking the derivative of Eq. (11.3.8) and setting it equal to zero. The resulting expression for x_m becomes

$$x_m^2 = \frac{1}{16\pi N_d(W - x_m)}. \tag{11.3.9}$$

Typically, x_m is much smaller than the depletion width W. With this approximation, x_m becomes

$$x_m \approx \sqrt{\frac{1}{16\pi N_d W}}. \tag{11.3.10}$$

The potential barrier lowering is simply equal to the potential energy at $x = x_m$. When approximation (11.3.10) is substituted for x_m and the resulting term in

$1/W$ is neglected, the barrier lowering is

$$\Delta\phi_B \approx -\frac{q^2}{\epsilon_s}\sqrt{\frac{WN_d}{4\pi}}. \tag{11.3.11}$$

The dielectric constant in the above result is that of the dielectric constant of the semiconductor. The above result implies that in a metal–semiconductor Schottky barrier, the effective barrier height is less than that predicted by Eq. (11.3.1).

Alternatively, an ohmic contact can be formed between a metal and a semiconductor. An ohmic contact forms if the work function of the metal is less than the work function of the semiconductor:

$$\phi_m < \phi_s. \tag{11.3.12}$$

For example, an ohmic contact can be formed between a metal and an n-type semiconductor, as shown in Figure 11.3.4. Note that ϕ_m is less than ϕ_s in this case. From our simple rules for junction formation, when the two materials are put into contact, the Fermi levels must align. Therefore there is a net flow of electrons from the metal into the semiconductor that produces a net negative charge in the semiconductor. Consequently, the electron energies in the semiconductor are raised and the bands bend up away from the interface. Note that no depletion layer forms as in the Schottky barrier and that the potential barrier to electron flow is small. Hence the barrier is easily overcome by the application of a small voltage, leading to ohmic behavior.

Figure 11.3.4. Energy-band diagram for an ohmic contact in equilibrium. a. Metal and semiconductor before to contact. b. Equilibrium band diagram after contact.

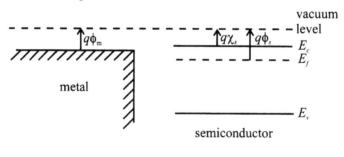

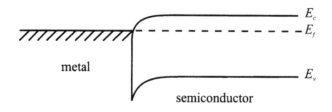

Often it is difficult to fabricate a true ohmic contact. Generally, finding an appropriate metal that when placed onto a semiconductor doped to various degrees will have no potential barrier is difficult. The metal used for the contacts must be such that it will not peel off the semiconductor. Since finding a contact metal is not always possible, what is typically done instead is to use various techniques to reduce the effective barrier height. The most common technique is to produce a narrow barrier that carriers can tunnel through on the application of a bias. These tunnel junctions, provided the barrier is sufficiently thin, yield nearly linear current-voltage characteristics, thus providing a good ohmic contact.

11.4 **Metal-Insulator–Semiconductor Junctions in Equilibrium**

The last junction type that we will consider is the metal-insulator–semiconductor, or MIS, junction. This structure forms the basis of many important semiconductor devices and is sufficiently different to warrant a separate treatment from the previous junction types. The most important MIS structure of use presently is the metal-oxide–semiconductor (MOS) system formed by aluminum, silicon dioxide, and silicon. Throughout this section, most of the emphasis is on this system, which is commonly referred to as the MOS system. The main reason why the MOS system is so commonly exploited is that a native oxide grows naturally on silicon with few defects. The technological maturity along with the natural compatibility of the silicon semiconductor and its native oxide, silicon dioxide, makes the MOS system highly useful for devices. In this book, we will briefly discuss MOS capacitors in Chapter 12 and transistors in Chapter 14. The reader is referred to the many excellent books, papers, etc., on MOS devices available, some of which are listed in the references.

In this section, we develop the basic theory of MIS junctions in equilibrium. Both ideal and nonideal systems are discussed. The most important difference between MIS structures and the previous junction types is that no dc current flows under bias in MIS structures because of the presence of the insulator layer. Hence all MIS structures are capacitive in nature.

The basic device structure considered here is sketched in Figure 11.4.1, three layers formed of a metal, insulator, and semiconductor. We first consider an ideal system with the following properties:

1. The metal–semiconductor work function difference ϕ_{ms} is zero at zero applied bias.
2. The insulator is perfect; it has zero conductivity, $\sigma = 0$.
3. No interface states located at the oxide–semiconductor interface are assumed to exist.
4. The semiconductor is uniformly doped.
5. There is a field-free region between the semiconductor and the back contact; there is no voltage drop within the bulk semiconductor.

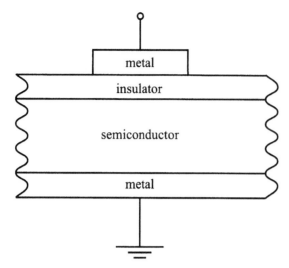

Figure 11.4.1. Sketch of a MIS system.

6. The structure is essentially one dimensional.
7. The metal gate can be treated as an equipotential surface.

Let us now consider the band diagram of an ideal MIS system in equilibrium. As must always be the case in equilibrium, the Fermi levels align throughout the entire structure, as shown in Figure 11.4.2. From assumption 1 above, the vacuum level is flat throughout the structure since the work functions are the same. Of course, this is in general not the case, but for now we assume its validity for simplicity. The symbols shown in Figure 11.4.2 are defined as the following: ϕ_m is the metal work function, ϕ_s is the semiconductor work function, χ is the electron affinity, and ψ_B is the potential difference between the Fermi level and the intrinsic level. The work function difference ϕ_{ms} for an n-type semiconductor can be determined from inspection of Figure 11.4.2 as

$$\phi_{ms} = 0 = \phi_m - \left[\chi + \frac{E_g}{2q} - \psi_B \right]. \tag{11.4.1}$$

Figure 11.4.2. Sketch of an ideal MIS system with an n-type semiconductor showing the energy bands and Fermi levels under equilibrium conditions.

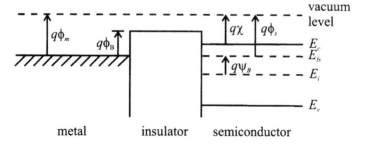

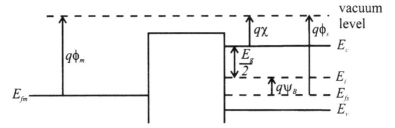

Figure 11.4.3. Ideal MIS system with a *p*-type semiconductor showing the energy bands and Fermi levels under equilibrium conditions.

The band diagram of a *p*-type semiconductor MIS system in equilibrium is sketched in Figure 11.4.3. Again, the metal–semiconductor work function difference is assumed to be zero and the Fermi level is flat everywhere throughout the junction. Inspection of Figure 11.4.3 yields an expression for ϕ_{ms} of

$$\phi_{ms} = 0 = \phi_m - \left[\chi + \frac{E_g}{2q} + \psi_B \right]. \tag{11.4.2}$$

In the above analysis, we have made many simplifying assumptions, many of which are invalid in most situations. In particular, it was assumed that the metal-semiconductor work function difference ϕ_{ms} is zero at zero applied bias and that no interface states or impurities are present within the system.

First, consider a MIS system in which the metal-semiconductor work function difference is not zero. Such a system is sketched in Figure 11.4.4.a. It is important to realize that flat band (condition in which the conduction and the valence bands have zero slope) does not occur at zero bias applied to the metal (typically referred to as zero gate bias). In other words, a voltage must be applied to the gate in order to attain the flat-band condition. This means then that in equilibrium, the energy bands within the semiconductor are bent.

To draw the energy-band diagram in equilibrium, we first note that, as in all the previous examples, the Fermi levels must align. For the specific example of Figure 11.4.4.a, the *p*-type semiconductor has a work function larger than that of the metal, $\phi_m < \phi_s$. As can be seen from the figure, the Fermi levels are not at the same level in the metal and semiconductor when they are held apart. Of course, once the MIS structure is formed, the Fermi levels must align. The Fermi levels align through charge transfer in the external circuit. Effectively, the metal has a net positive charge arising from the positive potential on the gate that lowers the electron energies at the insulator/semiconductor interface. Therefore the bands bend down as the interface is approached from within the bulk semiconductor material, as shown in Figure 11.4.4.b. The presence of a net positive charge on the metal in equilibrium results in an electric field that points from the metal into the semiconductor. This gives rise to a tilt in the conduction band within the insulator, as shown in the figure.

In the example chosen, in which a *p*-type semiconductor is used, the effective positive potential on the metal repels the majority carrier holes from

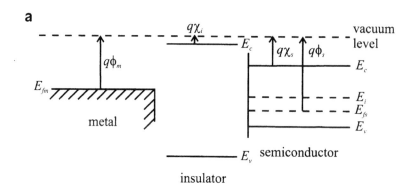

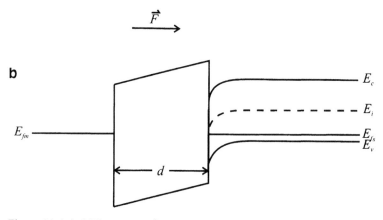

Figure 11.4.4. MIS structure for a nonideal system where $\phi_m < \phi_s$. a. Band diagram of the system before contact. b. Band diagram of the structure in equilibrium.

the interface. As a consequence, a depletion layer forms in the semiconductor. Therefore, in order to achieve flat-band conditions, a negative gate voltage equal to the work-function difference must be applied.

The presence of charge, because of either surface states or impurities within the insulator layer, which can be either fixed or mobile, alters the equilibrium band diagram of the system. Let us now consider each of these contaminants and their effect on the band diagram in succession.

In silicon/silicon dioxide structures, mobile charges due to sodium ion contamination are the most common source of impurities. In the processing of silicon/silicon dioxide devices it is easy to inadvertently introduce sodium, Na^+, ions into the oxide since Na^+ contamination readily occurs from human touch (sweat has Na^+ ions in large quantities) or through glassware and chemical reagents. Na^+ ions have a high mobility in SiO_2 at temperatures below $250°C$ at which most processing occurs. The presence of mobile charge in the insulator is particularly disruptive since Na^+ ions move around as a function of gate bias. Under the application of a negative gate bias, the Na^+ ions migrate to the metal-oxide interface, partially screening the gate bias with respect to the underlying semiconductor layer. In this way, the threshold voltage for strong inversion changes dramatically.

The change in voltage due to mobile ion charge, ΔV_m, is readily derived from Gaussian law as

$$\Delta V_m = \frac{Q_m}{C_i}, \tag{11.4.3}$$

where Q_m is the effective net charge of the mobile ions per unit area at the interface and C_i is the insulator capacitance. This can be reformulated in terms of the volume charge density of mobile charge, $\rho_m(x)$, as

$$\Delta V_m = \frac{1}{C_i} \left[\frac{1}{d} \int_0^d x \rho_m(x) \, dx \right], \tag{11.4.4}$$

where d is the thickness of the oxide. Note that the integral is weighted by x to give the effective charge at the interface per unit area.

In addition to mobile charge within the insulator, fixed charge can arise at either the interface or within the insulator itself. The interface charge is typically produced by dangling bonds between the underlying semiconductor and the covering insulator layer. In the case of the silicon/silicon dioxide system, the dangling bonds are due to trivalent silicon (excess silicon), excess oxygen, or impurities. The number of filled interface traps depends on the probability that a trap is filled (given in equilibrium by a Fermi–Dirac distribution function), which depends on the position of the Fermi level with respect to the interface trap state energy E_t. An example of this situation is shown in Figure 11.4.5.a. Note that the Fermi level is above some of the interface trap states, while some states lie above the Fermi level. For the most part, the interface states below the Fermi level are all filled while those above the Fermi level are empty. If a bias is applied to a MOS capacitor, the Fermi level remains fixed of course (since no particle exchange occurs across the insulator, that is, no dc current flows in the structure), but the bands bend in accordance with the applied bias. As a result of the band bending the interface trap energy E_t moves up or down depending on the nature of the applied bias, as shown in Figure 11.4.5. It is important to note that the bias alters the occupation of the interface states. As is readily seen from the diagram, application of a positive gate bias pulls the interface states below the Fermi level. As a result, the interface states become filled. Application of a negative bias pushes the interface states to energies above the Fermi level. As a result, these levels empty out. Clearly, the status of the interface changes with bias, affecting the behavior of the underlying semiconductor. We discuss the effects of bias on MIS structures in Section 11.8.

11.5 **Nonequilibrium Conditions: Current Flow in Homojunctions**

Now that we have discussed all the basic junctions in equilibrium it is necessary to address the nonequilibrium conditions for each junction type. Rather than

a Equilibrium

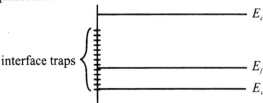

interface traps

b Positive gate bias

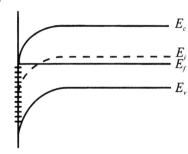

Figure 11.4.5. Sketch of a metal-insulator p-type semiconductor system including interface trap states. a. Equilibrium condition. b. Positive gate bias condition. c. Negative gate bias condition.

c Negative gate bias

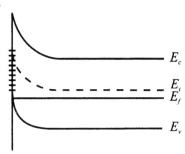

attempt a full discussion of current transport in junctions, only the salient issues that are of greatest usage in our later discussion of devices are addressed. A full discussion of junctions here would be impractical. In this section, we examine current flow in a p–n homojunction diode.

We start with the simplest such formulation for the current flow, called the Shockley equation. The Shockley equation can be derived under the following assumptions:

1. The depletion region width W is assumed to be abrupt.
2. Nondegenerate conditions hold throughout (the Boltzmann distribution approximation to the carrier densities holds on both the p and the n sides).
3. Low carrier injection.
4. No carrier generation occurs within W.

As in Section 10.8, Box 10.8.1, we define the electrostatic potential as $\psi = -E_i/q$, ϕ as the Fermi potential, $-E_f/q$, ϕ_n as the electron quasi-Fermi level, and ϕ_p as the hole quasi-Fermi level.

Following the results given in Box 10.8.1, the electron and hole carrier concentrations in nonequilibrium can be written in terms of the quasi-Fermi levels and the potential as

$$n = n_i e^{\frac{q(\psi - \phi_n)}{kT}}, \qquad p = n_i e^{\frac{q(\phi_p - \psi)}{kT}}. \tag{11.5.1}$$

The current density $\vec{J}_n$ can be found from its definition and the nonequilibrium carrier concentrations as follows. From Chapter 7, the current density was determined as

$$\vec{J}_n = q\mu_n n \vec{F} + q D_n \vec{\nabla}_n. \tag{11.5.2}$$

The field is given as the gradient of the potential

$$\vec{F} = -\vec{\nabla}\psi. \tag{11.5.3}$$

Substituting in the expressions for the electric field and the nonequilibrium carrier concentrations into the current density and using the Einstein relation to relate the mobility and diffusivity, we find that $\vec{J}_n$ becomes

$$\vec{J}_n = -q\mu_n n \vec{\nabla}\psi + q\mu_n \frac{kT}{q} \vec{\nabla}\left[n_i e^{\frac{q(\psi-\phi_n)}{kT}} \right] \tag{11.5.4}$$

$$= -q\mu_n n \vec{\nabla}\psi + q\mu_n n_i (\vec{\nabla}\psi - \vec{\nabla}\phi_n) e^{\frac{q(\psi-\phi_n)}{kT}} \tag{11.5.5}$$

$$= -q\mu_n n \vec{\nabla}\psi + q\mu_n n \vec{\nabla}\psi - q\mu_n n \vec{\nabla}\phi_n, \tag{11.5.6}$$

$$\vec{J}_n = -q\mu_n n \vec{\nabla}\phi_n. \tag{11.5.7}$$

The current density is found to be directly proportional to the gradient of the quasi-Fermi level.

Equation (11.5.7) is generally true and can be applied to any semiconducting system in nonequilibrium. In many applications the use of Eq. (11.5.7) is limited since the current density cannot conveniently be determined in this manner. For the present situation, current flow in a p–n homojunction, the current density can be found in a different way, as follows.

Our goal is to solve the one-dimensional current continuity equation subject to the boundary conditions on the carrier concentrations at the depletion region edges. First, let us determine the electron carrier concentration on the p side of the junction at the edge of the depletion region, $n_p(-x_p)$. The electron concentration on the p side is called the minority carrier concentration. The voltage drop V occurs completely across the depletion region; the doping concentrations are assumed sufficiently large that no voltage drop occurs across the undepleted n or p regions. The product of the nonequilibrium electron and hole concentrations at $x = -x_p$ yields

$$p_p(-x_p)n_p(-x_p) = n_i^2 e^{\frac{qV}{kT}}, \tag{11.5.8}$$

where we have used the result from Box 10.8.1 that the difference in the quasi-Fermi levels is equal to the bias V. Thus,

$$n_p(-x_p) = \frac{n_i^2 e^{\frac{qV}{kT}}}{p_p(-x_p)}. \tag{11.5.9}$$

The above result can be further simplified through use of the third assumption stated above, that of low-level injection. Implicit in this assumption is that the excess majority carrier concentrations Δn_n and Δp_p are small with respect to the equilibrium majority carrier concentrations, n_{n0} and p_{p0}. Therefore the electron carrier concentration at x_n is given as

$$n(x_n) = n_{n0} + \Delta n_n \sim n_{n0}. \tag{11.5.10}$$

The hole concentration at $-x_p$ is similarly determined as

$$p(-x_p) = p_{p0} + \Delta p_p \sim p_{p0}. \tag{11.5.11}$$

The law of mass action on the p side of the junction gives

$$\frac{n_i^2}{p_{p0}} = n_{p0}. \tag{11.5.12}$$

Clearly if $p_p(-x_p) \sim p_{p0}$ then

$$\frac{n_i^2}{p_p(-x_p)} = \frac{n_i^2}{p_{p0}} = n_{p0}. \tag{11.5.13}$$

The minority electron carrier concentration on the p side given by Eq. (11.5.9) can be rewritten as

$$n_p(-x_p) = n_{p0} e^{\frac{qV}{kT}}. \tag{11.5.14}$$

Similarly, the minority hole carrier concentration on the n side of the junction p_n is found as

$$p_n(x_n) = \frac{n_i^2}{n_{n0}} e^{\frac{qV}{kT}} = p_{n0} e^{\frac{qV}{kT}}. \tag{11.5.15}$$

Next we solve the one-dimensional current continuity equation subject to the above boundary conditions. The current continuity equation, given by Eq. (10.2.3), is

$$\frac{dn_n}{dt} = G_n - R_n + \frac{1}{q} \vec{\nabla} \cdot \vec{J}_n. \tag{11.5.16}$$

Substituting Eq. (11.5.2) in for the electron current density, we find that the current continuity equation becomes

$$\frac{dn_n}{dt} = G_n - R_n + \frac{1}{q}\vec{\nabla}\cdot[q\mu_n n\vec{F} + qD_n\vec{\nabla}n_n].$$
(11.5.17)

Assuming steady-state conditions, no generation processes present, and that a one-dimensional approximation is valid, we simplify the continuity equation

$$0 = -R_n + \mu_n n_n\frac{dF}{dx} + \mu_n F\frac{dn_n}{dx} + D_n\frac{d^2n_n}{dx^2}.$$
(11.5.18)

Similarly, for holes,

$$0 = -R_p + \mu_p p_n\frac{dF}{dx} + \mu_p F\frac{dp_n}{dx} + D_p\frac{d^2p_n}{dx^2}.$$
(11.5.19)

If we further assume that the electron and hole recombination rates within the depletion region W are zero, then the electron and hole current densities must be constant throughout the depletion region. Consequently, the current density must be the same within the depletion region as that at the depletion region edges, $x = x_n$ and $x = -x_p$. From Kirchoff's current law, the total current density throughout the structure must be constant. Therefore, if J is determined at $x = x_n$ or $x = -x_p$, then it is known throughout the structure.

The current density at the depletion region edges can be found as follows. Multiplying Eq. (11.5.18) by $\mu_p p_n$ yields

$$0 = -R_n\mu_p p_n + \mu_n n_n\mu_p p_n\frac{dF}{dx} + \mu_n\mu_p p_n F\frac{dn_n}{dx} + \mu_p p_n D_n\frac{d^2n_n}{dx^2}.$$
(11.5.20)

A similar result is found for the holes as

$$0 = -R_p\mu_n n_n + \mu_n n_n\mu_p p_n\frac{dF}{dx} - \mu_n n_n\mu_p F\frac{dp_n}{dx} + \mu_n n_n D_p\frac{d^2p_n}{dx^2}.$$
(11.5.21)

Within the undepleted or bulk n and p regions, the electric field is assumed to be zero. In essence, it is assumed that there is no voltage drop in the bulk regions. Therefore, at the edges of the depletion regions, x_n and $-x_p$, the electric field goes to zero. The terms that involve F in Eqs. (11.5.20) and (11.5.21) both vanish at either x_n or $-x_p$. With these assumptions, Eqs. (11.5.20) and (11.5.21) can be added together to get

$$0 = -R(\mu_p p_n + \mu_n n_n) + \mu_p p_n D_n\frac{d^2n_n}{dx^2} + \mu_n n_n D_p\frac{d^2p_n}{dx^2}.$$
(11.5.22)

The electron and hole mobilities can be expressed in terms of the diffusion constants D_n and D_p through use of the Einstein relations:

$$\mu_n = \frac{q}{kT} D_n, \qquad \mu_p = \frac{q}{kT} D_p. \tag{11.5.23}$$

When the Einstein relations are used, Eq. (11.5.22) becomes

$$0 = -R(\mu_p p_n + \mu_n n_n) + \frac{q}{kT} D_n D_p \left[p_n \frac{d^2 n_n}{dx^2} + n_n \frac{d^2 p_n}{dx^2} \right]. \tag{11.5.24}$$

It is reasonable to assume that the excess carrier concentrations on the same side of the junction are roughly the same. Thus

$$(n_n - n_{n0}) \sim (p_n - p_{n0}). \tag{11.5.25}$$

Taking the derivative of approximation (11.5.25) with respect to x yields

$$\frac{dn_n}{dx} - \frac{dn_{n0}}{dx} \sim \frac{dp_n}{dx} - \frac{dp_{n0}}{dx}. \tag{11.5.26}$$

Since the derivatives of the constant terms n_{n0} and p_{n0} are zero, it follows that

$$\frac{d^2 n_n}{dx^2} \sim \frac{d^2 p_n}{dx^2}. \tag{11.5.27}$$

With this substitution and with the ambipolar diffusion constant defined as

$$D_a = \frac{(n_n + p_n)}{\left(\frac{p_n}{D_n} + \frac{n_n}{D_p} \right)}, \tag{11.5.28}$$

Eq. (11.5.24) becomes

$$D_a \frac{d^2 p_n}{dx^2} - R = 0. \tag{11.5.29}$$

D_a is the general expression for the diffusion coefficient that is, useful when both carrier species contribute to the current.

The recombination rate, as discussed in Chapter 10, can be expressed as the ratio of the excess carrier concentration divided by the average lifetime. With an ambipolar lifetime given as τ_a, the recombination rates can be written in terms of the excess carrier concentrations as

$$R = \frac{(p_n - p_{n0})}{\tau_a} = \frac{(n_n - n_{n0})}{\tau_a}. \tag{11.5.30}$$

With the above substitutions, the expression for the minority hole carrier concentration p_n outside the depletion region on the n side of the junction becomes

$$D_a \frac{d^2 p_n}{dx^2} - \frac{(p_n - p_{n0})}{\tau_a} = 0. \tag{11.5.31}$$

Next we solve for p_n under the assumption of low-level injection, that is, $p_n \ll n_n$. The ambipolar diffusion constant can be simplified under this assumption to

$$D_a = \frac{(n_n + p_n)}{\left(\frac{p_n}{D_n} + \frac{n_n}{D_p}\right)} \sim \frac{n_n}{\left(\frac{n_n}{D_p}\right)} = D_p. \qquad (11.5.32)$$

Similarly, $\tau_a \sim \tau_p$. Equation (11.5.31) becomes then

$$\frac{d^2 p_n}{dx^2} - \frac{(p_n - p_{n0})}{D_p \tau_p} = 0. \qquad (11.5.33)$$

The product $D_p \tau_p$ in the denominator in Eq. (11.5.33) has units of length squared. We define the diffusion length, which physically corresponds to the average distance a carrier will diffuse before recombining, as

$$L_p = \sqrt{D_p \tau_p}. \qquad (11.5.34)$$

Finally, the equation for p_n is

$$\frac{d^2(p_n - p_{n0})}{dx^2} - \frac{(p_n - p_{n0})}{L_p^2} = 0, \qquad (11.5.35)$$

where we note that the second derivative of p_{n0} with respect to x is equal to zero and can be added to the left-hand side of Eq. (11.5.35) without altering its value.

We next solve Eq. (11.5.35) subject to the boundary conditions:

1. As x approaches ∞, p_n approaches p_{n0}, the equilibrium hole concentration on the n side.
2. At $x = x_n$, $p_n = p_{n0} e^{(qV/kT)}$.

The solution of Eq. (11.5.35) is readily found as

$$(p_n - p_{n0}) = A e^{\frac{x}{L_p}} + B e^{-\frac{x}{L_p}}. \qquad (11.5.36)$$

When boundary condition 1 is applied at $x = +\infty$, $(p_n - p_{n0}) = 0$. Therefore the coefficient A must be zero. Applying the second boundary condition at $x = x_n$, we have

$$(p_n - p_{n0})|_{x=x_n} = B e^{-\frac{x_n}{L_p}} = p_{n0}\left(e^{\frac{qV}{kT}} - 1\right). \qquad (11.5.37)$$

Therefore B is found as

$$B = p_{n0}\left(e^{\frac{qV}{kT}} - 1\right) e^{\frac{x_n}{L_p}}. \qquad (11.5.38)$$

Finally, $(p_n - p_{n0})$ is found to be

$$(p_n - p_{n0}) = p_{n0}\left(e^{\frac{qV}{kT}} - 1\right) e^{\frac{-(x - x_n)}{L_p}}. \qquad (11.5.39)$$

Using Expression (11.5.39) for the excess minority carrier concentration, we can find the current density J_p from Eq. (6.3.69). At $x = x_n$, the electric field F is zero. Therefore substituting Eq. (11.5.39) into Eq. (6.3.69) yields

$$J_p|_{x=x_n} = -q D_p \frac{d}{dx} \left[p_{n0} \left(e^{\frac{qV}{kT}} - 1 \right) e^{\frac{-(x-x_n)}{L_p}} + p_{n0} \right]. \tag{11.5.40}$$

Evaluating the derivative yields

$$J_p = \frac{q D_p}{L_p} p_{n0} \left(e^{\frac{qV}{kT}} - 1 \right) e^{\frac{-(x-x_n)}{L_p}} \bigg|_{x=x_n}. \tag{11.5.41}$$

The hole current density at the edge of the depletion region as well as throughout the depletion region is then

$$J_p = \frac{q D_p}{L_p} p_{n0} \left(e^{\frac{qV}{kT}} - 1 \right). \tag{11.5.42}$$

A similar analysis applied to the electron current density yields

$$J_n = \frac{q D_n}{L_n} n_{p0} \left(e^{\frac{qV}{kT}} - 1 \right). \tag{11.5.43}$$

The total current density J is determined from the sum of J_n and J_p. When Eqs. (11.5.42) and (11.5.43) are combined, J is found as

$$J = \left(\frac{q D_p p_{n0}}{L_p} + \frac{q D_n n_{p0}}{L_n} \right) \left(e^{\frac{qV}{kT}} - 1 \right), \tag{11.5.44}$$

which is the Shockley equation. We readily find the corresponding current as a function of voltage by multiplying Eq. (11.5.44) by the cross-sectional area of the junction A. The current becomes

$$I = J A = q A \left(\frac{D_p p_{n0}}{L_p} + \frac{D_n n_{p0}}{L_n} \right) \left(e^{\frac{qV}{kT}} - 1 \right). \tag{11.5.45}$$

Under forward bias the exponential term dominates and the current increases exponentially with increased bias. In reverse bias, the applied potential V is less than zero. Therefore the exponential term $e^{qV/kT}$ can be neglected with respect to 1. The reverse current is simply given then as

$$I = J A = -q A \left(\frac{D_p p_{n0}}{L_p} + \frac{D_n n_{p0}}{L_n} \right). \tag{11.5.46}$$

The current is plotted as a function of voltage in Section 11.6.

11.6 **Interpretation and Modifications of the Shockley Equation**

In this section, we discuss the meaning of the Shockley equation that has been derived in Section 11.5. Additionally, we discuss modifications to the Shockley equation needed to capture some of the nonidealities neglected in its formulation. Inspection of the Shockley equation shows that the current-voltage characteristic is asymmetric with the applied voltage V. Before we discuss the deviations from the ideal behavior specified by the Shockley equation, it is necessary first to examine the basic physics implicit in its description.

Under forward bias the current is dominated by diffusion that increases exponentially with increasing bias. In reverse bias the diffusion current is effectively choked off, leaving only the drift current. To understand the effect that the bias has on the diffusion current, it is useful to examine the potential diagram of a p–n homojunction under equilibrium, forward, and reverse biases. The potential and energy-band diagrams are shown in Figure 11.6.1. Recall that in equilibrium a built-in potential appears across the junction because of the space-charge layer formed between the p and the n layers. Consequently, the p side lies at a lower potential than the n side, as shown in Figure 11.6.1.a. If a forward bias is applied to the junction, the p side is biased positively with respect to the n side. Therefore the potential barrier between the two sides of the junction is lowered. Under reverse bias, the potential barrier across the junction is increased since the n side lies at a higher potential than the p side.

Figure 11.6.1. Potential and energy-band diagrams for a p–n homojunction under a. equilibrium, b. forward bias, c. reverse bias conditions.

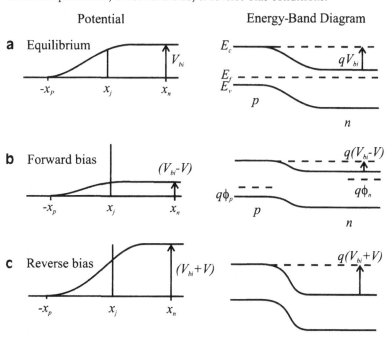

The question we seek to answer is, how does the size of the potential barrier influence the current flow? From the discussion in Chapter 5, the equilibrium energy distribution function $f(E)$ yields the probability of finding a carrier at energy E. Since the junction is formed between two weakly coupled subsystems, each separately near equilibrium, the carriers in each system can be described by an equilibrium distribution function. This implies that the carriers on either side of the junction can be described in terms of a Boltzmann or Fermi-like distribution function characterized by a quasi-Fermi level, ϕ_n or ϕ_p. A typical distribution function for the carriers on either side of the junction (assuming Fermi–Dirac statistics) is shown in Figure 11.6.2.a. The energy corresponding to the equilibrium value of the built-in voltage is marked in Figure 11.6.2.a as qV_{bi}. Those electrons whose forward-directed kinetic energies are less than qV_{bi} have insufficient energy to overcome the potential barrier and cannot contribute to the diffusion current. It is important to recognize that the forward-directed kinetic energy, not just the kinetic energy, needs to be greater than the potential barrier. Forward directed means that if the z direction is perpendicular to the potential barrier, then $1/2mv_z^2$ must be greater than the potential barrier height. Otherwise the carriers cannot surmount the potential barrier. These carriers are in essence trapped on their original side of the junction. However, the electrons within the high-energy tail of the distribution, those with forward-directed kinetic energies greater than the potential barrier, can diffuse from one side of the junction to the other. These are the carriers that contribute to the diffusion current. In equilibrium, the total diffusion current is precisely balanced by the total drift current, leading to a net zero current within the junction. The drift current comprises carriers that are generated within the depletion region or within a diffusion length of the depletion region and are swept out of the junction by the action of the built-in field. Since the field points from the n to the p side of the junction, the electrons drift toward the n side and the holes drift toward the p side, opposite to the direction of the diffusion currents.

When a forward bias is applied to the junction, the potential difference between the two sides of the junction is reduced, as shown in Figure 11.6.1.b. As the potential barrier decreases, the diffusion current increases exponentially. The exponential increase in the diffusion current can be understood from consideration of Figure 11.6.2. Since either carrier distribution is essentially Fermi–Dirac, lowering the potential barrier through the application of a forward bias leads to an exponential increase in the number of carriers whose energy exceeds the potential barrier, as shown in Figure 11.6.2.b. As can be seen from Figure 11.6.2.b, the carriers that can contribute to the diffusion current are those under the tail of the distribution from an energy of $q(V_{bi} - V)$ to ∞, which is essentially the area under the curve in this region. (Note that, for simplicity, it is assumed here that the distribution shown represents those electrons whose energy is only forward directed.) Compared with the area under the curve for the equilibrium case, there is an exponential increase in the number of carriers that can surmount

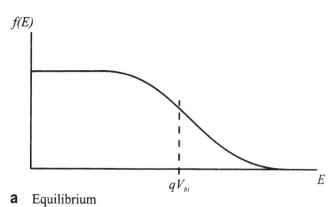

a Equilibrium

Figure 11.6.2. Carrier distribution functions and the corresponding potential barriers for a. equilibrium, b. forward and reverse bias conditions.

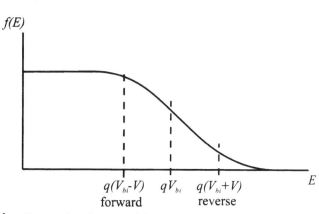

b Forward and reverse bias

the potential barrier. Therefore there is an exponential increase in the diffusion current.

Under reverse bias, the effective potential barrier is increased, leading to an exponential decrease in the number of carriers whose energy is greater than the potential barrier. The diffusion current then is choked off, leaving only the drift current. The drift current depends on the number of carriers generated within the depletion region or within a diffusion length of the depletion region edges. To first order, the drift current is independent of the magnitude of the bias. We can understand this readily by recognizing that the drift current is limited by the number of carriers that are produced within the depletion region, not by how fast the carriers are swept out of this region, which is of course related to the magnitude of the electric field. The rate at which carriers are generated within the depletion region is only weakly dependent on the magnitude of the electric field, as will be discussed below.

In summary, the Shockley equation can be understood qualitatively as follows. A forward bias acts to lower the potential barrier of the junction such that the carriers can more readily diffuse across it. A reverse bias acts to increase the potential barrier, thereby choking off the diffusion current. The drift current is basically independent of bias since it depends on the number of carriers generated within or near the depletion region. Under very large reverse bias, the depletion region becomes wider. Therefore more carriers can be generated within it, leading to a greater drift current. Nevertheless, this increase is not very great.

At this point, it is useful to examine modifications to the Shockley equation. In its derivation, a number of approximations are made, as stated above in Section 11.5. Additionally, numerous physical effects are omitted from the simple theory of p–n junctions. The critical factors omitted from the Shockley equation that most affect the current description are:

1. generation–recombination within the depletion region W,
2. high current injection under forward bias,
3. reverse bias breakdown through avalanche multiplication,
4. carrier tunneling in reverse bias, and
5. diode series resistance (voltage drop within the bulk regions).

Let us first consider the effect of generation–recombination processess within the depletion region when the junction is reverse biased. From Eq. (11.1.30), the depletion region width in equilibrium is found to be

$$W = \sqrt{\frac{2\epsilon_s V_{\text{bi}}}{q} \frac{(N_a + N_d)}{N_a N_d}}. \tag{11.6.1}$$

Under bias the depletion region width becomes

$$W = \sqrt{\frac{2\epsilon_s (V_{\text{bi}} - V)}{q} \frac{(N_a + N_d)}{N_a N_d}}. \tag{11.6.2}$$

As the bias becomes more negative, the depletion region width increases. The generation–recombination current will also increase as W increases since the probability of an electron–hole pair's being generated within the depletion region is greater the larger W is.

In reverse bias, the p–n product, as given by Eq. (10.3.28), becomes less than n_i^2. This implies minority carrier extraction from each side of the junction, since the minority carrier concentrations are less than the corresponding equilibrium carrier concentrations.

The current density in reverse bias is simply given by the drift current component plus the generation–recombination current within the depletion region. For simplicity the generation–recombination mechanisms will be assumed to be through trap states. As such, the current density can be written as

$$J = -\left(\frac{q D_p p_{n0}}{L_p} + \frac{q D_n n_{p0}}{L_n} \right) + \int_0^W q R_n \, dx, \tag{11.6.3}$$

where R_n is the net electron recombination rate through traps, as discussed in Chapter 10. Remember that R_n is determined from the difference in the trap capture and emission rates $R_{cn} - R_{en}$. If a spatially independent generation–recombination rate within the junction is assumed, the last term in Eq. (11.6.3) can be readily integrated to yield

$$\int_0^W q R_n \, dx = q R_n W. \tag{11.6.4}$$

From Chapter 10, R_n was determined to be

$$R_n = \frac{N_t C_n C_p (np - n_i^2)}{[C_n(n + n_1) + C_p(p + p_1)]}, \tag{11.6.5}$$

where N_t is the total number of traps and n_1 and p_1 are equilibrium electron and hole concentrations, respectively, defined as

$$n_1 = n_i e^{\frac{(E_t - E_i)}{kT}}, \qquad p_1 = n_i e^{\frac{(E_i - E_t)}{kT}}. \tag{11.6.6}$$

With these substitutions R_n becomes

$$R_n = \frac{N_t C_n C_p (np - n_i^2)}{\left\{ C_n \left[n + n_i e^{\frac{(E_t - E_i)}{kT}} \right] + C_p \left[p + n_i e^{\frac{(E_i - E_t)}{kT}} \right] \right\}}. \tag{11.6.7}$$

Under reverse bias, $np \ll n_i^2$, so n and p can be neglected with respect to the terms involving n_i above. Therefore R_n simplifies to

$$R_n = \frac{-N_t C_n C_p n_i^2}{\left\{ C_n \left[n_i e^{\frac{(E_t - E_i)}{kT}} \right] + C_p \left[n_i e^{\frac{(E_i - E_t)}{kT}} \right] \right\}}. \tag{11.6.8}$$

The mean recombination lifetime can be defined as

$$\frac{1}{\tau_e} = \frac{N_t C_n C_p}{C_n e^{\frac{(E_t - E_i)}{kT}} + C_p e^{\frac{(E_i - E_t)}{kT}}}. \tag{11.6.9}$$

With this definition, R_n becomes

$$R_n = -\frac{n_i}{\tau_e}. \tag{11.6.10}$$

The negative sign present in Eqs. (11.6.8) and (11.6.10) indicates that carrier generation dominates the generation–recombination process in reverse bias. This is readily understood since the number of free carriers present is less than the number of free carriers in equilibrium under reverse bias. Therefore, to restore the system to equilibrium, the net generation rate must exceed the net recombination rate.

Substituting into Eq. (11.6.3) the result for the recombination rate, we find that the expression for the reverse current density becomes

$$J_{\text{reverse}} = -\left(\frac{q D_p p_{n0}}{L_p} + \frac{q D_n n_{p0}}{L_n} \right) - \frac{q n_i W}{\tau_e}, \tag{11.6.11}$$

where the first term is the reverse saturation current density and the last term is the depletion layer generation current density. Note that J_{reverse} is not constant, independent of the reverse bias, because of the inclusion of the depletion layer current term. J_{reverse} depends on the reverse bias from the voltage dependence of the depletion region width.

Equation (11.6.11) can be simplified under the special case of a p^+–n junction. In this situation, the p side is much more highly doped than the n side of the junction. Consequently, $p_{n0} \gg n_{p0}$. The saturation current density can thus be simplified to

$$J_{\text{sat}} \sim \frac{q D_p p_{n0}}{L_p} = \frac{q D_p n_i^2}{N_d} \frac{1}{\sqrt{D_p \tau_p}}. \tag{11.6.12}$$

With the above substitution, the reverse saturation current density in a p^+–n junction can be written as

$$J_{\text{reverse}} = -q \sqrt{\frac{D_p}{\tau_p}} \frac{n_i^2}{N_d} - \frac{q n_i W}{\tau_e}. \tag{11.6.13}$$

Under large forward bias, a significant recombination current adds to the usual diffusion current. Under forward bias, the junction has a net injection of minority carriers. For the junction to recover equilibrium if the bias is removed, there would necessarily have to be a larger recombination rate compared with the generation rate. Hence the recombination current dominates the generation current under forward bias and is given as

$$R_n = \frac{N_t C_n C_p n_i^2 \left(e^{\frac{qV}{kT}} - 1 \right)}{C_n \left[n + n_i e^{\frac{(E_t - E_i)}{kT}} \right] + C_p \left[p + n_i e^{\frac{(E_i - E_t)}{kT}} \right]}. \tag{11.6.14}$$

If for simplicity it is further assumed that the intrinsic and the trap levels align and that the electron and hole capture cross sections are equal, then R_n becomes

$$R_n = \frac{N_t n_i^2 C \left(e^{\frac{qV}{kT}} - 1 \right)}{(n + p + 2n_i)}. \tag{11.6.15}$$

We can find the net recombination–generation current density by performing the integration in Eq. (11.6.4) with the expression for R_n given by Eq. (11.6.15) above. This is left as an exercise, problem 2, at the end of this chapter. As discussed in the problem statement, a simplified result for the current density can be obtained if it is assumed that the intrinsic and the trap levels are coincident and if the voltage is large with respect to kT. Note that the expression for R_n above has a positive sign, indicating that the net recombination rate exceeds the net generation rate. The net current density under forward bias including the recombination current is

$$J_{\text{forward}} = q\left(\frac{D_p p_{n0}}{L_p} + \frac{D_n n_{p0}}{L_n}\right)\left(e^{\frac{qV}{kT}} - 1\right) + \frac{q}{2}W\,CN_t n_i e^{\frac{qV}{2kT}}. \qquad (11.6.16)$$

For a p^+–n junction the forward current density can be simplified to

$$J_{\text{forward}} = q\sqrt{\frac{D_p}{\tau_p}}\,\frac{n_i^2}{N_d}e^{\frac{qV}{kT}} + \frac{q}{2}W\,CN_t n_i e^{\frac{qV}{2kT}}. \qquad (11.6.17)$$

The third effect that we consider for a more complete description of a p–n homojunction is reverse bias breakdown arising from avalanche multiplication. The topic of avalanche multiplication has been presented in Chapter 10. Further information will be presented in Chapter 12 when we discuss avalanche photodiodes, which are useful for light-detection applications. Here we discuss how the presence of avalanche multiplication alters the performance of a simple diode.

Under large reverse bias, the electric field within the depletion region can become sufficiently large that the electrons and holes drifting within it can be accelerated to high energies such that impact ionization events can occur. As discussed in Chapter 10, impact ionization is a threshold process; the initiating carriers must have sufficient energy to produce an electron–hole pair by means of a collision with the lattice. Either electrons or holes can initiate an impact ionization event, provided they are heated through the action of the electric field within the structure to energies greater than or equal to the bandgap energy. On attaining sufficient energy, a carrier can experience a collision with the lattice where it transfers energy to an electron initially within the valence band, leading to the promotion of that electron into the conduction band. As a result, two electrons are produced within the conduction band and a hole is left behind within the valence band. In this way, additional carriers are created, leading to an increase in the current.

The onset voltage for reverse bias breakdown is called the breakdown voltage V_{br}. When a diode is biased just at V_{br}, avalanche multiplication begins to occur. Further increase in reverse bias leads to a near exponential increase in the current, as shown in Figure 11.6.3. Eventually the multiplication rate increases to the point at which the device becomes unstable, which is called the avalanche breakdown condition. Generally, avalanche breakdown occurs when the probability of a carrier's impact ionizing within the depletion region becomes 100%. More will be said about the nature of avalanche breakdown in a diode in Chapter 12.

Avalanche breakdown is fully reversible; when the reverse bias voltage is lowered, the diode behaves as before. However, if the p–n junction is reverse biased well beyond the avalanche breakdown point an additional mechanism, called secondary breakdown, occurs. Secondary breakdown is a nonreversible process and leads to catastrophic failure of the junction.

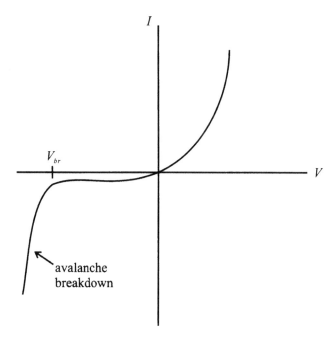

Figure 11.6.3. Plot of the current-voltage characteristic of a *p–n* homojunction illustrating reverse breakdown.

An additional mechanism, called Zener tunneling, also becomes important under reverse bias conditions, particularly in diodes made from narrow-bandgap materials. The band bending resulting from the application of a reverse bias electric field, as shown in Figure 11.6.4, can lead to tunneling of electrons from the valence band into the conduction band. The valence electrons experience a triangular-shaped barrier potential. The electrons tunnel from the valence band into unfilled conduction-band states, as shown in Figure 11.6.4.b.

The tunneling probability can be determined from use of the WKB approximation that was presented in Chapter 2. The tunneling probability T_t was found to be approximated as

$$T_t \sim e^{-\frac{2}{\hbar} \int |p(x)|\, dx}, \tag{11.6.18}$$

where the integral is defined over the potential barrier and is evaluated between the points $-x_1$ and x_2. The electron momentum can be readily determined from the kinetic energy as

$$\frac{\hbar^2 k^2}{2m} = KE = E - U. \tag{11.6.19}$$

Following the approach of Kane (1957), the difference between the energy and the potential energy can be estimated as

$$E - U = -\frac{\left(\frac{E_g}{2}\right)^2 - E_c^2}{E_g}. \tag{11.6.20}$$

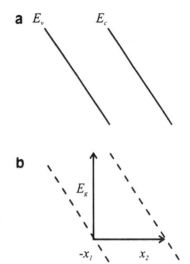

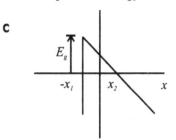

Figure 11.6.4. a. Energy-band diagram under a large reverse bias. b. Triangular potential barrier formed in the junction. c. Corresponding potential-energy diagram.

The energy can be subdivided into two components, $E_\perp$ and $E_\parallel$, where $E_\perp$ is the energy associated with the momentum perpendicular to the tunneling direction and $E_\parallel$ is the energy associated with the momentum in the tunneling direction. The tunneling probability can then be written as

$$T_t \sim e^{-2\int \frac{2m}{\hbar^2}\sqrt{\frac{\left(\frac{E_g^2}{4}-E_c^2\right)}{E_g}+E_\perp}\,dx}. \tag{11.6.21}$$

We can evaluate the tunneling probability assuming that $E_c = qFx$, where F is an applied reverse bias field leading to the conduction-band bending, as

$$T_t = e^{-\frac{E_g}{4E}}e^{-\frac{E_\perp}{E}}, \qquad E = \sqrt{2}\frac{qF\hbar}{2\pi\sqrt{mE_g}}, \tag{11.6.22}$$

(see Moll 1964, p. 250). T_t is the probability of tunneling for an electron. The derivation of Eqs. (11.6.22) is left as an exercise for the student.

The tunneling current can readily be determined from the following considerations. Since the energy can be decomposed into $E_\parallel$ and $E_\perp$, the wave vector

k can be subdivided also in terms of perpendicular and parallel components. $E_\perp$ has values then from 0 to total energy E. The number of electrons contributing to the tunneling flux per unit volume in a ring with perpendicular wave vector $k_\perp$ to $k_\perp + dk_\perp$ is given as

$$\text{flux} = q(\text{velocity of } k \text{ space})(\text{density of } k \text{ space})(\text{area of ring})$$
$$\times \; (\text{occupancy of states}).$$

The velocity in k space is simply dk/dt or $qF/\hbar$. The density of states in k space is $1/4\pi^3$ and the occupancy factor is the Fermi distribution. The area of the ring is $2\pi k_\perp dk_\perp$. With the assumption of parabolic energy bands,

$$E_\perp = \frac{\hbar^2 k_\perp^2}{2m^*}, \tag{11.6.23}$$

the flux becomes

$$\frac{q^2 F m^* dE_\perp F(E)}{2\pi^2 \hbar^3}. \tag{11.6.24}$$

We can now determine the tunneling current by taking the product of the incident flux, the tunneling probability, and the volume $A\,dx$. The differential dx is equal, though, to dE/qF. Additionally, it is necessary to account for the possibility that the conduction band is degenerate. Therefore the tunneling rate from the valence band into the conduction band depends on the product of $f_v(E)$ and $[1 - f_c(E)]$, where $f_c(E)$ and $f_v(E)$ are the conduction- and the valence-band Fermi–Dirac distribution functions. With these substitutions, the differential current dI is

$$dI = A \frac{q m^*}{2\pi^2 \hbar^3} e^{-\frac{E_g}{4E}} e^{-\frac{E_\perp}{E}} [\, f_v(E) - f_c(E)]\, dE_\perp dE, \tag{11.6.25}$$

where A is the area. Expression (11.6.25) allows for tunneling in both directions.

Finally, the expression for the tunneling current can be obtained by integration of Eq. (11.6.25) over the energy. The bounds are from 0 to E for $dE_\perp$. Then if it is assumed that $f_c(E) \sim 0$ and $f_v(E) \sim 1$ and that $E_\perp \ll E$, the tunneling current is given as

$$I = A \frac{q^3 \sqrt{2m^*} F V_a}{4\pi^3 \hbar^2 \sqrt{E_g}} e^{-\frac{\pi \sqrt{m^*} E_g^{3/2}}{2\sqrt{2}qF\hbar}}. \tag{11.6.26}$$

The last second-order effect present in a p–n homojunction considered here is that of the diode series resistance. Generally, there is a nonzero resistance within the bulk regions of the p–n junction diode. Hence an additional voltage drop will occur across these regions in addition to that across the depletion region. The presence of the series resistance results in a smaller current flow with applied voltage, as shown in Figure 11.6.5.

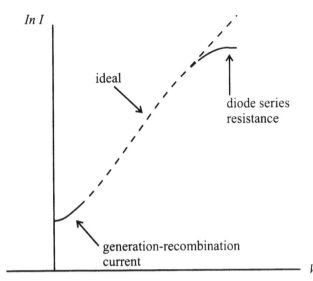

Figure 11.6.5. Logarithm of the current versus voltage of a forward biased *p–n* homojunction illustrating the influence of the diode series resistance.

11.7 **Current Flow in a Schottky Barrier Diode**

In previous sections, we have discussed current-flow processes within *p–n* homojunctions. The basic current-flow mechanisms in a Schottky barrier, although qualitatively similar, arise from different physical processes. The most important current-flow mechanism present in Schottky barriers is thermionic emission.

To understand the nature of thermionic emission, let us first review the shape of the energy bands in equilibrium in a Schottky barrier. The energy-band diagrams in equilibrium were presented in Figure 11.3.1. It is important to recognize that potential barriers to electron flow arise on both sides of the junction if $\phi_m > \phi_s$. When a bias is applied to the junction, then to first order the potential barrier ϕ_B remains unchanged. ϕ_B is lowered by the image-force lowering, as discussed in Section 11.3, but this effect is significantly less than the change in the potential difference on the semiconductor side of the junction. In other words, although the Schottky barrier ϕ_B is slightly affected by an applied bias, the built-in voltage is drastically changed by an applied bias. From inspection of Figure 11.3.1, the potential barrier looking into the metal from the semiconductor side depends on the band bending within the semiconductor, which varies strongly with bias.

In a rectifying metal–semiconductor contact in equilibrium, the electrostatic potential is higher in the semiconductor than in the metal since electrons transfer from the semiconductor into the metal in order to align the Fermi levels. Hence ionized donors are left behind within the semiconductor, forming a positive space-charge region. As a result, the semiconductor is at a higher potential than the metal. The energy bands bend down in the semiconductor away from the junction.

If a positive voltage is applied to the metal with respect to the semiconductor, the Schottky barrier becomes forward biased. The potential barrier looking into

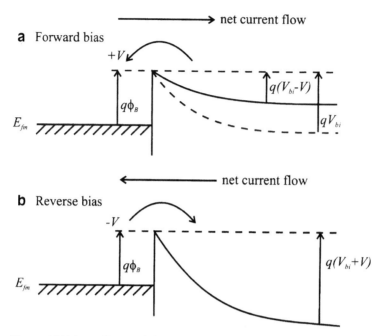

Figure 11.7.1. a. Forward, b. reverse biased Schottky barrier diode energy-band diagrams showing the current-flow direction and the effect on the potential barrier of the bias voltage.

the metal is thereby lowered, as shown in Figure 11.7.1.a. If a negative voltage is applied to the metal with respect to the semiconductor, the junction becomes reverse biased. The potential barrier on the semiconductor side is increased, as shown in Figure 11.7.1.b.

How does the current behave under these conditions? It is important to note that a Schottky barrier is essentially a unipolar current device. Electron motion dictates the current flow almost entirely. Although a hole current can be considered, typically it is completely negligible under most conditions. For simplicity, we will omit any further discussion of the hole component of the current in a Schottky barrier, recognizing that little loss of generality occurs. We define two electron current components. These are

1. J_n^-: the electron current due to electron transport from the metal into the semiconductor,
2. J_n^+: the electron current due to electron transit from the semiconductor into the metal.

The electron flux that comprises the current density J_n^+ must overcome the potential barrier on the semiconductor side of magnitude, $V_{bi} - V$, under forward biased conditions. The current density J_n^- consists of electrons that move from the metal into the semiconductor. These electrons need to overcome the potential barrier on the metal side ϕ_B, which remains generally unchanged under bias. Recall that the current moves in the direction opposite to the flux. Clearly, the current density J_n^+ increases dramatically with increasing forward bias since

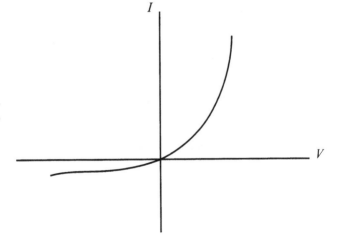

Figure 11.7.2. Plot of the current-voltage characteristic for a Schottky barrier diode.

the potential barrier height is lowered. Similarly, under reverse bias conditions, J_n^+ is greatly decreased since the potential barrier height is raised. Since J_n^- is essentially voltage independent, then the total current, the sum of both J_n^+ and J_n^-, is asymmetric with respect to the applied voltage, as shown in Figure 11.7.2.

The question, though, is, what is the physical cause of the asymmetry in the current-voltage characteristic? The answer is basically the same as that given for a *p–n* homojunction; the current arises from just those carriers whose forward-directed kinetic energies exceed the potential barrier height as seen from either side of the junction. In equilibrium, the net current is, of course, zero. J_n^+ is exactly balanced on average by J_n^-; the net flow of electrons from the metal to the semiconductor is equal to the net flow of electrons from the semiconductor to the metal. In the case of Schottky barriers, as opposed to that of the *p–n* homojunction, the current mechanism is thermionic emission or diffusion rather than solely diffusion. Let us concentrate next on determining the nature of the thermionic emission current.

In the context of a Schottky barrier, the thermionic emission current arises from electron emission from the metal into the semiconductor layer. Since the metal is essentially a perfect conductor, no diffusion current can flow within it. Nevertheless, carriers are transported from the metal into the semiconductor. To describe the current flux from the metal to the semiconductor, it is important to introduce a new mechanism. This is called thermionic emission.

Thermionic emission is a majority carrier current process in which only high-energy electrons, those whose energies exceed the potential barrier height, can be emitted from one side of the junction to the other. Strictly speaking, thermionic emission is typically associated with the escape of electrons from a hot surface or cathode. Therefore it is present in a system, like that of a Schottky barrier, only in which a metal contact or equivalently, a cathode, is present. In a *p–n* homojunction, the current is mainly diffusive since there is no cathode present on either side of the junction.

The question is, though, what is the nature of the current flow from the semiconductor back into the metal? There are two possibilities to consider. The first case is one in which the depletion region is sufficiently small that the free path length of the electrons is large with respect to the thickness of the potential barrier, such as is the situation for high-mobility materials. In this case, the electron flux from the semiconductor into the metal can be adequately described by thermionic emission alone. We discuss this case in detail below. The second case is that in which the depletion region is sufficiently large that the electrons suffer many collisions so that their transport to the metal interface is described best by diffusion rather than thermionic emission. In the latter case, the current flow from the semiconductor side to the metal is best thought of as a combination of thermionic emission and diffusion. A diffusion current flows from the edge of the depletion region to the metal-semiconductor interface. At the interface, the system can be modeled again as a thermionic emitter. So one can picture the semiconductor–metal flux as two currents in series, a diffusion current followed by a thermionic emission current. The net current is of course limited by the smaller of these two currents.

As discussed above, the potential barrier as seen from the semiconductor side of the junction is most sensitive to changes in bias. As the potential barrier height decreases (forward bias) or increases (reverse bias), the number of electrons with forward-directed kinetic energy sufficiently high to overcome the potential barrier at the junction either increases or decreases, respectively. Since the electron-energy distributions on either side of the junction are quasi-equilibrium distributions, the number of carriers contributing to the thermionic emission current changes exponentially with applied bias, as shown in Figure 11.7.3. Again, for simplicity, we assume that the distribution shown in Figure 11.7.3 is restricted to only those electrons whose kinetic energies are forward directed. Then under forward bias, there is an exponential increase in the number of carriers with forward-directed kinetic energies in excess of the potential barrier at the junction. All the carriers under the curve from $q(V_{bi} - V)$ to infinity can contribute to the current. This results in an exponential increase in the current density. Alternatively, under reverse bias, there is an exponential decrease in the number of

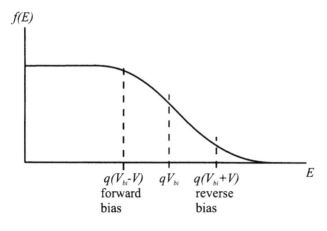

Figure 11.7.3. Energy distribution function showing the fraction of carriers capable of surmounting the potential barrier formed at the junction under equilibrium and forward and reverse biased conditions.

carriers with forward-directed kinetic energies in excess of the potential barrier at the junction. Only those carriers at the very high end of the distribution, ranging in energy from $q(V_{bi} + V)$ to infinity, can contribute to the current. Since the tail is relatively sparsely populated, the resulting current density is quite small.

Let us now consider the current flow as dictated by thermionic emission quantitatively. To determine the semiconductor–metal thermionic emission current density, it is necessary to find the number of electrons of sufficiently high energy in the forward direction that can overcome the potential barrier at the junction. J_n^+ can then be found from the net flux of electrons perpendicular to the potential barrier whose kinetic energy exceeds the potential height. If the x direction is chosen to be perpendicular to the interface, then J_n^+ becomes

$$J_n^+ = \int_{n(\text{bottom})}^{n(\infty)} q v_x \, dn, \tag{11.7.1}$$

where n (bottom) is the electron concentration at an energy equal to the barrier height and $n(\infty)$ is the concentration at $E = \infty$. Note that the integral is taken over the differential carrier concentration dn, which can be expressed in terms of the energy as

$$dn = D(E) f(E) \, dE. \tag{11.7.2}$$

Substituting Eq. (11.7.2) into Eq. (11.7.1) yields

$$J_n^+ = \int_{E_c}^{\infty} q v_x D(E) f(E) \, dE. \tag{11.7.3}$$

In most situations, the distribution function $f(E)$ can be approximated as a Maxwellian distribution:

$$f(E) \sim e^{-\frac{(E - E_f)}{kT}}. \tag{11.7.4}$$

When $q V_n$ is defined as $E_c - E_f$, $f(E)$ can be expressed as

$$f(E) = e^{-\frac{(E - E_c + q V_n)}{kT}} = e^{\frac{-q V_n}{kT}} e^{-\frac{(E - E_c)}{kT}}, \tag{11.7.5}$$

where $(E - E_c)$ is the kinetic energy of the electron. Remember that $E - E_c$ is the amount of energy above the band edge that the electron has, which is all kinetic. Therefore $E - E_c$ can be expressed in terms of the carrier velocity v as

$$E - E_c = \frac{1}{2} m^* v^2. \tag{11.7.6}$$

dE can then be found as

$$dE = m^* v \, dv. \tag{11.7.7}$$

The three-dimensional density-of-states function $D(E)$ is given by Eq. (5.1.11). For the case of interest here, the energy is measured with respect to the conduction-band edge E_c. Redefining E as $E - E_c$, we can write the density of states as

$$D(E)\,dE = \frac{\sqrt{2}}{\pi^2}\frac{(m^*)^{3/2}}{\hbar^3}\sqrt{E - E_c}\,dE. \tag{11.7.8}$$

But $(E - E_c)^{1/2} = (m^*/2)^{1/2}v$. The density of states $D(E)$ then becomes

$$D(E)\,dE = 2\left(\frac{m^*}{h}\right)^3 4\pi v^2\,dv. \tag{11.7.9}$$

Substituting Eq. (11.7.5) and Eq. (11.7.9) into Eq. (11.7.2) we obtain for dn

$$dn = 2\left(\frac{m^*}{h}\right)^3 e^{-\frac{qV_n}{kT}} e^{-\frac{1/2m^*v^2}{kT}} 4\pi v^2\,dv. \tag{11.7.10}$$

In three dimensions, v^2 is simply

$$v^2 = v_x^2 + v_y^2 + v_z^2, \tag{11.7.11}$$

and the differentials transform as

$$4\pi v^2\,dv = dv_x\,dv_y\,dv_z. \tag{11.7.12}$$

With the above substitutions, J_n^+ becomes

$$J_n^+ = \int qv_x\,dn = \int 2q\left(\frac{m^*}{h}\right)^3 e^{-\frac{qV_n}{kT}} v_x e^{-\frac{1/2m^*v^2}{kT}}\,dv_x\,dv_y\,dv_z. \tag{11.7.13}$$

Separating the integral into three separate integrals yields

$$J_n^+ = 2q\left(\frac{m^*}{h}\right)^3 e^{-\frac{qV_n}{kT}} \int_{v_{0x}}^{\infty} v_x e^{-\frac{1/2m^*v_x^2}{kT}}\,dv_x$$
$$\times \int_{-\infty}^{\infty} e^{-\frac{1/2m^*v_y^2}{kT}}\,dv_y \int_{-\infty}^{\infty} e^{-\frac{1/2m^*v_z^2}{kT}}\,dv_z. \tag{11.7.14}$$

But the integrals can be evaluated easily since

$$\int_{-\infty}^{+\infty} e^{-a^2x^2}\,dx = 2\int_0^{+\infty} e^{-a^2x^2}\,dx = \frac{\sqrt{\pi}}{a}. \tag{11.7.15}$$

Therefore the integrals over dv_y and dv_z are easily integrated to give

$$\int_{-\infty}^{\infty} e^{-\frac{1/2m^*v_y^2}{kT}}\,dv_y \int_{-\infty}^{\infty} e^{-\frac{1/2m^*v_z^2}{kT}}\,dv_z = \frac{2kT}{m^*}\pi. \tag{11.7.16}$$

The integral over v_x can be evaluated with

$$u \equiv \frac{m^* v_x^2}{2kT}, \qquad du = \frac{m^*}{kT} v_x \, dv_x, \tag{11.7.17}$$

which then yields

$$\frac{kT}{m^*} e^{-\frac{m^* v_{0x}^2}{2kT}}. \tag{11.7.18}$$

Therefore the current density J_n^+ becomes

$$J_n^+ = \frac{4\pi q m^* k^2 T^2}{h^3} e^{-\frac{q V_n}{kT}} e^{-\frac{m^* v_{0x}^2}{2kT}}. \tag{11.7.19}$$

The minimum kinetic energy needed to overcome the potential barrier, $1/2 m^* v_{0x}^2$, is simply equal to $q(V_{bi} - V)$, as is readily seen from inspection of Figure 11.7.1. J_n^+ can be expressed then as

$$J_n^+ = \frac{4\pi q m^* k^2 T^2}{h^3} e^{-\frac{q(V_n + V_{bi})}{kT}} e^{\frac{qV}{kT}}. \tag{11.7.20}$$

We can recast expression (11.7.20) for J_n^+ in a more convenient form by recognizing that the potential barrier height ϕ_B is simply equal to $V_n + V_{bi}$. This follows from inspection of Figure 11.7.1 and the definition of $q V_n = E_c - E_f$. J_n^+ finally becomes

$$J_n^+ = A^* T^2 e^{-\frac{q\phi_B}{kT}} e^{\frac{qV}{kT}}, \tag{11.7.21}$$

where A^* is defined as

$$A^* = \frac{4\pi q m^* k^2}{h^3}. \tag{11.7.22}$$

A^* is called the Richardson constant.

Equation (11.7.22) gives the current density in a metal–semiconductor junction due to electron flow from the semiconductor to the metal. The complete expression for the current density in a Schottky barrier is given as the sum of J_n^+ and J_n^-, the current density due to electron flow from the metal into the semiconductor. In equilibrium, J_n^- must equal $-J_n^+$. Therefore J_n^- can be found from

$$J_n^- = -J_n^+(V = 0) = -A^* T^2 e^{-\frac{q\phi_B}{kT}}. \tag{11.7.23}$$

Since the Schottky barrier height is relatively bias independent, J_n^- remains roughly constant with changing bias. Hence J_n^- is given by Eq. (11.7.23) over a wide range of reverse bias voltages.

The net current density due to electron transport J_n in a Schottky barrier is simply equal to the sum of J_n^+ and J_n^-. J_n is given then as

$$J_n = J_n^+ + J_n^- = A^* T^2 e^{-\frac{q\phi_B}{kT}} \left(e^{\frac{qV}{kT}} - 1 \right). \tag{11.7.24}$$

J_n can be reexpressed as

$$J_n = J_{st} \left(e^{\frac{qV}{kT}} - 1 \right), \tag{11.7.25}$$

where J_{st} is defined as

$$J_{st} = A^* T^2 e^{-\frac{q\phi_B}{kT}}. \tag{11.7.26}$$

The net current density is clearly asymmetric with bias. Under forward bias, the thermionic emission current of electrons from the semiconductor into the metal dominates and the current increases exponentially [the $e^{qV/kT}$ term clearly dominates the 1 term in Eq. (11.7.26) for J_n]. Under reverse bias the exponential term approaches zero and the constant term $-J_{st}$ gives rise to the total current flow. Hence the current-voltage characteristic is asymmetric, in agreement with our qualitative discussion above.

As discussed above, it is often important to determine both the thermionic emission current density as well as the diffusion current density in a Schottky barrier diode, especially in low-mobility materials. The net current flow in the junction is limited by the slower of these two processes. The diffusion velocity is associated with the transport of carriers from the edge of the depletion region W to the potential-energy maximum x_m. Physically, current flow within the region between $x = W$ and $x = x_m$ depends on electron diffusion. At the interface, in the region between $x = x_m$ and $x = 0$, which is typically very small ($\sim$200 nm), if the electron distribution is Maxwellian and if no electrons return from the metal other than those associated with the metal–semiconductor current density, then the semiconductor acts as a thermionic emitter. The metal–semiconductor junction can be pictured as being composed of two different regions: the interface region in which the current mechanism is basically thermionic emission, and the depletion region in which the current flows by diffusion. Since these currents flow in series with one another, the smaller of the two limits the total current.

Which of these mechanisms limits the total current can be understood as follows. If the diffusion velocity is much less than the recombination velocity, the current is limited by the diffusion process. Under this condition, the current flow in the Schottky barrier can best be described in terms of diffusion theory. If, however, the diffusion velocity is much less than the recombination velocity, the recombination process at the interface limits the net current flow. Under these conditions, the junction acts as a thermionic emitter and the current is thermionically limited.

Another way to view this is as follows. Either process, thermionic emission over the barrier or diffusion from the edge of the depletion region to the potential-energy maximum, can cause a bottleneck in the current flow. If the diffusion velocity is much less than the recombination velocity, the current is limited by the time it takes for the electrons to reach the interface and not by the time it takes for the electrons to recombine there. Therefore the junction is diffusion limited, and the current flow can be described by diffusion theory. In contrast, if the recombination velocity is much less than the diffusion velocity, the current flow is recombination limited; it takes relatively no time for the electrons to reach the interface after which it takes much longer for them to recombine, giving rise to current flow. Either theory, thermionic emission or diffusion, applies, depending on the nature of the junction.

11.8 **Metal-Insulator–Semiconductor Junctions in Nonequilibrium**

To close this chapter, let us discuss the operation of MIS structures under the application of a bias. The most important difference between an MIS structure and a p–n junction is that no dc current flows under bias in these devices because of the presence of the insulator layer. Hence all MIS devices act as capacitors.

Let us first consider the operation of an n-type ideal MIS junction under bias. Our choice of an ideal system is made simply for simplicity. Inclusion of nonidealities is reasonably easy to do by following the discussion in Section 11.4. The first condition we consider is a positive voltage applied to the metal gate with respect to the semiconductor layer. The positive bias on the metal attracts electrons within the semiconductor to the semiconductor–insulator interface. Hence an accumulation layer of electrons, the majority carriers in the n-type semiconductor, forms at the semiconductor/insulator interface. Because of the presence of the positive potential on the metal, commonly called the gate, the electron energies are lowered near the interface and the bands bend down, as shown in Figure 11.8.1. Sufficient electrons accumulate near the interface to balance the applied positive gate voltage. Subsequently, away from the interface, there is no net space charge resulting in flat-band conditions.

The band diagram sketched in Figure 11.8.1 can be understood in another way. The positive bias applied to the gate effectively deposits positive charge onto the metal. Sufficient negative charge, because of the accumulation of free, mobile electrons within the semiconductor, is induced at the insulator–semiconductor interface to maintain space-charge neutrality. The finite conductivity of the semiconductor requires that the negative charge accumulates within a layer adjacent to the interface. The presence of the insulator acts to separate the positive and the negative charges, producing an electric field across it, leading to the band bending shown in the conduction band of the insulator. The semiconductor bands bend down toward the interface since the electron energies are lowered by the positive charge on the metal. An easy way to remember which way the bands bend is to recall that electrons roll downhill in energy diagrams. The positive potential on the metal will act to attract electrons. Therefore an electron placed

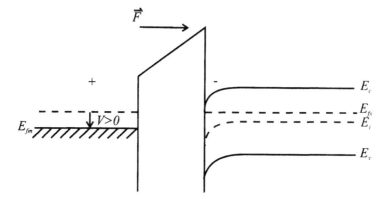

Figure 11.8.1. Sketch of an ideal MIS n-type structure in accumulation.

within the semiconductor will roll toward the interface, so the bands must bend down toward the interface.

There is no current flow within an MIS structure, and thus the Fermi levels are flat throughout the system. The bands generally bend when a bias is applied. The difference between the Fermi level and the intrinsic level then is different at different locations within the system. The electron concentration at the interface can be determined from the extent of the band bending by use of Eq. (10.3.11):

$$n_s = n_i e^{\frac{E_{fs} - E_i}{kT}}. \tag{11.8.1}$$

The electron concentration at the surface n_s is therefore much larger than in the bulk since the difference between the Fermi level E_{fs} and the intrinsic level E_i is greater at the surface (interface between the insulator and semiconductor) than in the bulk region. This case, application of a positive gate bias to an n-type MIS device, is called accumulation.

If a small negative gate bias is applied, the electrons within the n-type semiconductor are repelled from the surface of the semiconductor. A depletion region forms then at the insulator–semiconductor interface. The negative bias applied to the metal is equivalent to depositing negative charge on its surface. Therefore, to preserve space-charge neutrality, an equal amount of positive charge is induced within the semiconductor layer underneath through depletion. The presence of the negative charge on the metal raises the electron energies within the semiconductor at the interface. Hence the energy bands bend up toward the interface, as shown in Figure 11.8.2.

We can determine the direction by which the bands bend in Figure 11.8.2 by again using the fact that electrons roll downhill in energy-band diagrams. Since the metal is negatively charged, the electrons in the semiconductor will naturally be repelled from the surface. If an electron is placed at the surface of the semiconductor, it will move away from the oxide. Therefore the bands must bend such that the electrons roll away from the semiconductor–insulator interface.

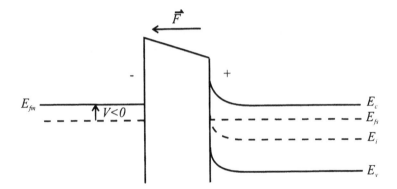

Figure 11.8.2. Ideal n-type MIS structure in depletion.

As can be seen from Figure 11.8.2, there again is a tilt in the oxide conduction band due to the potential difference between the metal and the semiconductor. In this case, the semiconductor is at a higher potential than the metal.

The electron concentration at the surface of the semiconductor can be determined from use of Eq. (11.8.1) again:

$$n_s = n_i e^{\frac{E_{fs} - E_i}{kT}}. \tag{11.8.2}$$

In this case, the electron concentration is less near the interface than in the bulk since $E_{fs} - E_i$ is smaller near the junction than within the bulk, as can be seen from Figure 11.8.2. So clearly the semiconductor region is depleted and the structure is said to be in depletion.

If the negative gate bias is sufficiently large, minority holes are attracted to the insulator–semiconductor interface, leading to an inversion of the surface. The bands bend strongly upward at the junction, as shown in Figure 11.8.3.

Figure 11.8.3. Ideal n-type MIS structure in inversion.

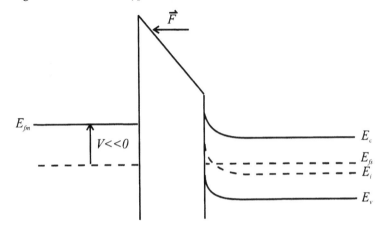

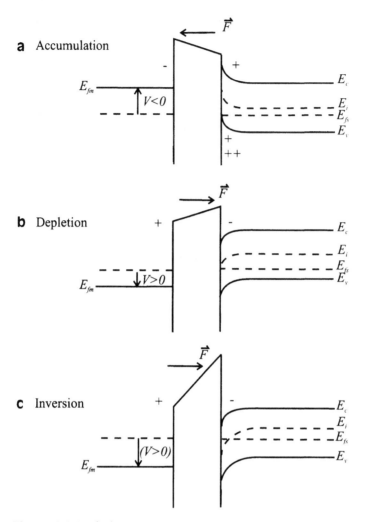

Figure 11.8.4. Ideal *p*-type MIS structure in a. accumulation, b. depletion, c. inversion.

The band bending is sufficient that the intrinsic level crosses the Fermi level near the surface. According to Eq. (11.8.1), the electron concentration becomes less than the intrinsic concentration n_i, while the hole concentration becomes greater than n_i near the surface. Therefore the semiconductor surface acts like a *p*-type material while the bulk remains *n* type. The semiconductor layer is said then to be inverted with a hole concentration of

$$p = n_i e^{\frac{E_i - E_{fs}}{kT}}. \tag{11.8.3}$$

The corresponding band diagrams for a *p*-type semiconductor MIS system under different bias conditions are presented in Figure 11.8.4. The behavior of the junction is the same as that given above except the signs of the bias voltages are of course reversed.

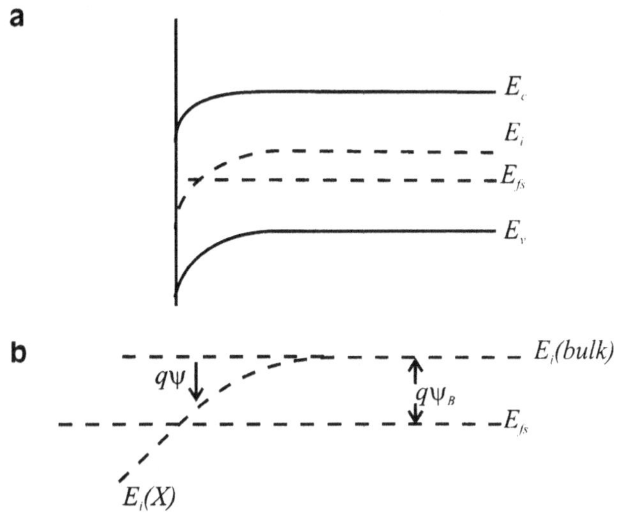

a

Figure 11.8.5. Band structure of an ideal *p*-type MIS structure a. biased in inversion, b. expanded view near the interface to show the behavior of the bands to facilitate the calculation of the surface charge performed in the text.

b

We next derive a quantitative expression for the surface-charge concentration as a function of applied gate bias. For convenience, we choose a *p*-type semiconductor. The results for an *n*-type semiconductor are, of course, easily obtained from those for a *p*-type system. ψ is defined as the electrostatic potential. The potential is assumed to be zero within the bulk region. All the voltage drop is assumed to occur at the interface, across both the oxide and the semiconductor depletion regions.

From inspection of Figure 11.8.5, $q\psi$ is found from the difference between the bulk intrinsic level $E_i(\text{bulk})$ and the position-dependent intrinsic level $E_i(x)$ as

$$q\psi = E_i(\text{bulk}) - E_i(x). \tag{11.8.4}$$

$q\psi_B$ is defined as the electrostatic potential difference between $E_i(\text{bulk})$ and E_{fs}, as shown in the diagram. The electron and hole concentrations can easily be obtained from

$$n_p = n_i e^{\frac{E_{fs}-E_i}{kT}} = n_i e^{\frac{E_{fs}-E_i(\text{bulk})}{kT}} e^{\frac{E_i(\text{bulk})-E_i}{kT}}, \tag{11.8.5}$$

$$p_p = n_i e^{\frac{E_i-E_{fs}}{kT}} = n_i e^{\frac{E_i(\text{bulk})-E_{fs}}{kT}} e^{\frac{E_i-E_i(\text{bulk})}{kT}}. \tag{11.8.6}$$

When the definition of ψ given above is used, the electron and hole concentrations in the *p*-type semiconductor, n_p and p_p, respectively, become

$$n_p = n_{p0} e^{\frac{q\psi}{kT}}, \qquad p_p = p_{p0} e^{\frac{-q\psi}{kT}}, \tag{11.8.7}$$

where

$$n_{p0} \equiv n_i e^{\frac{E_{fs}-E_i(\text{bulk})}{kT}}, \qquad p_{p0} \equiv n_i e^{\frac{E_i(\text{bulk})-E_{fs}}{kT}}. \tag{11.8.8}$$

n_{p0} and p_{p0} are the electron and hole equilibrium concentrations in the bulk material, respectively. The carrier concentrations at the interface, n_s and p_s, are given then as

$$n_s = n_{p0}e^{\frac{q\psi_s}{kT}}, \qquad p_s = p_{p0}e^{\frac{-q\psi_s}{kT}}, \qquad (11.8.9)$$

where ψ_s is the electrostatic potential at the surface.

Inspection of Eqs. (11.8.9) indicates that if the surface potential is less than zero, $\psi_s < 0$, then hole accumulation occurs at the surface, $p_s > p_{p0}$. This follows readily from the fact that a negative gate bias will attract majority holes to the surface of the semiconductor. At zero bias, $\psi = 0$, flat band (no band bending) occurs in an ideal MIS structure. In flat band, the surface carrier concentrations n_s and p_s are then simply equal to their equilibrium bulk values. Depletion occurs when the surface potential ψ_s is greater than zero but less than ψ_B. The onset of inversion begins when the electron and hole concentrations at the surface are equal, $n_s = p_s = n_i$. This occurs when $\psi_s = \psi_B$ since

$$n_s = n_{p0}e^{\frac{q\psi_s}{kT}} = n_i e^{\frac{E_f - E_i(\text{bulk})}{kT}} e^{\frac{q\psi_s}{kT}}. \qquad (11.8.10)$$

If $q\psi_s = q\psi_B$, then

$$n_s = n_i e^{\frac{E_f - E_i(\text{bulk})}{kT}} e^{\frac{E_i(\text{bulk}) - E_f}{kT}} = n_i. \qquad (11.8.11)$$

Finally, strong inversion occurs when the surface potential ψ_s exceeds ψ_B.

We can find the position-dependent potential $\psi(x)$ solving the one-dimensional Poisson equation,

$$\frac{d^2\psi}{dx^2} = -\frac{\rho(x)}{\epsilon_s}, \qquad (11.8.12)$$

where the charge density is given as

$$\rho(x) = q[N_d^+ - N_a^- + p_p - n_p]. \qquad (11.8.13)$$

Within the bulk, charge neutrality holds so $\rho(x) = 0$. The ionized donor and acceptor concentrations, N_d^+ and N_a^-, are equal to the bulk equilibrium electron and hole carrier concentrations since it is assumed that the semiconductor is uniformly doped. Therefore

$$N_d^+ - N_a^- = n_{p0} - p_{p0}. \qquad (11.8.14)$$

$p_p - n_p$ can be expressed as

$$p_p - n_p = p_{p0}e^{\frac{-q\psi}{kT}} - n_{p0}e^{\frac{q\psi}{kT}}. \qquad (11.8.15)$$

Substituting Eqs. (11.8.13), (11.8.14), and (11.8.15) into Poisson's equation yields

$$\frac{d^2\psi}{dx^2} = -\frac{q}{\epsilon_s}\left[p_{p0}\left(e^{\frac{-q\psi}{kT}} - 1\right) - n_{p0}\left(e^{\frac{q\psi}{kT}} - 1\right)\right].$$
(11.8.16)

Multiply both sides of Eq. (11.8.16) by $d\psi/dx$ and integrate with respect to x as

$$\int_0^F \frac{d\psi}{dx}\frac{d^2\psi}{dx^2}dx = \int_0^\psi -\frac{q}{\epsilon_s}\left[p_{p0}\left(e^{\frac{-q\psi}{kT}} - 1\right) - n_{p0}\left(e^{\frac{q\psi}{kT}} - 1\right)\right]\frac{d\psi}{dx}\,dx,$$
(11.8.17)

where F is the electric field. We can evaluate Eq. (11.8.17) by making the assignments

$$u \equiv \frac{d\psi}{dx}, \qquad du \equiv \frac{d^2\psi}{dx^2},$$
(11.8.18)

as

$$\int_0^F \frac{d\psi}{dx}d\left(\frac{d\psi}{dx}\right)$$
$$= \int_0^\psi -\frac{q}{\epsilon_s}\left[p_{p0}\left(e^{\frac{-q\psi}{kT}} - 1\right) - n_{p0}\left(e^{\frac{q\psi}{kT}} - 1\right)\right]d\psi.$$
(11.8.19)

The integrations can be readily performed to yield an expression for the square of the electric field F^2 as

$$F^2 = \frac{2q}{\epsilon_s}\frac{kT}{q}p_{p0}\left[\left(e^{\frac{-q\psi}{kT}} + \frac{q\psi}{kT} - 1\right) + \frac{n_{p0}}{p_{p0}}\left(e^{\frac{q\psi}{kT}} - \frac{q\psi}{kT} - 1\right)\right].$$
(11.8.20)

When the Debeye length L_D is defined as

$$L_D \equiv \sqrt{\frac{kT\epsilon_s}{p_{p0}q^2}},$$
(11.8.21)

the expression for the electric field becomes

$$F = \pm\frac{\sqrt{2}}{L_D}\frac{kT}{q}\sqrt{\left[\left(e^{\frac{-q\psi}{kT}} + \frac{q\psi}{kT} - 1\right) + \frac{n_{p0}}{p_{p0}}\left(e^{\frac{q\psi}{kT}} - \frac{q\psi}{kT} - 1\right)\right]},$$
(11.8.22)

where the $+$ sign holds for $\psi > 0$ and the $-$ sign applies to $\psi < 0$. The surface charge Q_s can be readily determined from the surface electric field by use of Gaussian law:

$$Q_s = -\epsilon_s F_s.$$
(11.8.23)

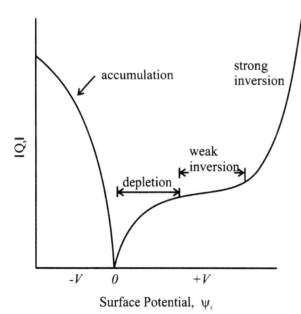

Figure 11.8.6. Plot of the space-charge density as a function of the surface potential to show the different regions of operation, that is, accumulation, depletion, week inversion, and inversion.

The surface electric field is determined from Eq. (11.8.22) with the surface potential ψ_s in place of ψ. The result is plotted in Figure 11.8.6.

It is interesting to analyze each of the four states of the MIS system in terms of Figure 11.8.6.

1. In accumulation, the surface potential ψ_s is less than zero and Q_s is positive. The expression for the field is dominated by the first term under the square root, $e^{-q\psi/kT}$. Hence

$$Q_s \sim e^{\frac{q|\psi_s|}{2kT}}. \tag{11.8.24}$$

2. Flat band occurs when the surface potential is zero. In this case, $Q_s = 0$, no excess surface charge is present.
3. Depletion occurs when $\psi_B > \psi_s > 0$. $Q_s \sim -\sqrt{\psi_s}$.
4. When the surface potential is very much larger than ψ_B, the surface is in strong inversion. Under these conditions, ψ_s is positive and the expression for the electric field is dominated by the fourth term under the square root:

$$Q_s \sim e^{\frac{q|\psi_s|}{2kT}}. \tag{11.8.25}$$

It is important to recognize that a similar plot to that shown in Figure 11.8.6 needs to be constructed for the particular semiconductor system of interest.

Before we conclude this section, it is useful to discuss the capacitance of an MIS structure. As mentioned above, no dc current flows in a MIS device. Hence these devices are all capacitive in nature. The capacitance-voltage characteristic of an ideal MIS system can be determined as follows. Charge neutrality requires

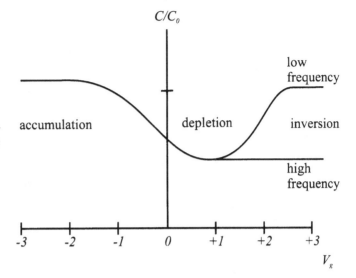

Figure 11.8.7. Capacitance-voltage characteristic for a p-type MIS structure.

that the amount of charge deposited on the metal, Q_m, be balanced by an opposite but equal charge in the semiconductor, Q_s. The charge in the semiconductor, though, can be a mixture of free charge (either mobile electrons or holes) and fixed ionic charge (arising from depletion). Therefore the charge balancing can be written as

$$Q_m = Q_s = Q_n + q N_a W, \tag{11.8.26}$$

where Q_n is the inversion layer charge and $q N_a W$ is the depletion layer charge. Clearly the MIS structure behaves like a capacitor with equal but opposite charges on either side of the insulator layer.

The capacitance-voltage characteristic for a p-type MIS capacitor is sketched in Figure 11.8.7. The actual capacitance depends on the state of the system, that is, whether it is biased into accumulation, depletion, or inversion. As discussed above, a negative gate bias leads to accumulation of holes at the interface. In this case, the capacitance is approximately equal to that of a parallel-plate capacitor with an insulator dielectric between the metal and the semiconductor plates. The capacitance is dominated by the oxide and has magnitude C_0, given as

$$C_0 = \frac{\epsilon_{ox} A_G}{d}, \tag{11.8.27}$$

where A_G is the gate area and d is the oxide thickness.

When a small positive gate bias is applied, the underlying semiconductor layer becomes depleted. Therefore the MIS structure behaves as two capacitors in series, one due to the oxide layer and the other due to the depletion layer. The net capacitance in this case is simply

$$C = \frac{C_0 C_s}{(C_0 + C_s)}, \tag{11.8.28}$$

where C_s is the semiconductor depletion layer capacitance:

$$C_s = \frac{\epsilon_s A_G}{W}. \tag{11.8.29}$$

The oxide capacitance is given by Eq. (11.8.27).

Finally, the capacitance of the device when biased into inversion depends on the frequency with which the voltage is applied. At low frequencies, the minority carrier response, in this case electrons, can readily follow the signal. Therefore the electrons can accumulate or dissipate at the oxide–semiconductor interface with the change in gate bias. Consequently, the minority carrier charge is added or subtracted at the interface sufficiently fast relative to the changing signal such that the system always is in a near-parallel-plate capacitive state. The total capacitance of the system is then simply C_0; the device behaves as a parallel-plate capacitor.

In the high-frequency limit, the minority carriers cannot follow the change in the applied bias. Therefore the minority carrier concentration remains fixed at roughly its dc value, and the depletion region width W fluctuates in size to satisfy charge neutrality. This situation is similar to that of two parallel-plate capacitors in series with one another. The inversion layer capacitance in this case is given as

$$C = \frac{C_0 C_s}{(C_0 + C_s)}. \tag{11.8.30}$$

Therefore the capacitance-voltage characteristic has two branches in the inversion condition, as shown in Figure 11.8.7.

PROBLEMS

1. Determine an expression for the electrostatic potential due to space charge within the depletion region of a Schottky barrier diode. Assume that the device can be modeled much like an abrupt p^+–n junction diode.

2. Determine an expression for the generation–recombination current in a forward bias p–n junction diode. Make the following assumptions:

   a. The intrinsic and trap levels are coincident.

   b. The trap coefficients for electrons and holes are the same.

   c. The maximum value of the recombination rate occurs at midgap.

   d. The voltage is large with respect to kT.

3. Calculate the small-signal capacitance ($C = dQ/dV$) at zero dc bias at 300 K for an ideal Schottky barrier formed between platinum ($\phi = 5.3$ eV) and silicon doped

with $N_d = 1 \times 10^{16}$ cm^{-3}. The area of the diode is 10^{-5} cm^2. $\epsilon_s = 11.9 \times \epsilon_0$, $\chi_s = 4.0$ eV, and $E_g = 1.1$ eV.

4. A silicon sample with an equilibrium electron concentration of $n_0 = 1 \times 10^{14}$ cm^{-3} is optically pumped such that 1×10^{13} electron-hole pair/cm^3 are created in every microsecond. If $\tau_n = \tau_p = 2$ μs, determine the optically induced voltage. Assume that $n_i = 1.5 \times 10^{10}$ cm^{-3} and $kT = 0.0259$ eV.

5. Assume that a reverse bias p–n junction diode can be treated as a parallel-plate capacitor. Derive an expression for the capacitance of the structure for an abrupt junction.

6. For a simple bipolar junction pnp transistor, make the following assumptions:

   a. Drift is negligible in the base region of the device.

   b. The emitter injection efficiency is 100%.

   c. The device is in steady state.

   d. The current flow in the base region is one dimensional and the undepleted base width is W.

   e. The EB junction is strongly forward biased and the CB junction is strongly reverse biased.

   f. The excess hole concentration at the edge of the EB junction in the base is:
   $$\Delta p_E = p_n\left(e^{qV/kT} - 1\right),$$
   where V is the EB voltage.

   g. The excess hole concentration at the edge of the CB junction in the base is:
   $$\Delta p_C = p_n\left(e^{qV/kT} - 1\right),$$
   where V is the CB voltage.

   h. There is no generation throughout the base region.

   i. The equilibrium hole concentration in the base region is negligible.

   Determine the excess hole concentration as a function of position within the base.

7. Consider a p^+–n junction diode. At $t = 0$ the device is forward biased and a forward current flows. Determine an expression for the total forward-injected hole charge in the n region of the forward bias diode. This is the charge stored from the forward injection. Assume that the diode has infinite length; the current at the edge of the depletion region on the n side is due only to holes.

8. An ideal metal–semiconductor contact is formed on n-type silicon doped at 5.0×10^{15} cm^{-3}. The metal has a work function of 3.8 eV and silicon has an affinity of 4.0 eV. Determine the difference in the work functions of the two materials. Draw the equilibrium band diagram and determine whether the contact is ohmic or rectifying. $E_g = 1.1$ eV, $n_i = 1.5 \times 10^{10}$ cm^{-3}, and $kT = 0.0259$ eV.

9. Consider an abrupt p–n junction with N_a acceptors on the p side and N_d donors on the n side. Calculate the space charge per unit area $Q(x)$ in the depletion region on the p side.

12

Semiconductor Photonic Detectors

Perhaps the most important application of compound semiconductor materials to date is in optoelectronic devices. The types of optoelectronic devices that are discussed here can be classified into two main categories, photonic detectors and emitters. In this section, we discuss photonic semiconductor detectors. In Chapter 13 we will discuss emitters.

Many compound semiconductors, such as GaAs, InP, GaInAs, GaN, ZnS, etc., are direct-bandgap materials. As discussed in Chapter 10, optical absorption and emission processes occur to first order in direct-gap systems. Therefore these materials are extremely useful as both detectors and emitters of electromagnetic radiation. Depending on the bandgap energy, different semiconductors can be used to detect radiation from the far infrared to the ultraviolet portion of the spectrum.

In this chapter, we discuss photonic detectors, concentrating on MIS structures (particularly charge-coupled devices, or CCDs), photoconductors, photodiodes, and avalanche photodiodes (APDs). To begin with, some issues in detection are discussed; later it is shown how each of the above-mentioned device types operate.

12.1 Basic Issues in Photonic Detection

The fundamental purpose of any photonic detector is to convert an input photonic signal into an electrical signal. The application in which the detector is used greatly affects the performance criterion of the detector. For example, a detector can be used for imaging. In these applications, a good detector is defined as one that provides a high degree of spatial resolution, gray-scale resolution (the ability to distinguish different shades on a totally white to totally black contrast scale), etc. Additionally, such a detector may have to operate at both low and high light levels. At low input light levels, the detector may need to provide gain, whereas at high input light levels the detector may have to be designed to avoid image smearing. Detectors used in imaging typically need to be two dimensional as well.

Detectors can also be used in modern communications systems. For example, in modern optical fiber communications networks, semiconductor detectors are used to capture the output optical signals from optical fibers. In this application, the detector must be designed to ensure signal recognition, avoid intersymbol interference, etc. The speed of response of the detector in these applications is crucial to its performance. In contrast, in most imaging applications, speed of response is relatively unimportant. Therefore the design and performance of a photonic detector depends critically on its planned application.

Let us first consider an imaging application familiar in our everyday lives. Consider the performance of a camcorder. These devices are essentially electronic cameras. As we see below, they use a device called a charge-coupled device (CCD) to capture photons and convert them into an electronic signal. Before we discuss the operation of a CCD let us first examine the basic issues that confront a designer of an electronic camera.

Essentially, one would like the camcorder to produce an image as sharp and clear as possible, approximating reality perhaps as well as photographic film does today. In a 35-mm camera, light enters through an aperture and is focused through a lens onto a planar surface where chemical film is placed. In a camcorder, an electronic chip is placed onto the focal plane instead. Depending on the image, different portions of the film or chip are exposed to different light levels, giving rise to regions of variable brightness across the detector. The film's or chip's ability to resolve sharp spatial variations in illumination and hence produce a sharp image depends to a great extent on the graininess or density of independent photosensitive elements within the detector. In chemical film, high resolution is obtained when the graininess of the film is reduced. However, this comes at the expense of reducing the film speed; 100 ASA film is significantly less grainy than 400 ASA film, although 100 ASA cannot be used as effectively as 400 ASA film for high-speed photography at reduced light levels. In electronic imaging, the graininess or spatial resolution depends on the density of independent detector elements. Each detector element is called a pixel. The density of pixels is critical in determining the spatial resolution of a camcorder.

To obtain some cursory understanding of the problems in image capture, it is instructive to do a simple numerical exercise. Let us consider a simple, although crude, example of how many photons are captured and then converted into electrons within a high-definition electronic television camera. Consider an input monochromatic light level of 200 lx (lx is the abbreviation for lux). 200 lx corresponds to the approximate light level of standard room lighting. A lux is defined as 1 lm/m^2 (lm is the abbreviation for lumen). A lumen is a unit of optical power and is defined at a fixed wavelength 555 nm. For our purposes, let us assume that the incident light is 555 nm. 200 lx of monochromatic, incident light of 555-nm wavelength can be converted as

$$200 \text{ lx} = 200 \left(\frac{\text{lm}}{\text{m}^2} \right) \left(\frac{1 \text{ W}}{680 \text{ lm}} \right) = 0.29 \frac{\text{W}}{\text{m}^2}. \tag{12.1.1}$$

The input optical power P_{opt} is obtained then as

$$P_{\text{opt}} = \frac{0.29 \text{ W}}{\text{m}^2} \frac{A \text{ m}^2}{1} \frac{\lambda}{hc} = \frac{\text{number of photons}}{\text{s}}, \tag{12.1.2}$$

where A is the area of the aperture through which the light passes. To calculate the number of electrons that are photogenerated within one pixel of a high-definition television electronic camera we need to make some assumptions:

1. Fill factor of $\sim 90\%$: this means that 90% of the area of the detector chip is active.
2. Quantum efficiency of 30%: quantum efficiency is defined as the probability that an impinging photon onto the detector is converted into an electron.
3. 2×10^6 pixels are present.
4. Shutter speed of 1/30 s.
5. Aperture radius of 0.5 cm.

With the above assumptions, the number of photons incident upon the detector chip per second is given as

$$\frac{0.29\,\mathrm{W}}{\mathrm{m}^2}\frac{7.8 \times 10^{-5}\,\mathrm{m}^2}{1}\frac{555 \times 10^{-7}\,\mathrm{cm}}{1.242 \times 10^{-4}\,\mathrm{eV\,cm}}\frac{1\,\mathrm{eV}}{1.602 \times 10^{-19}\,\mathrm{J}}. \tag{12.1.3}$$

Multiplying the above numbers together yields 6.31×10^{13} photons/s passing through the aperture. To determine the number of photons that are focused onto active pixel areas, multiply by the fill factor of 0.90, which yields 5.68×10^{13} photons/s. We find the number of photons incident upon a single detector by dividing by the number of pixels. This is

$$\frac{5.68 \times 10^{13}\,\mathrm{photons/s}}{2 \times 10^6\,\mathrm{pixels}} = 2.84 \times 10^7\frac{\mathrm{photons}}{\mathrm{s\,pixel}}. \tag{12.1.4}$$

If the quantum efficiency is 30%, the number of photogenerated electrons per pixel per second is 8.52×10^6. Given an integration time of 1/30 s, then the number of photogenerated electrons/frame per pixel is

$$\frac{8.52 \times 10^6}{30} = 2.84 \times 10^5\frac{\mathrm{electrons}}{\mathrm{frame\,pixel}} \tag{12.1.5}$$

at an incident light level of 200 lx, or approximate standard room lighting conditions.

From the above calculation, it is seen that far less than 1×10^6 electrons are produced per frame per pixel under typical conditions within a high-resolution electronic camera. Therefore an electronic imager must be able to count relatively small numbers of electrons in order to function properly. If the input light level is scaled down to 1 lx, which corresponds to approximately the limit of most current camcorders and the light level at night during a full moon, the number of electrons generated would be ~ 1400, a somewhat remarkably small number.

The above calculation brings us to an important aspect of detection. Since the signal levels can be relatively weak, meaning that few electrons are produced, the detector itself must not generate a large number of spurious electrons that would mask the detected signal. In other words, the detector must not produce noise (randomly generated carriers that are not attributed to the signal) that is greater or comparable with the signal level itself. Otherwise the detecting system cannot distinguish between the input signal and its own internal noise.

In some situations the input signal level for a detector is necessarily low. This is the case in night-vision applications, as we will see in Section 12.2, or in astronomical observations. In low-light-level imaging, either the noise of the detector must be suppressed below the input signal level or the detector must boost the signal level by providing gain. Gain within the detector acts to amplify the input detected signal above the noise level. Throughout this chapter we discuss detectors that provide gain and the different gain mechanisms available.

Aside from imaging applications, solid-state detectors are of increasing importance as detectors for light-wave communications systems. In these applications, it is essential that the detector be capable of distinguishing between a series of pulses arriving in rapid succession. A representative input optical pulse is shown in Figure 12.1.1.a. The input pulse shape is assumed to be Gaussian for illustrative purposes. In Figure 12.1.1.b two representative output pulse shapes are shown. The first pulse shape is sharper and more closely matches the input pulse shape. It is desirable that the output pulse shape be as close as possible to the input pulse shape. Sometimes, as will be discussed in Section 12.4, the output pulse shape is highly distorted, as shown by the second shape in Figure 12.1.1.b. In this case, there is a much longer decay of the pulse, called the fall time, than was present initially. A substantially increased fall time in the output pulse can lead to intersymbol interference; adjacent pulses in a pulse train begin to interfere with one another. Intersymbol interference can lead to an increase

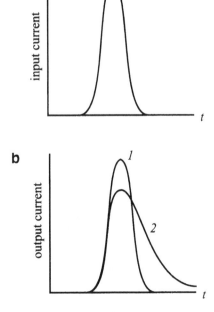

Figure 12.1.1. Representative input and output pulses showing possible fall time delays. a. Sketch of the input current as a function of time in the detector. b. Sketch of the output current as a function of time for two different conditions: 1, fast fall time, 2, slow fall time, which can lead to intersymbol interference.

in the bit error rate of the detector. In other words, an input 1 may be interpreted as a 0 or vice versa. Such a condition is, of course, undesirable. In essence then, a photodetector applied to high-speed communications systems must itself operate at high speed or high bandwidth.

High-bandwidth operation of a photodetector typically also requires gain. As mentioned above, gain in a photodetector boosts the input detected signal above the noise floor of the following electronics within the detector system. As the bandwidth of the detector increases, the bandwidth of the following electronics must increase as well. As is well known, the noise increases in an electronic circuit with increasing bandwidth. Currently there are silicon-based circuits that have very low noise levels. However, most of these circuits are restricted to bandwidths in the few megahertz range. To detect signals in the gigahertz frequency range, which is necessary for communications systems, the noise in the following electronics is currently too high to discriminate out the signal. Therefore front-end gain within the detector is necessary to overcome the system noise at gigahertz frequency operation. For these reasons, many high-frequency, high-gain detectors are being investigated for light-wave communications systems. In Section 12.5 we will discuss APDs, which are useful in communications systems for high-bandwidth, low-noise, high-gain operation.

Perhaps the most critical performance figure of merit for photodetectors in analog imaging applications is the signal-to-noise ratio. To characterize the signal-to-noise ratio, it is first necessary to define what is meant by noise. Noise arises simply from variations from average behavior. In other words, noise in a current arises from fluctuations from the mean value of the current. If the detected signal is completely deterministic, no fluctuations from the mean behavior occur; the detector has no noise.

Quantitatively, the noise of a signal can be determined from the current fluctuation in the output of the detector. We assume that the current measured during the time interval τ is simply

$$i = \frac{qn}{\tau},\tag{12.1.6}$$

where n is the total charge in the photodetector. The average photocurrent I can then be expressed as

$$I = \langle i \rangle = \frac{\langle n \rangle q}{\tau},\tag{12.1.7}$$

where $\langle n \rangle$ is the mean number of photogenerated carriers and τ is the time interval in which the current is measured.

If the photogeneration rate is strictly Poisson, meaning that the events are all independent and completely random (there are no correlations between events), then the mean is equal to the variance of the distribution. The variance is defined as

$$\sigma^2 = \mathrm{Var}(i_n) = \langle i_n^2 \rangle = \langle (i - \langle i \rangle)^2 \rangle = \langle i^2 \rangle - \langle i \rangle^2.\tag{12.1.8}$$

Applying Eq. (12.1.8) to the current yields

$$\langle i_n^2 \rangle = \langle (i - \langle i \rangle)^2 \rangle = \left(\frac{q}{\tau} \right)^2 \langle (n - \langle n \rangle)^2 \rangle. \tag{12.1.9}$$

If the quantity n is Poisson distributed, then the variance is simply

$$\langle (n - \langle n \rangle)^2 \rangle = \langle n \rangle. \tag{12.1.10}$$

Consequently, the mean-square noise current is

$$\langle i_n^2 \rangle = \frac{q^2 \langle n \rangle}{\tau^2}, \tag{12.1.11}$$

which can be rewritten as

$$\langle i_n^2 \rangle = \frac{q^2 \langle n \rangle}{\tau^2} = \left(\frac{q}{\tau} \right)^2 I \frac{\tau}{q} = \frac{qI}{\tau}. \tag{12.1.12}$$

The bandwidth of the detector, which is related to the response speed of the device, is given as

$$B = \frac{1}{2\tau}, \tag{12.1.13}$$

where τ is defined as the sampling time. Solving for τ in terms of B in Eq. (12.1.13) and substituting the result into Eq. (12.1.12) yields, for the mean-square noise current,

$$\langle i_n^2 \rangle = 2qIB. \tag{12.1.14}$$

This noise is due to fluctuations in the carrier arrival rates and is commonly called shot noise. There is shot noise in all semiconductor devices since it accompanies any randomness in the collected current.

An important figure of merit in detectors is the noise-equivalent power (NEP). The NEP is defined as the output signal power necessary to yield a signal-to-noise ratio of 1. The signal-to-noise power ratio can be determined from the ratio of the signal current i_s to the noise current i_n. The signal current is given as

$$i_s = \frac{q\eta P_s}{h\upsilon}, \tag{12.1.15}$$

where P_s is the input optical signal power, η is the quantum efficiency (defined as the probability that an electron–hole pair will be created from the incident photon), and $h\upsilon$ is the photon energy. The mean-square noise current $\langle i_n^2 \rangle$ is

$$\langle i_n^2 \rangle = 2qi_s B = \frac{2q^2 \eta P_s B}{h\upsilon}. \tag{12.1.16}$$

The signal-to-noise power ratio at the output is then given by the ratio of the squares of the currents as

$$\left(\frac{S}{N}\right)_p = \frac{i_s^2}{i_n^2} = \frac{\frac{q^2 \eta^2 P_s^2}{(h\upsilon)^2}}{\frac{2q^2 \eta P_s B}{h\upsilon}} = \frac{\eta P_s}{2h\upsilon B}. \tag{12.1.17}$$

The NEP can be found from Eq. (12.1.17). The NEP is defined as the signal power P_s necessary to yield a signal-to-noise power ratio of 1. Clearly the NEP is simply

$$\text{NEP} = \frac{2h\upsilon B}{\eta}. \tag{12.1.18}$$

The physical meaning of the NEP can be ascertained by substituting Eq. (12.1.13) in for B. The NEP at unit quantum efficiency is then

$$\text{NEP} = \frac{h\upsilon}{\tau}, \tag{12.1.19}$$

which implies that, on average, one signal photon can be detected per unit measurement time. This is an average detection capability of the detector.

It is important to consider the operation of the detector in the presence of background radiation in addition to that of the signal. If the background radiation is assumed to be constant or at least does not fluctuate at the same frequencies as that of the signal, the mean-square noise current is simply

$$\langle i_n^2 \rangle = 2q\langle i \rangle B = \frac{2\eta q^2 (P_s + P_B) B}{h\upsilon}, \tag{12.1.20}$$

where P_B is the background power. The signal-to-noise power ratio is then

$$\left(\frac{S}{N}\right)_p = \frac{\eta P_s^2}{2h\upsilon B (P_s + P_B)}. \tag{12.1.21}$$

The NEP in the presence of background radiation, called the background-limited NEP, can be readily obtained from Eq. (12.1.21). If the signal power is very much less than the background power such that the signal power can be neglected in the denominator, the $(\text{NEP})_\text{BL}$ can be approximated as

$$(\text{NEP})_\text{BL} = \sqrt{\frac{2h\upsilon B P_B}{\eta}}. \tag{12.1.22}$$

Finally, the detectivity of the detector is defined as the inverse of the NEP at a bandwidth of 1 Hz. The detectivity D is defined then as

$$D \equiv \frac{1}{\text{NEP}_\text{BL}}. \tag{12.1.23}$$

When normalized to a detector of a specific collecting area and for an application in which the background radiation is a limiting noise source, D serves as an absolute upper limit on the sensitivity of the system. The detectivity is most important in infrared imaging applications in which the background can overwhelm the signal.

12.2 **Low-Light-Level Imaging: Image Intensification**

One of the key developments in modern military operations is the capability to operate under limited visual conditions such as darkness, fog, rain, smoke, etc. Two basic different approaches have been developed for sensing in the dark. The first method, which is the prinicipal subject of this section, operates by amplifying the ambient optical radiation to produce an image that can be seen with the unaided eye. This technique is called image intensification, and the devices that are used to perform this function are called image intensifiers. The second method relies on thermal imaging. In this technique the heat emitted by an object is collected and converted into an image (thermal contrast is converted into visual contrast on a display). Thermal imaging relies on the collection of far longer wavelength radiation than image intensification and can be used during both day and night.

Thermal imagers provide far better imaging through smoke, fog, and rain than do image intensifiers. This is due to the fact that thermal radiation is emitted at much longer wavelengths, ~ 3–14 μm, longer typically than the dimensions of smoke, fog, or rain particles within the air, and is consequently not strongly scattered by the particles. Additionally, thermal imagers can detect objects at far greater distances than image intensifiers can. Subsequently, for long-range, full day/night/foul weather applications, a thermal imaging system is superior to an image intensifier. However, image intensification is of primary importance in piloting, maneuvering, and reconnaissance during which the operator needs to be able to distinguish among objects with the same temperature signature. Image intensifiers amplify the ambient background light.

Table 12.2.1 illustrates the range of light levels that the human eye responds to and representative sources and illuminance. As can be seen from Table 12.2.1, an image intensifier would need to operate over many orders of magnitude if it responded only to optical radiation originating from either the Moon or starlight. From the chart, mean starlight provides between 10^{-3}- and 10^{-4}-lx illuminance. The response range of the image intensifier can be reduced if it takes advantage of near-infrared radiation as well as optical radiation. Even though in darkness with no Moon, where there is little ambient optical radiation present, there is still near-visible, near-infrared radiation. The sources of this radiation are zodiacal light and airglow. Zodiacal light originates from solar radiation scattered by residual interplanetary dust within the orbital plane. Airglow is caused by excitation of atoms and molecules in the upper atmosphere that are heated by the Sun during the day and then that release this energy at night. While the radiation produced by airglow and zodiacal light cannot be seen by

Table 12.2.1
Low Light Sources

Illuminance (lux)	Source	Typical Seeing Conditions
10^6		Too bright
10^5	Glare	Full sun
10^4	Hazy sun	
10^3	Partial cloud	Optimum performance
10^2	Room lighting	
10^1	Candlelight	Reduced color and texture
1	Full moon	Inaccurate color and texture
10^{-1}	Half moon	Discrete objects discernible
10^{-2}	Thin moon	Limit of color perception
10^{-3}	Thin moon	Some outline perception
10^{-4}	No moon; starlight	Some contrast perception
10^{-5}	Overcast; no moon	Limit of light perception

the unaided eye, this radiation can be used by image intensifiers as a source of illumination. Typically, airglow and zodiacal radiation provide an illuminance between 10^{-2} and 10^{-3} lx, an order of magnitude higher than starlight alone. As will be discussed below, existing image intensifiers have high response to these wavelengths, typically of the order of 0.4 to 0.9 μm.

In general, any low-light-level detector must be able to distinguish the input signal of interest from noise. If the noise level of the detector and following electronics is extremely low, then low-light-level detection can proceed without front-end amplification. For example, at high input signal levels, front-end gain is unnecessary in a well-designed detector since the converted input signal is sufficiently above the noise floor that the signal can be readily discriminated from the noise. As the signal level diminishes in magnitude, it ultimately approaches the noise floor, making signal discrimination difficult. At low input signal levels then, improved detection can be achieved either by the lowering of the noise of the system or by an increase in the signal strength. It is difficult to lower the detector noise significantly without cooling well below room temperature, which is undesirable because of the concomitant cost, weight, and size constraints. Alternatively, by providing gain, the detector boosts the signal level well above the noise floor of the detector. Therefore for image intensification the device typically must provide front-end gain.

Image intensifiers were first introduced in the 1940s. All of these designs used tubes, as do most of the current image intensifiers. The primary advantage of tubes is their ability to yield very high gain at very low additional noise, thus greatly improving the signal–noise ratio of the detector. Existing image intensifier tubes, such as the GEN III, consist of three main components, a photocathode, a microchannel plate (MCP), and an output light phosphor, as shown in Figure 12.2.1. The tube is, of course, in high vacuum and requires high voltage for operation. Gain is provided within the tube by the MCP. Depending on the

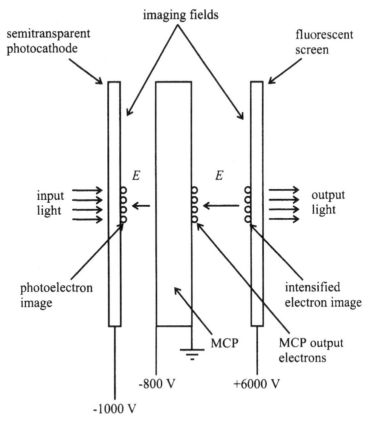

Figure 12.2.1. Basic tube anatomy for GEN II and GEN III image intensifier tubes.

applied voltage, a MCL can deliver several orders of magnitude gain, providing excellent signal-to-noise operation. GEN III tubes can operate in light levels as low as 10^{-7} lx and still enable the user to see well enough to manuever.

The photocathode is typically made from GaAs. Light falls incident upon the GaAs photocathode, where it is absorbed, forming electron–hole pairs. The choice of GaAs provides spectral coverage up to a cutoff of ~ 900 nm. This spectral coverage does not fully capture the night sky. As can be seen from Figure 12.2.2, the night sky peaks at longer wavelengths than the cutoff wavelength of GaAs. Therefore a GaAs photoabsorber is transparent to much of the night-sky irradiance. Nevertheless, GaAs is still the most commonly used photocathode mostly because of its relative technological maturity. Once the carriers are generated, because of the negative electron affinity of GaAs, they have sufficient kinetic energy to escape. Consequently, the electrons are emitted from the GaAs into vacuum and then accelerated into the MCP. As mentioned above, the MCP provides gain. A MCP is a compact aggregation of parallel microscopic glass tubes aligned parallel to the electron flow. One electron at the input can give rise to $\sim 10^4$ electrons at the output, depending on the applied voltage across the tube. The incident electron, called the primary, collides with the sides of the tube

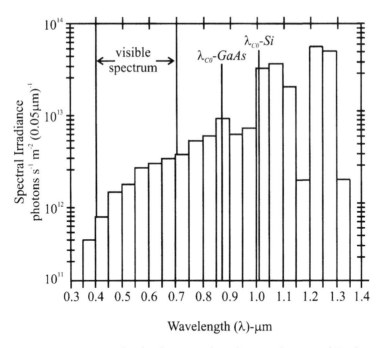

Figure 12.2.2. Natural night-sky spectral irradiance on horizontal Earth's surface. Corresponds to an illuminance of 1.1×10^{-2} lx. Data are from the *RCA Electro-Optics Handbook*.

while being accelerated by the high electric field applied across the tube. After colliding with the sides of the tube, the excess kinetic energy of the electron is transferred to the tube, leading to the production of secondary electrons. These secondary electrons provide the gain of the device.

Although GEN III tubes offer excellent low-light-level performance, they are primarily useful in direct-view applications, situations in which the user wears a helmet-mounted system. This is because GEN III tubes have photonic outputs rather than electronic. Helmet-mounted systems are impractical for commercial applications like automobile or truck driving and airplane piloting. In these situations, image-intensified data will be exploited much like a rearview mirror as a source of secondary information to aid the driver or the pilot. Consequently, it is important that the image-intensified data be digitized such that it can be projected onto a display and merged with other information, such as global-positioning-system and roadmap data. In this way, the driver will be furnished with maximum information to ensure safety and convenience.

The most obvious means of improving intensifier tubes is to adapt them such that they can provide a direct electronic output that can then be digitized and redisplayed. Attempts have been made to merge a GEN III tube with a CCD. In Section 12.3 we will discuss the basic operation of MIS detectors and CCDs.

A CCD comprises a two-dimensional matrix of MOS structures, each called a pixel. Charge is collected underneath each pixel and is clocked out sequentially across the array by the application of a sequence of voltages to each pixel.

Nearly perfect charge transfer between pixels is required for accurate image reconstruction, which in turn limits the readout speed of the device.

The CCD can be directly applied to low-light-level imaging, but this typically requires cryogenic cooling, slow-scan readout rates, and extended integration times. The main limitation in using CCDs directly in low light levels is that in their simplest implementation, they do not provide any front-end gain. Subsequently, to collect sufficient carriers to overcome readout noise, etc., the integration time of the CCD must be extended, resulting in slower readout rates. In fact, for very low light levels, insufficient carriers are collected to overcome the noise floor in a typical room-temperature-operated CCD. For these reasons, the direct usefulness of CCDs is restricted in low-light-level applications. One ready approach is to cool a CCD well below room temperature. Cooling greatly reduces the noise and dark current within a CCD, enabling low-light-level operation. Thermoelectric coolers can provide cooling to temperatures at which CCDs can operate for low light levels. However, cooling is often not acceptable because of cost, weight, and low power requirements. Alternatively, CCDs are often combined with image intensifier tubes to provide electronic output.

Two of the most promising means of using CCDs for image intensification are either optically coupling the CCD to the phosphor screen of an existing tube by an objective or a fiber-optic plate or inserting the CCD in place of the phosphor screen and directly bombarding it with electrons. Usage of CCDs in these modes does not circumvent the limitations of the tubes in terms of cost, size, and weight, since these devices are again vacuum devices that incorporate all or some of the tube components. Nevertheless, these CCD designs provide an electronic output that can be digitized to provide data integration.

The use of CCD designs in the electron-bombarded mode eliminates the MCP, phosphor screen, and output optics of the tube designs. Electrons emitted from the photocathode and accelerated by an electric field applied in vacuum impinge directly onto the CCD. Gain is achieved from the multiplication process that occurs when very-high-energy electrons (several kilo-electron-volts) are stopped within a semiconductor material. On average, an electron–hole pair is produced for every $\sim$3.6 eV of energy in silicon. Consequently, a gain of several thousand is achievable. The relatively low noise level of the CCD readout at room temperature (typically 10–100 electrons/pixel frame) coupled with this high gain ensures that the CCD can detect individual photoelectrons. The CCD must be backside bombarded to avoid damaging the front side connections as well as to optimize carrier production. Backside bombardment requires in turn wafer thinning down to $\sim$10 μm. Electron-bombarded CCDs require a radiation-hardened device. Damage to gate structures from kilo-electron-volt x rays and ballistic electrons can result in increased dark current as well as a reduction in pixel well capacity.

Optical coupling into a CCD is typically accomplished with a fiber-optic coupler. This can be accomplished without harming the array by use of a non-permanent oil coupling or direct bonding. The particular advantage of this approach is that it can be operated with existing image intensifiers. In addition, the optically coupled CCDs have good resolution performance.

CCDs can also be directly used if they are modified to provide front-end gain. When an APD array is inserted on top, a CCD can be effective in low-light-level imaging. An APD is a photodiode sufficiently reverse biased such that carrier multiplication by means of impact ionization occurs. Incident light is absorbed within the photodiode, and the subsequently photogenerated carriers are multiplied through impact ionization events. The carrier multiplication provides gain. The multiplied carriers are collected within the depletion layer of the APD and are transferred into the CCD wells for readout after a suitable integration time. If the amplification of the APD array is combined with the CCD readout, low-light-level imaging might also be possible. We discuss the detailed operation of APDs in Section 12.3.

12.3 Metal-Insulator–Semiconductor Detectors and Charge-Coupled Devices

The most widely used device in two-dimensional imaging applications is the metal-insulator–semiconductor (MIS) detector. A typical MIS detector array is a two-dimensional grid of MIS devices linked together through a CCD. Each MIS device is typically biased initially into deep depletion. Minority carriers are photogenerated by incident photons that pass through the transparent gate electrodes. The minority carriers are then collected within the potential well formed at the insulator–semiconductor interface. The amount of photogenerated minority carriers collected within an individual MIS device, or pixel, is proportional to the input light intensity onto that pixel. Therefore a two-dimensional grid of pixels can record a complete image.

There are in general two different types of MIS devices, surface-channel and buried-channel devices. Representative devices of both types are sketched in Figure 12.3.1. As can be seen from the figure, in the surface-channel device minority carriers are collected at the insulator–semiconductor interface. In the buried-channel device, minority carriers are collected away from the interface in a potential well formed within the semiconductor layer. The particular advantage of the buried-channel device is that the collected carriers are isolated spatially from the interface. As we discussed in Chapter 11, the insulator–semiconductor interface can have traps and surface states. These imperfections can trap electrons, thereby limiting the response of the MIS detector. In many materials systems, such as GaAs and InP, excellent control of the insulator–semiconductor interface cannot be uniformly maintained. Consequently, it is extremely difficult to make reproducible surface-channel detectors from these materials. For this reason, buried-channel devices are often used. However, the charge capacity of a buried-channel device is up to three times smaller than that of a comparable surface-channel device. Also, buried-channel structures can have a higher dark or leakage current than surface-channel devices. Excellent control of silicon and its oxide, SiO_2, can be readily achieved, leading to nearly perfect interfaces with

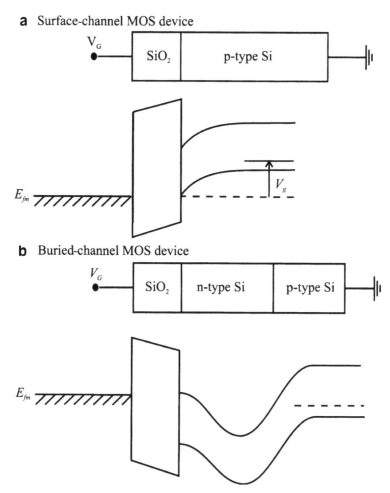

Figure 12.3.1. Sketches of different MOS devices for light detection within a CCD. a. Surface-channel MOS device. b. Buried-channel MOS device.

few, if any, surface states or impurities. Owing to the inherent advantages of surface-channel devices and the fact that silicon-based surface-channel structures can be reliably made, most silicon MIS detectors use surface-channel collection.

Because surface-channel MIS structures are the most common, we now restrict our discussion to these devices. Charge storage within the potential well formed under the MIS structure is controlled by the recombination time of the minority carriers and the maximum surface potential at the interface. During the readout phase of the detector, charge is transferred from one MIS device to another, as in a CCD. The electronic signal is fed into an electronic processer for amplification and image reconstruction.

To understand the operation of an MIS detector, consider an *n*-type MIS structure biased initially into deep depletion, as shown in Figure 12.3.2.a. If the device is illuminated by light of energy greater than or equal to the semiconductor bandgap energy, electron–hole pairs are photogenerated. The minority carrier

a

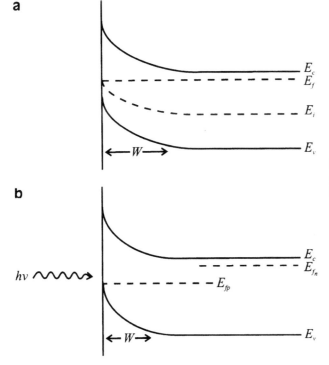

Figure 12.3.2. MIS detector with an
n-type semiconductor. a. Without illumination. b. With illumination.

b

holes are attracted to the surface by the negative gate potential, leading to a
buildup of positive charge. The presence of the holes in the potential well acts
to neutralize some of the excess negative charge on the gate. Therefore the
depletion region width decreases as the well fills with holes, as shown in Figure
12.3.2.b. Finally, the depletion region collapses entirely when the collected hole
charge balances the negative gate bias. Under this condition, no further charge is
collected. The MIS structure appears as a parallel-plate capacitor; the negative
charge on the metal gate is balanced by the positive charge collected in the
potential well at the semiconductor–insulator interface.

One of the key issues in the effectiveness of an MIS detector is its ability
to collect mainly charge photogenerated by the signal. In other words, for the
detector to give an accurate accounting of the signal strength, it should collect
only charge that has been generated by the absorption of the incident signal and
not charge that is generated spuriously, that is, randomly generated charge from
Shockley–Read–Hall centers, for example. Of course, the collection of some
spurious charge cannot be avoided, but the amount of such charge collected
must be minimized.

The total amount of signal charge collected by the detector is limited by
the storage time of the device τ_{st}, which is defined as the length of time before
which the potential well is filled by charge generated from spurious sources. If the
storage time is small, little signal charge will be collected before readout; the well
fills with charge due mostly to sources other than the signal. If the storage time is

large, the detector will fill very slowly with charge due to sources other than the signal. Therefore it is more likely to accumulate charge arising primarily from the signal, which, of course, is the desirable situation. It is necessary to design MIS detectors such that the storage time is large, particularly if weak signals need be detected. In the case of a weak input signal, the detector must focus on the source for a relatively long time in order to collect sufficient signal charge. The length of time that the detector stares at the target is often referred to as the integration time. If the storage time is less than the integration time, then the detector will collect much more noise than signal, and poor signal discernment will occur. Only if the integration time is much less than the charge storage time will the detector clearly distinguish the signal.

The storage time of an MIS device can be expressed in general as

$$\tau_{st} = \frac{C_{ox} V}{J_d}, \tag{12.3.1}$$

where C_{ox} is the oxide capacitance, $C_{ox} V$ gives the well charge storage capacity, and J_d is the dark current that arises because of spurious sources of generated charge. Clearly, as the dark current increases, the storage time decreases, implying that the well fills with spurious charge rather than signal charge. In order then to increase the storage time, the dark current must be minimized.

The dark current draws its name from the fact that it represents the current due to charge-generated independent of any input signal, hence in the dark. Sources of the dark current are

1. thermal generation of minority carriers within the bulk,
2. minority carrier generation within the depletion region,
3. generation of carriers out of fast surface states,
4. background flux generation,
5. tunneling current.

Each of these sources of dark current can be understood as follows.

The total dark current density can be quantitatively expressed as

$$J_d = q n_i \left[\frac{n_i L_p}{n_0 \tau_p} + \frac{W}{2\tau_0} + \frac{v_r}{2} \right] + q\eta\phi_B + J_{tunnel}, \tag{12.3.2}$$

where the first term is the diffusion current due to minority carriers generated within the bulk regions; the second term is the drift current within the depletion region; the third term is the surface recombination current due to generation–recombination processes by means of interface states; the fourth term is the background current due to photogenerated carriers arising from the incident background flux; and the fifth term is the tunneling current due to electrons tunneling out of the valence band into the conduction band.

The first component of the dark current mentioned above, thermal generation of carriers within the bulk, depends on the diffusion of the minority carriers and is hence characterized in terms of a diffusion current. The diffusion current depends on the number of carriers generated within a diffusion length of the depletion region. The diffusion current is due to minority carriers generated within the bulk region diffusing into the depletion region from which they are collected into the potential well. The expression for the diffusion current follows readily from the Shockley equation in reverse bias and by consideration of only the holes for n-type material:

$$J = \left(\frac{q D_p p_0}{L_p} + \frac{q D_n n_0}{L_n} \right) \left(e^{\frac{qV}{kT}} - 1 \right),$$

$$J \sim - \left(\frac{q D_p p_0}{L_p} + \frac{q D_n n_0}{L_n} \right),$$

$$J_{\text{diff}} = \frac{q D_p p_0}{L_p} = q p_0 \sqrt{\frac{D_p}{\tau_p}} = \frac{q p_0 L_p}{\tau_p} = \frac{q n_i^2 L_p}{n_0 \tau_p}. \tag{12.3.3}$$

Carriers generated within the depletion region are swept out by the electric field and thus constitute a depletion or drift current. The depletion region current, or drift current, arises from hole generation through Shockley–Read–Hall centers within the depletion region. In general the depletion region current is given by the generalization of the generation–recombination term within Eq. (11.6.11) as

$$J_{\text{depl}} = \frac{q W n_i^2}{(n_1 \tau_{po} + p_1 \tau_{no})}, \tag{12.3.4}$$

where n_1 and p_1 are defined in Eqs. (11.6.6). If the trap energy is coincident with the intrinsic level, which is a reasonable assumption for deep-level traps, then J_{depl} simplifies to

$$J_{\text{depl}} = \frac{q W n_i}{2 \tau_0}, \tag{12.3.5}$$

where the lifetimes are assumed to be equal.

The third term is due to the generation of minority carriers out of interface states. It is characterized by v_r, the recombination velocity. The recombination velocity enables the description of the current density due to a net recombination or generation of carriers as

$$J_{\text{rec}} = q(n - n_0) v_r, \tag{12.3.6}$$

where n is the nonequilibrium carrier concentration and n_0 is the equilibrium carrier concentration. In Eq. (12.3.2), v_r is the maximum surface recombination velocity.

The current generated by the incident background flux is simply given by the product of the electronic charge q, the quantum efficiency η, and the background photon flux density Φ_B as

$$J = q \eta \phi_B. \tag{12.3.7}$$

The quantum efficiency η is defined as in Section 12.1 as the probability that an electron–hole pair will be produced from an incident photon. Obviously the highest quantum efficiency is 100%; every incident photon produces an electron–hole pair. Equation (12.3.7) represents the limiting current for a detector. The detector cannot perform better than the background-limiting performance. The background is typically unimportant unless the detector is operating at relatively long wavelengths, generally into the infrared. At infrared wavelengths, the observer is most interested in some target that is embedded in an environment. The detector needs to be capable of discerning the target from the background. The background flux is given by Eq. (12.3.7). To see a target against the background, the target must be brighter than the background flux given above.

Finally, there can be an additional dark current term due to tunneling. Tunneling is most important in narrow-bandgap materials. In these materials, under sufficient bias, electrons can tunnel out of the valence band into the conduction band, thereby leaving holes behind in the valence band and producing electrons in the conduction band. The holes produced in the valence band are subsequently trapped in the potential well formed at the interface. The tunneling process is sketched in Figure 12.3.3. As can be seen from the figure, there are three main tunneling channels that can occur. These are

1. Direct interband tunneling. In this case, electrons tunnel directly from the valence band into the conduction band.
2. Thermal excitation to a deep-level trap state followed by a tunneling transition into the conduction band.

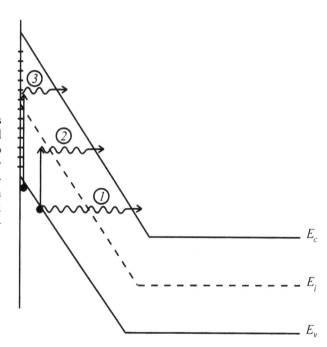

Figure 12.3.3. Tunneling processes in a MIS device: 1, direct interband tunneling; 2, thermal excitation to a deep-level trap state followed by a tunneling transition into the conduction band; 3, thermal excitation into a surface state followed by a tunneling transition into the conduction band.

E_c

E_i

E_v

3. Thermal excitation into a surface state followed by a tunneling transition into the conduction band.

The tunnel current density can be approximated as that due to tunneling from a triangular barrier as in a reverse biased *p-n* junction diode. The expression for the tunneling current, in this approximation, can be quantitatively expressed with Eq. (11.6.26) as

$$ J = \frac{q^3 F V_a \sqrt{2 m^*}}{4\pi^3 \hbar^2 \sqrt{E_g}} e^{-\frac{\pi \sqrt{m^*} E_g^{3/2}}{2\sqrt{2} q F \hbar}}, \tag{12.3.8} $$

where F is the field associated with the potential barrier and V_a is the surface potential.

Typically, MIS detectors are arranged in a two-dimensional array and are coupled into a CCD to enable readout of the photogenerated and collected charge. A CCD is essentially an array of closely spaced MIS capacitive devices. Charge is accumulated (in photodetection applications, the accumulated charge corresponds to photogenerated carriers), stored, and transferred from one capacitor to another to perform useful functions.

A CCD is most often compared with a bucket brigade. Charge is transferred out of one potential well into an adjacent well through variations in the gate potential applied to each device. Charge sloshes from one well into the next sequentially through a line through the entire array. This is similar to water being poured from one bucket into the next in an old-fashioned bucket brigade used for fighting fires. As in a bucket brigade, care must be exercised in a CCD such that most of the charge is successfully transferred from one device to the next so that at the output there is sufficient charge to be useful.

The basic operation of a CCD is quite simple. Typical CCDs are made with a *p*-type semiconductor substrate such that negative charge can be accumulated in the normal operation of the device. A CCD is subdivided into a series of unit cells consisting of two, three, or four different MIS devices. The gates of each adjacent MIS device are separately contacted by different electrodes so as to provide various clock voltages used in transferring charge from one device to the next.

As an example, let us consider the operation of a three-cell CCD device as shown in Figure 12.3.4. Let us assume that a charge packet is initially stored under the first gate, marked 1 in Figure 12.3.4. To hold the charge, a positive voltage is applied to the metal gate. The other gates in the cell, marked 2 and 3, are unbiased during this phase of operation of the device. To transfer the charge from gate 1 into the adjacent gate 2, the voltage on gate 1 must be slowly ramped to zero while the voltage on gate 2 is abruptly switched positive. The gradual ramping of the voltage on gate 1 is necessary for efficient charge transfer out of the first potential well into the second. Eventually, all the charge is transferred out of the first potential well, which now disappears, and into the second adjacent well. The sequence repeats itself in transferring the charge into the third well. In

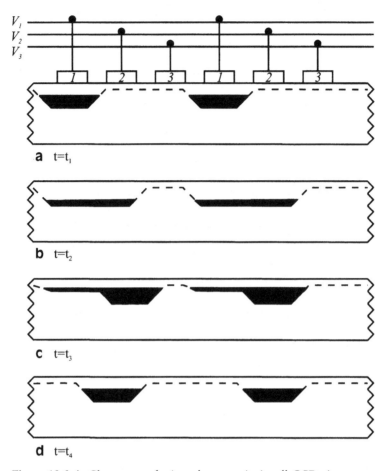

Figure 12.3.4. Charge transfer in a three-gate/unit-cell CCD. At $t = t_1$, charge is assumed to be collected under gate 1. At a later time t_2, the voltage on gate 2 is abruptly turned on (becomes positively charged) while the voltage on gate 1 starts to ramp to zero. During time t_3 the charge flows out of 1 into 2 because of the slow ramping of the voltage on gate 1. Finally, at time t_4 the charge is collected under the second gate.

this case the potential on the third gate is abruptly switched positive while the second gate potential is slowly ramped to zero. Once the charge is completely transferred to under the third gate, the full sequence is repeated until the charge is transferred out of the device. Alternative schemes in which two and four MIS structures per unit cell are used are also commonly used.

It is critical that the adjacent MIS structures be closely spaced. This is necessary to enable control of the surface potential and hence the flow of charge from one region to the next. Obviously, in the case for which the gates are widely spaced, the surface potential in the gap between adjacent gates depends on surface states and impurities that cannot be externally controlled.

Although CCDs are relatively mature, are used in many camera products, and can be adapted for low-light-level applications, they have some important limitations. The most serious limitation of CCDs is their need for nearly perfect charge-transfer efficiency. Charge-transfer efficiency η is defined as the fraction of charge transferred from one potential well to the next. The fraction of charge left behind is the transfer loss, or transfer inefficiency, ϵ. Only ~13% of the collected electrons are available at the output of the array if the charge-transfer efficiency η is 0.999 for a 1024 × 1024 array (see Fossum (1993)). This means that only 1 electron in 1000 can be lost. Since a typical charge packet in a CCD contains ~1000 electrons, then only 1 electron can be lost per transfer. At an η of 0.99999, where a far better fraction of electrons are available at the output, 98%, this translates into the loss of only one electron after one hundred transfers! Because a single broken bond in the semiconductor crystal can capture a signal electron, then nearly perfect semiconductor crystalline material is required to maintain the required η. This is the primary problem with CCDs and the principal reason why extension of CCDs to other material systems is technologically difficult.

The charge-transfer process from one device to another is due to several mechanisms: self-induced drift, thermal diffusion, and field fringing drift. Self-induced drift dominates the charge-transfer process at relatively large charge densities. Therefore, during the initial stages, when the charge concentration within the original potential well is high, self-induced drift drives the charge-transfer process. Self-induced drift arises from the nonuniform distribution of electrons under adjacent electrodes, which forms during the transfer process. The potential on the electrodes is also typically the same at this point. After most of the charge has transferred, the remaining charge moves by means of thermal diffusion and fringe-field drift. Thermal diffusion is, of course, much slower than either drift process and leads to a slow exponential decay of the remaining charge under the transferring gate. In most CCDs thermal diffusion is avoided when the device is designed such that fringe-field drift dominates the tail end of the transfer process. Fringe-field drift is due primarily to the potential difference between adjacent electrodes.

The maximum charge-transfer efficiency depends primarily on how fast the charge is transferred between adjacent wells and how much charge is trapped during the transfer process. Since the tail end of the charge transfers by means of a relatively slow diffusion process, transferring the last remaining charge can take a considerable amount of time. For this reason, high-speed operation of a CCD is often frustrated. As mentioned above, fringe fields can supply a drift component to the diffusion current, improving the transfer speed. Currently, usage of parallel outputs, circumventing charge transfer across the entire device and along each of its columns, can greatly improve CCD operating speed without compromising charge-transfer efficiency.

The presence of interface traps limits the charge-transfer efficiency. The speed with which the traps empty determines whether they influence the charge-transfer efficiency or not. If the traps emit electrons rapidly, which is the case for traps near the conduction-band edge, then they do not influence the transfer inefficiency if the emission time is much faster than the clock cycle time of the CCD. If the trap emission time is much larger than the CCD clock cycle time, then the transfer efficiency suffers. One method that is commonly used to combat deep-level trap state collection is to fill each well with a fixed level of charge, generally 10%–20% of saturation. Each well then has a zero offset voltage. The fixed stored charge is called a fat zero.

As discussed above, the dark current limits the performance of a MIS detector and hence a CCD. In terms of a CCD, the dark current presents three limitations on performance. These are finite storage time, fixed-pattern noise, and temporal noise. In Section 12.6 we will discuss noise in more detail; the subject is briefly introduced here. Noise, generally, describes fluctuations from average behavior. Fixed-pattern noise arises from the fact that the dark current is not uniform throughout the device. As a result, each device will exhibit a different dark current signature that leads to a fluctuation in the dark current from element to element. The temporal noise arises from the fact that the dark current is generated randomly in time. Hence there is a fluctuation from the average in time due to variations in the generation rate.

There are other sources of noise within a CCD other than those due to the dark current. These sources can be categorized into three types: noise associated with the input, noise associated with charge integration and transfer, and noise associated with the output. The input noise arises from fluctuations in the incident photon flux and fluctuations in the number of electrons injected into the potential well to produce a fat zero. The integration and transfer noise arises from fluctuations in charge trapping and reemitting, and from the dark current. Finally, the output noise is due to reset noise and noise in the accompanying electronics.

Many other issues are of importance in the design and utilization of CCDs for imaging. For example, the means by which the charge is read out and integrated is quite complicated and different approaches are required in different applications. Rather than attempt an exhaustive discussion of CCDs, I refer the reader to the references for more information.

12.4 **Photoconductors**

Although MIS detectors are of great importance since they can be integrated into large-scale two-dimensional arrays, one of their primary limitations is that they do not provide internal gain. As we discussed in Section 12.2, internal gain is necessary in many detection applications in order to overcome the noise floor of the detection system. Gain is particularly necessary in detecting weak signals or at high frequencies where the electronic amplifier noise is greatest. Subsequently, in light-wave communications systems, detectors that capture high-frequency, weak signals need to deliver internal gain. There are generally two different methods by which solid-state photodetectors can deliver gain. These are by photoconductive or multiplicative gain. In this section, we discuss photoconductive gain.

The gain derived from a photoconductor occurs from space-charge neutrality requirements. On photogeneration of an electron–hole pair, each carrier species

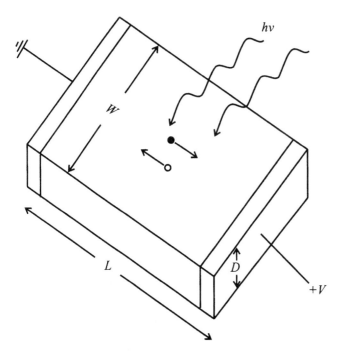

Figure 12.4.1. Sketch of a photo-conductive device.

drifts under the application of an applied electric field to its respective collecting contact, as shown in Figure 12.4.1. The electron moves toward the lower contact in Figure 12.4.1 while the hole drifts toward the upper contact. If the transit times of the electron and hole are vastly different, which is often the case since the low-field hole velocity is typically much less than the electron velocity in III–V compound semiconductors, then the electron traverses the device in far less time than the hole. The electron exits the semiconductor before the hole does. Subsequently, the semiconductor has a net positive charge. To maintain charge neutrality, an additional electron is injected from the external circuit. The new electron again drifts across the device. If the second electron traverses the device and is collected before the original hole is recombined, then the device has yet again a net positive charge. Once again, an electron is injected into the device to maintain space-charge neutrality. This process continues until the hole recombines.

It is important to note that photoconductive gain depends on a difference in the electron and hole transit times. Let us consider an extreme case, in which holes cannot freely propagate through the device yet electrons can. Following a photogeneration event, the electron propagates through the device under the influence of the electric field due to the applied potential on the contacts. As we require, the hole cannot propagate, but remains localized in the structure. The electron ultimately is collected at the high-potential contact. This results in the device's having a net positive charge because of the localized hole. To maintain space-charge neutrality, an electron is injected into the device from the low-potential contact. Again, the electron drifts toward the high-potential contact and is collected. In the meantime, the hole does not move since it is localized. This procedure continues until the hole finally recombines. Note that many electrons can traverse the device in the time it takes the hole to recombine. For example, a typical drift time for an electron in a 1-μm-long device is ~10 ps. The typical lifetime for hole recombination is at least 10 ns. There is ~3 orders of magnitude difference between the electron drift time and the hole recombination time. Subsequently, on average, 1000 electrons will traverse the device before the hole recombines. This leads to a gain in the collected current of 1000, called photoconductive gain. Note that the photoconductive gain is equal to the ratio of the hole recombination time τ to the electron transit time t as τ/t.

An example structure in which holes are highly localized but electrons are free to propagate is the effective mass filter (see Capasso, Sen, and Cho 1987). An effective-mass filter is created by use of a series of quantum wells. As was discussed in Chapter 2, if a series of quantum wells are brought into close contact, the highest-energy states within the wells interact. This produces a series of extended states that form a miniband within the multiquantum-well (MQW) stack. Electrons can propagate through the miniband and hence move through the entire MQW system. The lower-energy states typically remain localized. The wave functions do not overlap, and hence the electrons in these states cannot propagate through the system. Within the valence band, because of the much larger hole effective mass than the electron mass, even the highest-energy states can remain decoupled. The holes are localized everywhere within the system. Consequently electrons can propagate through the MQW stack but the holes are localized. This effective-mass filter can produce a high photoconductive gain. A typical structure is sketched in Figure 12.4.2.

The principal limitation of photoconductors is that the frequency response or bandwidth of these devices decreases with increasing gain. The actual device response time depends inversely on the hole lifetime. The longer a hole survives, the longer it takes the signal to decay. However, the device gain is directly proportional to the hole lifetime. The longer the hole lives, the greater the gain. There is thus a trade-off in the gain–bandwidth product of a photoconductor. The device is said then to be gain–bandwidth limited; the higher the gain, the lower the bandwidth and vice versa.

Let us quantitatively describe the operation of a photoconductor. Consider the photoconducting slab shown in Figure 12.4.1. A photoconductor is simply a light-sensitive semiconductor material contacted on both ends. The conductivity of a semiconductor depends on the carrier concentrations and mobilities. This follows readily from the expression derived in Chapter 6

$$\sigma = q\mu_n n. \tag{12.4.1}$$

If both carrier species are present, then the conductivity is given as

$$\sigma = q(\mu_n n + \mu_p p). \tag{12.4.2}$$

On illumination the conductivity of the semiconductor increases because of the production of excess free electrons and holes.

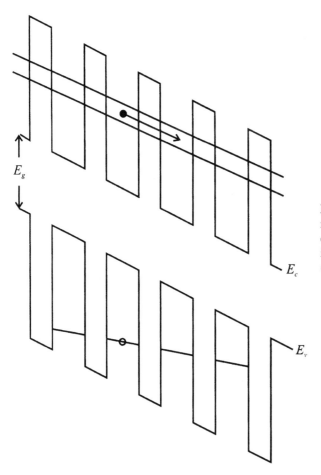

Figure 12.4.2. Sketch of an effective-mass filter showing a miniband formed in the conduction band and localized states only in the valence band. (See Capasso et al. 1987)

Photoconductive gain arises in the following way. Define δn_0 as the excess majority carrier concentration per unit volume at $t = 0$. At a later time t, the number of excess majority carriers is expressed as $\delta n(t)$. From the discussion in Chapter 10, $\delta n(t)$ decays exponentially to zero with a characteristic lifetime τ, the recombination lifetime in the absence of further pumping. The photogeneration rate of free carriers within the semiconductor material G depends on the incident optical power P_{opt}, the quantum efficiency η, and the photon energy $h\upsilon$, as

$$G = \frac{P_{\mathrm{opt}}}{h\upsilon} \frac{\eta}{V}, \tag{12.4.3}$$

where V is the volume of the material. In steady state, the net recombination rate must equal the net generation rate:

$$\frac{P_{\mathrm{opt}}}{h\upsilon} \frac{\eta}{V} = \frac{\delta n}{\tau}. \tag{12.4.4}$$

Therefore the excess electron carrier concentration is

$$\delta n = \frac{\eta P_{\mathrm{opt}} \tau}{h\upsilon V}. \tag{12.4.5}$$

If a uniform electric field is applied across the semiconductor, the steady-state drift velocity of each electron will be equal to the ratio of the distance traveled L to the transit time t as

$$v_d = -\frac{L}{t}. \tag{12.4.6}$$

The photocurrent I can be written then as

$$I = -qv_d\delta n A = q A \frac{L}{t} \frac{P_{\text{opt}}\eta\tau}{h\upsilon V}. \tag{12.4.7}$$

The photogenerated current I_{ph} is simply

$$I_{\text{ph}} = \frac{q\eta P_{\text{opt}}}{h\upsilon}. \tag{12.4.8}$$

The photoconductive gain g can now be found from the ratio of the photocurrent that flows in the external circuit I to the photogenerated current I_{ph} as

$$g = \frac{I}{I_{\text{ph}}} = \frac{\frac{\eta P_{\text{opt}}}{\upsilon}\frac{q\tau}{t}}{\frac{q\eta P_{\text{opt}}}{h\upsilon}}, \tag{12.4.9}$$

which reduces to

$$g = \frac{\tau}{t}. \tag{12.4.10}$$

The photoconductive gain is given as the ratio of the minority carrier lifetime τ to the transit time t as we had argued earlier qualitatively.

When the minority carrier lifetime, typically that of a hole, is increased, a very large gain can be attained. However, as the gain increases, the bandwidth invariably decreases since the response speed of the photoconductor diminishes in proportion to the hole lifetime. Therefore photoconductors are gain–bandwidth limited owing to the fundamental trade-off in gain versus speed of the device. For this reason, photoconductors are somewhat limited in high-bandwidth applications that also require gain, principally in light-wave communications systems.

One of the most promising photodetector designs that can also act as a photoconductor is a metal–semiconductor–metal (MSM) photodetector. A representative device is sketched in Figure 12.4.3. Light is incident upon the top of the device, as shown in the diagram. The Schottky barrier contacts form two sets of interdigitated fingers on the top surface of the device. One set is typically grounded while the other set of fingers is reverse biased. Electron–hole pairs are photogenerated within the bulk region of the device. The application of a bias to the metallic fingers creates an electric field within the underlying semiconductor that acts to sweep the photogenerated carriers out of the device. If a sufficiently large reverse bias is applied, the semiconductor layer becomes

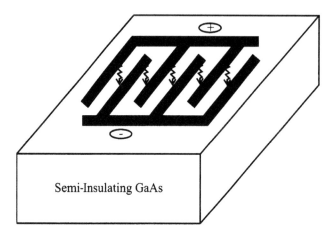

Figure 12.4.3. Typical MSM photo-detector device structure.

Semi-Insulating GaAs

depleted between the two contacts and into the depth of the device. The applied potential for which the two contact depletion regions touch is called the reach-through voltage, V_{RT}. When the device is operated at the reach-through voltage, the semiconductor layer is fully depleted between the contacts. As such the current flow is drift dominated as opposed to a slower diffusion process. How fast the carriers are collected and how many of them actually survive to be collected at the contacts within a particular time determine the speed and the responsivity, respectively, of the detector. Carriers generated deep within the semiconductor must traverse a greater distance before they are collected at the contacts compared with those generated near the surface. Depending on the magnitude of the electric field within the semiconductor, the time needed to collect those carriers generated deep in the device can vary drastically. Under low bias conditions, $\sim 5\text{--}10$ V, which is typical for most integrated circuit applications, this collection time can be prohibitively long in high-speed applications due to the fact that the semiconductor region is not fully depleted. As a result, the carriers must first diffuse into the depletion region, which is a relatively slow process.

One method by which the time response of an MSM detector can be improved is through the insertion of a double-heterostructure layer, as shown in Figure 12.4.4. As shown in the figure, an AlGaAs layer is inserted within the photoabsorptive GaAs layer, forming a double heterostructure. The double heterostructure acts to block those carriers generated deep within the device structure. As a result, only those carriers photogenerated within the top absorption layer are collected, leading to a fast overall response. However, the responsivity depends principally on the number of photogenerated carriers collected at the contacts. A high responsivity, especially at low input power levels, dictates that most of the photogenerated carriers be collected. Since many of the photogenerated carriers are produced deep within the semiconductor layer, the insertion of the double-heterostructure layer to improve the speed of the device necessarily reduces its responsivity as well. Therefore there is a fundamental trade-off

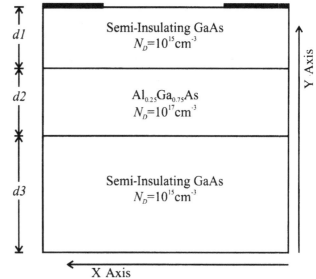

Figure 12.4.4. Cross-sectional view of a double-heterostructure GaAs-based MSM device.

between the speed of response and responsivity of a heterostructure MSM detector (see Salem, Smith, and Brennan 1994).

As an example of how the speed of response varies with the location of the double heterostructure, we examine quantitatively the impulse response of the device shown in Figure 12.4.4. Various thicknesses of the AlGaAs layer, d_2 in the diagram, and the depth of the top absorbing layer, d_1, are considered. The response is calculated with the Poisson equation coupled with the electron and hole drift-diffusion equations derived in Section 6.3. The contacts are assumed to be Schottky barriers with barrier heights equal to 0.7 eV. The choice of Schottky barriers is important in order to reduce the dark current of the MSM device.

As described above, the primary function of the buried heterostructure layers is to block the collection of the carriers photogenerated deep within the GaAs substrate by preventing them from diffusing back into the active layer and toward the contacts. In addition, at low applied bias, the barrier between the active and the AlGaAs layers acts to confine the photogenerated carriers within the active region. At higher applied bias, the heterostructure blockage of the photogenerated carriers becomes less effective because of the much greater band bending present in the device. Subsequently, the location of the AlGaAs layer greatly affects the charge-collection attributes of the device, depending on the field distribution and the applied bias. In the structures examined here, the background doping is n type, implying that the primary photogenerated carriers collected are holes.

The calculated impulse responses at different active layer thicknesses, along with the corresponding rise and fall times of the signal, are displayed in Figure 12.4.5. Five different structures are examined: four double-heterostructure devices and one GaAs bulk device 6 μm in width. The double-heterostructure devices consist of a top GaAs layer, d_1, ranging in width from 0.5 to 4 μm,

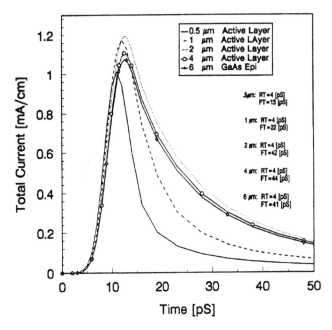

Figure 12.4.5. Calculated impulse response at different active layer thicknesses for the double-barrier MSM heterostructure device at an applied bias of -5 V. The doping levels are 10^{15} cm^{-3} for GaAs and 10^{17} cm^{-3} for the AlGaAs. RT, rise time; FT, fall time.

a 1-μm AlGaAs layer d_2, followed by a GaAs epilayer d_3, ranging in width from 4.5 to 1 μm. Note that the total width of all three layers combined remains constant at 6 μm. The rise and the fall times are defined as the time it takes the output signal to go from 10% to 90% and from 90% to 10% of its maximum, respectively. The voltage applied to the device is -5 V. As can be seen from Figure 12.4.5, the fastest response occurs for the device configuration with a 0.5-μm active layer thickness. This is obvious from both the curve corresponding to the 0.5-μm device as well as from its corresponding fall time. However, the maximum output signal magnitude for the 0.5-μm device is significantly less than for the other cases. This is as expected, since speed is achieved at the expense of lower output signal magnitude because the slower carriers, those generated deep within the device, are blocked from being collected by the heterojunction barrier. As the active layer thickness d_1 increases to 1 μm, more carriers are generated within the top, active GaAs region. As a result, a greater number of photogenerated carriers are collected, producing a higher output signal current. Although the $d_1 = 1$ μm device does not show as rapid a collection of the photogenerated carriers as the $d_1 = 0.5$ μm device, the field is sufficiently strong and the carriers are still relatively close to the collecting contacts that a reasonably high speed of response is retained; a fall time of 22 ps is achieved, compared with 13 ps for the 0.5-μm device. If the active layer thickness is increased further to 2 μm, the same trend is observed; more carriers are collected from the bottom of the active layer, resulting in a longer fall time, $\sim$42 ps, and slower speed of response. Interestingly, the maximum output signal current ultimately decreases with increasing d_1. The maximum signal current is plotted as a function of active layer width in Figure 12.4.6. As can be seen from

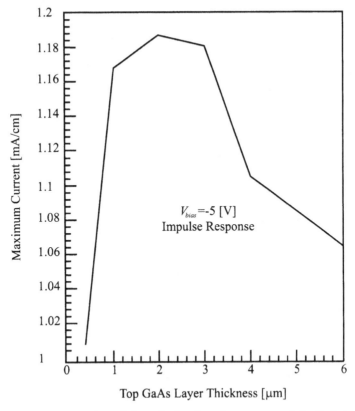

Figure 12.4.6. Plot of the calculated peak response of the GaAs/AlGaAs MSM detector as a function of the top GaAs layer thickness.

Figure 12.4.6, the output signal current reaches a maximum for an active layer width of $d_1 = 2 \ \mu$m. The maximum signal current is significantly less for a device with $d_1 = 4 \ \mu$m than with $d_1 = 2 \ \mu$m, although the fall times are comparable. Clearly, at an applied bias of -5 V, there is an optimal thickness, in terms of speed of response and collection efficiency, for layer d_1.

The MSM devices can also exhibit photoconductive gain. The devices are typically not designed to deliver much gain. Nevertheless, because of the somewhat uncontrollable difference in the electron transit time and hole lifetime, MSM devices show some gain in the output signal. Of course, photoconductive gain comes at the expense of bandwidth. Since the primary function of MSM devices is to provide high-bandwidth detection on a planar surface, gain is typically unimportant in their design.

12.5 **Photodiodes and Avalanche Gain**

A diode can be used as a photodetector. Such devices are called photodiodes and are simply a reverse biased *p-n* or more often a *p-i-n* junction diode. As discussed in Chapter 11, the reverse bias current is driven by generation–recombination processes. Generation of excess carriers within the depletion region greatly enhances the reverse bias current. Therefore, if a reverse biased junction is illuminated with light of energy in excess of the bandgap, the reverse bias current is significantly increased because of the presence of the photogenerated carriers.

The reverse bias aids the operation of the device in other ways. The depletion region in which the light is absorbed is enlarged by the action of the reverse bias, thereby enhancing the absorption. The photogenerated electron–hole pairs are separated by the action of the drift field. In addition, the field is typically large enough that high carrier drift velocities are attained. If the reverse bias is sufficiently large, impact ionization can occur. In this case, carrier multiplication and current gain occur in the device. A photodiode that operates in the multiplication mode is called an avalanche photodiode (APD). We will discuss APDs in some detail in Section 12.6.

Photodiodes are attractive in detection applications because they provide high quantum efficiency and very high bandwidth. The bandwidth, which is inversely proportional to the temporal response of the device, depends on three factors: the carrier diffusion to the depletion region edge, drift time within the depletion region, and the capacitance of the depletion region. Each of these components adds to the time delay of the device. Those carriers photogenerated outside of the depletion region must first diffuse to the depletion region edges, after which they are swept out of the device by the action of the reverse bias field. The diffusion process is in general slow and can take a relatively long time, depending on the width of the absorbing layer. Once the carriers reach the depletion region edge they are swept out by the action of the reverse bias field. Electrons drift toward the n contact while the holes drift toward the p contact. The time delay due to the carrier drift within the depletion region depends on both the width of the depletion region W and the relative speeds of the carriers. As mentioned above, in order to absorb the incident radiation fully, it is necessary to make the depletion region wide. However, as W increases, the transit time across the junction increases, thereby reducing the bandwidth of the device.

The speed of response also depends on the junction capacitance, that associated with the reverse biased junction. The device resembles a planar capacitor of capacitance:

$$C = \frac{\epsilon_s A}{W}.$$
(12.5.1)

Substituting Eq. 11.6.2 for the depletion region width W yields

$$C = \frac{\epsilon_s A}{\sqrt{\frac{2\epsilon_s}{q} \frac{(V_{bi}-V)(N_a+N_d)}{N_d N_a}}}.$$
(12.5.2)

If C_0 is defined as C (at $V = 0$), then for an abrupt junction the capacitance can be written as

$$C = \frac{C_0}{\sqrt{1 - \frac{V}{V_{bi}}}}. \tag{12.5.3}$$

In general, the capacitance of the junction is

$$C = \frac{C_0}{\left(1 - \frac{V}{V_{bi}}\right)^m}, \tag{12.5.4}$$

where m has values between 1/3 and 1/2 inclusive. $m = 1/3$ corresponds to a graded junction, while $m = 1/2$ holds for an abrupt junction.

As mentioned above, if a photodiode is sufficiently reverse biased, carrier multiplication by means of impact ionization can occur. These device types are called APDs. The current response in the external circuit due to one photogenerated electron–hole pair depends in part on the number of secondary electron–hole pairs created from the original carriers. Typically, an APD is illuminated from only one side, usually the p side. Therefore the photogenerated hole is immediately swept out of the device and the photogenerated electron, often called the parent electron or initiating electron, is accelerated through the depletion region by the action of the reverse bias field, as shown in Figure 12.5.1. Under sufficiently large reverse bias, the electron can attain threshold and impact ionize, thus increasing the current. The secondary electron–hole pair, daughter carriers, produced in the impact ionization event can also impact ionize as they move toward their respective collecting contacts. In this way, a cascade of secondary carriers is produced from each photogenerated electron–hole pair.

The total electron and hole currents in an APD can be determined in the following way. The electron current flows in the direction opposite to that of the electrons. We assume that the electrons are injected at $x = 0$ and move toward $x = W$, as shown in Figure 12.5.2. Therefore the electron current increases with increasing x while the hole current decreases with x. The variation of the electron

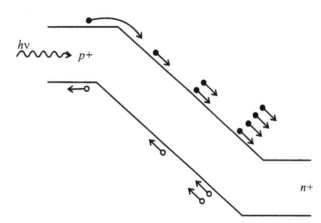

Figure 12.5.1. Schematic drawing illustrating impact ionization events in a reverse biased junction.

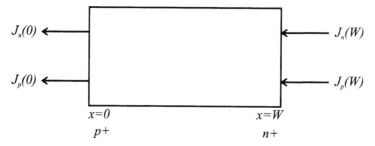

Figure 12.5.2. Current densities in a reverse biased junction. $x = 0$ corresponds to the p^+ region/depletion region boundary while $x = W$ is at the n^+/depletion region edge.

and hole currents with x are given as

$$\frac{dJ_n(x)}{dx} = \alpha(x)J_n(x) + \beta(x)J_p(x) + qG(x), \tag{12.5.5}$$

$$\frac{-dJ_p(x)}{dx} = \alpha(x)J_n(x) + \beta(x)J_p(x) + qG(x), \tag{12.5.6}$$

where α and β are the electron and the hole ionization rates, respectively. The total current density is constant in steady state. It is simply equal to the sum of the electron and hole current densities as

$$J = J_n(x) + J_p(x). \tag{12.5.7}$$

Using the expression for the total current density, we find that the variation of the electron and hole current densities becomes

$$\frac{dJ_n(x)}{dx} = [\alpha(x) - \beta(x)]J_n(x) + \beta(x)J + qG(x), \tag{12.5.8}$$

$$\frac{dJ_p(x)}{dx} = [\alpha(x) - \beta(x)]J_p(x) - \alpha(x)J - qG(x). \tag{12.5.9}$$

The total current density can be determined by integration of Eqs. (12.5.8) and (12.5.9) over x from 0 to W (the depletion region width). By using the integrating factor,

$$e^{-\int(\alpha-\beta)\,dx} \equiv e^{-\phi(x)}, \tag{12.5.10}$$

We can solve Eqs. (12.5.8) and (12.5.9) (see Stillman and Wolfe 1977, and Box 12.5.1)

$$J = \frac{J_p(W) + J_n(0)e^{\phi(W)} + qe^{\phi(W)}\int_0^W G(x)e^{-\phi(x)}\,dx}{1 - \int_0^W \beta(x)e^{\int_x^W(\alpha-\beta)\,dx'}\,dx} \tag{12.5.11}$$

$$J = \frac{J_p(W)e^{-\phi(W)} + J_n(0) + q\int_0^W G(x)e^{-\phi(x)}\,dx}{1 - \int_0^W \alpha(x)e^{-\int_x^W(\alpha-\beta)\,dx'}\,dx}. \tag{12.5.12}$$

Box 12.5.1 Calculation of the Multiplication Formulas

To see the origin of the multiplication formulas, Eqs. 12.5.13 and 12.5.14, start with the expression for the electron current density, Eq. 12.5.8, modified as

$$\frac{dJ_n}{dx} + (\beta - \alpha)J_n = \beta J + qG.$$

Notice that $e^{-\phi(x)}$ is an integrating factor. This can be seen as follows. Let u be defined as

$$u = J_n e^{\int_0^x (\beta - \alpha)\,dx'} = J_n e^{-\phi(x)}.$$

du/dx becomes then

$$\frac{du}{dx} = \frac{dJ_n}{dx} e^{\int_0^x (\beta - \alpha)\,dx'} + (\beta - \alpha)J_n\, e^{\int_0^x (\beta - \alpha)\,dx'}.$$

Notice that du/dx is simply equal to the product of the integrating factor and the left-hand side of the modified version of Eq. 12.5.8. Therefore,

$$\frac{d}{dx}\big[J_n e^{\int_0^x (\beta - \alpha)\,dx'}\big] = [\beta J + qG] e^{\int_0^x (\beta - \alpha)\,dx'}.$$

Integrating the above with respect to x from 0 to W obtains,

$$\int_0^W dx \frac{d}{dx}\big(J_n e^{\int_0^x (\beta - \alpha)\,dx'}\big) = \int_0^W [\beta J + qG] e^{\int_0^x (\beta - \alpha)\,dx'}\, dx.$$

The left-hand side above is an exact differential. Applying the following boundary conditions

$$x = 0 \quad J_n = J_{n0} = J_n(0); \quad x = W \quad J_p = J_{p0} = J_p(W),$$

the left-hand side becomes

$$J_n e^{\int_0^x (\beta - \alpha)\,dx'}\Big|_0^W = J_n(W) e^{\int_0^W (\beta - \alpha)\,dx'} - J_n(0).$$

Noticing that $J_{n0} = J_n(0)$ and $J_p(W) = J_{p0}$, obtains,

$$(J - J_{p0}) e^{\int_0^W (\beta - \alpha)\,dx'} - J_{n0} = \int_0^W e^{\int_0^x (\beta - \alpha)\,dx'} [\beta J + qG]\, dx.$$

For M_p: $J_{n0} \ll J_{p0}$; $J = M_p J_{p0}$ and assuming $G = 0$, the above expression becomes

$$J_{p0}(M_p - 1)z(W) = \int_0^W \beta z(x) M_p J_{p0}\, dx,$$

where $z(W)$ and $z(x)$ are defined as

$$z(W) = e^{\int_0^W (\beta - \alpha)\,dx'}; \quad z(x) = e^{\int_0^x (\beta - \alpha)\,dx'}.$$

Therefore, M_p is given as

$$M_p = \frac{z(W)}{z(W) - \int_0^W \beta z(x)\, dx},$$

which is equivalent to

$$M_p = \frac{z(W)}{1 - \int_0^W \alpha z(x)\, dx}.$$

The above is just Eq. 12.5.15 in the text. A similar analysis results in the other expressions given in the text.

If the generation rate $G(x)$ is assumed to be zero and single carrier injection conditions apply, only one carrier type is injected at any one time. Under these conditions, the electron or hole carrier density at the edge of the depletion region vanishes, depending on which carrier type is injected. The expressions for the hole and electron multiplication factors, M_p and M_n, can be found from Eqs. (12.5.11) and (12.5.12) by requiring that $J_n(0) = 0$ for pure hole injection or $J_p(W) = 0$ for pure electron injection in either of the two equations. M_p and M_n become

$$M_p = \frac{J}{J_p(W)} = \frac{1}{1 - \int_0^W \beta(x) e^{\int_x^W (\alpha - \beta) \, dx'} \, dx}, \qquad (12.5.13)$$

or equivalently, when Eq. (12.5.12) is used,

$$M_p = \frac{J}{J_p(W)} = \frac{e^{-\int_0^W (\alpha - \beta) \, dx'}}{1 - \int_0^W \alpha(x) e^{-\int_0^x (\alpha - \beta) \, dx'} \, dx}. \qquad (12.5.14)$$

Similarly, the electron multiplication rate is given as

$$M_n = \frac{J}{J_n(0)} = \frac{e^{\int_0^W (\alpha - \beta) \, dx'}}{1 - \int_0^W \beta(x) e^{\int_x^W (\alpha - \beta) \, dx'} \, dx}, \qquad (12.5.15)$$

or

$$M_n = \frac{J}{J_n(0)} = \frac{1}{1 - \int_0^W \alpha(x) e^{-\int_0^x (\alpha - \beta) \, dx'} \, dx}. \qquad (12.5.16)$$

It is important to note that the electron and hole multiplication factors have singularities when the denominators vanish. This condition is called avalanche breakdown and the device becomes unstable. Operation of an APD is confined to regions below avalanche breakdown for most applications.

It is instructive to analyze Eqs. (12.5.13)–(12.5.16) under two different conditions: electron ionization only occurs and both the electron and hole ionization coefficients are equal. In the first case, the hole ionization rate is zero, so $\beta = 0$. Therefore Eq. (12.5.15) reduces, when $\beta = 0$, to

$$M_n = e^{\int_0^W \alpha \, dx}. \qquad (12.5.17)$$

If the electron ionization rate α is independent of x, M_n becomes

$$M_n = e^{\alpha W}. \qquad (12.5.18)$$

Under these conditions the multiplication rate never diverges: avalanche breakdown does not occur. In principle, any gain is possible.

The second case, equal ionization rates, $\alpha = \beta$, implies that the multiplication factors are the same, $M_n = M_p$. Putting $\alpha = \beta$ into Eqs. (12.5.13) and (12.5.16) yields

$$M_n = M_p = \frac{1}{1 - \int_0^W \alpha \, dx} = \frac{1}{1 - \alpha W}. \tag{12.5.19}$$

In this case, avalanche breakdown occurs when $\alpha W = 1$. Physically, this means that when each injected carrier on the average generates an electron–hole pair during its transit through the depletion region, breakdown occurs.

12.6 **Noise Properties of Multiplication Devices**

The multiplication process adds an additional noise component to a device. In the previous sections we have discussed several types of multiplication processes. In vacuum devices, like a MCP or a photomultiplier tube (PMT), gain occurs by means of multiplication. In these devices gain occurs through secondary emission of electrons from metallic plates (PMTs) or from the sides of metallic glasses (MCPs) following the high-energy collision of an incident electron. Secondary electrons are emitted from the plates and are accelerated through vacuum until they strike another plate, as shown in Figure 12.6.1. In each such event, a cascade of secondary electrons is emitted for each incident-electron collision. The number of secondary electrons emitted is typically extremely large, of the order of ~ 1000. The large number of secondary carriers produced from a collision provides an extremely high gain in the device.

Figure 12.6.1. Sketch of a PMT showing the high gain of the device. Each of the cross-sectional circles represents a dynode or metallic plate at which carriers are produced because of secondary electronic collisions.

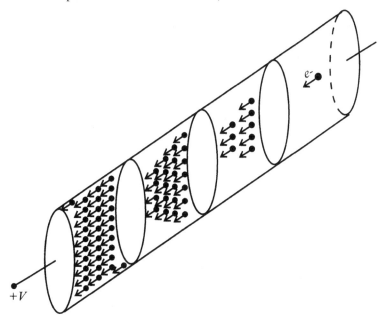

In APDs, on the other hand, gain arises from the production of secondary electron–hole pairs following an impact ionization event. In virtually all cases, only one secondary electron–hole pair is born per event, unless the initiating carrier is injected into the material with an extremely high energy, which we neglect here. Therefore the gain per event in an APD is very much less than that for a PMT or a MCP.

In all these devices, APDs, MCPs, and PMTs, randomness in the multiplication process produces an additional noise component. This additional noise is termed the excess noise. The excess noise occurs in addition to the usual noise sources present in an electronic device, such as the shot noise discussed above. As mentioned in Section 12.1, noise arises from fluctuations from the mean. In the case of shot noise, fluctuations from the mean in the arrival times of carriers produces noise. If no statistical fluctuations in the carrier velocities from the average ensemble behavior occur, then the carrier arrival rate, and hence the current, is completely deterministic in time. No shot noise is thus produced. Of course in practice, since devices operate at room temperature and electrons suffer collisions during their flights, there is a significant amount of shot noise in any semiconductor device.

In an APD, not only are there fluctuations in the carrier arrival times, but the number of collected carriers varies as well. The variation in the number of carriers arises from randomness in the multiplication rate; each initial photogenerated carrier gives rise to a random number of secondary electron–hole pairs. The additional randomness due to the multiplicative gain process leads to additional noise, which is the excess noise. APDs, MCPs, and PMTs all exhibit excess noise due to randomness in the multiplication processes in these devices.

The question is then, since the multiplication process adds an additional noise component to the system, of what use is the multiplicative gain? As discussed in Section 12.2, gain is of use in a photodetector only to boost the signal over the noise floor of the following electronics. If the gain comes at the expense of additional noise, then it is questionable if the gain is of any benefit. The advantage of the gain provided by an APD in terms of how it alters the signal-to-noise power ratio can be understood as follows.

The steady-state photocurrent in the presence of a mean multiplicative gain $\langle M \rangle$ is

$$I_p = \frac{q \eta P_{\mathrm{opt}}}{h \upsilon} \langle M \rangle; \tag{12.6.1}$$

all the variables are as defined previously. In addition to the input signal current, there is a current due to the background radiation I_B and one due to thermally generated carriers within the depletion region of the diode I_D, the dark current. The mean-square shot-noise current within the photodiode is then

$$\langle i_s^2 \rangle = 2q B (I_p + I_B + I_D) \langle M^2 \rangle, \tag{12.6.2}$$

where again B is the bandwidth.

We define the excess noise factor as F_e. F_e physically represents the contribution of the randomness in the multiplication process to the noise of the device. The excess noise factor is defined as

$$F_e = 1 + \frac{\text{Var}(M)}{\langle M \rangle^2}, \tag{12.6.3}$$

where $\text{Var}(M)$ is the variance in the multiplication rate. Note that if there is no variance in the multiplication rate, then the excess noise factor is simply equal to 1. From the definition of the variance given by Eq. (12.1.8), the excess noise factor can be rewritten as

$$F_e = \frac{\langle M^2 \rangle}{\langle M \rangle^2}. \tag{12.6.4}$$

The mean-square shot noise in the presence of excess noise is simply multiplied by the excess noise factor. The resulting mean-square noise current is

$$\langle i_s^2 \rangle = 2qB(I_p + I_B + I_D)F_e \langle M \rangle^2. \tag{12.6.5}$$

The thermal noise of the device can be obtained from the lumped contribution of the junction resistance R_j, the external load resistance R_L, and the input resistance of the following amplifier R_i. The thermal noise is then

$$\langle i_T^2 \rangle = \frac{4kTB}{R_{\text{eq}}}, \tag{12.6.6}$$

where $1/R_{\text{eq}} = 1/R_i + 1/R_L + 1/R_j$.

The signal-to-noise power ratio can now be determined from the ratio of the signal current squared to the sum of the mean-square shot-noise current and the mean-square thermal current, yielding

$$\left(\frac{S}{N} \right)_{\text{power}} = \frac{i_p^2}{i_s^2 + i_T^2}. \tag{12.6.7}$$

Substituting Eqs. (12.6.1), (12.6.5), and (12.6.6) into Eq. (12.6.7) for i_p, i_s, and i_T yields

$$\left(\frac{S}{N} \right)_{\text{power}} = \frac{\frac{1}{2}\left(\frac{q\eta P_{\text{opt}}}{h\upsilon}\right)^2 \langle M \rangle^2}{2q(I_p + I_B + I_D)\langle M \rangle^2 F_e B + \frac{4kTB}{R_{\text{eq}}}}, \tag{12.6.8}$$

where the extra factor of 1/2 in the numerator arises if the rms value of the input optical signal is chosen. Dividing through by $\langle M \rangle^2$, we find that Eq. (12.6.8) becomes

$$\left(\frac{S}{N} \right)_{\text{power}} = \frac{\frac{1}{2}\left(\frac{q\eta P_{\text{opt}}}{h\upsilon}\right)^2}{2q(I_p + I_B + I_D)F_e B + \frac{4kTB}{\langle M \rangle^2 R_{\text{eq}}}}. \tag{12.6.9}$$

From inspection of Eq. (12.6.9), it is apparent that as the mean gain increases, the thermal noise becomes less important. At sufficiently high gain, the thermal noise of the load resistor can be made negligible. Therefore the internal gain mechanism amplifies the signal current without increasing the thermal noise of the detector. However, the excess noise typically increases with increasing gain. This leads then to an optimum value of the multiplication rate that produces the maximum signal-to-noise ratio at a given optical input power.

Let us now examine how the excess noise factor varies with the gain of the device. We will take two different examples, a PMT and an APD. In a PMT many carriers are produced per collisional excitation. Typically, one to several thousand carriers are produced following each event. The average multiplication for an m-stage device is given by (see Teich, Matsuo, and Saleh 1986)

$$\langle M \rangle = \prod_{k=1}^{m} \langle \delta_k \rangle, \tag{12.6.10}$$

where $\langle \delta_k \rangle$ is the mean gain at the kth stage. At each stage, a random number of electrons are produced from a single electron collision. The variable δ_k represents the random number of secondary electrons produced. The excess noise arises from the fact that there is a variation in the multiplication rate from the average value. In the case of a PMT, for every stage there is a fluctuation in the number of carriers produced from the mean gain per stage. The excess noise factor in general is defined by Eq. (12.6.3). Therefore the excess noise factor for an m-stage PMT is found from the sum of the variations from each stage as

$$F_e = 1 + \frac{\text{Var}(\delta_1)}{\langle \delta_1 \rangle^2} + \frac{\text{Var}(\delta_2)}{\langle \delta_1 \rangle \langle \delta_2 \rangle^2} + \cdots + \frac{\text{Var}(\delta_m)}{\langle \delta_1 \rangle \langle \delta_2 \rangle \cdots \langle \delta_m \rangle^2}. \tag{12.6.11}$$

Note that again, if there are no variations from the average behavior, the excess noise factor is simply 1 and no additional noise arises from the gain process.

Typically, it can be assumed in a PMT that each stage is equivalent. Then the gain of the PMT is given as

$$\langle M \rangle = \langle \delta \rangle^m. \tag{12.6.12}$$

In this situation the excess noise factor reduces to

$$F_e = 1 + \frac{\text{Var}(\delta)}{\langle \delta \rangle} \left[\frac{1}{\langle \delta \rangle} + \cdots + \frac{1}{\langle \delta \rangle^m} \right]. \tag{12.6.13}$$

But we can sum expression (12.6.13) by using the geometric progression sum defined as

$$a_n = a_1 r^{n-1}, \qquad s_n = a_1 \frac{1 - r^n}{1 - r}, \tag{12.6.14}$$

where s_n is the sum of n terms. Applying the above relationship to the excess noise factor yields

$$F_e = 1 + \frac{\text{Var}(\delta)}{\langle \delta \rangle (\langle \delta \rangle - 1)} \left[1 - \frac{1}{\langle \delta \rangle^m} \right]. \tag{12.6.15}$$

In many PMTs, it is reasonable to assume that the multiplication process is Poisson. Any Poisson process has the property that the mean is equal to the variance, $\text{Var}(\delta) = \langle \delta \rangle$. With the definition of the gain as $\langle M \rangle = \langle \delta \rangle^m$ and the assumption of Poisson statistics, the excess noise factor can be written as

$$F_e = 1 + \frac{1}{\langle \delta \rangle - 1} \left[1 - \frac{1}{M} \right]. \tag{12.6.16}$$

If it is further assumed that the gain is 1000–10,000, which is typically the case, then F_e becomes

$$F_e = \frac{\langle \delta \rangle}{\langle \delta \rangle - 1}. \tag{12.6.17}$$

Note that the excess noise factor is very close to 1. Nearly noise-free multiplication occurs. This result is generally the case for a PMT. Owing to the large number of secondary carriers produced from a collisional excitation, the noise of a PMT is extremely low. For this reason, PMTs are used in many applications that require ultralow noise detection despite their large sizes, high voltage requirements, and weight.

The noise properties of an APD are quite different from those of a PMT. APDs can be broken down into two different categories, a conventional APD and a superlattice APD. The conventional APD is a reverse biased p-i-n junction where the intrinsic region is simply bulk semiconductor material. Most APDs are of the conventional type. In a conventional APD, an impact ionization event can occur virtually anywhere within the intrinsic region. There is no spatially preferred location in which an ionization event will occur. The superlattice APD consists of a series of stages after which impact ionization events occur preferentially. An illustrative example of a superlattice APD is shown in Figure 12.6.2. In the next section, we will discuss superlattice APDs in detail and discuss various types. For now, let us consider the noise properties of APDs.

One of the primary differences between a PMT and an APD that affects their noise properties is that in an APD both carrier species, electrons and holes, are produced. In a PMT, there is only one carrier species, electrons, present. The bipolar nature of the impact ionization process in APDs greatly affects the noise properties of the device. The other major difference between a PMT and an APD is that in a PMT many carriers are produced following an impact excitation event while in an APD only one electron–hole pair is produced per event. Again, this has an important effect on the noise properties of the device.

Consider the operation of the superlattice APD, as shown in Figure 12.6.2. Light is incident upon the p^+ contact. If the constituent material of the APD is a direct-gap semiconductor, the absorption coefficient is relatively high. This

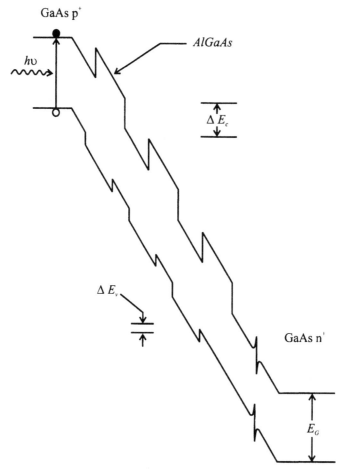

GaAs p⁺

$h\upsilon$

AlGaAs

ΔE_c

ΔE_v

GaAs n⁺

E_G

Figure 12.6.2. Energy-band diagrams for the simple multiquantum-well APD structure under bias.

results in strong absorption near the surface of the material. As such, most of the light will be absorbed within the p^+ region of the diode. The absorbed light within the p^+ region produces electron–hole pairs. The action of the reverse bias on the diode sweeps the photogenerated electrons into the intrinsic region where they begin their flights toward the n^+ collecting contact. The photogenerated holes on the other hand are recombined within the p^+ region or the p^+ contact. Therefore only electrons will initiate the impact ionization process under these conditions. These are called single carrier injection conditions. As the electrons travel through the intrinsic region, they are accelerated by the action of the reverse bias field to sufficiently high energies, on which they may suffer an impact ionization collision. Following an impact ionization event, as discussed in Chapter 10, an electron–hole pair is produced. The secondary hole travels backward in the device toward the p^+ contact while the secondary electron along with the primary electron continue toward the n^+ contact.

If the hole impact ionization rate is comparable with the electron ionization rate, the secondary hole may impact ionize during its flight toward the p^+

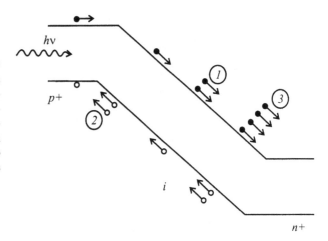

Figure 12.6.3. Sketch of an APD showing the regenerative process when both carrier species have comparable ionization rates. 1, first ionization event due to a primary electron, 2, first hole-initiated ionization event due to a secondary hole, 3, two electron initiated ionization events occurring later in time.

contact. If the secondary hole does ionize, then another electron–hole pair is produced. Now the secondary electron travels toward the n^+ contact while the secondary and the primary holes travel toward the p^+ contact. Note that if both carrier species again ionize, as shown in Figure 12.6.3, many secondary carriers are produced that can ultimately result in avalanche breakdown. As mentioned in Section 12.5, in avalanche breakdown the gain effectively is infinite and the device is unstable. As can easily be seen from Figure 12.6.3, if avalanche breakdown occurs, the device never shuts off because of the continuous regenerative process of both carriers ionizing.

If the secondary hole ionization rate is zero, then the device shuts off after the last hole reaches the p^+ contact. In this case, there is no feedback from hole ionization, as shown in Figure 12.6.4, and the device does not undergo avalanche breakdown. This condition is called single carrier ionization.

In the presence of secondary ionization, the multiplication noise greatly increases. The origin of the enhanced excess noise can be understood qualitatively as follows. At fixed gain there are fewer carriers in the active region of the device when the electron α and the hole β ionization rates are comparable than when

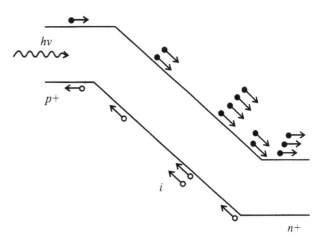

Figure 12.6.4. Sketch of an APD showing carrier births when only the electrons can ionize. Note in this case that there is no feedback from secondary hole ionization.

the hole ionization rate vanishes. This is obvious since each secondary carrier contributes equally to the gain when both carriers ionize. Secondary holes do not ionize when the hole ionization rate β is equal to zero. In this case, all the gain arises from electron ionization events alone. Clearly, many more carriers are needed within the active region at any given time, under steady-state conditions, to produce the same gain as when the carrier ionization rates are equal. Inspection of Figures 12.6.3 and 12.6.4 bears this expectation out. Statistical fluctuations become more significant as the particle number decreases. Therefore the case when $\alpha = \beta$ is inherently noisier than the case when $\beta = 0$.

The noise properties of the conventional and the superlattice APDs are somewhat different. In a superlattice APD, the ionization events are spatially localized in each stage, while in a conventional APD there is continuous ionization. For convenience and to provide a better comparison with the PMT, our discussion here of the noise properties is restricted to the case of a superlattice APD. The gain and the excess noise factor in a superlattice avalanching device when both carrier species ionize are given as (see Teich et al. 1986)

$$\langle M \rangle = \frac{(1 + P)^m (1 - k_s)}{[(1 + k_s P)^{m+1} - k_s (1 + P)^{m+1}]}, \tag{12.6.18}$$

$$F_e = 1 + \frac{(1 - \frac{1}{\langle M \rangle})(1 - k_s)}{2 + P(1 + k_s)} \left[-P + 2 \frac{(1 - k_s P^2)}{(1 + k_s P)} \right], \tag{12.6.19}$$

where P is the probability that an electron will ionize in each stage, Q is the probability that a hole will ionize in each stage, k_s is the ratio of Q to P, and m is the number of stages within the device.

The gain at a fixed number of stages is plotted as a function of P with Q as a parameter in Figure 12.6.5. The maximum gain the device delivers is limited by the onset of avalanche breakdown when Q is not zero. Curves 3 and 4 of Figure 12.6.5 show that avalanche breakdown occurs in a ten-stage device with $Q = 0.005$ when $P = 0.4$. As Q decreases, larger values of P are possible at fixed m before avalanche breakdown occurs. In addition, the excess noise factor strongly increases with increasing Q, as shown in Figure 12.6.6. In fact, for the $Q = 0$ case, in addition to avoiding avalanche breakdown, the excess noise factor approaches 1 at large values of P. In principle then, when $Q = 0$, any gain can be achieved at the lowest possible excess noise, depending of course on the number of stages in the device. Therefore it is clear that the optimal device structure, as measured by its noise performance, would be one in which $Q = 0$ and P is as close to 1 as possible.

Before we discuss novel APD device structures, specifically various types of superlattice APDs, let us consider the time response or bandwidth of APDs. The bandwidth is determined in part by how quickly the current pulse, generated by an incident pulse of photons, decays. If an APD exhibits considerable time dispersion, its ability to handle a high-speed data train is severely compromised. Adjacent pulses bleed into one another, causing what is known as intersymbol

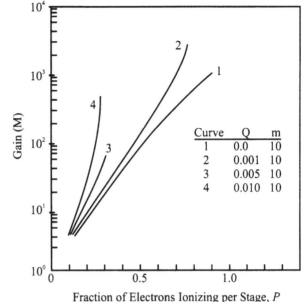

Figure 12.6.5. Mean multiplication for a ten-stage device as a function of the electron ionization probability P.

Curve	Q	m
1	0.0	10
2	0.001	10
3	0.005	10
4	0.010	10

Gain (M)

Fraction of Electrons Ionizing per Stage, P

interference. Consequently, the APD will fail to discern separate pulses, which leads to detection errors.

The time response of an APD is shortest in the absence of secondary hole ionization. Under these conditions, the time response depends only on the combined transit times of the electrons and the holes. However, when the secondary holes (those produced from electron-initiated impact ionization events) impact ionize, the impulse response shows considerable broadening.

The temporal response of the device is defined as the time course of the current delivered by the device to the external circuit in response to a single photogenerated carrier pair. The total current is the superposition of the electron and hole currents generated in the circuit from within the semiconductor, which are instantaneously proportional to the carrier velocities. Each initially

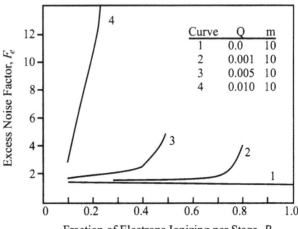

Figure 12.6.6. Excess noise factor for a ten-stage device as a function of electron ionization probability P at various values of hole ionization probability Q.

Curve	Q	m
1	0.0	10
2	0.001	10
3	0.005	10
4	0.010	10

Excess Noise Factor, F_e

Fraction of Electrons Ionizing per Stage, P

injected photogenerated electron traverses the entire length of the device and therefore delivers a complete charge q to the external circuit. Each secondary electron–hole pair created within the device, from an impact ionization event, also delivers a net charge q to the external circuit. Therefore each individual electron created within the depletion region of the APD induces only a portion of the total charge q in the external circuit, such that the total charge arising from the motion of the electron and hole taken together is q. Another way of understanding this is to recall that current flow is defined as the passage of charge through a complete circuit. Within the semiconductor, the total path is partly traversed by the electron and partly by the hole, such that the total distance traveled together is equal to the total device length W.

As mentioned above, in the absence of secondary ionization events, the signal duration depends basically on the width of the depletion region. The slowest response will then occur when an initiating carrier ionizes at the end of the depletion region. The pulse dies out when the subsequent secondary carrier completes its drift back to the edge of the depletion region from which the initiating carrier arose.

The bandwidth can be expressed in closed form when only one carrier species ionizes. If we assume that the device consists of several equal stages or partitions, like the device shown in Figure 12.6.2, then the bandwidth under single carrier injection, single carrier multiplication (SCISCM) conditions is simply

$$B = \frac{1}{2\pi} \frac{1}{m(\tau_e + \tau_h)}, \tag{12.6.20}$$

where m is the number of stages and τ_e and τ_h are the electron and the hole transit times per stage, respectively. Alternatively, in the presence of secondary ionization events, the pulse duration is significantly longer. Secondary ionization events can repeatedly occur throughout the device, leading to very long delays.

At the present time, no closed-form expression for the bandwidth exists under double-carrier ionization conditions. The time pulse is considerably broadened in the presence of secondary hole ionization. Once the hole ionization probability becomes greater than ~1%, the time response greatly increases. Therefore it is obvious that for high-speed operation it is necessary that an APD exhibit SCISCM conditions.

12.7 Superlattice Avalanche Photodiode Structures

Most of the compound semiconducting materials exhibit nearly equal electron and hole ionization rate ratios. For example, in bulk GaAs and InP, the electron and hole ionization rates are roughly comparable over the full range of breakdown fields. Therefore, based on the discussion in Section 12.6, the use of GaAs, InP, or other compound semiconductors in simple *p-i-n* APD structures provides unsatisfactory device performance. An important exception to this is silicon. In bulk silicon the electron ionization rate is significantly higher than the hole ionization rate, by a factor of nearly 20 at some field strengths.

For this reason, silicon APDs are attractive devices. However, silicon is an indirect-bandgap semiconductor, and, as such, it has a much lower absorption coefficient than GaAs or InP. Additionally, silicon cannot be used for most mid-wavelength to long-wavelength infrared detectors owing to its relatively large bandgap of 1.1 eV, which is far wider than what is necessary for most infrared detection applications. In these applications, other materials such as InAs, HgCdTe, and InSb have been applied.

Ideally, one would like to find a material that has a large difference in the impact ionization rates ratio and has a direct bandgap. Such a material would exhibit a high quantum efficiency and could be used in an APD to provide ultralow noise and high-bandwidth detection. Unfortunately, no material has been identified to date that has these characteristics and also spectrally matches the short infrared to optical part of the spectrum.

Alternatively, work has been recently performed to artificially alter materials to change their impact ionization rates ratio and hence their noise and bandwidth performance. Chin et al. (1980) first proposed a means of artificially altering the electron-to-hole ionization coefficient ratio through use of the MQW structure shown in Figure 12.6.2. This structure, which we refer to as the simple MQW device, consists of an intrinsic region formed by alternating layers of narrow- and wide-bandgap materials, such as GaAs and AlGaAs or GaInAs and AlInAs, respectively. As the electrons move through the depletion region, they alternately cross from one layer to the next. On transferring from the wide-gap layer to the narrow-gap layer, the carriers gain a kinetic-energy boost due to the conduction-band-edge discontinuity ΔE_c formed at the heterointerface. Owing to the nonlinear properties of the impact ionization rate and the existence of a threshold energy, the electron ionization rate is locally enhanced over its corresponding bulk rate. The origin of the enhancement can be understood as follows.

Since the ionization process has a threshold, if the carriers are heated above the threshold locally (in the case of the simple MQW APD immediately after crossing over the step on entering the narrow-gap material) ionization events can occur. Even if the overall applied field in an MQW APD is less than that necessary to cause impact ionization events within the constituent bulk material, impact ionization can still occur in the vicinity of the step owing to the additional kinetic energy transferred to the electrons from the potential discontinuity. In this way, the discontinuity acts to supply sufficient energy periodically to cause impact ionization that otherwise would not occur.

The total field acting on the carriers as they move through the MQW structure is the superposition of the overall constant bias field F_0 and the field due to the periodic potential $V(z)$ of the MQW structure. With the results from Chapter 10.5, Eqs. (10.5.13) and (10.5.16), the ionization rate α is given as the ratio of the probability of a carrier traveling time t_0 without scattering to the mean free distance for ballistic impact ionization L_b as

$$\alpha = \frac{P(t_0)}{L_b} = \frac{P(t_0) q F(z)}{E_{th}},$$

(12.7.1)

where the field $F(z)$ is no longer constant but depends on z. $P(t_0)$ is again given by Eq. (10.5.11) as

$$P(t_0) = e^{-\int_0^{t_0} \frac{dt'}{\tau(E)}}. \tag{12.7.2}$$

It is convenient to replace the integral over t to an integral over energy E by using $vt = d$, where the velocity v is simply

$$v = \frac{1}{\hbar} \frac{dE}{dk} \tag{12.7.3}$$

and the distance traveled in the field direction d is

$$d = \frac{dE}{qF(z)}. \tag{12.7.4}$$

Therefore t becomes

$$t = \frac{dE}{qF(z)} \hbar \left(\frac{dE}{dk}\right)^{-1}. \tag{12.7.5}$$

With this substitution, the integral for $P(t_0)$ becomes

$$P(t_0) = e^{-\frac{\hbar}{qF(z)} \int_0^{E_{th}} \left(\frac{dE}{dk}\right)^{-1} \frac{dE}{\tau(E)}}. \tag{12.7.6}$$

The impact ionization rate α_z is then

$$\alpha_z = \frac{qF_0}{\hbar} e^{-\frac{\hbar}{qF(z)} \int_0^{E_{th}} \left(\frac{dE}{dk}\right)^{-1} \frac{dE}{\tau(E)}}, \tag{12.7.7}$$

where we neglect the z dependence of the field in the prefactor term.

We can evaluate expression (12.7.7) for α_z by expanding the periodic potential in terms of a Fourier series. If a square-well potential structure is assumed, as is the case for a simple MQW APD, the potential can be expressed as

$$V(z) = \frac{V_0}{2} + \frac{2V_0}{\pi} \cos \frac{\pi z}{L} - \frac{2V_0}{3\pi} \cos \frac{3\pi z}{L} + \cdots, \tag{12.7.8}$$

where $2L$ is the superlattice period and V_0 is the conduction-band-edge discontinuity.

The average impact ionization rate is found by integrating the expression for α_z with respect to z. We first substitute $F_0 - dV(z)/dz$ for $F(z)$. For one half the superlattice period this yields

$$\alpha_z = \frac{q}{LE_{th}} \int_0^L F_0 e^{-\frac{\hbar}{qF(z)} \int_0^{E_{th}} \left(\frac{dE}{dk}\right)^{-1} \frac{1}{\tau(E)} dE} \, dz. \tag{12.7.9}$$

If only the first two Fourier components in the expansion of $F(z)$ are retained, α_z becomes

$$\alpha_z = \frac{1}{L}\frac{q}{E_{\text{th}}}\int_0^L F_0 e^{-\frac{C}{F_0 + \frac{2V_0}{L}\sin\frac{\pi z}{L}}}\, dz, \tag{12.7.10}$$

where C is defined as

$$C \equiv \frac{\hbar}{q}\int_0^{E_{\text{th}}}\left(\frac{dE}{dk}\right)^{-1}\frac{1}{\tau(E)}\, dE, \tag{12.7.11}$$

which is taken independently of z. We can determine C readily by first fitting α to the bulk experimental ionization rate. The exponential term in the integral can be simplified by a Taylor series expansion. Equation (12.7.10) becomes then

$$\alpha_z = \frac{1}{L}\frac{q}{E_{\text{th}}} F_0 e^{-\frac{C}{F_0}}\int_0^L e^{\frac{2V_0 C}{LF_0^2}\sin\frac{\pi z}{L}}\, dz, \tag{12.7.12}$$

which can be further simplified by a change of variables to

$$\alpha_z \sim \frac{1}{L}\frac{q}{E_{\text{th}}} F_0 e^{-\frac{C}{F_0}}\frac{2}{\pi}e^\beta \int_0^{\frac{\pi}{2}} e^{-\frac{\beta\gamma^2}{2}}\, d\gamma, \tag{12.7.13}$$

which finally can be integrated to

$$\alpha_z \sim \frac{1}{L}\frac{q F_0}{E_{\text{th}}} e^{-\frac{C}{F_0}}\sqrt{\frac{2}{\pi\beta}}e^\beta, \tag{12.7.14}$$

where β is defined as

$$\beta \equiv \frac{2V_0 C}{LF_0^2}. \tag{12.7.15}$$

It is important to note that the impact ionization rate depends exponentially on the conduction-band discontinuity V_0. This result is in accordance with the original model of impact ionization in a MQW structure, as proposed by Chin et al. In this theory, the ionization rate is found to vary exponentially with V_0, which is equivalent to either subtracting or adding V_0 to the bulk exponential factor.

Alternatively, if we evaluate the integral directly in Eq. (12.7.12), the ionization rate can be expressed in closed form as

$$\alpha_z = \frac{q}{E_{\text{th}}} F_0 e^{-\frac{C}{F_0}} I_0(\beta), \tag{12.7.16}$$

where $I_0(\beta)$ is the modified Bessel function of zero order. $I_0(\beta)$ is always greater than 1 for positive argument. Therefore it is clear that the impact ionization rate is always enhanced within a periodic potential structure over the corresponding bulk rate.

For the above reasons, there has been much research performed on superlattice or MQW APDs. From the above analysis it appears that the ionization rates can be enhanced from the bulk rates by use of a periodic potential structure. Additionally, it may be possible to alter the ionization rate ratio and thereby improve the noise and bandwidth performance of the device.

The first device structure proposed that exploits the discontinuity within the conduction band for altering the ionization rates was the simple MQW APD. The first proposed structure comprised alternating layers of GaAs and AlGaAs to form a MQW structure. In these materials the magnitude of the potential band-edge discontinuity depends on the relative aluminum concentration. At an aluminum concentration of 45%, GaAlAs becomes indirect. At this composition the conduction-band-edge discontinuity is currently believed to be $\sim$0.34 eV.

Numerical calculations, based on the direct solution of the Boltzmann equation, have indicated that an enhancement of roughly seven is possible in a GaAs/Al$_{0.45}$Ga$_{0.55}$As MQW APD for certain geometries and electric field strengths (see Brennan and Wang 1988). The electron impact ionization rate calculated with an ensemble Monte Carlo method (which is a stochastic approach in which the trajectory of each carrier is directly simulated) along with the corresponding bulk GaAs ionization rate is presented in Figure 12.7.1. At low electric-field strengths, the net ionization rate in the MQW structure significantly exceeds the bulk rate, as can be seen from the figure. As the electric field increases, the net rate in the MQW structure becomes less than the bulk rate.

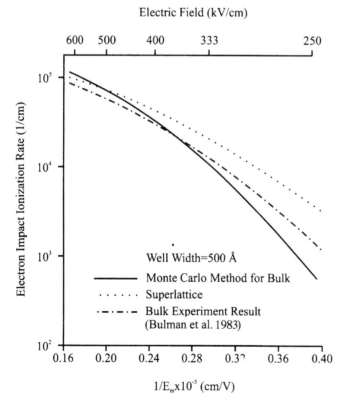

Figure 12.7.1. Calculated electron ionization rate in a simple GaAs/AlGaAs MQW APD.

This is because, at high field strengths, the carriers gain by far the largest fraction of the energy needed to reach threshold from the field rather than from the potential step. The carriers are already heated by the action of the field to energies near threshold. Therefore, at high electric-field strengths, the potential step only weakly contributes to the ionization rate.

At low applied fields, the potential step significantly alters the ionization rate by providing a larger proportion of the energy needed to reach threshold. If the field becomes too low, carrier trapping at the second interface (formed at the end of the well) can occur, leading to a space-charge buildup.

Although theory and experimental measurements indicate that some improvement in the ionization rate and possibly the ionization rate ratio is possible in a simple MQW device, the enhancement is relatively weak in GaAs/AlGaAs and would have little positive effect in an actual device and system. For this reason, alternative device structures have been pursued.

Numerous other approaches have been devised. Chief among these methods is the use of a built-in *p-i-n* or *p-n* minijunction in conjunction with an asymmetric unit cell. Examples of this type of device are shown in Figure 12.7.2. In

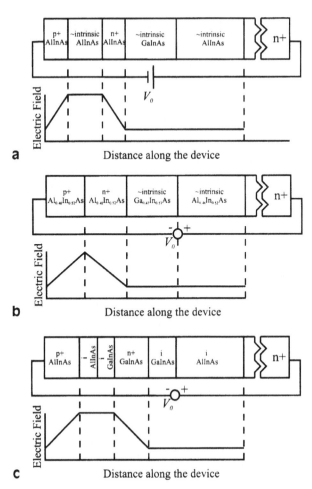

Figure 12.7.2. Sketch of doped barrier/well device schemes and their electric-field profiles.

the figure, the unit cells and corresponding electric-field profiles are sketched for three different types of designs. In each device, the unit cells are repeated within the intrinsic region of the structure. The entire device consists then of a p^+ region followed by several stages of unit cells forming the intrinsic region and then completed by an n^+ region of the diode. It has been predicted that these devices, collectively called the doped barrier and well MQW APDs, can offer roughly 4 orders of magnitude enhancement in the electron-to-hole ionization rate ratio (see Blauvelt, Margalit, and Yariv 1982; Brennan 1986; Brennan 1987a; and Brennan 1987b).

The basic unit cell of the doped-barrier quantum-well device consists of five separate layers made from two different bandgap materials, as shown in Figure 12.7.2a. The materials system chosen in Figure 12.7.2 is the $Ga_{0.47}In_{0.53}As/Al_{0.48}In_{0.52}As$ system. These materials are lattice matched at only these compositions. As the concentrations are altered, the two alloys are no longer lattice matched. When the layers are doped as shown in the figure, a minijunction is produced within the AlInAs layer. If the device is reverse biased so as to deplete the minijunction fully, the electric-field profile sketched in Figure 12.7.2a results. The device is typically operated at sufficient reverse bias to deplete the minijunction, and any residual bias produces an overall constant electric field throughout the structure. It is important to note further that the AlInAs layers have a much greater bandgap than the GaInAs layer. Therefore impact ionization is more likely to occur within the GaInAs layer.

As before, for electron-initiated ionization conditions, the device is illuminated from the p^+ side. As a result, under the reverse bias, the photogenerated electrons are injected into the intrinsic region. Therefore they enter each unit cell from the p^+ side. The electrons are heated by the high field formed by the minijunction in the AlInAs layers and the residual overall bias field before their injection into the narrower-bandgap GaInAs layer. The combined action of the high field and subsequent heterobarrier injection acts to superheat the electron distribution to energies at which impact ionization readily occurs. As a result, the electrons undergo a high rate of impact ionization within the GaInAs layers. Following the impact ionization events, a new electron–hole pair is produced. The secondary electron produced from the event drifts along with the primary toward the n^+ collecting contact. The secondary hole produced, however, drifts backward toward the p^+ contact. As such, the hole moves in the reverse direction within the unit cell to the electron. The hole then enters the unit cell from the intrinsic AlInAs layer. Since the field is relatively low in this layer, the holes are much cooler than the electrons when they are injected into the small-gap GaInAs layer. The hole temperature is then very much lower than the electron temperature within the GaInAs layer. As a result, few, if any, hole-initiated ionization events occur. The holes, of course, also drift through the high-field p-i-n region but instead of entering the GaInAs layer, they enter the much wider-bandgap, low-field AlInAs layer. Since the field is much less in the AlInAs layer, the holes cool considerably before they enter the GaInAs layer. A judicious choice of layer thicknesses and doping concentrations can provide substantial electron ionization at negligible hole ionization.

There are several independent parameters in each of the three designs shown in Figure 12.7.2. These parameters are the overall bias field, the junction doping concentration and width, the high-field intrinsic layer width, the narrow-gap intrinsic layer width, and the low-field, wide-gap layer width. In principle, the electron and hole ionization probabilities per stage, P and Q, respectively, can be expressed as functions of these six parameters. The device could then be optimized from maximizing the P-to-Q ratio, which provides the highest gain at the lowest noise, by means of the use of Lagrange multipliers. However, a simple analytical expression for P and Q is not at present practical since the dependence of these variables on the device parameters, the band structure, and phonon-scattering rates combined is most complicated. Instead, Brennan (1987a, 1987b) presented a numerical means by which these devices can be optimized.

The numerical approach used by Brennan is based on an ensemble Monte Carlo calculation of the electron and hole impact ionization rates in the device structure. The full details of the conduction and valence bands and phonon-scattering mechanisms are included in his simulation. Brennan used the Keldysh formula, discussed in Section 10.5, to determine the impact ionization transition rate. In order to determine the value of the two parameters p and E_{th}, which appear in the Keldysh formula, comparison of the calculated bulk ionization rates must be made to experiment. As a control on the calculations then, Brennan first determined the carrier impact ionization rates within the bulk constituent materials and compared these results with experiment. This provided reliable values of the Keldysh parameters. Brennan's model is thus calibrated to known experimental data for the constituent device materials and can then be used to reliably study the behavior of devices for which the ionization rates are unknown.

Brennan has exhaustively examined how each of the different device parameters influences the electron and hole impact ionization rates. Performance of the devices depends on two standards. First, the electron temperature within the narrow-gap layer must be high enough to affect a very high impact ionization rate. Second, the hole temperature within the narrow-gap layer must be low enough to inhibit impact ionization. The hole temperature within the narrow-gap layer depends, as does the electron temperature, on the doping concentration and high-field layer widths, but more importantly on the width of the low-field, wide-gap layer. The holes must cool within the low-field, wide-gap layer to a temperature low enough that, on reinjection into the narrow-gap layer, they do not suffer an impact ionization event. If the low-field, wide-gap layer is made too narrow, the holes reenter the narrow-gap layer with energy sufficiently large to cause an impact ionization. On the other hand, if this layer is very wide, $> \sim 2000$ Å, then the bandwidth of the device will suffer unnecessarily. Subsequently, there are many trade-offs in the design of the doped-barrier and doped-well devices.

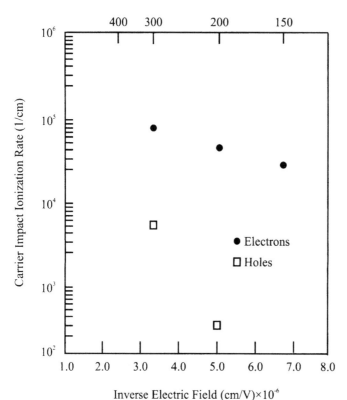

Figure 12.7.3. Calculated electron and hole impact ionization rates in a representative *p-i-n* doped device as a function of inverse applied electric-field strength.

From these calculations it appears that a significant enhancement of the electron ionization rate over the hole ionization rate is possible in all three devices. In fact, it is possible that a device can be made in which the hole ionization rate is completely suppressed while the electron ionization rate can be over 10^4 1/cm, essentially 4 orders of magnitude in difference. Calculations of the electron and hole impact ionization rates α and β for a representative *p-i-n* volume doped device are presented in Figure 12.7.3. Note that the hole ionization rate increases strongly with increasing bias field. However, at an applied field of 150 kV/cm, the hole ionization rate completely vanishes while the electron ionization rate is still very high, $>10^4$ cm^{-1}.

Recently experimental investigations of the performance of the *p-i-n*, *p-n* homojunction APDs have been made (see Aristin et al. 1992). Representative devices were grown by use of molecular-beam epitaxy (MBE) and fabricated. The experimentally measured breakdown voltage for a sample doped-barrier device for two different light source intensities is plotted in Figure 12.7.4. As can be seen from Figure 12.7.4, the breakdown voltage of the device is measured to be between 6.5 and 9 V, more than an order of magnitude lower than that of conventional APD structures. The low-voltage operation is due to the localized breakdown's arising from the fully depleted *p-i-n* regions within each

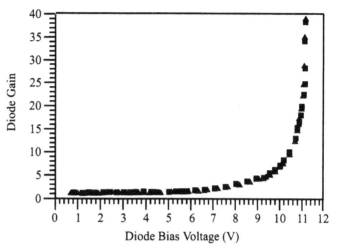

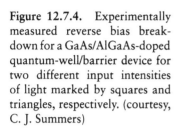

Figure 12.7.4. Experimentally measured reverse bias breakdown for a GaAs/AlGaAs-doped quantum-well/barrier device for two different input intensities of light marked by squares and triangles, respectively. (courtesy, C. J. Summers)

unit cell. **It should be noted that this is the first experimentally measured device that offers ultralow-voltage operation for interband impact ionization in GaAs/AlGaAs.** Other measurements made for a doped-well device again indicate low-voltage breakdown. In the doped-well devices, breakdown occurs from 9 to 14 V, somewhat higher than in the doped-barrier devices because of the choice of a lower doping concentration, but still approximately an order of magnitude lower than in conventional devices. As a consequence of the low-voltage operation, the doped-barrier/well structures are far more efficient devices than conventional APDs and as such are better suited to on-chip, low-power, low-dissipation environments.

The experimentally measured excess noise factor as a function of diode gain for the doped-barrier device examined above is plotted in Figure 12.7.5. The excess noise factor for bulk GaAs, for the case of equal electron and hole impact ionization rates (denoted by the curve marked $k = 1$ in the diagram), is also plotted in Figure 12.7.5 for comparison. The solid diamonds correspond to measurements made on the GaAs/AlGaAs-doped-barrier devices. As can be seen from the figure, the excess noise factor for the doped-barrier device is significantly less than that for $k = 10$ at low gain, <5, and is everywhere far less than that in bulk GaAs. **It is important to recognize that this is the lowest-noise GaAs-based interband avalanching semiconductor APD ever demonstrated to date.** From the noise measurements the electron and hole ionization rates can be evaluated. It is found that the electron-to-hole ionization rate ratio in this device has values between 50 and 12.5 for gains up to 5, and $\sim$5 for gains greater than 5.

A complete set of evaluations has also recently been performed for the doped-well devices (see Aristin et al. 1992). In the GaAs-doped-well devices, an electron-to-hole ionization rate ratio between 10 and 50 has been measured at low gain $M < 5$ and $\sim$5–10 for $M > 5$. Again, a relatively low excess noise factor approaching that of bulk silicon has been obtained at low gain. However, as in the doped-barrier device, the excess noise factor begins to increase significantly for gains greater than 5.

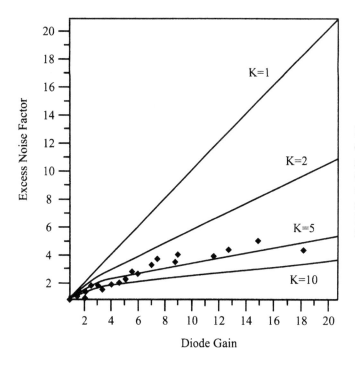

Figure 12.7.5. Plot of the excess noise factor versus diode gain. The diamonds represent experimentally measured excess noise factor data for a representative GaAs/AlGaAs-doped-barrier/well APD. (courtesy, C. J. Summers)

12.8 **Bipolar Phototransistors**

In this section, we discuss the application of bipolar transistors as photodetectors. The reader is referred to the review article by Campbell (1985) as well as the paper by Moriizumi and Takahashi (1972) for a full discussion of these devices. Only a very brief review of the operation of a bipolar transistor is provided here. Readers are referred to other books for a full discussion of the properties of bipolar transistors. A full discussion of the operation and properties of bipolar transistors is impractical in the current book.

Let us briefly review the very basics of bipolar transistor operation. Recall that the current in a reverse biased *p-n* junction is due primarily to the rate at which minority carriers are generated within a diffusion length of the junction. Generally, the reverse current is independent of the magnitude of the electric field and depends on only how often minority carriers are generated, not on how fast a carrier is transported across the junction. It is possible to increase the reverse current by an increase in the carrier generation rate. This is the situation for a photodiode. Recall that light illumination of a reverse biased diode leads to carrier generation. As a result, the reverse current of a diode is greatly increased.

Alternatively, the reverse current of a diode can be increased through electrical carrier injection. If carriers are electrically injected into a reverse biased junction, a similar characteristic as that obtained for optical injection (as shown in Figure 12.8.1) is obtained.

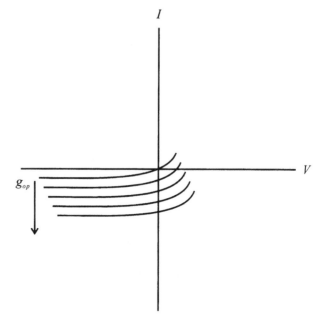

Figure 12.8.1. Current-voltage characteristic showing the effect of increased optical illumination g_{op} on the reverse saturation current of a diode.

How, though, can electrical carrier injection be accomplished? A forward biased *p-n* junction diode electrically injects carriers. Holes are injected from the *p* region into the *n* region and vice versa. Therefore, if a forward biased junction is attached to a reverse biased junction, the forward biased junction will inject carriers into the reverse biased *p-n* junction. Depending on how the two junctions are arranged, either holes or electrons can be injected.

This is the basic idea behind a bipolar junction transistor. The basic scheme is shown in Figure 12.8.2.a. The particular structure shown in the figure is a *p-n-p* bipolar transistor. The first *p* layer is called the emitter, the *n* layer the base, and the second *p* layer the collector. For the *p-n-p* bipolar transistor, holes are injected from the first *p* layer into the *n* region; hence the name emitter. If the *n* region or base is sufficiently thin, these holes survive to reach the depletion region formed between the *n* region and the second *p* region. The action of the reverse bias acts to sweep the holes out into the second *p* region. For this reason, it is called the collector. Subsequently, holes are electrically injected into the reverse biased junction.

A simple biasing scheme for a *p-n-p* transistor is shown in Figure 12.8.2.b. The transistor is biased into what is called the active mode and is said to be in the common-emitter configuration. For most of the holes that are injected into the base to be collected, the hole diffusion length must be much larger than the base width. In other words, the hole lifetime τ_p must be large with respect to the base transit time in order that most of the holes survive to be collected. These collected holes comprise the collector current I_c. The emitter current I_e is due primarily to the injection of the holes from the emitter into the base.

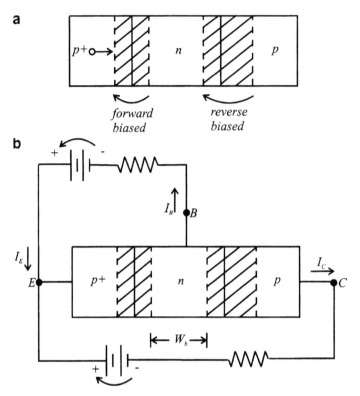

Figure 12.8.2. a. Basic scheme of a *p-n-p* bipolar transistor. b. Active biasing mode for common-emitter configuration.

There is a current in the base lead I_B. The base current arises from the following sources:

1. Recombination of injected holes with electrons stored in the base. The electrons lost through recombination must be restored because of space-charge neutrality requirements. The electrons are restored through the base contact by the current I_B.
2. Electron injection from the forward biased *p-n* junction emitter–base diode. The electrons lost (those injected from the base into the emitter as a consequence of the forward bias) must be supplied also by I_B.
3. Electrons are swept into the base by the reverse biased base–collector junction. These electrons reduce I_B.

In the common-emitter configuration, the output current is I_c while the input current is I_B. Changes in I_B are reflected and amplified in I_c. The bipolar transistor provides gain in this way. As discussed below, a small change in the base current leads to a large change in the collector current. Provided the frequency by which the base current is modulated is sufficiently low enough that the collector current can track it, small changes in the base current are amplified within the collector current. Therefore a signal is input to the base, and the resulting collector current, now substantially amplified, is used as the output current. The current amplification factor in the common-emitter configuration, β, is defined then as

$$\beta = \frac{i_c}{i_B}. \tag{12.8.1}$$

The physical basis of the gain in the common-emitter mode can be understood as follows. Let there be a small change in the base current, one that leads to a slight increase in the electron concentration in the base. Such a change can represent a small-signal variation on the input base current. The addition of extra negative charge in the base results in a deviation from space-charge neutrality. To restore space-charge neutrality, sufficient holes are injected from the emitter to balance the excess negative charge within the base. Since the hole lifetime in the base is relatively long and the transit time is small (because of the small base width) most of the injected holes simply transit the base and enter into the base–collector depletion region from which they are swept out into the collector to produce the collector current. The base remains with a net negative charge. Therefore additional holes are injected to balance again the negative charge within the base. Most of the injected holes also transit the base and become collected. This process continues until the excess electrons injected by the base current are ultimately all recombined. The net gain of the bipolar device is found from the ratio of the carrier lifetime τ to the transit time t as (see Box 12.8.1)

$$\beta = \frac{i_c}{i_B} = \frac{\tau}{t}. \tag{12.8.2}$$

Box 12.8.1   dc Current Gain in a Bipolar Junction Transistor

The dc current gain in a BJT can be determined using the charge control model. The excess minority carrier charge in the base region of the BJT, Q_B, can be determined by integrating the excess minority carrier concentration, $\delta n(x)$ (where we assume electrons are the minority carrier as is the case in a *npn* device), over the base volume. Q_B becomes

$$Q_B = \int_0^{W_B} q\, A\, \delta n(x)\, dx,$$

where A is the area and W_B is the base width. The collector current, I_c, arises from the flow of this injected charge. It is relatively easy to show that the injected current from a *p-n* junction can be expressed in general as

$$\frac{dQ}{dt} = I - \frac{Q}{t}.$$

Assuming steady-state conditions, the collector current can be expressed in terms of a mean base transit time, t, as

$$I_c = \frac{Q_B}{t},$$

continued

Box 12.8.1, cont.

where Q_B is the injected charge. The steady-state base current, I_B, is in turn equal to the amount of recombination current in the base. This is given as

$$I_B = \frac{Q_B}{\tau}.$$

Therefore, the dc current gain, I_c/I_B, becomes

$$\beta = \frac{I_c}{I_B} = \frac{\frac{Q_B}{t}}{\frac{Q_B}{\tau}} = \frac{\tau}{t},$$

which is Eq. 12.8.2.

This is precisely the same ratio as was found for photoconductive gain. It is important to note that physically both processes, photoconductive gain and bipolar transistor common-emitter gain, arise from the same physical process, that is, the basic requirement of space-charge neutrality.

Note that since the collector current is very much larger than the base current, the small fluctuations in the base current lead to fluctuations in a large collector current. The change in the collector current ΔI_c is very much larger than ΔI_B, leading to current amplification. Therefore if a small ac current is added to the dc bias current at the base, a corresponding ac current is produced at the collector. However, the ac collector current is amplified by the factor β.

As in a photoconductor, the gain of a bipolar transistor configured in the common-emitter mode is gain–bandwidth limited. As the carrier lifetime increases, the gain increases. However, the frequency response of the device will decrease since it takes longer for an initial signal to decay. Conversely, if the base transit time increases, the speed of the transistor increases but at the expense of decreased gain.

There is much more that can be said about bipolar junction transistors. Since this book focuses more on optoelectronic applications, we will refrain from discussing the many intricacies of bipolar transistors except for their application as photodetectors. The above is meant as only a short review of the subject before we discuss the usage of bipolar transistors for photodetection.

A common type of phototransistor is the heterojunction phototransistor. In this structure, the emitter comprises a wide-gap material compared with the base and the collector regions. Light is incident upon the device and photogenerates carriers in the base and the collector regions. The photogenerated electrons within the base, base–collector depletion region, and within a diffusion length of the depletion region within the collector accumulate within the base. As discussed above, the addition of extra negative charge within the base, in this case caused by photogeneration rather than electrical injection, results in a departure of the base from space-charge neutrality. Consequently, holes are injected from the emitter into the base to balance the excess negative charge produced by the photogenerated carriers. If the lifetime of the holes injected into the base is longer than the base transit time, most of the holes survive their flight through the base

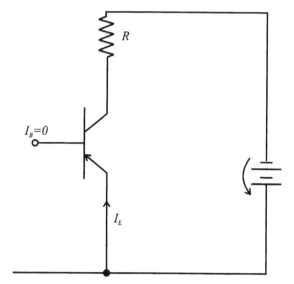

Figure 12.8.3. Biasing configuration of a *p-n-p* phototransistor with an open base.

and are swept out by the reverse bias field in the collector–base junction to form the collector current. This process continues until the excess photogenerated electrons are ultimately all recombined. Note that again the device exhibits gain, a small photogenerated input signal acts to produce a large output collector current. As in the case of electrical injection, if the frequency of the light input signal is sufficiently low that the collector current can track the input, then the phototransistor can amplify a small ac input light signal.

Operation of a bipolar transistor as a photodetector is typically accomplished when the device is configured in the common emitter mode and the base is open circuited, as shown in Figure 12.8.3. In this circuit configuration, the base current is zero. From the current node law, the emitter current is related to the collector current as

$$I_E = -I_c. \tag{12.8.3}$$

The energy-band diagram for the phototransistor is sketched in Figure 12.8.4. For simplicity, we make the following assumptions to calculate more easily the

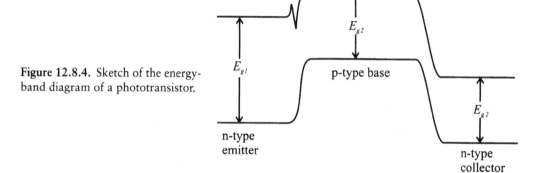

Figure 12.8.4. Sketch of the energy-band diagram of a phototransistor.

gain of the device. These are

1. Generation–recombination of carriers within the collector and the emitter depletion layers is negligible. Also the widths of the depletion regions with respect to the bulk regions are negligible.
2. There is no electric field within the bulk regions of the device.
3. A one-dimensional analysis is satisfactory.
4. The absorption coefficient and the quantum efficiencies are the same within each region of the device.

The generation rate of holes within the base is given by

$$G_1 = \eta \alpha F e^{-\alpha x}, \tag{12.8.4}$$

where η is the quantum efficiency, α is the absorption coefficient, and F is the incident-light flux (simply equal to the number of photons) at the surface of the emitter. The resulting current density due to the diffusion of holes into the emitter junction J_{p1} is determined by the solution of the one-dimensional diffusion equation subject to the boundary conditions (see problem 8). The result is

$$J_{p1} = g_1 F_1 - J_{op1}\left(e^{\frac{V_e}{V_T}} - 1\right), \tag{12.8.5}$$

where V_T is the thermal voltage, V_e is the emitter bias, and g_1 is defined as

$$g_1 = q\eta \frac{\alpha L_1}{1-\alpha^2 L_1^2}\left[\left(\alpha L_1 + \frac{E_2}{E_1}\right)e^{-\alpha W_1} - \frac{1}{E_1}\left(\alpha L_1 + \frac{s_1 L_1}{D_1}\right)\right]. \tag{12.8.6}$$

J_{op1} is defined as

$$J_{op1} = \frac{qD_1}{L_1}\frac{E_2}{E_1}p_1. \tag{12.8.7}$$

In Eqs. (12.8.6) and (12.8.7) the subscripts 1, 2, or 3 represent quantities within the emitter, base, or collector regions, respectively. Additionally, L is the minority carrier diffusion length, s is the recombination velocity, D is the diffusion coefficient, W is the width of the bulk regions of the device, and E_1 and E_2 are defined as

$$E_1 = \cosh\frac{W_1}{L_1} + \frac{s_1 L_1}{D_1}\sinh\frac{W_1}{L_1}, \tag{12.8.8}$$

$$E_2 = \frac{s_1 L_1}{D_1}\cosh\frac{W_1}{L_1} + \sinh\frac{W_1}{L_1}. \tag{12.8.9}$$

The electron current density flowing into the emitter from the base (recall the electron motion is from the emitter into the base, but the current moves in the opposite direction) can be found in a similar manner. It is

$$J_{n2e} = g_2 F_1 f_{n2e} - J_{0n2} \left[\left(e^{\frac{V_e}{V_T}} - 1 \right) \cosh \frac{W_2}{L_2} + 1 \right], \qquad (12.8.10)$$

where g_2 is defined as

$$g_2 = \frac{q}{\sinh \frac{W_2}{L_2}} \frac{\eta \alpha L_2 e^{-\alpha W_1}}{1 - \alpha^2 L_2^2} \qquad (12.8.11)$$

and f_{n2e} is defined as

$$f_{n2e} = \cosh \frac{W_2}{L_2} - \alpha L_2 \sinh \frac{W_2}{L_2} - e^{-\alpha W_2} \qquad (12.8.12)$$

and J_{0n2} is

$$J_{0n2} = \frac{q D_2}{L_2} \frac{1}{\sinh \frac{W_2}{L_2}} n_2. \qquad (12.8.13)$$

The electron current density flowing into the collector from the base is given as

$$J_{n2c} = - \left[J_{0n2} \left(e^{\frac{V_e}{V_T}} - 1 + \cosh \frac{W_2}{L_2} \right) - g_2 F f_{n2c} \right], \qquad (12.8.14)$$

where f_{n2c} is

$$f_{n2c} = 1 - \left[\cosh \frac{W_2}{L_2} + \alpha L_2 \sinh \frac{W_2}{L_2} \right] e^{-\alpha W_2}. \qquad (12.8.15)$$

To complete the description of the device, it remains for us to calculate the hole current density flowing into the collector. We find the solution by solving the diffusion equation for holes subject to the boundary conditions on the excess hole concentration at the collector junction. If it is assumed that the collector region width W_3 is very much larger than the hole diffusion length in the collector region L_3, then the hole current density can be expressed as

$$J_{p3} = -(g_3 F + J_{0p3}), \qquad (12.8.16)$$

where g_3 is defined as

$$g_3 = -q\eta \frac{\alpha L_3}{1 + \alpha L_3} e^{-\alpha W_1 - \alpha W_2}, \qquad (12.8.17)$$

and J_{0p3} is

$$J_{0p3} = -\frac{q D_3}{L_3} p_3. \qquad (12.8.18)$$

The optical gain of the transistor can now be evaluated. Since the base terminal is open circuited, the sum of the current densities flowing into the base region must be zero. Therefore

$$J_{n2e} + J_{p1} - J_{n2c} - J_{p3} = 0. \tag{12.8.19}$$

The optical gain is defined as the ratio of the increase in the total collector current due to the photogeneration of carriers to the photocurrent corresponding to the incidence number of photons. The optical gain can be expressed as

$$G = \frac{1}{q} \frac{J_{T,\text{opt}}}{F}, \tag{12.8.20}$$

where $J_{T,\text{opt}}$ is the optical component of the collector current. We can calculate the change in the total current due to illumination $J_{T,\text{opt}}$ from the total current by examining the components that depend on the illumination. The total current density is

$$
\begin{aligned}
J_T &= J_{n2c} + J_{p3} \\
&= J_{0n2}\left(e^{\frac{V_e}{V_T}} - 1 + \cosh\frac{W_2}{L_2}\right) - g_2 F f_{n2c} + g_3 F + J_{0p3}.
\end{aligned} \tag{12.8.21}
$$

The change in the total current due to the illumination arises from the change in the emitter bias V_e and the illumination terms involving F. We find the change in the emitter bias due to the illumination by substituting Eq. (12.8.13) and Eq. (12.8.18) for each current in Eq. (12.8.21), solving for the exponential term involving V_e and isolating the terms that depend on only F. The dependence of V_e on the illumination can be written as

$$\Delta\left(e^{\frac{V_e}{V_T}}\right) = \frac{F[g_1 + g_2(f_{n2e} - f_{n2c}) + g_3]}{J_{0p1} + J_{0n2}\left[\cosh\frac{W_2}{L_2} - 1\right]}. \tag{12.8.22}$$

The resulting increase in the total current density due to the illumination is equal to $J_{T,\text{opt}}$ which is

$$J_{T,\text{opt}} = J_{0n2}\Delta\left(e^{\frac{V_e}{V_T}}\right) + (g_3 - g_2 f_{n2c})F. \tag{12.8.23}$$

The optical gain can be determined now with Eqs. (12.8.20) and (12.8.23). If it is further assumed that the ratio

$$\frac{J_{0n2}}{J_{0p1}} \equiv \gamma \tag{12.8.24}$$

is very large, then the optical gain can be approximated as

$$G \sim \frac{\beta}{q}\left[g_1 + g_2\left(f_{n2e} - f_{n2c}\cosh\frac{W_2}{L_2}\right) + g_3\cosh\frac{W_2}{L_2}\right], \tag{12.8.25}$$

where β is the dc common-emitter current gain.

Finally, a simple expression for the optical gain can be obtained if some assumptions are made for the device geometry. Specifically, if it is assumed that

1. $W_1 \gg L_1$ or, equivalently, $\alpha L_1 \ll \alpha W_1 \ll 1$,
2. $s_1 = 0$,
3. $\alpha_2 W_2 \gg 1$,
4. $W_2 \ll L_2$,

the gain reduces to

$$G \sim \frac{\beta}{q}[q\eta\alpha L_1 + q\eta] \sim \beta\eta. \tag{12.8.26}$$

If it is not assumed that the quantum efficiencies of each layer are equal, the gain becomes

$$G \sim \beta\eta_2, \tag{12.8.27}$$

where η_2 is the quantum efficiency in the base region.

In conclusion, the optical gain of a phototransistor is equal to the product of the dc current gain and the quantum efficiency of the base region. The particular advantage of a phototransistor is that its gain can be uniform with bias. However, the bandwidth of most phototransistors is quite poor because of the relatively long lifetimes of the minority carriers generated within the base. For this reason, phototransistors have limited usefulness in high-frequency detector applications.

PROBLEMS

1. Determine the magnitude of the reverse bias necessary to cause Zener breakdown in a silicon p-n junction diode if the critical field is 10^6 V/cm. Assume that the junction is abrupt, has equal acceptor and donor doping concentrations on the p and the n sides, respectively, of 4.0×10^{18} cm^{-3}, and that breakdown occurs when the peak field is equal to the critical field. Assume that the diode is at room temperature. $n_i = 1.5 \times 10^{10}$ cm^{-3}, $\epsilon_0 = 8.85 \times 10^{-14}$ F/cm, $\epsilon_s = 11.8$.

2. Consider an n-type doped CdS rectangular bar with dimensions of length $L = 1$ cm, width $w = 0.5$ mm, and thickness $t = 5.0$ μm. Determine the change in resistance of the bar from dark to illuminated conditions if the optical excitation $g_{op} = 10^{21}$ EHP/cm^3 s. $N_d = 10^{14}$ cm^{-3}, $\mu_n = 250$ cm^2/V s, $\mu_p = 15$ cm^2/V-s. Assume that $n_i \ll N_d$, and $\tau_n = \tau_p = 10^{-6}$ s.

3. Consider a p^+-n-p silicon bipolar transistor. Given that the transistor has area $A = 2 \times 10^{-4}$ cm^2, base width $W_b = 1.0$ μm, emitter doping of 10^{18} cm^{-3}, base doping of 10^{16} cm^{-3}, hole lifetime in the base of 1 μs, and mobility of 400 cm^2/V s, determine the value of the base transport factor B, which is defined as the ratio of the collector to injected emitter currents. Assume that the emitter is long compared with the electron diffusion length and that the electron lifetime in the emitter is 0.1 μs. Assume also that Δp_c is negligible.

4. Determine an expression for the stored charge in the base region of a *p-n-p* bipolar transistor as a function of time $Q_b(t)$ if the base current is driven from I_B to zero at $t = 0$.

5. It is desired to use *n*-type silicon as an infrared MIS detector for 1.1-μm radiation. If diffusion is assumed to be the dominant source of the dark current, estimate the charge storage time of the device. Assume that $t_{ox} = 0.1\ \mu\text{m}, \epsilon_{ox} = 3.9$, applied voltage $= 1$ V, $n_i = 1.5 \times 10^{10}\ \text{cm}^{-3}, \mu_p = 450\ \text{cm}^2/\text{V s}, \tau_p = 1.0$ ms, and $n_0 = 1.0 \times 10^{14}\ \text{cm}^{-3}$.

6. Consider a GaAs-based APD that detects incident light of 1.42 eV. Determine the output current of the device if the input radiative power is 1.0 μW and the device has a gain of 10 and a quantum efficiency of 0.80.

7. Determine the gain of a GaAs APD that detects incident light of 0.87 μm if the output current of the device is 10 μA, the input radiative power is 1.0 μW, and the quantum efficiency is 0.80.

8. Derive the expression for the hole emitter current density J_{p1} for a bipolar heterojunction transistor under illumination. Solve the hole diffusion equation given as

$$\frac{d^2 \delta p}{dx^2} - \frac{\delta p}{L_1^2} = \frac{-\eta \alpha F e^{-\alpha x}}{D}$$

subject to the boundary conditions

$$D\frac{d\delta p}{dx}\bigg|_{x=0} = s\delta p, \qquad \delta p(W) = p\big(e^{\frac{V_e}{V_T}} - 1\big) = \Delta p_e,$$

where W is the width of the emitter, p is the hole concentration in the emitter, s is the surface recombination velocity at the surface of the emitter, $x = 0$ is the surface of the emitter, $x = W$ is the emitter/base junction, η is the quantum efficiency, L is the hole diffusion length, F is the input optical intensity, and α is the absorption coefficient.

13

Optoelectronic Emitters

As discussed in Chapter 10, a direct-gap semiconductor is a far more efficient emitter and detector of optical radiation since radiative transitions in these materials proceed to first order. In indirect-gap semiconductors, such as silicon and germanium, radiative transitions cannot proceed to first order but require a second process, phonon absorption or emission, etc., to occur. One of the most important characteristics of the compound semiconductors is that many of them are direct-gap semiconductors. For this reason, many of the compound semiconductors are used for light emitters such as light-emitting diodes (LEDs) and semiconductor diode lasers. In this chapter, we discuss optoelectronic semiconductor devices that emit photons.

13.1 Light-Emitting Diodes

We begin our discussion of optoelectronic emitters. Perhaps the simplest semiconductor light emitter is the LED. LEDs have become pervasive because of their low cost, high efficiency, wide spectral capabilities, relatively simple drive circuitry, high reliability, and very long lifetime. LEDs are most familiar as displays and indicator lamps. However, the relatively low cost, high efficiency, long lifetime, and reliability make them attractive candidates to replace incandescent bulbs in many applications, that is, particularly in those applications in which it is difficult to replace the lamp (like automobile dashboards). Future lighting systems in automobiles, traffic lights, outdoor lighting fixtures, etc., may also utilize LEDs for these reasons.

As discussed in Chapter 10, there are two basic types of radiative transitions, spontaneous and stimulated. Stimulated emission requires the presence of an electromagnetic field to induce a transition. Spontaneous emission, as the name implies, occurs without any external influence. LEDs utilize spontaneous emission processes.

In Chapter 10 it was further noted that there are several recombination mechanisms present within a semiconductor. These mechanisms involve band-to-band or band-to-bound transitions. LEDs typically exploit band-to-band transitions within a direct-gap semiconductor. An LED is simply a forward biased p–n junction. As we discussed in Chapter 11, under forward bias, minority carriers are electrically injected from one side into the other, that is, holes into the n side and electrons into the p side. Following injection, the excess minority carriers diffuse away from the junction, continuously recombining with the majority carriers. In a direct-gap semiconductor, most of the injected minority carriers recombine radiatively with the majority carriers. The radiative recombination events lead to photon emission at an energy near the bandgap. A fraction of the minority carriers recombine nonradiatively, typically through thermal channels. Hence the radiative efficiency of the junction is not 100%, but is typically less.

Radiative transitions can proceed either through band-to-band, band-to-bound, donor-to-acceptor, or excitonic processes. All these processes have been discussed earlier in Chapter 10, except excitonic transitions. Following a photon absorption event, the resulting electron–hole pair can become bound. This bound electron–hole pair is called an exciton. The binding energy of an exciton is relatively weak, since it arises from a screened Coulomb interaction. When the effective-mass approximation is used, the exciton can be described as a hydrogenlike atom, that is, an electron bound to a hole. This problem has been exhaustively examined in Chapter 3. The energy of an electron in a hydrogenlike atom is given as (see Section 3.2)

$$E = -\frac{1}{2n^2}\frac{Zq^4 m}{\hbar^2} \text{ (cgs)}, \qquad = -\frac{1}{2n^2}\frac{Zq^4 m}{[4\pi\epsilon_0\epsilon_s\hbar]^2} \text{ (mks).} \qquad (13.1.1)$$

In the above formula for the electron energy, it is assumed that the mass of the nucleus is infinite. This is a reasonable assumption, since the mass of even the lightest hydrogenlike nucleus, the proton, is several orders of magnitude larger than that of an electron. Therefore it can be readily assumed that the center of mass of the system is at the center of the proton. However, when the masses of the constituent particles are similar, as is the case for an exciton, the mass in Eq. (13.1.1) must be replaced by the reduced mass, given as

$$\mu = \frac{m_e m_h}{m_e + m_h}, \qquad (13.1.2)$$

where m_e and m_h are the electron and the hole effective masses, respectively. The energy of an exciton can thus be determined with the reduced-mass expression and Eq. (13.1.1) as

$$E = -\frac{1}{2n^2}\frac{q^4 \mu}{\hbar^2} \text{ (cgs)}, \qquad = -\frac{1}{2n^2}\frac{q^4 \mu}{[4\pi\epsilon_0\epsilon_s\hbar]^2} \text{ (mks).} \qquad (13.1.3)$$

Excitons produce sharp resonance peaks just below the energy gap E_g. As such, discrete lines appear in the absorption spectrum just below the absorption edge. Therefore the production of excitons leads to below-energy-gap absorption.

Excitons are formed in pure semiconductors at low temperatures. In bulk material, excitons dissociate quite easily. Even a relatively small electric field can ionize an exciton.

Band-to-bound transitions can involve deep levels, as discussed in Chapter 10; however, for radiative transitions, the bound states can also be donor or acceptor states. Donor-to-acceptor state transitions can occur that do not involve any band states. Alternatively, a band-to-bound transition can occur that involves a deep-level state. The deeper in energy an impurity state is, the

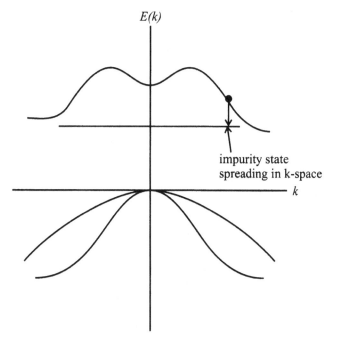

Figure 13.1.1. Sketch of the impurity state spreading in wave-vector space (*k* space), leading to a possible transition.

more localized an electron becomes when trapped into that state. As a result, its momentum, due to the Heisenberg Uncertainty Principle, becomes more uncertain and spread out. Therefore the impurity state spreads out in *k* space. If the impurity is introduced into an indirect-gap semiconductor, as shown in Figure 13.1.1, the spreading of the state in *k* space can enable a transition without the assistance of phonons. Sometimes deep-level impurities are added to indirect-gap semiconductors to enhance their radiative efficiency. In a junction LED, all the above mechanisms can play a significant role in light emission. Since a junction comprises doped materials, impurity-related transitions in particular play an important role.

   The choice of materials for an LED is dictated primarily by the spectral requirements of the intended application as well as the technological maturity of the material, whether it can be doped *p* and *n* types. The most commonly used materials for LEDs are GaAs and GaP and their related ternary compound $GaAs_{1-x}P_x$. GaP is indirect with an ~2.3-eV bandgap. Because GaP is indirect, impurities are typically added to aid radiative recombination. Nitrogen forms a donorlike trap state in GaP close to the conduction band. The nitrogen trap provides a radiative recombination channel by means of excitonic emission. The corresponding light is emitted in the green portion of the spectrum. Other colors can be achieved through use of alternative dopants in GaP, such as zinc and oxygen, or by use of $GaAs_{1-x}P_x$. $GaAs_{1-x}P_x$ changes from direct to indirect with increasing phosphorous composition. When different phosphorous compositions are chosen, direct-bandgap emission can occur. Red emission can be

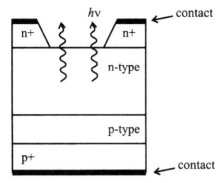

a surface-emitter LED

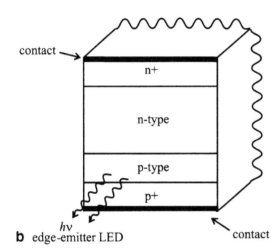

b edge-emitter LED

Figure 13.1.2. Sketches of a. surface-emitting, b. edge-emitting LEDs.

obtained through direct-bandgap emission at 40% phosphorous composition. Other colors, orange and yellow, can be obtained by the addition of nitrogen to indirect compositions of GaAsP. Blue emission can be obtained with InGaN alloys and are currently becoming commercially available.

An LED must be constructed such that the light emitted by the radiative recombination events can escape the structure. LEDs can be designed as either surface or edge emitters, as shown in Figure 13.1.2. Surface-emitting LEDs can be tailored such that the bottom edge reflects radiation back toward the top surface to enhance the output intensity. The main advantage of edge-emitter LEDs is that the emitted radiation is relatively direct. Subsequently, edge-emitting LEDs have a higher efficiency in coupling to an optical fiber.

The key figures of merit that define LEDs are the center wavelength, the spectral width, optical yield (optical output power per electrical input power), luminosity, and forward voltage. The spectral width is small for direct band-to-band radiative processes but much wider for indirect materials. The optical yield depends on the injection, recombination, and extraction efficiencies. The optical yield in many ways is the most important figure of merit for an LED since it describes the basic function of the device, that is, how efficient it is at converting an electrical current into usable light energy. Basically the optical yield is a bookkeeping tool that describes the efficiency with which injected carriers are converted into extractable photons.

Under forward bias, in a highly asymmetrically doped junction virtually only one minority carrier species is injected. As a result, the injection efficiency is large. Therefore a well-designed p–n junction diode can exhibit excellent injection efficiency.

The radiative recombination efficiency is defined as the ratio of emitted photons to injected electrons. The radiative efficiency depends on the relative strength of the radiative recombination channel compared with the other recombination mechanisms present, that is, thermal, Shockley–Read–Hall, surface state, or trap-assisted Auger rate. If the radiative recombination rate is comparable with or larger than the competing recombination mechanisms, as it typically is in a direct-bandgap semiconductor, the radiative recombination efficiency can be quite high, $\sim 50\%$.

Three different loss mechanisms reduce the photon extraction efficiency of an LED. These are 1, absorption within the LED material; 2, Fresnel loss; and 3, critical angle loss. Photons can be lost within the LED material itself, depending on the choice of material and substrate. If the substrate is of the same material as the diode itself, then a substantial number of photons will be reabsorbed, greatly reducing the photon extraction efficiency. Use of a wider-bandgap substrate than the LED material can greatly reduce the amount of reabsorbed photons.

Fresnel loss is due to the difference in the indices of refraction between the LED and air. In transmission line theory, if the impedances of two lines are different there will be a partially reflected wave and a transmitted wave. A similar finding was determined in Chapter 2 for quantum-mechanical particles incident upon a potential barrier; there is a probability of the particle's being reflected or transmitted even if its kinetic energy exceeds the potential barrier height. Fresnel loss arises from the same type of effect. In this case the indices of refraction are different, and hence there is reflected power at the interface. Essentially the two mediums are not impedance matched.

Critical angle loss is due to total internal reflection caused by the change in the index of refraction from the LED material to the air. The index of refraction of the LED material is greater than that of air. As a result, there is a critical angle for total internal reflection. Much of the light will strike the semiconductor–air interface at angles greater than the critical angle for total internal reflection and hence become trapped within the device.

The luminous intensity of an LED follows a Lambertian angular distribution:

$$I(\theta) = I_0 \cos \theta, \tag{13.1.4}$$

where I_0 is the intensity normal to the active layer and θ is the angle with respect to the normal direction. Clearly the luminous intensity decreases with increasing off-axis viewing angle. Instead of simply using an air–semiconductor interface,

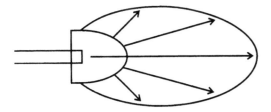

a Narrow radiation pattern design

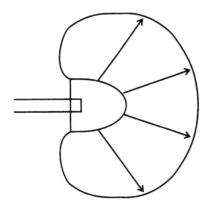

Figure 13.1.3. Sketches of a. narrow, b. wide radiation pattern LED designs.

b Wide radiation pattern design

the LED can be packaged in a plastic dome with a lower index of refraction. The viewing angle depends on the manner in which the LED is packaged. A LED positioned near the top of the plastic dome package produces a wider viewing angle than one positioned near the bottom of the package, as shown in Figure 13.1.3. The wide viewing angle design has a lower luminous on-axis intensity than the narrow viewing angle design at comparable bias.

13.2 Semiconductor Laser Physics

High-efficiency semiconductor lasers are of importance in communications, surgery, and optical recording. Semiconductor lasers are of critical importance to optical communication systems because they feature both small size and high-speed direct modulation. Recent developments of new semiconductor lasers make long-distance optical communications possible since these lasers operate at long wavelengths where optical fiber transmission losses are low. Semiconductor lasers are also of immense importance in video disk systems, laser printers, digital–audio disk systems, and eye surgery. Much of this growth is due to the development of reliable, efficient GaAs/AlGaAs laser devices.

In Chapter 10, we discussed both optical absorption and emission in semiconductors. In Section 10.8, spontaneous and stimulated emission were discussed in detail and the Einstein relations were developed. As was mentioned at the end of Section 10.8, the spontaneous emission rate is much less than the stimulated emission rate. The spontaneous emission rate is generally negligible compared with the stimulated emission rate and is typically neglected when the net rate is calculated with both processes present. A laser is a device in which the light emitted arises from stimulated emission. In fact, the word laser is an acronym standing for light amplification by the stimulated emission of radiation. The key then to understanding how a semiconductor laser works is first to determine the conditions under which a semiconductor will emit stimulated emission, and then to determine the resonance condition for amplification.

A laser is essentially a light amplifier. As is well known from elementary circuit theory, an oscillator requires gain and positive feedback. In a light amplifier, gain is achieved if the stimulated emission rate exceeds the absorption rate. Clearly, if the absorption rate is greater than the emission rate, there will be no net increase in the emitted power, but a loss of radiative power instead. On the other hand, if the stimulated emission rate exceeds the absorption rate, then the system will have a net increase in the emitted power. So a laser has to be designed such that the stimulated emission rate exceeds the absorption rate.

To achieve positive gain, the net emission rate must exceed the net absorption rate. Of course, the system is in nonequilibrium since there is a net difference in the absorption and emission rates. Neglecting the spontaneous emission rate and by using Eqs. (10.8.3) and (10.8.5), we find that the condition that the stimulated emission rate must exceed the absorption rate is given as

$$B_{21} f_2 (1 - f_1) P(E_{21}) > B_{12} f_1 (1 - f_2) P(E_{21}), \tag{13.2.1}$$

where $B_{21} = B_{12}$, as shown in Section 10.8, and f_1 and f_2 are nonequilibrium distribution functions defining the occupancy of the conduction and the valence bands, respectively. When the equality of the Einstein coefficients and the optical powers is used, inequality (13.2.1) simplifies to

$$f_2 (1 - f_1) > f_1 (1 - f_2), \tag{13.2.2}$$

or, equivalently,

$$f_2 > f_1. \tag{13.2.3}$$

Substituting the Fermi-like distributions with quasi-Fermi levels defined in Box 10.8.1 for the nonequilibrium distribution functions yields

$$\frac{1}{e^{\frac{E_2 - q\phi_2}{kT}} + 1} > \frac{1}{e^{\frac{E_1 - q\phi_1}{kT}} + 1}, \tag{13.2.4}$$

which further simplifies to

$$e^{\frac{q(\phi_2 - \phi_1)}{kT}} > e^{\frac{E_2 - E_1}{kT}} \tag{13.2.5}$$

or

$$q(\phi_2 - \phi_1) > (E_2 - E_1). \tag{13.2.6}$$

Inequality (13.2.6) gives the condition for net stimulated emission to occur, that is, the separation of the quasi-Fermi levels must exceed the photon emission energy. In other words, the downward stimulated emission rate will exceed the upward absorption rate, leading to net stimulated emission of radiation if the quasi-Fermi levels are separated in energy by more than the emitted photon energy. Therefore a semiconductor device must be devised such that the separation of the quasi-Fermi levels exceeds the energy gap in order to achieve a net stimulated emission rate and thus exhibit positive gain.

In a gas laser, the net stimulated emission rate will exceed the net absorption rate if the population of electrons in a higher-energy state exceeds that in a lower-energy state. This is called population inversion and is often cited as one of the primary conditions for lasing action in a gas laser. The condition that the difference in the quasi-Fermi levels exceed the energy gap in a semiconductor laser is essentially the same population inversion condition as in a gas laser. In any event, lasing can be achieved only if the net stimulated emission rate exceeds the absorption rate. Applying this requirement to a gas system leads to population inversion between two different levels while in a semiconductor it leads to the population inversion condition as specified by the difference in the quasi-Fermi levels.

Let us examine physically the meaning of the condition on the quasi-Fermi levels. Figure 13.2.1 shows a sketch of the conduction and the valence bands in a semiconductor and the electron and the hole populations when $F_n - F_p > E_g$ where F_n and F_p are $q\phi_2$ and $q\phi_1$ respectively. As can be seen from the figure, the

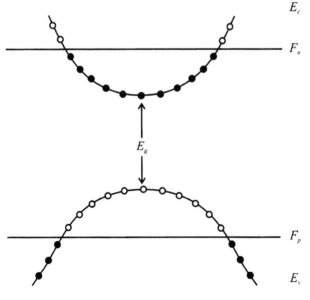

Figure 13.2.1. Sketch of the conduction and the valence bands of a semiconductor and the electron (solid circles) and hole (open circles) populations when $F_n - F_p > E_g$.

electron population is large in the conduction band while the hole population is large in the valence band. This is essentially inverted from the usual condition, wherein the populations of electrons in the conduction band and holes in the valence band are small. To achieve the condition pictured in Figure 13.2.1, the system can be optically pumped by a photon source of energy $h\upsilon > F_n - F_p$. Since the relaxation processes are ~3 orders of magnitude faster than the recombination rates, then under high illumination a significant population inversion can be obtained.

Alternatively, population inversion can be obtained through the forward biasing of a degenerately doped p–n junction diode. A system is said to be degenerately doped if it is highly doped with impurities such that the Fermi level lies within 3 kT of the conduction or the valence bands, as shown in Figure 13.2.2.a. Essentially, a material is said to be nondegenerate if the Boltzmann approximation can be made for the equilibrium distribution function. As we discussed in Chapter 5, if $E - E_F \gg kT$, then the Fermi–Dirac distribution

Figure 13.2.2. a. Degeneracy condition for a semiconductor. Note that the Fermi level lies close to or above the conduction-band edge for n-type material. For p-type material the Fermi level lies close to or below the valence-band edge. b. Forward bias degenerately doped p–n junction diode.

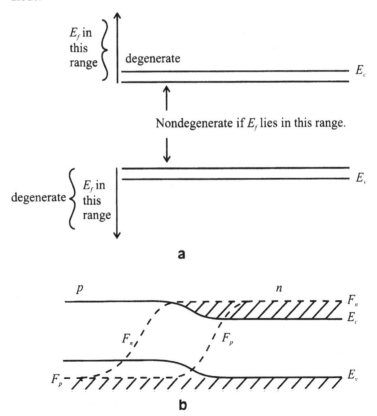

can be simplified to the Maxwell–Boltzmann distribution. As a result, the equilibrium concentration of electrons in the conduction band or holes in the valence band can be calculated with the forms derived in Section 5.8. If $E - E_f$ is comparable with or less than kT, then the Boltzmann approximation fails and the system is degenerate. Under these conditions, the full Fermi–Dirac distribution must be used to calculate the equilibrium carrier concentrations.

A degenerately doped p–n junction diode under forward bias is sketched in Figure 13.2.2.b. As can be seen from the figure, under forward bias within the transition region of the degenerate p–n junction diode, the electron and the hole concentrations are substantial. The populations can become inverted, provided the bias is sufficiently high. As a result, the condition for net stimulated emission can be met and positive gain can be achieved in the device.

An expression for the gain coefficient, or equivalently, the absorption coefficient, can be obtained with the results derived in Section 10.7. In Section 10.7 the transition rate for optical absorption was derived. The total absorption transition rate per unit volume for a direct transition in a semiconductor is obtained by use of Eq. (10.7.33). The transition rate is summed over all possible values of k, assuming a monochromatic source. The total optical transition rate for a band-to-band absorption event is given as

$$P = \frac{2V}{(2\pi)^3} \int |W_{\text{abs}}| f_v (1 - f_c) \, d^3k, \tag{13.2.7}$$

where W_{abs} is given by Eq. (10.7.33) and a factor of 2 accounts for a possible change of spin during the transition. The factors f_v and $(1 - f_c)$ are the valence- and the conduction-band occupancy probabilities. Clearly, for an absorption event to occur, there must be a vacancy in the conduction band and an electron within the valence band. The factor $1 - f_c$ is the probability that a vacancy exists within the conduction band and f_v is the probability that an electron exists within the valence band.

The absorption coefficient depends on the net absorption rate. The net absorption rate is equal to the difference between the absorption rate and the emission rate. The stimulated emission rate is essentially given by Eq. (13.2.7), except that the distribution functions are reversed. Therefore the stimulated emission rate is given as

$$P = \frac{2V}{(2\pi)^3} \int |W_{\text{abs}}| f_c (1 - f_v) \, d^3k. \tag{13.2.8}$$

The net absorption rate $r_{12}(\text{net})$ is given then as the difference between the absorption and the stimulated emission rates as

$$r_{12}(\text{net}) = \frac{2V}{(2\pi)^3} \int |W_{\text{abs}}| (f_v - f_c) \, d^3k. \tag{13.2.9}$$

The net absorption rate can also be expressed as the product of the absorption coefficient and the incident photon flux $\Phi(E)$. The photon flux is simply equal to the product of the incidence density of photons of energy E, $P(E)$, and the group velocity v_g. Therefore the net absorption rate can be written as

$$r_{12}(\text{abs}) = \alpha(E_{21})\Phi(E), \tag{13.2.10}$$

where $\alpha(E_{21})$ is the absorption coefficient. Using Eq. (13.2.10) for the net absorption rate, we can write the absorption coefficient as

$$\alpha(E_{12}) = \frac{\frac{2V}{(2\pi)^3} \int |W_{\text{abs}}|(f_v - f_c)\,d^3k}{\Phi(E)}. \tag{13.2.11}$$

In most situations, the difference between f_v and f_c is positive since the probability that a valence-band state is occupied is much greater than the probability that a conduction-band state is occupied. However, if the system is inverted, $f_c - f_v > 0$. Under population inversion the absorption coefficient becomes negative since $f_v - f_c < 0$. In this case, the system is said to have gain, and the absorption coefficient is called the gain coefficient. Note that the gain coefficient is negative.

In summary, a positive value of the absorption coefficient $\alpha(E_{21})$ represents net absorption while a negative value of $\alpha(E_{21})$ means that the stimulated emission rate exceeds the absorption rate and the system has gain. For a system to lase, it must have a negative value for the absorption coefficient; there must be a net output of radiation arising from stimulated emission. As we derived in Eq. (13.2.6), the net stimulated emission rate exceeds the absorption rate when the difference in the quasi-Fermi levels is greater than the photon energy. Therefore, if $\phi_2 - \phi_1 > h\upsilon = (E_2 - E_1)$, then the system has a positive gain coefficient.

Lasing occurs in a system if gain exceeds the total losses and the optical radiation is resonantly amplified. Losses in a laser arise from such sources as free-carrier absorption, scattering, mirror imperfections, and absorption. To determine whether a system will lase, it is necessary to calculate the gain coefficient and compare it with the total loss rate in the system. Additionally, the system has to be designed such that resonant amplification of the radiation can be achieved. Let us first consider what is required for resonant amplification.

In addition to exhibiting gain, a device must be operated in resonance in order to lase. The system can be driven into resonance by reflecting surfaces at the ends of the gain medium in order to form what is called a Fabry–Perot resonator. The light is successively reflected at each mirror, thus providing feedback of the radiation into the gain medium. The oscillation threshold is found from the requirement that the round-trip gain must exceed unity. In other words, oscillation occurs when the gain coefficient exceeds the loss over one round trip. The system will then lase on a narrow band of frequencies. The question is then, what is the oscillation condition?

The oscillation condition can be derived with Figure 13.2.3, following the approach of Yariv (1976). A laser can be envisioned as a cavity bounded on either side by two mirrors. Light is assumed incident from the left-hand side

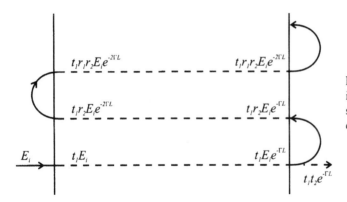

Figure 13.2.3. Schematic drawing of a Fabry–Perot resonantor showing the fields throughout the cavity.

with amplitude E_i, as shown in Figure 13.2.3. Each mirror surface is assumed to be partially reflecting with reflectivities r_1 and r_2, respectively. Within the cavity, the electromagnetic fields have a complex propagation constant Γ. The propagation constant is complex in general. The imaginary component of the propagation constant results in either loss or gain. The ratio of transmitted to incident fields at the left and the right sides of the cavity are defined as t_1 and t_2, respectively. The power reflectivities are defined as the modulus squared of the reflectivities r_1 and r_2:

$$R_1 = |r_1|^2, \qquad R_2 = |r_2|^2. \tag{13.2.12}$$

We can determine the resonance or oscillation condition by relating the transmitted field to the incident field. The field immediately to the right of the left boundary is simply

$$E = t_1 E_i, \tag{13.2.13}$$

while the field just inside the right boundary is

$$E = t_1 E_i e^{-\Gamma L}, \tag{13.2.14}$$

where L is the cavity length. The exponential term $e^{-\Gamma L}$ in Eq. (13.2.14) accounts for the cavity loss since Γ is a complex number; it leads to both propagating and evanescent components. The first portion of the field transmitted at the right-hand side is

$$t_1 t_2 E_i e^{-\Gamma L}, \tag{13.2.15}$$

while that reflected at the right is

$$t_1 r_2 E_i e^{-\Gamma L}. \tag{13.2.16}$$

In a like fashion, the field reflected back to the left boundary is given as

$$t_1 r_2 E_i e^{-2\Gamma L}. \tag{13.2.17}$$

The reflected part of this wave is then (see Figure 13.2.3)

$$t_1 r_1 r_2 E_i e^{-2\Gamma L}. \tag{13.2.18}$$

We can then find the net transmitted field to the right by summing up each of the components as

$$E_t = t_1 t_2 E_i e^{-\Gamma L} \left[1 + r_1 r_2 e^{-2\Gamma L} + r_1^2 r_2^2 e^{-4\Gamma L} + \cdots \right]. \tag{13.2.19}$$

The series can be readily summed to

$$E_t = \frac{E_i t_1 t_2 e^{-\Gamma L}}{1 - r_1 r_2 e^{-2\Gamma L}}. \tag{13.2.20}$$

When the denominator vanishes in expression (13.2.20), the system is in resonance. Hence the oscillation or resonance condition is simply

$$1 = r_1 r_2 e^{-2\Gamma L}. \tag{13.2.21}$$

Generally, the propagation constant Γ is a complex number. For convenience, we write Γ as

$$\Gamma = j\beta k_0 - \alpha. \tag{13.2.22}$$

With this substitution, the resonance condition becomes

$$1 = r_1 r_2 e^{-2(j\beta k_0 - \alpha)L} = r_1 r_2 e^{-2j\beta k_0 L} e^{2\alpha L}. \tag{13.2.23}$$

The quantity α can be expressed as the difference between the gain g and all the loss terms α_i:

$$\alpha = g - \alpha_i \tag{13.2.24}$$

Below we qualitatively describe the different loss mechanisms that comprise α_i. We can rewrite the resonance condition by making use of Eq. (13.2.24) as

$$1 = r_1 r_2 e^{-2j\beta k_0 L} e^{2(g - \alpha_i)L}. \tag{13.2.25}$$

Note that the phase term $e^{-2j\beta k_0 L}$ does not change the amplitude of the field since it has unity amplitude. Therefore the amplitude of the field is governed by the product of the reflectivities and the exponential term $e^{2(g - \alpha_i)L}$, which must equal unity. Therefore the amplitude requirement is given as

$$1 = r_1 r_2 e^{2(g - \alpha_i)L}. \tag{13.2.26}$$

Solving for the gain g yields

$$g = \alpha_i + \frac{1}{2L} \ln\left(\frac{1}{r_1 r_2}\right). \tag{13.2.27}$$

The last term in Eq. (13.2.27) represents the mirror reflectivity loss. For the system to achieve resonance, the gain g must equal the sum of all the losses α_i and the mirror reflectivity loss. If the gain is less than the sum of these losses, the amplitude requirement is not met and the amplitude will decay with time. Under these conditions, the system will not lase.

As seen from Eq. (13.2.27), if the loss terms are large, a large gain must be achieved in order to reach resonance. The larger the loss terms, the larger the minimum gain required for making the device lase. The gain in a semiconductor laser depends in turn on the current density injected into the junction diode. There is then a minimum current density necessary for a semiconductor device to lase. The threshold current density is defined as the current density necessary to make the device gain equal to the sum of the device internal and mirror reflective losses and emitted radiation. Laser efficiency increases as the threshold current density decreases since stable laser operation then occurs at a progressively lower input current. The threshold current density depends on both the semiconductor material parameters and device geometry. Radiation and carriers not confined to the active region of the laser can increase the threshold current density and thus reduce the device efficiency since they do not directly add to the output-beam power. The threshold current density can be found from the sum of the current densities corresponding to the loss terms and to the output signal produced.

In general, there are significant losses in a semiconductor laser. These loss mechanisms include both internal losses, denoted above as α_i and the mirror reflectivity loss. The internal losses consist of contributions from several different loss mechanisms. Among these are free-carrier absorption in both the active and the cladding layers, scattering loss, and coupling loss. Free-carrier absorption, denoted as α_k, is a major unavoidable loss mechanism in both the active and the cladding layers. The active layer is defined as the layer in which the population inversion exists, that is, the gain medium. The cladding layers are the surrounding semiconductor regions. Within the cladding layers no gain occurs. The free-carrier absorption in the active region can be simply expressed as the product of the optical confinement factor γ, which is defined as the fraction of the field confined within the active region, and the loss term α_k. The free-carrier loss within the cladding layers is equal then to the product of the fraction of the field within these layers $(1 - \gamma)$ and the cladding layer loss term $\alpha_{k,x}$. The free-carrier absorption loss arises physically from free carriers' absorbing photons out of the radiation field, thereby lowering the output-beam intensity. Scattering loss α_s is due to the scattering of radiation out of the optical waveguide by either nonplanar heterojunction interfaces or imperfections in the dielectric layers. Finally, coupling losses α_c occur when the optical field spreads beyond the wide-bandgap confining layers. All these losses can be combined into one total internal loss term α_i, as

$$\alpha_i = \gamma \alpha_k + (1 - \gamma)\alpha_{k,x} + \alpha_s + \alpha_c. \tag{13.2.28}$$

Equation (13.2.28) for the internal loss terms can be substituted into Eq. (13.2.27) to give a general expression for the gain. The minimum gain necessary for lasing is equal then to the sum of all the collective loss terms and the mirror reflection loss term as

$$g = \gamma \alpha_k + (1 - \gamma)\alpha_{k,x} + \alpha_s + \alpha_c + \frac{1}{2L} \ln\left(\frac{1}{r_1 r_2}\right). \qquad (13.2.29)$$

We can write the mirror loss terms in terms of the power reflectivities R_1 and R_2 by using Eqs. (13.2.12). If $R_1 = R_2$, then the gain can be rewritten as

$$g = \gamma \alpha_k + (1 - \gamma)\alpha_{k,x} + \alpha_s + \alpha_c + \frac{1}{L} \ln\left(\frac{1}{R}\right). \qquad (13.2.30)$$

The total loss in a semiconductor laser is then found from the sum of the internal loss terms and by the mirror loss term.

Typically, the threshold current density is determined empirically. As mentioned above, it depends greatly on the losses present in the device. To reduce the threshold current density J_{th}, it is necessary to improve both the electrical and the optical characteristics of the device structure. In other words, it is important to confine the optical modes within the gain region and reduce the amount by which the optical modes leak into the cladding layers and substrate. Therefore it is desirable that the optical confinement of the modes should be as close to one as possible. Careful material growth acts to reduce the scattering losses α_s by producing fine interfaces. Therefore the internal and the mirror loss terms can be reduced through improvement of the optical confinement and facet reflectivity, as well as through careful device fabrication. The threshold current density can also be effectively reduced by an improvement in carrier confinement, thus increasing the efficiency in generating stimulated emission events.

We can describe the threshold current density quantitatively by summing the current densities corresponding to the loss terms and to the output signal produced. In steady state the net input current must be balanced by the net recombination rate. The continuity equation gives then

$$\frac{dn}{dt} = \frac{J}{qd} - \frac{n}{\tau} = 0, \qquad (13.2.31)$$

where τ is the recombination lifetime, d is the width of the active region of the laser, and n is the electron concentration within the active region. Therefore, at threshold, the current density is given as

$$J_{th} = \frac{qdn_{th}}{\tau}. \qquad (13.2.32)$$

The very minimum threshold current density would occur if there were no losses in the device. Assuming no losses in the structure, the minimum current density occurs when the stimulated emission rate balances the absorption rate. Under such conditions, there is no gain in the medium. The corresponding charge concentration is typically referred to as the nominal concentration n_{nom}. The threshold current density with no losses is then

$$J_{\text{th}}^0 = \frac{qdn_{\text{nom}}}{\tau}. \tag{13.2.33}$$

If the losses are included, then the minimum threshold current density must be greater than the sum of J_{th}^0 and the current density that corresponds to the loss terms. The resulting expression is

$$J_{\text{th}} = J_{\text{th}}^0 + \frac{d}{\eta\beta\Gamma}\left[\alpha_i + \frac{1}{L}\ln\left(\frac{1}{R}\right)\right], \tag{13.2.34}$$

where η is the quantum efficiency, β is defined as the gain constant, and Γ is the optical confinement factor.

In Section 13.3 we will discuss various device geometries that reduce the threshold current density. Analysis of Eq. (13.2.34) indicates that as the optical confinement factor increases, the threshold current density decreases. Also, if the loss terms, both internal and mirror reflective, are reduced, the threshold current density is further reduced. Several semiconductor laser designs have been advanced to reduce the threshold current density. Of chief importance to this end are the quantum-well lasers, single-quantum-well lasers, multiquantum-well lasers, and graded-index confinement lasers.

13.3 Semiconductor Laser Structures

Simple homojunction semiconductor lasers have very large threshold current densities since both carrier and optical confinements within the active region are minimal. Carriers injected by the action of the forward bias into the active region readily diffuse outward and thus do not contribute to the lasing process. In addition, the optical fields leak considerably into the surrounding inactive semiconductor layers bordering the gain medium. Both of these effects conspire to greatly increase the threshold current density for lasing.

Heterostructures were first suggested by Kroemer as a means of providing both carrier and optical confinements within the active region with a subsequent decrease in the threshold current density. Double-heterostructure systems, consisting of a narrow-bandgap material bordered on either side by a larger bandgap material, are the most successful of these laser types. The presence of the heterostructure potential barriers on either side of the narrow-gap layer, which forms the active region of the device, as shown in Figure 13.3.1, provides both optical and electrical confinements. The enhanced optical and electrical confinements provided by the heterostructure substantially reduce the threshold current density of the laser.

Optical confinement is achieved from the waveguiding nature of the double-heterostructure geometry. Because of the difference in energy bandgaps between

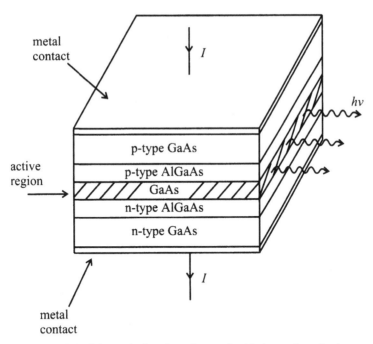

Figure 13.3.1. Schematic drawing of a *p–n* double-heterojunction laser.

the active and the cladding layers, the dielectric constants and hence indices of refraction are significantly different. This produces significant reflection at the heterointerfaces, leading to zig-zagging of the wave front through the active region. In this way, the optical confinement is greatly enhanced.

The band-edge discontinuities at each edge of the active region serve to confine the injected electrons and holes. The greater confinement leads to a greater percentage of carriers that recombine by means of stimulated emission, adding to the optical output power. Fewer carriers need be injected in a heterostructure device in order to achieve the same output power; therefore the required threshold current density is less.

Further improvement in semiconductor laser performance has been achieved through use of single-quantum-well (SQW) and multiple-quantum-well (MQW) structures. Specifically, these devices, in particular MQW structures, exhibit very low threshold current densities, providing for extended continuous-wave operation, excellent linearity in light–current characteristics, stable narrow-beam divergence, and decreased temperature dependence of the threshold current.

The energy-band diagram of a MQW laser is sketched in Figure 13.3.2. The quantum wells are formed by the sandwiching of a narrow-bandgap material, GaAs in the case shown by Figure 13.3.2, between two wide-bandgap layers, $Al_xGa_{1-x}As$ for example. The stimulated emission arises from electronic transitions between the confined quantum states within the wells for the electrons and the holes. As discussed in Chapter 2, the confined-quantum-state energies depend on the well width and depth and the electron and hole effective masses. If the

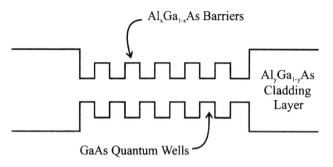

Figure 13.3.2. Sketch of the conduction and the valence bands for a MQW laser.

square wells formed are approximated as infinite, the confined-quantum-state energies for the electrons and the holes are given by the closed-form expressions derived in Chapter 2 as

$$E_n = \frac{n^2 \pi^2 \hbar^2}{2m^* L^2},$$ (13.3.1)

where n is an integer specifying the quantum state, m^* is the carrier effective mass, and L is the width of the well. Generally, the eigenenergies of the electrons and the holes in the system have an additional component due to the motion in the nonquantized directions (assumed here as x and y). The total eigenenergies of the carriers are of the form

$$E(n, k_x, k_y) = E_n + \frac{\hbar^2 (k_x^2 + k_y^2)}{2m^*},$$ (13.3.2)

where E_n is approximated by Eq. (13.3.1). Neglecting any change in the kinetic energy in the nonquantized directions, the energy of a photon emitted following an electronic transition between two confined quantum states, one in the conduction band and the other in the valence band, is easily seen from Figure 13.3.3 to be

$$h\upsilon = E_g + \frac{n^2 \pi^2 \hbar^2}{2m_c^* L^2} + \frac{n^2 \pi^2 \hbar^2}{2m_\upsilon^* L^2}.$$ (13.3.3)

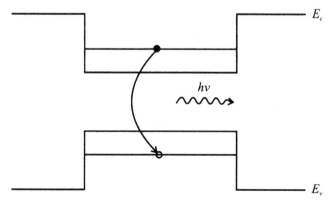

Figure 13.3.3. Sketch of the energy-band structure of a quantum-well laser showing a confined-quantum-state transition.

The result given by Eq. (13.3.3) is simply the sum of the energy-bandgap of the narrow-bandgap material and the confined-quantum-state energies of the electron and the hole, respectively (see Example 2.3.2 in Chapter 2). It is important to note that the emitted photon in the quantum-well device has an energy greater than the bandgap.

A GaAs homojunction laser lases at $\sim$0.90 μm, which is below the visible in the near-infrared portion of the spectrum. For optical mass-memory storage, laser printing, and video disk playback applications, lasers operating in the visible portion of the spectrum are preferred. Subsequently, either a wider-bandgap material can be used for the lasing medium or a quantum-well device can be devised in which the confined states match the desired lasing frequency.

Another important application of semiconductor lasers is for optical sources operating in the 1.3–1.65-μm range for optical fiber communications systems. The primary materials utilized in these applications are $Ga_xIn_{1-x}As_yP_{1-y}/InP$ and $Ga_xIn_{1-x}As/InP$, quaternary and ternary compounds epitaxially grown onto InP. In these lasers, GaInAsP or GaInAs is typically used as the active layer while InP forms the cladding layers in a double-heterostructure device.

It is interesting to compare the operation of double heterostructure lasers, in which no spatial quantization effects occur, with quantum-well lasers. In a quantum-well device, the density of states is quite different from that in a double-heterostructure system. As discussed in Chapter 5, the density of states of a three-dimensional system, which is the case for a double-heterostructure laser, increases as $\sqrt{E}$. The density of states for a two-dimensional system is independent of E (see Problem 4 in Chapter 8) and is constant for each quantum state. As a result, the density of states has a steplike structure, as shown by Figure 13.3.4. For lasing to occur, the gain must exceed the total losses

Figure 13.3.4. Density of states as a function of energy for three-dimensional (dashed curve) and two-dimensional (solid curve) systems.

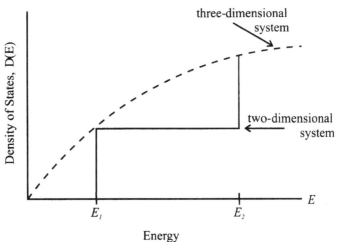

in the device. Generally, the losses are approximately the same in comparable double-heterostructure and quantum-well devices. Therefore the primary factor influencing the threshold current density is the difference in the density of states of the two devices. It is easier to saturate a two-dimensional system by band filling since, for any energy above the band minimum, there are fewer states than in a three-dimensional system. The lower density of states of the two-dimensional system in the quantum-well laser requires fewer carriers for obtaining population inversion. As a consequence, the threshold current density should be less in a quantum-well laser than in a comparable double-heterostructure laser.

Although a quantum-well laser typically has a lower threshold current density than a heterostructure laser, this is not always the case. In the case of a SQW laser, the threshold current density increases when the well width decreases below 100 Å. As in the case of confined-state impact ionization (see Section 10.6), for the injected carriers to be collected within the wells, they must first suffer an energy-relaxing collision while over the well such that its energy becomes less than the barrier height. Repeated scatterings can ultimately result in the thermalization of the carriers within the well. As the quantum well becomes smaller and approaches the electron-scattering mean-free-path length, many of the electrons will suffer insufficient scatterings to be collected. As a result, many of the injected electrons pass over the well without being captured. Subsequently, for small-quantum-well-width devices, many electrons must be injected to produce sufficiently thermalized carriers for lasing action. This of course results in a higher threshold current density.

The use of a MQW instead of a SQW enhances the probability of carrier capture into the wells. We can understand this by recognizing that in a MQW structure there is a greater chance of a carrier's being collected by a well since the total collision cross section of a carrier with a well is obviously larger the greater the number of wells present. In the single-well device, there is a reasonably high probability that an injected carrier will transit the device without being captured within the quantum well, particularly as the well width decreases in width. The use of MQWs increases the odds that an injected carrier will be collected within the well and hence contribute to the stimulated emission. For this reason, MQW devices have been observed to have lower threshold current densities than SQW devices.

The reduction of the threshold current density in a MQW structure is limited by two major problems. The presence of multiple wells alters the effective refractive index of the active region to roughly the spatial average of the composite layers. If the cladding layers have the same composition as the barriers, the refractive-index step is greatly reduced, leading to significant leakage of the optical fields into the surrounding layers. Hence the optical confinement factor is greatly reduced, thus increasing J_{th}, the threshold current density. In addition, the presence of barriers reduces the carrier collection efficiency since carrier injection into the various wells is hindered somewhat by the large potential barriers formed. These problems can be circumvented by use of a higher aluminum content in the composition of the cladding layers than that of the quantum-well

a

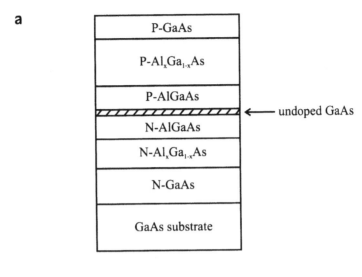

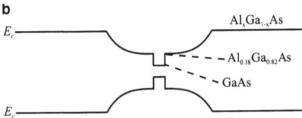

Figure 13.3.5. a. Layers, b. band diagram for a graded-index separate confinement heterostructure laser.

barriers. This improves both the carrier collection and the optical confinement factor. However, the use of a higher aluminum content in the cladding layers leads to wider beam divergence.

Another approach to reducing the threshold current density is to embed a SQW within a graded funnellike region, as shown in Figure 13.3.5. The device sketched in Figure 13.3.5 is called the graded-index separate-confinement heterostructure (GRIN-SCH) laser (see Tsang 1982). As is shown in the figure, the graded region surrounding the quantum well provides a means by which the injected carriers can be funneled into the well, greatly reducing the threshold current density. Note that the bandgap grading results in an electric field that directs the electrons toward the quantum well from either side. As a result, injected electrons tend to roll within the funnel region until they suffer sufficient scatterings to thermalize. Fewer electrons injected from the n side can diffuse into the p side without being collected by the well. Consequently, carrier confinement is much improved over both the SQW and the MQW structures without sacrificing optical confinement. In addition to improving the electrical confinement of the injected carriers, the surrounding graded region enables independent control of the optical and carrier collections separate from the well. Subsequently, far smaller wells can be made than is practical in simple SQW devices.

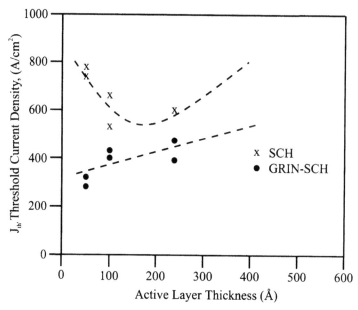

Figure 13.3.6. Plot of the experimental measurements of the threshold current density as a function of well thickness for both SQW and GRIN-SCH lasers. Data taken from the paper by Hersee et al. (1984). "Reprinted with permission from APPLIED PHYSICS LETTERS. Copyright 1984 American Institute of Physics."

The threshold current density of GRIN-SCH lasers has been found to be well below that of comparable SQW or MQW lasers. Figure 13.3.6 shows experimental data from Hersee et al. (1984), which clearly illustrates the dramatic improvement in the current-density threshold in the GRIN-SCH structure over other structures. The improvement stems from the fact that the carriers thermalize more efficiently into the quantum well owing to the presence of the funnellike region surrounding the well.

The reader is referred to the references for more information about semiconductor lasers and laser theory. We next turn our attention in the ensuing sections to other emitters, useful predominantly for flat-panel displays.

13.4 Flat-Panel Displays: An Overview

One of the most important emerging markets for photonic emitters is in flat-panel display technology. The applications of flat-panel displays are becoming ubiquitous. The reader is keenly aware of the overwhelming importance of cathode ray tube (CRT) displays currently in use for most personal computers, workstations, etc. One of the primary limitations of extending personal computers and related information technology platforms (driving aids, navigation aids, etc.) to more diverse applications is the fact that CRT displays are relatively large, heavy, fragile, expensive, and have extensive power requirements. These aspects conspire to make CRTs essentially useless in portable systems or those for which weight, size, or fragility become important (for example, in airplane cockpits or automobile dashboards). Currently flat-panel displays are manufactured and marketed primarily for laptop computer applications. However, there are additional applications in which flat-panel displays can be utilized, provided they overcome existing limitations in price and performance. Some of the emerging applications, aside from lightweight laptop computers, are information appliances (machines that have reduced complexity from personal computers and are specialized and easy to use), telephone systems with Internet e-mail functions, automobile driving aids, airplane navigation, and function displays, etc. In this section, we discuss some of the competing technologies utilized for flat-panel displays, specifically those somewhat related to semiconductors and semiconductor devices. In Section 13.5 we examine one of these, electroluminescent (EL) devices, in some detail.

As discussed above, the vast majority of computer terminal displays are presently made with CRTs. The principal types of flat-panel displays are EL displays, liquid-crystal displays (LCDs), field-emission displays (FEDs), and plasma displays. Each of these emerging technologies, in turn, has its own limitations. LCDs, although they provide high-quality resolution at low-power requirements, emit in only a single color. Plasma displays can be made to emit in different colors, but at present these displays are prohibitively expensive. Present EL devices consume much power and therefore require high-voltage drive sources. However, these devices can be made to display all the colors (red, green, and blue), necessary to give full color representation, can be fabricated as large-area flat panels, and have exhibited long lifetimes under adverse environments.

It is useful to review the different requirements that flat-panel displays must achieve. The primary engineering figures of merit for flat-panel displays are

1. Luminous efficiency. The luminous efficiency is defined as the ratio of the luminous output flux divided by the input electrical power. Of course, it is desirable that the luminous efficiency be as high as possible.
2. Viewing angle. The viewing angle is the maximum angle from the normal for a viewer before the contrast decays sufficiently as to be unreadable.
3. Color and resolution. The display must achieve accurate representations of color and provide high resolution, rendering clear and sharp images.
4. Portability and power. In many applications, power consumption and portability are of overwhelming importance. For example, in digital watches, miniature televisions, automotive clocks, and camcorder viewfinders, displays must operate at a very low input power with a very low battery drain.

Let us examine briefly some of the types of flat-panel display technologies available today. The first device type we consider is the LCD. A LCD does not generate any light of its own. Basically the device creates an image by controlling light that either passes through it or is reflected off of it.

As has been discussed throughout this text, in crystalline solids the atoms are generally fixed in both position and orientation. The atoms can undergo only small thermal motions about their fixed-equilibrium lattice sites. In liquids, on the other hand, the atoms comprising the liquid can move freely and their orientations can move in random directions. Liquid crystals have some intermediate order between these two extremes. For example, the simplest situation is one in which molecules are generally fixed in space but can undergo free rotation as well as vibration. Generally, in liquid crystals there is some positional and orientational order, but far less than that within solids. The structure of the liquid-crystal phases depends principally on the arrangement of the molecules and their interaction.

There are two general types of liquid crystals. These are thermotropic and lyotropic crystals. Thermotropic liquid crystals comprise only one molecular species. Two general types of thermotropic liquid crystals exist, called calamitic and discotic. Calamitic liquid crystals are formed of identical molecules in which the length along one axis is much longer or shorter than that along the other two axes. Therefore calamitic liquid crystals are rodlike molecules. Discotic liquid crystals, as the name implies, are formed of identical disk-shaped molecules. Lyotropic liquid crystals are multicomponent systems consisting of two or more constituents. In lyotropic liquid crystals, concentration dictates the behavior of the system while in thermotropic liquid crystals temperature changes transform the system from one phase to another.

Calamitic liquid-crystal molecules can be categorized into one of three types. These are nematic, smectic, and chiral or cholesteric. The nematic phase is characterized by a nearly parallel arrangement of the molecules. The long axes of the molecules tend to point along a certain direction as the molecules move from one point to another. It is important to recognize that within one plane or sheet of a nematic liquid crystal the molecules are all nearly aligned. Between different sheets, the molecules may be aligned differently from one sheet to another. This is of great importance, as we will see below, for the most popular LCDs.

The smectic phase has several different configurations. However, the two most important smectic phases are the smectic-A and smectic-C phases. These phases differ in their orientation. There is a preferred direction for the long axis of the molecules. This direction is called the director. In the smectic-A phase, as shown in Figure 13.4.1.a, the director is parallel to the layer thickness. In the smectic-C phase, as shown in Figure 13.4.1.b, the director is oriented at an angle θ with respect to the layer thickness.

The chiral phase, often referred to as the cholesteric phase, is one in which the director rotates along a direction perpendicular to the director, for example along a helical axis, as shown in Figure 13.4.2. Alternatively, the director can rotate about a cone.

The key importance of liquid crystals in device applications arises from their anisotropic electrical and optical properties. Basically owing to the anisotropy of the molecules and the macroscopic structure of the liquid crystal, the electric and

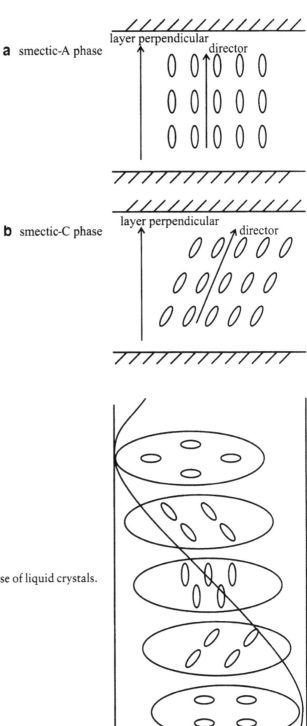

Figure 13.4.1. Sketches of the smectic-*A* and the smectic-*C* phases of liquid crystals.

Figure 13.4.2. Sketch of the chiral phase of liquid crystals.

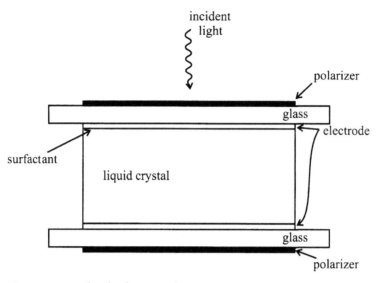

Figure 13.4.3. Sketch of a twisted nematic display.

the magnetic properties are different along the director direction and directions perpendicular to the director. Additionally, the anisotropy is responsible for a linear birefringence. Light polarized parallel to the direction of the director propagates with a different index of refraction than light polarized perpendicular to the direction of the director.

The most important configuration of LCDs is the twisted nematic display, as shown in Figure 13.4.3. The liquid crystal is encapsulated between two glass planes onto which thin electrodes are deposited on the same side as the liquid crystal. Contacting the liquid crystal itself are layers of a surfactant that provide a preferential direction by which the liquid crystal is oriented. The surfactant has a preferred direction, which in turn aligns the liquid-crystal molecules. The preferential directions of the surfactants on the top and the bottom surfaces are rotated by 90° with respect to one another. Therefore the director of the liquid crystal is rotated by 90° from the top to the bottom of the liquid crystal.

Polarizers are placed on the top and the bottom glass layers on the other side from the liquid crystal. The polarizers are arranged such that they pass light of parallel polarization to that of the corresponding surfactant on the other side of the glass layer. Light incident upon the top of the device is polarized along the direction of the top surfactant layer. If the thickness of the liquid crystal is chosen appropriately, the polarization of the light rotates with the director as it passes through the liquid crystal. As mentioned above, the molecules within a single sheet or plane of a nematic liquid crystal are all nearly aligned. The angle of alignment, though, need not be identical between adjacent sheets. In the display structure, since the boundaries fix the alignment of the molecules, the directors within the planes of the molecules distributed between the top and the bottom boundaries must be slowly rotated to provide the full 90° rotation. As such, as the light propagates through the device, it slowly rotates in polarization so that it is aligned with the bottom polarizing layer. Therefore, in its inactive state, the device is fully transparent.

If a voltage is applied to the electrodes, the situation is quite different. The applied voltage orients the director in the liquid crystal parallel to the electric field. Except for a thin boundary layer adjacent to each glass plate, the director within each sheet of liquid crystal is oriented parallel to the direction of propagation of the light. The polarization direction of the incident light is no longer rotated as it passes through the liquid crystal. When the light passing through the device intersects the second polarizer, its polarization is perpendicular to that of the polarizer sheet. As a result, the light is absorbed and the device is now opaque. This arrangement of the twisted nematic liquid crystals is called the normally white mode of operation because the display is bright when off.

The device can be used in either a transmissive or a reflective mode. If a reflector is added below the bottom polarizer, the device can be used in the reflective mode. In this mode, dark areas are seen for cells in which a voltage has been applied, while gray areas correspond to those cells for which no voltage has been supplied. The reflective mode is most commonly used in watches and calculator displays. The transmissive mode requires a light behind the display. The light is then blocked for those electrodes that are biased, while devices for which no voltage has been applied appear transparent.

Much more, of course, can be said about liquid crystals and their applications. The interested reader is directed to the references. Although there are several technologies of importance for flat-panel displays, in this book, we will treat only two additional topics from LCDs, since they are most closely related to semiconductors and semiconductor devices. In Section 13.5, we discuss EL devices. These devices operate with a high-energy, high electric field much like APDs, and other devices already studied in this book. For this reason, we will examine them separately. To complete this section, we briefly examine field-emission displays or FEDs.

A FED is a vacuum microelectronic device. A FED can be thought of as something like a micro-CRT. As in a CRT, electrons are emitted from a cathode and travel through vacuum to hit a target phosphor, producing luminescence. However, a FED has several advantages over CRTs and vacuum tubes in general. The principal advantages of FEDs are as follows.

1. They provide much higher current densities than thermionic vacuum tubes.
2. They do not require heating of the electron source to provide electron emission as in a CRT. In thermal emission, heat is used to provide the necessary kinetic energy to the electrons to overcome the work function of the emitter material. In a FED device, electrons are emitted from the cathode by means of field emission. In field emission, the sharp tip of the cathode produces a sufficiently high electric field that can overcome the work function of the emitter on the application of an applied bias.
3. FEDs can be made compact and operate at lower voltages than CRTs. Additionally, FEDs can be packaged into thin units, provide a high luminous

efficiency, and have massive redundancy. Each pixel can have up to several thousand cathodes, so that variations between cathode tips and failures become unimportant.

In addition, FEDs have important advantages over semiconductor devices stemming from the fact that they are vacuum devices. These can be summarized as follows.

1. No power dissipation occurs within the transport region of a FED since no collisions occur in vacuum. In contrast, within a semiconductor, substantial collisions (i.e., phonons, impurities, etc.) occur as electrons move through the material, thus leading to energy dissipation.
2. Operation is essentially temperature independent within a FED. Semiconductors are highly sensitive to temperature variations, and can change from semi-insulating at low temperatures to conducting at high temperatures.
3. FEDs are radiation hard; these devices are impervious to radiation damage. Semiconducting devices, on the other hand, are highly sensitive to alpha particle or other radiation damage, which leads to spurious charge generation.

The workings of a FED device can be understood as follows. Field emission occurs from an applied electric field supplied to a sharp tip of a cathode in vacuum. As a result, the potential barrier at the surface of the cathode is greatly reduced, which in turn triggers quantum-mechanical tunneling of the electron into the vacuum. These electrons then travel through vacuum and impinge on a phosphor, producing phosphorescence. A schematic sketch of the structure is shown in Figure 13.4.4.

For use in flat-panel displays, the emitting cathode array is typically fabricated on an insulating baseplate. Each pixel contains three color elements.

Figure 13.4.4. Sketch of a FED emitter.

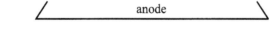

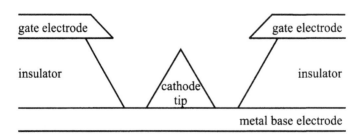

Three metal gate electrodes are used in each pixel, one for each color. A pixel is activated first by the application of a baseline voltage up to the threshold for emission. To vary intensity, a variable bias is applied to the gate electrode. Electrons are then emitted from the cathode tip in an amount proportional to the electrode bias. On being emitted from the cathode tip, the electons are then accelerated through the vacuum toward a distant anode, onto which they impinge. It should be noted that the electric field between the tip and the electrode is essentially radial. Therefore the emitted electrons can obtain a sizable transverse velocity from the action of this radial field. This transverse velocity in turn leads to an angular spreading of the electrons once they reach the distant collecting anode. In principle, however, a second electrode with a coaxial hole can be arranged so as to focus the electrons into parallel beams. An alternative biasing scheme can be used in which the intensity is modulated by control of the length of time in which a bias is applied. In such a scheme, each pixel must be addressed separately.

Once the electrons are emitted from the cathode, they are accelerated through the vacuum by the positively biased anode. The high-energy electrons excite the anode phosphor, which produces luminescence.

Although FEDs certainly offer many attractive features, they suffer from some important limitations as well. Specifically, because of vacuum breakdown and the resulting arcing between the anode and cathode tips, there is a minimum screen separation distance that must be maintained. In turn, spreading of the emitted electrons can present an important limitation in the resolution of the display. Much work is presently in progress toward refining FED technology. The interested reader is directed to the references listed at the end of this book for further information.

13.5   **Electroluminescent Devices**

EL displays essentially consist of a thin semiconducting film that acts as the host for luminescent centers from which optical radiation is produced. Devices operating with either applied dc or ac voltages exist. A typical ac EL display is shown in Figure 13.5.1. The device is a symmetrical insulator–semiconductor–insulator structure that produces light under sufficiently high applied bias. The active region of the device comprises a semiconductor layer that includes intentionally doped deep-level centers that provide luminescence. The semiconductor layer is sandwiched between two insulators that are then contacted by means of one opaque and one transparent electrode. As shown in the diagram, the transparent electrode is placed above the glass substrate. The emitted light is viewed from the substrate side of the device.

The device operation can be understood through examination of Figure 13.5.2. At high electric-field strengths, the semiconductor energy bands bend strongly, as shown in the figure. The high band bending leads to a reduced energy barrier for electrons trapped within interface states at the first insulator–semiconductor boundary. As can be seen from Figure 13.5.2, the large band

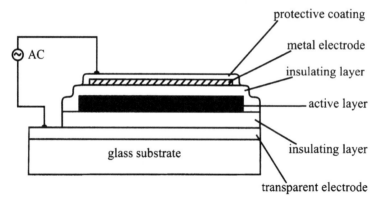

Figure 13.5.1. Schematic of an ac thin-film EL device.

bending results in a triangular potential barrier. The greater the band bending, the smaller the triangular potential barrier becomes. At some point, electrons trapped within the interface states can begin to tunnel out of these states into the conduction band of the semiconductor.

Operation of the EL device breaks down into several different physical processes. Following carrier injection from interface trap tunneling, the free carriers are accelerated by the electric field within the semiconductor layer, often referred to as a phosphor, since it is the light-emitting layer. The injected electrons traverse the phosphor under the action of the applied electric field. The carriers are heated to very high energies on average. As was discussed in Section 10.5, the competing

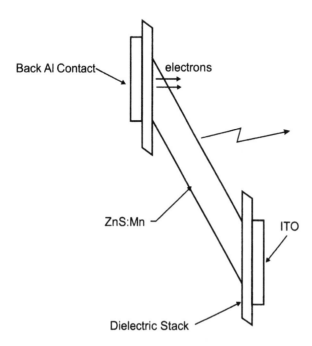

Figure 13.5.2. Illustration of the operation of a thin-film EL device, depicting hot-electron and impact excitation processes. (courtesy, B. K. Wagner)

processes of field heating and energy relaxation through phonon scattering shape the carrier distribution and determine the population of the carriers in the high-energy tail. Depending on the relative strength of the phonon-scattering rate, a fraction of the electrons reach sufficient energies that they can impact excite the luminescent centers introduced into the phosphor. In the impact excitation process, the high-energy, or hot, electron transfers its excess kinetic energy through a collision to an electron bound to the luminescent center, much like the case of impact ionization. However, in the luminescent mechanism, the bound electron is excited to another bound electronic state within the same atom. This is different from impact ionization in which the initially bound electron is freed by the impact excitation collision. In any event, both impact ionization and impact excitation require that the initiating carrier reach a relatively high energy. The excited state of the center ultimately decays back to its ground state. The decay may occur radiatively or by other means, generally thermally. If the decay occurs radiatively, depending on the nature of the excitation, an output photon of a particular wavelength is produced.

The vast majority of electrons injected into the semiconductor layer do not contribute to the output brightness by means of impact exciting a luminescent center. Most of the carriers either do not achieve sufficient energy to initiate an impact excitation event or simply do not strike a luminescent center. The situation is similar to that for impact ionization. In Section 10.5, it was found that the impact ionization rate depended on two different factors, that is, the probability that a carrier reaches sufficient energy to initiate an ionization event (called the threshold energy) and the probability that an ionization event will occur once a carrier has attained threshold. The probability that an event occurs once the carrier has attained threshold depends of course on the relative strength of the ionizing collision (the transition rate) compared with the energy-relaxing phonon-scattering rate. The ionization transition rate depends in turn on the scattering cross section of the collision.

Similarly, the impact excitation rate depends on two separate factors, the probability that a carrier will reach sufficient energy to cause an event and that an excitation event will occur. The probability that a carrier will reach the threshold energy for the event is found in the same way as for impact ionization. As discussed in Section 10.5, attainment of the threshold energy depends to a great extent on the scattering rate. Two different models were invoked, the Shockley model and the Wolff model. In the Shockley model, only lucky carriers reached threshold. A lucky carrier is one that survives its flight to threshold without suffering any collisions. Alternatively, in the Wolff model, the carriers are assumed to suffer many collisions, which leads to the development of a well-defined distribution function. Only those carriers within the high-energy tail of the distribution contribute to the ionization process. It was also argued that a more realistic picture of the process combines features of both the Shockley and the Wolff models. The Baraff and numerical theories predict that, at low applied fields, the nonequilibrium distribution function resembles that predicted by the Shockley model while at high electric fields the distribution function is more like that of Wolff. These same results apply as well to the impact excitation problem as they did to the impact ionization problem.

Although the probability that a carrier will achieve threshold is the same for both the impact ionization and impact excitation mechanisms, the probability that the carrier suffers an impact excitation collision is quite different from that for an impact ionization event. In Section 10.4, the impact ionization transition rate was formulated. The ionization transition rate for impact excitation is found in a similar manner, but the physical process is somewhat different.

The first thing to recognize is that the target is much different in the impact excitation process than in the impact ionization process. The target in the impact excitation scheme is a single atom, a deep center impurity embedded within the phosphor material. For a photon of visible wavelength to be emitted, it is necessary that the impact excitation transition occur between the ground and a high-energy excited state. The most common center is Mg^{2+}, which can be active in several host phosphors. The actual calculation of the impact excitation cross section is extremely difficult to perform, since even within a vacuum, the calculation of cross sections is quite difficult. The addition of a host crystal adds further complexity. The matrix element for the collision is governed by a screened Coulomb interaction, as in Section 10.4. The primary difference in the calculations, though, is in the initial and the final states of the target atom's electron. Owing to the indistinguishability of the electrons, the n-valence-shell electrons within the target atom are described by an n-electron collective wave function. In essence, the collision is treated as an excitation of the entire shell. The state vectors used to describe the ground and the excited states of the atom must be functions of the $3n$ space and n spin coordinates. It should be noted that beyond writing down the Hamiltonian and the transition-rate formulation, little progress has currently been made in evaluating expressions for the collision cross section owing to the complexity of evaluating the matrix element, including multielectron states. For this reason, a quantitative treatment of the impact excitation mechanism is omitted. In any case, the impact excitation transition rate is substantially smaller than the corresponding impact ionization transition rate. The basic reason why the impact excitation rate is smaller than the ionization rate is that the targets are far fewer in number. Recall that the target atoms are simply deep-level impurities whose concentration rarely exceeds $\sim 10^{18}$ cm^{-3}. In contrast, the target atoms for the impact ionization process are the constituent atoms of the host semiconductor itself whose concentration is $\sim 10^{23}$ cm^{-3}, at least 5 orders of magnitude greater. On the basis of this observation alone, the impact ionization transition rate is expected to be substantially greater than the impact excitation rate. It should be noted, though, that the impact excitation mechanism becomes active at lower carrier energies than the impact ionization rate in most host phosphors. The host phosphors are generally selected such that they have bandgap energies greater than the optical radiation energy desired.

Since the collision cross section for impact excitation is small and only those carriers within the high-energy tail of the distribution can contribute to the process, the net result is that few injected carriers contribute to exciting the luminescent centers. Therefore most of these carriers become collected on the other side of the structure, leading to a space-charge buildup that further affects electronic injection into the device. If the applied bias is reversed, as in an ac device, these electrons and new ones trapped near the second electrode are accelerated back into the active semiconductor layer. Through use of an ac bias, continuous illumination of the display can be achieved. Since the source of charge carriers in an ac display is interface states at the semiconductor–insulator boundary, there are relatively few carriers present at any given time within the active semiconductor layer. Therefore, ac EL displays, owing to the limited number of electrons available to impact excite the centers, operate at relatively low brightness levels.

EL devices can also be operated with dc voltages. These structures consist of MIS, metal–insulator–metal, or Schottky barriers. In dc EL devices, the source of free-charge carriers is the metal electrode. Through application of a reverse bias in a Schottky barrier device, electrons tunnel from the metal into the semiconductor. Similar tunneling of electrons through a narrow insulator layer in MIS devices leads to significant carrier injection. The principal advantage of dc EL devices over their ac counterparts is due to the much larger free-carrier concentration within the semiconductor layer available to impact excite a luminescent center, typically Mn or comparable rare-earth elements.

Direct-current EL devices are operated in two different modes, continuous and pulsed. Pulsed operation typically results in longer device lifetimes at the same brightness levels compared with continuous operation. Nevertheless, performance deterioration with time is a critical disadvantage of dc EL devices. The deterioration in continuous-mode dc devices is due to the increase in resistivity, which can be partially offset through pulsed operation. Resistive deterioration does not occur in ac devices since they are capacitive structures. Nevertheless, dc devices still retain several advantages over ac devices in addition to large free-carrier concentrations. These advantages are lower power consumption, more versatility in shape and size, and ease of construction.

At present, all EL devices suffer from poor power efficiencies, that is, $\sim 0.1\%$ for dc EL devices. The poor device efficiencies appear to be due to several factors: difficulty in heating sufficient carriers to energies high enough to cause an impact excitation event, the small collisional cross sections for excitation, and limited center concentrations due to quenching effects. The successful operation of an EL device depends on the probability of an injected electron impact exciting a luminescent center, which depends in turn upon the collision cross section, the density of centers, and the probability that an electron will achieve the threshold energy for excitation and then actually excite a center through a collision. As discussed above, the combined effect of all these issues greatly reduces the number of injected carriers that successfully impact excite a center. For this reason, the efficiency of an EL device is quite poor.

The main requirements for effective luminescence of a center are that it should have inner-shell transitions well shielded from the applied electric field, be isovalent or neutral to avoid drift-aided diffusion of the centers (this is of greatest

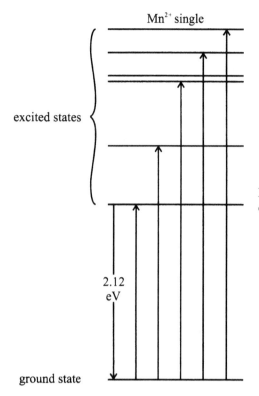

Mn^{2+} single

excited states

ground state

2.12
eV

Figure 13.5.3. Energy-level scheme for ground and excited states of Mn^{2+} in the ZnS lattice.

importance in dc devices since the field always points in the same direction), have a large cross section for impact excitation, have a high solubility within the host material, have a high radiative efficiency within the center, and emit within the visible part of the spectrum. To date, the center identified with the most favorable qualities is manganese within either ZnS or ZnSe. The transitions excited in the luminescence of a manganese center are shown in Figure 13.5.3. The principal transition is at 2.12 eV, which corresponds to yellow light.

Because of the many restrictions on the nature of the center listed above, it is difficult to improve on the centers themselves. Recall that the efficiency depends on the combined effects of the carriers' achieving threshold energy and then exciting a center to produce a radiative transition. Since the choice of centers is highly restricted, there is not much that can be done to improve efficiency regarding the center. Consequently, efficiency enhancement can best be achieved by an increase in the number of hot electrons available within the semiconductor layer.

The most obvious means of increasing the hot-electron concentration is by application of a high electric field across the semiconductor layer. In steady state, the carrier heating from the field is balanced, on average, by inelastic-scattering processes, as discussed in Section 10.5. The principal scattering processes that act to relax the energy of the distribution are phonon scattering, impact ionization, and, in the case of EL devices, impact excitation. Both impact ionization and excitation are threshold processes. In both mechanisms, the carriers must achieve a threshold energy in order to initiate the process. A center must be excited to the energy of the desired photon that is to be emitted. For visible radiation, the center must be excited to energies significantly higher than 2 eV. Therefore the initiating electron must be heated to at least these energies to cause an impact excitation event.

The maximum electric field that can be applied, though, is limited by the onset of drift-aided diffusion of the excitation centers, avalanche breakdown resulting from impact ionization, and ultimately by dielectric breakdown of the semiconductor itself. Therefore few electrons survive from field heating alone to high energies at which impact excitation can occur, resulting in very low efficiency operations of existing EL devices.

The competing processes of field heating and phonon cooling result in only a very small fraction of the injected electrons attaining high energies. Ensemble Monte Carlo studies, of the kind described briefly in Section 10.5, have been performed for electron transport in ZnSe and ZnS (see Brennan 1988). The Monte Carlo calculated electron number density function is plotted as a function of energy in Figure 13.5.4. As can be seen from the figure, the high-energy tail of the electron number density function is relatively small, even at an applied electric field of 2000 kV/cm. Much of the efficiency lost in an EL device comes from the relative difficulty of heating the electron distribution.

Clearly, much remains to be done to improve the performance and operation of EL devices. Higher efficiency operation will require improved materials,

Figure 13.5.4. Monte Carlo calculated energy distribution function in ZnS at an electric-field strength of 2 MV/cm.

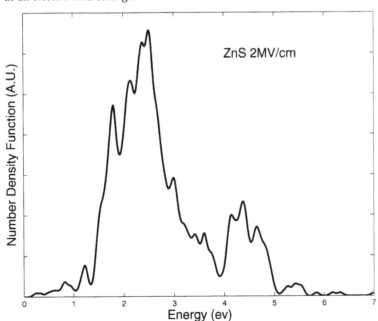

possibly new impact centers, or perhaps even new device designs. It is expected that refinement and improvement of EL devices will remain an area of critical importance for some time since they have a direct impact on the already large and potentially enormous market of flat-panel displays.

PROBLEMS

1. If it is assumed that the electron and the hole concentrations are equal and that equal transition rates exist between the carriers, determine the minimum carrier concentration for population inversion in InP. Assume that $n_i \sim 10^7$ cm^{-3} and $E_g = 1.35$ eV.

2. Assume that the net gain coefficient, defined as the difference between the gain and the loss terms at threshold, of a semiconductor laser is equal to 0.01 cm^{-1}. If the mirror reflectances are assumed to be equal and the cavity length is 20 cm long, what is the value of the reflectance?

3. Determine the cavity length of a semiconductor laser when the collective loss terms equal the mirror loss terms. Assume that the mirror reflectances are both equal to 0.50 and that the collective loss terms equal 30 cm^{-1}.

14

Field-Effect Devices

One of the most important physical mechanisms of importance to semiconductor devices is the field effect. Several important devices exploit this effect in their operation, such as metal-oxide–semiconductor field-effect transistors (MOSFETs), metal–semiconductor field-effect transistors (MESFETs), and junction field-effect transistors (JFETs). In fact, the field effect transistor (FET) is arguably the most important innovation that has fueled the computer and information revolution. In this chapter, the fundamentals of the FET operation are presented; the reader is directed to the references for a more comprehensive study.

14.1 The Field Effect

The field effect can be simply defined as the modulation of the conductivity of an underlying semiconductor layer by the application of an electric field to a gate electrode on the surface. As we learned in Chapter 11, the application of a bias to a MIS structure results in a modulation in the carrier concentration within the underlying semiconductor layer. If the semiconductor is naturally n type and a positive gate bias is applied, electrons accumulate at the semiconductor–insulator interface. Conversely, if a negative gate bias is applied to the same structure, the electrons are repelled from the interface and, depending on the magnitude of the bias, the underlying semiconductor layer is either depleted or inverted. If the semiconductor becomes inverted, the carrier type changes.

The conductivity of the semiconductor will vary as the carrier concentration is changed. In Chapter 6 we found that the electrical conductivity can be expressed as the product of the charge density, charge, and carrier mobility as

$$\sigma = qn\mu. \tag{14.1.1}$$

Therefore, when the carrier concentration in the underlying semiconductor layer is changed through the action of the gate bias, the conductivity of the semiconductor is altered. When the applied gate bias is such that the semiconductor is depleted, the free-carrier concentration is greatly reduced. As a result, the resistivity of the semiconductor increases or, equivalently, the conductivity decreases. When the gate voltage is such that the semiconductor is biased into inversion, a highly conducting channel of the opposite carrier type to that of the bulk semiconductor material forms at the semiconductor–insulator interface.

The field effect can thus be used as a gating mechanism. By a change in the applied gate bias, the underlying semiconductor can be made either highly conducting or resistive, thereby altering the current-carrying properties of the layer. When the gate is biased such that the semiconductor layer is conducting, a current can flow through the layer. A conducting channel is opened at the semiconductor–insulator interface. Alternatively, if the gate is biased such that the semiconductor layer is depleted, only a small reverse leakage current can flow, which is typically negligible. Under this condition, no conducting channel exists at the semiconductor–insulator interface. Therefore the field effect can be used to open and close a conducting channel at the surface of the semiconductor. As such, a MIS structure can be used as a gate to control current flow between two reservoirs of charge, which are in turn connected to an external circuit by ohmic contacts. This is the underlying principle behind a FET.

Different types of field-effect structures exists. There are three main types. These are a junction field-effect structure, a metal–semiconductor field-effect structure, and a metal-insulator–semiconductor field-effect structure. In each type of device structure, the basic operating principle is the same, that is, the conductivity of the underlying semiconductor region is modulated by the application of a gate bias. The only real difference among these device types is how the conductivity changes in each case. In this chapter, we discuss each of these field-effect structures and show how each is used as a gating mechanism to provide switching action.

14.2 Qualitative Operation of the JFET and the MESFET

The simplest application of the field effect in a transistor structure is that of a JFET. A simple JFET structure is sketched in Figure 14.2.1. Such a device consists of two islands of highly doped material, typically an n-type doped semiconductor, separated by a lightly doped n-type layer, called the channel, which is bordered on either side by p^+ layers contacted to an external bias by a metal gate electrode. As the voltage on the gate is changed, the extent of the depletion region is altered. Under a forward bias, that is, application of a positive voltage to the gate electrode, the width of the underlying depletion region decreases. Under reverse bias the depletion region increases in size. If a negative bias is applied to each electrode, the underlying p^+n junctions become reverse biased, leading to an increase in the depletion region widths. The depletion region then intrudes into the n-channel region, thereby constricting its width. Since the resistance of the channel depends on the channel area A as

$$R = \frac{\rho L}{A}, \tag{14.2.1}$$

the resistance increases as the width of the channel and hence its area decreases.

If a positive voltage is applied to the drain contact (marked D in Figure 14.2.1) then electrons will flow from the source contact (marked S in Figure 14.2.1) to the drain contact. Hence the current flows from the drain to the source, opposite to the direction of the electron flow. As the depletion region is constricted by the application of a negative gate bias, the channel resistance increases. Therefore

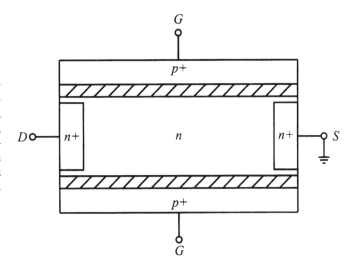

Figure 14.2.1. Simplified version of a JFET device in equilibrium. The contacts marked D, S, and G are the drain, source, and gate, respectively, in the device. The cross-hatched region represents the depletion region between the p and the n regions.

if the drain–source voltage is kept constant, as the resistance increases the drain current decreases. Further increase in the gate bias ultimately results in the two depletion regions under each gate touching, which is commonly called punch through. When this occurs, the resistance of the channel reaches its peak value and the drain current will not increase with any increase in the drain–source voltage.

The workings of a JFET can be qualitatively understood with the help of Figure 14.2.2. For the purposes of this discussion, it is assumed that there is a negligible voltage drop between the contacts and the corresponding ends of the channel. This implies that there is zero voltage drop between the drain contact and the drain region as well as zero voltage drop between the source contact and the source region. Two cases of operation need to be considered. These correspond to zero and nonzero gate voltages. It is interesting to consider the operation of the device when a drain voltage is applied under both conditions on the gate voltage.

Consider first the case of zero gate voltage, as shown in Figure 14.2.2. Since the source is grounded, the voltage drop between the source and the gate is then simply zero. If a positive voltage is applied to the drain, the drain end of the device becomes reverse biased. As we discussed in detail in Chapter 11, the depletion region of a p^+n junction increases in width with increasing reverse bias. In addition, most of the depletion region extends within the lightly doped side of the junction. For the present device structure, since the n region is lightly doped with respect to the gate p^+ regions, the reverse bias applied to the drain acts to deplete out the n regions near the drain end of the device.

At low drain voltage, electrons begin to flow from the source to the drain through the undepleted n region sandwiched between the two p^+n junctions forming the gate on either side of the device. The n region is called the channel. At small drain voltages and zero gate voltage, the channel behaves as a simple resistor with a small resistance. Therefore the variation of the drain current

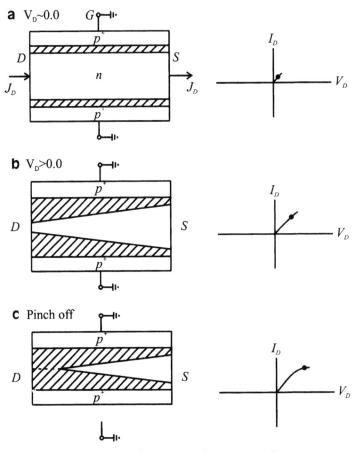

Figure 14.2.2. Behavior of a JFET under zero gate bias at various drain voltages. a. Near-zero drain voltage. The drain current I_D versus drain voltage V_D is linear in this range. b. $V_D > 0.0$. The drain current departs from a linear relationship because of the increased channel resistance. c. Pinch-off condition. I_D saturates at this point.

I_D with drain voltage V_D is linear, as shown in the accompanying plot of the current-voltage characteristic in Figure 14.2.2.a.

As the drain voltage increases, the voltage difference between the drain and the grounded gate increases proportionately. Because the source and the gate are grounded, at least for this configuration of the device, as the drain–source voltage increases, the voltage drop along the channel increases. Therefore the effective reverse bias of the drain to gate increases, causing the depletion region to increase in width near the drain end of the channel. As a result, the channel is constricted by the encroaching depletion regions from both the top and the bottom gates. Since the potential is much lower near the source side of the device, the p^+n junctions near the source are less reverse biased. Therefore the effective channel is wider near the source than near the drain, as shown in Figure 14.2.2.b. Under these conditions the drain-current versus drain-voltage characteristic departs from a linear relationship.

As the drain voltage is further increased, the channel region becomes even more constricted near the drain end. Ultimately, the drain potential is increased to the point where the top and the bottom depletion regions extend sufficiently that they both touch near the center of the device. When this occurs, the original conducting channel between the source and the drain has now been effectively closed and the channel is said to be pinched off. When the device is in pinch off, no conduction path exists between the source and the drain. Instead, part of the path is now fully depleted. As was discussed in Chapter 11, the free-carrier concentration is very small within a fully depleted layer and as a result the region has a high resistance. The resistance of the device then greatly increases in pinch off. The current saturates with increasing drain voltage after the pinch-off point is reached. Further increase in the drain voltage does not result in greater current flow.

The physical basis for the current saturating after pinch off can be understood as follows. As the drain voltage is increased above the point at which the channel pinches off, called $V_{D, sat}$, the pinched-off region of the channel increases from just a point to a region of finite length ΔL. On the source side of this extended region, the voltage changes from zero (at the source) to $V_{D, sat}$ at the edge of ΔL. The applied drain voltage in excess of the saturation voltage, $V_D - V_{D, sat}$, is dropped across ΔL since the voltage on the drain side is V_D. If $\Delta L \ll L$, then the source to pinch-off point separation remains essentially unchanged. Hence exactly the same voltages as were present at the start of saturation are retained and the drain current remains the same. This follows since the shape of the conducting region and the potential applied across it do not change. If $\Delta L \sim L$, then $V_{D, sat}$ is dropped across a much smaller channel segment, $L - \Delta L$, and the drain current will increase as the drain voltage increases above the saturation value. Hence, under these conditions, the current-voltage characteristic may not be completely flat beyond the pinch-off voltage.

The second case is that of a gate voltage applied to the device. If a negative bias is applied to the gates, the gate p^+n junctions are reverse biased, even if the drain voltage is zero. Consequently, the two depletion regions increase in width and the net channel width decreases. Therefore the channel resistance is large for a given drain voltage when the gates are reverse biased ($V_G < 0$) rather than when the gates are unbiased ($V_G = 0$). The channel becomes pinched off at a lower source–drain voltage since it is narrower to begin with. The effect of the gate bias on the channel width is illustrated in Figure 14.2.3. As is readily seen from the figure, the behavior of the device is essentially the same under an applied gate bias as it is without a gate bias except that the pinch-off condition is reached at a lower applied drain bias. In fact, the device can be biased into pinch off even at zero drain bias. The applied gate bias that causes pinch off at zero drain bias is denoted as V_p.

Before we develop a quantititive description of the JFET in Section 14.3, let us examine another device type that is similar to a JFET. This structure is called a metal–semiconductor field-effect transistor or MESFET. A typical MESFET structure is sketched in Figure 14.2.4. The basic structure consists of an n-type

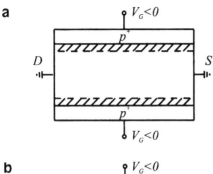

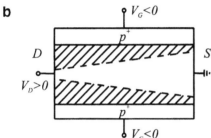

Figure 14.2.3. Simplified sketches of a JFET with an applied gate voltage. In each drawing, the shaded area represents the depletion region. a. Channel shape under negative gate bias and zero drain bias. b. Channel shape at small positive drain voltage. c. Pinch-off condition.

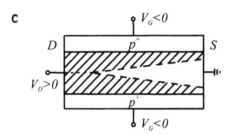

layer epitaxially grown on top of a semi-insulating substrate. The epitaxial layer is typically only ~0.3–0.6 μm thick. We create source and drain regions by highly doping the epitaxial layer by using either ion implantation or diffusion. The source and the drain are separated by a variable-length region in which the conductivity is controlled by the field effect with a Schottky barrier. At zero gate bias and small drain bias, a conducting channel connects the source

Figure 14.2.4. Sketch of an *n*-channel MESFET.

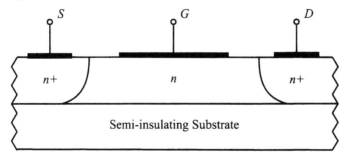

and the drain through the top epitaxial layer. Since the bottom layer is semi-insulating, it is highly resistive and virtually no current can flow through it. Therefore all the current between the source and the drain must flow within the top epitaxial layer. Application of a reverse bias to the gate contact acts to deplete the underlying semiconductor region, thus constricting the conducting channel between the encroaching top depletion region and the bottom semi-insulating layer. As the gate voltage increases, the depletion region extends deeper into the top epitaxial region. Ultimately, with sufficient reverse bias applied to the gate, the top epitaxial region becomes fully depleted between the metal gate and the substrate, closing off the conducting channel between the source and the drain.

The MESFET behaves similarly to the JFET; however, the channel conductivity is modulated by different effects. In the JFET, the channel conductivity is altered by two encroaching depletion regions formed by reverse biased p^+n junctions. In the MESFET, the channel conductivity is modulated by the action of a reverse biased Schottky barrier contact.

As in the JFET, two bias cases need to be considered. These are zero and nonzero gate bias conditions. Consider first the case in which there is no bias applied to the Schottky gate contact. In this case, the semiconductor layer beneath the gate is undepleted and remains n type and thus conducting. If a small positive bias is applied to the drain, a small current will flow from the drain to the source (electron flux from the source to the drain). As before in the JFET, the current is directly proportional to the applied drain bias, resulting in a linear I-V characteristic. As the drain bias increases, the Schottky barrier becomes reverse biased on the drain side. As a result, the depletion region is larger near the drain than near the source, as shown in Figure 14.2.5. This is analogous to the JFET under low-drain-bias conditions.

At a higher drain bias, the depletion region ultimately extends to contact the semi-insulating layer, as shown in Figure 14.2.6. The conducting channel no longer connects the source and the drain regions, and the device is said to be pinched off. As in the case of the JFET, the current saturates.

The operation of a MESFET under a gate bias is again virtually the same as that for a JFET. The application of a gate bias acts to deplete the region underneath the Schottky contact. Therefore, the channel will become pinched off at a lower drain bias. As in the JFET, the channel can be pinched off by application of the gate bias alone. If the gate is sufficiently reverse biased, the region underneath the gate becomes fully depleted and the channel connecting the source and the drain is pinched off. The gate voltage at which this occurs is called the pinch-off voltage, V_p.

The JFETs and MESFETs discussed above are called depletion-mode devices. A depletion-mode device is defined as a structure wherein the application of a gate bias acts to pinch off the channel, closing the conducting path between the source and the drain. In essence, the conducting channel of the device is turned off. Alternatively, one can design a device in which the application of a gate bias creates the conducting channel. In this case, the gate bias turns the

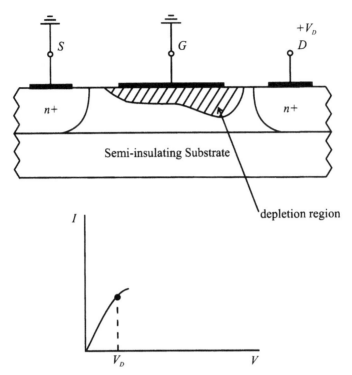

Figure 14.2.5. Sketch of a MESFET and its corresponding current-voltage characteristic at low drain bias.

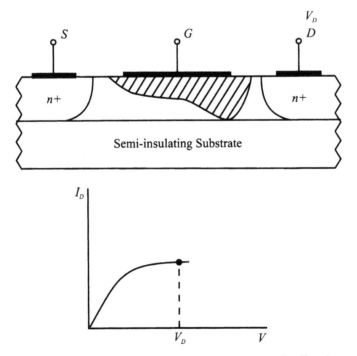

Figure 14.2.6. Sketch of a MESFET biased into pinch off and its corresponding current-voltage characteristic.

conducting channel of the device on. These structures are called enhancement-mode devices. In enhancement-mode devices, the device is nonconducting at zero gate bias; the region under the gate is fully depleted. To make the device conducting, a forward gate bias must now be applied. At some forward gate bias, the channel is pinched off at a single point. Further forward bias on the gate will open the channel. The gate–source voltage necessary to bring the device just to the onset of the channel opening is called the threshold voltage V_T. The applied forward gate voltage must be kept close to the threshold voltage since a substantial gate leakage current will flow with significant forward bias of the gate Schottky contact.

MESFETs are generally much easier to fabricate than JFETs since they can readily be made by use of a planar process. In addition, MESFETs require only the formation of ohmic and Schottky metal contacts as well as only one doping type. Because of the simplicity of the device and its relative ease of fabrication it is particularly attractive for use in many different materials systems. MESFETs are particularly important for the compound semiconductors such as GaAs and InP. As we will see in Section 14.6 later in this chapter, an alternative FET structure is the MOSFET or MISFET. In these devices, an insulator layer is formed below the gate contact to reduce the gate leakage current. However, many materials do not form native oxides, and, as a result, it is difficult to grow a highly controlled and defect-free insulator onto these materials. Most compound semiconductors suffer from this limitation. For this reason, MESFETs are typically used since no insulator layer is necessary, although performance usually does not match that of a MISFET.

In Section 14.3 quantitative descriptions of the JFET and, equivalently, the MESFET, are presented. The basic formulas for these devices are very much the same.

14.3 **Quantitative Description of JFETs and MESFETs**

In this section, we develop a quantitative description of the behavior of a JFET and a MESFET. As discussed in Section 14.2, these two device structures operate in effectively the same manner. In both device types, the current flow is essentially within the bulk regions of the device. The only real difference is that the gating action is controlled by p^+n junctions within a JFET and a Schottky barrier in a MESFET. For this reason, it is possible to formulate expressions describing the current-voltage characteristic that apply to both device types.

The geometry of the device to be considered is sketched in Figure 14.3.1. For simplicity, we will work with a JFET device structure. The channel is assumed to be along the y direction and the width of the device is taken to be Z. The equilibrium width of the channel is $2a$ and is measured in the x direction. The voltage as a function of distance along the channel is $V(y)$. $W(y)$ is the junction depletion region width along the channel. As discussed in Section 14.2, the depletion region extends predominately within the n region since the doping is very much less than that within the p^+ region.

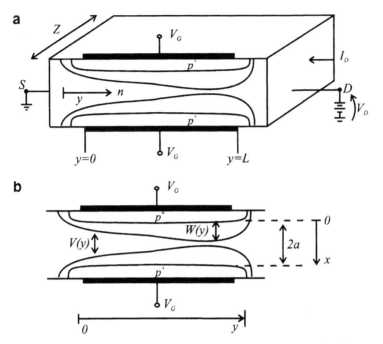

Figure 14.3.1. a. JFET device geometry used in deriving the drain-current expression. b. Cross-sectional view of the device.

In deriving an expression for the drain current it is useful to make some simplifying assumptions:

1. The p^+n junctions are assumed to be abrupt and are uniformly doped.
2. The device is symmetric about the plane $x = 2a$.
3. The current flow occurs within only the undepleted portion of the channel.
4. Breakdown conditions are avoided within each junction.
5. The voltage drop between the contacts and between the source region and $y = 0$ and the drain region and $y = L$ are negligible.

First consider the operation of the device if both the gate and the drain voltages are such that the device is biased below pinch off. The total current density consists of only electron flow since the JFET is a unipolar device. This is also true for a MESFET. In both device structures, the current flow between the source and the drain contacts flows from one carrier type, typically electrons. Therefore the total current density can be expressed solely in terms of the electron current density J_n as

$$\vec{J}_n = q\mu_n n \, \vec{F} + q D_n \vec{\nabla} n. \tag{14.3.1}$$

The electron concentration within the conducting channel is essentially equal to the doping concentration within the channel N_d, assuming all the dopants are ionized. If we neglect the diffusion component in Eq. (14.3.1) the magnitude of

J_n becomes

$$J_n = q\mu_n N_d F_y = -q\mu_n N_d \frac{dV}{dy}, \qquad (14.3.2)$$

where V is the potential drop from one end of the channel to the other along the source–drain direction. Since there are no sources or sinks of charge, the current flowing through any cross-sectional plane within the channel must be equal to the drain current in steady state, I_D. Therefore the drain current can be determined by integration of the expression for the current density over the cross-sectional area of the device:

$$I_D = -\iint J_n \, dx \, dz, \qquad (14.3.3)$$

which is negative because the current density is an electron flux. Equation (14.3.3) can be integrated over z since the current density is assumed to be uniform along the z direction. The drain current is found then when Eq. (14.3.2) is substituted into Eq. (14.3.3) to get

$$I_D = -Z \int_{W(y)}^{2a-W(y)} J_n \, dx = 2Z \int_{W(y)}^{a} q\mu_n N_d \frac{dV}{dy} \, dx. \qquad (14.3.4)$$

The integration along the x direction can be performed if the mobility and the carrier concentrations are both assumed to be independent of x. The assumption that the mobility is independent of x is effectively equivalent to assuming that it is field independent. Note that x is the direction along the depletion region widths in which the field changes with distance. It is reasonable to assume a field-independent or constant mobility if the field changes slowly or is very low. However, if the field changes strongly, such an assumption may not be valid. In Section 14.4, we will relax the assumption of a constant mobility and numerically calculate the current-voltage characteristic with a field-dependent mobility. Using the constant-mobility assumption, the integration over x can be performed as

$$I_D = 2q Z \mu_n N_d \frac{dV}{dy} x \Big|_{W(y)}^{a},$$

$$I_D = 2q Z \mu_n N_d \frac{dV}{dy} a \left[1 - \frac{W(y)}{a} \right]. \qquad (14.3.5)$$

Since the steady-state current is everywhere the same along y, the drain current can be integrated with respect to y to give

$$\int_0^L I_D \, dy = I_D L = 2q Z \mu_n N_d a \int_{V(0)\sim 0}^{V(L)\sim V_D} \left[1 - \frac{W(V)}{a} \right] dV,$$

$$I_D = \frac{2q Z \mu_n N_d a}{L} \int_0^{V_D} \left[1 - \frac{W(V)}{a} \right] dV. \qquad (14.3.6)$$

To proceed further, we need to find an expression for $W(V)$.

An additional assumption is now required. This assumption is called the gradual-channel approximation. It assumes that the rate of change of the potential is small in the y direction with respect to the x direction. In other words, the rate of change of the potential in the channel is much less along the y or channel direction than along the x direction. This assumption is reasonable provided that the channel length L, is very much larger than half the channel width a. Note that if the gate voltage is zero, the same potential drop occurs from 0 to L along the channel as from the center of the channel region to the p^+ contact. Therefore the depletion region width can be approximated with a one-dimensional model that varies only in the x direction. Hence the depletion region width as a function of voltage is given as

$$W(V) = \sqrt{\frac{2\epsilon_s(V_{bi} - V)}{qN_d}}, \tag{14.3.7}$$

where it is assumed that the acceptor concentration is very much larger than the donor concentration $N_a \gg N_d$, and ϵ_s is the dielectric constant of the semiconductor.

The applied voltage across the p^+n junctions, V, can be reexpressed in terms of the gate potential and the potential in the channel $V(y)$, as shown in Figure 14.3.2, as

$$V = V_G - V(y). \tag{14.3.8}$$

The expression for $W(V)$ given by Eq. (14.3.7) can be rewritten then as

$$W(V) = \sqrt{\frac{2\epsilon_s[V_{bi} + V(y) - V_G]}{qN_d}}. \tag{14.3.9}$$

It is important to realize that the depletion region width approaches a, half the channel width, as the gate voltage approaches V_p at zero applied drain voltage.

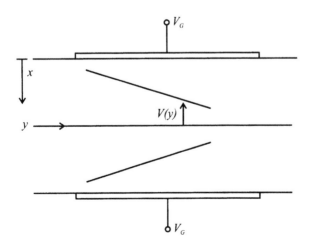

Figure 14.3.2. Diagram used for relating the applied potential drop appearing across the p^+n junctions to the gate and the channel voltages.

When the drain voltage vanishes, the voltage along the channel $V(y)$ is simply zero. With these realizations, we can obtain an expression for half the channel width a as

$$W(V = 0) = a = \sqrt{\frac{2\epsilon_s(V_{bi} - V_p)}{qN_d}}. \tag{14.3.10}$$

Dividing Eq. (14.3.9) by Eq. (14.3.10) yields

$$\frac{W(V)}{a} = \sqrt{\frac{V_{bi} + V(y) - V_G}{V_{bi} - V_p}}. \tag{14.3.11}$$

Therefore the drain current can be expressed as

$$I_D = \frac{2qZN_d a}{L} \int_0^{V_D} \left[1 - \sqrt{\frac{V_{bi} + V(y) - V_G}{V_{bi} - V_p}} \right] dV. \tag{14.3.12}$$

The integral in Eq. (14.3.12) can be evaluated directly since it is of the form

$$\int 1 - \sqrt{A + Bu}\, du: \quad A \equiv \frac{V_{bi} - V_G}{V_{bi} - V_p}, \quad B \equiv \frac{1}{V_{bi} - V_p}. \tag{14.3.13}$$

The integration yields

$$\int 1 - \sqrt{A + Bu}\, du = u + \frac{2}{3B}\sqrt{(A + Bu)^3}. \tag{14.3.14}$$

Therefore the drain current becomes

$$I_D = \frac{2q\mu_n ZN_d a}{L} \left\{ V_D + \frac{2}{3}(V_{bi} - V_p) \left[\left(\frac{V_{bi} - V_G}{V_{bi} - V_p} + \frac{V_D}{V_{bi} - V_p} \right)^{\frac{3}{2}} \right. \right.$$
$$\left. \left. - \left(\frac{V_{bi} - V_G}{V_{bi} - V_p} \right)^{\frac{3}{2}} \right] \right\}, \tag{14.3.15}$$

which is valid for biases below pinch off. When Eq. (14.3.15) is simplified, the drain current becomes

$$I_D = \frac{2q\mu_n ZN_d a}{L} \left\{ V_D - \frac{2}{3}(V_{bi} - V_p) \left[\left(\frac{V_{bi} - V_G + V_D}{V_{bi} - V_p} \right)^{\frac{3}{2}} \right. \right.$$
$$\left. \left. - \left(\frac{V_{bi} - V_G}{V_{bi} - V_p} \right)^{\frac{3}{2}} \right] \right\}, \tag{14.3.16}$$

which holds for bias voltages on the drain and the gate such that the gate–drain voltage difference is less than that needed to pinch off the channel.

On reaching pinch off, the current saturates. Therefore the drain current I_D becomes equal to the saturation drain current $I_{D, sat}$. It is important to realize that the pinch-off condition is attained at different drain voltages for different gate biases. Therefore the difference between the gate voltage and the saturation drain voltage, $V_G - V_{D, sat}$, is equal to the pinch-off voltage at zero drain bias V_p. This follows readily from the fact that, after pinch off, the width of the depletion region becomes equal to a at a channel bias $V = V_{D, sat}$. Therefore a can be found as

$$a = \sqrt{\frac{2\epsilon_s(V_{bi} + V_{D, sat} - V_G)}{q N_d}}. \tag{14.3.17}$$

Comparing Eq. (14.3.17) for a with that given in Eq. (14.3.10) for a yields

$$V_{D, sat} = V_G - V_p. \tag{14.3.18}$$

An expression for the saturation drain current can now be obtained as

$$I_{D, sat} = \frac{2q\mu_n Z N_d a}{L} \left\{ V_{D, sat} - \frac{2}{3}(V_{bi} - V_p) \right.$$
$$\left. \times \left[\left(\frac{V_{D, sat} + V_{bi} - V_G}{V_{bi} - V_p} \right)^{\frac{3}{2}} - \left(\frac{V_{bi} - V_G}{V_{bi} - V_p} \right)^{\frac{3}{2}} \right] \right\}, \tag{14.3.19}$$

which simplifies to

$$I_{D, sat} = \frac{2q\mu_n Z N_d a}{L} \left\{ V_G - V_p - \frac{2}{3}(V_{bi} - V_p) \left[1 - \left(\frac{V_{bi} - V_G}{V_{bi} - V_p} \right)^{\frac{3}{2}} \right] \right\}. \tag{14.3.20}$$

The pinch-off voltage can be determined as follows. Pinch off occurs at the drain end of the channel when the depletion region width becomes equal to half of the channel width. The depletion region can be determined from the formulas for a p^+n junction. Neglecting the built-in voltage with respect to the gate to drain voltage, we find that the width of the depletion region under the gate is given as

$$W = \sqrt{\frac{2\epsilon_s |V_G|}{q N_d}}. \tag{14.3.21}$$

Therefore the pinch-off voltage V_p is the gate voltage at which $W = a$, the half-width of the channel region. V_p is then

$$V_p = \frac{q a^2 N_d}{2\epsilon_s}. \tag{14.3.22}$$

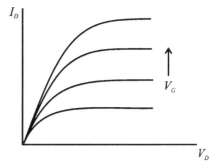

Figure 14.3.3. Sketch of the drain current I_D, versus drain voltage V_D, for an n-channel JFET device.

In the saturation regime, the current-voltage characteristic can be modeled by the following relationship, which is found to hold empirically:

$$I_{D,\,sat} = I_{D,0}\left[1 - \frac{V_G}{V_p}\right]^2, \qquad (14.3.23)$$

where $I_{D,0}$ is the drain current at zero applied gate voltage, in other words, when the gate is shorted to the source. It has been found experimentally that Eq. (14.3.23) approximates the saturation characteristic quite well in most JFET devices. The resulting current-voltage characteristic for an n-channel JFET is sketched in Figure 14.3.3.

The situation for a MESFET is similar to that derived above. Basically, the same expressions can be used to describe the behavior of a MESFET as for a JFET. However, the above formulas often do not provide highly accurate agreement with experimental measurements. Some of the approximations made in deriving the above expressions are not suitable for many actual device structures. For example, it was assumed that the mobility is field independent. This is typically a poor approximation, especially for materials such as GaAs. In addition, it was assumed that there is a negligible voltage drop between the active channel and the source and the drain contacts. However, in practice, there can be a significant voltage drop in these regions. As a means of improving the agreement with experiment, source and drain resistances are introduced into the model, as shown in Figure 14.3.4. The voltage at the source end of the channel now becomes $V = I_D R_s$, where R_s is the source resistance. Similarly, the voltage at the drain end of the channel is given as $V = V_D - I_D R_D$. In Section 14.4 we discuss numerical models of MESFETs in which nonideal effects are considered.

Before we finish this section it is important to introduce an important quantity called the transconductance

Figure 14.3.4. Simplified MESFET device sketch including source and drain resistances.

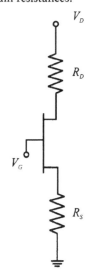

of the FET. The transconductance is defined as

$$g_m \equiv \left.\frac{\partial I_D}{\partial V_G}\right|_{V_D = \text{constant}}. \tag{14.3.24}$$

The transconductance is of importance in the small-signal model of the device. Basically, it relates the small-signal drain current to the small-signal gate voltage under saturation conditions. If the device is operated above pinch off, the small-signal drain current i_d is given as

$$i_d = g_m v_g, \tag{14.3.25}$$

where v_g is the small-signal gate voltage. In Section 14.18 we will discuss the small-signal models of JFETs and MESFETs and return to this discussion.

14.4 Numerical Analysis of MESFETs

In Section 14.3, we developed an analytical expression for the current-voltage characteristic for a MESFET and a JFET. Several assumptions were made in its derivation. In state-of-the-art MESFET structures, many of the assumptions made in deriving the analytical expression for the current are no longer valid. Principally, the gradual-channel approximation and the assumption that the mobility is field independent are suspect in many present device structures. As was mentioned in Section 14.3, in the gradual-channel approximation it is assumed that the rate of change of the potential is much less along the channel direction than perpendicular to the channel direction. This assumption is reasonable provided that the channel length is much larger than half of the channel width. In state-of-the-art MESFET structures, the channel length is typically less than 0.5 μm. In addition, owing to the relatively high electric-field strength within the channel, usage of a constant mobility is questionable. Finally, in the derivation of Eqs. (14.3.16) and (14.3.21), it was assumed that breakdown conditions do not occur. Again, as the channel length decreases, the field strength increases, which in turn can result in breakdown.

As an alternative to the analytical model presented in Section 14.3, detailed numerical models have been developed to study FETs and MESFETs in particular. Most of these models are based on the solution of the drift-diffusion equations along with the Poisson equation. The details of how these equations are obtained were presented in Section 6.3. As was discussed in that section, the drift-diffusion equations are derived directly from the Boltzmann equation by taking its first two moments. As such, the drift-diffusion equations do not provide an exact solution of the Boltzmann equation. Instead they provide only an approximate solution. Nevertheless, the drift-diffusion equations, when solved numerically coupled to the Poisson equation, provide a much more comprehensive solution than that provided by analytical solutions for the current.

Other numerical techniques besides the drift-diffusion approach are available. The ensemble Monte Carlo method, for example, provides an exact solution of the Boltzmann equation. The Monte Carlo method requires extensive computational resources since it solves the Boltzmann equation through the direct simulation of the carriers' trajectories. Because of its large computational expense, the Monte Carlo method is not well suited to studying device structures with large spatial dimensions, device structures that operate over many orders of magnitude different time scales, or devices in which thermal effects are important. For these reasons, other techniques, principally the drift-diffusion approach, are used in semiconductor device simulation.

The particular advantage of the drift-diffusion formulation is that it is computationally efficient, relatively easy to implement, and requires relatively little parameterization. Hence the drift-diffusion formulation is used extensively in semiconductor device simulation. However, the drift-diffusion solution is highly approximate since, by truncating the solution of the Boltzmann equation, it neglects many important physical effects such as thermal gradients, nonstationary transport effects, and energy-dependent phenomena. These effects, although of little importance in most large-dimensional structures and low-power devices, are of increasing importance in state-of-the-art submicrometer feature-length devices. For simplicity, we concentrate here on the usage of only numerical, drift-diffusion solutions; the reader is directed to the references for more advanced techniques.

In this section, we examine the numerical solution of the current-voltage characteristic for a GaAs-based MESFET by using the drift-diffusion equations. The model can be summarized as follows. The basic set of semiconductor equations used are determined following the procedure discussed in Section 6.3. We obtain the drift-diffusion solution by taking the first two moments of the Boltzmann equation. From the first moment, the electron and hole continuity equations can be obtained. The second moment provides the charge flux. These equations, coupled with the Poisson equation, constitute the drift-diffusion solution. The continuity equations are

$$\frac{\partial n}{\partial t} + \vec{\nabla} \cdot (n\vec{v}_n) = G - R \tag{14.4.1}$$

for electrons and

$$\frac{\partial p}{\partial t} + \vec{\nabla} \cdot (p\vec{v}_p) = G - R \tag{14.4.2}$$

for holes, where $n v_n$ and $p v_p$ are the electron and the hole fluxes, respectively.

The charge-flux equations are determined from the second moment of the Boltzmann equation. These are

$$n\vec{v}_n = \mu_n k_B T_e \vec{\nabla} n + n\mu_n \left[\vec{\nabla}(E_c - \chi) - k_B \vec{\nabla} T_e + \frac{3}{2} k_B T_e \frac{\vec{\nabla} m_e}{m_e} \right], \tag{14.4.3}$$

$$p\vec{v}_p = \mu_p k_B T_h \vec{\nabla} p + p\mu_p \left[\vec{\nabla}(E_v - \chi - E_g) - k_B \vec{\nabla} T_h + \frac{3}{2} k_B T_h \frac{\vec{\nabla} m_h}{m_h} \right], \tag{14.4.4}$$

where T_e and T_h are the electron and the hole temperatures, respectively, and χ is the affinity.

In addition to Eqs. (14.4.3) and (14.4.4) and the Poisson equation, auxiliary equations describing the mobilities and the generation–recombination mechanisms must be provided. Typically, all the pertinent generation–recombination mechanisms are included based on the results discussed in Chapter 10. The primary mechanisms for generation and recombination arise from thermal effects, three-particle Auger recombination, optical effects, impact ionization, surface recombination, and tunneling. Formulas for each of these effects have been discussed throughout this book.

The Shockley–Read–Hall mechanism is included by use of Eq. (10.3.35). The Auger formulation is given by

$$R_{\text{auger}} = (np - n_i^2)(C_n n + C_p p), \tag{14.4.5}$$

where the C's are the Auger coefficients. Impact ionization is included by use of the ionization rate coefficients determined in Chapter 10. The generation rate due to impact ionization can be written as

$$G_{ii} = \alpha \frac{|J_n|}{q} + \beta \frac{|J_p|}{q}, \tag{14.4.6}$$

where α and β are the electron and the hole ionization rate coefficients, respectively. Expressions for α and β were presented in Chapter 10.

In addition to the above formulations for the generation–recombination mechanisms, mobility models must also be included. Semiempirical relationships for the carrier mobilities are typically used. For electrons in GaAs, the most commonly used mobility model is given as

$$\mu_{\text{high}} = \frac{\mu_{\text{low}} + v_{\text{sat}} \frac{F^3}{F_{\text{crit}}^4}}{1 + \left(\frac{F}{F_{\text{crit}}}\right)^4}, \tag{14.4.7}$$

where μ_{low} is the low field mobility, which is usually assumed to be constant, F_{crit} is the critical electric field for intervalley transfer, and v_{sat} is the electron saturation velocity. For holes, the field-dependent mobility is given as

$$\mu_{\text{high}} = \frac{\mu_{\text{low}}}{\left[1 + \left(\frac{\mu_{\text{low}} F}{v_{\text{sat}}}\right)^\beta\right]^{\frac{1}{\beta}}}, \tag{14.4.8}$$

where β is a fitting parameter.

The continuity equations, Poisson equation, and carrier-flux equations are discretized and solved on a two-dimensional grid. The resulting system of

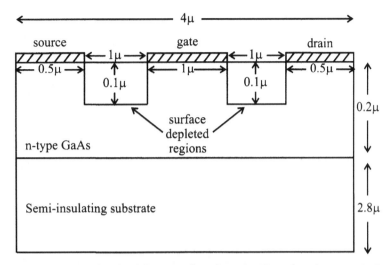

Figure 14.4.1. Sketch of the GaAs-based MESFET used in the numerical calculations.

equations is linearized by a multidimensional Newton's method and is solved with standard mathematical techniques. For a detailed discussion of the method of solving a system of differential equations for semiconductor device design the reader is referred to the book by Snowden (1988).

As an example of the method and as a means of examining the relevance of numerical simulation, a representative GaAs-based MESFET is simulated. The device structure is typical of those used in practice. The device structure is sketched in Figure 14.4.1. As can be seen from the figure, the gate length chosen is 1.0 μm and the source and drain to gate spacings are both 1.0 μm. The width of the n-type GaAs layer in which all the transport occurs is only 0.2 μm. For this structure, the gate length is substantially larger than half the channel width, which enables some comparison with the analytical formulation. Underneath the free surfaces between the source and the gate and the drain and the gate, the semiconductor region is assumed to be depleted. This depletion is called surface depletion. It often occurs in GaAs-based devices because the semiconductor surface typically contains significant surface states. These surface states will trap free electrons, thus depleting the semiconductor layer. For simplicity, it is assumed here that surface depletion regions of well-defined geometry occur. The entire structure is grown onto a semi-insulating substrate in which, of course, negligible conduction occurs. Therefore electron transport from the source to the drain occurs through the narrow channel formed between the gate and surface depletion regions and the substrate.

As discussed in Sections 14.2 and 14.3, the gate controls the conducting channel. Application of a gate bias in this device turns the device off. At zero gate bias then, the conducting channel exists between the source and the drain. The calculated drain current as a function of drain voltage at zero gate bias is

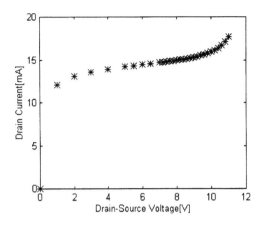

Figure 14.4.2. Calculated drain current as a function of applied drain voltage for the device shown in Figure 14.4.1. The increase in the drain current seen at voltages above ~11 V is due to breakdown.

shown in Figure 14.4.2. As can be seen from the figure, the current has a slight slope within the saturation region that arises from the source and the drain resistances. At ~11 V, the drain current begins to increase strongly with bias. At this voltage, the onset of avalanche multiplication occurs, leading to a significant increase in the drain current.

It is interesting to determine where within the device carrier multiplication is occurring. Multiplication, of course, depends strongly on the magnitude of the electric field. However, knowledge of the electric-field strength is, by itself, inadequate to determine where within a structure breakdown occurs. This is because the generation rate, which specifies the actual number of carriers that are produced, depends on not only the ionization rate but also on the current density as reflected by Eq. (14.4.6). Although the electric field can be sufficiently strong in a region to cause carrier multiplication, it does not guarantee that the device is in breakdown. This is because carrier multiplication requires, in addition to a high electric-field strength, initiating carriers to enter the region to trigger the avalanching. If no carriers enter the high-field region, carrier multiplication and hence breakdown cannot occur. Therefore, to identify where within a device breakdown occurs, it is necessary to examine both the electric-field profile and the current density. Both of these quantities for the MESFET structure sketched in Figure 14.4.1 are plotted in Figure 14.4.3.

As can be seen from Figure 14.4.3, the electric field peaks at the surface in two positions, near the gate and near the drain. The large increase in the electric field at these two points is due to the rather abrupt change in the potential from the metal gate to the semiconductor surface. Again, a high electric field is present at the surface of the semiconductor between the drain contact and the semiconductor region. This is also due to the rather abrupt change in potential in moving from the metal to the semiconductor surface.

In contrast, the current density peaks only near the drain. The current is much higher near the drain contact than at the gate since most of the carriers exit the device at this point. In other words, current crowding occurs at the drain edge on the gate side of the contact. Coupled with the high electric-field

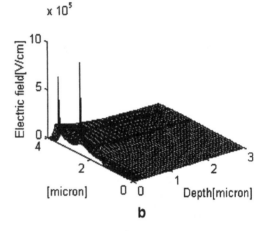

Figure 14.4.3. a. Plot of the current density, b. electric-field strength in the device structure sketched in Figure 14.4.1 at the onset of breakdown. Note that the peak electric field occurs at the surface of the device on the drain side of the gate region.

strength, the maximum generation rate should occur at only one point, near the drain at the surface of the device. As can be seen from Figure 14.4.4., the generation rate peaks at precisely this point. Therefore breakdown within the MESFET occurs at the surface of the device, not within the bulk. Typically, most MESFET structures suffer surface breakdown rather than bulk breakdown.

Figure 14.4.4. Calculated impact ionization generation rate within the MESFET device of Figure 14.4.1. Note that the generation rate is largest near the drain contact because of the combined effects of the large electric field and current density.

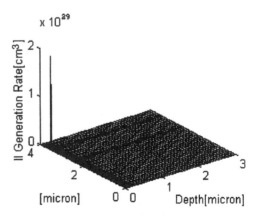

Usage of a numerical model, such as the one presented here, provides an accurate means of determining the breakdown voltage and its location within a MESFET. Numerical models greatly extend the range of standard theoretical tools, such as the analytical models developed in Section 14.3. Numerical device simulation, combined with theory and experiment, provides a comprehensive means of understanding, evaluating, and optimizing semiconductor devices.

14.5 A Brief Introduction to HEMTs

As the reader is well aware, most transistor devices are of the MOSFET type. MOSFETs are discussed in Section 14.6. However, before we discuss MOSFETs, it is useful to complete our introduction to MESFETs by discussing one of thin derivatives that has become of great importance in compound semiconductor applications.

As was discussed in Sections 11.4 and 12.3, the principal virtue of the MOS system is that nearly perfect material, with an extremely low number of interface states, can be prepared. For this reason, MOSFET devices can be manufactured with an extremely high degree of reliability and reproducibility. This makes MOSFET devices the overwhelming choice for device types in present and most likely future integrated circuit applications.

Although MOS devices are currently the key component in the vast majority of commercial semiconductor device products, they are potentially less attractive than other devices for $>\sim 5$ GHz device applications. The emerging industry of cellular telephones will require higher bandwidths as the usage of cellular phones increases. To provide higher bandwidths, the frequency operation of cellular phones will have to progress to higher frequencies. In addition, to extend the spatial operating distances of cellular phones and reduce the number of substations needed, higher operating power is also desirable. To achieve the dual goals of higher-power delivery and higher-frequency operation, transistor amplifiers must be improved.

Typically, higher-frequency operation of a transistor can be attained through either a reduction in the device dimensions or by an increase in the operating speed of the device itself. There is of course a limit to which one can reduce the dimensions of a device. Although short-channel MOSFET devices can operate at fairly high frequencies, they fall short of the performance of FETs made from other materials such as GaAs or InP. A simple, although somewhat crude argument for why GaAs and InP FETs can achieve a higher frequency of operation than silicon-based FETs is that the electron mobility is substantially higher in GaAs and InP than in silicon. The mobility is higher in GaAs and InP than in silicon since the effective mass of the electrons is substantially lower and the phonon-scattering rate is somewhat lower in these materials. From the simple formula derived for the mobility in Section 6.2,

$$\mu = \frac{q\tau}{m^*},$$ (14.5.1)

where we have used the effective mass m^* in place of the free-electron mass, it is clear that the mobility is higher in materials for which both the effective mass and the scattering rate (inversely proportional to τ) are smaller.

The problem that one encounters when using GaAs- or InP-based FETs is that of finding an acceptable insulator on which to place the gate contact. For silicon, silicon dioxide is a native oxide and thus can be grown to produce an excellent and highly controllable and reproducible insulating layer on which to place the gate. Unfortunately, there is no insulating material that can be readily grown onto either GaAs or InP that offers anywhere near the degree of excellence of silicon dioxide on silicon. What is typically done instead is to use a Schottky barrier as the gate contact allowing the fabrication of a MESFET. As was discussed in the previous sections, the current flow in a MESFET is within the bulk material. Therefore the mobility of the carriers contributing to the current flow is that corresponding to the bulk material.

As mentioned above, the low field bulk mobility of GaAs and InP is quite large, significantly higher than in silicon, at least for intrinsic material. If the material is doped, the mobility of course decreases because of the increase in impurity scattering and the concomitant reduction in τ. In order for the MESFET to deliver a sizable current, there must be free carriers in the device. The source of these free carriers comes from doping the material. Subsequently, a trade-off occurs. Increased doping of the GaAs or InP increases the current-carrying capacity of the FET but at the expense of a decrease in the mobility and hence the speed of operation of the device. Therefore one cannot in general make a MESFET that will have a very high operating frequency and deliver a high current.

Ideally, one desires a device with a high mobility and a high carrier concentration. The high-electron-mobility transistor (HEMT) provides such a structure.

HEMTs utilize modulation doping to increase the free-electron carrier concentration without increasing the scattering rate. As discussed in Section 11.2, if a wide-bandgap material, such as AlGaAs, is a highly doped n type forming a heterojunction with a narrow gap, intrinsic material, such as GaAs, then electrons will move from the wide-bandgap material to the narrow-gap material to equilibrate the Fermi level. The resulting energy-band diagram for a representative GaAs/AlGaAs modulation-doped structure is shown in Figure 14.5.1. As can be seen from Figure 14.5.1, a notch forms in the conduction-band edge at the GaAs/AlGaAs interface. The transferred electrons from the AlGaAs layer accumulate within this notch and are spatially separated from the donor atoms within the AlGaAs layer. As a result, the electron concentration within the narrow-gap material near the interface is very much increased. The increase in the electron concentration occurs without the introduction of impurities and hence, scattering centers, within the narrow-gap layer. Therefore the high mobility of the intrinsic GaAs is retained while the free-electron concentration is enhanced. Modulation doping produces a high carrier concentration without sacrificing the carrier mobility.

The most common heterojunction system used for HEMT devices is the GaAs/AlGaAs system. A typical HEMT structure is sketched in Figure 14.5.2.

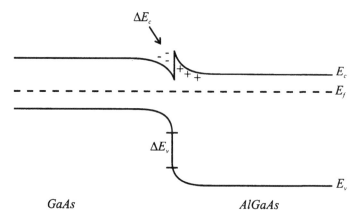

Figure 14.5.1. Sketch of the energy-band diagram for a GaAs/AlGaAs modulation-doped structure. Note that the electrons accumulated in the conduction-band notch within the GaAs layer are spatially separated from the donors in the AlGaAs layer.

As can be seen from Figure 14.5.2, a thin intrinsic AlGaAs layer is sandwiched between the doped AlGaAs layer, which provides the electrons for the modulation doping, and the intrinsic GaAs layer in which the transport occurs. By insertion of the intrinsic AlGaAs layer, the donor ions are separated more from the charge carriers, further reducing the effects of ionized impurity scattering. As a consequence, the carrier mobility approaches that of intrinsic GaAs.

As in a standard FET device, the electron flux is from the source to the drain under standard biasing conditions. The conducting channel formed between the source and the drain exists at the AlGaAs/GaAs interface. By the modulation-doping effect, the GaAs layer becomes highly conducting and thus produces a highly conducting channel between the source and the drain. A simplified HEMT device structure is sketched along with the corresponding energy-band

Figure 14.5.2. Sketch of a GaAs/AlGaAs HEMT device.

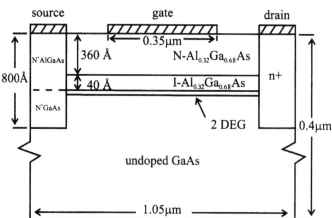

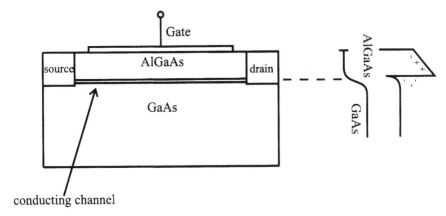

Figure 14.5.3. Sketch of a HEMT device structure along with the energy-band structure.

diagram of the device in Figure 14.5.3. As can be seen from Figures 14.5.3 and 14.5.1, the band bending at the GaAs/AlGaAs interface results in a narrow notch within the conduction band. Since the dimensions of the notch are typically comparable with the de Broglie wavelength of an electron, spatial quantization effects can occur. As discussed in Section 11.2, the notch can be approximated by a triangular well and the spatial quantization levels can be determined. Much of the transport occurs within these quantized levels.

The primary advantage of the HEMT over a MESFET is that it can operate with a high carrier concentration without sacrificing the high carrier mobility of intrinsic GaAs. It should be noted however, that the high carrier mobility is important only near the source contact of the device, where the electric field is low. As the carriers near the drain end of the device, hot-electron effects dominate the transport physics and dictate the speed of response of the device. For a full description of the workings of this device, the reader should consult the references.

14.6 MOSFETs: Long-Channel Devices

Arguably, the most important semiconductor device is the MOSFET. The key device within an integrated circuit is the MOSFET. As the reader is well aware, the development and the application of the integrated circuit have lead to the computer or information-age revolution. Virtually all aspects of modern society are affected by the integrated circuit and hence by the MOSFET.

The MOSFET is a three-terminal device, much like the JFET and MESFET, in which the drain current is controlled through use of the field effect. As described in the previous sections, the field effect provides a means by which the conductivity of the underlying semiconductor material can be modulated by the application of an electric field to the gate electrode. In the previous sections, a p^+n junction or a Schottky barrier formed the key structure in JFETs and MESFETs, respectively, through which the field effect was used. In a MOSFET,

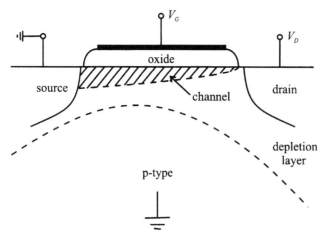

Figure 14.6.1. Schematic drawing of an *n*-channel MOSFET device.

the gating action is provided by a MIS structure, wherein the semiconductor is silicon and the insulator is its native oxide, silicon dioxide. In Section 11.8, the operation under bias of a MIS structure was examined in detail. As was discussed, the semiconductor layer can be biased into one of three different states, accumulation, depletion, or inversion, depending on the nature of the applied gate bias. The MOSFET is designed in a manner similar to that of the MESFET in that two islands of highly doped *n*-type or *p*-type material (both are of course the same in an individual device) constituting the source and the drain are separated by a semiconductor material of different doping concentration. Depending on the applied gate bias, the region separating the source and the drain can be inverted such that a low-resistance channel opens, connecting the source and the drain contacts. A typical MOSFET device is sketched in Figure 14.6.1.

The device sketched in Figure 14.6.1 is an *n*-channel MOSFET. When the gate bias is such that the underlying semiconductor region is biased into inversion, the source and the drain are connected electrically by the inversion region. Under these conditions, an ohmic current flow arises; the drain current versus drain–source voltage characteristic is linear.

As in MESFETs, there are two general types of devices, enhancement- and depletion-mode structures. Enhancement-mode devices are ones in which the channel does not exist at zero applied gate bias. In other words, no low-resistance path connecting the source and the drain regions exists in equilibrium. Depletion-mode devices, in contrast, are structures in which the channel exists in equilibrium and under low drain bias. A gate bias must be applied in order to turn the device off.

There are also two different device types in terms of carrier type. The MOSFET, like the JFET and MESFET, is a unipolar device; only one carrier species principally contributes to its operation. Therefore either *n*- or *p*-channel devices can be made in which electrons or holes, respectively, act as the current-carrying species.

The actual performance of a MOSFET device depends on the existence of a conducting channel being formed under the gate contact. Since the source and the drain contact regions are of a different doping type than the bulk semi-conductor region, in order to form a conducting channel the semiconductor layer must be biased into inversion. There is a minimum gate voltage, called the threshold voltage, V_T, for which the semiconductor becomes inverted. The threshold voltage is simply equal to the sum of the flat band voltage, the voltage necessary first to make the bands flat and the amount of voltage necessary to achieve strong inversion $2\psi s$, as defined in Section 11.8. The threshold voltage can be expressed as

$$V_T = \phi_{\mathrm{ms}} - \frac{Q_i}{C_i} - \frac{Q_d}{C_i} + 2\psi_s, \tag{14.6.1}$$

where ϕ_{ms} is the metal–semiconductor work function difference, Q_i/C_i is the cumulative effect of oxide, traps, and impurity charge, and Q_d/C_i is the voltage needed to accommodate the depletion region. Q_d is the doping concentration within the channel region.

Let us restrict our discussion to the Al/Si material system for simplicity. Although polysilicon is more often used in state-of-the-art integrated circuits, the workings of the MOSFET remain virtually the same. In the Al/Si system, the metal–semiconductor work function difference is always negative. In addition, the impurity charge Q_i is always positive since it is due mostly to ionized sodium atoms. Hence the second term in Eq. (14.6.1) is also always negative. The last two terms in Eq. (14.6.1) can be either positive or negative, depending on the device. If the device is an n-channel structure, then Q_d is negative (ionized acceptors are formed within the channel region because of depletion of the p-type doping in the bulk semiconductor material). The term Q_d/C_i is then positive in Eq. (14.6.1). Conversely, if the device is a p-channel structure, then Q_d is positive (ionized donors are formed within the channel region because of depletion of the n-type doping in the bulk semiconductor material). Finally, the last term in the threshold voltage expression is positive if the device is an n-channel device and is negative if the device is a p-channel device.

A p-channel MOS device always operates in the enhancement mode independent of the doping concentration. This follows readily from Eq. (14.6.1) since all the terms for the threshold voltage are always negative when applied to the Al/n-Si material system. Hence the threshold voltage is always negative, which implies that a negative gate bias is necessary in order to induce a conducting channel underneath the gate. An n-channel device can be operated in either the depletion or the enhancement mode depending on the value of the threshold voltage, which depends in turn on the acceptor doping concentration.

Let us consider the current flow in an enhancement-mode MOSFET. As is the situation in a JFET, at a small drain–source bias the current flow is linear with the applied drain voltage at fixed gate bias, provided the semiconductor is inverted. As the drain bias increases, it eventually reaches a point at which the channel near the drain end is no longer inverted. As in a JFET device, the channel pinches off near the drain under different drain and gate bias conditions. This

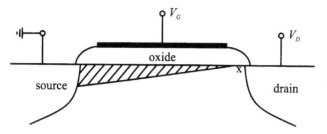

Figure 14.6.2. MOSFET structure showing the pinch-off point, marked as X in the drawing.

can be easily understood from consideration of Figure 14.6.2. The potential difference at the drain end of the channel, marked as point X in the figure, is small if the applied gate voltage V_G and drain–source voltage V_D are comparable. Consequently, point X is not sufficiently positive with respect to the substrate to remain inverted. Hence the conducting channel is pinched off at this point. Similar to the JFET operation, after pinch off, further increase in the drain–source voltage fails to increase the current. The number of carriers arriving at point X, the pinch-off point, remains the same since the voltage at the pinch-off point does not change; the pinch-off point only moves closer to the source. Of course, the actual current increases in saturation somewhat since the effective channel length decreases from L to L' with increased V_D.

It is important to realize that most of the source–drain voltage drop occurs across the pinch-off point. This follows since the pinch-off point is highly resistive because the conductive channel is choked off in this region. Therefore the source–drain region can be modeled as two resistors in series. The first region, the conductive channel, has a very small resistance, while the second region, the noninverted pinch-off point, is highly resistive. Hence the voltage divider rule predicts that the voltage drop across the pinch-off point is very large. As a consequence, the electric field within the pinch-off region is very large.

The current-voltage characteristic for an ideal MOSFET can be derived in several ways. We first make some assumptions to simplify the analysis:

1. The gate is an ideal MOS.
2. The drift current is the only relevant current channel.
3. The mobility of the carriers in the inversion layer is constant.
4. The channel is uniformly doped.
5. The leakage current is negligible.
6. The gradual channel approximation holds; the electric field in the direction perpendicular to the MOS surface is much larger than the longitudinal electric field.

The basic device structure under different drain bias conditions is sketched in Figure 14.6.3. In Figure 14.6.3.a, the gate bias exceeds the threshold voltage while the drain bias is close to zero. Hence the current-voltage characteristic is essentially linear. As the drain voltage is increased, the depletion region spreads deeper into the bulk semiconductor material (often called the substrate region) until pinch off is attained. When the drain voltage exceeds $V_G - V_T$, the

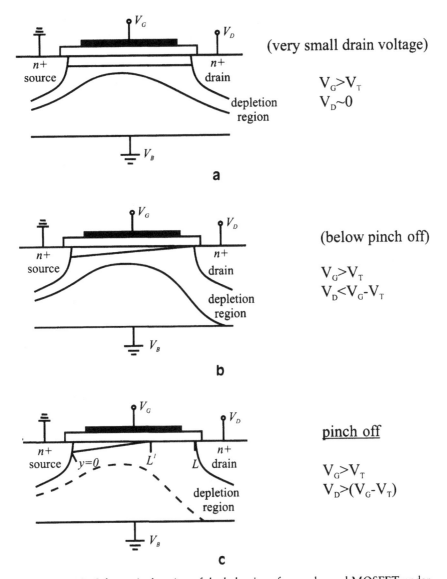

(very small drain voltage)

$V_G > V_T$
$V_D \sim 0$

a

(below pinch off)

$V_G > V_T$
$V_D < V_G - V_T$

b

pinch off

$V_G > V_T$
$V_D > (V_G - V_T)$

c

Figure 14.6.3. Schematic drawing of the behavior of an *n*-channel MOSFET under different biasing conditions, a. Small applied drain voltage. b. Below pinch off. c. Pinch off.

difference in the gate and the threshold voltages, the channel becomes pinched off. This is because the voltage difference between the gate and the underlying semiconductor region near the drain is no longer greater than the threshold voltage. Hence the region marked as L' to L in the figure is no longer inverted but is only depleted. As mentioned above, once the layer is no longer inverted, it becomes highly resistive.

The easiest manner by which the current-voltage characteristic can be computed is to use charge-control analysis. The first condition we consider is that of small applied source–drain voltage V_D, but with the channel region biased into strong inversion. In this case, the current is simply given by the quotient of the channel charge Q_n and the carrier transit time t_{tr} as

$$I_D = -\frac{Q_n}{t_{tr}}. \qquad (14.6.2)$$

If only a drift-current component exists, which is a reasonable assumption below pinch off, then the carrier transit time can be expressed as

$$t_{tr} = \frac{L}{v_d}, \qquad (14.6.3)$$

where L is the channel length and v_d is the drift velocity. The drift velocity can be reexpressed in terms of the channel mobility μ_n as

$$v_d = -\mu_n F = \mu_n \frac{V_D}{L}. \qquad (14.6.4)$$

The transit time becomes then

$$t_{tr} = \frac{L^2}{\mu_n V_D} \qquad (14.6.5)$$

and the channel charge is

$$Q_n = -C_i(V_G - V_T)WL. \qquad (14.6.6)$$

Therefore the drain current is

$$I_D = -\frac{Q_n}{t_{tr}} = C_i(V_G - V_T)WL\mu_n \frac{V_D}{L^2}, \qquad (14.6.7)$$

which simplifies to

$$I_D = C_i \frac{W}{L}(V_G - V_T)\mu_n V_D. \qquad (14.6.8)$$

Clearly, if V_D is small, the drain current increases linearly with the drain voltage V_D.

As the drain voltage increases, it changes the channel bias near the drain. The voltage near the drain end of the device relative to the gate decreases. As a result, the voltage difference of the gate electrode and the underlying semiconductor region, that is, the channel, becomes less, reducing the amount of inversion layer charge Q_n. Let the average voltage above threshold between the gate and the channel be $V_G - V_D/2$. The channel charge Q_n becomes then

$$Q_n = -C_i\left(V_G - V_T - \frac{V_D}{2}\right)WL, \qquad (14.6.9)$$

which yields for the drain current

$$I_D = \mu_n \frac{W}{L} C_i \left(V_G - V_T - \frac{V_D}{2} \right) V_D. \tag{14.6.10}$$

Finally, as the drain voltage increases further, the channel becomes pinched off. The region $L' - L$ is no longer inverted but is in depletion instead. Hence the resistance of this region is much greater than within the channel. Most of the voltage drop between the source and the drain then occurs across the pinch-off region. Therefore the electric field within the pinch-off region is very high.

The current depends not on how fast the electrons are swept out into the drain, but on the rate at which the carriers arrive at the edge of the depletion region formed at the pinch-off point. This situation is similar to that for a reverse biased junction wherein the current flow occurs through a depletion region. To first order, the carrier arrival rate is insensitive to the drain voltage V_D. Consequently the drain current saturates for $V_{D,\,sat} = V_G - V_T$. Under this condition, the drain current becomes

$$I_D = \frac{\mu W C_i}{2L} (V_G - V_T)^2, \tag{14.6.11}$$

which approximates the saturated drain current, often quite well.

A more precise formulation of the current-voltage characteristic was first provided by Pao and Sah (1966). This formulation includes both the drift and the diffusion currents. The drain-current expressions derived above, wherein only the drift current is considered, is valid as long as the device is not in saturation. However, once the device is pinched off, it is essential to include the diffusion current in order to obtain the drain current. In the above, only an approximate formulation for the saturated drain current was obtained. With the Pao and Sah model, a more exact solution is possible.

The drain current can be determined by integration of the two-dimensional current density over the cross-sectional area of the device. If the y direction is chosen from the source to the drain and the x direction is from the top of the device into the substrate, the current can be given as

$$I_D = \int_0^x J(x, y) Z \, dx, \tag{14.6.12}$$

where Z is the device width. For an n-channel device, the current density can be approximated as an electron current only, J_n. The electron current density is given as

$$\vec{J}_n = q \mu_n n \vec{F}_y + q D_n \vec{\nabla} n, \tag{14.6.13}$$

where F_y is the electric field along the source–drain direction, that is, along y. Using the expression for the nonequilibrium electron concentration given by

Eqs. (11.5.1), we find that

$$n = n_i e^{\frac{q(\psi - \phi_n)}{kT}}.$$

(14.6.14)

Taking the gradient of n and simplifying yields

$$\vec{\nabla} n = \frac{qn}{kT}[\vec{\nabla}\psi - \vec{\nabla}\phi_n].$$

(14.6.15)

But the gradient of ψ is simply given as

$$-\vec{\nabla}\psi = +\vec{F}_y.$$

(14.6.16)

With the above substitutions, the diffusion current can be reexpressed as

$$q D_n \vec{\nabla} n = -q\mu_n n \vec{F}_y - \frac{q^2 D_n}{kT} n \vec{\nabla}\phi_n.$$

(14.6.17)

When expression (14.6.17) for the diffusion current is combined with the drift current, the electron current density becomes

$$\vec{J}_n = -q\mu_n n \vec{\nabla}\phi_n.$$

(14.6.18)

Substituting Eq. (14.6.18) for the current density into Eq. (14.6.12) and integrating over y, we find that the drain current is given as

$$\int_0^L I_D \, dy = \int_0^L \int_0^x q\mu_n Z n(x, y) \frac{d\phi_n}{dy} \, dx \, dy,$$

(14.6.19)

where it is assumed that the quasi-Fermi level varies only in the y direction. Evaluating the integral on the left-hand side yields finally

$$I_D = \frac{1}{L} \int_0^L q\mu_n Z \frac{d\phi_n}{dy} dy \int_0^x n(x, y) \, dx.$$

(14.6.20)

The above result can be integrated numerically. It provides an improved estimate of the drain current in a long-channel MOSFET. Of course, it is necessary to provide some estimate of the electron carrier concentration as well as the variation of the quasi-Fermi level.

Pao and Sah evaluated the current density numerically for a long-channel MOSFET. They have shown that this solution is valid for the entire range of drain–source voltages.

To complete this section, it is useful to provide an expression for the current density in the subthreshold regime. When the gate voltage is below the threshold voltage in an enhancement-mode device and the surface is in weak inversion, the analytical expressions for the drain current as a function of drain bias derived above no longer hold. As the reader may recall, the analytical expressions for the drain current were obtained assuming that the surface was inverted. When the surface is not inverted and a drain voltage is applied, the drain current is not necessarily equal to zero. The current that flows between the source and the drain under this condition is called the subthreshold current.

The subthreshold current is important in digital circuit design. The MOSFET in digital circuits is switched into cutoff in order to turn the drain current off. Cutoff corresponds to operation within the subthreshold region. Ideally no drain current flows with drain bias when the device is biased into cutoff. However, in general a small subthreshold current flows. This has the unfortunate effect of drawing current when the device is switched off, leading to significant power consumption in integrated circuits.

The subthreshold current can be estimated as follows. In weak inversion, the current flow is dominated by diffusion. This is because the channel resistance is large. The current density is given then by

$$J_{\text{diff}} = q D_n \frac{n(0) - n(L)}{L}, \tag{14.6.21}$$

where $n(0)$ and $n(L)$ are the electron concentrations at the source and the drain ends of the channel, respectively. The current is obtained when J_{diff} is multiplied by the cross-sectional area that is given by the product of the width of the transistor and the effective channel thickness. The channel thickness is assumed to be given by the depth into the bulk material at which the potential changes by kT/q. Let F_s be the surface field; the channel thickness x_{ch} is given as

$$x_{\text{ch}} = \frac{kT}{q F_s}. \tag{14.6.22}$$

The surface electric field can be found from the solution of Poisson's equation under the depletion approximation to be

$$F_s = \sqrt{\frac{2 \psi_s q N_B}{\epsilon_s}}, \tag{14.6.23}$$

where ψ_s is the surface potential and N_B is the doping concentration. The electron concentration at the source end of the channel is given as

$$n(0) = n_i e^{\frac{q(\psi_s - V_{SB} - \psi_B)}{kT}}, \tag{14.6.24}$$

while at the drain it is

$$n(L) = n_i e^{\frac{q(\psi_s - V_{DB} - \psi_B)}{kT}}, \tag{14.6.25}$$

where V_{SB} and V_{DB} are the source–substrate and drain–substrate voltages and $q\psi_B$ is the energy difference between the intrinsic level and the Fermi level, as defined in Section 11.8. Substituting Eqs. (14.6.24) and (14.6.25) into Eq. (14.6.21) and using Eqs. (14.6.22), (14.6.23), and the Einstein relation, I_D becomes

$$I_D = W V_t \sqrt{\frac{\epsilon_s}{2 q N_B \psi_s}} \frac{q \mu V_t n_i}{L} \left[e^{\frac{\psi_s - V_{SB} - \psi_B}{V_t}} - e^{\frac{\psi_s - V_{DB} - \psi_B}{V_t}} \right], \tag{14.6.26}$$

where V_t is the thermal voltage, kT/q. Recognizing that

$$V_{DB} = V_{DS} + V_{SB} \tag{14.6.27}$$

and that

$$N_B = n_i e^{\frac{\psi_B}{V_t}}, \tag{14.6.28}$$

we find that the expression for the drain current becomes

$$I_D = \frac{q\,W\mu\,V_t^2 n_i}{L} \sqrt{\frac{\epsilon_s}{2qN_B\psi_s}}\, e^{\frac{\psi_s - V_{SB} - \psi_B}{V_t}} \left[1 - e^{\frac{-V_{DS}}{V_t}} \right], \tag{14.6.29}$$

or, equivalently,

$$I_D = \frac{q\,W\mu\,V_t^2 n_i}{L} \sqrt{\frac{\epsilon_s}{2qn_i\psi_s}}\, e^{\frac{\psi_s - V_{SB} - 1.5\psi_B}{V_t}} \left[1 - e^{\frac{-V_{DS}}{V_t}} \right]. \tag{14.6.30}$$

At this point, the surface potential ideally should be solved as a function of V_G. However, an explicit solution cannot be determined. Therefore different approximations have been made in the literature to complete the expression for the drain current. One such relationship, given by Grotjohn and Hoefflinger (1984) is

$$I_D = \frac{\mu\,WC_{ox}}{LA} V_t^2 e^{\frac{A(V_G - V_T)}{V_t}} \left[1 - e^{\frac{-V_{DS}}{V_t}} \right], \tag{14.6.31}$$

where V_T is the threshold voltage, C_{ox} is the oxide capacitance, V_G is the gate voltage, and A is given as

$$\frac{1}{A} = \frac{C_{ox} + C_B + C_{FS}}{C_{ox}}, \tag{14.6.32}$$

where C_B is the depletion capacitance and C_{FS} is the capacitance due to fast surface states.

14.7 Short-Channel Effects in MOSFETs

In Section 14.6 we derived expressions for the drain current based on the assumption that the channel is long. As the channel length decreases, the use of the gradual-channel approximation becomes suspect. The gradual-channel approximation assumes that the effect of the longitudinal electric field is negligible with respect to the transverse field. As the channel length decreases, the depletion layer widths become comparable with the channel length and the field becomes effectively two dimensional. Short-channel effects are quite important in present, state-of-the-art MOSFET devices, since feature lengths are near $0.3\ \mu\text{m}$.

There are many effects that alter device performance at short-channel lengths. These effects arise from three primary sources: 1, the electric-field profile becomes two dimensional; 2, the electric-field strength in the channel becomes very high; and 3, a decrease in the physical separation between the source and the drain. The most important short-channel effects can be classified into one of these sources as follows:

I. Two-dimensional potential profile.
1. threshold voltage reduction; depletion layer charge is effectively lowered.
2. mobility reduction by gate-induced surface fields.
II. High electric fields present within the channel.
1. carrier-velocity saturation
2. impact ionization near the drain
3. gate oxide charging
4. parasitic bipolar effect
III. Decrease in the physical separation between the source and drain.
1. punchthrough
2. channel-length modulation

In this section, we discuss each of the above effects, its origin, and its effect on the device operation.

The two-dimensionality of the potential profile alters the device performance since a considerable amount of the bulk charge Q_B is no longer controlled by the gate voltage but depends on the drain–source voltage. The close proximity of the source and the drain regions cause a fraction of the bulk charge directly under the channel to have field lines terminated at the source and the drain rather than the channel. Consequently, as the channel length decreases, a larger fraction of the bulk charge under the channel has field lines terminated at the source and the drain junctions. The bulk charge appears in the expression given for the threshold voltage, Eq. (14.6.1) (for the specific case considered in Section 14.6, Q_d is the bulk charge). As can be seen from inspection of Eq. (14.6.1), the threshold voltage increases with increasing Q_d. Therefore, since the gate voltage controls less depletion layer charge, the threshold voltage is effectively reduced. In addition, the surface potential is increased, resulting in an increase in the subthreshold current. This degradation is often referred to as drain induced barrier lowering, DIBL.

The transverse electric fields, which become of increasing importance as the channel length decreases, alter the carrier mobility. It has been found that the action of the gate field, normal to the channel in an MOS transistor, degrades the carrier mobility. The mobility reduction is associated, to a some extent, with enhanced surface scattering at the MOS interface. The reduction in the surface mobility can be modeled as

$$\mu = \frac{\mu_0}{1 + \theta(V_{GS} - V_T)}, \tag{14.7.1}$$

where μ_0 is the mobility at the threshold voltage and θ is the mobility reduction factor.

The second category of effects arise from the high electric field present in the channel. As the channel length decreases, if the voltage is not reduced, the channel electric field increases substantially. The first effect we consider is carrier-velocity saturation. Typically, in silicon the carrier drift velocity increases linearly with the applied electric field. However, both the electron and the hole velocities saturate at applied electric fields in excess of ~100 kV/cm. In short-channel devices, the electric field near the drain can attain values in excess of 400 kV/cm. In general, the velocity saturation is given by

$$v = \frac{\mu_0 F_y}{\left[1 + \left(\frac{F_y}{F_c}\right)^\alpha\right]^{\frac{1}{\alpha}}}, \tag{14.7.2}$$

where F_c is the critical electric field and F_y is the channel field. α is a parameter. Typically, α has a value near 2.

When the drain–source voltage is high, the electric field near the drain end of the device can be large enough to cause impact ionization of carriers moving from the source. The impact ionization rate near the drain has a strong field dependence. If the electric field near the drain end of the device exceeds 150 kV/cm, electron-initiated impact ionization can be appreciable. The generated electrons are swept into the drain, increasing the drain–source current, while the generated holes are swept into the substrate, as is seen from Figure 14.7.1. As can be seen from the figure, carrier multiplication at the drain end of the device leads to hole injection into the substrate. The injected holes create what is called the parasitic bipolar effect, since the holes act to forward bias the source/substrate junction, leading to additional electron injection from the source into the channel. The net result is a further increase in the drain current.

Figure 14.7.1. Sketch of a MOSFET showing parasitic bipolar operation.

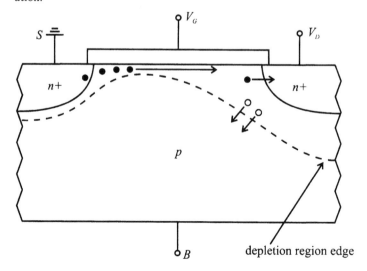

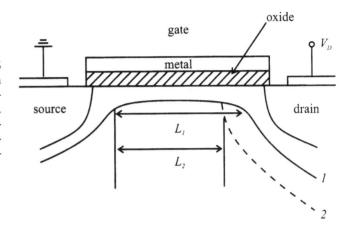

Figure 14.7.2. Sketch of an MOS device showing channel-length modulation. Curve 1 is the depletion region edge for $V_{D1} < V_{D2}$. Note that as the voltage increases to V_{D2}, the depletion region expands, reducing the effective channel length.

The last effect we consider that can be categorized as a high-field effect is gate-oxide charging. The very high channel electric field heats the carriers to very high kinetic energy near the drain. The carriers, typically electrons, can reach sufficient energy such that they can transfer from the silicon into the gate oxide. The actual amount of energy that the electrons must attain is quite high. Nevertheless, barrier lowering and quantum-mechanical broadening can act to reduce the SiO_2 barrier. Carrier injection results in oxide charging, which leads to threshold voltage shifts. It should be noted that threshold voltage shifts due to gate-oxide charging occur over time. With continued use, the electron concentration within the SiO_2 increases, ultimately reaching the level at which the threshold voltage can be affected. Gate-oxide charging then is a long-term degradation mechanism in a MOSFET.

The last category of short-channel effects includes those due to the decrease in the physical separation of the source and the drain. The most important effects in this category are channel-length modulation and punchthrough. The channel-length modulation effect is caused by the shortening of the effective channel length of the transistor because of an increase in the drain depletion region as the drain voltage is increased. The effect is sketched in Figure 14.7.2. As can be seen from the figure, the effective channel length is reduced. The resulting channel length is simply equal to the metallurgical channel length minus the source and the drain depletion region widths. To a good approximation, the depletion layer width at the source end of the device can be neglected with respect to the drain end. Channel-length modulation is modeled as

$$I_D = \frac{I_o L}{L - L_D}, \tag{14.7.3}$$

where L_D is a geometrically reduced channel length, L is the original channel length, and I_o is the original drain current. Note that the drain current increases as the channel length decreases. This results in an output conductance, defined as a nonzero slope of the drain current versus drain voltage curve for the device.

Punchthrough arises when the channel length is very small. Under this condition, the source and the drain depletion regions can touch, resulting in a large increase in the drain current.

Some of the limitations of short-channel MOSFETs can be overcome if the applied voltage is scaled with the gate length. As mentioned above, as the dimensions become small, at fixed drain voltage, the electric-field magnitude increases strongly. By the scaling of the applied voltage with the channel length, some of the above short-channel effects, principally those associated with a high electric field, are reduced. There of course is a limit as to how low the voltage can be reduced before noise becomes of importance.

At this writing, the ultimate limits to MOSFET devices are still being debated. As the dimensions continue to shrink, ultimately quantum-mechanical effects will alter the operation of the device. Much work is currently underway toward understanding the effects quantum phenomena will have on ultrasubmicrometer MOSFETs. For more information on this topic, the reader is directed to the references, particularly the book by Ferry (1991).

14.8 Alternating-Current Operation of FETs

In this section, a brief introduction to the ac operation of JFETs and MOSFETs is presented. Let us first discuss the ac operation of a JFET and then extend this analysis to that of a MOSFET. At low-frequency operation, capacitive effects can be neglected. If the JFET is modeled as a two-port network, the input is across the gate–source while the output is across the drain–source terminals. In a JFET, between the gate and the source is a reverse biased p^+n junction diode. To first order, if the reverse saturation current in this diode can be neglected, then the gate to source acts as an open circuit; no current flows between the two terminals. The input can then be modeled as an open circuit.

At the output, the dc drain current I_D is a function of both the dc source–drain voltage V_D and the dc gate voltage V_G. The total drain current, including the small-signal drain current i_D and the dc drain current I_D, is then a function of the total drain and gate voltages, including the dc biases and the small-signal biases, v_D and v_G. The small-signal drain current can then be written as

$$i_D = I_D(V_D + v_D, V_G + v_G) - I_D(V_D, V_G). \tag{14.8.1}$$

Expanding Eq. (14.8.1) to first order yields

$$I_D(V_D + v_D, V_G + v_G) = I_D(V_D, V_G) + \left.\frac{\partial I_D}{\partial V_D}\right|_{V_G} v_D + \left.\frac{\partial I_D}{\partial V_G}\right|_{V_D} v_G. \tag{14.8.2}$$

When expression (14.8.2) is substituted for the total current into Eq. (14.8.1) for the small-signal drain current, i_D can be written as

$$i_D = \left.\frac{\partial I_D}{\partial V_D}\right|_{V_G} v_D + \left.\frac{\partial I_D}{\partial V_G}\right|_{V_D} v_G. \tag{14.8.3}$$

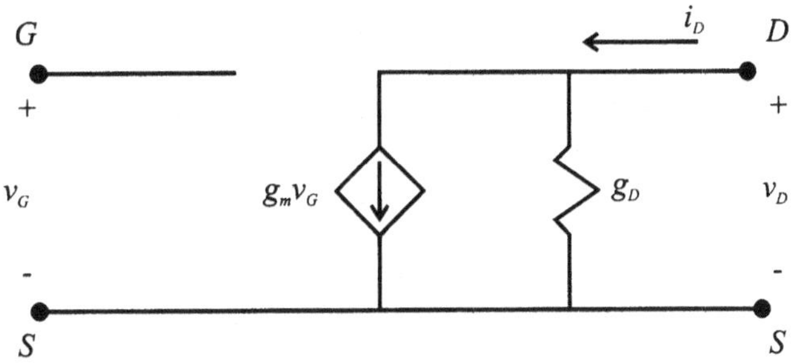

Figure 14.8.1. Sketch of the small-signal equivalent circuit for a JFET.

If we make the following definitions,

$$g_D \equiv \left. \frac{\partial I_D}{\partial V_D} \right|_{V_G}, \qquad g_m \equiv \left. \frac{\partial I_D}{\partial V_G} \right|_{V_D}, \tag{14.8.4}$$

the small-signal drain current becomes

$$i_D = g_D v_D + g_m v_G. \tag{14.8.5}$$

The quantity g_D is called the drain or channel conductance while g_m is the transconductance or mutual conductance. Therefore the small-signal drain current can be written as the sum of two components, one based on the drain conductance and the other involving the transconductance.

Using these parameters, we can construct a small-signal model of the JFET as shown in Figure 14.8.1. As discussed above, the input is simply an open circuit. Therefore the gate–source contacts are sketched in the figure as an open circuit. The output small-signal drain current is equal to the sum of the drain conductance and the transconductance terms, and hence the small-signal drain current can be written as their parallel combination.

We can readily see the importance of the transconductance by examining the operation of the device after pinch off. To an excellent approximation, after pinch off the dc drain current is invariant with increasing drain voltage. Therefore the derivative of the drain current with respect to the drain voltage is zero after pinch off. As a result, the channel conductance vanishes and the small-signal drain current is given then as

$$i_D = g_m v_G. \tag{14.8.6}$$

Knowledge of the transconductance enables determination of the small-signal drain current as a function of the input gate voltage after pinch off.

Although JFETs have been applied in some ac applications, other structures have found greater application in high-frequency amplification. As will be discussed below, the MESFET is the most attractive device candidate for high-power, high-frequency amplifiers compared with bipolar or MOSFET devices. The primary advantages of MESFET devices for high-frequency operation are the fact that they can be readily realized in many different materials systems and that the channel thickness can be accurately controlled. The ideal semiconductor for high-power, high-frequency device applications has simultaneously a large mobility, a large maximum drift velocity, and a large avalanche breakdown field. Since silicon has a relatively low electron mobility and a low breakdown field, it is less attractive than other wide-bandgap materials for high-power, high-frequency applications.

Wide-bandgap semiconductors form the most attractive alternative to silicon for high-temperature, high-power applications because of several inherent material advantages. Among these advantages are high thermal conductivities, high saturation drift velocities, small dielectric constants, high breakdown voltages, and very low thermally generated leakage currents. For high-power, high-temperature operation, it is desirable to maximize the breakdown voltage as well as the drift region conductance. Maximization of the drift region conductance increases the power handling of the device. At high-temperature operation, the thermal properties of the semiconductor become important, particularly its thermal conductivity. A high thermal conductivity considerably eases heat-sinking problems, which also improves packing densities of high-power devices. However, in practice, the heat-sinking capability of devices may be limited by the substrate. This is clearly the case for GaN grown on sapphire.

At high operating temperatures, moderate-bandgap materials like silicon become intrinsic. Wide-bandgap semiconductors can operate at temperatures as high as 600 °C, considerably higher than silicon or GaAs, without becoming intrinsic. Very low thermal leakage currents can therefore be expected if defect densities can be held at acceptably low levels. Wide-bandgap semiconductors typically have the additional feature of a small dielectric constant. As a result of a lower dielectric constant, devices made from these materials have lower parasitic capacitances, thereby providing higher-frequency operation than silicon at comparable dimensions. The combination of all these features, that is, high saturated drift velocity, high thermal conductivity, reduced capacitances, high breakdown voltages, and low leakage currents, makes the wide-bandgap semiconductors very attractive for high-power, high-frequency devices and for high-temperature device applications.

For high-power, high-frequency operation, the device type of choice is the FET. The principal advantages of the FET over bipolar junction transistors (BJTs) are

1. high input impedance (gate drive circuitry is simplified),
2. excellent safe operating area,
3. better output characteristics for paralleling,
4. high-speed or high-frequency operation is better because FETs are unipolar devices. The slow recovery due to minority carrier transport in BJTs is eliminated in FETs.

Of the FET devices, MOSFETs and MESFETs are typically more useful than JFETs for high-frequency operation. Between these structures, silicon-

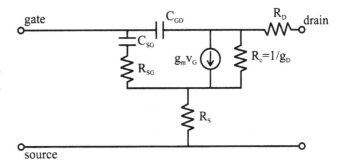

Figure 14.8.2. Sketch of the small-signal ac equivalent circuit for a MESFET.

power MOSFETs have a much higher on-resistance. The presence of a high on-resistance leads to a high power loss at low frequencies in high-voltage devices. The high on-state resistance of silicon MOSFETs limits their application to below 1000 V at 1 Amp. Wide-bandgap semiconductor MESFETs such as SiC devices have 2 orders of magnitude lower on-resistances, leading to much higher power tolerance. Diamond devices are projected to have roughly 3 orders of magnitude lower on-resistances than conventional silicon devices and 1 order of magnitude improvement over SiC. Therefore a much lower power loss can be expected with diamond than either SiC or silicon-based devices.

To assess the ac performance of a MESFET it is useful to construct an equivalent-circuit model of the device. A simplified equivalent circuit for a MESFET structure is shown in Figure 14.8.2. The model is somewhat more complicated than that used for the JFET since the gate–source terminals are no longer assumed to be a simple open circuit. Instead the gate to source is modeled as a capacitor C_{SG} and a resistor R_{SG} in series. Source R_S and drain R_D resistances are included in the model to take account of the finite resistance of the source and the drain regions on either side of the channel. A capacitance between the gate and the drain is also included. The quantity R_C is the channel resistance that is specified by the inverse of the channel conductance. In more sophisticated models, additional resistances are typically added to account for the resistance of the source, gate, and drain contacts and an additional capacitance between the drain and the source is used to represent the substrate capacitance. These quantities are associated with the extrinsic properties of the device. The model shown in Figure 14.8.2 is often called an intrinsic model, albeit a somewhat simplified one.

The key figures of merit for microwave operation of a MESFET are the frequency at unity current gain f_t and the maximum frequency f_{max}. f_t reflects the total time delay of the device independent of the circuit, while f_{max} sets an upper limit for the frequency at which the device can operate. Generally a structure is designed such that f_{max} is significantly higher than f_t. If f_{max} is much more than 2 times greater than f_t, then there is a large impedance mismatch between the input and the output of the device, making its circuit implementation difficult. Conversely, if f_{max} is less than f_t, then the short delay of the device cannot be profitably exploited in a real circuit application.

f_t is often referred to as the unity short-circuit current gain frequency since it is obtained from the ratio of the magnitudes of the ideal output drain current, or equivalently the short-circuited output drain current, to the input current. The short-circuited output current is

$$I_D = g_m V_G, \tag{14.8.7}$$

while the magnitude of the input current is

$$|I_{\text{input}}| = \omega C_{GS} V_G. \tag{14.8.8}$$

f_t is found then from the ratio of the output to the input current magnitudes and by setting that ratio to unity. This results in

$$f_t = \frac{g_m}{2\pi C_{GS}}. \tag{14.8.9}$$

Another way of viewing the frequency performance of a MESFET is to consider the delay associated with the device operation. In general, the total device delay depends principally on the combined effect of the transit time and the charging time of the associated device capacitances. As mentioned above, f_t is associated with the total time delay through the device including the delays associated with the carrier transport as well as the charging and the discharging of the parasitic capacitances outside of the device. Another expression for f_t can be obtained as

$$\frac{1}{2\pi f_t} = \frac{L}{V_{\text{sat}}} + \tau_{\text{ch}} + \tau_{\text{dr}} + \tau_{\text{fringe}} + \tau_{\text{pad}}, \tag{14.8.10}$$

where L is the gate length, V_{sat} is the saturated carrier velocity, τ_{ch} is the time required for charging and discharging the nonvelocity-saturated part of the channel, τ_{dr} is the time taken to cross the drain depletion region, and τ_{fringe} and τ_{pad} are the times required for charging and discharging the gate fringes and pad capacitances. From this expression it is clear that in order to boost the frequency performance of a MESFET the transit time as well as the capacitive charging and discharging times must be minimized.

The high-frequency limitations of MESFETs depend on the material and the device geometry. The most important geometric effect on the frequency performance is the gate length. Very short channel lengths improve the frequency performance of a device in several ways. Decreasing the gate length of the device decreases the gate–source capacitance C_{GS} and increases the transconductance. From inspection of Eq. (14.8.10), decreasing the gate length reduces the carrier transit time. Therefore the value of f_t is increased, providing higher-frequency device operation.

In addition to the above effects, the carrier velocity can be increased from its saturated value by a reduction in the gate length, thereby further reducing the total delay. This effect is called velocity overshoot. High electric fields can

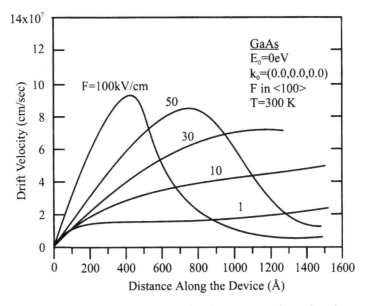

Figure 14.8.3. Average electron drift velocity versus device length at various electric-field strengths in GaAs at zero launching energy and at 300 K.

produce velocity overshoot over small distances by driving the electrons to velocities above the corresponding steady-state velocity. Figure 14.8.3 shows the calculated transient electron velocity in GaAs as a function of distance for various applied field strengths. In each of the calculations, the electrons are assumed initially to be at zero kinetic energy. As can be seen from Figure 14.8.3, at low applied electric fields, 1 kV/cm, the velocity does not overshoot its steady-state value significantly. As the field is increased, the velocity overshoots its steady state value by a large amount. This can be seen for GaAs at fields of 10 and 30 kV/cm. The electron transit time at these fields will be substantially reduced by the overshoot compared with the case in which they travel with their steady-state velocity. However, as the applied field is increased more, the overshoot dramatically decreases because of the transferred electron effect. On transferring from the central, low-mass gamma valley into the large effective-mass satellite valleys, the electron drift velocity decreases sharply. Clearly, there is a limited window of applied fields and conditions that will lead to a significant velocity overshoot in GaAs.

Alternative materials from GaAs and silicon can be used to enhance further the high-frequency performance of a MESFET. Materials that have large intervalley separation energies as well as low effective masses within the central valley are particularly attractive candidates for high-frequency device operation. Such materials offer the possibility of shorter transit times under the gate and a higher velocity overshoot. The most promising candidates are GaInAs/AlInAs-based devices. Work is currently underway on developing these materials for use as MESFET and HEMT devices. Future wireless communications systems and satellite relay systems may incorporate these types of devices in order to operate at higher frequencies.

PROBLEMS

1. An Al gate p-channel MOS transistor is made on an n-type silicon substrate with $N_d = 1 \times 10^{16}$ cm^{-3}. The SiO$_2$ thickness is 150 nm in the gate region and the effective interface charge $Q_i = 2 \times 10^{11}$ qC/cm^2. Note: $\phi_{ms} = -0.25$V; $kT = 0.0259$ eV; $n_i = 1.5 \times 10^{10}$cm^{-3}; $\epsilon_{ox} = 3.9$; $\epsilon_{si} = 11.9$; $\epsilon_0 = 8.85 \times 10^{-14}$ F/cm.

   a. Determine the maximum depletion region width.

   b. Determine the threshold voltage for the device.

2. One can transform an n-channel depletion-mode MOSFET into an n-channel enhancement-mode MOSFET by implanting a layer of ionized boron. Find the dosage of ionized boron required for forming an n-channel enhancement-mode MOSFET with a threshold voltage of +2V. Assume that $N_a = 1 \times 10^{16}$ cm^{-3}; $\epsilon_{ox} = 3.9$; $\epsilon_{si} = 11.9$; $\epsilon_0 = 8.85 \times 10^{-14}$ F/cm; t_{ox} (oxide thickness) $= 10^{-5}$ cm; $\phi_m = 3.2$ eV; $\chi_s = 3.25$ eV; $Q_i = 5.0 \times 10^{11}$ qC/cm^2.

3. Calculate $\partial V_D / \partial I_D$ for a JFET below saturation. What does the simple theory imply about this resistance beyond pinch off?

4. Consider a MOS system 3×15 μm in area composed of an Al gate, SiO$_2$ insulator, and p-type Si substrate. Data: t_{ox} (oxide thickness) $= 120$ nm; Q_{ox} (surface-state oxide charge) $= 3 \times 10^{10}$qC/cm^2; $\epsilon_{ox} = 3.9$; ρ (Si resistivity) $= 12.5$ Ω cm; D_p (Si) $= 13$ cm^2/s; E_g (Si) $= 1.12$ eV; ϕ_m (Al) $= 3.20$ eV; χ (Si) $= 3.25$ eV; surface mobility $= 1/2$ bulk mobility.

   a. Determine the total oxide capacitance.

   b. Determine the surface potential at strong inversion.

   c. Determine the threshold voltage.

References

Chapters in which sources have been consulted are indicated in parentheses at the end of the reference.

Allam, J., Capasso, F., Alavi, K., and Cho, A. Y. (1987). Near-single carrier-type multiplication in a multiple graded-well structure for a solid-state photomultiplier. IEEE Electron Dev. Lett. **EDL-8**, 4–6. (10)

Allen, G. (1953). Band structures of one-dimensional crystals with square-well potentials. Phys. Rev. **91**, 531–3. (8)

Aristin P., Torabi A., Garrison A. K., Harris H. M., and Summers C. J. (1992). New doped multiple-quantum-well avalanche photodiode: The doped barrier AlGaAs/GaAs multiple-quantum-well avalanche photodiode. Appl. Phys. Lett. **60**, 85–7. (12)

Ashcroft, N. W. and Mermin, J. D. (1976). *Solid State Physics*. New York: Holt, Rinehart & Winston. (5)

Baraff, G. M. (1962). Distribution functions and ionization rates for hot electrons in semiconductors. Phys. Rev. **128**, 2507–18. (10)

Barbe, D. F. (1975). Imaging devices using the charge-coupled concept. Proc. IEEE **63**, 38–67. (12)

Bassani, F. and Pastori Parravicini, G. (1977). *Electronic States and Optical Transitions in Solids*. Oxford: Pergamon. (7)

Baym, G. (1969). *Lectures on Quantum Mechanics*. Reading, MA: Benjamin. (4)

Bhatnagar, M. and Baliga, B. J. (1993). Comparison of 6H-SiC, 3C-SiC, and Si for power devices. IEEE Trans. Electron. Dev. **40**, 645–55. (14)

Bhattacharya, P. (1994). *Semiconductor Optoelectronic Devices*. Englewood Cliffs, NJ: Prentice-Hall. (12)

Blauvelt, H., Margalit, S., and Yariv, A. (1982). Single-carrier-type dominated impact ionization in multilayered structures. Electron. Lett. **18**, 375–6. (12)

Bohm, D. (1951). *Quantum Theory*, Parts 1 and 2. Englewood Cliffs, NJ: Prentice-Hall. (1)

Bohm, D. and Pines, D. (1953). A collective description of electron interactions: III. Coulomb interactions in a degenerate electron gas. Phys. Rev., **92**, pp. 609–625. (9)

Brennan, K. (1986). Theory of the doped quantum well superlattice APD: a new solid-state photomultiplier. J. Quantum Electron. **QE-22**, 1999–2016. (12)

Brennan, K. (1987a). The *p-n* junction quantum well APD: a new solid state photodetector for lightwave communications systems and on-chip detector applications. IEEE Trans. Electron. Dev. **ED-34**, 782–92. (12)

Brennan, K. (1987b). The *p-n* heterojunction quantum well APD: a new high-gain low-noise high-speed photodetector suitable for lightwave communications and digital applications. IEEE Trans. Electron. Dev. **ED-34**, 793–803. (12)

Brennan, K. (1988). Theory of high field electronic transport in bulk ZnS and ZnSe. J. Appl. Phys. **64**, 4024–30. (13)

Brennan, K. and Hess, K. (1986). A theory of enhanced impact ionization due to the gate field and mobility degradation in the inversion layer of MOSFETs. IEEE Electron. Dev. Lett. **EDL-7**

Brennan, K. and Wang, Y. (1988). Field and spatial geometry dependencies of the electron and hole ionization rates in GaAs/AlGaAs multiquantum well APDs. IEEE Trans. Electron. Dev. **35**, 634–41. (12)

Brennan, K. F. and Mansour, N. S. (1991). Monte Carlo calculation of electron impact ionization in bulk InAs and HgCdTe. J. Appl. Phys. **69**, 7844–7. (10)

Brodie, I. and Schwoebel, P. R. (1994). Vacuum microelectronic devices. Proc. IEEE **82**, 1006–34. (13)

Bude, J. and Hess, K. (1992a). Thresholds of impact ionization in semiconductors. J. Appl. Phys. **72**, 3554–6. (10)

Bude, J., Hess, K., and Iafrate, G. J. (1992b). Impact ionization in semiconductors: effects of high electric fields and high scattering rates. Phys. Rev. B **45**, 10958–64. (10)

Bulman, G. E., Robbins, V. M., Brennan, K. F., Hess, K., and Stillman, G. E. (1983). Experimental determination of impact ionization coefficients in (100) GaAs. IEEE Electron Device Lett. **EDL-4**, 181–5. (12)

Burt, M. G. (1985). An alternative expression for the impact ionisation coefficient in a semiconductor derived using lucky drift theory. J. Phys. C **18**, L477-L481. (10)

Burt, M. G., Brand, S., Smith C., and Abram, R. A. (1984). Overlap integrals for Auger recombination in direct bandgap semiconductors: calculations for conduction and heavy-hole bands in GaAs and In P. J. Phys. C **17**, 6385–401. (10)

Callaway, J. (1991). *Quantum Theory of the Solid State*, 2nd ed. Boston: Academic. (9)

Campbell, J. C. (1985). Phototransistors for lightwave communications, In *Semiconductors and Semimetals*, Vol. 22, Part D, eds., R. K. Willardson and A. C. Beer, pp. 389–447. New York: Academic. (12)

Capasso, F. (1985). Physics of avalanche photodiodes. In *Semiconductors and Semimetals*, Vol. 22, Part D, eds., R. K. Willardson and A. C. Beer, pp. 1–172. New York: Academic. (10)

Capasso, F., Tsang, W. T., Hutchinson, A. L., and Williams, G. F. (1982). Enhancement of the electron impact ionization rate in a superlattice: a new avalanche photodiode with large ionization rates ratio. Appl. Phys. Lett. **40**, 38–40. (12)

Capasso, F., Allam, J., Cho, A. Y., Mohammed, K., Malik, R. J., Hutchinson, A. L., and Sivco, D. (1986). New avalanche multiplication phenomenon in quantum well superlattices: evidence of impact ionization across the band-edge discontinuity. Appl. Phys. Lett. **48**, 1294–6. (10)

Capasso, F., Sen, S., and Cho, A. Y. (1987). Resonant tunneling: physics, new transistors and superlattice devices. Proc. SPIE **792**, 10–17. (12)

Cappy, A., Carnez, B., Fauquembergues, R., Salmer, G., and Constant, E. (1980). Comparative potential performance of Si, GaAs, GaInAs, InAs submicrometer-gate FETs. IEEE Trans. Electron Dev. **ED-27**, 2158–60. (14)

Casey, Jr., H. C. and Panish, M. B. (1978). *Heterostructure Lasers, Part A: Fundamental Principles*. Orlando, FL: Academic. (13)

Chin, R., Holonyak, Jr., N., Stillman, G. E., Tang, J. Y., and Hess, K. (1980). Impact ionization in multilayered heterojunction structures. Electron. Lett. **16**, 467–9. (12)

Chuang, S. L. and Hess, K. (1986). Impact ionization across the conduction band edge discontinuity of quantum well heterostructures. J. Appl. Phys. **59**, 2885–94. (10)

Cohen, M. L. and Bergstresser, T. K. (1966). Band structures and pseudopotential form factors for fourteen semiconductors of the diamond and zinc-blende structures. Phys. Rev. **141**, 789–96. (8)

Colclaser, R. A. and Diehl-Nagle, S. (1985). *Materials and Devices for Electrical Engineers and Physicists*. New York: McGraw-Hill. (8)

Collings, P. J. and Patel, J. S. (1997). Introduction to the science and technology of liquid crystals. In *Handbook of Liquid Crystal Research*, eds., P. J. Collings and J. S. Patel. New York: Oxford U. Press. (13)

Conwell, E. and Weisskopf, V. F. (1950). Theory of impurity scattering in semiconductors. Phys. Rev. **77**, 388–90. (9)

Demus, D. (1990). Types and classification of liquid crystals. In *Liquid Crystals: Applications and Uses*, Vol. 1, ed., B. Bahadur. Singapore: World Scientific. (13)

Dirac, P. A. M. (1958). *The Principles of Quantum Mechanics*. Oxford: Clarendon. (3)

Eisberg, R. and Resnick, R. (1974). *Quantum Physics of Atoms, Molecules, Solids, Nuclei, and Particles*. New York: Wiley. (2)

Esaki, L. and Chang, L. L. (1994). New transport phenomenon in a semiconductor "superlattice." Phys. Rev. Lett. **33**, 495–8. (2)

Fawcett, W., Boardman, A. D., and Swain, S. (1970). Monte Carlo determination of electron transport in gallium arsenide. J. Phys. Chem. Solids **31**, 1963–90. (9)

Ferry, D. K. (1991). *Semiconductors*. New York: MacMillan. (9, 11, 14)

Feynman, R. P., Leightov, R. B., and Sands, M. L. (1965). *The Feynman Lectures on Physics*, Addison-Wesley. (5)

Fichtner, W. and Poetzl, H. W. (1979). MOS modelling by analytical approximations. I. Subthreshold current and threshold voltage. Int. J. Electron. **46**, 33–55. (14)

Fossum, E. R. (1993). Active pixel sensors: Are CCDs dinosaurs? Proc. SPIE **1900**, 2–14. (12)

French, A. P. (1968). *Special Relativity*. New York: Norton. (3)

Frisch, D. H. and Smith, J. H. (1963). Measurement of the relativistic time-dilation using Mu-mesons. Am. J. Phys. **31**, 342–355. (3)

Gaylord, T. K. (1970). High electric field conduction anisotropies in semiconductors. Ph.D. dissertation, Rice University, Houston, TX. (8)

Gaylord, T. K. and Brennan, K. F. (1988). Semiconductor superlattice electron wave interference filters. Appl. Phys. Lett. **53**, 2047–9. (2)

Giuliani, G. F. and Quinn, J. J. (1982). Lifetime of a quasiparticle in a two-dimensional electron gas. Phys. Rev. B **26**, 4421–8. (9)

Grotjohn, T. and Hoefflinger, B. (1984). A parametric short-channel MOS transistor model for subthreshold and strong inversion current. IEEE Trans. Electron Dev. ED-31, 234–46. (14)

Gurnett, K. W. (1996). The light emitting diode (LED) and its application. Microelectron. J. **27**, 37–41. (13)

Harrison, W. A. (1979). *Solid State Theory*. New York: Dover. (8)

Hersee, S. D., De Cremoux, B., and Duchemin, J. P. (1984). Some characteristics of the GaAs/GaAlAs graded-index separate-confinement heterostructure quantum well laser structure. Appl. Phys. Lett. **44**, 476–8. (13)

Hess, K. (1988). *Advanced Theory of Semiconductor Devices*. Englewood Cliffs, NJ: Prentice Hall. (10)

Holonyak, Jr., N. (1985). Quantum-well semiconductor lasers (review). Sov. Phys. Semicond. **19**, 943–58. (13)

Holonyak, Jr., N. (1987). Semiconductor alloy lasers – 1962. IEEE J. Quantum Electron. **QE-23**, 684–91. (13)

Holonyak, Jr., N., Kolbas, R. M., Dupuis, R. D., and Dapkus, P. D. (1980). Quantum well heterostructure lasers. IEEE J. Quantum Electron. **QE-16**, 170–85. (2)

Hsu, F.-C., Ko, P.-K., Tam, S., Hu, C., and Muller, R. S. (1982). An analytical breakdown model for short-channel MOSFETs. IEEE Trans. Electron. Dev. **ED-29**, 1735–40. (14)

Jackson, J. D. (1975). *Classical Electrodynamics*. 2nd ed. New York: Wiley. (1)

Jacoboni, C. and Lugli, P. (1989). *The Monte Carlo Method for Semiconductor Device Simulation*. Vienna: Springer-Verlag. (9, 10)

Jones, W. and March, N. H. (1973). *Theoretical Solid State Physics*, Vol. 1, *Perfect Lattices in Equilibrium*. New York: Dover. (8)

Jordan, T. F. (1969). *Linear Operators for Quantum Mechanics*. New York: Wiley. (1)

Kane, E. O. (1967). Electron scattering by pair production in silicon, Phys. Rev. **159**, 624–631. (10)

Kane, E. O. (1957). Band structure of Indian Antionide. J. Phys. Chem Solids **1** 249–261. (11)

Keldysh, L. V. (1965). Concerning the theory of impact ionization in semiconductors. Sov. Physics, JETP, **21**, 1135–1144. (10)

Kinch, M. A. (1981). Metal-insulator-semiconductor infrared detectors. In *Semiconductors and Semimetals*, Vol. 18, eds., R. K. Willardson and A. C. Beer, pp. 313–378. New York: Academic. (12)

Kittel, C. (1963). *Quantum Theory of Solids*. New York: Wiley. (9)

Kittel, C. (1976). *Introduction to Solid State Physics*, 5th ed. New York: Wiley. (6)

Kittel, C. and Kroemer, H. (1980). *Thermal Physics*. 2nd ed. San Francisco: Freeman. (5)

Kosonocky, W. F. and Sauer, D. J. (1975). The ABCs of CCDs. Electron. Des. **23**, 58–63. (12)

Kroemer, H. (1963). A proposed class of heterojunction injection lasers. Proc. IEEE **51**, 1782. (13)

Landau, L. D. and Lifshitz, E. M. (1977). *Quantum Mechanics (Nonrelativistic Theory)*. Oxford: Pergamon. (2)

Landau, L. and Lifshitz, E. M. (1980). *Statistical Physics*, 3rd ed., Part 1. Oxford: Pergamon. (5)

Landsberg, P. T. (1978). *Thermodynamics and Statistical Mechanics*. New York: Dover. (5)

Leonard, W. F. and Martin, Jr., T. L. (1980). *Electronic Structure and Transport Properties of Crystals*. Huntington, NY: Krieger. (7)

Levine, B. F., Choi, K. K., Bethea, C. G., Walker, J., and Malik, R. J. (1987). Quantum well avalanche multiplication initiated by 10 μm intersubband absorption and photoexcited tunneling. Appl. Phys. Lett. **51**, 934–6. (10)

Liboff, R. L. (1992). *Introductory Quantum Mechanics*, 2nd ed. Reading, MA: Addison-Wesley. (4)

Liechti, C. A. (1976). Microwave field-effect transistors – 1976. IEEE Trans. Microwave Theory and Technol. **MTT-24**, 279–300. (14)

Lundstrom, M. (1990). *Modular Series on Solid State Devices, Fundamentals of Carrier Transport*, Vol. X. Reading, MA: Addison-Wesley. (6)

Madelung, O. (1978). *Introduction to Solid-State Theory*. Berlin: Springer-Verlag. (9)

Mandl, F. (1971). *Statistical Physics*. London: Wiley. (5)

Marsland, J. S. (1987). A lucky drift model, including a soft threshold energy, fitted to experimental measurements of ionization coefficients. Solid-State Electron. **30**, 125–32. (10)

Masuda, H., Nakai, M., and Kubo, M. (1979). Characteristics and limitation of scaled-down MOSFETs due to two-dimensional field effect. IEEE Trans. Electron. Dev. **ED-26**, 980–6. (14)

Matsuo, K., Teich, M. C., and Saleh, B. E. A. (1985). Noise properties and time response of the staircase avalanche photodiode. IEEE Trans. Electron. Dev. **ED-32**, 2615–23. (12)

McGervey, J. D. (1971). *Introduction to Modern Physics*. New York: Academic. (5)

McKelvey, J. P. (1966). *Solid State and Semiconductor Physics*. Malabar, FL: Krieger. (5)

Merzbacher, E. (1970). *Quantum Mechanics*, 2nd ed. New York: Wiley. (1)

Moll, J. (1964). *Physics of Semiconductors*. New York: McGraw-Hill. (11)

Moriizumi, T. and Takahashi, K. (1972). Theoretical analysis of heterojunction phototransistors. IEEE Trans. Electron Dev. **ED-19**, 152–9. (12)

Muller, R. S. and Kamins, T. I. (1977). *Device Electronics for Integrated Circuits*. New York: Wiley. (14)

Pao, H. C. and Sah, C. T. (1966). Effects of diffusion current on characteristics of metal-oxide (insulator) – semiconductor transistors. Solid-State Electron. **9**, 927–37. (14)

Pearce, C. W. and Yaney, D. S. (1985). Short-channel effects in MOSFETs. IEEE Electron. Dev. Lett. **EDL-6**, 326–8. (14)

Philips, E. N. (1980). Electro-optics – image intensifiers. ITT Internal Rep., pp. 1–19, Doc. JSI/100. Roanoke, VA: ITT Defense. (12)

Pierret, R. F. (1983). *Modular Series on Solid State Devices*, Vol. 1, *Semiconductor Fundamentals*. Reading, MA: Addison-Wesley. (10)

Pierret, R. F. (1996). *Semiconductor Device Fundamentals*. Reading, MA: Addison-Wesley. (11)

Pollock, C. R. (1995). *Fundamentals of Optoelectronics*. Chicago: Irwin. (13)

Pucel, R. A., Haus, H. A., and Statz, H. (1975). Signal and noise properties of GaAs microwave field-effect transistors. In *Advances in Electronics and Electron Physics*, Vol. 38, pp. 195–265. New York: Academic. (14)

Purcell, E. M. (1963). *Electricity and Magnetism*. New York: McGraw-Hill. (3)

Quinn, J. J. (1962). Range of excited electrons in metals. Phys. Rev. **126**, 1453–7. (9)

Rack, P. D., Naman, A., Holloway, P. H., Sun, S.-S., and Tuenge, R. T. (1996). Materials used in electroluminescent displays," MRS Bull., **21**, 49–58. (13)

Reif, F. (1965). *Fundamentals of Statistical and Thermal Physics*. New York: McGraw-Hill. (1, 5)

Resnick, R. (1972). *Basic Concepts in Relativity and Early Quantum Theory*. New York: Wiley.

Ridley, B. K. (1983). A model for impact ionisation in wide-gap semiconductors. J. Phys. C **16**, 4733–51. (10)

Ridley, B. K. (1988). *Quantum Processes in Semiconductors*, 2nd ed. Oxford: Clarendon. (10)

Ridley, B. K. (1993). *Quantum Processes in Semiconductors*, 3rd ed. Oxford: Clarendon. (11)

Saleh, B. E. A. and Teich, M. C. (1991). *Fundamentals of Photonics*. New York: Wiley. (12)

Salem, A. F., Smith, A. W., and Brennan, K. F. (1994). Theoretical study of the effect of an AlGaAs double heterostructure on metal-semiconductor-metal photodetector performance. IEEE Trans. Electron Dev. **41**, 1112–1119. (12)

Sano, N. and Yoshii, A. (1994). Impact ionization rate near thresholds in Si. J. Appl. Phys. **75**, 5102–5105. (10)

Saxon, D. S. (1968). *Elementary Quantum Mechanics*. San Francisco, CA: Holden-Day. (7)

Scheffer, T. and Nehring, J. (1990). Twisted nematic and supertwisted nematic mode LCDs. In *Liquid Crystals: Applications and Uses*, Vol. 1, ed., B. Bahadur. Singapore: World Scientific. (13)

Schiff, L. (1955). *Quantum Mechanics*. New York: McGraw-Hill. (1)

Schiff, L. I. (1968). *Quantum Mechanics*, 3rd ed. New York: McGraw-Hill. (7)

Selberherr, S. (1984). *Analysis and Simulation of Semiconductor Devices*. Vienna, New York: Springer-Verlag. (6)

Shenai, K., Scott, R. S., and Baliga, B. J. (1989). Optimum semiconductors for high-power electronics. IEEE Trans. Electron. Dev. **36**, 1811–23. (14)

Shichijo, H. (1980). Theoretical studies of high field transport in III-V semiconductors. Ph.D. dissertation, University of Illinois, Urbana, IL. (9)

Shichijo, H. and Hess, K. (1981). Band-structure-dependent transport and impact ionization in GaAs. Phys. Rev. B **23**, 4197–207. (10)

Shockley, W. (1961). Problems related to p-n junctions in silicon. Solid-State Electron. **2**, 35–67. (10)

Singh, J. (1993). *Physics of Semiconductors and Their Heterostructures*. New York: McGraw-Hill. (9)

Smith, A. (1992). Light confinement and hydrodynamic modelling of semiconductor structures by volumetric methods. Ph.D. dissertation, Georgia Tech., Atlanta, Ga. (6)

Smith, R. A. (1979). *Semiconductors*, 2nd ed. Cambridge: Cambridge U. Press. (11)

Smith, J. S., Chin, L. C., Margalit, S., and Yariv, A. (1983). A new infrared detector using electron emission from multiple quantum wells. J. Vac. Sci. Technol. B **1**, 376–8. (10)

Snowden, C. M. (1988). *Semiconductor Device Modelling*. London: Peregrinus. (14)

Stillman, G. E. and Wolfe, C. M. (1977). Avalanche photodiodes. In *Semiconductors and Semimetals*, Vol. 12, eds., R. K. Willardson and A. C. Beer, pp. 291–393. New York: Academic. (12)

Streetman, B. G. (1980). *Solid State Electronic Devices*, 2nd ed. Englewood Cliffs, NJ: Prentice-Hall. (10)

Streetman, B. G. (1995). *Solid State Electronic Devices*, 4th ed. Englewood Cliffs, NJ: Prentice-Hall. (11)

Sze, S. M. (1981). *Physics of Semiconductor Devices*, 2nd ed. New York: Wiley. (5)

Taylor, E. F. and Wheeler, J. A. (1963). *Spacetime Physics*. San Francisco: W. H. Freeman.

Teich, M. C., Matsuo, K., and Saleh, B. E. A. (1986). Excess noise factors for conventional and superlattice avalanche photodiodes and photomultiplier tubes. IEEE J. Quantum Electron. **QE-22**, 1184–93. (12)

Titchmarsh, E. C. (1955). *Eigenfunction Expansions Associated with Second-Order Differential Equations*. Oxford: Oxford University Press. (1)

Tsang, W. T. (1982). Extremely low threshold (AlGa)As graded-index waveguide separate-confinement heterostructure lasers grown by molecular beam epitaxy. Appl. Phys. Lett. **40**, 217. (13)

Tsang, W. T. (1984). Heterostructure semiconductor lasers prepared by molecular beam epitaxy. IEEE J. Quantum Electron. **QE-20**, 1119–32. (12)

Tsividis, Y. (1987). *Operation and Modeling of the MOS Transistor*. New York: McGraw-Hill. (14)

Tsu, R. and Esaki, L. (1973). Tunneling in a finite superlattice. Appl. Phys. Lett. **22**, 562–4. (2)

Uyemura, J. P. (1984). Course Notes: EE4350. Atlanta, GA: Georgia Institute of Technology. (9)

Van Duzer, T. and Turner, C. W. (1981). *Principles of Superconductive Devices and Circuits*. New York: Elsevier. (6)

Wada, T. and Frey, J. (1979). Physical basis of short-channel MESFET operation. IEEE Trans. Electron. Dev. **ED-26**, 476–89. (14)

Wada, Y. and Tomizawa, M. (1988). Drain avalanche breakdown in gallium arsenide MESFETs. IEEE Trans. Electron Dev. **35**, 1765–70. (14)

Wang, Y. and Brennan, K. F. (1990a). Theoretical study of a classically confined solid-state photomultiplier. IEEE J. Quantum Electron. **26**, 1838–44. (10)

Wang, Y., Park, D. H., and Brennan, K. F. (1990b). Theoretical analysis of confined quantum state GaAs/AlGaAs solid-state photomultipliers. IEEE J. Quantum Electron. **26**, 285–95. (10)

Wang, Y. and Brennan, K. F. (1994). Semiclassical study of the wave vector dependence of the internal impact ionization rate in bulk silicon. J. Appl. Phys. **75**, 313–9.

Warriner, R. A. (1977). Computer simulation of gallium arsenide field-effect transistors using Monte Carlo methods. Solid-State and Electron. Dev. **1**, 105–10. (14)

Webb, P. P., McIntyre, R. J., and Conradi, J. (1974). Properties of avalanche photodiodes. RCA Rev. **35**, 2334–477. (12)

Williams, C. K., Glisson, T. H., Hauser, J. R., Littlejohn, M. A., and Abusaid, M. F. (1985). Two-dimensional Monte Carlo simulation of a submicron GaAs MESFET with a nonuniformly doped channel. Solid-State Electron. **11**, 1105–9. (14)

Wolfe, C. M., Holonyak, Jr., N., and Stillman, G. E. (1989). *Physical Properties of Semiconductors*, Englewood Cliffs, NJ: Prentice-Hall. (10)

Wolff, P. A. (1954). Theory of electron multiplication in silicon and germanium. Phys. Rev. **95**, 1415–20. (10)

Yariv, A. (1976). *Introduction to Optical Electronics*, 2nd ed. New York: Holt. (13)

Yariv, A. (1982). *An Introduction to Theory and Applications of Quantum Mechanics*. New York: Wiley. (5)

Young, J. F. and Kelly, P. J. (1993). Many-body treatment of hot-electron scattering from quasiequilibrium electron-hole plasmas and coupled plasmon-longitudinal-optic-phonon modes in GaAs. Phys. Rev. B **47**, 6316–29. (9)

Ziman, J. M. (1960). *Electrons and Phonons: The Theory of Transport Phenomena in Solids*. Oxford: Oxford U. Press. (9)

Ziman, J. M. (1969). *Elements of Advanced Quantum Theory*. Cambridge: Cambridge U. Press. (6)

Ziman, J. M. (1972). *Principles of the Theory of Solids*. Cambridge: Cambridge U. Press. (8)

Index